HIGHER ALGEBRA

[Classical and Modern]

For BSc [General] Course of studies of all Indian Universities, covering, in particular the syllabi of the Universities of Calcutta, Burdwan, North Bengal, Kalyani, Viswa Bharati and Vidyasagar University. Also suitable text for the Technical Colleges.

HIGHER ALGEBRA

[Classical and Modern]

Ram Krishna Ghosh

Formerly Dean of Science, Professor and Head of the Department of Mathematics
St. Xavier's College, Kolkata
Sometime Senior Lecturer in Mathematics, University of Cape Coast, Ghana
Sometime Part-time Lecturer in Mathematics, University of Kalyani, Kalyani

Kantish Chandra Maity

Formerly Professor of Mathematics, St. Xavier's College, Kolkata
Formerly Principal, Raja Peary Mohan College, Uttarpara
Formerly Head of the Dept of Mathematics, Asutosh College, Kolkata

To be had of

New Central Book Agency (P) Ltd

LONDON
HYDERABAD ERNAKULAM BHUBANESWAR
NEW DELHI KOLKATA PUNE GUWAHATI

NCBA

REGD OFFICE

8/1 Chintamoni Das Lane, Kolkata 700 009, India
email: ncbapvtltd@eth.net

OVERSEAS

NCBA (UK) Ltd, 149 Park Avenue North, Northampton, NN3 2HY, UK
email: ncbauk@yahoo.com

EXPORT

NCBA Exports Pvt. Ltd, 212 Shahpur Jat, New Delhi 110 049
email: ncbaexports@ncbapvtltd.com

BRANCHES

2/27 Ansari Road, 1st Floor
Daryaganj, **New Delhi** 110 002
email: ncbadel@ncbapvtltd.com

Shop Nos. 3 & 4, Vinayak Towers
681/B Budhwar Peth, Appa Balwant Chowk
Pune 411 002
email: ncbapune@ncbapvtltd.com

Shop No. 15, Bharati Towers, Ground Floor
Forest Park
Bhubaneswar 751 001
email: ncbabub@ncbapvtltd.com

House No. 3-1-315, 1st Floor
Nimboliadda, Kachiguda, **Hyderabad** 500 027
email: ncbahydb@ncbapvtltd.com

GSS Shopping Complex, 1st Floor
Opposite College Post Office, Convent Road
Ernakulam 682 035
0484-4021054 / 53

Radhamayee Complex (Second Floor)
Opp. Axis Bank, Panbazar, Jaswanta Road
Guwahati 781 001
email: ncbaguwahati@ncbapvtltd.com

HIGHER ALGEBRA ● Ram Krishna Ghosh, Kantish Chandra Maity

First Published: 1980, Fifth Edition: 1999
Reprinted: 2001, 2002, Completely revised Sixth Edition: 2005
Reprinted: 2011, 2013, 2014, 2015, 2018

PUBLISHER AND TYPESETTER
Central Educational Enterprises (P) Ltd
54B Patuatola Lane, Kolkata 700 009

PRINTER
New Central Book Agency (P) Ltd
Web-Offset Division, Dhulagarh, Sankrail, Howrah

PROJECT SUPERVISOR
Abhijit Sen

ISBN: 978 81 7381 104 3

Price ₹ 415.00

Preface to the Revised Sixth Edition

As advised by the University Grants Commission, the Indian Universities are gradually trying to introduce a uniform syllabi for the Three-Year Degree Courses. The University of Calcutta has accordingly revised the syllabuses of B.Sc. (both Honours and General) courses in Mathematics effective from the academic session 2003-2004.

The present text on **Higher Algebra** was first published in 1980 and currently running through six editions. The authors have already introduced most of the recommendations of the U.G.C. in their fourth edition (1995).

In the present edition, the authors have carefully reviewed each and every chapter and introduced innovative techniques of presentation as expected by the U.G.C. The students were always kept in their minds and the syllabuses of the University of Calcutta (as revised) have been followed very religiously. Some topics under the heading "**Additional Topics**" are meant to cover the syllabus of all the Indian Universities.

The authors do hope that the presentation will be beneficial to the students and will be appreciated by the teachers. The authors are thankful to the teachers and the students for their continued support for the last twenty five years.

Suggestions for further improvement will be gratefully received. In spite of the best efforts some errors might have gone unnoticed. The authors will be thankful if such errors be brought to their notice.

The authors wish to express sincere thanks to the publisher and their entire staff, who were responsible for the careful publication of the book.

Janmastami
27 August 2005

R.K. Ghosh
K.C. Maity

Preface to the First Edition

This text book has been written to include those topics of Algebra which are mentioned in the Mathematics syllabuses of the Universities of Calcutta, Burdwan, Jadavpur, North Bengal and Kalyani.

The book consists mainly of two sections—one containing topics of Classical Algebra and the other containing those of Abstract Algebra. Infinite sequences and series have been discussed under a separate section. This topic has been included by some Universities (like North Bengal and Burdwan) in the Algebra Syllabus.

In the preparation of the book the authors have aimed at producing a treatise in which the subject is developed logically, complete as far as it goes, and the spirit of the current trends in modern algebra is maintained. A careful reader of this book will not miss such trends even in classical topics like *complex numbers, complex functions* and *Theory of Equations.* The underlying concept of linearity has been stressed upon while discussing *Matrix Algebra* and *Finite difference equations.* In presenting the *fundamentals of Abstract Algebra* the main considerations were not so much the depth but the motivation, the logical order and the utilitarianism of the new concepts. Algebra serves as an introduction to modern analysis and accordingly the inclusion of a chapter on *Infinite sequence* and *series* under Algebra may be useful to the students who wish to study Analysis.

As in the case of other texts written by the present authors the needs of the students were constantly kept in mind. Basic ideas have been explained through varied examples. Theorems are proved with expected amount of rigour. Carefully selected examples are worked out. A Set of well-graded Review exercises (along with Hints, for solving comparatively harder problems) are given at the end of each chapter. It is hoped that the book will be found really useful to all section of students.

The authors gratefully acknowledge the assistance and inspiration that they have received from a large number of colleagues as well as former students in the preparation of this book.

In particular, the authors are greatly indebted to Professor Pratul Kumar Bagchi, Head of the Department of Mathematics, Vidyasagar College, Calcutta for his invaluable suggestions, criticism and constant inspiration.

Thanks are also due to the publishers and the staff who had to work against heavy odds in bringing out the present treatise.

The authors will be extremely grateful to any one who will bring to their notice any necessary corrections and give suggestions for improvement of the present book.

Authors

Calcutta
1st February, 1980

Contents

Part I

Classical Algebra

(CHAPTERS 1–5)

Chapter 1

De Moivre's Theorem and Its Applications

1.	Complex Number Defined: Associated Terms (Recapitulations)
2.	De Moivre's Theorem: Its applications

1.1 Introduction

In the present chapter we shall discuss **De Moivre's Theorem and its Applications.**

It will be, therefore, assumed that our student—readers are already acquainted with the basic knowledge of complex numbers.[1] However, for their convenience we briefly recapitulate some basic definitions and notations:

Definitions of Complex Numbers and Associated Terms

I. Definition of a complex number and elementary operations with complex numbers.

A complex number z may be defined by an expression of the form $a+ib$, where a and b are real numbers; and we denote

$$z = a + ib \ (a, b \text{ real}).$$

It has been found more logical to define a complex number z as an ordered pair (a, b) of real numbers and write $z = (a, b)$, where a and b are real:

$$\begin{aligned} a &= \text{Real part of } z = \text{Rl}\, z \\ \text{and} \quad b &= \text{Imaginary part of } z = \text{Im}\, z. \end{aligned}$$

We assume the statements given below: **(1–7).**

1. **Equality:** $(a, b) = (c, d) \Rightarrow a = c, b = d$,

   i.e., two complex numbers are equal if their real parts are equal and their imaginary parts are equal. This technique is called *"Equating real and imaginary parts"*

[1]See present authors' **Higher Secondary Mathematics, Vol. I**

2. Addition: We add two complex numbers by the rule:

$$(a, b) + (c, d) = (a + c, b + d)$$

i.e., the sum of two complex numbers is a complex number since $a + c, b + d$ *are both real.*

3. Multiplication: We can also multiply two complex numbers by the rule:

$$(a, b) \cdot (c, d) = (ac - bd, ad + bc).$$

Clearly, *The product of two complex numbers is also a complex number.*

4. Definition of i: We shall write a real number a in the complex form $(a, 0)$.

A complex number $(0, b)$ is called a **Purely Imaginary Number.**

In particular, $1 = (1, 0)$ *is a real number* and $(0, 1)$ is a purely imaginary number denoted by i. The complex number $(0, 0)$, denoted by $\underset{\sim}{0}$ is the zero of the *complex number, system.*

By the rule of multiplication,

$$\begin{aligned} i^2 &= i \cdot i = (0, 1) \cdot (0, 1) = (-1, 0) = -1; \\ i^3 &= i^2 \cdot i = -i; \\ i^4 &= i^2 \cdot i^2 = 1, \text{ etc.} \end{aligned}$$

A purely imaginary number $(0, b) = (0, 1)(b, 0) = ib$

$$\therefore (a, b) = (a, 0) + (0, b) = a + ib.$$

Henceforth, in place of (a, b) **we can write** $a + ib$ **and add or multiply two complex numbers by the ordinary rule of Algebra—only point to remember "Replace** i^2 **by** -1", whenever it occurs.

▶ Example 1.1.1
$$\begin{aligned} (3 + 2i)(6 - 8i) &= 3.6 + 3 \cdot (-8i) + (2i) \cdot 6 + (2i) \cdot (-8i) \\ &= 18 - 24i + 12i - 16i^2 \\ &= 18 - 12i + 16 \; (\because i^2 = -1) \\ &= 34 - 12i. \end{aligned}$$

▶ Example 1.1.2
$$\begin{aligned} \frac{i^5 + i^9 + i^{14}}{i^6 + i^{11} + i^{13}} &= \frac{i^4 \cdot i + i^8 \cdot i + i^{14}}{i^6 + i^{10} \cdot i + i^{12} \cdot i} \\ &= \frac{(i^2)^2 \cdot i + (i^2)^4 \cdot i + (i^2)^7}{(i^2)^3 + (i^2)^5 \cdot i + (i^2)^6 \cdot i} \\ &= \frac{i + i + (-1)}{-1 + (-i) + i} \\ &= \frac{-1 + 2i}{-1} = 1 - 2i. \end{aligned}$$

In the following discussions we shall take a, b, c, d, m—all real numbers:

5. Multiply a complex number by a real number:

$$m(a, b) = m(a + ib) = ma + imb = (ma, mb).$$

6. Subtraction:

$$\begin{aligned}(a, b) - (c, d) &= (a, b) + (-1)(c, d)\\ &= (a, b) + (-c, -d)\\ &= \{(a - c), (b - d)\},\\ \text{i.e., } (a + ib) - (c + id) &= (a - c) + i(b - d).\end{aligned}$$

7. Division: We can divide a complex number (a, b) by another complex number (c, d), provided $(c, d) \neq (0, 0)$.

$$\begin{aligned}\frac{(a, b)}{(c, d)} &= \frac{a + ib}{c + id}(c \neq 0, d \neq 0)\\ &= \frac{(a + ib)(c - id)}{(c + id)(c - id)} = \frac{(ac + bd)}{c^2 + d^2} + \frac{i(bc - ad)}{c^2 + d^2}\\ &= \left(\frac{ac + bd}{c^2 + d^2}, \frac{bc - ad}{c^2 + d^2}\right).\end{aligned}$$

II. Terms Associated with a Complex Number.

1. Conjugate of a complex number: If $z = (a, b) = a + ib$ be a complex number, then the complex number $(a, -b) = a - ib$ is called the *conjugate of* z, denoted by $\bar{z}$. It is easy to verify:

(a) $(\bar{\bar{z}}) = z$ (z is any complex number).

(b) $\overline{z_1 + z_2} = \bar{z}_1 + \bar{z}_2$ (z_1, z_2 are any two complex numbers).

(c) $(\overline{z_1 \cdot z_2}) = \bar{z}_1 \cdot \bar{z}_2$.

(d) $\left(\overline{\dfrac{z_1}{z_2}}\right) = \dfrac{\bar{z}_1}{\bar{z}_2}$, where z_2 is a non-zero complex number.

(e) $z + \bar{z} = 2\,\mathrm{Rl}\, z$ and hence $(z + \bar{z})^2 \geq 0$.

(f) $z - \bar{z} = 2\,\mathrm{Im}\, z$ and hence $(z - \bar{z})^2 \leq 0$.

See that when we divided $a+ib$ by $c+id$, we have multiplied both Numerator and Denominator by the conjugate of the Denominator, e.g.,

$$\frac{3 + 5i}{2 + 3i} = \frac{(3 + 5i)(2 - 3i)}{(2 + 3i)(2 - 3i)} = \frac{21 + i}{13} = \frac{21}{13} + i\frac{1}{13} = \left(\frac{21}{13}, \frac{1}{13}\right).$$

Note 1. $(a + ib)(a - ib) =$ *a* ***positive real number*** $a^2 + b^2$ *unless both* $a = 0, b = 0$.

2. Modulus of a complex number: We shall define the Modulus of a complex number $z = a + ib$ by the positive square root of $a^2 + b^2$ and denote it by $|z|$, thus

$$|z| = |a + ib| = +\sqrt{a^2 + b^2}.$$

It is useful to remember:

$$\boxed{z\bar{z} = |z|^2}$$

With regard to **modulus the following laws hold** (they can be easily verified).

(a) $|z_1 \cdot z_2| = |z_1||z_2|$.

(b) $\left|\dfrac{z_1}{z_2}\right| = \dfrac{|z_1|}{|z_2|}$, where z_2 is a non-zero complex number.

(c) $|z| = |\bar{z}|$.

(d) A Result which is most frequently used:

$$|z_1 + z_2| \leq |z_1| + |z_2| \text{ (known as } \textbf{Triangle Inequality}\text{)}$$

Proof. We observe that

$$\begin{aligned}
|z_1 + z_2|^2 &= (z_1 + z_2) \cdot (\overline{z_1 + z_2}) = (z_1 + z_2) \cdot (\bar{z}_1 + \bar{z}_2) \\
&= z_1 \cdot \bar{z}_1 + z_1 \cdot \bar{z}_2 + z_2 \cdot \bar{z}_1 + z_2 \cdot \bar{z}_2 \\
&= |z_1|^2 + z_1\bar{z}_2 + z_2\bar{z}_1 + |z_2|^2.
\end{aligned}$$

Also we see that $(\overline{z_1 \cdot \bar{z}_2}) = \bar{z}_1 \cdot (\bar{\bar{z}}_2) = \bar{z}_1 \cdot z_2$, i.e., $z_1 \cdot \bar{z}_2$ and $z_2 \cdot \bar{z}_1$ are two conjugate complex numbers. Clearly their difference is a purely imaginary number or zero, i.e., $(z_1\bar{z}_2 - z_2\bar{z}_1)^2 \leq 0$.

Now

$$\begin{aligned}
(z_1\bar{z}_2 + z_2\bar{z}_1)^2 &= (z_1\bar{z}_2 - z_2 \cdot \bar{z}_1)^2 + 4z_1\bar{z}_2 \cdot z_2\bar{z}_1 \\
&= (z_1\bar{z}_2 - z_2 \cdot \bar{z}_1)^2 + 4|z_1|^2|z_2|^2 \\
&\leq 4|z_1|^2|z_2|^2 \text{ [since } (z_1\bar{z}_2 - z_i \cdot \bar{z}_1)^2 \leq 0].
\end{aligned}$$

Hence

$$\begin{aligned}
|z_1 + z_2|^2 &= |z_1|^2 + |z_2|^2 + z_1\bar{z}_2 + z_2 \cdot \bar{z}_1 \\
&\leq |z_1|^2 + |z_2|^2 + 2|z_1||z_2|, \\
\text{i.e.,} \quad &\leq (|z_1| + |z_2|)^2 \\
\text{or,} \quad |z_1 + z_2 &\leq |z_1| + |z_2|.
\end{aligned}$$

(e) Extension of the Rule (d) above:

$$|z_1 + z_2 + \cdots + z_n| \leq |z_1| + |z_2| + \cdots + |z_n|$$

(f) $-|z| \leq \text{Rl} \cdot z \leq |z|$.

$((\text{Rl} \cdot z)^2 = a^2 < a^2 + b^2 < |z|^2$. Now use the fact

$$a^2 < b^2 \Rightarrow -|b| < a < |b|).$$

III. Complex Plane: Argand Diagram: Modulus—Amplitude Form of a Complex Number

Geometrically, a real number is represented by a **point on a line.** But a complex number $a + ib$ or (a, b) is represented by a **point on a plane.** We take a plane and through a given point O on the plane we draw two mutually perpendicular lines X′OX and Y′OY. A point, whose co-ordinates with reference to these two lines as axes are (a, b), represents the complex number (a, b) or $a + ib$. With each complex number $z = a + ib$ we associate a point P on the plane. Such a plane is called a **complex plane** or a **Gaussian Plane.** A diagram on this plane is known as **Argand Diagram.** [See **Fig. 1.1**]

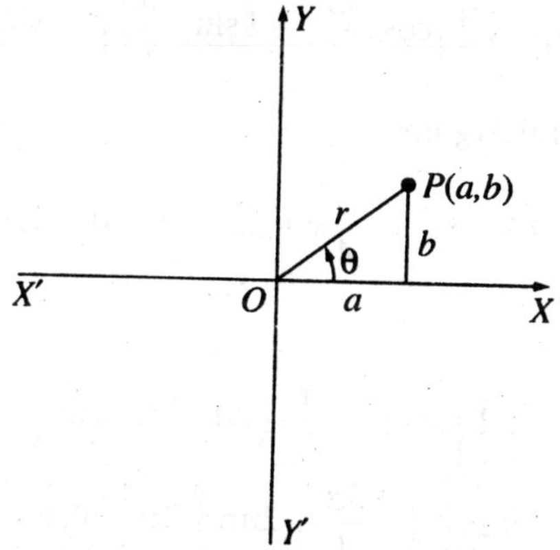

Fig. 1.1: Complex plane. Point P represents $z = a + ib$.

Let (r, θ) be the polar co-ordinates of P with reference to O as pole and OX as the initial line. See **Fig. 1.1.**

$$OP = r = \sqrt{a^2 + b^2} = \text{modulus of } z = a + ib.$$

θ = the angle which OP makes with the positive direction of X′OX (also called **Real Axis**).

= **Amplitude of z** (or **Argument of z**)

Clearly, $a = r\cos\theta$ and $b = r\sin\theta (r > 0)$
$\therefore$ we write $z = a + ib = r(\cos\theta + i\sin\theta)$.

This form of writing a complex number is known as the **Modulus-Amplitude Form** of writing a complex number $a + ib$.

We shall take θ so that $-\pi < \theta \leq \pi$. Then θ is called the *Principal value of the argument.*

We write θ = **arg z** or **amp z.** The angle θ which OP makes with the positive direction of Real axis X′OX may be taken as

$$\mathbf{arg\, z + 2n\pi\ (n = 0, \pm 1, \pm 2, \ldots).}$$

In this case we write Arg z = **arg z** + $2n\pi$ (where $-\pi < \arg z \leq \pi$).

A short form of writing
$\cos\theta + i\sin\theta$ is **cis θ**;
$\cos\theta - i\sin\theta = \cos(-\theta) + i\sin(-\theta)$ = **cis** $(-\theta)$.

▶ **Example 1.1.3** *Write the complex number* $-1-i=(-1,-1)$ *in Modulus-Amplitude form.*

Solution: Put -1 = real part $= r\cos\theta$, -1 = Imaginary part $= r\sin\theta$.

Then $r=+\sqrt{2}$ = modulus of the complex number $-1-i$.

We obtain θ, which satisfies $\cos\theta=-1/\sqrt{2}$, $\sin\theta=-1/\sqrt{2}$ and $-\pi<\theta\le\pi$.

Since both $\cos\theta$ and $\sin\theta$ are negative, θ must be in the third quadrant. But for the principal argument, $-\pi<\theta\le\pi$, the angle θ should be taken as $-\frac{3\pi}{4}=\arg(-1-i)$ (Not $\pi+\pi/4$)

$$\therefore\quad -1-i=\sqrt{2}\left[\cos\frac{-3\pi}{4}+i\sin\frac{-3\pi}{4}\right]=\sqrt{2}\,\text{cis}\left(\frac{-3\pi}{4}\right)$$

Note that the General Argument

$$\textbf{Arg}\,(-1-i)=-\frac{3}{4}\pi+2n\pi\ (n=0,\pm1,\pm2,\ldots)$$

Thus we may write:

$$\begin{aligned}-1-i &= \sqrt{2}\left[\cos\left(\frac{-3\pi}{4}+2n\pi\right)+i\sin\left(\frac{-3\pi}{4}+2n\pi\right)\right]\\ &= \sqrt{2}\,\text{cis}\left(-\frac{3\pi}{4}+2n\pi\right)\quad (n=0,\pm1,\pm2,\ldots)\end{aligned}$$

Important Results to Remember:

It is easy to check **(Principal arguments are taken)**

$$\begin{aligned}1 &= 1+i\cdot 0=1\{\cos 0^\circ+i\sin 0^\circ\}=\text{cis}\,0^\circ\\ -1 &= -1+i\cdot 0=1\{\cos\pi+i\sin\pi\}=\text{cis}\,\pi\\ i &= 0+i\cdot 1=1\left\{\cos\frac{\pi}{2}+i\sin\frac{\pi}{2}\right\}=\text{cis}\,\frac{\pi}{2}\\ -i &= 0+i(-1)=1\left\{\cos\left(-\frac{\pi}{2}\right)+i\sin\left(-\frac{\pi}{2}\right)\right\}=\text{cis}\left(-\frac{\pi}{2}\right)\\ 1+i &= 1+i\cdot(1)=\sqrt{2}\left\{\cos\frac{\pi}{4}+i\sin\left(\frac{\pi}{4}\right)\right\}=\sqrt{2}\,\text{cis}\,\frac{\pi}{4}\\ 1-i &= 1+i(-1)=\sqrt{2}\left\{\cos\left(-\frac{\pi}{4}\right)+i\sin\left(-\frac{\pi}{4}\right)\right\}=\sqrt{2}\,\text{cis}\left(-\frac{\pi}{4}\right)\\ -1+i &= -1+i(1)=\sqrt{2}\left\{\cos\left(\frac{3\pi}{4}\right)+i\sin\left(\frac{3\pi}{4}\right)\right\}=\sqrt{2}\,\text{cis}\left(\frac{3\pi}{4}\right)\\ -1-i &= -1+i(-1)=\sqrt{2}\left\{\cos\left(-\frac{3\pi}{4}\right)+i\sin\left(-\frac{3\pi}{4}\right)\right\}=\sqrt{2}\,\text{cis}\left(-\frac{3\pi}{4}\right)\end{aligned}$$

Regarding Moduli of complex numbers it is true:

Modulus of the product of two complex numbers = Product of their moduli.

Modulus of the quotient of two complex numbers = Quotient of their moduli.

Thus in case of Product, we obtain

$$\begin{aligned} & r_1(\cos\theta_1 + i\sin\theta_1)\cdot r_2(\cos\theta_2 + i\sin\theta_2) \\ = \; & r_1 r_2\{\cos(\theta_1+\theta_2) + i\sin(\theta_1+\theta_2)\}. \end{aligned}$$

Again in case of Quotient, we obtain

$$\frac{r_1(\cos\theta_1 + i\sin\theta_1)}{r_2(\cos\theta_2 + i\sin\theta_2)} = \frac{r_1}{r_2}\{\cos(\theta_1-\theta_2) + i\sin(\theta_1-\theta_2)\}$$

But with regard to the arguments it is to be noted:

Argument of the Product of two complex numbers = Sum of their Arguments.

Argument of the Quotient of two complex numbers = Difference of their Arguments.

(These results are true only if General Arguments are considered),

$$\text{i.e.,}\quad \begin{aligned} \operatorname{Arg}(z_1 z_2) &= \operatorname{Arg} z_1 + \operatorname{Arg} z_2 \\ \operatorname{Arg}(z_1/z_2) &= \operatorname{Arg} z_1 - \operatorname{Arg} z_2 \quad (z_2 \neq 0). \end{aligned}$$

These results are **not always true** if the Principal Arguments of the complex numbers are considered. See the following example:

► Example 1.1.4 *Let* $z_1 = 1 - i$, $z_2 = i$ *so that* $z_1 z_2 = (1-i)i = 1 + i$.

$$\textit{Principal argument of } z_1 = -\frac{\pi}{4};$$

$$\textit{Principal argument of } z_2 = +\frac{\pi}{2};$$

$$\textit{Principal argument of } z_1\cdot z_2 = +\frac{\pi}{4}.$$

So, it is true here that $\arg z_1\cdot z_2 = \arg z_1 + \arg z_2 \left(\frac{\pi}{4} = -\frac{\pi}{4} + \frac{\pi}{2}\right)$.

Again, let $z_2 = i$ and $z_3 = -1 + i$ so that $z_2\cdot z_3 = i(-1+i) = -1 - i$.

Principal arguments of z_2, z_3 and $z_2\cdot z_3$ are respectively $\frac{\pi}{2}, \frac{3\pi}{4}; -\frac{3\pi}{4}$.

Here, $\arg z_2 + \arg z_3 \neq \arg z_2 z_3$

$$\left(\because \frac{\pi}{2} + \frac{3\pi}{4} \neq -\frac{3\pi}{4}\right)$$

Correct Interpretation of the statement: $\operatorname{Arg}(z_1\cdot z_2) = \operatorname{Arg} z_1 + \operatorname{Arg} z_2$.

Some value of $\operatorname{Arg} z_1$ **+ some value of** $\operatorname{Arg} z_2$ **= One of the values of** $\operatorname{Arg}(z_1\cdot z_2)$

One should remember: General argument of z has many values = Principal value of the argument of $z + 2n\pi$ $(n = 0, \pm 1, \pm 2, \ldots)$

▶ Example 1.1.5 *If the ratio* $\frac{z-i}{z-1}$ *be purely imaginary, where z is any complex number, then show that z lies on a circle whose centre is the point* $\frac{1}{2}+i\frac{1}{2}$ *and radius is* $\frac{1}{\sqrt{2}}$. **[C.P. 2001]**

Solution: Take $z = x + iy$. Then,

$$\begin{aligned}\frac{z-i}{z-1} &= \frac{x+i(y-1)}{(x-1)+iy} = \frac{\{x+i(y-1)\}\{(x-1)-iy\}}{(x-1)^2+y^2}\\ &= \frac{x(x-1)+y(y-1)}{(x-1)^2+y^2} + \frac{i\{(x-1)(y-1)-y(x)\}}{(x-1)^2+y^2}.\end{aligned}$$

Since, by the given condition, $(z-i)(z-1)$ is purely imaginary, so we write $x(x-1)+y(y-1)=0$

$$\text{or} \quad x^2+y^2-x-y=0 \quad \text{or} \quad \left(x-\frac{1}{2}\right)^2+\left(y-\frac{1}{2}\right)^2=\left(\frac{1}{\sqrt{2}}\right)^2.$$

Hence z his on a circle where centre is $\frac{1}{2}+i\frac{1}{2}$ and radius $=\frac{1}{\sqrt{2}}$.

▶ Example 1.1.6 *Prove that* $\arg z_1 \sim \arg z_2$ *is an even multiple of* π, *if*

$$|z_1+z_2| = |z_1|+|z_2|.$$

Solution: Let

$$\begin{aligned} z_1 &= r_1(\cos\theta_1 + i\sin\theta_1);\\ z_2 &= r_2(\cos\theta_2 + i\sin\theta_2).\end{aligned}$$

Then

$$z_1+z_2 = (r_1\cos\theta_1 + r_2\cos\theta_2) + i(r_1\sin\theta_1 + r_2\sin\theta_2).$$

Given

$$|z_1+z_2| = |z_1|+|z_2|,$$

which gives

$$|z_1+z_2|^2 = |z_1|^2+|z_2|^2+2|z_1||z_2|,$$

i.e., $(r_1\cos\theta_1 + r_2\cos\theta_2)^2 + (r_1\sin\theta_1 + r_2\sin\theta_2)^2 = r_1^2+r_2^2+2r_1r_2$

$\Rightarrow$ $\cos(\theta_1 \sim \theta_2) = 1$ and hence $\theta_1 \sim \theta_2 = 2k\pi$, or, $\arg z_1 \sim \arg z_2$ $= 2k\pi =$ an even multiple of π]

▶ Example 1.1.7 *Prove that* $\arg z - \arg(-z) = \pm\pi$, *according as* $\arg z$ *is positive or negative.*

Solution: Let

$$z = r(\cos\theta + i\sin\theta);$$

mod $z = r$, principal amp. $z = \arg z = \theta$, where $-\pi < \theta \leq \pi$.

Then

$$-z = -r(\cos\theta + i\sin\theta) = r[\cos(\theta+\pi) + i\sin(\theta+\pi)],$$

i.e., argument of $-z$ is $\theta+\pi$ (not necessarily the principal value of the argument.

Case 1. *Suppose* $\arg z$ *is positive, i.e.,* $0 < \theta \leq \pi$.

Then $\pi < \theta + \pi \leq 2\pi$, i.e., $\theta + \pi$ is not the P.V. of $\arg \cdot (-z)$.
$\text{Arg}(-z) = \theta + \pi + 2k\pi$ and P.V. should be between $-\pi$ and 0. This is possible if $k = -\pi$ and P.V. of $\text{Arg}(-z) = \theta - \pi$.
In that case $\arg z - \arg(-z) = \theta - (\theta - \pi) = +\pi$

Case 2. *Suppose* $\arg z$ *is negative, i.e.,* $-\pi < \theta < 0$, *then* $0 < \theta + \pi < \pi$.

$$\therefore \text{P.V. of } \text{Arg}(-z) = \theta + \pi.$$

Hence $\arg z - \arg(-z) = \theta - (\theta + \pi) = -\pi$ hence etc.]

▶ **Example 1.1.8** *If* z_1 *and* z_2 *be a pair of conjugate complex numbers and* z_3, z_4 *be another pair of conjugate complex numbers, then prove that*

$$amp.\left(\frac{z_1}{z_4}\right) = amp.\left(\frac{z_3}{z_2}\right)$$ [C.P. 2003]

Solution: Let

$$\begin{aligned}
z_1 &= r_1(\cos\theta_1 + i\sin\theta_1),\ z_2 = r_1(\cos\theta_1 - i\sin\theta_1) \\
z_3 &= r_2(\cos\theta_2 + i\sin\theta_2),\ z_4 = r_2(\cos\theta_2 - i\sin\theta_2) \\
\frac{z_1}{z_4} &= \frac{r_1}{r_2}\frac{\cos\theta_1 + i\sin\theta_1}{\cos\theta_2 - i\sin\theta_2} = \frac{r_1}{r_2}\frac{(\cos\theta_1 + i\sin\theta_1)(\cos\theta_2 + i\sin\theta_2)}{(\cos\theta_2 - i\sin\theta_2)(\cos\theta_2 + i\sin\theta_2)} \\
&= \frac{r_1}{r_2}(\cos\theta_1 + i\sin\theta_1)(\cos\theta_2 + i\sin\theta_2)\ (\because \text{Denr.} = \cos^2\theta_2 + \sin^2\theta_2 = 1) \\
&= \frac{r_1}{r_2}\{\cos\theta_1\cos\theta_2 - \sin\theta_1\sin\theta_2 + i(\sin\theta_1\cos\theta_2 + \cos\theta_1\sin\theta_2)\} \\
&= \frac{r_1}{r_2}\{\cos(\theta_1 + \theta_2) + i\sin(\theta_1 + \theta_2)\};\ \text{i.e., amp.}\left(\frac{z_1}{z_4}\right) = \theta_1 + \theta_2.
\end{aligned}$$

Similarly obtain amp. $\left(\frac{z_3}{z_2}\right) = \theta_1 + \theta_2$; hence etc.

▶ **Example 1.1.9** *For two complex numbers* z_1 *and* z_2, *prove that*

$$|z_1 + z_2|^2 + |z_1 - z_2|^2 = 2(|z_1|^2 + |z_2|^2).$$

Solution:

$$\begin{aligned}
|z_1 + z_2|^2 &= (z_1 + z_2)\cdot(\overline{z_1 + z_2}) = (z_1 + z_2)\cdot(\bar{z}_1 + \bar{z}_2) \\
&= z_1\bar{z}_1 + z_1\cdot\bar{z}_2 + z_2\cdot\bar{z}_1 + z_2\cdot\bar{z}_2 \\
&= |z_1|^2 + |z_2|^2 + z_1\bar{z}_2 + z_2\bar{z}_1.
\end{aligned}$$

Similarly

$$|z_1 - z_2|^2 = |z_1|^2 + |z_2|^2 - z_1\cdot\bar{z}_2 - z_2\cdot\bar{z}_1.$$

On addition, the result follows.

1.2 De Moivre's Theorem: Statement

In its most general form, De Moivre's Theorem states:

For all real values of n, $\cos n\theta + i \sin n\theta$ is one of the values of

$$(\cos \theta + i \sin \theta)^n.$$

More explicitly,

(i) if n is a positive integer, $(\cos\theta + i\sin\theta)^n$ **has only one value, namely**

$$\cos n\theta + i \sin n\theta.$$

(ii) if n is a negative integer, $(\cos\theta + i\sin\theta)^n$ has also only one value, namely

$$\cos n\theta + i \sin n\theta;$$

(iii) if n is a positive fraction of the form $\frac{p}{q}$ in its lowest terms, p and q being positive integers $(q \neq 0)$, then $(\cos\theta + i\sin\theta)^{p/q}$ has q values of which one value is

$$\left(\cos \frac{p}{q}\theta + i \sin \frac{p}{q}\theta\right);$$

(iv) if n is a negative fraction $-\frac{p}{q}$ in its lowest terms, p and q being positive integers $(q \neq 0)$, then $(\cos\theta + i\sin\theta)^{-p/q}$ has q values of which one value is

$$\cos\left(-\frac{p}{q}\theta\right) + i \sin\left(-\frac{p}{q}\theta\right), \text{ or, } \cos\frac{p\theta}{q} - i \sin\frac{p\theta}{q};$$

(v) if n be any **irrational number**, then $(\cos\theta + i\sin\theta)^n$ has infinite number of values, of which one value is $\cos n\theta + i\sin n\theta$.

We shall prove all the cases except the last case where n is an irrational number. In all the above cases, **θ is a real number.**

1.3 Proof of De Moivre's Theorem for integral indices

We shall prove the theorem in two steps:

Step 1. *De Moivre's Theorem for Positive Integral Indices.*

For all positive integers n and for any real θ, $(\cos\theta + i\sin\theta)^n$ has only one value, namely $\cos n\theta + i \sin n\theta$.

Proof. (i) For $n = 1$, it is easy to see that

$$(\cos\theta + i\sin\theta)^1 = \cos\theta + i\sin\theta,$$

i.e., the theorem is true for $n = 1$.

Even for $n = 2$, we see that

$$\begin{aligned}(\cos\theta + i\sin\theta)^2 &= (\cos\theta + i\sin\theta)(\cos\theta + i\sin\theta)\\ &= (\cos^2\theta - \sin^2\theta) + i\cdot 2\sin\theta\cos\theta\\ &= \cos 2\theta + i\sin 2\theta,\end{aligned}$$

i.e., the theorem is true for $n = 2$.

(ii) Let us assume that the theorem is true for a certain positive integer k, i.e., let us assume that

$$(\cos\theta + i\sin\theta)^k = \cos k\theta + i\sin k\theta.$$

Multiply both sides by $(\cos\theta + i\sin\theta)$. Then

$$\begin{aligned}\text{L.H.S.} &= (\cos\theta + i\sin\theta)^{k+1}\\ \text{and R.H.S.} &= (\cos k\theta + i\sin k\theta)(\cos\theta + i\sin\theta)\\ &= (\cos k\theta\cos\theta - \sin k\theta\sin\theta) + i(\sin k\theta\cos\theta + \cos k\theta\sin\theta)\\ &= \cos(k+1)\theta + i\sin(k+1)\theta,\end{aligned}$$

i.e., the theorem is true for $n = k + 1$, if we assume it to be true for $n = k$.

(iii) Now in **(i)** we have observed that the theorem is true for $n = 1$ or 2.

Hence, by the principle of mathematical induction, the theorem is TRUE for all positive integers n.

Step 2. *De Moivre's Theorem for Negative Integral Indices.*

For all negative integers n and for any real θ, $(\cos\theta + i\sin\theta)^n$ has only one value, namely, $\cos n\theta + i\sin n\theta$.

Proof. Let n be a negative integer $= -m$ (say), where m is a positive integer. Then

$$\begin{aligned}(\cos\theta + i\sin\theta)^n &= (\cos\theta + i\sin\theta)^{-m} = \frac{1}{(\cos\theta + i\sin\theta)^m}\\ &= \frac{1}{(\cos\theta + i\sin\theta)^m}\quad\left[\begin{array}{l}\text{assuming Step 1, m being}\\ \text{a positive integer}\end{array}\right]\\ &= \frac{\cos m\theta - i\sin m\theta}{(\cos m\theta + i\sin m\theta)(\cos m\theta - i\sin m\theta)}\\ &= \frac{\cos m\theta - i\sin m\theta}{\cos^2 m\theta + \sin^2 m\theta} = \cos m\theta - i\sin m\theta\\ &= \cos(-n\theta) - i\sin(-m\theta)\ (\because\ m = -n)\\ &= \cos n\theta + i\sin n\theta.\end{aligned}$$

This completes the proof of De Moivre's theorem for integral indices.

1.4 Alternative Method of Proving De Moivre's theorem for Positive Integral Indices

Step 1. *We prove, by the method of induction*

$$(\cos\theta_1 + i\sin\theta_1)(\cos\theta_2 + i\sin\theta)\cdots \textit{ to n factors}$$
$$= \cos(\theta_1 + \theta_2 + \cdots + \theta_n) + i\sin(\theta_1 + \theta_2 + \cdots + \theta_n). \qquad (1.4.1)$$

(i) Since

$$\begin{aligned}(\cos\theta_1 + i\sin\theta_1)(\cos\theta_2 + i\sin\theta_2)\\ = \cos\theta_1\cos\theta_2 - \sin\theta_1\sin\theta_2 + i(\sin\theta_1\cos\theta_2 + \cos\theta_1\sin\theta_2)\\ = \cos(\theta_1+\theta_2) + i\sin(\theta_1+\theta_2),\end{aligned}$$

i.e., the result (1.4.1) is True for $n = 2$.

(ii) Suppose that the result (1.4.1) is true for a certain positive integer k (say), i.e., we assume that

$$\begin{aligned}(\cos\theta_1 + i\sin\theta_1)(\cos\theta_2 + i\sin\theta_2)\cdots \text{ to } k \text{ factors}\\ = \cos(\theta_1+\theta_2+\cdots+\cdots+\theta_k) + i\sin(\theta_1+\theta_2+\cdots+\theta_k).\end{aligned}$$

We multiply both sides by $\cos\theta_{k+1} + i\sin\theta_{k+1}$. Then

$$\begin{aligned}\text{L.H.S.} &= (\cos\theta_1 + i\sin\theta_1)(\cos\theta_2 + i\sin\theta_2)\cdots \text{ to } k+1 \text{ factors}\\ \text{and R.H.S.} &= \{\cos(\theta_1+\theta_2+\cdots+\theta_k) + i\sin(\theta_1+\theta_2+\cdots+\theta_k)\}\times\\ &\qquad (\cos\theta_{k+1} + i\sin\theta_{k+1})\\ &= \cos(\theta_1+\theta_2+\cdots+\theta_k)\cos\theta_{k+1}\\ &\quad -\sin(\theta_1+\theta_2+\cdots+\theta_k)\sin\theta_{k+1}\\ &\quad +i\{\sin(\theta_1+\theta_2+\cdots+\theta_k)\cos\theta_{k+1}\\ &\quad +\cos(\theta_1+\theta_2+\cdots+\theta_k)\sin\theta_{k+1}\}\\ &= \cos(\theta_1+\theta_2+\cdots+\theta_k+\theta_{k+1})\\ &\quad +i\sin(\theta_1+\theta_2+\cdots+\theta_k+\theta_{k+1}),\end{aligned}$$

i.e., the result (1.4.1) is true for $n = k+1$ if we assume it to be true for $n = k$.

But by Step (i) above, it is true for $n = 2$; hence it must be true for $n = 3$ and so on, i.e., by induction the result (1.4.1) is true for all positive integers n.

Deduction. We have proved that

$$\begin{aligned}(\cos\theta_1 + i\sin\theta_1)(\cos\theta_2 + i\sin\theta_2)\cdots \text{ to } n \text{ factors}\\ = \cos(\theta_1+\theta_2+\cdots+\theta_n) + i\sin(\theta_1+\theta_2+\cdots+\theta_n).\end{aligned}$$

Put $\theta_1 = \theta_2 = \theta_3 = \cdots = \theta_n = \theta$ (say). Then we get

$$(\cos\theta + i\sin\theta)^n = \cos n\theta + i\sin n\theta$$

which is **De Moivre's theorem for positive integral index.**

1.5 Proof of De Moivre's Theorem for Rational Indices

Step 1. *Let n be any positive fraction $\frac{p}{q}$ in its lowest terms, p and q being positive integers ($q \neq 0$) and let θ be any real numbers. We shall prove that*

$$\left(\cos\frac{p}{q}\theta + i\sin\frac{p}{q}\theta\right)$$

is one of the values of $(\cos\theta + i\sin\theta)^{p/q}$.

Proof. Now, by De Moivre's theorem for a positive integral index q,

$$\begin{aligned}\left(\cos\frac{p}{q}\theta + i\sin\frac{p}{q}\theta\right)^q &= \cos\left(\frac{p}{q}\theta\cdot q\right) + i\sin\left(\frac{p}{q}\theta\cdot q\right)\\ &= \cos p\theta + i\sin p\theta\end{aligned}$$

$\therefore$ $\cos\frac{p}{q}\theta + i\sin\frac{p}{q}\theta$ **is one of the qth roots of** $\cos p\theta + i\sin p\theta$.

[**Our definition.** If $z^q = a^p$ (z, a are complex numbers), then we say, z is one of the qth roots of a^p.]

or, $\cos\frac{p}{q}\theta + i\sin\frac{p}{q}\theta$ **is one of the qth roots of** $(\cos\theta + i\sin\theta)^p$

Thus for a positive fractional index $\frac{p}{q}$,

$\cos\frac{p}{q}\theta + i\sin\frac{p}{q}\theta$ **is one of the values of** $(\cos\theta + i\sin\theta)^{p/q}$.

Step 2. *Let* $n = -\frac{p}{q}$ *is its lowest terms,* p, q *being positive integers* $(q \neq 0)$.

$$\left\{\cos\left(-\frac{p}{q}\theta\right) + i\sin\left(-\frac{p}{q}\theta\right)\right\}^q = \cos(-p\theta) + i\sin(-p\theta)$$

(*by De Moivre's theorem for positive integral index* q).

$\therefore$ $\cos\left(-\frac{p\theta}{q}\right) + i\sin\left(-\frac{p\theta}{q}\right)$ **is one of the qth roots of** $\cos(-p\theta) + i\sin(-p\theta)$

or, $\cos\left(-\frac{p\theta}{q}\right) + i\sin\left(-\frac{p\theta}{q}\right)$ **is one of the qth roots of** $(\cos\theta + i\sin\theta)^{-p}$.

$\therefore$ $\cos\left(-\frac{p\theta}{q}\right) + i\sin\left(-\frac{p\theta}{q}\right)$ **is one of the values of** $(\cos\theta + i\sin\theta)^{-\frac{p}{q}}$.

We have thus proved that for all rational values of n and for any real θ,

$$\cos n\theta + i\sin n\theta$$

is one of the values of $(\cos\theta + i\sin\theta)^n$.

The result is true for all irrational values of n—the proof is beyond the scope of this Text.

1.6 All Possible Values of $(\cos\theta + i\sin\theta)^{p/q}$

If p *is any integer (positive or negative) and* q *is a positive integer* (p *and* q *are prime to each other*), *then there are* q *values of*

$$(\cos\theta + i\sin\theta)^{p/q}$$

and these values are

$$\cos\left(\frac{p\theta}{q} + \frac{2r\pi}{q}\right) + i\sin\left(\frac{p\theta}{q} + \frac{2r\pi}{q}\right) \quad (r = 0, 1, 2, \ldots, q-1)$$

The value $\cos\frac{p\theta}{q}+i\sin\frac{p\theta}{q}$ *(corresponding to* $r=0$*) is called the principal value of* $(\cos\theta+i\sin\theta)^{p/q}$.

We have

$$\left\{\cos\left(\frac{p\theta}{q}+\frac{2r\pi}{q}\right)+i\sin\left(\frac{p\theta}{q}+\frac{2r\pi}{q}\right)\right\}^q$$

$$=\cos\left(\frac{p\theta+2r\pi}{q}\times q\right)+i\sin\left(\frac{p\theta+2r\pi}{q}\times q\right)$$

(using De Moivre's theorem for the positive integral index q)

$$=\cos(p\theta+2r\pi)+i\sin(p\theta+2r\pi)$$

$$=\cos p\theta+i\sin p\theta,\text{ when } r \text{ is any integer}$$

$$=(\cos\theta+i\sin\theta)^p \text{ (using De Moivre's theorem for any integral index)}$$

$$\therefore\ \cos\left(\frac{p\theta}{q}+\frac{2r\pi}{q}\right)+i\sin\left(\frac{p\theta}{q}+\frac{2r\pi}{q}\right)$$

is **one** of the qth roots of $(\cos\theta+i\sin\theta)^p$, i.e., **one** of the values of

$$(\cos\theta+i\sin\theta)^{p/q},$$

where r is any positive or negative integer.

But the angles $\frac{p\theta}{q}+\frac{2r\pi}{q}$ where r is given the values $0, 1, 2, 3, \ldots, (q-1)$ are all different and no two have at the same time equal cosines and equal sines.

$$\therefore\ \cos\left(\frac{p\theta}{q}+\frac{2r\pi}{q}\right)+i\sin\left(\frac{p\theta}{q}+\frac{2r\pi}{q}\right)$$

has q distinct values for these q integers.

Also by putting r equal to any other integer, this expression repeats one or other of these q values.

$\therefore$ $(\cos\theta+i\sin\theta)^{p/q}$ has q distinct values

$$\cos\left(\frac{p\theta}{q}+\frac{2r\pi}{q}\right)+i\sin\left(\frac{p\theta}{q}+\frac{2r\pi}{q}\right)\ [r=0,1,2,3,\ldots,(q-1).]$$

Q.1. What are the qth roots of $(\cos\theta+i\sin\theta)$?

Ans. The qth roots of $(\cos\theta+i\sin\theta)$, i.e., the q values of $\sqrt[q]{\cos\theta+i\sin\theta}$ are

$$\frac{\cos(\theta+2r\pi)}{q}+i\sin\frac{(\theta+2r\pi)}{q}\text{ where } r=0,1,2,\ldots,q-1.$$

The **principal value** is $\cos\frac{\theta}{q}+i\sin\frac{\theta}{q}$ (when $r=0$).

Q.2. What can you say about $(\sin\theta+i\cos\theta)^n$? [C.P. 1993]

Ans.
$$(\sin\theta+i\cos\theta)^n=\left\{\cos\left(\frac{\pi}{2}-\theta\right)+i\sin\left(\frac{\pi}{2}-\theta\right)\right\}^n$$

$$=\cos n\left(\frac{\pi}{2}-\theta\right)+i\sin n\left(\frac{\pi}{2}-\theta\right)$$

$$=\cos\left(\frac{n\pi}{2}-n\theta\right)+i\sin\left(\frac{n\pi}{2}-n\theta\right).$$

1.7 Solved Problems

► Example 1.7.1 *Express* $\dfrac{(\cos\theta + i\sin\theta)^6}{(\sin\theta + i\cos\theta)^5}$ *in the form* $a + ib$. **[C.P. 1997]**

Solution: We write

$$\begin{aligned}
\frac{(\cos\theta + i\sin\theta)^6}{(\sin\theta + i\cos\theta)^5} &= (\cos\theta + i\sin\theta)^6 \cdot (\sin\theta + i\cos\theta)^{-5} \\
&= (\cos\theta + i\sin\theta)^6 \cdot \left\{\cos\left(\frac{\pi}{2} - \theta\right) + i\sin\left(\frac{\pi}{2} - \theta\right)\right\}^{-5} \\
&= (\cos 6\theta + i\sin 6\theta) \cdot \left\{\cos\left(-\frac{5\pi}{2} + 5\theta\right) + i\sin\left(-\frac{5\pi}{2} + 5\theta\right)\right\} \\
&\quad \text{(using De Moivre's Theorem for } n = 6, n = -5\text{)} \\
&= \cos\left\{6\theta - \frac{5\pi}{2} + 5\theta\right\} + i\sin\left\{6\theta - \frac{5\pi}{2} + 5\theta\right\} \\
&= \cos\left(-\frac{5\pi}{2} + 11\theta\right) + i\sin\left(-\frac{5\pi}{2} + 11\theta\right) \\
&= \cos\left(\frac{5\pi}{2} - 11\theta\right) - i\sin\left(\frac{5\pi}{2} - 11\theta\right) \\
&= \cos\left(2\pi + \frac{\pi}{2} - 11\theta\right) - i\sin\left(2\pi + \frac{\pi}{2} - 11\theta\right) \\
&= \cos\left(\frac{\pi}{2} - 11\theta\right) - i\sin\left(\frac{\pi}{2} - 11\theta\right) \\
&= \sin 11\theta - i\cos 11\theta \\
&= a + ib, \text{ where } a = \sin 11\theta,\ b = -\cos 11\theta.
\end{aligned}$$

► Example 1.7.2 *Prove that* $\dfrac{(\cos\frac{\pi}{8} + i\sin\frac{\pi}{8})^4}{(\cos\frac{\pi}{4} + i\sin\frac{\pi}{4})^2} = 1.$ **[C.P. 1990]**

Solution:

$$\begin{aligned}
\text{L.H.S.} &= \left(\cos\frac{\pi}{8}\cdot 4 + i\sin\frac{\pi}{8}\cdot 4\right) \cdot \left(\cos\frac{\pi}{4} + i\sin\frac{\pi}{4}\right)^{-2} \\
&= \left(\cos\frac{\pi}{2} + i\sin\frac{\pi}{2}\right) \cdot \left\{\cos(-2)\cdot\frac{\pi}{4} + i\sin(-2)\cdot\frac{\pi}{4}\right\} \\
&= \left(\cos\frac{\pi}{2} + i\sin\frac{\pi}{2}\right) \cdot \left\{\cos\left(-\frac{\pi}{2}\right) + i\sin\left(-\frac{\pi}{2}\right)\right\} \\
&= \left(\cos\frac{\pi}{2} + i\sin\frac{\pi}{2}\right) \cdot \left(\cos\frac{\pi}{2} - i\sin\frac{\pi}{2}\right) \\
&= i\cdot(-i) = +1 = \text{R.H.S. } \textbf{(Proved)}
\end{aligned}$$

[In the first line we have used De Moivre's theorem for $n = 4$ and in the second line we have used De Moivre's theorem for $n = -2$.]

► Example 1.7.3 *Prove that*

$$\left(\frac{1 + \cos\theta + i\sin\theta}{1 + \cos\theta - i\sin\theta}\right)^n = (\cos\theta + i\sin\theta)^n = \cos n\theta + i\sin n\theta$$

where n *is a positive integer and* $\theta \neq (2k+1)\pi$.

Solution:

$$\frac{1+\cos\theta+i\sin\theta}{1+\cos\theta-i\sin\theta} = \frac{2\cos^2\frac{\theta}{2}+2i\sin\frac{\theta}{2}\cos\frac{\theta}{2}}{2\cos^2\frac{\theta}{2}-2i\sin\frac{\theta}{2}\cos\frac{\theta}{2}} = \frac{2\cos\frac{\theta}{2}\left(\cos\frac{\theta}{2}+i\sin\frac{\theta}{2}\right)}{2\cos\frac{\theta}{2}\left(\cos\frac{\theta}{2}-i\sin\frac{\theta}{2}\right)}$$

$$= \frac{\cos\frac{\theta}{2}+i\sin\frac{\theta}{2}}{\cos\frac{\theta}{2}-i\sin\frac{\theta}{2}} \quad \left[\begin{array}{l}\text{Cancelling } \cos\frac{\theta}{2} \text{ from both} \\ \text{numerator and denominator} \\ \text{since } \theta \neq (2k+1)\pi, \cos\frac{\theta}{2}\neq 0\end{array}\right]$$

$$= \frac{\left(\cos\frac{\theta}{2}+i\sin\frac{\theta}{2}\right)^2}{\cos^2\frac{\theta}{2}+\sin^2\frac{\theta}{2}} \quad \left[\begin{array}{l}\text{Multiplying both numer-} \\ \text{ator and denominator by} \\ \cos\frac{\theta}{2}+i\sin\frac{\theta}{2}\end{array}\right]$$

$$= \left(\cos\frac{\theta}{2}+i\sin\frac{\theta}{2}\right)^2 = \cos\frac{\theta}{2}\cdot 2+i\sin\frac{\theta}{2}\cdot 2$$

$$= \cos\theta+i\sin\theta.$$

Hence,

$$\left(\frac{1+\cos\theta+i\sin\theta}{1+\cos\theta-i\sin\theta}\right)^n = (\cos\theta+i\sin\theta)^n = \cos n\theta+i\sin n\theta. \textbf{ (Proved)}$$

▶ **Example 1.7.4** *If m is a positive integer, prove that*

$$\left(\frac{1+\sin\theta+i\cos\theta}{1+\sin\theta-i\cos\theta}\right)^m = (\sin\theta+i\cos\theta)^m$$

$$= \left\{\cos\left(\frac{\pi}{2}-\theta\right)+i\sin\left(\frac{\pi}{2}-\theta\right)\right\}^m$$

$$= \cos\left(\frac{m\pi}{2}-m\theta\right)+i\sin\left(\frac{m\pi}{2}-m\theta\right).$$

Solution:

$$\frac{1+\sin\theta+i\cos\theta}{1+\sin\theta-i\cos\theta} = \frac{(\sin^2\theta+\cos^2\theta)+(\sin\theta+i\cos\theta)}{1+\sin\theta-i\cos\theta}$$

$$= \frac{(\sin^2\theta-i^2\cos^2\theta)+(\sin\theta+i\cos\theta)}{1+\sin\theta-i\cos\theta}$$

$$= \frac{(\sin\theta+i\cos\theta)\{\sin\theta-i\cos\theta+1\}}{\sin\theta-i\cos\theta+1}$$

$$= \sin\theta+i\cos\theta$$

$$\therefore \left(\frac{1+\sin\theta+i\cos\theta}{1+\sin\theta-i\cos\theta}\right)^m = (\sin\theta+i\cos\theta)^m$$

$$= \left\{\cos\left(\frac{\pi}{2}-\theta\right)+i\sin\left(\frac{\pi}{2}-\theta\right)\right\}^m$$

$$= \cos\left(m\frac{\pi}{2}-m\theta\right)+i\sin\left(\frac{m\pi}{2}-m\theta\right) \textbf{ (Proved)}$$

▶ **Example 1.7.5** *Prove that*

$$\frac{\left(\cos\frac{\pi}{22}+i\sin\frac{\pi}{22}\right)^{11}\left(\cos\frac{\pi}{21}-i\sin\frac{\pi}{21}\right)^7}{\left(\cos\frac{\pi}{36}+i\sin\frac{\pi}{36}\right)^{12}} = \frac{1}{2}\left(\sqrt{3}-i\right)$$

Solution: Using De Moivre's theorem for $n = 11, n = 7$ and $n = 12$, we write the L.H.S.

$$= \frac{\left(\cos\frac{\pi}{22}\cdot 11 + i\sin\frac{\pi}{22}\cdot 11\right)\left(\cos\frac{\pi}{21}\cdot 7 - i\sin\frac{\pi}{21}\cdot 7\right)}{\cos\frac{\pi}{36}\cdot 12 + i\sin\frac{\pi}{36}\cdot 12}$$

$$\left[\text{Note that } \left(\cos\frac{\pi}{21} - i\sin\frac{\pi}{21}\right)^7 = \left\{\cos\left(-\frac{\pi}{21}\right) + i\sin\left(-\frac{\pi}{21}\right)\right\}^7\right.$$

$$= \cos\left(-\frac{\pi}{21}\cdot 7\right) + i\sin\left(-\frac{\pi}{21}\cdot 7\right)$$

$$\left.= \cos\frac{\pi}{21}\cdot 7 - i\sin\frac{\pi}{21}\cdot 7\right]$$

$$= \frac{\left(\cos\frac{\pi}{2} + \sin\frac{\pi}{2}\right)\cdot\left(\cos\frac{\pi}{3} - i\sin\frac{\pi}{3}\right)}{\cos\frac{\pi}{3} + i\sin\frac{\pi}{3}}$$

$$= \left(\cos\frac{\pi}{2} + i\sin\frac{\pi}{2}\right)\left\{\cos\left(-\frac{\pi}{3}\right) + i\sin\left(-\frac{\pi}{3}\right)\right\}\left\{\cos\frac{\pi}{3} + i\sin\frac{\pi}{3}\right\}^{-1}$$

$$= \left(\cos\frac{\pi}{2} + i\sin\frac{\pi}{2}\right)\left\{\cos\left(-\frac{\pi}{3}\right) + i\sin\left(-\frac{\pi}{3}\right)\right\}\left\{\cos\left(-\frac{\pi}{3}\right) + i\sin\left(-\frac{\pi}{3}\right)\right\}$$

$$= \cos\left(\frac{\pi}{2} - \frac{\pi}{3} - \frac{\pi}{3}\right) + i\sin\left(\frac{\pi}{2} - \frac{\pi}{3} - \frac{\pi}{3}\right)$$

[using $(\cos\theta_1 + i\sin\theta_1)(\cos\theta_2 + i\sin\theta_2)(\cos\theta_3 + i\sin\theta_3)$
$= \cos(\theta_1 + \theta_2 + \theta_3) + i\sin(\theta_1 + \theta_2 + \theta_3)$]

$$= \cos\left(-\frac{\pi}{6}\right) + i\sin\left(-\frac{\pi}{6}\right) = \cos\frac{\pi}{6} - i\sin\frac{\pi}{6} = \frac{\sqrt{3}}{2} - i\frac{1}{2}$$

$$= \frac{1}{2}\left(\sqrt{3} - i\right) = \text{R.H.S. (Proved)}$$

▶ **Example 1.7.6** *Express* $\left(\sqrt{3} + i\right)$ *in the form* $r(\cos\theta + i\sin\theta)$, *and hence find* $\left(\sqrt{3} + i\right)^6$.

Solution: Let $\sqrt{3} = r\cos\theta$ and $1 = r\sin\theta$.

Then $r = 2$, and θ is such that $\cos\theta = \frac{\sqrt{3}}{2}$ and $\sin\theta = \frac{1}{2}$ so that $\theta = \frac{\pi}{6}$ $(-\pi < \theta \le \pi)$.

$$\therefore\ \sqrt{3} + i = 2\left[\cos\frac{\pi}{6} + i\sin\frac{\pi}{6}\right]$$

$$\text{and hence } \left(\sqrt{3} + i\right)^6 = 2^6\left[\cos\frac{\pi}{6} + i\sin\frac{\pi}{6}\right]^6$$

$$= 2^6\left\{\cos\frac{\pi}{6}\cdot 6 + i\sin\frac{\pi}{6}\cdot 6\right\}$$

$$= 2^6(\cos\pi + i\sin\pi)$$

$$= -2^6 \ (\because\ \cos\pi = -1 \text{ and } \sin\pi = 0).$$

▶ **Example 1.7.7** *Prove that if* α, β, γ *are reals and* $\cos\alpha + \cos\beta + \cos\gamma = 0$, *and* $\sin\alpha + \sin\beta + \sin\gamma = 0$,

then
$$\cos 3\alpha + \cos 3\beta + \cos 3\gamma = 3\cos(\alpha + \beta + \gamma),$$
$$\sin 3\alpha + \sin 3\beta + \sin 3\gamma = 3\sin(\alpha + \beta + \gamma).$$

Solution: Let
$$\begin{aligned} a &= \cos\alpha + i\sin\alpha \\ b &= \cos\beta + i\sin\beta \\ c &= \cos\gamma + i\sin\gamma. \end{aligned}$$

Then, by using the given conditions we find that $a+b+c=0$.

But
$$\begin{aligned} a^3+b^3+c^3-3abc &= (a+b+c)(a^2+b^2+c^2-ab-ac-bc) \\ &= 0 \ (\because\ a+b+c=0) \end{aligned}$$

$$\therefore\ a^3+b^3+c^3 = 3abc \tag{1}$$

Now $a^3 = (\cos\alpha + i\sin\alpha)^3 = \cos 3\alpha + i\sin 3\alpha$

Similarly, $b^3 = \cos 3\beta + i\sin 3\beta$ and $c^3 = \cos 3\gamma + i\sin 3\gamma$.

Also
$$\begin{aligned} abc &= (\cos\alpha + i\sin\alpha)(\cos\beta + i\sin\beta)(\cos\gamma + i\sin\gamma) \\ &= \cos(\alpha+\beta+\gamma) + i\sin(\alpha+\beta+\gamma) \end{aligned}$$

Hence from (1) we can write

$$\begin{aligned} &(\cos 3\alpha + i\sin 3\alpha) + (\cos 3\beta + i\sin 3\beta) + (\cos 3\gamma + i\sin 3\gamma) \\ &= 3[\cos(\alpha+\beta+\gamma) + i\sin(\alpha+\beta+\gamma)] \end{aligned}$$

Equating real and imaginary part we get

$$\left.\begin{aligned} \cos 3\alpha + \cos 3\beta + \cos 3\gamma &= 3\cos(\alpha+\beta+\gamma) \\ \sin 3\alpha + \sin 3\beta + \sin 3\gamma &= 3\sin(\alpha+\beta+\gamma) \end{aligned}\right\} \textbf{(Proved)}$$

▶ Example 1.7.8 *Prove that*

$$\frac{(\cos\theta - i\sin\theta)^6}{\cos\theta + i\sin\theta)^{12}} = \cos 22\theta - i\sin 22\theta.$$

Try yourself.

1.7.1 Application of De Moivre's Theorem for Integral Indices

Expressing $\sin n\theta$ and $\cos n\theta$ in powers of $\sin\theta$ and $\cos\theta$, n being a positive integer.

▶ Example 1.7.9 *Prove that:*

(i) $\cos 3\theta = 4\cos^3\theta - 3\cos\theta$; **[C.P. 1999]**

$\sin 3\theta = 3\sin\theta - 4\sin^3\theta$.

(ii) $\cos 5\theta = 16\cos^5\theta - 20\cos^3\theta + 5\cos\theta$;

$\sin 5\theta = 16\sin^5\theta - 20\sin^3\theta + 5\sin\theta$.

(iii) $\cos 7\theta = 64\cos^7\theta - 112\cos^5\theta + 56\cos^3\theta - 7\cos\theta$;

$\sin 7\theta = -64\sin^7\theta + 112\sin^5\theta - 56\sin^3\theta + 7\sin\theta$.

Note that in these examples we have expressed $\cos n\theta$ *in powers of* $\cos\theta$, *where* n *is an odd positive integer and* θ *is any real number.*

Solution:

(i) $(\cos\theta + i\sin\theta)^3 = \cos^3\theta + 3\cos^2\theta(i\sin\theta) + 3\cos\theta(i^2\sin^2\theta) + (i\sin\theta)^3$

(Using Binomial theorem for positive integral index 3)

L.H.S. $= \cos 3\theta + i\sin 3\theta$,

(Using De Moivre's theorem for positive integral index 3)

R.H.S. $= \cos^3\theta - 3\cos\theta\sin^2\theta + i(3\cos^2\theta\sin\theta - \sin^3\theta)$.

Equating Real and Imaginary parts we get

(Real parts) $\cos 3\theta = \cos^3\theta - 3\cos\theta\sin^2\theta = \cos^3\theta - 3\cos\theta(1-\cos^2\theta)$.

Thus,

$$\boldsymbol{\cos 3\theta = 4\cos^3\theta - 3\cos\theta \text{ (in powers of } \cos\theta)}$$

(Imaginary parts) $\sin 3\theta = 3\cos^2\theta\sin\theta - \sin^3\theta = 3(1-\sin^2\theta)\sin\theta - \sin^3\theta$

i.e., $\sin 3\theta = 3\sin\theta - 4\sin^3\theta$ (in powers of $\sin\theta$).

(ii) $(\cos\theta + i\sin\theta)^5 = \cos^5\theta + 5\cos^4\theta(i\sin\theta) + 10\cos^3\theta(i\sin\theta)^2$
$+10\cos^2\theta(i\sin\theta)^3 - 5\cos\theta(i\sin\theta)^4 + (i\sin\theta)^5$

(Binomial expansion with index 5)

i.e., $\cos 5\theta + i\sin 5\theta = \cos^5\theta - 10\cos^3\theta\sin^2\theta + 5\cos\theta\sin^4\theta$
$+i[5\cos^4\theta\sin\theta - 10\cos^2\theta\sin\theta + \sin^5\theta]$

Equating

$$\begin{aligned}
\textbf{(Real parts)},\ \cos 5\theta &= \cos^5\theta - 10\cos^3\theta\sin^2\theta + 5\cos\theta\sin^4\theta \\
&= \cos^5\theta - 10\cos^3\theta(1-\cos^2\theta) \\
&\quad +5\cos\theta(1-\cos^2\theta)^2 \\
&= \cos^5\theta - 10\cos^3\theta + 10\cos^5\theta \\
&\quad +5\cos\theta(1-2\cos^2\theta+\cos^4\theta) \\
&= 5(1-2\sin^2\theta+\sin^4\theta)\sin\theta \\
&\quad -10(1-\sin^2\theta)\sin^3\theta + \sin^5\theta,
\end{aligned}$$

i.e., $\cos 5\theta = 16\cos^5\theta - 20\cos^3\theta + 5\cos\theta$ (in powers of $\cos\theta$)

$$\begin{aligned}
\textbf{(Imaginary parts)},\ \sin 5\theta &= 5\cos^4\theta\sin\theta - 10\cos^2\theta\sin^3\theta + \sin^5\theta \\
&= 5(1-\sin^2\theta)^2\sin\theta \\
&\quad -10(1-\sin^2\theta)\sin^3\theta + \sin^5\theta \\
&= 5(1-2\sin^2\theta+\sin^4\theta)\sin\theta \\
&\quad -10(1-\sin^2\theta)\sin^3\theta + \sin^5\theta,
\end{aligned}$$

i.e., $\sin 5\theta = 16\sin^5\theta - 20\sin^3\theta + 5\sin\theta$ (in powers of $\sin\theta$)

(iii) Expanding $(\cos\theta + i\sin\theta)^7$ by Binomial theorem and writing

$$(\cos\theta + i\sin\theta)^7 = \cos 7\theta + i\sin 7\theta$$

and then equating real and imaginary parts we get, as before.

$$\begin{aligned}
\boldsymbol{\cos 7\theta} &= \boldsymbol{64\cos^7\theta - 112\cos^5\theta + 56\cos^3\theta - 7\cos\theta} \\
&\quad \textbf{(in powers of } \boldsymbol{\cos\theta}\textbf{)} \\
\textbf{and } \boldsymbol{\sin 7\theta} &= \boldsymbol{-64\sin^7\theta + 112\sin^5\theta - 56\sin^3\theta + 7\sin\theta} \\
&\quad \textbf{(in powers of } \boldsymbol{\sin\theta}\textbf{)}.
\end{aligned}$$

▶ Example 1.7.10

(i) *Assuming the results obtained in Ex. 1.7.9 above, express*

$$\frac{\cos 3\theta}{\cos\theta}, \frac{\cos 5\theta}{\cos\theta} \text{ and } \frac{\cos 7\theta}{\cos\theta} \text{ in powers of } \sin\theta.$$

(ii) *Also express*

$$\frac{\sin 3\theta}{\sin\theta}, \frac{\sin 5\theta}{\sin\theta} \text{ and } \frac{\sin 7\theta}{\sin\theta} \text{ in powers of } \cos\theta.$$

Solution:

(i) We have already proved that $\cos 3\theta = 4\cos^3\theta - 3\cos\theta$.

$$\therefore \frac{\cos 3\theta}{\cos\theta} = 4\cos^2\theta - 3$$ (provided θ is not those values which make $\cos\theta = 0$, i.e., $\theta \neq (2k+1)\frac{\pi}{2}$, k is any integer)

$$= 4(1-\sin^2\theta) - 3,$$

$$\text{i.e., } \frac{\cos 3\theta}{\cos\theta} = 1 - 4\sin^2\theta \text{ (in powers of } \sin\theta).$$

⇒ We have already proved that $\cos 5\theta = 16\cos^5\theta - 20\cos^3\theta + 5\cos\theta$.

$$\begin{aligned}\therefore \frac{\cos 5\theta}{\cos\theta} &= 16\cos^4\theta - 20\cos^2\theta + 5\\ &= 16(1-\sin^2\theta)^2 - 20(1-\sin^2\theta) + 5\\ &= 16(1 - 2\sin^2\theta + \sin^4\theta) - 20 + 20\sin^2\theta + 5\\ &= 16\sin^4\theta - 12\sin^2\theta + 1 \text{ (in powers of } \sin\theta).\end{aligned}$$

Similarly, from $\cos 7\theta = 64\cos^7\theta - 112\cos^5\theta + 56\cos^3\theta - 7\cos\theta$, we obtain

$$\frac{\cos 7\theta}{\cos\theta} = -64\sin^6\theta + 80\sin^4\theta - 24\sin^2\theta + 1 \text{ (in powers of } \sin\theta)$$

(ii) We have already obtained $\sin 3\theta = 3\sin\theta - 4\sin^3\theta$.

$$\therefore \frac{\sin 3\theta}{\sin\theta} = 3 - 4\sin^2\theta = 3 - 4(1-\cos^2\theta) = 4\cos^2\theta - 1 \text{ (in powers of } \cos\theta).$$

Similarly

$$\frac{\sin 5\theta}{\sin\theta} = 16\cos^4\theta - 12\cos^2\theta + 1 \text{ (in powers of } \cos\theta)$$

Similarly

$$\frac{\sin 7\theta}{\sin\theta} = 64\cos^6\theta - 80\cos^4\theta + 24\cos^2\theta - 1 \text{ (in powers of } \cos\theta)$$

Note. When n is an odd positive integer, we can express

$$\cos n\theta \text{ in powers of } \cos\theta,$$
$$\sin n\theta \text{ in powers of } \sin\theta,$$

and further we can express

$$\frac{\cos n\theta}{\cos\theta} \text{ in powers of } \sin\theta$$

$$\text{and} \quad \frac{\sin n\theta}{\sin\theta} \text{ in powers of } \cos\theta$$

▶ Example 1.7.11 (n is an even positive integer) *Prove that*

(i) $\dfrac{\sin 4\theta}{\cos\theta} = -8\sin^3\theta + 4\sin\theta$ (*in powers of* $\sin\theta$);

(ii) $\cos 4\theta = 8\cos^4\theta - 8\cos^2\theta + 1$ (*in power of* $\cos\theta$);

(iii) $\dfrac{\sin 8\theta}{\cos\theta} = -128\sin^7\theta + 192\sin^5\theta - 80\sin^3\theta + 8\sin\theta$ (*in powers of* $\sin\theta$);

(iv) $\cos 8\theta = 128\cos^8\theta - 256\cos^6\theta + 160\cos^4\theta - 32\cos^2\theta + 1$ (*in powers of* $\cos\theta$).

We shall prove (i) *and* (ii) *and leave the others* ***as Exercises for the students.***

Solution:

$$\begin{aligned}(\cos\theta + i\sin\theta)^4 &= \cos^4\theta - 6\cos^2\theta\sin^2\theta + \sin^4\theta \\ &\quad + i(4\cos^3\theta\sin\theta - 4\cos\theta\sin^3\theta) \\ \text{or } \cos 4\theta + i\sin 4\theta &= \cos^4\theta - 6\cos^2\theta\sin^2\theta + \sin^4\theta \\ &\quad + i(4\cos^3\theta\sin\theta - 4\cos\theta\sin^3\theta).\end{aligned}$$

Equating real and imaginary parts we get

$$\begin{aligned}\cos 4\theta &= \cos^4\theta - 6\cos^2\theta\sin^2\theta + \sin^4\theta \\ &= \cos^4\theta - 6\cos^2\theta(1-\cos^2\theta) + (1-\cos^2\theta)^2 \\ &= \cos^4\theta - 6\cos^2\theta + 6\cos^4\theta + 1 + \cos^4\theta - 2\cos^2\theta.\end{aligned}$$

$\therefore$ $\mathbf{\cos 4\theta = 8\cos^4\theta - 8\cos^2\theta + 1}$ **(in powers of $\cos\theta$).**

And $\sin 4\theta = 4\cos^3\theta\sin\theta - 4\cos\theta\sin^3\theta$.

$$\therefore \frac{\sin 4\theta}{\sin\theta} = 4(1-\sin^2\theta)\sin\theta - 4\sin^3\theta,$$

i.e., $\dfrac{\sin 4\theta}{\cos\theta} = \mathbf{-8\sin^3\theta + 4\sin\theta}$ **(in powers of $\sin\theta$)**

Expressing $\cos^n\theta$ in terms of cosines of multiples of θ, when n is a positive integer.

Let us put $\cos\theta + i\sin\theta = x$. Then $\cos\theta - i\sin\theta = x^{-1}$.
Then we have

$$\cos n\theta + i\sin n\theta = x^n$$
$$\text{and } \cos n\theta - i\sin n\theta = x^{-n}.$$

We can then write

$$x + x^{-1} = 2\cos\theta; \quad x - x^{-1} = 2i\sin\theta;$$
$$x^n + x^{-n} = 2\cos n\theta; \quad x^n - x^{-n} = 2i\sin n\theta.$$

▶ **Example 1.7.12** *Expand $\cos^6\theta$ in terms of cosines of multiples of θ.*

Solution:

$$\begin{aligned}(2\cos\theta)^6 &= \left(x + x^{-1}\right)^6 \\ &= x^6 + 6x^5\left(x^{-1}\right) + 15x^4\cdot\left(x^{-1}\right)^2 + 20x^3\cdot\left(x^{-1}\right)^3 \\ &\quad + 15x^2\cdot\left(x^{-1}\right)^4 + 6x^1\cdot\left(x^{-1}\right)^5 + \left(x^{-1}\right)^6 \\ &= \left(x^6 + x^{-6}\right) + 6\left(x^4 + x^{-4}\right) + 15\left(x^2 + x^{-2}\right) + 20 \\ &= 2\cos 6\theta + 6(2\cos 4\theta) + 15(2\cos 2\theta) + 20.\end{aligned}$$

$$\therefore \cos^6\theta = \frac{1}{2^5}[\cos 6\theta + 6\cos 4\theta + 15\cos 2\theta + 10].$$

▶ **Example 1.7.13** *Prove that*

$$2^6\cos^3\theta\sin^4\theta = \cos 7\theta - \cos 5\theta - 3\cos 3\theta + 3\cos\theta.$$

Solution:

$$(2\cos\theta)^3 = \left(x + x^{-1}\right)^3$$
$$(2i\sin\theta)^4 = \left(x - x^{-1}\right)^4.$$

$$\begin{aligned}\therefore (2\cos\theta)^3(2i\sin\theta)^4 &= \left(x + x^{-1}\right)^3\left(x - x^{-1}\right)^4 \\ &= \left(x + x^{-1}\right)^3\cdot\left(x - x^{-1}\right)^3\cdot\left(x - x^{-1}\right) \\ &= \left(x^2 - x^{-2}\right)^3\cdot\left(x - x^{-1}\right) \\ &= \left(x^6 - 3x^2 + 3x^{-2} - x^{-6}\right)\cdot\left(x - x^{-1}\right) \\ &= \left(x^7 + x^{-7}\right) - 3\left(x^3 + x^{-3}\right) \\ &\quad + 3\left(x^1 + x^{-1}\right) - \left(x^5 + x^{-5}\right) \\ &= 2\cos 7\theta - 3(2\cos 3\theta) + 3(2\cos\theta) - (2\cos 5\theta).\end{aligned}$$

$$\begin{aligned}\therefore 2^6\cos^3\theta\sin^4\theta &= \cos 7\theta - 3\cos 3\theta + 3\cos\theta - \cos 5\theta \quad (\because i^4 = 1) \\ &= \cos 7\theta - \cos 5\theta - 3\cos 3\theta + 3\cos\theta.\end{aligned}$$

► Example 1.7.14 *For any positive integer n and for any real θ, prove that*

(i) $\cos n\theta = \cos^n\theta - \frac{n(n-1)}{1\cdot 2}\cos^{n-2}\theta\sin^2\theta + \cdots$

(ii) $\sin n\theta = n\cos^{n-1}\theta\sin\theta - \frac{n(n-1)(n-2)}{1\cdot 2\cdot 3}\cos^{n-3}\theta\sin^3\theta + \cdots$

(iii) $\tan n\theta = \dfrac{n\tan\theta - \frac{n(n-1)(n-2)}{1\cdot 2\cdot 3}\tan^3\theta + \cdots}{1 - \frac{n(n-1)}{1\cdot 2}\tan^2\theta + \frac{n(n-1)(n-2)(n-3)}{1\cdot 2\cdot 3\cdot 4}\tan^4\theta + \cdots}$

Solution:

$$\begin{aligned}
(\cos\theta + i\sin\theta)^n &= \cos^n\theta + n\cos^{n-1}\theta(i\sin\theta) \\
&\quad + \frac{n(n-1)}{1\cdot 2}\cos^{n-2}\theta(i^2\sin^2\theta) \\
&\quad + \frac{n(n-1)(n-2)}{1\cdot 2\cdot 3}\cos^{n-3}\theta(i\sin\theta)^3 + \cdots \\
\therefore\ \cos n\theta + i\sin n\theta &= \left\{\cos^n\theta - \frac{n(n-1)}{1\cdot 2}\cos^{n-2}\theta\sin^2\theta + \cdots\right\} \\
&\quad i\left\{n\cos^{n-1}\theta\sin\theta - \frac{n(n-1)(n-2)}{1\cdot 2\cdot 3}\right. \\
&\quad \left.\cos^{n-3}\theta\sin^3\theta + \cdots\right\}.
\end{aligned}$$

Equating real and imaginary parts, we get (i) and (ii).
On division,

$$\tan n\theta = \frac{\sin n\theta}{\cos n\theta} = \frac{n\cos^{n-1}\theta\sin\theta - \frac{n(n-1)(n-2)}{1\cdot 2\cdot 3}\cos^{n-3}\theta\sin^3\theta + \cdots}{\cos^n\theta - \frac{n(n-1)}{1\cdot 2}\cos^{n-2}\theta\sin^2\theta + \cdots}$$

Dividing both numerator and denominator by $\cos^n\theta$, we get

$$\tan n\theta = \frac{n\tan\theta - \frac{n(n-1)(n-2)}{1\cdot 2\cdot 3}\tan^3\theta + \cdots}{1 - \frac{n(n-1)}{1\cdot 2}\tan^2\theta + \frac{n(n-1)(n-2)(n-3)}{1\cdot 2\cdot 3\cdot 4}\tan^4\theta - \cdots}$$

1.7.2 Applications of De Moivre's Theorem for Fractional Indices

► Example 1.7.15 *Cube roots of unity: $\sqrt[3]{1}$, i.e., solve for z, where $z^3 = 1$.*

Solution: We write

$$\begin{aligned}
z &= (1)^{1/3} = (\cos 0^\circ + i\sin 0^\circ)^{1/3} \\
&= \{\cos(0^\circ + 2r\pi) + i\sin(0^\circ + 2r\pi)\}^{1/3} \\
&= \cos\frac{2r\pi}{3} + i\sin\frac{2r\pi}{3} \text{ (using De Moivre's Theorem)}
\end{aligned}$$

Since the index is $\frac{1}{3}$ with denominator 3.

We get three distinct values of z, namely

When $r = 0$, $z = \cos 0° + i \sin 0° = 1$ **(principal value of $\sqrt[3]{1}$)**.

When $r = 1$, $z = \cos\frac{2\pi}{3} + i\sin\frac{2\pi}{3} = \omega$ (say).

When $r = 2$, $z = \cos\frac{4\pi}{3} + i\sin\frac{4\pi}{3} = \omega^2$.

Thus $\sqrt[3]{1}$ has three values $1, \omega, \omega^2$ of which one value is real (namely 1—this is the **principal value** of $\sqrt[3]{1}$) and the other two conjugate complex numbers—if one is denoted by ω, then the other is ω^2.

It is easy to verify that

$$\begin{aligned} 1 + \omega + \omega^2 = 0 \text{ and } \omega^3 &= \left(\cos\frac{2\pi}{3} + i\sin\frac{2\pi}{3}\right)\left(\cos\frac{4\pi}{3} + i\sin\frac{4\pi}{3}\right) \\ &= \cos\left(\frac{2\pi}{3} + \frac{4\pi}{3}\right) + i\sin\left(\frac{2\pi}{3} + \frac{4\pi}{3}\right) \\ &= \cos 2\pi + i\sin 2\pi \\ \text{i.e., } \omega^3 &= 1. \end{aligned}$$

See that (i) $\omega^n + \omega^{2n} + 1 = 3$ if $n = 3k$. [C.P. 2005]

(ii) $(1-\omega)(1-\omega^2)(1-\omega^4)(1-\omega^8) = 9$. [C.P. 2003]

For, L.H.S. $= \{(1-\omega)(1-\omega^2)\}^2 = \{1 - \omega - \omega^2 + \omega^3\}^2 = \{1+1+1\}^2 = 9$.

▸ Example 1.7.16 *Fourth roots of unity: $\sqrt[4]{1}$, i.e., solve: $z^4 = 1$.*

Solution: We write

$$\begin{aligned} z &= (1)^{1/4} = (\cos 0° + i\sin 0°)^{1/4} = \{\cos 2r\pi + i\sin 2r\pi\}^{1/4} \\ &= \cos\frac{2r\pi}{4} + i\sin\frac{2r\pi}{4}, \text{ where } r = 0, 1, 2, 3. \end{aligned}$$

When $r = 0$, we get the principal value of $\sqrt[4]{1}$, namely $\cos 0° + i\sin 0° = 1$.

When $r = 1$, we get another value of

$$\sqrt[4]{1} = \cos\frac{2\pi}{4} + i\sin\frac{2\pi}{4} = i \left(\because \cos\frac{\pi}{2} = 0, \sin\frac{\pi}{2} = 1\right).$$

When $r = 2$, we get the third value of $\sqrt[4]{1}$, namely

$$\cos\frac{4\pi}{4} + i\sin\frac{4\pi}{4} = -1 \ (\because \cos\pi = -1, \sin\pi = 0).$$

When $r = 3$, we get the fourth value of $\sqrt[4]{1}$, namely

$$\begin{aligned} \cos\frac{6\pi}{4} + i\sin\frac{6\pi}{4} &= \cos\frac{3\pi}{2} + i\sin\frac{3\pi}{2} = -i \\ &\left(\because \cos\frac{3\pi}{2} = 0, \ \sin\frac{3\pi}{2} = -1\right), \end{aligned}$$

i.e., the four fourth roots of unity are $1, -1, i, -i$.

▶ **Example 1.7.17** *nth roots of unity, $\sqrt[n]{1}$, i.e., to find z such that $z^n = 1$.*

Solution:

$$\begin{aligned} z &= (1)^{1/n} = (\cos 2r\pi + i\sin 2r\pi)^{1/n} \\ &= \cos\frac{2r\pi}{n} + i\sin^2\frac{r\pi}{n} \quad (r = 0, 1, 2, \ldots, n-1), \end{aligned}$$

i.e., there are n nth roots of unity.

Case 1. Let $n =$ an even positive integer $2m$ (say).

$$\left.\begin{aligned} \text{For} \quad r = 0, \ \sqrt[n]{1} &= \cos 0^\circ + i\sin 0^\circ = 1 \\ r = m = \tfrac{n}{2}, \ \sqrt[n]{1} &= \cos\pi + i\sin\pi = -1 \end{aligned}\right\} \textbf{Two real values.}$$

The other $n-2$ values are pairs of conjugate complex numbers. Corresponding to $r = k$ (other than $r = 0, r = n/2$)

$$\sqrt[n]{1} = \cos\frac{2k\pi}{n} + i\sin\frac{2k\pi}{n} = \cos\frac{2k\pi}{2m} + i\sin\frac{2k\pi}{2m}.$$

Corresponding to $r = 2m - k$, we get

$$\begin{aligned} \sqrt[n]{1} &= \cos^2\frac{(2m-k)\pi}{2m} + i\sin\frac{2(2m-k)\pi}{2m} \\ &= \cos\left(2\pi - \frac{2k\pi}{2m}\right) + i\sin\left(2\pi - \frac{2k\pi}{2m}\right) \\ &= \cos\frac{2k\pi}{2m} - i\sin\frac{2k\pi}{2m}, \end{aligned}$$

i.e., $r = k$, $r = 2m - k$; we get a pair of conjugate complex numbers

$$\cos\frac{2k\pi}{2m} \pm i\sin\frac{2k\pi}{2m}.$$

∴ when n is an even positive integer $\sqrt[n]{1}$ has two real values $1, -1$ and other $n-2$ values are

$$\cos\frac{2r\pi}{n} \pm i\sin\frac{2r\pi}{n} \quad \left[r = 1, 2, 3, \ldots, \left(\frac{n}{2} - 1\right)\right]$$

Case 2. Let n be an odd positive integer, we have only one real value 1, corresponding to $r = 0$ and the other $n - 1$ values give pairs of conjugate complex numbers, namely

$$\cos\frac{2r\pi}{n} \pm i\sin\frac{2r\pi}{n} \quad \left(r = 1, 2, 3, \ldots, \frac{n-1}{2}\right).$$

▶ **Example 1.7.18**

(i) *Find all the six values of $\sqrt[6]{1}$.*

(ii) *Find all the seven values of $\sqrt[7]{1}$.*

Solution:

(i) $$(1)^{1/6} = (\cos 2r\pi + i\sin 2r\pi)^{1/6}$$
$$= \cos\frac{2r\pi}{6} + i\sin\frac{2r\pi}{6} \quad (r = 0, 1, 2, 3, 4, 5).$$

$$\left.\begin{array}{l}\text{Corresponding to } r = 0,\ \sqrt[6]{1} = \cos 0^\circ + i\sin 0^\circ = 1\\ \text{Corresponding to } r = 3,\ \sqrt[6]{1} = \cos\pi + i\sin\pi = -1\end{array}\right\}.$$

$$\left.\begin{array}{rcl}\text{Corresponding to } r = 1,\ \sqrt[6]{1} &=& \cos\frac{2\pi}{6} + i\sin\frac{2\pi}{6}\\ r = 5,\ \sqrt[6]{1} &=& \cos\frac{10\pi}{6} + i\sin\frac{10\pi}{6}\\ &=& \cos\left(2\pi - \frac{2\pi}{6}\right) + i\sin\left(2\pi - \frac{2\pi}{6}\right)\\ &=& \cos\frac{2\pi}{6} - i\sin\frac{2\pi}{6}\end{array}\right\}.$$

Similarly, corresponding $r = 2$ and $r = 4$, we get

$$\sqrt[6]{1} = \cos\frac{4\pi}{6} + i\sin\frac{4\pi}{6} \text{ and } \cos\frac{4\pi}{6} - i\sin\frac{4\pi}{6}.$$

(ii) $$(1)^{1/7} = (\cos 2r\pi + i\sin 2r\pi)^{1/7}$$
$$= \cos\frac{2r\pi}{7} + i\sin\frac{2r\pi}{7} \quad (r = 0, 1, 2, 3, 4, 5, 6).$$

Corresponding to $r = 0$, $\sqrt[7]{1} = \cos 0^\circ + i\sin 0^\circ = 1$.

Corresponding to $(r = 1, r = 6)$; $(r = 2, r = 5)$; $(r = 3, r = 5)$, we get three pairs of values

$$\cos\frac{2\pi}{7} \pm i\sin\frac{2\pi}{7};\ \cos\frac{4\pi}{7} \pm i\sin\frac{4\pi}{7};\ \cos\frac{6\pi}{7} \pm i\sin\frac{6\pi}{7}.$$

▶ **Example 1.7.19**

(i) *Find all the values of* $\left(\sqrt{3} + i\right)^{1/6}$. **[C.P. 1990]**

(ii) *Find all the values of* $(i - 1)^{1/6}$. **[C.P. 1998]**

Solution:

(i) In $\sqrt{3} + i$, put $\sqrt{3} = r\cos\theta$, $1 = r\sin\theta \Rightarrow r = +2$, θ is such that $\cos\theta = \frac{\sqrt{3}}{2}$, and $\sin\theta = \frac{1}{2}$, i.e., $\theta = \frac{\pi}{6}$ (Principal amplitude).

so that

$$\sqrt{3} + i = 2\left(\cos\frac{\pi}{6} + i\sin\frac{\pi}{6}\right) = 2\left[\cos\left(\frac{\pi}{6} + 2r\pi\right) + i\sin\left(\frac{\pi}{6} + 2r\pi\right)\right].$$

$$\therefore\ \left(\sqrt{3}+i\right)^{1/6} = 2^{1/6}\left[\cos\left(\frac{\pi}{6}+2r\pi\right)+i\sin\left(\frac{\pi}{6}+2r\pi\right)\right]^{1/6}$$
$$= 2^{1/6}\left[\cos\left(\frac{\pi}{36}+\frac{2r\pi}{6}\right)+i\sin\left(\frac{\pi}{36}+\frac{2r\pi}{6}\right)\right]$$
where $r = 0,1,2,3,4,5$ **(six values).**

The Principal value of $(\sqrt{3}+i)^{1/6}$ is $\cos\frac{\pi}{36}+i\sin\frac{\pi}{36}$ corresponding to $r=0$.

(ii) Similar problems

$$-1+i = \sqrt{2}\left(-\frac{1}{\sqrt{2}}+\frac{1}{\sqrt{2}}i\right) = \sqrt{2}\left[\cos\frac{3}{4}\pi+i\sin\frac{3}{4}\pi\right]$$
$$= \sqrt{2}\left\{\cos\left(\frac{3}{4}\pi+2r\pi\right)+i\sin\left(\frac{3}{4}\pi+2r\pi\right)\right\}.$$

$\therefore$ 6 values of $(i-1)^{1/6}$ are given by

$$\left(\sqrt{2}\right)^{1/6}\left\{\cos\frac{\frac{3\pi}{4}+2r\pi}{6}+i\sin\frac{\frac{3\pi}{4}+2r\pi}{6}\right\}\ (r=0,1,2,3,4,5).$$

▶ **Example 1.7.20** *Solve for z in* $(z-1)^n = z^n$*, where* n *is a positive integer.*

Solution: We write $(z-1)^n = z^n$ in the form

$$\left(\frac{z-1}{z}\right)^n = 1,\quad \text{or,}\quad \frac{z-1}{z} = (1)^{1/n} = \cos\frac{2r\pi}{n}+i\sin\frac{2r\pi}{n}$$
$$(r = 0,1,2,3,\ldots,n-1),$$
$$\text{or,}\quad 1-\frac{1}{z} = \cos\frac{2r\pi}{n}+i\sin\frac{2r\pi}{n},$$
$$\text{or,}\quad \frac{1}{z} = \left(1-\cos\frac{2r\pi}{n}\right)-i\sin\frac{2r\pi}{n}$$
$$= 2\sin^2\frac{r\pi}{n}-i\left(2\sin\frac{r\pi}{n}\cos\frac{r\pi}{n}\right)$$
$$= 2\sin\frac{r\pi}{n}\left[\sin\frac{r\pi}{n}-i\cos\frac{r\pi}{n}\right],$$
$$\text{or,}\quad z = \frac{1}{2}\left(\frac{1}{\sin\frac{r\pi}{n}}\right)\left[\frac{1}{\sin\frac{r\pi}{n}-i\cos\frac{r\pi}{n}}\right]$$
$$= \frac{1}{2}\frac{1}{\sin\frac{r\pi}{n}}\left[\frac{\sin\frac{r\pi}{n}+i\cos\frac{r\pi}{n}}{\sin^2\frac{r\pi}{n}+\cos^2\frac{r\pi}{n}}\right]$$
$$= \frac{1}{2}\left[1+i\cot\frac{r\pi}{n}\right],\quad r = 0,1,2,\ldots,(n-1).$$

▶ Example 1.7.21 *Solve:* (i) $x^7 = 1$, (ii) $x^7 + 1 = 0$. [C.P. 1982]

Solution:

(i) $x^7 = 1$ gives $x = (\cos 2r\pi + i\sin 2r\pi)^{1/7}$,

$$\text{or, } x = \cos\frac{2r\pi}{7} + i\sin\frac{2r\pi}{7} \quad (r = 0,1,2,3,4,5,6)$$

(ii) $x^7 = -1$ gives
$$\begin{aligned} x &= (\cos\pi + i\sin\pi)^{1/7} \\ &= \{\cos(\pi + 2r\pi) + i\sin(\pi + 2r\pi)\}^{1/7} \\ &= \cos(2r+1)\frac{\pi}{7} + i\sin(2r+1)\frac{\pi}{7} \quad (r = 0,1,2,3,4,5,6). \end{aligned}$$

The seven values of $\sqrt[7]{-1}$ are

$$\cos\frac{\pi}{7} + i\sin\frac{\pi}{7},\ \cos\frac{3\pi}{7} + i\sin\frac{3\pi}{7},\ \cos\frac{5\pi}{7} + i\sin\frac{5\pi}{7},\ \cos\pi + i\sin\pi,$$
$$\cos\frac{9\pi}{7} + i\sin\frac{9\pi}{7},\ \cos\frac{11\pi}{7} + i\sin\frac{11\pi}{7},\ \cos\frac{13\pi}{7} + i\sin\frac{13\pi}{7},$$

These seven values can be written as

$$\cos\frac{\pi}{7} + i\sin\frac{\pi}{7};\ \cos\frac{3\pi}{7} + i\sin\frac{3\pi}{7};\ \cos\frac{5\pi}{7} + i\sin\frac{5\pi}{7};\ \cos\pi + i\sin\pi = -1;$$
$$\cos\frac{9\pi}{7} + i\sin\frac{9\pi}{7} = \cos\left(2\pi - \frac{5\pi}{7}\right) + i\sin\left(2\pi - \frac{5\pi}{7}\right) = \cos\frac{5\pi}{7} - i\sin\frac{5\pi}{7};$$
$$\cos\frac{11\pi}{7} + i\sin\frac{11\pi}{7} = \cos\left(2\pi - \frac{3\pi}{7}\right) + i\sin\left(2\pi - \frac{3\pi}{7}\right) = \cos\frac{3\pi}{7} - i\sin\frac{3\pi}{7};$$
$$\cos\frac{13\pi}{7} + i\sin\frac{13\pi}{7} = \cos\left(2\pi - \frac{\pi}{7}\right) + i\sin\left(2\pi - \frac{\pi}{7}\right) = \cos\frac{\pi}{7} - i\sin\frac{\pi}{7}.$$

We write these seven values of $\sqrt[7]{-1}$ as

$$-1,\ \cos(2r+1)\frac{\pi}{7} \pm i\sin(2r+1)\frac{\pi}{7} \quad (r = 0,1,2)$$

1.8 Miscellaneous Types of Solved Problems

▶ Example 1.8.1 *Find the general value of* θ (θ *real*) *if*

$$(\cos\theta + i\sin\theta)(\cos 2\theta + i\sin 2\theta)\cdots(\cos n\theta + i\sin n\theta) = 1$$

(*n is a given positive integer*)

Solution: L.H.S. can be written as

$$\begin{aligned} &\cos(\theta + 2\theta + 3\theta + \cdots + n\theta) + i\sin(\theta + 2\theta + 3\theta + \cdots + n\theta) \\ =\ &\cos\frac{n(n+1)}{2}\theta + i\sin\frac{n(n+1)}{2}\theta. \end{aligned}$$

Thus, given,

$$\cos\frac{n(n+1)}{2}\theta + i\sin\frac{n(n+1)}{2}\theta = 1.$$

Hence, equating real and imaginary parts we get

$$\cos\frac{n(n+1)}{2}\theta = 1 \text{ and } \sin\frac{n(n+1)}{2} = 0.$$

Both the relations are satisfied by

$$\frac{n(n+1)}{2}\theta = 2k\pi \ (k = 0, \pm 1, \pm 2, \ldots)$$

$$\text{or,} \quad \theta = \frac{4k\pi}{n(n+1)} \ (k = 0, \pm 1, \pm 2, \ldots) \ \textbf{Answer}$$

▶ **Example 1.8.2** *If* $x + \frac{1}{x} = 2\cos\theta$, *then prove that* $x^n + \frac{1}{x^n} = 2\cos n\theta$. [C.P. 1992]

Solution: $x + \frac{1}{x} = 2\cos\theta \;\Rightarrow\; x^2 - 2x\cos\theta + 1 = 0$

$$\Rightarrow x = \frac{2\cos\theta \pm \sqrt{4\cos^2\theta - 4}}{2} = \cos\theta \pm i\sin\theta.$$

(i) If $x = \cos\theta + i\sin\theta$, then

$$x^{-1} = (\cos\theta + i\sin\theta)^{-1} = \cos(-\theta) + i\sin(-\theta) = \cos\theta - i\sin\theta.$$

$$x^n = \cos n\theta + i\sin\theta,\ x^{-n} = \cos n\theta - i\sin n\theta \Rightarrow x^n + \frac{1}{x^n} = 2\cos n\theta.$$

(ii) If $x = \cos\theta - i\sin\theta$, then $x^{-1} = \cos\theta + i\sin\theta$.

$$x^n = \cos n\theta - i\sin n\theta;\ x^{-n} = \cos n\theta + i\sin n\theta.$$

$$\therefore\ x^n + x^{-n} = 2\cos n\theta.$$

▶ **Example 1.8.3** *Show that the product of all the four values of* $\left(1 + i\sqrt{3}\right)^{3/4}$ *is* 8.

Solution: Let

$$1 + i\sqrt{3} = r(\cos\theta + i\sin\theta).$$

Then

$$r\cos\theta = 1$$
$$\text{and} \quad r\sin\theta = \sqrt{3}$$

$\Rightarrow\ r = 2$ and θ is such that $\cos\theta = \frac{1}{2}$ and $\sin\theta = \frac{\sqrt{3}}{2}$

$\Rightarrow\ r = 2, \theta = \frac{\pi}{3}$ (Principal amplitude).

$$\begin{aligned}\therefore\ \left(1 + i\sqrt{3}\right)^{3/4} &= 2^{3/4}\left\{\cos\left(\frac{\pi}{3} + 2r\pi\right) + i\sin\left(\frac{\pi}{3} + 2r\pi\right)\right\}^{3/4} \\ &= 2^{3/4}\left\{\cos\left(\frac{\pi}{3} + 2r\pi\right)\frac{3}{4} + i\sin\left(\frac{\pi}{3} + 2r\pi\right)\frac{3}{4}\right\}\end{aligned}$$

[r = 0, 1, 2, 3].

$\therefore$ the four values are

$$2^{3/4}\left[\cos\frac{\pi}{4}+i\sin\frac{\pi}{4}\right];\ (r=0)$$
$$2^{3/4}\left[\cos\frac{7\pi}{4}+i\sin\frac{7\pi}{4}\right];\ (r=1)$$
$$2^{3/4}\left[\cos\frac{13\pi}{4}+i\sin\frac{13\pi}{4}\right];\ (r=2)$$
$$2^{3/4}\left[\cos\frac{19\pi}{4}+i\sin\frac{19\pi}{4}\right];\ (r=3).$$

Their product

$$= 2^3\left[\cos\left(\frac{\pi}{4}+\frac{7\pi}{4}+\frac{13\pi}{4}+\frac{19\pi}{4}\right)+i\sin\left(\frac{\pi}{4}+\frac{7\pi}{4}+\frac{13\pi}{4}+\frac{19\pi}{4}\right)\right]$$
$$= 8(\cos 10\pi + i\sin 10\pi) = 8\ (\because\ \sin 10\pi = 0, \cos 10\pi = 1).$$

▶ Example 1.8.4 *If n be a positive integer, prove that*
$$(1+i)^n + (1-i)^n = 2^{\frac{n+2}{2}}\cos\frac{n\pi}{4}.$$ [C.P. 1996]

Solution:
$$1+i = \sqrt{2}\left(\cos\frac{\pi}{4}+i\sin\frac{\pi}{4}\right);$$
$$1-i = \sqrt{2}\left[\cos\left(-\frac{\pi}{4}\right)+i\sin\left(-\frac{\pi}{4}\right)\right].$$

$$\begin{aligned}(1+i)^n+(1-i)^n &= \left(\sqrt{2}\right)^n\left[\cos\frac{n\pi}{4}+i\sin\frac{n\pi}{4}\right] + \left(\sqrt{2}\right)^n\left[\cos\frac{n\pi}{4}-i\sin\frac{n\pi}{4}\right]\\ &= \left(\sqrt{2}\right)^n 2\cos\frac{n\pi}{4} = 2^{\frac{n}{2}+1}\cos\frac{n\pi}{4}\\ &= 2^{\frac{n+2}{2}}\cos\frac{n\pi}{4}.\end{aligned}$$

We also obtain
$$(1+i)^n-(1-i)^n = \left(\sqrt{2}\right)^n\cdot 2i\sin\tfrac{n\pi}{4} = 2^{n/2}\cdot 2i\sin\tfrac{n\pi}{4}$$

▶ Example 1.8.5 *If α and β be the two roots of $x^2-2x+4=0$, prove that*
$$\alpha^n+\beta^n = 2^{n+1}\cos\frac{n\pi}{3}.$$ [C.P. 1980]

Solution: $x^2-2x+4=0 \Rightarrow (x-1)^2=-3 \Rightarrow x = 1\pm i\sqrt{3}$
To find

$$\alpha^n+\beta^n = \left(1+i\sqrt{3}\right)^n + \left(1-i\sqrt{3}\right)^n$$
$$1+i\sqrt{3} = 2\left(\frac{1}{2}+i\frac{\sqrt{3}}{2}\right) = 2\left[\cos\frac{\pi}{3}+i\sin\frac{\pi}{3}\right]$$
$$1-i\sqrt{3} = 2\left[\cos\left(-\frac{\pi}{3}\right)+i\sin\left(-\frac{\pi}{3}\right)\right].$$

$$\therefore \left(1+i\sqrt{3}\right)^n = 2^n\left[\cos\frac{n\pi}{3}+i\sin\frac{n\pi}{3}\right]$$
$$\left(1-i\sqrt{3}\right)^n = 2^n\left[\cos\frac{n\pi}{3}-i\sin\frac{n\pi}{3}\right].$$

$$\therefore \alpha^n+\beta^n = 2^n\cdot 2\cos\frac{n\pi}{3} = 2^{n+1}\cos\frac{n\pi}{3}.$$

► Example 1.8.6 *Prove that the sum of the 99th powers of the roots of the equation* $x^7-1=0$ *is zero.*

Solution: $x^7-1=0 \Rightarrow x^7 = 1 = \cos 2r\pi + i\sin 2r\pi$

$$\Rightarrow x = (\cos 2r\pi + i\sin 2r\pi)^{1/7}$$
$$= \cos\frac{2r\pi}{7}+i\sin\frac{2r\pi}{7}\ (r=0,1,2,3,4,5,6).$$

When $r=0$, we get one root $= \cos 0° + i\sin 0° = 1$

$r=1$, we get the second root $= \cos\frac{2\pi}{7}+i\sin\frac{2\pi}{7} = \alpha$ (say)

$r=2$, we get the third root $= \cos\frac{4\pi}{7}+i\sin\frac{4\pi}{7} = \alpha^2$

$r=3$, we get the fourth root $= \cos\frac{6\pi}{7}+i\sin\frac{6\pi}{7} = \alpha^3$

....

$r=6$, we get the seventh root $= \cos\frac{12\pi}{7}+i\sin\frac{12\pi}{6} = \alpha^6$

$\therefore$ sum of the 99th powers of the roots of the equations

$$= (1)^{99}+\alpha^{99}+(\alpha^2)^{99}+(\alpha^3)^{99}+\cdots+(\alpha^6)^{99}$$

$$= \frac{1-(\alpha^{99})^7}{1-\alpha^{99}}\quad\left[\begin{array}{l}\text{Sum of a G.P., whose first term is}\\ \text{1 and common ratio is } \alpha^{99}\text{, the}\\ \text{number of terms being 7.}\end{array}\right]$$

$$= \frac{1-(\alpha^7)^{99}}{1-\alpha^{99}} = \frac{1-1}{1-\alpha^{99}} = 0\left[\begin{array}{rl}\because \alpha^7 &= \left(\cos\frac{2\pi}{7}+i\sin\frac{2\pi}{7}\right)^7\\ &= \cos 2\pi + i\sin 2\pi = 1\end{array}\right.$$

See that $\alpha^{99}\neq 1$ so that denominator $\neq 0$]

► Example 1.8.7 *If* $x+\frac{1}{x} = 2\cos\frac{\pi}{7}$*, then prove that* $x^7+\frac{1}{x^7} = -2.$ **[C.P. 2003]**

Solution: $x+\frac{1}{x} = 2\cos\frac{\pi}{7} \Rightarrow x^2-2x\cos\frac{\pi}{7}+1=0$

$$\Rightarrow x = \frac{2\cos\frac{\pi}{7}\pm\sqrt{4\cos^2\frac{\pi}{7}-4}}{2} = \cos\frac{\pi}{7}\pm i\sin\frac{\pi}{7}$$

If

$$x = \cos\frac{\pi}{7} + i\sin\frac{\pi}{7}, \text{ then } x^{-1} = \cos\frac{\pi}{7} - i\sin\frac{\pi}{7}.$$

$$\begin{aligned} \therefore x^7 &= \left(\cos\frac{\pi}{7} + i\sin\frac{\pi}{7}\right)^7 \\ &= \cos\pi + i\sin\pi \\ &= -1 \end{aligned} \qquad \begin{aligned} \therefore x^{-7} &= \left(\cos\frac{\pi}{7} - i\sin\frac{\pi}{7}\right)^7 \\ &= \cos\pi - i\sin\pi \\ &= -1 \end{aligned}$$

$$\therefore x^7 + x^{-7} = -2.$$

On the other hand we could take

$$x = \cos\frac{\pi}{7} - i\sin\frac{\pi}{7}, \text{ then } x^{-1} = \cos\frac{\pi}{7} + i\sin\frac{\pi}{7}.$$

Then as before

$$x^7 = \cos\pi - i\sin\pi \text{ and } x^{-7} = \cos\pi + i\sin\pi,$$

$$\text{i.e., } x^7 + \frac{1}{x^7} = 2\cos\pi = -2.$$

▶ **Example 1.8.8** *If* $a + \frac{1}{a} = 2\cos\alpha$ *and* $b + \frac{1}{b} = 2\cos\beta$, *then show that* $\cos(\alpha - \beta)$ *is one of the values of* $\frac{1}{2}\left(\frac{a}{b} + \frac{b}{a}\right)$. [C.P. 1988, '91]

Solution: As before

$$\begin{aligned} a + \frac{1}{a} &= 2\cos\alpha \Rightarrow a = \cos\alpha \pm i\sin\alpha \\ \text{and } b + \frac{1}{b} &= 2\cos\beta \Rightarrow b = \cos\beta \pm i\sin\beta. \end{aligned}$$

Let us take

$$\begin{aligned} a &= \cos\alpha + i\sin\alpha \\ \text{and } b &= \cos\beta + i\sin\beta. \end{aligned}$$

Then

$$\begin{aligned} a^{-1} &= \cos\alpha - i\sin\alpha \\ \text{and } b^{-1} &= \cos\beta - i\sin\beta. \end{aligned}$$

$$\begin{aligned} \therefore \frac{a}{b} + \frac{b}{a} &= a\cdot b^{-1} + b\cdot a^{-1} \\ &= (\cos\alpha + i\sin\alpha)(\cos\beta - i\sin\beta) + (\cos\beta + i\sin\beta)(\cos\alpha - i\sin\alpha) \\ &= (\cos\alpha + i\sin\alpha)\{\cos(-\beta) + i\sin(-\beta)\} \\ &\quad + (\cos\beta + i\sin\beta)\{\cos(-\alpha) + i\sin(-\alpha)\} \\ &= \{\cos(\alpha-\beta) + i\sin(\alpha-\beta)\} + \cos(\beta-\alpha) + i\sin(\beta-\alpha) \\ &= \{\cos(\alpha-\beta) + i\sin(\alpha-\beta)\} + \{\cos(\alpha-\beta) - i\sin(\alpha-\beta)\} \\ \text{or, } \frac{a}{b} + \frac{b}{a} &= 2\cos(\alpha-\beta). \end{aligned}$$

Next we take

$$a = \cos\alpha + i\sin\alpha, \qquad b = \cos\beta - i\sin\beta$$
$$a^{-1} = \cos\alpha - i\sin\alpha, \qquad b^{-1} = \cos\beta + i\sin\beta.$$

Then,

$$\begin{aligned}
\frac{a}{b}+\frac{b}{a} &= (\cos\alpha + i\sin\alpha)(\cos\beta + i\sin\beta) + (\cos\beta - i\sin\beta)(\cos\alpha - i\sin\alpha)\\
&= \{(\cos\alpha\cos\beta - \sin\alpha\sin\beta) + i(\sin\alpha\cos\beta + \cos\alpha\sin\beta)\}\\
&\quad + \{\cos\beta\cos\alpha - \sin\beta\sin\alpha - i(\sin\beta\cos\alpha + \sin\alpha\cos\beta)\}\\
&= \cos(\alpha+\beta) + i\sin(\alpha+\beta) + \cos(\beta+\alpha) - i\sin(\beta+\alpha)\\
&= 2\cos(\alpha+\beta).
\end{aligned}$$

Hence the statement that "$\cos(\alpha-\beta)$ is one of the values of $\frac{1}{2}\left(\frac{a}{b}+\frac{b}{a}\right)$" is true.

► Example 1.8.9 *If* $(1+x)^n = c_0 + c_1x + c_2x^2 + \cdots + c_nx^n$ (n *is a positive integer*), *then prove that:*

(i) $c_0 - c_2 + c_4 - \cdots = 2^{n/2}\cos\frac{n\pi}{4}$;

(ii) $c_1 - c_3 + c_5 - \cdots = 2^{n/2}\sin\frac{n\pi}{4}$;

(iii) $c_0 + c_4 + c_8 - \cdots = 2^{n-2} + 2^{\frac{n-2}{2}}\cos\frac{n\pi}{4}$;

(iv) $c_1 + c_5 + c_9 + \cdots = 2^{n-2} + 2^{\frac{n-2}{2}}\sin\frac{n\pi}{4}$. [C.P. 1980]

Solution: Put $x = i$ in the given identity

$$\begin{aligned}
(1+i)^n &= c_0 + c_1 i + c_2 i^2 + c_3 i^3 + \cdots + c_n i^n\\
&= (c_0 - c_2 + c_4 - c_6 + \cdots) + i(c_1 - c_3 + c_5 - c_7 + \cdots). \qquad (1)
\end{aligned}$$

Then put $x = -i$ in the given identity

$$\begin{aligned}
(1-i)^n &= c_0 + c_1(-i) + c_2(-i)^2 + c_3(-i)^3 + \cdots + c_n(-i)^n\\
&= (c_0 - c_2 + c_4 - c_6 + \cdots) - i(c_1 - c_3 + c_5 - \cdots). \qquad (2)
\end{aligned}$$

Adding,

$$(1+i)^n + (1-i)^n = 2(c_0 - c_2 + c_4 - c_6 + \cdots).$$

See Ex. 1.8.4 worked out above. We then obtain

$$2^{\frac{n+2}{2}}\cos\frac{n\pi}{4} = 2(c_0 - c_2 + c_4 - c_6 + \cdots)$$

$$\text{or,}\quad c_0 - c_2 + c_4 - c_6 + \cdots = \frac{2^{\frac{n+2}{2}}}{2}\cos\frac{n\pi}{4} = 2^{n/2}\cos\frac{n\pi}{4}.$$

This proves (i).

Subtracting (2) from (1) we also obtain

$$2i(c_1 - c_3 + c_5 - c_7 + \cdots) = 2i\cdot 2^{n/2} \sin\frac{n\pi}{4},$$

$$\text{or, } c_1 - c_3 + c_5 - c_7 + \cdots = 2^{n/2} \sin\frac{n\pi}{4}.$$

This proves (ii).

Now, put $x = 1, x = -1$ in the given identity. We then obtain

$$\left\{\begin{array}{rcl} 2^n &=& c_0 + c_1 + c_2 + c_3 + c_4 + c_5 + c_6 + \cdots + c_n \\ 0 &=& c_0 - c_1 + c_2 - c_3 + c_4 - c_5 + c_6 - \cdots + (-1)^n c_n \end{array}\right. \quad \text{(A)}$$

$$\text{Adding we get, } 2^n = 2(c_0 + c_2 + c_4 + c_6 + \cdots)$$

$$\text{or, } 2^{n-1} = c_0 + c_2 + c_4 + c_6 + \cdots$$

In (i) we already obtained

$$2^{n/2} \cos\frac{n\pi}{4} = c_0 - c_2 + c_4 - c_6 + \cdots$$

$$\text{(Adding)} \quad 2^{n-1} + 2^{n/2} \cos\frac{n\pi}{4} = 2(c_0 + c_4 + c_8 + \cdots)$$

$$\therefore \quad c_0 + c_4 + c_8 + \cdots = 2^{n-2} + 2^{n/2-1} \cos\frac{n\pi}{4}$$

This proves (iii).

From (A) by subtraction we get

$$2^n = 2(c_1 + c_3 + c_5 + c_7 + \cdots)$$

$$\text{or, } \quad 2^{n-1} = c_1 + c_3 + c_5 + c_7 + \cdots$$

Also from (ii) we write

$$2^{n/2} \sin\frac{n\pi}{4} = c_1 - c_3 + c_5 - c_7 + \cdots$$

$$\text{(Adding)} \quad 2^{n-1} + 2^{n/2} \sin\frac{n\pi}{4} = 2(c_1 + c_5 + c_9 + \cdots)$$

$$\therefore \quad c_1 + c_5 + c_9 + \cdots = 2^{n-2} + 2^{n/2-1} \sin\frac{n\pi}{4}.$$

This proves (iv).

▶ Example 1.8.10 *If* $\cos\alpha + \cos\beta + \cos\gamma = \sin\alpha + \sin\beta + \sin\gamma = 0$, *then prove that*

(i) $\cos 3\alpha + \cos 3\beta + \cos 3\gamma = 3\cos(\alpha + \beta + \gamma)$; **[C.P. 1990]**

(ii) $\sin 3\alpha + \sin 3\beta + \sin 3\gamma = 3\sin(\alpha + \beta + \gamma)$;

(iii) $\sin^2\alpha + \sin^2\beta + \sin^2\gamma = \cos^2\alpha + \cos^2\beta + \cos^2\gamma = \frac{3}{2}$; **[C.P. 1999]**

(iv) $\cos 4\alpha + \cos 4\beta + \cos 4\gamma = -2\sum \cos(2\beta + 2\gamma)$;

(v) $\sin 4\alpha + \sin 4\beta + \sin 4\gamma = -2\sum \sin(2\beta + 2\gamma)$.

(***Note that*** α, β, γ ***are reals***)

Solution: Let

$$\left.\begin{aligned} a &= \cos\alpha + i\sin\alpha, \\ b &= \cos\beta + i\sin\beta, \\ c &= \cos\gamma + i\sin\gamma. \end{aligned}\right\} \text{ See Ex. 1.7.7.}$$

We first prove (i) and (ii):
Using the given conditions, we get

$$\begin{aligned} a+b+c &= (\cos\alpha+\cos\beta+\cos\gamma)+i(\sin\alpha+\sin\beta+\sin\gamma) \\ &= 0+i0=0. \end{aligned}$$

Now we have the identity

$$a^3+b^3+c^3-3abc = (a+b+c)(a^2+b^2+c^2-ab-ac-bc).$$

$\therefore$ when $a+b+c=0$, we may write

$$a^3+b^3+c^3 = 3abc$$

or, $$(\cos\alpha+i\sin\alpha)^3+(\cos\beta+i\sin\beta)^3+(\cos\gamma+i\sin\gamma)^3$$
$$=3\{(\cos\alpha+i\sin\alpha)(\cos\beta+i\sin\beta)\times(\cos\gamma+i\sin\gamma)\}.$$

Using De Moivre's theorem

$$(\cos 3\alpha+i\sin 3\alpha)+(\cos 3\beta+i\sin 3\beta)+(\cos 3\gamma+i\sin 3\gamma)$$
$$=3\{\cos(\alpha+\beta+\gamma)+i\sin(\alpha+\beta+\gamma)\}$$

Equating real and imaginary parts, we obtain (i) and (ii).
For proving (iii):
We have

$$\frac{1}{a}+\frac{1}{b}+\frac{1}{c} = \frac{bc+ca+ab}{abc},$$

or, $$(\cos\alpha+i\sin\alpha)^{-1}+(\cos\beta+i\sin\beta)^{-1}+(\cos\gamma+i\sin\gamma)^{-1} = \frac{bc+ca+ab}{abc}$$

or, $$(\cos\alpha-i\sin\alpha)+(\cos\beta-i\sin\beta)+(\cos\gamma-i\sin\gamma) = \frac{bc+ca+ab}{abc}$$

or, $$(\cos\alpha+\cos\beta+\cos\gamma)-i(\sin\alpha+\sin\beta+\sin\gamma) = \frac{bc+ca+ab}{abc}.$$

From the given conditions (i.e., $\Sigma\cos\alpha = 0, \Sigma\sin\alpha = 0$) we observe that the L.H.S. = 0.

So we can write $bc+ca+ab=0$
We know the identity

$$a^2+b^2+c^2 = (a+b+c)^2-2(ab+bc+ca)$$
$$(\cos\alpha+i\sin\alpha)^2+(\cos\beta+i\sin\beta)^2+(\cos\gamma+i\sin\gamma)^2=0$$
$$(\because\ a+b+c=0 \text{ and } bc+ca+ab=0).$$

$$\therefore\ (\cos 2\alpha + i\sin 2\alpha) + (\cos 2\beta + i\sin 2\beta) + (\cos 2\gamma + i\sin 2\gamma) = 0 + i\cdot 0.$$

Equating real parts

$$\cos 2\alpha + \cos 2\beta + \cos 2\gamma = 0,$$
$$\text{or,}\quad (2\cos^2\alpha - 1) + (2\cos^2\beta - 1) + (2\cos^2\gamma - 1) = 0,$$
$$\text{or,}\quad \cos^2\alpha + \cos^2\beta + \cos^2\gamma = 3/2.$$

Also

$$(1 - 2\sin^2\alpha) + (1 - 2\sin^2\beta) + (1 - 2\sin^2\gamma) = 0,$$
$$\text{or,}\quad \sin^2\alpha + \sin^2\beta + \sin^2\gamma = 3/2.$$

This proves (iii).
For proving (iv) and (v) we use the identity

$$a^4 + b^4 + c^4 = (a^2 + b^2 + c^2)^2 - 2(a^2b^2 + b^2c^2 + c^2a^2)$$

We have already proved that $a^2 + b^2 + c^2 = 0$.

$$\therefore\quad a^4 + b^4 + c^4 = -2(a^2b^2 + b^2c^2 + c^2a^2),$$
$$\text{or,}\quad (\cos\alpha + i\sin\alpha)^4 + (\cos\beta + i\sin\beta)^4 + (\cos\gamma + i\sin\gamma)^4$$
$$= -2\sum(\cos\alpha + i\sin\beta)^2(\cos\beta + i\sin\beta)^2,$$
$$\text{or,}\quad \cos 4\alpha + \cos 4\beta + \cos 4\gamma + i(\sin 4\alpha + \sin 4\beta + \sin 4\gamma)$$
$$= -2\sum\{\cos(2\alpha + 2\beta) + i\sin(2\alpha + 2\beta)\}.$$

Equating real and imaginary parts we can easily obtain **(iv)** and **(v)**.

▶ Example 1.8.11 *If α and β are the roots of the equation*

$$x^2 - 2x\cos\theta + 1 = 0,$$

then find the equation whose roots are α^n and β^n, given that n is a positive integer. **[C.P. 1995]**

Solution: $x^2 - 2x\cos\theta + 1 = 0$ gives

$$x = \frac{2\cos\theta \pm \sqrt{4\cos^2\theta - 1}}{2} = \cos\theta \pm i\sin\theta$$

Let us call

$$\alpha = \cos\theta + i\sin\theta$$
$$\beta = \cos\theta - i\sin\theta.$$

Then

$$\alpha^n = \cos n\theta + i\sin n\theta$$
$$\text{and } \beta^n = \cos n\theta - i\sin n\theta,$$

so that

$$\begin{aligned} \alpha^n + \beta^n &= 2\cos n\theta \\ \text{and } (\alpha\beta)^n &= \{(\cos\theta + i\sin\theta)(\cos\theta - i\sin\theta)\}^n \\ &= (\cos^2\theta + \sin^2\theta)^n = 1. \end{aligned}$$

$\therefore$ the equation whose roots are α^n and β^n is

$$x^2 - (\alpha^n + \beta^n)x + \alpha^n\cdot\beta^n = 0,$$

$$\text{or,}\quad x^2 - 2x\cos n\theta + 1 = 0.$$

▶ Example 1.8.12 *If* $(a_1 + ib_1)(a_2 + ib_2)\cdots(a_n + ib_n) = A + iB$ $(a_1, a_2, a_3, \ldots, a_n, b_1, b_2, \ldots, b_n, A, B$ *are all reals*), *then prove that:*

(i) $\tan^{-1}\dfrac{b_1}{a_1} + \tan^{-1}\dfrac{b_2}{a_2} + \cdots + \tan^{-1}\dfrac{b_n}{a_n} = \tan^{-1}\dfrac{B}{A}$ *and*

(ii) $(a_1^2 + b_1^2)(a_2^2 + b_2^2)\cdots(a_n^2 + b_n^2) = A^2 + B^2$. **[C.P. 1995]**

Solution: We put
$a_1 = r_1\cos\theta_1, b_1 = r_1\sin\theta_1$, so that $a_1 + ib_1 = r_1(\cos\theta_1 + i\sin\theta_1)$
$a_2 = r_2\cos\theta_2, b_2 = r_2\sin\theta_2$, so that $a_2 + ib_2 = r_2(\cos\theta_2 + i\sin\theta_2)$
....
$a_n = r_n\cos\theta_n, b_n = r_n\sin\theta_n$, so that $a_n + ib_n = r_n(\cos\theta_n + i\sin\theta_n)$.
Also put $A = R\cos\theta, B = R\sin\theta$ so that $A + iB = R(\cos\theta + i\sin\theta)$.
Given condition then becomes

$$\begin{aligned} & r_1(\cos\theta_1 + i\sin\theta_1)\cdot r_2(\cos\theta_2 + i\sin\theta_2)\cdots r_n(\cos\theta_n + i\sin\theta_n) \\ & = R(\cos\theta + i\sin\theta) \\ \text{or,}\quad & r_1r_2\cdots r_n\{\cos(\theta_1 + \theta_2 + \cdots + \theta_n) + i\sin(\theta_1 + \theta_2 + \cdots + \theta_n\} \\ & = R(\cos\theta + i\sin\theta). \end{aligned}$$

Equating real and imaginary parts we get

$$\begin{aligned} r_1r_2\cdots r_n\cos(\theta_1 + \theta_2 + \cdots + \theta_n) &= R\cos\theta \\ r_1r_2\cdots r_n\sin(\theta_1 + \theta_2 - \cdots + \theta_n) &= R\sin\theta \end{aligned}$$

$$\begin{aligned} \therefore\quad & R^2 = r_1^2r_2^2r_3^2\cdots r_n^2, \\ \text{i.e.,}\quad & A^2 + B^2 = (a_1^2 + b_1^2)(a_2^2 + b_2^2)\cdots(a_n^2 + b_n^2) \\ \text{and}\quad & \tan(\theta_1 + \theta_2 + \cdots + \theta_n) = \tan\theta, \\ \text{i.e.,}\quad & \theta_1 + \theta_2 - \cdots + \theta_n = \theta \\ \text{or,}\quad & \tan^{-1}\frac{b_1}{a_1} + \tan^{-1}\frac{b_2}{a_2} + \cdots + \tan^{-1}\frac{b_n}{a_n} = \tan^{-1}\frac{B}{A}. \end{aligned}$$

▶ Example 1.8.13 (i) *Find the principal amplitudes of* $(\sin 50° + i\cos 50°)^6$.
(ii) *Find the modulus and principal value of the amplitude of* $(\cos 50° + i\sin 50°)^6$. **[C.P. 1996]**

Solution:

(i)
$$\begin{aligned}\sin 50^\circ + i\cos 50^\circ &= \sin(90^\circ - 40^\circ) + i\cos(90^\circ - 40^\circ)\\ &= \cos 40^\circ + i\sin 40^\circ \text{ (in De Moivre's form)}\end{aligned}$$

$$\begin{aligned}\therefore\ (\sin 50^\circ + i\cos 50^\circ)^6 &= (\cos 40^\circ + i\sin 40^\circ)^6\\ &= \cos 240^\circ + i\sin 240^\circ\\ &= \cos(360^\circ - 120^\circ) + i\sin(360^\circ - 120^\circ)\\ &= \cos 120^\circ - i\sin 120^\circ = \cos(-120^\circ) + i\sin(-120^\circ)\end{aligned}$$

and $-180^\circ < -120^\circ < 180^\circ$, therefore, the principal amplitude is -120°.

(ii)
$$\begin{aligned}(\cos 50^\circ + i\sin 50^\circ)^6 &= \cos 300^\circ + i\sin 300^\circ\\ &= \cos(360^\circ - 60^\circ) + i\sin(360^\circ - 60^\circ)\\ &= \cos 60^\circ - i\sin 60^\circ\\ &= \cos(-60^\circ) + i\sin(-60^\circ).\end{aligned}$$

$\therefore$ **modulus = 1, principal amplitude** $= -60^\circ$ ($\because\ -180^\circ < -60^\circ < 180^\circ$).

▶ **Example 1.8.14** *If* $x = \cos\alpha + i\sin\alpha$, $y = \cos\beta + i\sin\beta$, $z = \cos\gamma + i\sin\gamma$ *and* $x + y + z = xyz$, *then prove that* $\cos(\beta - \gamma) + \cos(\gamma - \alpha) + \cos(\alpha - \beta) = -1$.

[C.P. 2005]

Solution: $x + y + z = xyz$ gives $\cos\alpha + \cos\beta + \cos\gamma = \cos(\alpha + \beta + \gamma)$ and $\sin\alpha + \sin\beta + \sin\gamma = \sin(\alpha + \beta + \gamma)$. Squaring and adding, we get $3 + 2\sum\cos(\beta - \gamma) = 1$; hence etc.

Exercises on Chapter 1

1. **(Textual)**

   (a) State De Moivre's theorem for any real index.

   (b) Prove, by the principle of mathematical induction,

   $$\begin{aligned}&(\cos\theta_1 + i\sin\theta_1)(\cos\theta_2 + i\sin\theta_2)\cdots(\cos\theta_n + i\sin\theta_n)\\ &\quad = \cos(\theta_1 + \theta_2 + \cdots + \theta_n) + i\sin(\theta_1 + \theta_2 + \cdots + \theta_n)\end{aligned}$$

   and hence deduce De Moivre's theorem for a positive integral index.

   (c) State De Moivre's theorem for any rational index and use it to obtain all the three cube roots of unity.

2. Prove that
$$\frac{\left(\cos\frac{\pi}{8} + i\sin\frac{\pi}{8}\right)^4}{\left(\cos\frac{\pi}{4} - i\sin\frac{\pi}{4}\right)^2} = 1.$$

3. Prove that
$$\left(\frac{1 + \cos\theta + i\sin\theta}{1 + \cos\theta - i\sin\theta}\right)^n = \cos n\theta + i\sin n\theta.$$

4. Solve $x^5 = 1$. [*Ans.* $\cos\frac{2r\pi}{5} + i\sin\frac{2r\pi}{5}\,(r = 0, 1, 2, 3, 4)$]

5. Solve $x^7 + x^4 + x^3 + 1 = 0$.

[*Ans.* $\cos\left(\frac{\pi}{3} + \frac{2r\pi}{3}\right) + i\sin\left(\frac{\pi}{3} + \frac{2r\pi}{3}\right)$; $r = 0, 1, 2, 3$ and $\cos\left(\frac{\pi}{4} + \frac{2r\pi}{4}\right) + i\sin\left(\frac{\pi}{4} + \frac{2r\pi}{4}\right)$; $r = 0, 1, 2, 3$]

6. **If $x + iy = \frac{1}{2+\cos\theta+i\sin\theta}$ (x, y, θ are all reals), then prove that when θ varies, the point (x, y) moves in a circle of centre $(\frac{2}{3}, 0)$.** **[C.P. 1994, 2002]**

[*Hints*: $x + iy = \frac{2+\cos\theta-i\sin\theta}{(2+\cos\theta)^2+\sin^2\theta}$. Equate real and imaginary parts and eliminate θ. The eliminant becomes $(x - \frac{2}{3})^2 + y^2 = (\frac{1}{3})^2$; hence etc.]

7. If n be a positive integer, prove that

$$(1 + i)^n + (1 - i)^n = 2^{n/2+1}\cos\frac{n\pi}{4}.$$

8. Express $\left(\frac{\cos\theta + i\sin\theta}{\sin\theta + i\cos\theta}\right)^6$ in the form $a + ib$. [*Ans.* $(-\cos 12\theta) + i(-\sin 12\theta)$]

9. Use De Moivre's theorem to find a formula for $\cos 3\theta$.

10. Use De Moivre's theorem to prove that $\alpha^n + \beta^n = 2^{n/2+1}\cos\frac{n\pi}{4}$ where α, β are the roots of $x^2 - 2x + 2 = 0 = 0$ and n is a positive integer. **[C.P. 2004]**

11. If $2\cos\alpha = m + \frac{1}{m}$, $2\cos\beta = n + \frac{1}{n}$ and $2\cos\gamma = p + \frac{1}{p}$, then prove that

$$2\cos(\alpha + \beta + \gamma) = mnp + \frac{1}{mnp}.$$

12. Find the cube roots of (-1) using De Moivre's theorem **[C.P. 2002]**

$$\begin{aligned} x^3 &= -1 = \cos\pi + i\sin\pi = \cos(2r\pi + \pi) + i\sin(2r\pi + \pi), \\ \text{i.e., } x &= \cos\frac{(2r\pi + \pi)}{3} + i\sin\frac{(2r\pi + \pi)}{3}, \quad r = 0, 1, 2 \end{aligned}$$

13. **Express $\sin^4\theta\cos^4\theta$ in terms of sines and cosines of multiples of θ.** **[C.P. 2002]**

[*Ans.* $\frac{1}{2^7}(\cos 8\theta - 4\cos 4\theta + 3)$]

14. If $x + \frac{1}{x} = 2\cos\frac{\pi}{7}$, then show that $x^7 + \frac{1}{x^7} = -2$. **[C.P. 2003]**

15. **(a) Let $z = x + iy$ (x, y real). If $2|z - 1| = |z - 2|$, prove that on the complex plane z describes a circle $3(x^2 + y^2) = 4x$.**

(b) On the complex plane, let $P(z)$ be a variable point such that $|z + 3i| = 4$. Find the locus of P. **[C.P. 2005]**

[*Hints*: $x^2 + (y + 3)^2 = 4^2$ or $(\frac{x}{4})^2 + (\frac{y+3}{4})^2 = 1$, an ellipse.]

16. If the amplitude of $\frac{z-1}{z+1}$ is $\frac{\pi}{4}$, where $z = x + iy$ $(i^2 = -1)$, show that z lies on a circle with centre at i [i.e., at (0, 1)]. What is the radius? **[C.P. 2004]**

[*Hints*: $\frac{z-1}{z+1} = \frac{\{(x-1)+iy\}\{(x+1)-iy\}}{(x+1)^2+y^2}$; $\tan\theta = \frac{(x+1)y-y(x-1)}{x^2-1+y^2} = \tan\frac{\pi}{4} = 1$, i.e., $2y = x^2 - 1 + y^2$ or $x^2 + y^2 - 2y + 1 = 2$ or $x^2 + (y-1)^2 = (\sqrt{2})^2$.
$\therefore$ z lies on a circle with centre (0, 1), i.e., at i and radius $\sqrt{2}$.]

17. If $2\cos\alpha = m + \frac{1}{m}$, $2\cos\beta = n + \frac{1}{n}$, $2\cos\gamma = p + \frac{1}{p}$, prove that

$$2\cos(\alpha+\beta+\gamma) = mnp + \frac{P1}{mnp}.$$ **[C.P. 2001]**

Chapter 2

Functions of a Complex Variable (Exponential, Sine, Cosine and Logarithm of a Complex Number)

	Functions of a Complex Variable
1.	Exponential, Sine, Cosine and Logarithm of a Complex Number.
2.	Definition of $a^z (a \neq 0)$.
3.	Inverse Circular and Hyperbolic Functions.

2.1 Introduction: Definitions

In the present chapter we shall define a few important functions of a complex variable z—*Exponential functions, Trigonometric* and *Hyperbolic functions.* Their inverses —Logarithmic functions, Inverse Trigonometric and Inverse Hyperbolic functions will also be introduced.

Definitions

1. In real analysis we have proved that the infinite series

$$1 + \frac{x}{\lfloor\underline{1}} x + \frac{x^2}{\lfloor\underline{2}} + \frac{x^3}{\lfloor\underline{3}} + \cdots + \frac{x^n}{\lfloor\underline{n}} + \cdots$$

has a sum e^x (for all real values of x), called *exponential function* of the real variable x.

2. We have also proved that, for any real x

$$\begin{aligned} \sin x &= x - \frac{x^3}{\lfloor\underline{3}} + \frac{x^5}{\lfloor\underline{5}} - \frac{x^7}{\lfloor\underline{7}} + \cdots \\ \cos x &= 1 - \frac{x^2}{\lfloor\underline{2}} + \frac{x^4}{\lfloor\underline{4}} - \frac{x^6}{\lfloor\underline{6}} + \cdots \end{aligned}$$

3. On similar lines we define Exponential Functions of the complex variable z:

$$e^z = 1 + \frac{z}{\underline{|1}} + \frac{z^2}{\underline{|2}} + \frac{z^3}{\underline{|3}} + \cdots + \frac{z^n}{\underline{|n}} + \cdots$$

Instead of e^z, we often write Exp z or E(z).

4. We also define, for a complex variable z

$$\sin z = z - \frac{z^3}{\underline{|3}} + \frac{z^5}{\underline{|5}} - \frac{z^7}{\underline{|7}} + \cdots$$

$$\cos z = 1 - \frac{z^2}{\underline{|2}} + \frac{z^4}{\underline{|4}} - \frac{z^6}{\underline{|6}} + \cdots$$

5. We shall assume the truth of Addition Theorem:

$$E(z_1 + z_2) = E(z_1) \cdot E(z_2) \text{ [or, } e^{z_1 + z_2} = e^{z_1} \cdot e^{z_2}]$$

6. Assuming the definition of Exponential Function we can write

$$e^{iy} = 1 + \frac{iy}{\underline{|1}} + i^2 \frac{y^2}{\underline{|2}} + i^3 \frac{y^3}{\underline{|3}} + i^4 \frac{y^4}{\underline{|4}} + \cdots$$

$$= \left(1 - \frac{y^2}{\underline{|2}} + \frac{y^4}{\underline{|4}} - \cdots\right) + i\left(y - \frac{y^3}{\underline{|3}} + \frac{y^5}{\underline{|5}} - \cdots\right)$$

$$\left.\begin{aligned} \text{i.e., } e^{iy} &= \cos y + i \sin y \\ \text{Also } e^{-iy} &= \cos y + i \sin y \end{aligned}\right\} \text{ where } y \text{ is real.}$$

Even when z is complex we assume

$$\left.\begin{aligned} e^{iz} &= \cos z + i \sin z \\ e^{-iz} &= \cos z - i \sin z \end{aligned}\right\}, \textbf{ Euler's Exponential Values}$$

whence it also follows

$$\left.\begin{aligned} \cos z &= \frac{1}{2}(e^{iz} + e^{-iz}) \\ \sin z &= \frac{1}{2i}(e^{iz} - e^{-iz}) \end{aligned}\right\}.$$

We shall then define:

$$\tan z = \frac{\sin z}{\cos z}, \quad \cot z = \frac{\cos z}{\sin z}, \quad \sec z = \frac{1}{\cos z}, \quad \operatorname{cosec} z = \frac{1}{\sin z}.$$

These functions (sin z, cos z, tan z, cot z, sec z, cosec z) are called **Trigonometric** or **Circular** functions of the complex variable z.

7. We define **Hyperbolic functions** of z thus:

$$\textbf{Cosine hyperbolic } z = \cosh z = \frac{1}{2}(e^{z} + e^{-z})$$

$$\textbf{Sine hyperbolic } z = \sinh z = \frac{1}{2}(e^{z} - e^{-z}).$$

$$\tanh z = \frac{\sinh z}{\cosh z}, \quad \coth z = \frac{\cosh z}{\sinh z}, \quad \operatorname{sech} z = \frac{1}{\cosh z}, \quad \operatorname{cosech} z = \frac{1}{\sinh z}.$$

A very important relation:

$$\cosh^2 z - \sinh^2 z = 1.$$

More generally, any usual trigonometric relation can be replaced by the corresponding hyperbolic relation; **only any product of two sines should be changed into the product of two corresponding hyperbolic sines with a sign changed: e.g.,**

$$\begin{aligned} \cosh(z_1 + z_2) &= \cosh z_1 \cosh z_2 + \sinh z_1 \sinh z_2 \\ &\qquad (\text{cf. } \cos(x+y) = \cos x \cos y - \sin x \sin y) \\ \cosh 2z &= \cosh^2 z + \sinh^2 z \\ &\qquad (\text{cf. } \cos 2x = \cos^2 x - \sin^2 x) \end{aligned}$$

but see that

$$\begin{aligned} \sinh 2z &= 2 \sinh z \cosh z \\ \sinh(z_1 + z_2) &= \sinh z_1 \cosh z_2 + \cosh z_1 \sinh z_2, \text{ etc.} \end{aligned}$$

8. **Logarithmic functions**

Definition. If w and z are two complex numbers so related that $e^z = w$, then z is called a *logarithm* of w, written as

$$\begin{aligned} z &= \operatorname{Log} w, \text{ using capital letter L} \\ \text{i.e., } e^z &= w \Rightarrow z = \operatorname{Log} w. \end{aligned}$$

9. **Complex Numbers with Complex Exponents**

If a be a non-zero complex number (i.e., $a \neq 0 + i0$) and z is any complex number, then by a^z we shall mean $e^{z \operatorname{Log} a}$.

$$\text{i.e., } a^z = e^{z \operatorname{Log} a} \quad (a \neq 0, z \text{ is complex})$$

In the next article we shall give illustrative examples of all that we have defined above.

2.2 Solved Problems: Logarithmic, Exponential and Hyperbolic Function

▶ Example 2.2.1 *Express* e^z, $\text{Log}\, z$ *in the form* $A + iB$ $(A, B$ *real*$)$, *where* $z = x + iy$ $(x, y$ *real*$)$.

Solution:

$$\begin{aligned}\text{(i)}\quad e^z &= e^{x+iy} = e^x \cdot e^{iy} \text{ (Addition theorem)}\\ &= e^x(\cos y + i \sin y) \text{ (Euler's exponential values)}\\ &= e^x\{\cos(y + 2n\pi) + i\sin(y + 2n\pi)\},\ n = 0, \pm 1, \pm 2, \ldots\end{aligned}$$

Thus e^z has been expressed in modulus-amplitude form, with modulus e^x and amplitude $(y + 2n\pi)$.

The **principal value of the amplitude** of e^z is obtained by choosing the integer n such that

$$-\pi < 2n\pi + y \leq \pi.$$

(ii) Let $\text{Log}\, z = \alpha + i\beta$. $(\alpha, \beta$ real$)$.
Then, by definition say

$$\begin{aligned} e^{\alpha + i\beta} &= z = r(\cos\phi + i\sin\phi),\\ \text{or, } e^{\alpha}\cdot(\cos\beta + i\sin\beta) &= r(\cos\phi + i\sin\phi)\\ &= r[\cos(\phi + 2n\pi) + i\sin(\phi + 2n\pi)].\end{aligned}$$

Comparing modulus and amplitude we get

$$e^{\alpha} = r \text{ and } \beta = \phi + 2n\pi\cdot (n = 0, \pm 1, \pm 2, \ldots),$$
$$\text{or,}\quad \alpha = \log r \text{ and } \beta = 2n\pi + \phi.$$

$$\therefore\ \text{Log}\, z = \alpha + i\beta = \log r + i(\alpha + 2n\pi) = \mathbf{A + iB\ (form)},$$

i.e., $\text{Log}\, z$ is a many-valued function.

If ϕ_1 be the principal value of the amplitude of z, so that $-\pi < \phi_1 \leq \pi$ then we define the **principal value** of $\text{Log}\, z$ (written as $\log z$) by $\log z = \log r + i\phi_1$.

> **Remember**
> $\text{Log}\, z = \log|z| + i$ (Principal amplitude of $z + 2n\pi$) $n = 0, \pm 1, \pm 2, \ldots$
> Principal value of $z = \log z = \log|z| + i$ (Principal amplitude of z)
> We shall use this result as a **standard formula.**

▶ Example 2.2.2 *Express* $\text{Log}\, i$, $\text{Log}(-i)$, $\text{Log}\, 1$ *in the form* $A + iB$. *Write down the principal values in each case.*

Solution:

(i) By our definition

$$\begin{aligned}\text{Log}\, i &= \log|i| + i \text{ (principal amplitude of } i + 2n\pi)\\ &\quad (n = 0, \pm 1, \pm 2, \ldots)\\ &= \log 1 + i\left(\frac{\pi}{2} + 2n\pi\right) = i\left(\frac{\pi}{2} + 2n\pi\right).\end{aligned}$$

Principal value of Log i, written as

$$\log i = i\frac{\pi}{2}\ (n = 0).$$

(ii) $\text{Log}(-i) = \log|-i| + i\,(\text{principal amplitude of } -i + 2n\pi)$ $(n = 0, \pm1, \pm2, \ldots)$

$$= \log 1 + i\left(-\frac{\pi}{2} + 2n\pi\right) = i\left(-\frac{\pi}{2} + 2n\pi\right).$$

$\therefore\ \log(-i) = i\left(-\frac{\pi}{2}\right)\ (n = 0)$ [This is the principal value of Log(−i)].

(iii) $\text{Log } 1 = \log|1| + i\,(\text{principal amplitude of } 1 + 2n\pi)$ $(n = 0, \pm1, \pm2, \ldots)$

$$= i(0^\circ + 2n\pi) = 2n\pi i.$$

$\therefore\ \log 1 = 0$ (principal value of Log 1).

► Example 2.2.3 *Express* $\text{Log}(1+i), \text{Log}(-2), \text{Log}(-3-4i)$ *in the form* $A + iB$. *Write the principal values in each case.*

Solution:

(i) $\text{Log}(1+i) = \log|1+i| + i\,[\text{principal amplitude of } (1+i) + 2n\pi]$

$$= \log\sqrt{2} + i\left(\frac{\pi}{4} + 2n\pi\right)\ (n = 0, \pm1, \pm2, \ldots).$$

$$\therefore\ \log(1+i) = \frac{1}{2}\log 2 + i\frac{\pi}{4}.$$

We obtain Principal amplitude of 1 + i thus:
$1 + i = r(\cos\phi + i\sin\phi)$
$r\cos\phi = 1,\ r\sin\phi = 1$
$\therefore\ r = \sqrt{2}$, ϕ is determined in such a way
$\cos\phi = \frac{1}{\sqrt{2}}, \sin\phi = \frac{1}{\sqrt{2}}$ and $-\pi < \phi \le \pi$,
i.e., we find the principal amplitude of $1 + i$. Clearly $\phi = \frac{\pi}{4}$.

(ii) $\text{Log}(-2) = \text{Log}(-2 + i\cdot 0)$

$$= \log|-2| + i\,(\text{principal amplitude of } -2 + 2n\pi)$$
$$= \log 2 + i(\pi + 2n\pi)$$
$$= \log 2 + i(2n+1)\pi\ (n = 0, \pm1, \pm2, \ldots).$$

$$\therefore\ \log(-2) = \log 2 + i\pi.$$

To obtain Principal amplitude of −2, we write
$-2 + i\cdot 0 = r(\cos\phi + i\sin\phi)$
$r\cos\phi = -2,\ r\sin\phi = 0$
$\therefore\ r = 2$, and ϕ is such that $\cos\phi = -1$,
$\sin\phi = 0$ and $-\pi < \phi \le \pi$ so that $\phi = \pi$.

(iii) By definition

$$\begin{aligned}\text{Log}(-3-4i) &= \log|-3-4i| + i(\phi + 2n\pi) \\ &= \log 5 + i(\phi + 2n\pi)\ (n = 0, \pm 1, \pm 2, \ldots),\end{aligned}$$

where ϕ is defined in such a way that

$$\cos\phi = -\frac{3}{5},\ \sin\phi = -\frac{4}{5} \text{ and } -\pi < \phi \leq \pi.$$

> **To obtain Principal amplitude of $-3-4i$:**
> Let $-3-4i = r(\cos\phi + i\sin\phi)$
> $r\cos\phi = -3,\ r\sin\phi = -4,\ r^2 = 25$, i.e., $r = 5$
> ϕ is such that $\cos\phi = \dfrac{-3}{5}$ and $\sin\phi = -\dfrac{4}{5}$ and $-\pi < \phi \leq \pi$

Principal value of $\text{Log}(-3-4i) = \log(-3-4i) = \log 5 + i\phi$ (taking $n = 0$).

▶ Example 2.2.4 *What is the Principal value of* $\text{Log}\{\cos\theta + i\sin\theta\}$ *where* $-\pi < \theta \leq \pi$?

Solution:

$$\begin{aligned}\text{Log}(\cos\theta + i\sin\theta) &= \log|\cos\theta + i\sin\theta| + i(\theta + 2n\pi) \\ &\qquad (n = 0, \pm 1, \pm 2, \ldots) \\ &= \log 1 + i(\theta + 2n\pi) = i(\theta + 2n\pi).\end{aligned}$$

$\therefore$ principal value of $\text{Log}(\cos\theta + i\sin\theta) = \log(\cos\theta + i\sin\theta) = i\theta$ (taking $n = 0$).

▶ Example 2.2.5 *Express* a^z *in the form* $A + iB$ (A, B *real*) *if*

$$a = r(\cos\theta + i\sin\theta)(-\pi < \theta \leq \pi) \text{ and } z = x + iy\ (x, y \text{ real}).$$

Solution: By definition

$$\begin{aligned}a^z &= e^{z\,\text{Log}\,a} = e^{(x+iy)\{\log r + i(\theta + 2n\pi)\}} \\ &= e^{x\log r - y(\theta+2n\pi)} \cdot e^{i\{y\log r + x(\theta+2n\pi)\}} \\ &= r^x \cdot e^{-y(\theta+2n\pi)} \cdot [\cos\{y\log r + x(\theta + 2n\pi)\} + i\sin\{y\log r + x(\theta + 2n\pi)\}] \\ &= A + iB\end{aligned}$$

$$\begin{aligned}\text{where } A &= r^x e^{-y(\theta+2n\pi)} \cos\{y\log r + x(\theta + 2n\pi)\}, \\ B &= r^x e^{-y(\theta+2n\pi)} \sin\{y\log r + x(\theta + 2n\pi)\}.\end{aligned}$$

> $$\begin{aligned}\text{Log}\,a &= \text{Log}\,r(\cos\theta + i\sin\theta) \\ &= \log r + i(\theta + 2n\pi)\ (n = 0, \pm 1, \pm 2, \ldots)\end{aligned}$$

The principal value of a^z is

$$r^x \cdot e^{-y\theta}[\cos\{y\log r + x\theta\} + i\sin\{y\log r + x\theta\}].$$ [C.P. 1989]

▶ Example 2.2.6

(i) *Express i^i in the form $A + iB$. What is its principal value?*

(ii) *Show that the values of i^i are real and are in G.P.* **[C.P. 1997]**

Solution:

(i) By definition

$$\begin{aligned} i^i &= e^{i\,\text{Log}\, i} \\ &= e^{i\{i(\frac{\pi}{2}+2n\pi)\}} \\ &= e^{-(\frac{\pi}{2}+2n\pi)} \\ &= e^{-(\frac{\pi}{2}+2n\pi)} + i\cdot 0 \\ &= A + iB\ (B = 0) \end{aligned}$$

$$\begin{aligned} \text{Log}\, i &= \log|i| + i\left(\tfrac{\pi}{2} + 2n\pi\right) \\ &= \log 1 + i\left(\tfrac{\pi}{2} + 2n\pi\right) \\ &= i\left(\tfrac{\pi}{2} + 2n\pi\right) \quad n = 0, \pm 1, \pm 2, \ldots \end{aligned}$$

(ii) See that the values of i^i are $e^{-\frac{\pi}{2}}, e^{-(\frac{\pi}{2}\pm 2\pi)}, e^{-(\frac{\pi}{2}\pm 4\pi)}, \ldots$. They are all real numbers and they form a G.P. with a common ratio $e^{-2\pi}$ or $e^{2\pi}$.

The principal value of i^i is $e^{-\frac{\pi}{2}}$.

▶ Example 2.2.7 *Express $(-i)^i$ in the form $A + iB$ and state its principal value.*

Solution: By definition

$$\begin{aligned} (-i)^i &= e^{i\,\text{Log}(-i)} \\ &= e^{i\cdot i(-\frac{\pi}{2}+2n\pi)} \\ &= e^{\frac{\pi}{2}-2n\pi} \\ &= e^{-(4n-1)\frac{\pi}{2}} \\ &\quad (n = 0, \pm 1, \pm 2, \ldots) \end{aligned}$$

$$\begin{aligned} \text{Log}(-i) &= \log|-i| + i\left(-\tfrac{\pi}{2} + 2n\pi\right) \\ &= \log 1 + i\left(-\tfrac{\pi}{2} + 2n\pi\right) \\ &= i\left(-\tfrac{\pi}{2} + 2n\pi\right) \\ &\quad n = 0, \pm 1, \pm 2, \ldots \end{aligned}$$

Principal value is $e^{\frac{\pi}{2}}$.

▶ Example 2.2.8 *Express $i^{\log(1+i)}$ in the form $A + iB$.*

Solution: We first express $1 + i$ in the modulus-amplitude form:

$$\begin{aligned} 1 + i &= \sqrt{2}\left(\frac{1}{\sqrt{2}} + i\frac{1}{\sqrt{2}}\right) \\ &= \sqrt{2}\left(\cos\frac{\pi}{4} + i\sin\frac{\pi}{4}\right) \\ &= \sqrt{2}\left\{\cos\left(\frac{\pi}{4} + 2n\pi\right) + i\sin\left(\frac{\pi}{4} + 2n\pi\right)\right\} \\ &\quad \{n = 0, \pm 1, \pm 2, \ldots\} \end{aligned}$$

so that $|1 + i| = \sqrt{2}$ and $\text{Amp}(1 + i) = \pi/4 + 2n\pi$

$$\begin{aligned} \text{Log}(1 + i) &= \log|1 + i| + i\left(\frac{\pi}{4} + 2n\pi\right) \\ &= \log\sqrt{2} + i\left(\frac{\pi}{4} + 2n\pi\right) \\ &= \log\sqrt{2} + i\left(\frac{\pi}{4} + 2n\pi\right). \end{aligned}$$

$$\therefore \log(1+i) = \text{Principal value of } \mathrm{Log}(1+i)$$
$$= \log\sqrt{2} + i\frac{\pi}{4}.$$

Now, by definition

$$\begin{aligned}
i^{\log(1+i)} &= e^{\log(1+i)\,\mathrm{Log}\,i} \\
&= e^{\left(\log\sqrt{2}+i\frac{\pi}{4}\right)i(4n+1)\frac{\pi}{2}} \\
&= e^{-\frac{\pi}{4}(4n+1)\frac{\pi}{2}+i\{\log\sqrt{2}\cdot(4n+1)\frac{\pi}{2}\}} \\
&= e^{-\frac{\pi^2}{8}(4n+1)}\cdot e^{i\{\log\sqrt{2}\cdot(4n+1)\frac{\pi}{2}\}} \\
&= e^{-\frac{\pi^2}{8}(4n+1)} \times \\
&\quad \left[\cos\left\{\log\sqrt{2}\cdot(4n+1)\frac{\pi}{2}\right\} + i\sin\left\{\log\sqrt{2}\cdot(4n+1)\frac{\pi}{2}\right\}\right] \\
&= A + iB,
\end{aligned}$$

$$\text{where } A = e^{-\frac{\pi^2}{8}(4n+1)}\cos\left\{\log\sqrt{2}\cdot(4n+1)\frac{\pi}{2}\right\}$$
$$B = e^{-\frac{\pi^2}{8}(4n+1)}\sin\left\{\log\sqrt{2}\cdot(4n+1)\frac{\pi}{2}\right\}.$$

The principal value of $i^{\log(1+i)}$ is, therefore,

$$= e^{-\frac{\pi^2}{8}}\left[\cos\left\{\log\sqrt{2}\cdot\frac{\pi}{2}\right\} + i\sin\left\{\log\sqrt{2}\cdot\frac{\pi}{2}\right\}\right]$$
$$= e^{-\frac{\pi^2}{8}}\left[\cos\left(\frac{\pi}{4}\log 2\right) + i\sin\left(\frac{\pi}{4}\log 2\right)\right].$$

► Example 2.2.9 ***What is a periodic function? What is its period? Prove that e^z is a periodic function of period $2n\pi i$.***

> ***A function $f(z)$ of a complex variable z is said to be a periodic function of z, if there exists k, real or complex, such that $f(z+k) = f(z)$ for all complex numbers z. k is then called the period of $f(z)$.***

We verify that e^z is a periodic function of z of period $2n\pi i$.

Solution: Let k be the period of $f(z) = e^z$. Then $f(z+k) = f(z)$, for all complex numbers z, i.e., $e^{z+k} = e^z$.

Put $z = 0$, $e^k = e^0 = 1$.

Let $k = \alpha + i\beta$ (α, β real). Then

$$e^k = e^{\alpha+i\beta} = e^\alpha \cdot e^{i\beta} = e^\alpha(\cos\beta + i\sin\beta).$$

$\therefore$ $e^k = 1$ gives $e^\alpha = 1$ and $\beta = 0 + 2n\pi$ ($n = 0, \pm1, \pm2, \ldots$), i.e., $\alpha = 0$ and $\beta = 2n\pi$ and $k = 0 + i\cdot 2n\pi = 2n\pi i$.

We now verify

$$\begin{aligned}
e^{z+2n\pi i} &= e^z \cdot e^{2n\pi i} \\
&= e^z(\cos 2n\pi + i\sin 2n\pi) \\
&= e^z \cdot 1 = e^z.
\end{aligned}$$

Thus $k = 2n\pi i$ is the period of the periodic function e^z.

Important relations that exist between Trigonometric and Hyperbolic Functions:

► **Example 2.2.10** *Prove that*

(i) $\sin(-z) = -\sin z$;

(ii) $\cos(-z) = \cos z$;

(iii) $\sin 0 = 0$;

(iv) $\cos 0 = 1$.

Solution:

(i) $\sin(-z) = \dfrac{e^{i(-z)} - e^{-i(-z)}}{2i} = \dfrac{1}{2i}\{-(e^{iz} - e^{-iz})\} = -\dfrac{e^{iz} - e^{-iz}}{2i} = -\sin z.$

(ii) $\cos(-z) = \dfrac{e^{i(-z)} + e^{-i(-z)}}{2} = \dfrac{e^{iz} + e^{-iz}}{2} = \cos z.$

(iii) $\sin 0 = \dfrac{e^{i\cdot 0} - e^{-i\cdot 0}}{2i} = \dfrac{1-1}{2i} = 0.$

(iv) $\cos 0 = \dfrac{e^{i\cdot 0} + e^{-i\cdot 0}}{2} = \dfrac{1+1}{2} = 1.$

► **Example 2.2.11** ***Express*** $\sin(x+iy)$, $\cos(x+iy)$ ***in the form*** $A + iB$ $(x, y$ ***are reals***$)$. **[C.P. 1982]**

Solution:

$$\begin{aligned}
\sin(x+iy) &= \frac{e^{i(x+iy)} - e^{-i(x+iy)}}{2i} \\
&= \frac{e^{-y+ix} - e^{y-ix}}{2i} = \frac{e^{-y}.e^{ix} - e^{y}.e^{-ix}}{2i} \\
&= \frac{e^{-y}.(\cos x + i\sin x) - e^{y}.(\cos x - i\sin x)}{2i} \\
&= \cos x \frac{e^{-y} - e^{y}}{2i} + i\sin x \frac{e^{-y} + e^{y}}{2i} \\
&= \frac{-1}{i}\cos x \cdot \frac{e^{y} - e^{-y}}{2} + \sin x \cdot \frac{e^{y} + e^{-y}}{2} \\
&= i\cos x \sinh y + \sin x \cosh y \quad (\because\ -\tfrac{1}{i} = i).
\end{aligned}$$

i.e., $\sin(x+iy) = \sin x \cosh y + i\cos x \sinh y$. **[C.P. 1982]**

Similarly obtain,

$$\cos(x+iy) = \cos x \cosh y - i\sin x \sinh y.$$

▶ Example 2.2.12 ***Prove that:***

$$\text{(i)}\quad \left.\begin{aligned}\sin(iz) &= i\sinh z\\ \cos(iz) &= \cosh z\end{aligned}\right\};$$

$$\text{(ii)}\quad \left.\begin{aligned}\sinh(iz) &= i\sin z\\ \cosh(iz) &= \cos z\end{aligned}\right\}.$$

Solution:

$$\text{(i) } \sin(iz) = \frac{e^{i\cdot(iz)} - e^{-i\cdot(iz)}}{2i} = \frac{e^{-z} - e^{z}}{2i^2}\cdot i = i\left(\frac{e^{z} - e^{-z}}{2}\right),$$

$$\text{i.e., } \sin(iz) = i\sinh z.$$

Similarly all the other relations can be obtained by using the definitions of $\cos z$, $\sinh z$, $\cosh z$.

We leave them as Exercises.

▶ Example 2.2.13 ***Prove that:***

(i) $\cosh^2 z - \sinh^2 z = 1$ (*but* $\cos^2 z + \sin^2 z = 1$);

(ii) $\cosh^2 z + \sinh^2 z = \cosh 2z$ (*but* $\cos^2 z - \sin^2 z = \cos 2z$).

Solution:

(i) To prove: $\cosh^2 z - \sinh^2 z = 1$.

$$\begin{aligned}\text{L.H.S.} &= \left(\frac{e^{z}+e^{-z}}{2}\right)^2 - \left(\frac{e^{z}-e^{-z}}{2}\right)^2\\ &= \frac{1}{4}\{(e^{z}+e^{-z})^2 - (e^{z}-e^{-z})^2\}\\ &= \frac{1}{4}\cdot 4e^{z}\cdot e^{-z} = e^{\circ} = 1 \textbf{ (proved).}\end{aligned}$$

To prove: $\cos^2 z + \sin^2 z = 1$.

$$\begin{aligned}\text{L.H.S.} &= \left\{\frac{e^{iz}+e^{-iz}}{2}\right\}^2 + \left\{\frac{e^{iz}-e^{-iz}}{2}\right\}^2\\ &= \frac{1}{4}\{(e^{iz}+e^{-iz})^2 - (e^{iz}-e^{-iz})^2\}\\ &= \frac{1}{4}\cdot 4e^{iz}\cdot e^{-iz} = e^{\circ} = 1 \textbf{ (proved).}\end{aligned}$$

(ii) Similarly obtain these results.

▶ Example 2.2.14 *If* $z = x + iy$ (x, y *real*) *then prove:*

(i) $|\sin z|^2 = \sin^2 x + \sinh^2 y$;

(ii) $|\cos z|^2 = \cos^2 x + \sinh^2 y.$

Solution:

$$\begin{aligned}\text{(i)}\ \sin z &= \sin(x+iy) = \sin x \cos iy + \cos x \sin iy\\ &= \sin x \cosh y + i \cos x \sinh y\\ &\quad (\because\ \cos iy = \cosh y; \sin iy = i \sinh y).\end{aligned}$$

$$\begin{aligned}\therefore |\sin z|^2 &= \sin^2 x \cosh^2 y + \cos^2 x \sinh^2 y\\ &= \sin^2 x(1+\sinh^2 y) + (1-\sin^2 x)\sinh^2 y\\ &= \sin^2 x + \sinh^2 y.\end{aligned}$$

Similarly the other result (ii) can be obtained.

▶ Example 2.2.15 ***Periods of*** $\sin z$ ***and*** $\cos z$: *For definition see Ex. 2.2.9 worked above.*

Check: $2k\pi$ (*k is any integer or zero*) *is a period of* $\sin z$ *or* $\cos z$.

Check: *The hyperbolic functions* $\sinh z$ *and* $\cosh z$ *are periodic functions, each of period* $2k\pi i$.

▷ Example 2.2.16 ***Prove the familiar results: (using definitions only)***

(i) $\cos(z_1 \pm z_2) = \cos z_1 \cos z_2 \mp \sin z_1 \sin z_2;$

(ii) $\sin(z_1 \pm z_2) = \sin z_1 \cos z_2 \pm \cos z_1 \sin z_2;$

(iii) $\sin 2z = 2 \sin z \cos z;$

(iv) $\cos 2z = \cos^2 z - \sin^2 z;$

(v) $\sin 3z = 3 \sin z - 4 \sin^3 z;$

(vi) $\cos 3z = 4 \cos^3 z - 3 \cos z;$

(vii) $\tan(z_1 + z_2) = \dfrac{\tan z_1 + \tan z_2}{1 - \tan z_1 \tan z_2};$

(viii) $\sin z_1 + \sin z_2 = 2 \sin \frac{1}{2}(z_1+z_2) \cos \frac{1}{2}(z_1 - z_2);$

(ix) $\cos z_1 + \cos z_2 = 2 \cos \frac{1}{2}(z_1+z_2) \cos\left(\dfrac{z_1 - z_2}{2}\right).$

where z, z_1, z_2 *are all complex numbers.*

All these results can be proved by using definitions only.
However, we prove, for instance

$$\begin{aligned}\cos(z_1+z_2) &= \cos z_1 \cos z_2 - \sin z, \sin z_2;\\ \sin(z_1+z_2) &= \sin z_1 \cos z_2 + \cos z_1 \sin z_2.\end{aligned}$$

[C.P. 1981]

Proof.

$$e^{i(z_1+z_2)} = \cos(z_1+z_2) + i\sin(z_1+z_2) \text{ [Euler's Exponential Values]}$$

$$\text{or, } e^{iz_1}\cdot e^{iz_2} = \cos(z_1+z_2) + i\sin(z_1+z_2)$$

$$\text{or, } (\cos z_1 + i\sin z_1)(\cos z_2 + i\sin z_2) = \cos(z_1+z_2) + i\sin(z_1+z_2)$$

$$\text{or, } (\cos z_1\cos z_2 - \sin z_1\sin z_2) + i(\sin z_1\cos z_2 + \cos z_1\sin z_2)$$
$$= \cos(z_1+z_2) + i\sin(z_1+z_2). \quad (1)$$

Replacing z_1 by $-z_1$ and z_2 by $-z_2$, we obtain

$$\{\cos(-z_1)\cos(-z_2) - \sin(-z_1)\sin(-z_2)\}$$
$$+ i\{\sin(-z_1)\cos(-z_2) + \cos(-z_1)\sin(-z_2)\}$$
$$= \cos\{-z_1 - z_2\} + i\sin\{-z_1 - z_2\}.$$

Using,

$$\sin(-z) = -\sin z,\ \cos(-z) = \cos z,$$

we get

$$\{\cos z_1\cos z_2 - \sin z_1\sin z_2\} + i\{-\sin z_1\cos z_2 - \cos z_1\sin z_2\}$$
$$= \cos(z_1+z_2) - i\sin(z_1+z_2). \quad (2)$$

Adding and subtracting (1) and (2) the formulas for $\cos(z_1+z_2)$ and $\sin(z_1+z_2)$ follow immediately.

Otherwise. The students can verify the truth of these results by writing down the definitions of $\cos z_1$, $\cos z_2$, $\sin z_1$, $\sin z_2$, etc. in terms of exponential functions.

▶ Example 2.2.17 ***Zeros of $\sin z$ and $\cos z$.***

Definition 2.2.1 *If for $z = \alpha$, $f(\alpha) = 0$, we say α is a **zero of** $f(z)$, or, we say that α is **a root of the equation** $f(z) = 0$.*

What are the zeros of $\sin z$ and $\cos z$?
Solution: Let $z = x + iy$ (x, y real).
Then

$$\sin z = 0 \Rightarrow \sin(x+iy) = 0$$
$$\Rightarrow \sin x\cosh y + i\cos x\sinh y = 0$$
$$\Rightarrow \sin x\cosh y = 0 \text{ and } \cos x\sinh y = 0.$$

Since x, y are real and $\cosh y \geq 1$ (i.e., $\cosh y$ cannot be zero), we conclude that $\sin x = 0$ and hence $x = n\pi (n = 0, \pm1, \pm2, \ldots)$. But for these values of x, $\cos x \neq 0$ and hence $\sinh y$ should be zero, i.e., y should be zero. Hence

$$\sin z = \sin(x+iy) = 0 \Rightarrow x = n\pi \text{ and } y = 0,$$
$$\text{i.e., } z = x + iy = n\pi\ (n = 0, \pm1, \pm2, \pm3, \ldots).$$

Similarly, prove that

$$\cos z = 0 \Rightarrow z = (2n+1)\frac{\pi}{2}\ (n = 0, \pm1, \pm2, \ldots).$$

► **Example 2.2.18** *If* $\tan(x+iy) = u+iv$ $(x, y, u, v$ *are all reals*), *prove that*

$$(u^2+v^2-1)\tan 2x + 2u = 0$$
$$(u^2+v^2+1)\tanh 2y - 2v = 0.$$

[C.P. 1992]

Solution: We use the fact that if $\tan(x+iy) = u+iv$, then $\tan(x-iy) = u-iv$. Then

$$\begin{aligned}\tan 2x &= \tan(x+iy+x-iy)\\ &= \frac{\tan(x+iy)+\tan(x-iy)}{1-\tan(x+iy)\cdot\tan(x-iy)}\\ &= \frac{(u+iv)+(u-iv)}{1-(u+iv)(u-iv)} = \frac{2u}{1-u^2-v^2},\end{aligned}$$

i.e., $(1-u^2-v^2)\cdot\tan 2x = 2u,$

or, $(u^2+v^2-1)\tan 2x + 2u = 0$ **(First part proved)**

and

$$\begin{aligned}\tan 2iy &= \tan[(x+iy)-(x-iy)]\\ &= \frac{\tan(x+iy)-\tan(x-iy)}{1+\tan(x+iy)\tan(x-iy)}\\ &= \frac{(u+iv)-(u-iv)}{1+(u+iv)(u-iv)} = \frac{2iv}{1+u^2+v^2},\end{aligned}$$

$$\text{or, } \frac{\sin(2iy)}{\cos(2iy)} = \frac{2iv}{1+u^2+v^2},$$

$$\text{or, } \frac{i\sinh 2y}{\cosh 2y} = \frac{2iv}{1+u^2+v^2} \quad [\because \sin(iz) = i\sinh z \text{ and } \cos(iz) = \cosh z]$$

$\therefore$ $(u^2+v^2+1)\tanh 2y - 2v = 0$ **(Second part proved)**

► **Example 2.2.19** *If* $\sin(x+iy) = u+iv$ $(x, y, u, v$ *are all reals*), *prove that*

$$u^2\operatorname{cosec}^2 x - v^2\sec^2 x = 1 \textit{ and } u^2\operatorname{sech}^2 y + v^2\operatorname{cosech}^2 y = 1.$$

Solution:

$$\begin{aligned}&\sin(x+iy) = u+iv\\ \Rightarrow\ &\sin x\cos iy + \cos x\sin iy = u+iv\\ \Rightarrow\ &\sin x\cosh y + i\cos x\sinh y = u+iv \quad \left[\because \begin{array}{l}\sin iy = i\sinh y,\\ \cos iy = \cosh y\end{array}\right]\\ \Rightarrow\ &u = \sin x\cosh y,\ v = \cos x\sinh y \qquad (1)\end{aligned}$$

$$\therefore \frac{u}{\sin x} = \cosh y, \quad \frac{v}{\cos x} = \sinh y$$

Now using $\cosh^2 y - \sinh^2 y = 1$, we get

$$\frac{u^2}{\sin^2 x} - \frac{v^2}{\cos^2 x} = 1,$$

$$\text{or,} \quad u^2 \operatorname{cosec}^2 x - v^2 \sec x = 1 \text{ (First part).}$$

Again, from (1) we write

$$\frac{u}{\cosh y} = \sin x \text{ and } \frac{v}{\sinh y} = \cos x$$

Since $\sin^2 x + \cos^2 x = 1$, it follows

$$\frac{u^2}{\cosh^2 y} + \frac{v^2}{\sinh^2 y} = 1$$

$$\text{or, } u^2 \operatorname{sech}^2 y + v^2 \operatorname{cosech}^2 y = 1 \text{ (Second part)}$$

▶ **Example 2.2.20** *If* α, β *are real and* $|\cos(\alpha + i\beta)| = 1$, *then prove that* $\sin^2 \alpha = \sinh^2 \beta$.

Solution:

$$\cos(\alpha + i\beta) = \cos\alpha \cos i\beta - \sin\alpha \sin i\beta = \cos\alpha \cosh\beta - i \sin\alpha \sinh\beta$$

$$\therefore \quad |\cos(\alpha + i\beta)^2| = \cos^2\alpha \cosh^2\beta + \sin^2\alpha \sinh^2\beta = 1$$

(by the given condition),

$$\text{or,} \quad (1 - \sin^2\alpha)\cosh^2\beta + \sin^2\alpha(\cosh^2\beta - 1) = 1,$$

$$\text{or,} \quad \cosh^2\beta - 1 = \sin^2\alpha, \text{ i.e., } \sin^2\alpha = \sinh^2\beta.$$

2.3 Inverse Functions: Trigonometric and Hyperbolic Functions

Definition 2.3.1 *Let z be a given complex number and suppose w is such a complex number that* $\sin w = z$. *Then w is called an* **inverse sine** *of z, or* $w = \operatorname{Sin}^{-1} z$.

Let us take an example:

▶ **Example 2.3.1 Find all those complex numbers z for which $\sin z = 2$.**

Solution:

$$\sin z = 2 \Rightarrow \frac{e^{iz} - e^{-iz}}{2i} = 2, \quad \text{or,} \quad e^{iz} - \frac{1}{e^{iz}} = 4i,$$

$$\text{or,} \quad (e^{iz})^2 - 4i(e^{iz}) - 1 = 0.$$

$$\therefore e^{iz} = \left(2 \pm \sqrt{3}\right) i.$$

(i) First take, $e^{iz} = \left(2 + \sqrt{3}\right) i$. Then $iz = \operatorname{Log}\left(2 + \sqrt{3}\right) i$

We know that $\operatorname{Log} z = \log|z| + i(\theta + 2n\pi)$, $n = 0, \pm1, \pm2$, etc. when θ is the principal amplitude of z, i.e., $-\pi < \theta \leq \pi$

$$\therefore\ iz = \log\left|(2+\sqrt{3})i\right| + i\left(\frac{\pi}{2}+2n\pi\right)$$

$$\text{i.e., } z = \frac{1}{i}\log\left(2+\sqrt{3}\right) + \left\{\frac{\pi}{2}+2n\pi\right\}.$$

$$\therefore\ z = -i\log\left(2+\sqrt{3}\right) + \frac{\pi}{2} + 2n\pi = 2n\pi + \left[\frac{\pi}{2} - i\log(2+\sqrt{3})\right].$$

(ii) Next take

$$e^{iz} = \left(2-\sqrt{3}\right)i,$$

$$\begin{aligned} \text{or, } iz &= \text{Log}(2-\sqrt{3})i = \log\left|(2-\sqrt{3})i\right| + i\left(\frac{\pi}{2}+2n\pi\right) \\ &= \log(2-\sqrt{3}) + i\left(\frac{\pi}{2}+2n\pi\right) = -\log(2+\sqrt{3}) + i\left(\frac{\pi}{2}+2n\pi\right), \\ \text{i.e., } z &= i\log(2+\sqrt{3}) + \frac{\pi}{2} + 2n\pi = (2n+1)\pi - \left\{\frac{\pi}{2} - i\log(2\sqrt{3})\right\}, \end{aligned}$$

i.e., values of z for which $\sin z = 2$ are

$$n\pi + (-1)^n\left\{\frac{\pi}{2} - i\log(2+\sqrt{3})\right\}.$$

We then write

$$\text{Sin}^{-1} 2 = n\pi + (-1)^n\left\{\frac{\pi}{2} - i\log(2+\sqrt{3})\right\}.$$

The particular value when $n = 0$, is the principal value of inverse sine of 2, denoted by $\sin^{-1} 2$.

$$\text{i.e., } \sin^{-1} 2 = \frac{\pi}{2} - i\log(2+\sqrt{3}).$$

Note 1. $\sin^{-1} 2 = n\pi + (-1)^n \sin^{-1} 2$.

▶ Example 2.3.2 *Express* $\text{Cos}^{-1} i$ *in the form of* $A + iB$ (A, B *real*).

Solution: Let $\text{Cos}^{-1} i = \alpha + i\beta$ (α, β real).
Then,

$$\cos(\alpha + i\beta) = i,$$

$$\text{or, } \cos\alpha\cosh\beta - i\sin\alpha\sinh\beta = i.$$

Equating real and imaginary parts we get

$$\cos\alpha\cosh\beta = 0$$

$$\text{and } \sin\alpha\sinh\beta = -1.$$

Since $\cosh\beta > 0$, $\cos\alpha = 0$ from the first relation and hence

$$\alpha = (2k+1)\frac{\pi}{2}.$$

(i) If $k = 2m$, then $\alpha = 2m\pi + \frac{\pi}{2}$ and $\sin\alpha = 1$ and hence

$$\begin{aligned}\sinh\beta = -1, \quad &\text{i.e.,}\quad e^{\beta} - e^{-\beta} = -2,\\ &\text{i.e.,}\quad e^{2\beta} + 2e^{\beta} - 1 = 0,\\ &\text{i.e.,}\quad e^{\beta} = \frac{-2 \pm \sqrt{8}}{2} = \sqrt{2} - 1 \;(\because\ e^{\beta} \neq \text{negative}),\\ &\text{or,}\quad \beta = \log(\sqrt{2} - 1).\end{aligned}$$

$$\begin{aligned}\therefore\ \operatorname{Cos}^{-1} i &= 2m\pi + \frac{\pi}{2} + i\log(\sqrt{2} - 1)\\ &= 2m\pi + \left\{\frac{\pi}{2} - i\log(\sqrt{2} + 1)\right\}.\end{aligned}$$

(ii) If $k = 2m - 1$, then $\alpha = 2m\pi - \frac{\pi}{2}$ and then $\sinh\beta = 1$ whence, as before $e^{\beta} = \sqrt{2} + 1$ (neglecting negative value for e^{β}), i.e., $\beta = \log(\sqrt{2} + 1)$.

$$\therefore\ \operatorname{Cos}^{-1} i = 2m\pi - \left\{\frac{\pi}{2} - i\log(\sqrt{2} + 1)\right\}.$$

$$\therefore\ \operatorname{Cos}^{-1} i = 2m\pi \pm \cos^{-1} i,$$

where $\cos^{-1} i$ = principal value of

$$\operatorname{Cos}^{-1} i = \frac{\pi}{2} - i\log(\sqrt{2} + 1).$$

Remember. $\operatorname{Cos}^{-1} Z = 2n\pi \pm \cos^{-1} z$, **for any complex number** z.

▶ Example 2.3.3 *Express* $\operatorname{Tan}^{-1}(1 + i)$ *in the form* $A + iB$ (A, B *real*).

Solution: Let

$$\operatorname{Tan}^{-1}(1 + i) = A + iB \ (A, B \text{ real}).$$

Then

$$\tan(A + iB) = 1 + i.$$

We can then write

$$\tan(A - iB) = 1 - i.$$

Now,

$$\begin{aligned}\tan 2A &= \tan(A + iB + A - iB)\\ &= \frac{\tan(A + iB) + \tan(A - iB)}{1 - \tan(A + iB)\tan(A - iB)}\\ &= \frac{1 + i + 1 - i}{1 - (1 + i)(1 - i)} = -2\end{aligned}$$

whence $2A = 2n\pi + [\pi + \tan^{-1}(-2)]$.

$$\therefore\ A = \frac{1}{2}[(2n + 1)\pi + \tan^{-1}(-2)].$$

Again

$$\begin{aligned}\tan 2iB &= \tan[(A+iB)-(A-iB)]\\ &= \frac{\tan(A+iB)-\tan(A-iB)}{1+\tan(A+iB)\tan(A-iB)}\\ &= \frac{(1+i)-(1-i)}{1+(1+i)(1-i)} = \frac{2i}{3},\end{aligned}$$

i.e., $i\tanh 2B = \dfrac{2i}{3}$, or, $\tanh 2B = \dfrac{2}{3}$,

or, $\dfrac{e^{2B}-e^{-2B}}{e^{2B}+e^{-2B}} = \dfrac{2}{3}$, or, $\dfrac{e^{4B}-1}{e^{4B}+1} = \dfrac{2}{3}$,

or, $e^{4B} = 5$ (Componendo and Dividendo) or, $B = \dfrac{1}{4}\log 5$.

$$\begin{aligned}\therefore\ \mathrm{Tan}^{-1}(1+i) &= A+iB\\ &= \frac{1}{2}[(2n+1)\pi + \tan^{-1}(-2)] + i\cdot\frac{1}{4}\log 5\\ &= n\pi + \left[\frac{\pi}{2} + \frac{1}{2}\tan^{-1}(-2)\right] + \frac{i}{4}\log 5.\end{aligned}$$

Principal value of

$$\begin{aligned}\mathrm{Tan}^{-1}(1+i) &= \frac{\pi}{2} + \frac{\tan^{-1}(-2)}{2} + \frac{i}{4}\log 5\\ &= \tan^{-1}(1+i).\end{aligned}$$

$$\therefore\ \mathrm{Tan}^{-1}\,1+i = n\pi + \tan^{-1}(1+i).$$

Remember. $\mathrm{Tan}^{-1} z = n\pi + \tan^{-1} z$, **$z$ being any complex number.**

▶ Example 2.3.4 *If* $\mathrm{Cos}^{-1}(x+iy) = u+iv$ $(x>0, y>0)$, *then prove that* $\cos^2 u$ *and* $\cosh^2 v$ *are the roots of the equation*

$$t^2 - t(1+x^2+y^2) + x^2 = 0.$$

Solution: Given: $\mathrm{Cos}^{-1}(x+iy) = u+iv$ (u, v real).

$$\begin{aligned}\therefore\ \cos(u+iv) = x+iy,\quad \text{i.e.,}\quad &\cos u\cos iv - \sin u \sin iv = x+iy,\\ \text{i.e.,}\quad &\cos u\cosh v - i\sin u\sinh v = x+iy\end{aligned}$$

and hence $\cos u\cosh v = x$ and $-\sin u\sinh v = y$.

$$\begin{aligned}\therefore\ 1+x^2+y^2 &= 1+\cos^2 u\cosh^2 v + \sin^2 u\sinh^2 v\\ &= 1+\cos^2 u(1+\sinh^2 v) + (1-\cos^2 u)\sinh^2 v\\ &= 1+\cos^2 u + \sinh^2 v\\ &= \cos^2 u + \cosh^2 v.\end{aligned}$$

Hence, $\cos^2 u$ and $\cosh^2 v$ are the two roots of the quadratic equation

$$t^2 - (\cos^2 u + \cosh^2 v)t + \cos^2 u\cosh^2 v = 0,$$

i.e., $t^2 - (1+x^2+y^2)t + x^2 = 0$ **(Proved).**

2.4 Laws of Logarithms

If z_1, z_2, z_3 are complex numbers, then

(a) $\operatorname{Log} z_1 + \operatorname{Log} z_2 = \operatorname{Log} z_1 z_2$.

(b) $\operatorname{Log} z_1 - \operatorname{Log} z_2 = \operatorname{Log}(z_1/z_2)$.

(c) $\operatorname{Log} \frac{1}{z} = -\operatorname{Log} z$.

(d) $\operatorname{Log} z^m = m \operatorname{Log} z$ (m is a rational number).

In each of the above cases it is to be remembered that the equality is **not always** TRUE. The interpretations are as follows:

(a) Every value $\operatorname{Log} z_1 + \operatorname{Log} z_2$ = **Some value** of $\operatorname{Log} z_1 z_2$. **Conversely, every value** of $\operatorname{Log} z_1 z_2$ **is some value** of $\operatorname{Log} z_1 + \operatorname{Log} z_2$. **Similar interpretations for others:** e.g., in **(d)** Every value of $m \operatorname{Log} z$ is some value of $\operatorname{Log} z^m$.

▶ Example 2.4.1

$$\operatorname{Log} i^2 = \operatorname{Log}(\cos\pi + i\sin\pi) = \log 1 + i(\pi + 2k\pi).$$

$$2\operatorname{Log} i = 2\operatorname{Log}\left(\cos\frac{\pi}{2} + i\sin\frac{\pi}{2}\right) = 2i\left(\frac{\pi}{2} + 2p\pi\right).$$

Here every value of $2\operatorname{Log} i$ is some value of $\operatorname{Log} i^2$; but only **some values** of $\operatorname{Log} i^2$ are values of $2\operatorname{Log} i$ [the others are values of $2\operatorname{Log}(-i)$]

▶ Example 2.4.2 *The results* (a) *and* (b) *are not necessarily true for principal values.*

Let $z_1 = \cos\frac{3\pi}{4} + i\sin\frac{3\pi}{4}$, $z_2 = \cos\frac{3\pi}{4} + i\sin\frac{3\pi}{4}$.
Thus,

$$\begin{aligned}\log z_1 + \log z_2 &= \log\left(\cos\frac{3\pi}{4} + i\sin\frac{3\pi}{4}\right) + \log\left(\cos\frac{3\pi}{4} + i\sin\frac{3\pi}{4}\right)\\ &= i\left(\frac{3\pi}{4}\right) + i\frac{3\pi}{4} = i\frac{3\pi}{2}\end{aligned}$$

but

$$\begin{aligned}\log(z_1 z_2) &= \log\left\{\left(\cos\frac{3\pi}{4} + i\sin\frac{3\pi}{4}\right)\left(\cos\frac{3\pi}{4} + i\sin\frac{3\pi}{4}\right)\right\}\\ &= \log\left[\cos\left(\frac{3\pi}{4} + \frac{3\pi}{4}\right) + i\sin\left(\frac{3\pi}{4} + \frac{3\pi}{4}\right)\right]\\ &= \log\left[\cos\left(-\frac{\pi}{2}\right) + i\sin-\left(\frac{\pi}{2}\right)\right] = i\left(-\frac{\pi}{2}\right),\end{aligned}$$

i.e., $\log z_1 + \log z_2 \neq \log z_1 z_2$ **(Principal values being taken).**

2.5 Solved Problems

▶ Example 2.5.1 *Express* $\cos(\alpha + i\beta)$, α, β *are real numbers in the form* $A + iB$ $(A, B$ *real*$)$. **[C.P. 1987]**

Solution:

$$\begin{aligned}\cos(\alpha + i\beta) &= \cos\alpha\cos i\beta - \sin\alpha\sin(i\beta)\\ &= \cos\alpha\cosh\beta - i\sin\alpha\sinh\beta\\ &= A + iB,\end{aligned}$$

where $A = \cos\alpha\cosh\beta$ and $B = -\sin\alpha\sinh\beta$.

Alternatively

$$\begin{aligned}\cos(\alpha + i\beta) &= \frac{e^{i(\alpha+i\beta)} + e^{-i(\alpha+i\beta)}}{2}\\ &= \frac{e^{-\beta}.e^{i\alpha} + e^{\beta}.e^{-i\alpha}}{2}\\ &= \frac{e^{-\beta}(\cos\alpha + i\sin\alpha) + e^{\beta}(\cos\alpha - i\sin\alpha)}{2}\\ &= \cos\alpha\left(\frac{e^{\beta} + e^{-\beta}}{2}\right) - i\sin\alpha\left(\frac{e^{\beta} - e^{-\beta}}{2}\right)\\ &= \cos\alpha\cosh\beta - i\sin\alpha\sinh\beta.\end{aligned}$$

▶ Example 2.5.2 *If* $\tan(\theta + i\phi) = \tan\beta + i\sec\beta (0 < \beta < \pi)$, *prove that* $2^{2\phi} = \cot\frac{\beta}{2}$.

Solution:

$$\tan(\theta + i\phi) = \tan\beta + i\sec\beta \text{ (given)}.$$

Then, it is also true,

$$\tan(\theta - i\phi) = \tan\beta - i\sec\beta.$$

$$\therefore \quad \tan(2i\phi) = \tan\{(\theta + i\phi) - (\theta - i\phi)\},$$

$$\text{or,} \quad \frac{\sin(2i\phi)}{\cos(2i\phi)} = \frac{\tan(\theta + i\phi) - \tan(\theta - i\phi)}{1 + \tan(\theta + i\phi)\tan(\theta - i\phi)},$$

$$\text{or,} \quad \frac{i\sinh 2\phi}{\cosh 2\phi} = \frac{(\tan\beta + i\sec\beta) - (\tan\beta - i\sec\beta)}{1 + (\tan\beta + i\sec\beta)(\tan\beta - i\sec\beta)},$$

$$\text{or,} \quad \tan(2i\phi) = \frac{2i\sec\beta}{1 + \tan^2\beta + \sec^2\beta} = \frac{i}{\sec\beta},$$

$$\text{or,} \quad i\tanh 2\phi = i\cos\beta, \text{ i.e., } \tanh 2\phi = \cos\beta,$$

$$\text{or,} \quad \frac{e^{2\phi} - e^{-2\phi}}{e^{2\phi} + e^{-2\phi}} = \cos\beta.$$

By componendo and dividendo we get

$$\frac{2e^{2\phi}}{2e^{-2\phi}} = \frac{1 + \cos\beta}{1 - \cos\beta} = \frac{2\cos^2\frac{\beta}{2}}{2\sin^2\frac{\beta}{2}},$$

$$\text{i.e.,} \quad e^{4\phi} = \text{col}^2\frac{\beta}{2} \Rightarrow e^{2\phi} = \cot\frac{\beta}{2} \textbf{ (Proved).}$$

▶ Example 2.5.3 *If* $\alpha + i\beta = \cot(x + iy)$, *prove that*

(i) $\alpha + \beta^2 - 2\alpha \cot 2x = 1$, *and*

(ii) $\dfrac{\alpha^2 + \beta^2 - 1}{\alpha^2 + \beta^2 + 1} = \dfrac{\cos 2x}{\cosh 2y}$.

Solution: Since $\alpha + i\beta = \cot(x + iy)$, we can also write $\alpha - i\beta = \cot(x - iy)$.

(i)
$$\begin{aligned}\cot 2x &= \cot\{(x + iy) + (x - iy)\}\\ &= \frac{\cot(x+iy)\cot(x-iy) - 1}{\cot(x+iy) + \cot(x-iy)}\\ &= \frac{(\alpha+i\beta)(\alpha-i\beta) - 1}{(\alpha+i\beta)+(\alpha-i\beta)}\\ &= \frac{\alpha^2+\beta^2-1}{2\alpha},\end{aligned} \tag{1}$$

or, $2\alpha \cot 2x = \alpha^2 + \beta^2 - 1$,

i.e., $\alpha^2 + \beta^2 - 2\alpha \cot 2x = 1$ **(Proved).**

(ii)
$$\cot(2iy) = \cot\{(x+iy) - (x-iy)\} = \frac{\cot(x-iy)\cot(x+iy) + 1}{\cot(x-iy) - \cot(x+iy)}$$

or, $\dfrac{\cos(2iy)}{\sin(2iy)} = \dfrac{\alpha^2+\beta^2+1}{-2i\beta}$, or, $\dfrac{\cosh 2y}{i \sinh 2y} = \dfrac{\alpha^2+\beta^2+1}{-2i\beta}$

i.e., $\dfrac{\cosh 2y}{\sinh 2y} = -\dfrac{\alpha^2+\beta^2+1}{2\beta}$. (2)

Now,

$$\begin{aligned}\cosh^2 2y &= \frac{\cosh^2 2y}{\cosh^2 2y - \sinh^2 2y} = \frac{1}{1 - \tanh^2 2y}\\ &= \frac{1}{1 - \frac{4\beta^2}{(\alpha^2+\beta^2+1)^2}} \text{ [from (2)]}\\ &= \frac{(\alpha^2+\beta^2+1)^2}{(\alpha^2+\beta^2+1)^2 - 4\beta^2} = \frac{(\alpha^2+\beta^2+1)^2}{(\alpha^2+\beta^2-1)^2 + 4\alpha^2}\end{aligned}$$

[See that we can write
$(\alpha^2+\beta^2+1)^2 - 4\beta^2 = (\alpha^2+\beta^2-1)^2 + 4\alpha^2$]

$$\text{or, } \cosh 2y = \frac{\alpha^2+\beta^2+1}{\sqrt{(\alpha^2+\beta^2-1)^2 + 4\alpha^2}} \tag{3}$$

From the relation (1)

$$\cos 2x = \frac{\alpha^2+\beta^2-1}{\sqrt{(\alpha^2+\beta^2-1)^2 + 4\alpha^2}}. \tag{4}$$

From (3) and (4) we get

$$\frac{\cos 2x}{\cosh 2y} = \frac{\alpha^2+\beta^2-1}{\alpha^2+\beta^2+1}. \text{ **(Proved).**}$$

▶ **Example 2.5.4** *If* $\tan\log(x+iy) = a+ib(a^2+b^2 \neq 1)$, *prove that*

$$\tan\log(x^2+y^2) = \frac{2a}{1-a^2-b^2}.$$ [C.P. 1993]

Solution: From the given condition

$$\begin{aligned} \log(x+iy) &= \tan^{-1}(a+ib). \\ \therefore\ \log(x-iy) &= \tan^{-1}(a-ib). \end{aligned}$$

$$\begin{aligned} \therefore\ \log(x+iy)(x-iy) &= \log(x+iy)+\log(x-iy) \\ &= \tan^{-1}(a+ib)+\tan^{-1}(a-ib) \\ &= \tan^{-1}\frac{a+ib+a-ib}{1-(a+ib)(a-ib)}, \\ \text{i.e., } \log(x^2+y^2) &= \tan^{-1}\frac{2a}{1-a^2-b^2}, \\ \text{or, } \tan\log(x^2+y^2) &= \frac{2a}{1-a^2-b^2}\ \textbf{(Proved).} \end{aligned}$$

▶ **Example 2.5.5** *Prove that*

$$\sin\left(i\log\frac{a-ib}{a+ib}\right) = \frac{2ab}{a^2+b^2}\ (a, b\ \textit{real}).$$ [C.P. 2001]

Solution: Put $a = r\cos\theta$, $b = r\sin\theta$, i.e., $\tan\theta = \frac{b}{a}$.

$$\therefore\ \frac{a-ib}{a+ib} = \frac{r(\cos\theta - i\sin\theta)}{r(\cos\theta + i\sin\theta)} = \frac{e^{-i\theta}}{e^{i\theta}} = e^{-2i\theta}.$$

Hence

$$-2i\theta = \log\frac{a-ib}{a+ib}, \text{ or, } 2\theta = i\log\frac{a-ib}{a+ib}.$$

$$\therefore\ \sin\left(i\log\frac{a-ib}{a+ib}\right) = \sin 2\theta = \frac{2\tan\theta}{1+\tan^2\theta} = \frac{2\frac{b}{a}}{1+b^2/a^2} = \frac{2ab}{a^2+b^2}\ \textbf{(Proved).}$$

▶ **Example 2.5.6** *Show that the equation*

$$\tan\left(i\log\frac{x-iy}{x+iy}\right) = 2$$

represents the hyperbola $x^2 - y^2 = xy$. [C.P. 1965]

Solution: Put $x = r\cos\theta$, $y = r\sin\theta$.
Then,

$$\begin{aligned} \frac{x-iy}{x+iy} &= \frac{\cos\theta - i\sin\theta}{\cos\theta + i\sin\theta} \\ &= \frac{(\cos\theta - i\sin\theta)^2}{(\cos\theta + i\sin\theta)(\cos\theta - i\sin\theta)} \\ &= \cos 2\theta - i\sin 2\theta. \end{aligned}$$

$$\text{Log}\left(\frac{x-iy}{x+iy}\right) = \text{Log}(\cos 2\theta - i\sin 2\theta)$$
$$= \log|(\cos 2\theta - i\sin 2\theta)| + i(-2\theta + 2k\pi).$$

$$\therefore \log\frac{x-iy}{x+iy} = -2i\theta \text{ (Principal Value).}$$

$$\therefore \tan\left(i\log\frac{x-iy}{x+iy}\right) = 2 \text{ gives}$$

$$\tan[i(-2i\theta)] = 2, \text{ or, } \tan 2\theta = 2, \text{ i.e., } \frac{2\tan\theta}{1-\tan^2\theta} = 2,$$
$$\text{or, } \frac{2\frac{y}{x}}{1-y^2/x^2} = 2,$$
$$\text{or, } x^2 - y^2 = xy \text{ (which represents a hyperbola).}$$

▶ **Example 2.5.7** *Prove that* $i\log\frac{x-i}{x+i} = \pi - 2\tan^{-1}x$. **[C.P. 1987]**
Similar to the previous ***Example.*** *Put* $x = r\cos\theta$ *:nd* $1 = r\sin\theta$ *and proceed.*

▶ **Example 2.5.8** *Solve for z, the equation* $\cos z = -2$. **[C.P. 1986]**

Solution: $\cos z = -2$ gives

$$e^{iz} + e^{-iz} = -4, \text{ or, } e^{2iz} + 4e^{iz} + 1 = 0,$$
$$\text{i.e., } e^{iz} = \frac{-4 \pm \sqrt{16-4}}{2} = -2 \pm \sqrt{3}.$$

When, $e^{iz} = -2 + \sqrt{3}$, $iz = \text{Log}(-2+\sqrt{3}) = \log(2-\sqrt{3}) + i(\pi + 2k\pi)$ $(k = 0, \pm1, \pm2, \ldots)$, i.e., $z = (2k+1)\pi - i\log(2-\sqrt{3})$.

But when $e^{iz} = -2 - \sqrt{3}$, $iz = \text{Log}(-2-\sqrt{3}) = \log(2+\sqrt{3}) + i(\pi + 2k\pi)$ $(k = 0, \pm1, \pm2, \ldots)$, i.e., $z = (2k+1)\pi - i\log(2+\sqrt{3}) = (2k+1)\pi + i\log(2-\sqrt{3})$

$$\left(\because \log(2+\sqrt{3}) = \log\frac{(2+\sqrt{3})(2-\sqrt{3})}{2-\sqrt{3}} = -\log(2-\sqrt{3})\right)$$

$$\therefore z = (2k+1)\pi \mp i\log(2-\sqrt{3}).$$

▶ **Example 2.5.9** *Express* $Log\{Log(e^{i\theta})\}$ *in the form* $A + iB$ $(-\pi < \theta \le \pi)$.

Solution:

$$\text{Log}\, e^{i\theta} = \text{Log}(\cos\theta + i\sin\theta)$$
$$= \log|\cos\theta + i\sin\theta| + i(2n\pi + \theta)$$
$$= \log 1 + i(2n\pi + \theta) \text{ (n is any integer)}$$
$$= i(2n\pi + \theta) \text{ (See Ex. 2.2.4).}$$

$$\therefore \text{Log}(\text{Log}\, e^{i\theta}) = \text{Log}\{i(2n\pi + \theta)\}$$
$$= \text{Log}(2n\pi + \theta)\left[\cos\left(2k\pi + \frac{\pi}{2}\right) + i\sin\left(2k\pi + \frac{\pi}{2}\right)\right]$$
$$= \log(2n\pi + \theta) + i\left[2k\pi + \frac{\pi}{2}\right] \text{ (in the form } A + iB).$$

▶ Example 2.5.10 *Express* $\log(1+\cos 2\theta + i\sin 2\theta)$, *where* $\frac{\pi}{2} < \theta < \frac{3\pi}{2}$ *in the form* $A + iB$ (A, B *real*).

Solution: We first find the general case:

$$\text{Log}(1+\cos 2\theta + i\sin 2\theta) = \text{Log}(2\cos\theta)[\cos\theta + i\sin\theta]. \quad \textbf{(1)}$$

For

$$\frac{\pi}{2} < \theta < \frac{3\pi}{2}, \quad \cos\theta < 0;$$

also

$$\frac{\pi}{2} - \pi < \theta - \pi < \frac{3\pi}{2} - \pi, \text{ i.e., } -\frac{\pi}{2} < \theta - \pi < \frac{\pi}{2}.$$

So we write (1) as

$$\begin{aligned} &= \text{Log}(-2\cos\theta)[\cos(\theta-\pi) + i\sin(\theta-\pi)] \\ &= \log(-2\cos\theta) + i(\theta - \pi + 2k\pi) \ (k = 0, \pm1, \pm2, \ldots). \end{aligned}$$

Putting

$$\begin{aligned} k &= 0 \text{ we get the principal value } \log(1+\cos 2\theta + i\sin 2\theta) \\ &= \log(-2\cos\theta) + i(\theta - \pi) \\ &= A + iB, \text{ where } A = \log(-2\cos\theta) \text{ and} \\ &\qquad B = \theta - \pi. \textbf{ (Proved).} \end{aligned}$$

▶ Example 2.5.11 *If* $u + iv = \tan(x+iy)$, *show that* $u^2 + v^2 + 2u\cot 2x = 1$. *Also prove that* $u^2 + v^2 + 1 - 2v\coth 2y = 0$. **[C.P. 1992, 2004]**

Solution: $u + iv = \tan(x+iy)$. Then, $u - iv = \tan(x - iy)$.

$$\begin{aligned} \therefore \tan 2x &= \tan\left(\overline{x+iy} + \overline{x-iy}\right) \\ &= \frac{\tan(x+iy) + \tan(x-iy)}{1 - \tan(x+iy)\tan(x-iy)} \\ &= \frac{u + iv + u - iv}{1 - (u+iv)(u-iv)} \\ &= \frac{2u}{1 - u^2 - v^2}. \end{aligned}$$

$$\therefore \cot 2x = \frac{1 - u^2 - v^2}{2u} \text{ or, } u^2 + v^2 + 2u\cot 2x = 1 \textbf{ (Proved first part)}$$

Again

$$\begin{aligned} \tan(2iy) &= \tan\{(x+iy) - (x-iy)\} \\ &= \frac{\tan(x+iy) - \tan(x-iy)}{1 + \tan(x+iy)\tan(x-iy)} \\ &= \frac{2iv}{1+u^2+v^2}, \text{ i.e., } i\tanh 2y = \frac{2iv}{1+u^2+v^2}, \text{ or, } \coth 2y = \frac{1+u^2+v^2}{2v}, \\ &\qquad \text{or, } u^2 + v^2 + 1 - 2v\coth 2y = 0 \textbf{ (Proved second part).} \end{aligned}$$

▶ **Example 2.5.12** *If* $\cosh^{-1}(x+iy)+\cosh^{-1}(x-iy)=\cosh^{-1}a$ *(x, y, a are real and $a>1$), prove that (x,y) lies on an ellipse.*

Solution: Let

$$\cosh^{-1}(x+iy)=\alpha+i\beta\ (\alpha, \beta \text{ real}).$$

Then

$$\cosh^{-1}(x-iy)=\alpha-i\beta.$$

$\therefore$ from the given relation it follows

$$\alpha+i\beta+\alpha-i\beta=\cosh^{-1}a, \text{ i.e., } \cosh 2\alpha=a. \tag{1}$$

From $\cosh^{-1}(x+iy)=\alpha+i\beta$, we write

$$\begin{aligned}
x+iy &= \cosh(\alpha+i\beta)\\
&= \cosh\alpha\cosh i\beta+\sinh\alpha\sinh i\beta\\
&= \cosh\alpha\frac{e^{i\beta}+e^{-i\beta}}{2}+\sinh\alpha\frac{e^{i\beta}-e^{-i\beta}}{2},\\
\text{i.e., } x+iy &= \cosh\alpha\cos\beta+i\sinh\alpha\sin\beta\\
\Rightarrow x &= \cosh\alpha\cos\beta \text{ and } y=\sinh\alpha\sin\beta.
\end{aligned}$$

From this it follows

$$\left(\frac{x}{\cosh\alpha}\right)^2+\left(\frac{y}{\sinh\alpha}\right)^2=1. \tag{2}$$

From (1),

$$2\cosh^2\alpha=1+a \text{ and } 1+2\sinh^2\alpha=a.$$

Hence, (2) becomes

$$\frac{x^2}{(1+a)/2}+\frac{y^2}{(a-1)/2}=1 \text{ (an ellipse since } a>1).$$

▶ **Example 2.5.13** *If* $y=\log\tan\left(\frac{\pi}{4}+\frac{x}{2}\right)$, *then prove that* $x=-i\,Log\tan\left(\frac{\pi}{4}+i\frac{y}{2}\right)$.

Solution: From the given relation we write

$$e^y=\tan\left(\frac{\pi}{4}+\frac{x}{2}\right)=\frac{1+\tan\frac{x}{2}}{1-\tan\frac{x}{2}}.$$

By componendo and dividendo

$$\frac{e^y-1}{e^y+1}=\tan\frac{x}{2}, \text{ or, } \frac{e^{y/2}-e^{-y/2}}{e^{y/2}+e^{-y/2}}=\tan\frac{x}{2}, \text{ i.e., } \tanh\frac{y}{2}=\tan\frac{x}{2}.$$

This gives $$\frac{\frac{e^{ix/2}-e^{-ix/2}}{2i}}{\frac{e^{ix/2}+e^{-ix/2}}{2}}=\frac{\sinh\frac{y}{2}}{\cosh\frac{y}{2}}=\frac{1}{i}\tan\frac{iy}{2}.$$

[**Remember.** $\sinh z = \frac{1}{i}\sin(iz)$ and $\cosh z = \cos(iz)$.]

$$\text{or,}\quad \frac{e^{ix}-1}{e^{ix}+1} = \tan i\frac{y}{2}.$$

Again, by componendo and dividendo

$$e^{ix} = \frac{1+\tan\frac{iy}{2}}{1-\tan\frac{iy}{2}} = \tan\left(\frac{\pi}{4}+i\frac{y}{2}\right).$$

Hence

$$ix = \operatorname{Log}\tan\left(\frac{\pi}{4}+i\frac{y}{2}\right),\ \text{or, } x = -i\operatorname{Log}\tan\left(\frac{\pi}{4}+i\frac{y}{2}\right)\ \textbf{(Proved)}.$$

▶ Example 2.5.14 *When x is real and $|x| \le 1$ and $-\frac{\pi}{4} \le \tan^{-1}x \le \frac{\pi}{4}$, deduce that*

$$\tan^{-1}x = x - \frac{x^3}{3} + \frac{x^5}{5} - \frac{x^7}{7} + \cdots$$

*This is called **Gregory's series.***

Solution: Let θ be real.

Then,

$$\begin{aligned} e^{i\theta} &= \cos\theta + i\sin\theta \\ \text{and } e^{-i\theta} &= \cos\theta - i\sin\theta. \end{aligned}$$

$$\therefore\ \frac{e^{i\theta}}{e^{-i\theta}} = \frac{\cos\theta+i\sin\theta}{\cos\theta-i\sin\theta} = \frac{1+i\tan\theta}{1-i\tan\theta}\ \left(\theta \ne (2n+1)\frac{\pi}{2}\right).$$

$$\therefore\ \operatorname{Log}\frac{1+i\tan\theta}{1-i\tan\theta} = 2i\theta \text{ or } \operatorname{Log}(1+i\tan\theta) - \operatorname{Log}(1-i\tan\theta) = 2i\theta$$

$$\Rightarrow\ \log(1+i\tan\theta) - \log(1-i\tan\theta) + 2n\pi i = 2i\theta$$

$$\Rightarrow\ \left(i\tan\theta - \frac{i^2\tan^2\theta}{2} + \frac{i^3\tan^3\theta}{3} - \cdots\right)$$

$$-\left(-i\tan\theta - \frac{i^2\tan^2\theta}{2} - \frac{i^3\tan^3\theta}{3} - \cdots\right)(-1 \le \tan\theta \le 1) + 2n\pi i = 2i\theta.$$

[Using the expansion of $\log(1+z)$ as the infinite series $z - \frac{z^2}{2} + \frac{z^3}{3} - \cdots$ ($|z| < 1$).]

We thus get

$$2i\theta = 2n\pi i + 2i\left[\tan\theta - \frac{\tan^3\theta}{3} + \frac{\tan^5\theta}{5} - \cdots\right].$$

$$\theta - n\pi = \tan\theta - \frac{\tan^3\theta}{3} + \frac{\tan^5\theta}{5} - \cdots\ (-1 \le \tan\theta \le 1).$$

Taking the principal value

$$\theta = \tan\theta - \frac{\tan^3\theta}{3} + \frac{\tan^5\theta}{5} - \cdots\ \left(-1 \le \tan\theta \le 1; -\frac{\pi}{4} \le \theta \le \frac{\pi}{4}\right).$$

Let $\tan\theta = x$.

$$\therefore\ \tan^{-1}x = x - \frac{x^3}{3} + \frac{x^5}{5}\cdots\ \left(-1 \le x \le 1; -\frac{\pi}{4} \le \tan^{-1}x \le \frac{\pi}{4}\right).$$

An Important Example

▶ **Example 2.5.15 (a)** *If* $\log\sin(\theta+i\phi)=\alpha+i\beta$ (θ,ϕ,α,β *are reals*), *then prove that* $2e^{2\alpha}=\cosh 2\phi-\cos 2\theta$. **[C.P. 2005]**

Solution: Since $\log\sin(\theta+i\phi)=\alpha+i\beta$, we can write

$$\sin(\theta+i\phi)=e^{\alpha+i\beta}=e^{\alpha}\cdot e^{i\beta}=e^{\alpha}(\cos\beta+i\sin\beta),$$

$$\text{or,}\quad \sin\theta\cos i\phi+\cos\theta\sin i\phi=e^{\alpha}(\cos\beta+i\sin\beta),$$

$$\text{or,}\quad \sin\theta\cosh\phi+i\cos\theta\sinh\phi=e^{\alpha}\cos\beta+ie^{\alpha}\sin\beta$$

$$(\because \cos i\phi=\cosh\phi,\ \sin i\phi=i\sinh\phi).$$

Equating real and imaginary parts, we get $\sin\theta\cosh\phi=e^{\alpha}\cos\beta$
$\cos\theta\sinh\phi=e^{\alpha}\sin\beta$.

Squaring and adding we obtain

$$\begin{aligned}
e^{2\alpha}(\cos^2\beta+\sin^2\beta) &= \sin^2\theta\cosh^2\phi+\cos^2\theta\sinh^2\phi,\\
\text{or, } 2e^{2\alpha} &= 2\sin^2\theta\cosh^2\phi+2\cos^2\theta\sinh^2\phi\\
&= \sin^2\theta(1+\cosh 2\phi)+\cos^2\theta(\cosh 2\phi-1)\\
&\qquad (\because 2\cosh^2\phi=1+\cosh 2\phi\\
&\qquad \text{and } 2\sinh^2\phi=\cosh 2\phi-1)\\
&= \sin^2\theta\cosh 2\phi+\cos^2\theta\cosh 2\phi-\cos^2\theta+\sin^2\theta\\
&= \cosh 2\phi(\sin^2\theta+\cos^2\theta)-(\cos^2\theta-\sin^2\theta)\\
&= \cosh 2\phi-\cos 2\theta \text{ (Proved).}
\end{aligned}$$

(b) *If* $\log\sin(\theta+i\phi)=\alpha+i\beta$ (θ,ϕ,α,β *are reals*), *then prove that*
(i) $2\cos 2\theta=e^{2\phi}+e^{-2\phi}-4e^{2\alpha}$; (ii) $\dfrac{\cos(\theta-\beta)}{\cos(\theta+\beta)}=e^{2\phi}$.

Solution: (ii) is same as the part (a) which we have already proved. See that

$$2\cos 2\theta=2\cosh 2\phi-4e^{2\alpha}\quad\left(\because \cosh 2\phi=\frac{e^{2\phi}+e^{-2\phi}}{2}\right),$$

$$\text{or,}\quad 2e^{2\alpha}=\cosh 2\phi-\cos 2\theta.$$

(ii) We have already observed that

$$\sin\theta\cosh\phi=e^{\alpha}\cos\beta$$
$$\text{and}\quad \cos\theta\sinh\phi=e^{\alpha}\sin\beta.$$

On division, we get

$$\tan\theta\coth\phi=\cot\beta,$$
$$\text{or,}\quad \tan\beta\tan\theta=\tanh\phi.$$

$\therefore$ By componendo and dividendo we get

$$\frac{1+\tan\beta\tan\theta}{1-\tan\beta\tan\theta}=\frac{1+\tanh\phi}{1-\tanh\phi},\quad\text{or,}\quad \frac{\cos\beta\cos\theta+\sin\beta\sin\theta}{\cos\beta\cos\theta-\sin\beta\sin\theta}=\frac{\cosh\phi+\sinh\phi}{\cosh\phi-\sinh\phi},$$

$$\text{or,}\quad \frac{\cos(\theta-\beta)}{\cos(\theta+\beta)}=\frac{e^{\phi}}{e^{-\phi}}=e^{2\phi}\ \text{(Proved).}$$

Complex Functions

Summary

1. **Exponential Function:**

$$e^z = 1 + \frac{z}{\lfloor 1} + \frac{z^2}{\lfloor 2} + \frac{z^3}{\lfloor 3} + \cdots + \frac{z^n}{\lfloor n} + \cdots \text{ (z is a complex variable).}$$

2. **Addition Theorem:** $e^{z_1} \cdot e^{z_2} = e^{z_1 + z_2}$ (z_1, z_2 are complex numbers).

3. **Trigonometric Functions:**

$$\operatorname{Sin} z = \frac{e^{iz} - e^{-iz}}{2i} \text{ and } \operatorname{Cos} z = \frac{e^{iz} + e^{-iz}}{2}.$$

4. **Euler's Exponential Values:**

$$e^{iz} = \cos z + i \sin z; e^{-iz} = \cos z - i \sin z.$$

5. **Hyperbolic Functions:**

$$\sinh z = \frac{e^z - e^{-z}}{2} \text{ and } \cosh z = \frac{e^z + e^{-z}}{2}$$

so that $\cosh^2 z - \sinh^2 z = 1$ and $\cosh^2 z + \sinh^2 z = \cosh 2z$.

6. **Logarithmic Functions:** Given a complex number z; if there exists a complex number w such that $e^w = z$, then $\operatorname{Log} z = w$.
If $|z| = r$ and principal amp. of $z = \theta$ $(-\pi < \theta \leq \pi)$ then

$$\operatorname{Log} z = \log r + i(\theta + 2n\pi); \; n = 0, \pm 1, \pm 2, \ldots$$

and P.V. of $\operatorname{Log} z = \log z = \log r + i\theta$.

7. $\operatorname{Log} z_1 z_2 = \operatorname{Log} z_1 + \operatorname{Log} z_2$ in the sense every value of $\operatorname{Log} z_1 + \operatorname{Log} z_2$ is some value of $\operatorname{Log} z_1 z_2$. (This formula may not be true for P.V.)

8. **Important Results:**

$$\left.\begin{aligned} \sin iz &= i \sinh z; & \cos iz &= \cosh z \\ \sinh(iz) &= i \sin z; & \cosh(iz) &= \cos z \end{aligned}\right\}$$

$$\sin(z_1 + z_2) = \sin z_1 \cos z_2 + \cos z_1 \sin z_2$$

$$\cos(z_1 + z_2) = \cos z_1 \cos z_1 - \sin z_2 \sin z_2$$

9. If a is a non-zero complex number and z is a complex number, then a^z is defined by $e^{z \operatorname{Log} a}$.

10. **Inverse Functions:** If z and w are complex numbers such that $\cos z = w$ then z is called **Inverse Cosine** of w i.e., $z = \cos^{-1} w$. Similar definitions for other Inverse Trigonometric Functions and also of Inverse Hyperbolic Functions.

Exercises on Chapter 2

All the Worked Out Problems are to be Carefully Revised

For $\cos\theta + i\sin\theta$ we shall often write $\text{cis}\,\theta$

1. Express in the form $A + iB$ (A, B real):

$$e^{i\frac{\pi}{2}};\ e^{i} + e^{-i};\ e^{\frac{1}{2}+i\frac{\pi}{4}};\ e^{\text{cis}\,\theta}(-\pi < \theta \le \pi).$$

2. Write $\sin(x \pm iy)$; $\cos(x \pm iy)$; $\text{cosec}(x + iy)$; $\tan(x + iy)$ in the form $A + iB$ (A, B real).

3. Express $\cosh(x+iy)$, $\tanh(x-iy)$, $\sec(x+iy)$, $\cot(x+iy)$ in the form $\alpha + i\beta$ (α, β real).

4. If α, β are real, prove that

$$|\sin(\alpha + i\beta)|^2 = \sin^2\alpha + \sinh^2\beta.$$

5. Simplify $\dfrac{\cos(x+iy)}{\cos(x-iy)} + \dfrac{\cos(x-iy)}{\cos(x+iy)}$ in the form $\dfrac{2(1+\cos 2x\cosh 2y)}{\cos 2x + \cosh 2y}$.

6. If $\sin(x+iy) = u + iv$, prove that

$$\begin{aligned} u^2\,\text{cosec}^2\,x - v^2\sec^2 x &= 1 \\ u^2\,\text{sech}^2\,y + v^2\,\text{cosech}^2 y &= 1\ (x, y, u, v \text{ are reals}) \end{aligned}$$

7. If $\tan(x+iy) = u + iv$, prove that

$$\begin{aligned} (u^2+v^2-1)\tan 2x + 2u &= 0 \\ (u^2+v^2+1)\tanh 2y - 2v &= 0 \end{aligned}$$

[C.P. 1992]

8. If α, β are real and $|\cos(\alpha + i\beta)| = 1$, then prove that

$$\sin^2\alpha = \sinh^2\beta.$$

9. Find the general values and principal values of $\text{Log}\,i$; $\text{Log}(1+i)$; $\text{Log}(-2)$; $\text{Log}(-3-4i)$; $\text{Log}\,\text{cis}\,\theta\ (-\pi < \theta \le \pi)$; i^{i}; $(1+i)^{1+i}$; $(-1)^{-i}$; $(1+i)^{-i}$.

10. Express in the form $A + iB$.

$$\log(1 + \cos 2\theta + i\sin 2\theta)\ \left(\frac{\pi}{2} < \theta < \frac{3\pi}{2}\right)$$

11. Prove that, for $0 < x < \pi$

$$\log\sin(x+iy) = \frac{1}{2}\log(\cosh^2 y - \cos^2 x) + i\tan^{-1}(\cot x \tanh y).$$

12. Express $\operatorname{Log}\{\operatorname{Log}(\cos\theta + i\sin\theta)\}$ in the form $A + iB$.

13. Show that the ratio of the principal value of

$$(1+i)^{1-i} \text{ and } (1-i)^{1+i} \text{ is } \sin(\log 2) + i\cos(\log 2).$$

14. Prove that

(a) $\sin\left(i\log\dfrac{a-ib}{a+ib}\right) = \dfrac{2ab}{a^2+b^2}$ (a, b real); [C.P. 2001]

(b) $\tan\left(i\log\dfrac{a-ib}{a+ib}\right) = \dfrac{2ab}{a^2-b^2}$ (a, b real). (See Ex. 2.5.6) [C.P. 2003]

15. Show that the equation

$$\tan\left(i\log\frac{x-iy}{x+iy}\right) = 2$$

represents the rectangular hyperbola $x^2 - y^2 = xy$.

16. Prove that

(a) $\sin(\log i^i) = -1$;

(b) $\tan(\alpha + i\beta) = \dfrac{2\sin 2\alpha + i(e^{2\beta} - e^{-2\beta})}{2\cos 2\alpha + e^{2\beta} + e^{-2\beta}}$.

17. (a) If $\operatorname{Sin}^{-1} i = \alpha + i\beta$, prove that $\alpha = 2n\pi$ and $\beta = -\log(\sqrt{2} - 1)$.

(b) Prove that the principal value of $\operatorname{Tan}^{-1}(\operatorname{cis}\theta)$ is

$$\frac{\pi}{4} + \frac{i}{2}\log\tan\left(\frac{\pi}{4} + \frac{\theta}{2}\right), -\frac{\pi}{2} < \theta \le \frac{\pi}{2}.$$

(c) Prove that $\operatorname{Sin}^{-1}(ix) = 2n\pi \pm i\log(\sqrt{1+x^2} + x)$.

18. Solve:

(a) $\cos z = 0, \cos z = c$ (c is a real number > 1);

(b) $\sin z = 1\frac{2}{3}; \cos z = -2.6$.

19. If $\operatorname{Cos}^{-1}(x+iy) = u + iv$, where $x > 0$ and $y > 0$, prove that $\cos^2 u$ and $\cosh^2 v$ are the roots of the equation

$$t^2 - t(1 + x^2 + y^2) + x^2 = 0.$$

20. If $\operatorname{Sin}^{-1}(\alpha + i\beta) = u + iv$, where $\alpha > 0$ and $\beta > 0$, prove that $\sin^2 u$ and $\cosh^2 v$ are the roots of the equation $t^2 - t(1 + \alpha^2 + \beta^2) + \alpha^2 = 0$.

21. Using Gregory's series prove that

$$\frac{\pi}{8} = \frac{1}{1.3} + \frac{1}{5.7} + \frac{1}{9.11} + \cdots$$

22. If $-\frac{\pi}{2} < x < \frac{\pi}{2}$, prove that

$$\tan^{-1}\frac{1-\cos\theta}{1+\cos\theta} = \tan^2\frac{\theta}{2} - \frac{1}{3}\tan^6\frac{\theta}{2} + \frac{1}{5}\tan^{10}\frac{\theta}{2} - \cdots$$

23. If $\log\sin(\theta + i\phi) = \alpha + i\beta$, prove that

$$\frac{\cos(\theta-\beta)}{\cos(\theta+\beta)} = e^{2\phi}.$$

24. Express $\tan^{-1}(\cos\theta + i\sin\theta)$ in the form $u + iv$ if

(a) $-\frac{\pi}{2} < \theta < \frac{\pi}{2}$;

(b) $\frac{\pi}{2} < \theta < \frac{3\pi}{2}$.

25. If $x + iy = c\cosh(\alpha + i\beta)$, prove that α=constant, represents a family of confocal hyperbolas and β=constant, represents a family of confocal ellipses.

26. If $\cos(x + iy) = \cos\theta + i\sin\theta$, then prove that

$$\cos 2x + \cosh 2y = 2.$$

27. If $\tan(\theta + i\phi) = \sin(\alpha + i\beta)$, prove that

$$\frac{\tan\alpha}{\tanh\beta} = \frac{\sin 2\beta}{\sinh 2\phi}.$$

Hints and Answers

1. i; $2\cos 1$ [Put $z = 1$ in the def. of $\cos z$]; $\frac{e^{1/2}}{\sqrt{2}}(1+i)$; $e^{\cos\theta}\,\text{cis}(\sin\theta)$.

2. $\sin x\cosh y \pm i\cos x\sinh y$; $\cos x\cosh y \mp i\sin x\sinh y$:

$$\begin{aligned}\operatorname{cosec}(x+iy) &= \frac{1}{\sin(x+iy)} = \frac{2\sin(x-iy)}{2\sin(x+iy)\sin(x-iy)}\\ &= \frac{2(\sin x\cosh y - i\cos x\sinh y)}{\cosh 2y - \cos 2x}.\\ \tan(x+iy) &= \frac{2\sin(x+iy)\sin(x-iy)}{2\cos(x+iy)\cos(x-iy)} = \frac{\sin 2x + i\sinh 2y}{\cos 2x + \cosh 2y}.\end{aligned}$$

3. Ans. $\cosh x\cos y + i\sinh x\sin y$; $\frac{\sinh 2x - i\sin 2y}{\cosh 2x + \cos 2y}$;

$$\begin{aligned}\sec(x+iy) &= \frac{1}{\cos(x+iy)} = \frac{2\cos(x-iy)}{2\cos(x+iy)\cos(x-iy)}\\ &= \frac{2\cos x\cosh y + i2\sin x\sinh y}{\cos 2x + \cosh 2y}\\ &= \frac{2\cos x\cosh y}{\cos 2x + \cosh 2y} + i\frac{2\sin x\sinh y}{\cos 2x + \cosh 2y}\\ \cot(x+iy) &= \frac{\sin 2x - i\sinh 2y}{\cosh 2y - \cos 2x}\\ &= \frac{\sin 2x}{\cosh 2y - \cos 2x} - i\frac{\sinh 2y}{\cosh 2y - \cos 2x} = \alpha - i\beta.\end{aligned}$$

4. Since $\sin(\alpha + i\beta) = \sin\alpha\cosh\beta + i\cos\alpha\sinh\beta$.

$$\begin{aligned} \therefore\ |\sin(\alpha+i\beta)|^2 &= \sin^2\alpha\cosh^2\beta + \cos^2\alpha\sinh^2\beta \\ &= \sin^2\alpha(1+\sinh^2\beta) + (1-\sin^2\alpha)\sinh^2\beta \\ &= \sin^2\alpha + \sinh^2\beta. \end{aligned}$$

6. $\sin x\cosh y + i\cos x\sinh y = u + iv$. Equate real and imaginary parts.

$$\begin{aligned} \text{Sin}^2 x &= \frac{u^2}{\cosh^2 y}, \cos^2 x = \frac{v^2}{\sinh^2 y} \text{ and } \sin^2 x + \cos^2 x = 1;\ \text{hence etc.} \\ \cosh^2 y &= \frac{u^2}{\sin^2 x} \text{ and } \sinh^2 y = \frac{v^2}{\cos^2 x} \text{ and } \cosh^2 y - \sinh^2 y = 1;\ \text{hence etc.} \end{aligned}$$

7. If $\tan(x+iy) = u + iv$, we may take $\tan(x - iy) = u - iv$.
   See Example 2.2.18.

8. $\cos^2\alpha\cosh^2\beta + \sin^2\alpha\sinh^2\beta = 2$. Simplify.

9. $i(\pi/2 + 2n\pi)$, p.v. $= i\pi/2$;
   $\frac{1}{2}\log 2 + i(\pi/4 + 2n\pi)$, p.v. $= \frac{1}{2}\log 2 + i\pi/4$;
   $\log 2 + i(\pi + 2n\pi)$, p.v. $= \log 2 + i\pi$;
   $\log 5 + i(\alpha + 2n\pi)$ where α is s.t. $\cos\alpha = -3/5$ and $\sin\alpha = -4/5$
   and $-\pi < \alpha \le \pi$, p.v. $= \log 5 + i\alpha$;

$$\begin{aligned} \text{Log}(\cos\theta + i\sin\theta) &= \log|\cos\theta + i\sin\theta| + i(\theta + 2n\pi)\ (\because\ -\pi < \theta \le \pi) \\ &= i(\theta + 2n\pi);\ \text{p.v.} = i\theta; \end{aligned}$$

$$\begin{aligned} i^i &= \exp\{i\,\text{Log}\,i\} = \exp\{i(i(\pi/2+2n\pi)\} = e^{-\pi/2-2n\pi}, \text{p.v.} = e^{-\pi/2}, \\ (1+i)^{1+i} &= \exp\{(1+i)\,\text{Log}(1+i)\} \\ &= \exp\left\{(1+i)\left[\frac{1}{2}\log 2 + i(\pi/4+2n\pi)\right]\right\} \\ &= e^{\frac{1}{2}\log 2}.\, e^{(\frac{\pi}{4}+2n\pi)}\left[\text{cis}\left(\frac{1}{2}\log 2 + \frac{\pi}{4} + 2n\pi\right)\right] \\ &= \sqrt{2}e^{-\frac{\pi}{4}+2n\pi}\left[\text{cis}\left(\frac{1}{2}\log 2 + \frac{\pi}{4} + 2n\pi\right)\right], \\ \text{p.v.} &= \sqrt{2}e^{-\pi/4}\,\text{cis}\left\{\frac{\pi}{4} + \frac{1}{2}\log 2\right\}; \\ (-i)^{-i} &= e^{\frac{1}{2}(4n-1)\pi}, \text{p.v.} = e^{-\pi/2}; \\ (1+i)^{-i} &= e^{2n\pi+\pi/4}\,\text{cis}\left(-\frac{1}{2}\log 2\right), \text{p.v.} = e^{\pi/4}\,\text{cis}\left(-\frac{1}{2}\log 2\right). \end{aligned}$$

10. See worked out Ex. 2.5.10.

11. If $\sin(x+iy) = r(\cos\theta + i\sin\theta)$, then

$$\log\{\sin(x+iy)\} = \frac{1}{2}\log r^2 + i\,\text{amp.}(\text{cis}\,\theta)$$

$$r^2 = \sin(x+iy)\cdot\sin(x-iy) = \sin^2 x + \sinh^2 y = \cosh^2 y - \cos^2 x$$

Also $r\cos\theta = \sin x \cosh y > 0$ for $0 < x < \pi$;

$\therefore\ \cos\theta > 0$ and $r\sin\theta = \cos x \sinh y$;

$$\text{amp.}(\text{cis}\,\theta) = \tan^{-1}(\tan\theta) = \tan^{-1}\left(\frac{\cos x \sinh y}{\sin x \cosh y}\right) = \tan^{-1}(\cot x \tanh h).$$

$$\therefore\ \log\sin(x+iy) = \frac{1}{2}\log(\cosh^2 y - \cos^2 x) + i\tan^{-1}(\cot x \tanh y).$$

Note that if it is mentioned that $\pi < x < 3\pi/2$ and $y > 0$ then

$$\text{amp. cis}\,\theta = -\pi + \tan^{-1}(\tan\theta) \ \ \text{[Why?]}$$

12. $\text{Log}(\cos\theta + i\sin\theta) = i(2n\pi + \theta)$.

$$\begin{aligned}\therefore\ \text{Log}\{\text{Log}(\cos\theta + i\sin\theta)\} &= \text{Log}\, i(2n\pi+\theta)\\ &= \text{Log}(2n\pi+\theta)\left\{\text{cis}\left(2k\pi+\frac{\pi}{2}\right)\right\}\\ &= \text{Log}(2n\pi+\theta) + i\left(2k\pi+\frac{\pi}{2}\right)\\ &= A + iB.\end{aligned}$$

See worked out Ex. 2.2.4.

13. $(1+i)^{1-i} = e^{(1-i)\,\text{Log}(1+i)} = e^{(1-i)\{\log\sqrt{2}+i(\pi/4+2n\pi)\}}$

$\therefore$ p.v. of $(1+i)^{1-i}$ is $e^{(1-i)\log\sqrt{2}+i(\pi/4)} = e^{\log\sqrt{2}-i(\pi/4-\log\sqrt{2})}$.

Similarly p.v. of $(1-i)^{1+i}$ is $e^{\log\sqrt{2}-i(\pi/4-\log\sqrt{2})}$.

Hence, their ratio $= e^{i\left(\frac{\pi}{2}-\log 2\right)}$

$$\begin{aligned}&= \cos\left(\frac{\pi}{2}-\log 2\right) + i\sin\left(\frac{\pi}{2}-\log 2\right)\\ &= \sin\log 2 + i\cos(\log 2).\end{aligned}$$

14. (a), (b) see **worked out Examples 2.5.5 and 2.5.6.**

18. (a) $z = n\pi + \dfrac{\pi}{2};\ z = 2n\pi \pm i\log\left(c+\sqrt{c^2-1}\right);$

(b) $z = 2n\pi + \dfrac{\pi}{2} \pm i\log 3;\ z = (2n+1)\pi \pm i\log 5.$

21. Put $x = 1$ in Gregory's Series and group the terms.

22. Prove that $0 \le \dfrac{1-\cos\theta}{1+\cos\theta} \le 1$ when $-\dfrac{\pi}{2} \le \theta \le \dfrac{\pi}{2}$.

Expand $\tan^{-1}\dfrac{1-\cos\theta}{1+\cos\theta}$ by Gregory's Series.

23. See **Ex. 2.5.15.**

24. (a) $\dfrac{\pi}{4} + \dfrac{1}{4}\log\dfrac{1+\sin\theta}{1-\sin\theta}$. (b) $\dfrac{-\pi}{4} + \dfrac{1}{4}\log\dfrac{1+\sin\theta}{1-\sin\theta}$

A Model Test Paper

1. Using Euler's exponential values of sin zand cos z, where z is a complex number, establish that $\cos 2z = 1 - 2\sin^2 z$.

2. If $\sin^{-1}(u + iv) = a + ib$, then prove that $1 + u^2 + v^2 = \sin^2 a + \cosh^2 b$.

3. Show that $\tan\left(i \log \dfrac{a - ib}{a + ib}\right) = \dfrac{2ab}{a^2 - b^2}$ (a, b real)

4. Prove that $\text{Sin}\left(i \log \dfrac{a - ib}{a + 2b}\right) = \dfrac{2ab}{a^2 + b^2}$ (a, b real)

5. If $\tan \log(x + iy) = a + ib$, where $a^2 + b^2 \neq 1$, prove that
$$\log(x^2 + y^2) = \frac{2a}{1 - a^2 - b^2}.$$

6. If $\log \sin(\theta + i\phi) = \alpha + i\beta$, prove that
$$2e^{2\alpha} = \cosh 2\phi - \cos 2\theta \ (\theta, \phi, \alpha, \beta \text{ are all real numbers}).$$

7. Show that all the values of i^i are real and are in G.P.

8. If $\tan(x - iy) = u - iv$, show that $u^2 + v^2 + 2u \cot 2x = 1$.

9. If $(a_1 + ib_1)(a_2 + ib_2) \cdots (a_n + ib_n) = A + iB$, prove that
$$\tan^{-1} \frac{b_1}{a_1} + \tan^{-1} \frac{b_2}{a_2} + \cdots + \tan^{-1} \frac{b_n}{a_n} = \tan^{-1} \frac{B}{A}.$$

10. Prove that
$$\log \frac{a + ib - x}{a + ib + x} = \frac{1}{2} \log \frac{(a - x)^2 + b^2}{(a + x)^2 + b^2} + i\left\{\tan^{-1} \frac{b}{a - x} - \tan^{-1} \frac{b}{a + x}\right\}$$

11. Express $\text{Log}(1 - \cos\theta - i\sin\theta)$ in the form $A + iB$ (A, B real) and write down the principal value.

Chapter 3

Theory of Polynomials

Theory of Polynomials	
1.	(*a*) **Polynomials:** Fundamental Theorem of Classical Algebra (*statement only*): Consequences. (*b*) **Polynomials with real coefficients:** The nth degree polynomial equations has exactly n roots. (*c*) Nature of roots of an equation (*Surd* or *Complex roots occur in pairs*).
2.	Relation between roots and coefficients. Symmetric functions of roots.
3.	**Transformations of equations:** Equation of the squared differences of a cubic: Cardan's method of solution of a cubic.
4.	(*a*) Statement of Descarte's Rule of Signs and its applications. (*b*) If the polynomial $f(x)$ has opposite signs for two real values of x, e.g., a and b, the equation $f(x) = 0$ has an odd number of real roots between a and b; if $f(a)$ and $f(b)$ are of same sign, either no real root or an even number of roots lies between a and b (*statements only*). (*c*) Rolle's Theorem and its direct applications in Polynomial Functions.

3.1 Introduction: Definitions

An expression of the form

$$f(x) = a_0x^n + a_1x^{n-1} + a_2x^{n-2} + \cdots + a_{n-1}x + a_n,$$

where $a_0 \neq 0$, n is a positive integer or zero and the coefficients $a_0, a_1, a_2, \ldots, a_n$ are independent of x, is called a *polynomial in x of degree n*. We shall usually take $a_0, a_1, a_2, \ldots$ real numbers and then $f(x)$ will be called a *polynomial in x with real coefficients.* The coefficients may be complex numbers also—we then call it a *polynomial with complex coefficients.*

a_0 is the **leading coefficient** and the term a_0x^n is the **leading term.** If all the coefficients are zero, then the polynomial is known as the **zero-polynomial** (no degree is associated with it). But if $a_n \neq 0$ but all other coefficients are zero, then $f(x)$ = a constant function a_n; in that case we call it a **polynomial of degree zero.**

▶ Example 3.1.1

$$\begin{aligned} f(x) &= a_0x + a_1, \\ \phi(x) &= a_0x^2 + a_1x + a_2 \\ \psi(x) &= a_0x^3 + a_1x^2 + a_2x + a_3, \\ g(x) &= a_0x^4 + a_1x^3 + a_2x^2 + a_3x + a_4, \text{ etc.} \end{aligned}$$

In all the above cases, we assume that $a_0 \neq 0$.

Then, $f(x)$ is a polynomial of **degree one or a linear expression in x.**

$\phi(x)$ is a polynomial of **degree two or a quadratic expression in x.**

$\psi(x)$ is a polynomial of **degree three or a cubic expression in x.**

$g(x)$ is a polynomial of **degree four or a biquadratic expression in x.**

If α be a value of x (real or complex) for which $f(\alpha) = 0$, then α is called **a zero** of the polynomial $f(x)$ or α is called **a root of the polynomial equation** $f(x) = 0$.

▶ Example 3.1.2 *If* $f(x) = x^2 - 6x + 5$ *be a polynomial of second degree (or a quadratic), then* $f(5) = 0$ *and* $f(1) = 0$ *and so* 5, 1 *are the two* ***zeros*** *of* $f(x)$; *they are also called the* ***roots*** *of* $f(x) = 0$.

Note that though the coefficients of a polynomial are all real numbers, the polynomial equation may possess complex roots.

Two polynomials $f(x)$ and $g(x)$ of same degree are said to be *identically equal* if, and only if the coefficients of the like powers are equal. We then write $f(x) \equiv g(x)$.

Given: $ax^3 + bx^2 + cx + d \equiv -3x^3 + 2x^2 + 5$.

We can then *conclude:* $a = -3, b = 2, c = 0, d = 5$.

We shall assume familiar rules of addition, subtraction and multiplication of two polynomials. In each case the result is another polynomial.

Addition and multiplication are both commutative and they are connected by the Distributive Laws. Moreover,

if $f(x) \cdot g(x) = 0$ and $f(x) \neq 0$, then $g(x) = 0$.

if $f(x) \cdot g(x) = f(x) \cdot h(x)$ and $f(x) \neq 0$, then $g(x) = h(x)$.

We shall discuss the theory of polynomials in four sections:

Section I: Division Algorithm: Synthetic Division. Fundamental theorem of Classical Algebra and its immediate consequences.

Section II: Relations between roots and coefficients of a polynomial equation $f(x) = 0$. Nature of the roots. Symmetric functions of the roots.

Section III: *Transformations:* equation of squared differences of a cubic. Cardan's Method of solving a cubic equation.

Section IV: Location of roots of a polynomial equation with real coefficients. Descartes' Rule of Signs. Applications of Rolle's Theorem.

SECTION I

3.2 Division Algorithm: Synthetic Division

Division Process can be stated as:

Let $f(x)$ be a polynomial of degree n.

Let $\phi(x)$ be another polynomial of degree $m < n$.

Then, by Division Algorithm (or *Division Process*) of dividing $f(x)$ by $\phi(x)$, we mean:

There exists a polynomial $Q(x)$ of degree $n - m$ and another polynomial $R(x)$ of degree lower than m such that

$$f(x) = \phi(x) \cdot Q(x) + R(x).$$

$Q(x)$ is called the quotient and $R(x)$, the remainder. $f(x)$ (*dividend*) and $\phi(x)$ (*divisor*).

In particular, if $f(x)$ be a polynomial of degree n and α is a number, *real* or *complex*, then we can divide $f(x)$ by $x - \alpha$ and obtain a polynomial $Q(x)$, quotient of degree $n - 1$ and a remainder R, which is either zero or a constant such that

$$f(x) = (x - \alpha)\, Q(x) + R.$$

The quotient Q and the remainder R can be obtained either by the usual **long division** process or by a short process called **Synthetic Division.**

We give an example:

Long Process

$$\begin{array}{r|l} & x^3 - 2x^2 - 2x \;\; - 6 \\ \hline x - 3 & x^4 - 5x^3 + 4x^2 + 6 \\ & x^4 - 3x^3 \\ \hline & \;\; - 2x^3 + 4x^2 + 6 \\ & \;\; - 2x^3 + 6x^2 \\ \hline & \qquad - 2x^2 + 0.x + 6 \\ & \qquad - 2x^2 + 6x \\ \hline & \qquad\quad - 6x + 6 \\ & \qquad\quad - 6x + 18 \\ \hline & \qquad\qquad - 12 \end{array}
\qquad
\left\{\begin{array}{l} \phi(x) = x - 3 \text{ (Divisor)} \\ f(x) = x^4 - 5x^3 + 4x^2 + 6 \text{ (Dividend)} \\ Q(x) = x^3 - 2x^2 - 2x - 6 \text{ (Quotient)} \\ R = -12 \text{ (Remainder)} \end{array}\right.$$

Short Process: Synthetic Division

3	1	−5	4	0	6
		3	−6	−6	−18
	1	−2	−2	−6	−12

The last entry in the third line (i.e., −12) is the Remainder and the third line is Q(x) of degree 3, i.e.,

$$Q(x) = x^3 - 2x^2 - 2x - 6.$$

Rules

1. Write the coefficient as detached coefficient (if a particular power of x be missing take zero as its coefficient.)
2. Write 3 to the left of the leading coefficient.
3. The leading coefficient 1 comes in the first position of the third line.
4. Multiply this leading coefficient by 3 and put it in the second line below −5; add and put the sum −2 in third line.
5. Multiply this −2 by 3 and put it below 4; add and put the sum −2 in the third line and so on.

We give below a few more examples of synthetic division:

► Example 3.2.1 *Divide* $x^4 - x^2 + 5x - 6$ *by* $x + 2$, *using synthetic division.*

Solution: Here, on the left of the leading coefficient take −2 (NOT +2).

−2	1	0	−1	5	−6
		−2	4	−6	2
	1	−2	3	−1	−4

Quotient $= x^3 - 2x^2 + 3x - 1$, **Remainder** $= -4$.

► Example 3.2.2 *Divide* $2x^4 - 6x^3 + 7x^2 - 5x + 1$ *by* $2x - 3$ (*synthetic division*).

Solution: Here first divide by $x - \frac{3}{2}$ [see that $2x - 3 = 2\left(x - \frac{3}{2}\right)$]

$\frac{3}{2}$	2	−6	7	−5	1
		3	$-\frac{9}{2}$	$-\frac{15}{4}$	$-\frac{15}{8}$
	2	−3	$\frac{5}{2}$	$-\frac{5}{4}$	$-\frac{7}{8}$

On division by $x - 3/2$, Remainder $= -7/8$ and Quotient $= 2x^3 - 3x^2 + \frac{5}{2}x - \frac{5}{4}$.
∴ on division by $2(x - 3/2)$.
Here **Quotient** $= \frac{1}{2}\left(2x^3 - 3x^2 + \frac{5}{2}x - \frac{5}{4}\right)$ and **Remainder** $= -\frac{7}{8}$.

[The reader may observe that if Q and R are respectively the quotient and remainder, when $f(x)$ is divided by $x - \frac{b}{a}$, then

$$\begin{aligned} f(x) &= \left(x - \frac{b}{a}\right) Q + R \\ &= \frac{1}{a}[(ax-b)Q] + R] \end{aligned}$$

▶ **Example 3.2.3** [If, on dividing $f(x)$ by $x - \alpha$ we find $R = 0$, then we say that $f(x)$ *is divisible by* $x - \alpha$ or we say that $x - \alpha$ is a **factor** of $f(x)$.]

Show that $2x^4 - 7x^3 + 8ax^2 - 3bx + 17$ *is divisible by* $x - 2$ *when* $32a - 6b - 7 = 0$. **[C.P. 1993]**

Solution: We first divide $2x^4 - 7x^3 + 8ax^2 - 3bx + 17$ by $x - 2$, using synthetic division.

2	2	−7	8a	−3b	17
		4	−6	16a − 12	32a − 6b − 24
	2	−3	8a − 6	16a − 3b − 12	32a − 6b − 7

In order that the given polynomial may be exactly divisible by $x - 2$, the remainder $32a - 6b - 7 = 0$. Hence, the problem.

▶ **Example 3.2.4** *Exercises for Practice: Find the Quotient and Remainder in the following division:*

1. *Divide* $x^5 - 3x^4 + x^3 - 5x - 12$ *by* $x + 3$;
2. *Divide* $2x^3 + 2x^2 - 3x - 4$ *by* $x - 5$;
3. *Divide* $x^3 + 2x^2 - 3x - 4$ *by* $2x - 1$;
4. *Divide* $x^4 - 1$ *by* $x + 2$. **[C.P. 1989]**

Try yourself. No answers are given.

▶ **Example 3.2.5** *Find the Remainder, by synthetic division, when the polynomial* $3x^4 - 4x^3 + 2x^2 - 9x + 1$ *is divided by* $2x + 1$. **[C.P. 1997]**

Solution:

$-\frac{1}{2}$	3	−4	2	−9	1
		$-\frac{3}{2}$	$\frac{11}{4}$	$-\frac{19}{8}$	$\frac{91}{16}$
	3	$-\frac{11}{2}$	$\frac{19}{4}$	$-\frac{91}{8}$	$\frac{107}{16}$

[Note that **Quotient** should be $\frac{1}{2}\left(3x^3 - \frac{11}{2}x^2 + \frac{19x}{4} - \frac{91}{8}\right)$] [*Ans.* Remainder = $\frac{107}{16}$.]

▶ **Example 3.2.6** *What is the relation between* a *and* b *if* $x^3 + 2x^2 + 2ax + b$ *is exactly divisible by* $x + 2$? **[C.P. 1996]**

[*Ans.* $-4a + b = 0$. See Ex. 3.2.3 above.]

▶ Example 3.2.7 *Find the Remainder when* $x^3 + 5x^2 + 1$ *is divided by* $x + 3$.

[C.P. 1995]

Solution:

-3	1	5	0	1
		-3	-6	18
	1	2	-6	19

[*Ans.* **Remainder = 19.**]

3.3 Fundamental Theorem of Classical Algebra

We first state **Remainder Theorem** and **Factor Theorem** and then state the **Fundamental Theorem.**

3.3.1 Remainder Theorem

Let

$$f(x) = a_0x^n + a_1x^{n-1} + a_2x^{n-2} + \cdots + a_{n-1}x + a_n \ (a_0 \neq 0)$$

be a polynomial of degree n with coefficients all real.

Let α be any number, real or complex.

We divide $f(x)$ by $x - \alpha$. By division algorithm there exist two polynomials $Q(x)$ and $R(x)$ such that

$$f(x) = (x - \alpha)\, Q(x) + R(x),$$

where $R(x)$ is either zero or some constant.

In this identity we put $x = \alpha$. We then have $f(\alpha) = R(\alpha)$, i.e., the remainder is the value of $f(x)$ at $x = \alpha$. This is what is known as **Remainder Theorem.**

▶ Example 3.3.1 *Find the Remainder when* $x^3 + 5x^2 + 1$ *is divided by* $x + 3$.

[C.P. 1995]

Solution: By Remainder Theorem, the required remainder is $f(-3)$ where

$$f(x) = x^3 + 5x^2 + 1.$$

[*Ans.* $f(-3) = (-3)^3 + 5(-3)^2 + 1 = -27 + 45 + 1 = 19$.]

3.3.2 Factor Theorem

If α *be a root of the polynomial equation* $f(x) = 0$, *then* $(x - \alpha)$ *is a factor of* $f(x)$. *Conversely, if* $(x - \alpha)$ *be a factor of* $f(x)$, *then* α *is a root of* $f(x) = 0$.

This result is known as the **Factor Theorem.**

This follows immediately from the Remainder Theorem: we know that if $f(x)$ be divided by $(x - \alpha)$ and if $Q(x)$ is the quotient, then $f(\alpha)$ is the remainder and we can write

$$f(x) = (x - \alpha)\, Q(x) + f(\alpha).$$

$\therefore$ if α be a root of $f(x) = 0$ [i.e., if $f(\alpha) = 0$], then

$$f(x) = (x - \alpha)\, Q(x).$$

Hence, $(x - \alpha)$ *is a factor of* $f(x)$.

Conversely, if $(x - \alpha)$ be a factor of $f(x)$, then the remainder obtained when $f(x)$ is divided by $(x - \alpha)$ should be zero and hence, $f(\alpha)$ should be zero or α should be a root of $f(x) = 0$.

► Example 3.3.2 *Show that* $f(x) = x^4 - 3x^3 + 2x^2 + 2x - 4$ *is divisible by* $x - 2$.

Solution: See that $f(2) = 2^4 - 3 \cdot 2^3 + 2 \cdot 2^2 + 2 \cdot 2 - 4 = 0$.

$\therefore$ by Factor Theorem, $f(x)$ is divisible by $x - 2$.

3.3.3 Fundamental Theorem of Classical Algebra

Statement: *Let* $f(x)$ *be a polynomial with coefficients, real or complex and degree* $n \geq 1$. *The Fundamental Theorem asserts that the equation* $f(x) = 0$ *has at least one root which may be real or complex.*

The proof of this theorem is beyond the scope of this book.

Consequences:

Theorem 3.3.1 *A polynomial* $f(x)$ *of nth degree* $(n \geq 1)$ *with its coefficients, real or complex, has exactly* n *roots.*

We shall break up this theorem into two parts. We shall prove that

(i) $f(x) = 0$ has at least n roots, and

(ii) $f(x) = 0$ cannot have more than n roots.

Then the result of the theorem will follow.

(i) *To prove that*

$$f(x) = a_0 + a_1 x + a_2 x^2 + \cdots + a_n x^n = 0 \; (a_n \neq 0,\; n \geq 1),$$

where $a_0, a_1, a_2, \ldots, a_n$ *are real or complex numbers, has at least* n *roots.*

Proof. By the Fundamental Theorem of Algebra, the equation $f(x) = 0$ has at least one root, say α_1 (real or complex).

$$\therefore\; (x - \alpha_1) \text{ must be a factor of } f(x). \text{ [Factor Theorem]}$$

Hence, if we divide $f(x)$ by $(x - \alpha_1)$ there is no remainder and the quotient (say, $f_1(x)$ is a polynomial of degree $(n - 1)$ with its leading coefficient a_n, which is not zero; thus,

$$f(x) = (x - \alpha_1) f_1(x).$$

The equation $f_1(x) = 0$ has, by the same Fundamental Theorem, at least one root, say α_2 (real or complex), so that as before $f_1(x) = (x - \alpha_2) f_2(x)$, where $f_2(x)$ is a polynomial (with its coefficient $= a_n$) of degree $(n - 2)$.

We proceed in a similar manner till we come to a polynomial $f_{n-1}(x)$ of degree one with its leading coefficient $a_n (\neq 0)$, say, $a_n x + b$. We may write

$$f_{n-1}(x) = a_n(x - \alpha_n).$$

Hence, it follows that

$$f(x) = a_n(x - \alpha_1)(x - \alpha_2) \cdot (x - \alpha_{n-1})(x - \alpha_n).$$

Each factor of the form $x - \alpha_r$ gives a root of $f(x) = 0$ and thus, this equation has *at least* n *roots.*

(ii) *To prove that* $f(x) = 0$ [*where* $f(x)$ *is a polynomial of degree* n *defined in* (i)] *cannot have more than* n *roots.*

If possible, let the equation $f(x) = 0$ have $(n+1)$ roots $\alpha_1, \alpha_2, \alpha_3, \ldots, \alpha_n, \alpha_{n+1}$. Then $f(x)$ has factors $(x - \alpha_1), (x - \alpha_2, (x - \alpha_3), \ldots, (x - \alpha_n), (x - \alpha_{n+1})$, which when multiplied out, becomes an expression of degree $(n + 1)$ while we have assumed that $f(x)$ is of degree n. This contradiction shows that $f(x)$ cannot have $(n + 1)$ roots and, as such, it cannot have more than n roots in any case.

Note: We have taken at least one coefficient of $f(x) \neq 0$. If all the coefficients of $f(x)$ are zero, $f(x)$ is said **to vanish identically** or **$f(x) \equiv 0$ is an identity.** e.g., $f(x) = (x - a)(b - c) + (x - b)(c - a) + (x - c)(a - b) = 0$ **is an identity.**

Important Observations

Let $f(x) = a_0 + a_1x + a_2x^2 + \cdots + a_nx^n$ $(a_n \neq 0)$ be a polynomial of degree n.

1. If $f(x)$ be a polynomial of degree n, then $f(x) = 0$ has exactly n roots, say $\alpha_1, \alpha_2, \ldots, \alpha_n$. But there is no reason why these n roots should be all distinct. Some of them may be equal. We call them **multiple roots.**

2. If the root α occurs m times, then $(x - \alpha)^m$ must be a factor of $f(x)$ and if $(x - \alpha)^{m+1}$ is not a factor, then α is a multiple root of $f(x) = 0$ of **multiplicity** m.

   Conversely, if α is a **root of multiplicity** m, then $(x - \alpha)^m$ must be a factor of $f(x)$ but $(x - \alpha)^{m+1}$ is not a factor, e.g., if $f(x) = 0$ has a double root 3, then $(x - 3)^2$ is a factor of $f(x)$.

3. Suppose $f(x) = 0$ has m_1 roots each equal to α
   m_2 roots each equal to β
   ....
   m_k roots of each equal to λ

   where $m_1 + m_2 + \cdots + m_k = n$ (degree of $f(x)$).

   Then we can write

$$f(x) = a(x - \alpha)^{m_1}(x - \beta)^{m_2} \ldots (x - \lambda)^{m_k}$$

   where a_n is the coefficient of x^n (highest power of x).

Theorem 3.3.2 *If two polynomials, neither of which is of degree higher than n, are equal for a greater number of values of the variable than n, then the coefficients of like powers of the variable are equal, and the equality of the two polynomials becomes an identity.*

Proof. Let

$$\begin{aligned} f(x) &= a_0 + a_1x + a_2x^2 + \cdots + a_nx^n \ (a_n \neq 0) \\ \phi(x) &= b_0 + b_1x + b_2x^2 + \cdots + b_nx^n \ (b_n \neq 0). \end{aligned}$$

Given that $f(x) = \phi(x)$ for more than n values of x.
Let us consider the polynomial

$$\begin{aligned} F(x) &= f(x) - \phi(x) \\ &= (a_0 - b_0) + (a_1 - b_1)x + (a_2 - b_2)x^2 + \cdots + (a_n - b_n)x^n. \end{aligned}$$

Now $F(x) = 0$ can have only n roots, but it is assumed to vanish for more than n values of x. This can be true only if the coefficients vanish, or

$$a_0 - b_0 = 0, \ a_1 = b_1, \ a_2 = b_2, \ \cdots, \ a_n = b_n.$$

Then $F(x) = 0$ is true for all values of x and, consequently, $f(x) = \phi(x)$ for all values of x (and hence, it is an identity).

▶ **Example 3.3.3** *Find the values of the literal coefficients in order that the equation*

$$x^3 + 3x^2 + 7 = ax^4 + (b-5)x^3 + (c+1)x^2 + (d-2)x + f$$

may be an identity.

Solution: Equating coefficients of like powers of x gives

$$\begin{aligned} a &= 0, \ b - 5 = 1, \ c + 1 = 3, \ d - 2 = 0, \ f = 7 \\ \text{or, } a &= 0, \ b = 6, \ c = 2, \ d = 2, \ f = 7. \end{aligned}$$

▶ **Example 3.3.4** *Find the values of* A, B, C *for which*

$$A(x-3)(x-1) + B(x+1)(x-1) + C(x+1)(x-3) = 6x - 10$$

is an identity.

Solution: We write

$$\begin{aligned} & A\,(x^2 - 4x + 3) + B\,(x^2 - 1) + C\,(x^2 - 2x - 3) = 6x - 10, \\ \text{or, } \ & (A + B + C)x^2 + (-4A - 2C - 6)x + 3A - B - 3C + 10 = 0. \end{aligned}$$

This is an identity if and only if

$$\begin{aligned} A + B + C &= 0, \\ -4A - 2C - 6 &= 0, \\ 3A - B - 3C &= -10. \end{aligned}$$

Solving, $A = -2, B = 1, C = 1$.

3.4 Nature of the Roots of an Equation: Surd or Complex Roots Occur in Pairs

3.4.1 Complex Roots

Theorem 3.4.1 *If* $f(x)$ *be a polynomial in* x *with real coefficients and if* $\alpha + i\beta$ $(\alpha, \beta$ *real and* $\beta \neq 0)$ *be one root of* $f(x) = 0$, *then the conjugate complex number* $\alpha - i\beta$ *is also a root of* $f(x) = 0$.

This result can also be stated as: *In a polynomial equation with real coefficients imaginary roots occur in pairs.*

Proof. We first observe that

$$\begin{aligned} \{x - (\alpha + i\beta)\} \times \{x - (\alpha - i\beta)\} &= \{(x - \alpha) - t\beta\} \times \{(x - \alpha) + i\beta)\} \\ &= (x - \alpha)^2 + \beta^2 \\ &= \text{a quadratic expression with real coefficients.} \end{aligned}$$

Let us divide the given polynomial $f(x)$ (whose coefficient are all real) by the quadratic expression $(x - \alpha)^2 + \beta^2$. The remainder, if not zero, can be at most of degree one. Let the remainder $= Rx + S$ and the quotient $= Q$. Then, by division algorithm,

$$f(x) = \left\{(x - \alpha)^2 + \beta^2\right\} Q + Rx + S. \tag{3.4.1}$$

Now, by hypothesis, $\alpha + i\beta$ is a root of $f(x) = 0$, i.e., $f(\alpha + i\beta) = 0$.
From the identity (3.4.1) substituting $\alpha + i\beta$ for x we get

$$f(\alpha + i\beta) = 0 \times Q + R(\alpha + i\beta) + S = (R\alpha + S) + iR\beta.$$

$$\therefore\ R\alpha + S = 0 \text{ and } R\beta = 0\ [\because\ f(\alpha + i\beta) = 0].$$

Now $\beta \neq 0$, hence, $R = 0$; consequently $S = 0$, i.e., the remainder $Rx + S = 0$.
Thus if $\alpha + i\beta$ be a root of $f(x) = 0$, then

$$f(x) = \left\{(x - \alpha)^2 + \beta^2\right\} Q.$$

$$\therefore\ f(\alpha - i\beta) = \left\{(\alpha - i\beta - \alpha)^2 + \beta^2\right\} Q = 0 \cdot Q = 0,$$

i.e., $\alpha - i\beta$ is a root of $f(x) = 0$ **(Proved).**

Note: If $\alpha + i\beta$ be a multiple complex root of multiplicity r, then $\alpha - i\beta$ must also be a multiple complex root of multiplicity r (the coefficients of $f(x)$ should be real).

***The theorem is not necessarily true if the coefficients are not all reals.**

3.4.2 Surd Roots

Theorem 3.4.2 *Let* $f(x)$ *be a polynomial whose coefficients are all rational and let* $\alpha + \sqrt{\beta}(\beta \neq 0, \sqrt{\beta}$ *is not rational) be one root of* $f(x) = 0$. *Then,* $\alpha - \sqrt{\beta}$ *must also be a root of* $f(x) = 0$.

[*In a polynomial equation with rational coefficients quadratic surd roots occur in pairs.*]

The proof is similar to that of the previous theorem.

Outline of the Proof: Here we shall divide $f(x)$ by

$$\left\{x-\left(\alpha+\sqrt{\beta}\right)\right\}\times\left\{x-\left(\alpha-\sqrt{\beta}\right)\right\}=(x-\alpha)^2-\beta.$$

Let $Rx+S$ be the remainder and let Q be the quotient. Then

$$\begin{aligned} f(x) &= \left\{(x-\alpha)^2-\beta\right\}Q+(Rx+S) \\ \therefore\ 0 &= f\left(\alpha+\sqrt{\beta}\right)=0\times Q+R\left(\alpha+\sqrt{\beta}\right)+S=(R\alpha+S)+R\sqrt{\beta} \\ &\Rightarrow\ R\alpha+S=0,\ R\sqrt{\beta}=0 \\ &\Rightarrow\ R=0\ (\because\ \beta\neq 0) \text{ and hence, } S=0, \text{ i.e., remainder } Rx+S=0 \end{aligned}$$

i.e., $f(x)$ is divisible by both $\left\{x-\left(\alpha+\sqrt{\beta}\right)\right\}$ and $\left\{x-\left(\alpha-\sqrt{\beta}\right)\right\}$.

Since $\left\{x-\left(\alpha-\sqrt{\beta}\right)\right\}$ is a factor of $f(x)$, $\alpha-\sqrt{\beta}$ is a root of $f(x)=0$.

Note: This theorem is also not necessarily true if the coefficients are not all rational numbers.

3.5 Solved Problems

► Example 3.5.1 ***Check:*** *$x=i$ is a root of $x^3+7x-6i=0$. Therefore, $x=-i$ must be another root of this equation.*

Answer is: *Not necessarily. Why?*

Solution: We easily check that $x=i$ is a root of $x^3+7x-6i=0$, but $x=-i$ is not a root. The reason is that in applying the theorem **(Theorem 3.4.1, Art. 3.4)** the coefficients of the polynomial must be all real; but here all the coefficients are not real.

► Example 3.5.2 ***Check:*** *$x=\sqrt{3}-i\sqrt{2}$ is a root of*

$$x^3+\left(1-2\sqrt{3}\right)x^2+\left(5-2\sqrt{3}\right)x+5=0.$$

Can we at once conclude that $\sqrt{3}+i\sqrt{2}$ must be another root of this equation?

Answer is: Yes. Because the coefficients are all real and we can apply **Theorem 3.4.1, Art. 3.4.**

► Example 3.5.3 *Criticize the validity of the following conclusion:*

$$\text{“}x^4+\left(1-2\sqrt{2}\right)x^3+\left(4-2\sqrt{2}\right)x^2+\left(3-4\sqrt{2}\right)x+1=0$$

has $x=-1+\sqrt{2}$ as a root; hence $x=-1-\sqrt{2}$ must be another root”.

Answer is $x=-1-\sqrt{2}$ is not necessarily a root, since not all the coefficients of the given equation are **rational.**

In fact, by putting $x=-1-\sqrt{2}$ it will be found that $x=-1-\sqrt{2}$ is not a root.

▶ **Example 3.5.4** *Write the equation of lowest degree with real coefficients having 2 and $1-3i$ as two of its roots.*

Solution: The required equation is

$$(x-2)\{x-(1-3i)\}\{x-(1+3i)\}=0,$$
$$\text{or,}\quad (x-2)\{(x-1)^2+9\}=0,$$
$$\text{or,}\quad (x-2)(x^2-2x+10)=0,$$
$$\text{or,}\quad x^3-4x^2+14x-20=0.$$

▶ **Example 3.5.5** *Form the equation of lowest degree with rational coefficients having $\sqrt{5}-1$ and -6 as two of its roots.*

Solution: The required equation is

$$(x+6)\left\{x-\left(\sqrt{5}-1\right)\right\}\left\{x-\left(-\sqrt{5}-1\right)\right\}=0.$$

Simplifying $x^3+8x^2+8x-24=0$.

▶ **Example 3.5.6** *Solve the equation*

$$x^4-7x^3+27x^2-47x+26=0,$$

given that one of its roots is $2+3i$.

Solution: Here the coefficients are all reals. Hence if one root of the given equation is $2+3i$, there exists another root $2-3i$. So

$$(x-2-3i)(x-2+3i)$$

is a factor of

$$x^4-7x^3+27x^2-47x+26,$$
$$\text{i.e.,}\quad (x-2)^2+9=x^2-4x+13 \text{ is a factor.}$$

Division of

$$x^4-7x^3+27x^2-47x+26$$

by

$$x^2-4x+13$$

gives another factor

$$x^2-3x+2=(x-2)(x-1),$$

i.e., it gives two roots 2, 1. Thus the given equation has four roots $2+3i, 2-3i, 2, 1$.

$$\begin{array}{r|l} & x^2-3x\ +2 \\ \hline x^2-4x+13 & x^4-7x^3+27x^2-47x+26 \\ & x^4-4x^3+13x^2 \\ \hline & \quad -3x^3+14x^2-47x+26 \\ & \quad -3x^3+12x^2-39x \\ \hline & \qquad\qquad 2x^2-8x+26 \\ & \qquad\qquad 2x^2-8x+26 \\ \hline \end{array}$$

▶ **Example 3.5.7** *Solve the equation*

$$x^4 - 3x^3 - 5x^2 + 9x - 2 = 0$$

if one root be $2 + \sqrt{3}$. **[C.P. 2004]**

Solution: The given equation is a polynomial equation with rational coefficients. Hence, if $2 + \sqrt{3}$ is a root, $2 - \sqrt{3}$ must also be another root.

$$\text{See that } \left\{x - \left(2 + \sqrt{3}\right)\right\} \times \left\{x - \left\{2 - \sqrt{3}\right\}\right\} = (x-2)^2 - 3 = x^2 - 4x + 1.$$

$\therefore$ $x^2 - 4x + 1$ must be a factor of $x^4 - 3x^3 - 5x^2 + 9x - 2$.

$$\begin{array}{r|l} & x^2 + x - 2 \\ \hline x^2 - 4x + 1 & x^4 - 3x^3 - 5x^2 + 9x - 2 \\ & x^4 - 4x^3 + x^2 \\ \hline & \quad x^3 - 6x^2 + 9x \\ & \quad x^3 - 4x^2 + x \\ \hline & \quad -2x^2 + 8x - 2 \\ & \quad -2x^2 + 8x - 2 \\ \hline \end{array}$$

Division by $x^2 - 4x + 1$ will show that $x^2 + x - 2$ is also a factor or, $(x-1)$, $(x+2)$ are factors, or $x = 1, -2$ are roots.

Thus, we get all the four roots of the given equation

$$x = 1, -2, 2 + \sqrt{3}, 2 - \sqrt{3}.$$

▶ **Example 3.5.8** *Show that the roots of the equation*

$$\frac{1}{x-a} + \frac{1}{x-b} + \frac{1}{x-c} = \frac{1}{x} \quad (a > b > c > 0)$$

are all real. **[C.P. 2003]**

Solution: The given problem can be restated as:

To prove: $\dfrac{x}{x-a} + \dfrac{x}{x-b} + \dfrac{x}{x-c} = 1$, has *all its roots real,*

i.e., To prove: $\left(1 + \dfrac{a}{x-a}\right) + \left(1 + \dfrac{b}{x-b}\right) + \left(1 + \dfrac{c}{x-c}\right) = 1$, has *all its roots real,*

i.e., To prove: $\dfrac{a}{x-a} + \dfrac{b}{x-b} + \dfrac{c}{x-c} = -2$, has *all its roots real.*

Proof. This equation is a polynomial equation in x of degree 3 with all its coefficients real.

Let $\alpha + i\beta$ (α, β real) be a root of this equation. Then $\alpha - i\beta$ must be another root of this equation. Hence we can write

$$\frac{a}{\alpha + i\beta - a} + \frac{a}{\alpha + i\beta - b} + \frac{a}{\alpha + i\beta - c} = 1$$

$$\text{and} \quad \frac{a}{\alpha - i\beta - a} + \frac{a}{\alpha - i\beta - b} + \frac{a}{\alpha - i\beta - c} = 1.$$

On subtraction,

$$\left(\frac{a}{\alpha+i\beta-a}-\frac{a}{\alpha-i\beta-a}\right)+\left(\frac{b}{\alpha+i\beta-b}-\frac{b}{\alpha-i\beta-b}\right)$$
$$+\left(\frac{c}{\alpha+i\beta-c}-\frac{c}{\alpha-i\beta-c}\right)=0$$

$$\text{or}\quad a\left\{\frac{-2i\beta}{(\alpha-a)^2+\beta^2}\right\}+b\left\{\frac{-2i\beta}{(\alpha-b)^2+\beta^2}\right\}+c\left\{\frac{-2i\beta}{(\alpha-c)^2+\beta^2}\right\}=0$$

$$\text{or}\quad -2i\beta\left[\frac{a}{(\alpha-a)^2+\beta^2}+\frac{b}{(\alpha-b)^2+\beta^2}+\frac{c}{(\alpha-c)^2+\beta^2}\right]=0.$$

The expression within the bracket [ ] cannot be zero since a, b, c are all positive. Hence only conclusion is $\beta = 0$. This proves that the equation cannot have a root $\alpha + i\beta$, where $\beta \neq 0$. In other words, the equation has all its real.

▶ Example 3.5.9 *Prove that the roots of the equation*

$$\frac{1}{x-1}+\frac{2}{x-2}+\frac{3}{x-3}=x \text{ are all real.}$$

Try yourself (Exactly similar to **Ex. 3.5.8.**)

▶ Example 3.5.10 *Show that the equation*

$$\frac{L^2}{x-\alpha}+\frac{M^2}{x-\beta}+\cdots+\frac{T^2}{x-\lambda}=x-p,$$

where $L, M, \ldots, T, \alpha, \beta, \ldots, \lambda$ *and* p *are all distinct real numbers, cannot have any imaginary root.*

Solution: The equation can be expressed as a polynomial in x with real coefficients.
If possible, let $m + in$ ($n \neq 0, m, n$ real) be an imaginary root of the given polynomial equation.
Then $m - in$ must also be a root of this equation.
Put $x = m + in$ and then $x = m - in$ in the given equation:

$$\frac{L^2}{(m-\alpha)+in}+\frac{M^2}{(m-\beta)+in}+\cdots+\frac{T^2}{(m-\lambda)+in} = (m-p)+in$$
$$\frac{L^2}{(m-\alpha)-in}+\frac{M^2}{(m-\beta)-in}+\cdots+\frac{T^2}{(m-\lambda)-in} = (m-p)-in.$$

Subtracting we get

$$-2in\left[\frac{L^2}{(m-\alpha)^2-n^2}+\frac{M^2}{(m-\beta)^2+n^2}+\cdots+\frac{T^2}{(m-\lambda)^2+n^2}\right]=2in.$$

Then either $n = 0$ or else

$$\frac{L^2}{(m-\alpha)^2+n^2}+\frac{M^2}{(m-\beta)^2+n^2}+\cdots+\frac{T^2}{(m-\lambda)^2+n^2}=-1.$$

The left side is the sum of number of positive real numbers and hence, positive, and it cannot be equal to -1.

$\therefore$ our supposition that an imaginary root $m + in$ is a root of the given equation leads to the only conclusion that $n = 0$, i.e., the given equation cannot have any imaginary root.

In other words, all the roots of the given equation are real.

▶ Example 3.5.11 *Find an equation of the fourth degree with rational coefficients, one root of which is* $\sqrt{2} + i\sqrt{3}$. **[C.P. 2002]**

Solution: Since one root of the equation is $\sqrt{2}+i\sqrt{3}$, another root will be $\sqrt{2}-i\sqrt{3}$, a third root should be $-\sqrt{2}+i\sqrt{3}$, fourth root should be $-\sqrt{2}-i\sqrt{3}$. Hence, the required equation will be

$$\left(x-\sqrt{2}-i\sqrt{3}\right)\left(x-\sqrt{2}+i\sqrt{3}\right)\left(x+\sqrt{2}-i\sqrt{3}\right)\left(x+\sqrt{2}+i\sqrt{3}\right)=0,$$

or, $$\left\{\left(x-\sqrt{2}\right)^2+3\right\}\left\{\left(x+\sqrt{2}\right)^2+3\right\}=0,$$

or, $$\left\{x^2-2\sqrt{2}x+5\right\}\left\{x^2+2\sqrt{2}x+5\right\}=0,$$

or, $$(x^2+5)^2-\left(2\sqrt{2}x\right)^2=0,$$

or, $x^4 + 2x^2 + 25 = 0$ (coefficients are all rational.)

3.6 H.C.F. and L.C.M. of Polynomials

A polynomial which is a divisor of two or more polynomials is called a *common divisor* of these polynomials.

A common divisor of two or more polynomials is called their *greatest common divisor* (g.c.d.) or *highest common factor* (h.c.f.) if it is divisible by every common divisor of these polynomials. Two polynomials are said to be **relatively prime** if their g.c.d. is 1.

A polynomial which is a multiple of two or more polynomials is a *common* multiple of these polynomials.

A common multiple of two or more polynomials is called their *lowest common multiple* (l.c.m.) if it is a divisor of every common multiple of these polynomials.

▶ Example 3.6.1

$$\begin{aligned} f(x) &= x^4+x^3-9x^2-3x+18=(x+3)(x-2)\left(x+\sqrt{3}\right)\left(x-\sqrt{3}\right) \\ \phi(x) &= x^5+6x^2-49x+42=(x+3)(x-2)(x-1)\left(x+\sqrt{7}i\right)\left(x-\sqrt{7}i\right). \end{aligned}$$

The common divisors are $x+3$, $x-2$, $(x+3)(x-2)$.

$$\therefore \text{g.c.d.} = (x+3)(x-2) = x^2+x-6.$$

▶ Example 3.6.2

$$\begin{aligned} f(x) &= x^3+x^2-2=(x-1)(x^2+2x+2) \\ \phi(x) &= x^3+2x^2-3=(x-1)(x^2+3x+3). \end{aligned}$$

The l.c.m. $= (x-1)(x^2+2x+2)(x^2+3x+3)$.

3.6.1 A Practical Method of finding the H.C.F. of Two Polynomials

▶ Example 3.6.3 *Find the H.C.F. of* $f(x) = x^4 - x^3 + x - 1$ *and* $g(x) = x^3 - 3x^2 + 3x - 2$.

Rules. (Euclidean Algorithm)

(i) Divide $f(x)$, *the polynomial of higher degree*
by $g(x)$, *the polynomial of lower degree.*

(ii) Divide $g(x)$ by the **Remainder** obtained in the previous division.
The process is continued till the Remainder is zero,
Last divisor is the required H.C.F.

Note: Before division a little modification (multiply or divide the Remainder by a non-zero real numbers) may be necessary.

$$
\begin{array}{r|l|l}
x^3 - 3x^2 + 3x - 2 & x^4 - x^3 + 0x^2 + x - 1 & x + 2 \\
 & \underline{x^4 - 3x^3 + 3x^2 - 2x} & \\
 & \quad 2x^3 - 3x^2 + 3x - 1 & \\
 & \quad \underline{2x^3 - 6x^2 + 6x - 4} & \\
 & \qquad 3x^2 - 3x + 3 & \\
\end{array}
$$

$$
\begin{array}{r|l|l}
\to x^2 - x + 1 & x^3 - 3x^2 + 3x - 2 & x - 2 \\
 & \underline{x^3 - x^2 + x} & \\
 & -2x^2 + 2x - 2 & \\
 & \underline{-2x^2 + 2x - 2} & \\
\end{array}
$$

The transition from $3x^2 - 3x + 3$ to $x^2 - x + 1$ is just to ensure that fractions do not occur at the next stage of division.
Required H.C.F. $= x^2 - x + 1$.

Short-cut Method (Synthetic Method)

$$
\begin{array}{r|rrrr|rrrrr|l}
1-2 & 1 & -3 & 3 & -2 & 1 & -1 & 0 & 1 & -1 & 1+2 \\
 & 1 & -1 & 1 & & 1 & -3 & 3 & -2 & & \\
\hline
 & & -2 & 2 & -2 & & 2 & -3 & 3 & -1 & \\
 & & -2 & 2 & -2 & & 2 & -6 & 6 & -4 & \\
\hline
 & & & & & & 3 & 3 & -3 & 3 & \\
 & & & & & \leftarrow & & 1 & -1 & 1 & \\
\end{array}
$$

Required H.C.F. $= x^2 - x + 1$.

▶ Example 3.6.4 **(Exercises for Practice)** *Find the H.C.F. of*

(i) $x^4 - 6x^2 - 8x - 3$ *and* $x^3 - 3x - 2$; [*Ans.* $x^2 + 2x + 1$]

(ii) $x^6+3x^5+6x^4+7x^3+6x^2+3x+1$ *and* $x^5+x^4-x^3-x^2-x-1$.

[*Ans.* $x^4+2x^3+3x^2+2x+1$]

The following theorem will be very useful for our later discussions.

Theorem 3.6.1 *A multiple root of order* r *of the polynomial equation* $f(x)=0$ *is multiple root of order* $(r-1)$ *of the derived equation* $f'(x)=0$, *where* $f'(x)$ *is the derivative of* $f(x)$ *with respect to* x.

Solution: Let α be the root of $f(x)=0$ of multiplicity r. Then $(x-\alpha)^r$ is a factor of $f(x)$ but $(x-\alpha)^{r+1}$ is not a factor.

$\therefore$ we can write $f(x)=(x-\alpha)^r\phi(x)$, where $\phi(x)$ does not contain the factor $(x-\alpha)$.

Then

$$\begin{aligned} f'(x) &= r(x-\alpha)^{r-1}\phi(x)+(x-\alpha)^r\phi'(x) \\ &= (x-\alpha)^{r-1}[r\phi(x)+(x-\alpha)\phi'(x)] \\ &= (x-\alpha)^{r-1}g(x), \end{aligned}$$

where $g(x)=r\phi(x)+(x-\alpha)\phi'(x)$ is not divisible by $(x-\alpha)$.

$\therefore$ α is a root of $f'(x)=0$ of multiplicity $(r-1)$.

Note: Clearly α is a root of $f''(x)=0$ of multiplicity $(r-2)$ and so on.

Common Root: Use of H.C.F. *If* $f(x)$ *and* $g(x)$ *be two polynomials of degrees* m *and* n *respectively, then their common roots are given by* $\phi(x)=0$, *where* $\phi(x)$ *is the h.c.f. of* $f(x)$ *and* $g(x)$. *Hence, obtain a rule for determining the multiple roots and the orders of their multiplicity.*

General Discussions. Let $\alpha_1, \alpha_2, \ldots, \alpha_p$ $(p\le m, p\le n)$ be common roots of $f(x)$ and $g(x)$. Then

$$\begin{aligned} f(x) &= (x-\alpha_1)(x-\alpha_2)\cdots(x-\alpha_p)\psi(x); \\ g(x) &= (x-\alpha_1)(x-\alpha_2)\cdots(x-\alpha_p)\theta(x); \end{aligned}$$

where $\psi(x)$ and $\theta(x)$ are polynomials of degrees $m-p$ and $n-p$ respectively, and there are no common factors between them.

$$\therefore\ \phi(x)=(x-\alpha_1)(x-\alpha_2)\cdots(x-\alpha_p)$$

is the h.c.f. of the two polynomials $f(x)$ and $g(x)$, and $\phi(x)=0$ gives the common roots.

To determine the multiple roots and the order of their multiplicity of $f(x)=0$, we are to find the h.c.f. of $f(x)$ and $f'(x)$ and if

$$(x-\alpha)^p(x-\beta)^q(x-\gamma)^rF(x)$$

be the h.c.f., then α, β, γ are multiple roots of $f(x)=0$ of multiplicity $p+1$, $q+1$, $r+1$ respectively.

▶ Example 3.6.5 **As a numerical example,** *suppose we are to find the multiple roots, if any, of*

$$f(x)=x^3+x^2-16x+20.$$

We find the h.c.f. of $f(x)=x^3+x^2-16x+20$ *and* $f'(x)=3x^2+2x-16$.

The h.c.f. is $x-2$. Hence, 2 is a double root of $f(x)=0$, i.e., x^2-4x+4 is a factor of $x^3+x^2-16x+20$. The other factor must be $x+5$. See the division shown under ***

Synthetic Process of finding the H.C.F.

***

$$\begin{array}{r|l}
 & x+5 \\ \hline
x^2-4x+4 & x^3+x^2-16x+20 \\
 & x^3-4x^2+4x \\ \hline
 & 5x^2-20x+20 \\
 & 5x^2-20x+20 \\ \hline
 & \cdots
\end{array}$$

$3+8$	$3+2-16$	$1+1-16+20$	$1+1$
	$3-6$	$\times 3$	
	$8-16$	$3+3-48+60$	
	$8-16$	$3+2-16$	
		$1-32+60$	
		$\times 3$	
		$3-96+180$	
		$3+2-16$	
		$-98 \mid -98+196$	
		$\leftarrow 1-2$	

h.c.f. $=x-2$.

Observe that $f(x)=(x-2)^2(x+5)$. Hence, the roots of $f(x)=0$ are $2, 2, -5$.

Further Solved Examples

► Example 3.6.6 *Find the condition that* $ax^3+3bx^2+3cx+d=0$ *may have two equal roots and if that condition is satisfied* (*i.e., if the equation has two equal roots*), *then prove that each equal root* $=\dfrac{1}{2}\dfrac{bc-ad}{ac-b^2}$.

Solution: Let

$$f(x)=ax^3+3bx^2+3cx+d=0$$

have two equal roots α, α. Then α must be a root of the derived equation

$$f'(x)=3ax^2+6bx+3c=0.$$

We, therefore, have

$$a\alpha^3+3b\alpha^2+3c\alpha+d = 0 \quad (3.6.1)$$

$$a\alpha^2+2b\alpha+c = 0 \quad (3.6.2)$$

Multiplying (3.6.2) by α we obtain

$$a\alpha^3+2b\alpha^2+c\alpha=0. \quad (3.6.3)$$

From (3.6.1) and (3.6.3), by subtraction

$$b\alpha^2+2c\alpha+d=0. \quad (3.6.4)$$

From (3.6.2) and (3.6.4), by the method of cross multiplication we get

$$\frac{\alpha^2}{2bd-2c^2}=\frac{\alpha}{bc-ad}=\frac{1}{2ac-2b^2},$$

$$\text{i.e., } \alpha = \frac{bc-ad}{2ac-2b^2} \quad \text{(2nd and 3rd ratio)}$$

$$= \frac{2(bd-c^2)}{bc-ad} \quad \text{(1st and 2nd ratio).}$$

$\therefore$ the condition for equal roots is

$$\frac{1}{2}\frac{bc-ad}{ac-b^2} = \frac{2(bd-c^2)}{bc-ad},$$

$$\text{or,}\quad (bc-ad)^2 = 4(ac-b^2)(bd-c^2)$$

and the equal roots $= \dfrac{bc-ad}{2(ac-b^2)}$ or, $\dfrac{2(bd-c^2)}{bc-ad}$.

▶ Example 3.6.7

(a) *How to express a polynomial* $f(x)$ *of degree* n *in powers of* $x-h$?

$$f(x) = a_0(x-h)^n + a_1(x-h)^{n-1} + \cdots + a_{n-1}(x-h) + a_n.$$

How to get a_n***?***

a_n *is the remainder when* $f(x)$ *is divided by* $x-h$ *[see that* $f(h) = a_n$*]*
Suppose $Q(x)$ *is the quotient when* $f(x)$ *is divided by* $x-h$.
$Q(x)$ *is a polynomial of degree* $n-1$.

How to get a_{n-1}***?***

a_{n-1} *is the remainder when* $Q(x)$ *is divided by* $x-h$ $[Q(h) = a_{n-1}]$.
Continuing this process we can find all the coefficients.

(b) *Express* $3x^3-4x^2+5x+6$ *as a polynomial in* $(x+1)$. **[C.P. 2003]**

Solution: Here

$$f(x) = 3x^3-4x^2+5x+6 = a_0(x+1)^3 + a_1(x+1)^2 + a_2(x+1) + a_3.$$

We use the process of synthetic division: here $h = -1$.

−1	3	−4	5	6
		−3	7	−12
	3	−7	12	$-6 = a_3$
		−3	10	
	3	−10	$22 = a_2$	
		−3		
	$3 = a_0$	$-13 = a_1$		

$$\therefore\ f(x) = 3x^3-4x^2+5x+6 = 3(x+1)^3 - 13(x+1)^2 + 22(x+1) - 6.$$

(c) *Express* $3x^3+2x^2+x+80$ *as a polynomial in* $(x+5)$.
Try yourself. [*Ans.* $(x+5)^3 - 13(x+5)^2 + 56(x+5)$.]

Exercises on Chapter 3(I)

1. Find, by *synthetic division*, the Quotient and the Remainder

   (a) if $x^4-4x^3+6x^2-4x+1$ is divided by $3x-2$.

   [*Ans.* $\frac{1}{3}x^3 - \frac{10}{9}x^2 + \frac{34}{27}x - \frac{40}{80}; \frac{1}{81}$]

(b) if $2x^3 + x^2 - 5x - 3$ is divided by $x - 2$. [*Ans.* $2x^2 + 5x + 5; 7$]

2. Use the process of synthetic division to find

(a) $f(-4)$ if $f(x) = 2x^4 - 5x^2 - 32x + 6$. [*Ans.* 566]

(b) $f(-2/3)$ if $f(x) = 6x^3 + 4x^2 + 3x + 2$. [*Ans.* 0]

3. If $4x^4 - (a-1)x^3 + ax^2 - 6x + 1$ is divisible by $2x - 1$, prove that $a = 13$.

4. Express $x^3 + 2x^2 + x + 80$ as a polynomial in

(a) $x + 1$; [*Ans.* $(x+1)^3 - (x+1) + 80$]

(b) $x + \frac{1}{3}$. [*Ans.* $\left(x + \frac{1}{3}\right)^3 + \left(x + \frac{1}{3}\right)^2 + \frac{2156}{27}$.]

5. Express $4x^5 - 6x^4 + 3x^3 + x^2 - x - 1$ in the form

$$4(x-1)^5 + 14(x-1)^4 + 19(x-1)^3 + 14(x+1)^2 + 6(x-1).$$

6. Use Remainder Theorem to prove that $x - 3$ is a factor of $x^4 - 6x^3 + ax + b$ if $3a + b = 81$.

7. Prove that $x^4 + 2x^3 - 13x^2 - 14x + 21$ is exactly divisible by $x + 4$.

8. If $x^3 + 3px + q$ has a factor of the form $(x - a)^2$, prove that $q^2 + 4p^3 = 0$.

9. Find a cubic in x which shall vanish when $x = 1$ and $x = -2$ and has the values 4 and 28 when $x = -1$ and $x = 2$ respectively. [C.P. 1997]

[*Hints*: Let the required cubic be $(x-1)(x+2)(ax+b)$. Obtain $a = 3, b = 1$.]

10. Find a quadratic in x which shall vanish when $x = -3/7$ and have values 24 and 62 when $x = 3$ and $x = 4$ respectively. [*Ans.* $7x^2 - 11x - 6$]

11. Solve the following equations, using the condition given

(a) $x^4 + 2x^3 - 5x^2 + 6x + 2 = 0$, given that one of the roots is $-2 + \sqrt{3}$;

[*Ans.* $-2 \pm \sqrt{3}$ and $1 \pm i$]

(b) $x^4 - 12x^3 + 72x^2 - 312x + 676 = 0$, given that one of the roots is $5 + i$;

[*Ans.* $5 \pm i$ and $1 \pm 5i$]

(c) $x^4 - 2x^3 - 22x^2 + 62x - 15 = 0$, given that one root is $2 + \sqrt{3}$;

[*Ans.* $2 \pm \sqrt{3}, 3, -5$]

(d) $2x^3 - 15x^2 + 46x - 42 = 0$, given that one root is $3 + \sqrt{-5}$;

[*Ans.* $3 \pm \sqrt{-5}$ and $3/2$]

For (a)–(d), See Ex. 3.5.7.

12. Find a cubic equation, two of whose roots are 1 and $3 + 2i$.

[*Ans.* $x^3 - 7x^2 + 19x - 13 = 0$]

13. Find all the roots of

$$2x^6 - 3x^5 + 5x^4 + 6x^3 - 27x + 81 = 0,$$

if it is known that one root is $\sqrt{2} + i$. [*Ans.* $\sqrt{2}+i, \sqrt{2}-i, -\sqrt{2}+i, -\sqrt{2}-i$]

14. (a) Solve the equation

$$x^6 - 4x^5 - 11x^4 + 40x^3 + 11x^2 - 4x - 1 = 0,$$

given that one root is $\sqrt{2} + \sqrt{3}$. [*Ans.* Six roots are $\pm\sqrt{2} \pm \sqrt{3}$ and $2 \pm \sqrt{5}$]

[*Hints*: The required equation

$$\left(x - \sqrt{2} - \sqrt{3}\right)\left(x + \sqrt{2} - \sqrt{3}\right)\left(x + \sqrt{2} + \sqrt{3}\right)\left(x - \sqrt{2} + \sqrt{3}\right)(x - a)(x - b) = 0.$$

To determine a, b, divide $f(x)$ by

$$\left[\left(x - \sqrt{2} - \sqrt{3}\right)\left(x + \sqrt{2} - \sqrt{3}\right)\left(x + \sqrt{2} + \sqrt{3}\right)\left(x - \sqrt{2} + \sqrt{3}\right)\right] = x^4 - 10x^2 + 1$$

and equate the quotient $x^2 - 4x - 1 = 0 \Rightarrow x = 2 \pm \sqrt{5}$.]

(b) Find the equation of the fourth degree with rational coefficients, one root of which is $\sqrt{2 + i\sqrt{3}}$. [C.P. 1996]

15. (a) Show that the roots of the equation

$$\frac{1}{x-a} + \frac{1}{x-b} + \frac{1}{x-c} = \frac{1}{x}$$

are all real (a, b, c are three distinct positive real numbers). [C.P. 1993]

(b) Show that the roots of the equation

$$\frac{1}{x-1} + \frac{2}{x-2} + \frac{3}{x-3} + \frac{4}{x-4} + \frac{5}{x-5} = 6$$

are all real.

(c) Show that the equation

$$\frac{A^2}{x-a} + \frac{B^2}{x-b} + \frac{C^2}{x-c} + \cdots + \frac{L^2}{x-l} = x - m,$$

where $a, b, c, \ldots, l, m, A, B, C, \ldots, L$ are all distinct real numbers, cannot have any imaginary root.

16. Find the H.C.F. of:

(a) $3x^4 - 4x^3 + 1, 4x^4 - 5x^3 - x^2 + x + 1$; [*Ans.* $(x-1)^2$]

(b) $2x^4 - 13x^2 + x + 15, 3x^4 - 2x^3 - 17x^2 + 12x + 9$; [*Ans.* $x^2 + x - 3$]

(c) $2x^5 - 5x^2 + 3, 3x^5 - 5x^3 + 2$. [*Ans.* $(x-1)^2$]

17. Solve the following equations after finding the multiple roots:

(a) $4x^3 + 20x^2 - 23x + 6 = 0$; [*Ans.* $\frac{1}{2}, \frac{1}{2}, -6$]

(b) $x^4 - 8x^3 + 24x^2 - 32x + 16 = 0$; [*Ans.* 2, 2, 2, 2]

(c) $x^3 + x^2 - 16x + 20 = 0.$ [*Ans.* 2, 2, −5]

18. Assuming that the equations

$$\begin{aligned} 4x^4 + 12x^3 - x^2 - 15x &= 0 \\ \text{and } 6x^4 + 13x^3 - 4x^2 - 15x &= 0 \end{aligned}$$

have common roots, find all possible values of x in the two equations.

[*Ans.* 0, 1, −3/2, −5/2; 0, 1, −3/2, −5/3]

19. Find the common roots of

$$\begin{aligned} x^3 - 3x^2 - 10x + 24 &= 0 \\ \text{and } x^3 - 6x^2 - 40x + 192 &= 0. \end{aligned}$$

[*Ans.* 4]

20. Find a biquadratic equation with rational coefficients, one root of which is $\sqrt{3} - \sqrt{5}$. [*Ans.* $x^4 - 16x^2 + 4 = 0$]

21. Find the equal roots of the equation

$$x^4 - 5x^3 - 9x^2 + 81x - 108 = 0.$$

[*Hints*: Check: H.C.F. of $f(x)$ and $f'(x)$ is $(x-3)^2$]

[*Ans.* Three equal roots 3, 3, 3]

22. Solve the equation $x^5 - 15x^3 + 10x^2 + 60x - 72 = 0$ by testing for equal roots.

[*Ans.* 2, 2, 2, −3, −3]

23. If $1, a, b, \ldots, k$ be the n roots of $x^n - 1 = 0$, prove that

$$(1-a)(1-b)(1-c)\cdots(1-k) = n$$

[*Hints*: $x^n - 1 = (x-1)(x-a)(x-b)\cdots(x-k)$. Differentiate. Then put $x = 1$.]

24. If $\alpha, \beta, \gamma, \ldots, \lambda$ be the roots of .

$$x^n + p_1x^{n-1} + p_2x^{n-2} + \cdots + p_{n-1}x + p_n = 0.$$

prove that

$$\begin{aligned} &(1+\alpha^2)(1+\beta^2)(1+\gamma^2)\cdots(1+\lambda^2) \\ &= (p_1 - p_3 + p_5 - \cdots)^2 + (1 - p_2 + p_4 - \cdots)^2. \end{aligned}$$

[*Hints*: $x^n + p_1x^{n-1} + p_2x^{n-2} + \cdots + p_{n-1}x + p_n = (x-\alpha)(x-\beta)\cdots(x-\lambda)$. Put $x = \pm i$. Multiply.]

25. If $2+3i$ be one of the roots of the equation $x^4 - 2x^3 + 6x^2 + 22x + 13 = 0$, find the other roots. **[C.P. 2001]**

[*Ans.* $2-3i, -1, -1$]

SECTION II

3.7 Relation between Roots and Coefficients

We have already discussed how to construct an equation with given roots. Thus an equation with three roots $\alpha_1, \alpha_2, \alpha_3$ can be written as

$$(x-\alpha_1)(x-\alpha_2)(x-\alpha_3)=0.$$

When multiplied out, this becomes

$$x^3-x^2(\alpha_1+\alpha_2+\alpha_3)+x(\alpha_2\alpha_3+\alpha_3\alpha_1+\alpha_1\alpha_2)-\alpha_1\alpha_2\alpha_3=0$$

In shorter form we write

$$x^3-x^2(\Sigma\alpha_1)+x(\Sigma\alpha_1\alpha_2)-\alpha_1\alpha_2\alpha_3=0.$$

More generally, if $\alpha_1, \alpha_2, \alpha_3, \ldots, \alpha_n$ be n roots of

$$f(x)=a_0x^n+a_1x^{n-1}+a_2x^{n-2}+\cdots+a_{n-1}x+a_n=0\ (a_0\neq 0).$$

Then we can write

$$f(x)=A(x-\alpha_1)(x-\alpha_2)(x-\alpha_3)\cdots(x-\alpha_n),$$

where A is independent of x since $f(x)$ is of degree n in x; by equating coefficients of x^n, it follows $A=a_0$.

$$\therefore\ (x-\alpha_1)(x-\alpha_2)(x-\alpha_3)\cdots(x-\alpha_n)$$
$$=x^n+\frac{a_1}{a_0}x^{n-1}+\frac{a_2}{a_0}x^{n-2}+\cdots+\frac{a_n}{a_0}.$$

But expanding the left side of this identity we get

$$x^n-\Sigma\alpha_1x^{x-1}+\Sigma\alpha_1\alpha_2x^{n-2}-\cdots+(-1)^n\alpha_1\alpha_2\cdots\alpha_n.$$

Equating coefficients of like powers of this identity we get

$$\Sigma\alpha_1=-\frac{a_1}{a_0},\ \Sigma\alpha_1\alpha_2=\frac{a_2}{a_0},$$

$$\Sigma\alpha_1\alpha_2\alpha_3=-\frac{a_3}{a_0}\cdots\alpha_1\alpha_2\alpha_3\cdots\alpha_n=(-1)^n\frac{a_n}{a_0}.$$

Remember. The relations established remain true even if any of the roots α_1, α_2, ... are equal to one another.
e.g., Suppose $ax^3+bx^2+cx+d=0$ has three roots α, β, β. Then

$$\alpha+\beta+\beta=-\frac{b}{a},\ \ \alpha\beta+\alpha\beta+\beta^2=\frac{c}{a};\ \ \alpha\beta^2=-\frac{d}{a}.$$

3.7.1 Symmetric Functions of Roots

We shall often get expressions containing roots, α_1, α_2, α_3, etc. of a polynomial equation $f(x) = 0$. Such expressions are called *symmetric functions of roots* if the expressions remain unaltered when any two of the roots are interchanged (the expression must, of course, contain all the roots), e.g., if α, β, γ be the three roots of a cubic equations, expressions like

$$\alpha^2 + \beta^2 + \gamma^2,$$
$$\frac{1}{\alpha} + \frac{1}{\beta} + \frac{1}{\gamma},$$
$$\alpha^2\beta + \alpha\beta^2 + \alpha\gamma^2 + \alpha^2\gamma + \beta^2\gamma + \beta\gamma^2$$

are all symmetric functions of the roots of the cubic.

3.7.2 Abbreviated Form

A symmetric function of roots which is the sum of a number of terms of the same type is often written in an abbreviated form, thus:

Choose any one of the terms and place the sign Σ (sigma) before it.
e.g., in the above expressions we write

$$\begin{aligned} \Sigma\alpha^2 &= \alpha^2 + \beta^2 + \gamma^2, \\ \Sigma\frac{1}{\alpha} &= \frac{1}{\alpha} + \frac{1}{\beta} + \frac{1}{\gamma}, \\ \Sigma\alpha^2\beta &= \alpha^2\beta + \alpha^2\gamma + \alpha\beta^2 + \alpha\gamma^2 + \beta^2\gamma + \beta\gamma^2. \end{aligned}$$

Our final assertion is: *With the help of the relations between roots and coefficients, any symmetric function of the roots of a polynomial equation can be expressed in terms of the coefficients.*

3.8 Solved Problems

Direct Applications of the Formulae Connecting Roots and Coefficients.

▶ Example 3.8.1 *If the sum of two roots of the cubic*

$$x^3 + a_1x^2 + a_2x + a_3 = 0$$

is zero, prove that $a_1a_2 = a_3$.

Solution: Let the three roots of the cubic be α, β, γ, where $\alpha + \beta = 0$.
$\alpha + \beta + \gamma = -a_1$ and so $\gamma = -a_1$ ($\because \alpha + \beta = 0$).
$\alpha\beta + \gamma(\alpha + \beta) = a_2$ and so $\alpha\beta = a_2$ ($\because \alpha + \beta = 0$).
$\alpha\beta\gamma = -a_3 \Rightarrow a_2(-a_1) = -a_3$, i.e., $a_1a_2 = a_3$.

▶ Example 3.8.2 *If* α, β, γ *be the roots of* $x^3 + x + 1 = 0$. *Prove that*

$$\alpha^2 + \beta^2 + \gamma^2 = -2.$$

Solution:

$$\alpha+\beta+\gamma = 0,$$
$$\alpha\beta+\beta\gamma+\gamma\alpha = 1,$$
$$\alpha\beta\gamma = -1.$$

Now

$$\alpha^2+\beta^2+\gamma^2 = (\alpha+\beta+\gamma)^2-2(\alpha\beta+\beta\gamma+\gamma\alpha) = 0-2.1=-2.$$

▶ Example 3.8.3 *The roots of the equation*

$$x^3-3x^2+kx+3=0$$

are in A.P. Find the value of k *and solve the equation.*

Solution: Let the roots be $\alpha-\beta$, α, $\alpha+\beta$ so that they are in A.P.
Now

$$(\alpha-\beta)+\alpha+(\alpha+\beta) = 3;$$
$$(\alpha-\beta)\alpha+\alpha(\alpha+\beta)+(\alpha-\beta)(\alpha+\beta) = k;$$
$$(\alpha-\beta)\,\alpha\,(\alpha+\beta) = -3.$$

From the first relation,

$$3\alpha=3, \text{ or, } \alpha=1.$$

From the last relation,

$$\alpha(\alpha^2-\beta^2)=-3, \quad \text{or,} \quad 1(1-\beta^2)=-3,$$
$$\text{or,} \quad \beta^2=4, \text{ i.e., } \beta=\pm 2.$$

$$\therefore\ k=3\alpha^2-\beta^2=3-4=-1.$$

Roots of the equation are

$$\left.\begin{array}{l} -1,\ 1,\ \ 3 \text{ (Taking } \beta=2) \\ 3,\ 1,\ -1 \text{ (Taking } \beta=-2) \end{array}\right\}$$

▶ Example 3.8.4 *Solve the equation*

$$2x^3-21x^2+42x-16=0,$$

given that the roots are in G.P.

Solution: Let the roots be $\alpha\beta$, α, $\frac{\alpha}{\beta}$ so that they are in G.P.

$$\alpha\beta+\alpha+\frac{\alpha}{\beta}=\frac{21}{2};$$
$$\alpha\beta\cdot\alpha+\alpha\beta\cdot\frac{\alpha}{\beta}+\alpha\cdot\frac{\alpha}{\beta}=21;$$
$$\alpha\beta\cdot\alpha\cdot\frac{\alpha}{\beta}=8.$$

The first two relations give

$$\left.\begin{aligned} \alpha\left(\beta+1+\tfrac{1}{\beta}\right) &= \tfrac{21}{2} \\ \text{and } \alpha^2\left(\beta+1+\tfrac{1}{\beta}\right) &= 21 \end{aligned}\right\} \text{ so that } \alpha = 2.$$

Also

$$\beta+1+\tfrac{1}{\beta} = \tfrac{21}{4}, \quad \text{or,} \quad 4\beta^2 - 17\beta + 4 = 0$$
$$\Rightarrow \beta = 4 \quad \text{or,} \quad \tfrac{1}{4}.$$

$\therefore$ the roots are $8, 2, \frac{1}{2}$ (Taking $\beta = 4$), or, $\frac{1}{2}$, 2, 8 (Taking $\beta = \frac{1}{4}$).

▶ **Example 3.8.5** *If the roots of*

$$x^4 + ax^3 + bx^2 + cx + d = 0$$

are in G.P., then prove that $c^2 = a^2d$. **[C.P. 2000]**

Solution: We take the four roots in the form $\frac{\alpha}{\beta^3}$, $\frac{\alpha}{\beta}$, $\alpha\beta$, $\alpha\beta^3$ so that they form a G.P.

$$\frac{\alpha}{\beta^3} + \frac{\alpha}{\beta} + \alpha\beta + \alpha\beta^3 = -a,$$
$$\text{or,} \quad \alpha\left(\frac{1}{\beta^3} + \frac{1}{\beta} + \beta + \beta^3\right) = -a. \tag{1}$$

$$\frac{\alpha}{\beta^3}\cdot\frac{\alpha}{\beta} + \frac{\alpha}{\beta^3}\cdot\alpha\beta + \frac{\alpha}{\beta^3}\cdot\alpha\beta^3 + \frac{\alpha}{\beta}\cdot\alpha\beta + \frac{\alpha}{\beta}\cdot\alpha\beta^3 + \alpha\beta\cdot\alpha\beta^3 = b. \tag{2}$$

$$\frac{\alpha}{\beta^3}\cdot\frac{\alpha}{\beta}\cdot\alpha\beta + \frac{\alpha}{\beta^3}\cdot\frac{\alpha}{\beta}\cdot\alpha\beta^3 + \frac{\alpha}{\beta^3}\cdot\alpha\beta\cdot\alpha\beta^3 + \frac{\alpha}{\beta}\cdot\alpha\beta\cdot\alpha\beta^3 = -c,$$
$$\text{or,} \quad \alpha^3\left(\frac{1}{\beta^3} + \frac{1}{\beta} + \beta + \beta^3\right) = -c. \tag{3}$$

$$\frac{\alpha}{\beta^3}\cdot\frac{\alpha}{\beta}\cdot\alpha\beta\cdot\alpha\beta^3 = d, \text{ or, } \alpha^4 = d. \tag{4}$$

From (1) and (3) we obtain

$$\alpha^2 = \frac{-c}{-a} = \frac{c}{a}.$$

$$\therefore \alpha^4 = \frac{c^2}{a^2}, \text{ i.e., } d = \frac{c^2}{a^2}, \text{ using (4)}.$$

$$\therefore c^2 = a^2d \textbf{ (Proved)}.$$

▶ **Example 3.8.6** *Solve the equation*

$$4x^3 + 16x^2 - 9x - 36 = 0,$$

where the sum of two roots is zero. **[C.P. 2003]**

Solution: Let the roots be α, β, γ, where $\alpha + \beta = 0$.

$$\alpha + \beta + \gamma = -4, \ \alpha\beta + \beta\gamma + \gamma\alpha = -\tfrac{9}{4}, \ \alpha\beta\gamma = 9.$$

From the first relation,

$$\gamma = -4 \ (\because \ \alpha + \beta = 0).$$

From the second relation,

$$\alpha\beta = -\tfrac{9}{4} \text{ and } \alpha = -\beta,$$

$$\textbf{or, } -\beta^2 = -\tfrac{9}{4}, \text{ i.e., } \beta = \pm\tfrac{3}{2}.$$

If $\beta = \frac{3}{2}$, $\alpha = -\frac{3}{2}$. But if $\beta = -\frac{3}{2}$, then $\alpha = \frac{3}{2}$.
$\therefore$ the required roots are $-\frac{3}{2}, \frac{3}{2}, -4$, **or,** $+\frac{3}{2}, -\frac{3}{2}, -4$.

▶ Example 3.8.7 *Solve* $x^3 - 4x^2 - 3x + 18 = 0$, *given that two of its roots are equal.*

Solution: Let the roots be α, α, β. Then

$$\begin{aligned} 2\alpha + \beta &= 4; \\ \alpha^2 + 2\alpha\beta &= -3; \\ \alpha^2\beta &= -18. \end{aligned}$$

Now we may write

$$\begin{aligned} \alpha^2 + 2\alpha\beta = -3, \quad &\text{or,} \quad \alpha^2 + 2\alpha(4 - 2\alpha) - 3 \ (\because \ 2\alpha + \beta = 4), \\ &\text{or,} \quad 3\alpha^2 - 8\alpha - 3 = 0, \\ &\text{or,} \quad (3\alpha + 1)(\alpha - 3) = 0. \end{aligned}$$

$$\therefore \ \alpha = -1/3, \text{ or, } 3.$$

If $\alpha = -\frac{1}{3}$, $\beta = 14/3$; but if $\alpha = 3$, then $\beta = -2$ (using $2\alpha + \beta = 4$).

We now verify that the relation $\alpha^2\beta = -18$ is satisfied by only the second pair of values of α and β but not by the first pair.

$\therefore$ **the required roots are 3, 3, −2.**

▶ Example 3.8.8 *Solve:* (a) $4x^3 - 24x^2 + 23x + 18 = 0$, (b) $x^3 - 3x^2 - 6x + 8 = 0$ *given that the roots are in A.P.* **[C.P. 2004]**

Solution: (a) It will be convenient if we take the three roots as

$$\alpha - \beta, \alpha, \alpha + \beta \text{ (common difference } = \beta).$$

Then,

$$(\alpha - \beta) + \alpha + (\alpha + \beta) = \tfrac{24}{4} = 6, \text{ i.e., } \alpha = 2.$$

Also

$$(\alpha - \beta)\alpha + \alpha(\alpha + \beta) + (\alpha - \beta)(\alpha + \beta) = \tfrac{23}{4},$$

$$\text{i.e.,} \quad 3\alpha^2 - \beta^2 = \tfrac{23}{4}, \text{ (putting } \alpha = 2). \quad \therefore \ \beta = \pm\tfrac{5}{2}.$$

Taking $\beta = +\frac{5}{2}$, the roots are $-\frac{1}{2}, 2, 4\frac{1}{2}$.
Taking $\beta = -\frac{5}{2}$, the roots become $4\frac{1}{2}, 2, -\frac{1}{2}$.
$\therefore$ the required roots can be written as $-\frac{1}{2}, 2, 4\frac{1}{2}$, or, $4\frac{1}{2}, 2, -\frac{1}{2}$.

(b) Exactly as above. See that the roots of this equation are $-2, 1, 4$ (or, $4, 1, -2$). **(Reqd. Answer)**

Practical Hints
1. For a Cubic Equation, if the roots are in A.P. assume the roots as: $\alpha - \beta, \alpha, \alpha + \beta$ (C.D. $= \beta$)   **For a Biquadratic Equation, if the roots are in A.P. assume the roots as:** $\alpha - 3\beta, \alpha - \beta, \alpha + \beta, \alpha + 3\beta$ (C.D. $= 2\beta$)
2. For a Cubic Equation, if the roots are in G.P., then take the roots as: $\alpha\beta, \alpha, \alpha/\beta$ $\left(\text{C.R.} = \frac{1}{\beta}\right)$ **For a Biquadratic Equation, if the roots are in G.P., then take the roots as:** $\frac{\alpha}{\beta^3}, \frac{\alpha}{\beta}, \alpha\beta, \alpha\beta^3$ (C.R. $= \beta^2$)

► Example 3.8.9 *If the roots of*

$$x^4 + px^3 + qx^2 + rx + s = 0,$$

are in G.P. find a relation between the coefficients p, q, r, s.

[We may call such a relation as a necessary condition for the roots to be in G.P.]

Solution: Let the roots be α/β^3, α/β, $\alpha\beta$ and $\alpha\beta^3$.

Then

$$\sum \frac{\alpha}{\beta^3} = \alpha\left(\frac{1}{\beta^3} + \frac{1}{\beta} + \beta + \beta^3\right) = -p. \tag{1}$$

$$\sum \frac{\alpha}{\beta^3}\cdot\frac{\alpha}{\beta} = \alpha^2\left(\frac{1}{\beta^4} + \frac{1}{\beta^2} + 2 + \beta^2 + \beta^4\right) = q. \tag{2}$$

$$\sum\left(\frac{\alpha}{\beta^3}\cdot\frac{\alpha}{\beta}\cdot\alpha\beta\right) = \alpha^3\left(\frac{1}{\beta^3} + \frac{1}{\beta} + \beta + \beta^3\right) = -r. \tag{3}$$

$$\frac{\alpha}{\beta^3}\cdot\frac{\alpha}{\beta}\cdot\alpha\beta\cdot\alpha\beta^3 = s. \tag{4}$$

From (4) $\alpha^4 = s$.
From (1) and (3) $\alpha^2 = r/p$ (on division)

$$\therefore \frac{r^2}{p^2} = s, \text{ or, } p^2 s = r^2, \text{ is the required relation.}$$

► Example 3.8.10 (a) *For the cubic equation* $x^3 + ax^2 + bx + c = 0$,*prove that the necessary condition for the roots to be in G.P. is that* $b^3 = a^3c$. **[C.P. 1988]**

(b) *If the roots of the equation* $x^3 - px^2 + qx - r = 0$ *are in G.P., show that* $p^3r = q^3$. [C.P. 2005]

Solution: (a) Let the roots be $\alpha\beta$, α, $\frac{\alpha}{\beta}$ so that they are in G.P.

Then

$$\alpha\beta + \alpha + \frac{\alpha}{\beta} = \alpha\left(\beta + 1 + \frac{1}{\beta}\right) = -a \quad (1)$$

$$\alpha\beta\cdot\alpha + \alpha\beta\cdot\frac{\alpha}{\beta} + \alpha\cdot\frac{\alpha}{\beta} = \alpha^2\left(\beta + 1 + \frac{1}{\beta}\right) = +b \quad (2)$$

$$\alpha\beta\cdot\alpha\cdot\frac{\alpha}{\beta} = \alpha^3 = -c. \quad (3)$$

From (1) and (2), on division, $\alpha = -\frac{b}{a}$.

Putting this value of α in (3) we get $\left(-\frac{b}{a}\right)^3 = -c$, or, $b^3 = a^3c$ (**Proved**).

(b) Either follow the method given in (a), **or**

Let α, β, γ be the roots. Since the roots are in G.P. $\beta^2 = \alpha\gamma$.

Now, $\alpha + \beta + \gamma = p$, $\alpha\beta + \beta\gamma + \alpha\gamma = q$, $\alpha\beta\gamma = r$.

Hence from the last relation $\beta^3 = r$ using $\alpha r = \beta^2$ and from the second relation $\beta(\alpha + \gamma + \beta) = q$ **or** $\beta p = q$ or $\beta^3 = q^3/p^3 = r$.

Hence $p^3r = q^3$.

▶ **Example 3.8.11** *Solve the equation*

$$6x^3 - 11x^2 - 3x + 2 = 0$$

whose roots are in Harmonic Progression (H.P.).

Solution: Let α, β, γ be the roots of the given equation. Since the roots are in H.P. we may write

$$\frac{1}{\alpha} + \frac{1}{\gamma} = \frac{2}{\beta}. \quad (1)$$

Remember. If α, β, γ are in H.P., then $\frac{1}{\alpha}$, $\frac{1}{\beta}$, $\frac{1}{\gamma}$ are in A.P.

From the relations between the roots and coefficients we write

$$\alpha + \beta + \gamma = \frac{11}{6}. \quad (2)$$

$$\alpha\beta + \alpha\gamma + \beta\gamma = -\frac{1}{2}. \quad (3)$$

$$\alpha\beta\gamma = -\frac{1}{3}. \quad (4)$$

From (1) we obtain,

$$2\alpha\gamma = \alpha\beta + \beta\gamma,$$

$$\text{i.e.,}\quad 3\alpha\gamma = \alpha\beta + \beta\gamma + \gamma\alpha = -\tfrac{1}{2}, \text{ using (3).}$$

Hence,

$$3\alpha\beta\gamma = -\tfrac{1}{2}\beta,$$

$$\text{or,}\quad -1 = -\tfrac{1}{2}\beta, \text{ from (4), i.e., } \beta = 2.$$

From (2) it then follows,

$$\alpha+\gamma=\tfrac{11}{6}-2=-\tfrac{1}{6} \tag{5}$$

and from (3) it also follows,

$$\beta(\alpha+\gamma)+\alpha\gamma=-\frac{1}{2}, \quad \text{or,} \quad 2\left(-\frac{1}{6}\right)+\alpha\gamma=-\frac{1}{2},$$

$$\text{or,} \quad \alpha\gamma=-\frac{1}{6}.$$

Now,

$$(\alpha-\gamma)^2=(\alpha+\gamma)^2-4\alpha\gamma=\left(-\frac{1}{6}\right)^2-4\left(-\frac{1}{6}\right)=\frac{25}{36}, \text{ i.e., } \alpha-\gamma=\pm\frac{5}{6}. \tag{6}$$

From (5) and (6)

$$\alpha+\gamma=-\frac{1}{6}$$

$$\text{and } \alpha-\gamma=\frac{5}{6}$$

whence, $\alpha=\frac{1}{3}$ and $\gamma=-\frac{1}{2}$,
or, $\alpha+\gamma=-\frac{1}{6}$ and $\alpha-\gamma=-\frac{5}{6}$,
whence, $\alpha=-\frac{1}{2}$ and $\gamma=\frac{1}{3}$.
$\therefore$ the required roots can be written as $\frac{1}{3},2,-\frac{1}{2}$, or, as $-\frac{1}{2},2,\frac{1}{3}$.

▶ Example 3.8.12 *The equation*

$$x^3-5x^2+4x+20=0$$

has two roots which are equal in magnitude and opposite in sign. Solve it.

[C.P. 1987]

Solution: Since two roots are equal in magnitude and opposite in sign, we can assume that the roots are $\alpha,-\alpha,\beta$.

From the relations between roots and coefficients,
$\alpha-\alpha+\beta=5$, i.e., $\beta=5$.
$\alpha(-\alpha)+\alpha\beta+(-\alpha)\beta=-4$, i.e., $\alpha^2=4$, or, $\alpha=\pm2$.
$\alpha\cdot(-\alpha)\cdot\beta=-20$, i.e., $\alpha^2\beta=20$ (satisfied by $\beta=5$ and $\alpha=\pm2$).
The required roots are $2,-2,5$.

▶ Example 3.8.13 *Solve*

$$x^4+4x^3+8x^2-120x+900=0,$$

given that two of its roots are in the form $a\pm ia$, and the other two roots are in the form $b\pm ib$ (a,b real).

Solution: Let the four roots be $a + ia$, $a - ia$, $b + ib$, $b - ib$.

Their sum $= 2(a+b) = -4$, or, $a + b = -2$.

Product $= (a^2 + a^2)(b^2 + b^2) = 900$, or, $a^2 \cdot b^2 = \frac{900}{4} = 225$.

So we have, $a + b = -2$ and $ab = \pm 15$.

$$\therefore\ (a-b)^2 = (a+b)^2 - 4ab \begin{array}{rl} = & -15, \quad \text{if } ab = +15 \\ = & 64, \quad \text{if } ab = -15. \end{array}$$

Thus, if $ab = +15$, a, b will not give real values. So we take $ab = -15$.

Solving $a + b = -2$, $a - b = \pm 8$, we get $a = 3$, $b = -5$, or; $a = -5$, $b = 3$.

Hence, the required roots are $3 \pm 3i$, $-5 \pm 5i$.

▶ Example 3.8.14 *The equation*

$$f(x) = x^4 + ax^3 + bx^2 + cx + d = 0$$

has three equal roots α, α, α and the fourth root is $\beta (\neq \alpha)$. Prove that

$$\alpha = \frac{6c - ab}{3a^2 - 8b}.$$

Solution: Since

$$\begin{aligned} f(x) &= 0 \text{ has three equal roots } \alpha, \alpha, \alpha, \\ f'(x) &= 0 \text{ has two equal roots } \alpha, \alpha, \\ f''(x) &= 0 \text{ has a roots } \alpha, \end{aligned}$$

$\therefore$ we have,

$$\alpha^4 + a\alpha^3 + b\alpha^2 + c\alpha + d = 0. \quad (1)$$

$$4\alpha^3 + 3a\alpha^2 + 2b\alpha + c = 0. \quad (2)$$

$$12\alpha^2 + 6a\alpha + 2b = 0, \text{ or, } 6\alpha^2 + 3a\alpha + b = 0. \quad (3)$$

From (2) and (3)

$$\begin{aligned} 12\alpha^3 + 9a\alpha^2 + 6b\alpha + 3c = 0 \ [\text{multiply(2)by3}] \\ \text{and} \quad 12\alpha^3 + 6a\alpha^2 + 2b\alpha = 0 \ [\text{multiply (3) by } 2\alpha] \end{aligned}$$

Hence, on subtraction, we get

$$3a\alpha^2 + 4b\alpha + 3c = 0. \quad (4)$$

From (3) and (4) by cross-multiplication

$$\frac{\alpha^2}{9ac - 4b^2} = \frac{\alpha}{3ab - 18c} = \frac{1}{24b - 9a^2}$$

whence

$$\alpha = \frac{3ab - 18c}{24b - 9a^2} = \frac{6c - ab}{3a^2 - 8b}.$$

Symmetric Functions of the Roots

► Example 3.8.15 *If α, β, γ, δ be the roots of the biquadratic*

$$x^4 - 15x^2 - 10x + 24 = 0,$$

prove that

$$\Sigma\alpha^2 = 30$$
$$\text{and } \Sigma\alpha^2\beta = -30.$$

Solution: From the given equation we write

$$\Sigma\alpha = 0, \Sigma\alpha\beta = -15, \Sigma\alpha\beta\gamma = 10, \alpha\beta\gamma\delta = 24.$$

(i) $\Sigma\alpha^2 = (\Sigma\alpha)^2 - 2\Sigma\alpha\beta = 0 - 2 \times (-15) = \mathbf{30}$;

(ii) $\Sigma\alpha^2\beta = \Sigma\alpha \times \Sigma\alpha\beta - 3\Sigma\alpha\beta\gamma = 0 - 3 \times 10 = \mathbf{-30}$.

[See that $\Sigma\alpha \times \Sigma\alpha\beta$ contains terms like $\alpha\cdot\alpha\beta$ occurring once only but terms like $\alpha\cdot\beta\gamma$ occurring three times.]

► Example 3.8.16 *If α, β, γ be the roots of the equation*

$$x^3 + px + q = 0,$$

express in terms of p and q the values of the symmetric functions:

(i) $\Sigma\alpha^2$; (ii) $\Sigma\alpha^3$; (iii) $\Sigma\alpha^5$.

Solution: From the relations between roots and coefficients we get

$$\Sigma\alpha = 0,\ \Sigma\alpha\beta = p,\ \alpha\beta\gamma = -q.$$

(i) $\Sigma\alpha^2 = \alpha^2 + \beta^2 + \gamma^2 = (\Sigma\alpha)^2 - 2\Sigma\alpha\beta = 0 - 2p = \mathbf{-2p}$;

(ii) See that

$$\begin{aligned}\alpha^3 + \beta^3 + \gamma^3 - 3\alpha\beta\gamma &= (\alpha + \beta + \gamma)(\alpha^2 + \beta^2 + \gamma^2 - \alpha\beta - \beta\gamma - \gamma\alpha) \\ &= 0\ (\because\ \Sigma\alpha = 0 \text{ here}).\end{aligned}$$

$$\therefore\ \Sigma\alpha^3 = 3\alpha\beta\gamma = \mathbf{-3q}.$$

(iii) Observe the following method carefully:

Since α is a root of

$$x^3 + px + q = 0.$$

$$\therefore\ \alpha^3 + p\alpha + q = 0.$$

We may obtain,

$$\alpha^5 + p\alpha^3 + q\alpha^2 = 0.$$

Similarly,

$$\begin{aligned}\beta^5 + p\beta^3 + q\beta^2 &= 0, \\ \gamma^5 + p\gamma^3 + q\gamma^2 &= 0.\end{aligned}$$

Adding

$$\Sigma\alpha^5 + p\Sigma\alpha^3 + q\Sigma\alpha^2 = 0,$$
$$\text{or,}\quad \Sigma\alpha^5 = -p\Sigma\alpha^3 - q\Sigma\alpha^2 = (-p)(-3q) - q(-2p) = 5pq.$$

For Problems of Symmetric Functions of Roots: Remember
1. $\Sigma\alpha^2 = (\Sigma\alpha)^2 - 2\Sigma\alpha\beta$.
2. $\Sigma\alpha^2\beta = \Sigma\alpha\cdot\Sigma\alpha\beta - 3\Sigma\alpha\beta\gamma$.
3. $\Sigma\alpha^3 = \Sigma\alpha\cdot\Sigma\alpha^2 - \Sigma\alpha^2\beta$.
4. $\Sigma\alpha^2\beta\gamma = \Sigma\alpha\cdot\Sigma\alpha\beta\gamma - 4\Sigma\alpha\beta\gamma\delta$.
5. $\Sigma\alpha^2\beta^2 = (\Sigma\alpha\beta)^2 - 2\Sigma\alpha^2\beta\gamma - 6\Sigma\alpha\beta\gamma\delta$.
6. $\Sigma\alpha^3\beta = \Sigma\alpha^2\cdot\Sigma\alpha\beta - \Sigma\alpha^2\beta\gamma$.
7. $\Sigma\alpha^4 = (\Sigma\alpha^2)^2 - 2\Sigma\alpha^2\beta^2$.

▶ **Example 3.8.17** *If $\alpha_1, \alpha_2, \alpha_3$ be three roots of the equation*

$$x^3 + x + 1 = 0,$$

then prove that $(\alpha_1^2+1)(\alpha_2^2+1)(\alpha_3^2+1) = 1.$ **[C.P. 2002]**

Solution: Since $\alpha_1, \alpha_2, \alpha_3$ are the three roots of $x^3 + x + 1 = 0$, we can write

$$x^3 + x + 1 = (x-\alpha_1)(x-\alpha_2)(x-\alpha_3).$$

Put $x = i$,

$$i^3 + i + 1 = (i-\alpha_1)(i-\alpha_2)(i-\alpha_3).$$

Put $x = -i$,

$$-i^3 - i + 1 = (-i-\alpha_1)(-i-\alpha_2)(-i-\alpha_3).$$

Multiply

$$(i^3+i+1)(-i^3-i+1) = -(i^2-\alpha_1^2)(i^2-\alpha_2^2)(i^2-\alpha_3^2),$$
$$\text{or,}\quad 1\cdot 1 = (1+\alpha_1^2)(1+\alpha_2^2)(1+\alpha_3^2)\ (\because\ i^3+i=0, -i^3-i=0).$$

Exercises on Chapter 3(II)

1. Construct equations whose roots are:

   (a) 1, 2, 3. (b) $\pm1, \pm2, 3$.

2. Solve the following cubic equations under the condition stated against each:

   (a) $x^3 - 6x^2 + 3x + 10 = 0$; the roots are in A.P. **[C.P. 1987]**
   (b) $2x^3 - 21x^2 + 42x - 16 = 0$; the roots are in G.P.
   (c) $6x^3 - 11x^2 + 6x - 1 = 0$; the roots are in H.P.
   (d) $x^3 - 7x^2 + 36 = 0$; one root is twice another root.

(e) $x^3 - 3x^2 - 4x + 12 = 0$; sum of two of its roots is zero. [C.P. 1990]

(f) $x^3 - 5x^2 - 2x + 24 = 0$; product of two of its roots is 12.

(g) $x^3 - 10x^2 + 33x - 36 = 0$; two of its roots are equal in magnitude and sign.

(h) $3x^3 - 7x^2 - 36x - 20 = 0$; sum of two of its roots is 3.

(i) $x^3 - 9x^2 + 14x + 24 = 0$; two of its roots are in the ratio 3 : 2. [See Hints]

(j) $4x^3 + 16x^2 - 9x - 36 = 0$; the sum of two of its roots being zero. [See Ex. 3.8.12] [C.P. 1980, 2003]

3. Solve the following biquadratic equations subject to the condition stated against each:

(a) $x^4 + x^3 - 16x^2 - 4x + 48 = 0$; product of two of its roots is 6.

(b) $x^4 + 15x^3 + 70x^2 + 120x + 64 = 0$; roots are in G.P.

(c) $x^4 + 4x^3 - 2x^2 - 12x + 9 = 0$; two pairs of equal roots.

(d) $4x^4 - 4x^3 - 13x^2 + 9x + 9 = 0$; sum of two of its roots is zero.

(e) $x^4 - 11x^3 + 28x^2 + 36x = 144$; product of two of its roots and the product of other two roots differ only in sign.

(f) $x^4 - 2x^3 - 3x^2 + 4x - 1 = 0$; product of two of its roots is 1.

(g) $x^4 - 8x^3 + 14x^2 + 8x - 15 = 0$; roots are in A.P.

4. Given below cubic equations and some conditions satisfying the roots: You are to deduce in each case a relation between the coefficients.

	Equation	Condition satisfying the roots	To deduce the following Relations among coefficients
(i)	$x^3 + px^2 + qx + r = 0$;	roots are in A.P. →	$2p^3 + 27r = 9pq$. [C.P. 2004]
(ii)	$x^3 + px + q = 0$;	two of its roots are equal →	$4q^3 + 27r^2 = 0$.
	[Prove that the repeated roots is given by → $3r/2q$]		
(iii)	$x^3 + px^2 + qx + r = 0$;	two of its roots differ only in sign →	$pq = r$. [C.P. 2003]
(iv)	$x^3 + 3px^2 + 3qx + r = 0$;	roots are in H.P. →	$2q^3 = r(3pq - r)$.
(v)	$x^3 + px^2 + qx + r = 0$;	two of its roots α, β are connected by $\alpha\beta = -1$ →	$1 + q + pr + r^2 = 0$.

5. (a) If the roots $\alpha, \beta, \gamma, \delta$ of the equation

$$x^4 + px^3 + qx^2 + rx + s = 0,$$

be connected by the relation $\alpha\beta + \gamma\delta = 0$, then show that

$$p^2s + r^2 = 4qs.$$

(b) If the sum of the two roots of

$$x^4 + px^3 + qx^2 + rx + s = 0,$$

be equal to the sum of its other two roots, then prove that

$$p^3 + 8r = 4pq.$$ **[See Hints]**

(c) $x^4 - px^2 + qx - r = 0$ has three roots equal, and the other distinct. Prove that $p^2 = 12r$ and $9q^2 = 32pr$; obtain the repeated roots in terms of p and q.

(d) If the sum of two of its roots of

$$x^4 + px^3 + px^2 + rx + s = 0$$

be zero, then prove that

$$p^2s - pqr + r^2 = 0 \ (s \neq 0).$$

[For nos. 6, 7 and 8, see Hints and Answers.]

6. Prove that if the equation

$$x^4 + ax^3 + bx^2 + cx + d = 0$$

has three equal roots α, α, α and the fourth root distinct, then

$$3(ab - 6c)^2 = (8b - 3a^2)(9ac - 4b^2)$$

and that $\alpha = \dfrac{6c - ab}{3a^2 - 8b}$.

7. If α, β, γ be the roots of

$$ax^3 + 3bx^2 + 3cx + d = 0,$$

justify the following statements:

(a) The condition that α, β, γ are in A.P. is

$$a^2d - 3abc + 2b^3 = 0.$$

(b) The condition that α, β, γ are in G.P. is $ac^3 = b^3d$. **[C.P. 1982]**

(c) The condition that α, β, γ are in H.P. is $ad^2 - 3bcd + 2c^3 = 0$.

8. Find the conditions that the roots of

$$x^4 + px^3 + qx^2 + rx + s = 0$$

may be in: (a) A.P.; (b) G.P. (c) H.P.

9. (a) If α, β, γ be the roots of

$$x^3 - 4x^2 + 2x - 1 = 0,$$

find the values of: (i) $\Sigma\alpha^2$; (ii) $\Sigma\alpha^3$; (iii) $\Sigma\frac{1}{\alpha}$.

(b) If α, β, γ be the roots of $x^3 - x^2 + 4x - 1 = 0$, find the values of

i. $\Sigma\alpha^2$;
ii. $\Sigma\alpha^3$;
iii. $(\alpha + 1)(\beta + 1)(\gamma + 1)$;
iv. $\Sigma(\alpha - \beta)^2$;
v. $\Sigma\alpha(\beta - \gamma)^2$.

(c) If $\alpha_1, \alpha_2, \alpha_3$ be the roots of

$$x^3 + px + q = 0,$$

find in terms of p and q the values of

i. $\Sigma\alpha_1^4$;
ii. $\Sigma\alpha_1^2\alpha_2$;
iii. $\Sigma\alpha_1^2\alpha_2^2$.

10. α, β, γ are the roots of

$$x^4 + px^3 + qx^2 + rx + s = 0;$$

find in terms of the coefficients the values of

(a) $\Sigma\alpha^2$;
(b) $\Sigma\frac{1}{\alpha}$;
(c) $\Sigma\alpha^2\beta$;
(d) $\Sigma\alpha^2\beta^2\gamma^2$;
(e) $\Sigma\alpha\beta\gamma(\alpha + \beta\gamma)$.

11. If $\alpha, \beta, \gamma, \delta$ be the roots of

$$x^4 - 3x^3 + 4x^2 - 5x + 6 = 0,$$

find the value of

$$(\alpha^2 + 3)(\beta^2 + 3)(\gamma^2 + 3)(\delta^2 + 3).$$

[C.P. 2000]

Hints and Answers

1. (a) $x^3 - 6x^2 + 11x - 6 = 0$.

(b) $x^5 - 3x^4 - 5x^3 + 15x^2 + 4x - 12 = 0$.

2. (a) $-1, 2, 5$;
(b) $\frac{1}{2}, 2, 8$;
(c) $1, \frac{1}{2}, \frac{1}{3}$;
(d) $-2, 3, 6$.
Let the roots be α, 2α, β.
(e) $3, 2, -2$;
(f) $-2, 3, 4$;
(g) $3, 3, 4$;
(h) $-2/3, -2, 5$;
(i) Let the roots be 3α, 2α, β. Then $3\alpha + 2\alpha + \beta = 9$, $3\alpha \cdot 2\alpha + 3\alpha \cdot \beta + 2\alpha \cdot \beta = 14$, $3\alpha \cdot 2\alpha \cdot \beta = -24$. Solve for α, β. [*Ans.* 6, 4, −1.]
(j) $\frac{3}{2}, -\frac{3}{2}, -4$.

3. In a **biquadratic equation** with four roots $\alpha, \beta, \gamma, \delta$ write 4 relations:

$$\begin{aligned} \Sigma\alpha = (\alpha+\beta) + (\gamma+\delta) &= ? \\ \Sigma\alpha\beta = \alpha\beta + \gamma\delta + (\alpha+\beta)(\gamma,\delta) &= ? \\ \Sigma\alpha\beta\gamma = \alpha\beta(\gamma+\delta) + \gamma\delta(\alpha+\beta) &= ? \\ \alpha\beta \cdot \gamma\delta &= ? \end{aligned}$$

(a) $\Sigma\alpha = -1$, $\Sigma\alpha\beta = -16$, $\Sigma\alpha\beta\gamma = 4$, $\alpha\beta\gamma\delta = 48$. Given $\alpha\beta = 6$. Since $\alpha\beta\gamma\delta = 48$, it follows that $\gamma\delta = 8$. The calculations become simpler if we adopt the following procedure:

Let $\alpha + \beta = l$, $\gamma + \delta = m$. See that the roots of $x^2 - lx + 6 = 0$ are α, β and the roots of $x^2 - mx + 8 = 0$ are γ, δ. We then write

$$x^4 + x^3 - 16x^2 - 4x + 48 = (x^2 - lx + 6)(x^2 - mx + 8).$$

Comparing the coefficients we easily obtain $l = 5$ and $m = -6$,

$$\begin{aligned} \text{i.e., } x^4 + x^3 - 16x^2 - 4x + 48 &= (x^2 - 5x + 6)(x^2 + 6x + 8) \\ &= (x-3)(x-2)(x+4)(x+2). \end{aligned}$$

$\therefore$ required roots are $3, 2, -2, -4$.

(b) Let the roots be a, ar, ar^2 and ar^3. Call them $\alpha, \gamma, \delta, \beta$. See that

$$\alpha\beta = a \cdot ar^3 \text{ and } \gamma\delta = ar \cdot ar^2, \text{ i.e., } \alpha\beta = \gamma\delta.$$

Since

$$\alpha\beta\gamma\delta = 64, \text{ i.e., } (\alpha\beta)^2 = 64, \alpha\beta = +8, \text{ or, } -8.$$

Let $\alpha + \beta = l$, $\gamma + \delta = m$. Take $\alpha\beta = \gamma\delta = 8$ (Not -8. Why?)
As in (a)

$$x^4 + 15x^3 + 70x^2 + 120x + 64 = (x^2 - lx + 8)(x^2 - mx + 8).$$

Comparing coefficients, we get

$$l + m = -15, lm = 54 \text{ whence } l = -9, m = -6.$$

$$\therefore f(x) = (x^2 + 9x + 8)(x^2 + 6x + 8) = (x+8)(x+1)(x+4)(x+2).$$

$\therefore$ roots are $-1, -2, -4, -8$.

(c) **Ans.** $1, 1, -3, -3$.

(d) Let the roots be $\alpha, \beta, \gamma, \delta$. Then

$$(\alpha+\beta)+(\gamma+\delta)=1 \tag{i}$$

$$(\alpha+\beta)+(\gamma+\delta)+\alpha\beta+\gamma\delta=-\frac{13}{4} \tag{ii}$$

$$\gamma\delta(\alpha+\beta)+\alpha\beta(\gamma+\delta)=-\frac{9}{4} \tag{iii}$$

$$\alpha\beta\cdot\gamma\delta=\frac{9}{4} \tag{iv}$$

Since the sum of two roots is zero, we take

$$\alpha+\beta=0. \tag{v}$$

From (i) and (v), we obtain

$$\gamma+\delta=1. \tag{vi}$$

From (iii) we now get

$$\alpha\beta=-\frac{9}{4}. \tag{vii}$$

Dividing (iv) by (vii) we have

$$\gamma\delta=-1. \tag{viii}$$

From (v) and (vii) we find $\alpha+\beta=0, \alpha\beta=-\frac{9}{4}$, i.e., α, β are the roots of

$$x^2-\frac{9}{4}=0, \text{ i.e., } \alpha, \beta=\frac{3}{2}, -\frac{3}{2}.$$

From (vi) and (viii) we find $\gamma+\delta=1, \gamma\delta=-1$, i.e., γ, δ are the roots of

$$x^2-x-1=0, \text{ i.e., } \gamma, \delta=\frac{1\pm\sqrt{5}}{2}.$$

$\therefore$ required roots are $\frac{3}{2}, -\frac{3}{2}, \frac{1+\sqrt{5}}{2}, \frac{1-\sqrt{5}}{2}$.

(e) Ans. $-2, 3, 4, 6$ (Here take $\alpha\delta=-\gamma\beta$).

(f) Ans. $\frac{1}{2}\left(3+\sqrt{5}\right), \frac{1}{2}\left(3-\sqrt{5}\right), \frac{1}{2}\left(-1+\sqrt{5}\right), \frac{1}{2}\left(-1-\sqrt{5}\right)$
(Take $\alpha\beta=1$).

(g) Ans. $5, 3, 1, -1$. Let the roots be $\alpha-3\beta, \alpha-\beta, \alpha+\beta, \alpha+3\beta$.

$$\Sigma(\alpha-3\beta)=8, \text{ i.e., } 4\alpha=8, \text{ i.e., } \alpha=2.$$

$$\Sigma(\alpha-3\beta)(\alpha-\beta)=14,$$

$$\begin{aligned}\text{i.e., } (\alpha^2-4\alpha\beta+3\beta^2) &+ (\alpha^2-2\alpha\beta-3\beta^2)\\ &+ (\alpha^2-9\beta^2)+(\alpha^2-\beta^2)\\ &+ (\alpha^2+2\alpha\beta-3\beta^2)\\ &+ (\alpha^2+4\alpha\beta+3\beta^2)=14,\end{aligned}$$

or, $6\alpha^2-10\beta^2=14$, i.e., $\beta^2=1$ (since $\alpha=2$), **or,** $\beta=\pm1$.

$\therefore$ required roots are $2-3, 2-1, 2+1, 2+3$ (Taking $\beta=1$)

or, $2+3, 2+1, 2-1, 2-3$ (Taking $\beta=-1$)

5. (b)

$$\begin{aligned}(\alpha+\beta)+(\gamma+\delta) &= -p &(1)\\ (\alpha+\beta)(\gamma+\delta)+\alpha\beta+\gamma\delta &= q &(2)\\ \alpha\beta(\gamma+\delta)+\gamma\delta(2+\beta) &= -r &(3)\\ \alpha\beta\cdot\gamma\delta &= 8. &(4)\end{aligned}$$

Since the sum of the two roots=the sum of the other two, let

$$\alpha+\beta=\gamma+\delta. \qquad (5)$$

From (1) and (5)

$$\alpha+\beta=\gamma+\delta=-\frac{p}{2}.$$

From (5) it then follows

$$\alpha\beta+\gamma\delta=-\frac{r}{-p/2}=\frac{2r}{p}.$$

From (2) we then have

$$\left(-\frac{p}{2}\right)\left(-\frac{p}{2}\right)+\frac{2r}{p}=q, \quad \text{or,} \quad \frac{p^2}{4}+\frac{2r}{p}=q,$$

$$\text{or,} \quad p^3+8r=4pq.$$

6. If α be a triple root of $f(x)=0$, there it is a double root of $f'(x)=0$ and a single root of $f''(x)=0$.

$\therefore$ we have

$$\alpha^4+a\alpha^3+b\alpha^2+c\alpha+d=0. \qquad (1)$$

$$4\alpha^3+3a\alpha^2+2b\alpha+c=0, \qquad (2)$$

$$\left.\begin{aligned}12\alpha^2+6a\alpha+2b &= 0,\\ \text{i.e., } 6\alpha^2+3a\alpha+b &= 0\end{aligned}\right\}. \qquad (3)$$

Multiply (3) by 2α and (2) by 3 and subtract

$$3a\alpha^2+4b\alpha+3c=0. \qquad (4)$$

Solving (3) and (4) by the method of cross multiplication

$$\frac{\alpha^2}{9ac-4b^2}=\frac{\alpha}{3ab-18c}=\frac{1}{24b-9a^2}.$$

$$\frac{\alpha^2}{\alpha}=\frac{9ac-4b^2}{3ab-18c} \text{ and } \frac{\alpha}{1}=\frac{3ab-18c}{24b-9a^2},$$

i.e., condition for three equal roots is

$$\begin{aligned}(3ab-18c)^2 &= (9ac-4b^2)(24b-9a^2),\\ \text{or, } 3(ab-bc)^2 &= (8b-3a^2)(9ac-4b^2).\end{aligned}$$

From the last two ratios the value of α is

$$\frac{3ab-18c}{24b-9a^2}=\frac{6c-ab}{3a^2-8b}.$$

7. (a) Here $2\alpha = \beta + \gamma$. But $\Sigma\alpha = -\frac{3b}{a}$. Hence, $\alpha = -\frac{b}{a}$. Put this value of α in the given equation. The required condition will follow.

   (b) Take the roots: $\alpha\beta, \alpha, \frac{\alpha}{\beta}$. Then their product $= \alpha^3 = -\frac{d}{a}$. Put this value of α in the given equation.

   (c) Put $x = \frac{1}{y}$ and obtain

$$dy^3 + 3cy^2 + 3by + a = 0$$

   whose roots are $\frac{1}{\alpha}, \frac{1}{\beta}, \frac{1}{\gamma}$. The condition for the roots of the given equation to be in H.P. is the same as to find the condition. for the roots of this new equation to be in A.P. Hence, etc.

8. (a) Let the roots be $\alpha - 3\beta, \alpha - \beta, \alpha + \beta, \alpha + 3\beta$.
   Using the relations between roots and coefficients we deduce

$$\alpha = -p/4;$$
$$\beta^2 = (3p^2 - 8q)/80;$$
$$4\alpha(\alpha^2 - 5\beta^2) = -r;$$
$$(\alpha^2 - \beta^2)(\alpha^2 - 9\beta^2) = s.$$

   Put the values of α and β^2 in the last two relations:
   We get then two conditions:

$$p^3 - 4pq + 8r = 0$$
$$\text{and} \quad (36q - 11p^2)(4q + p^2) = 1600s.$$

   These are the required conditions.

   (b) Take the roots $\frac{\alpha}{\beta^3}, \frac{\alpha}{\beta}, \alpha\beta, \alpha\beta^3$. Proceed as in (i) and obtain two relations involving the coefficients.

   (c) Write $\frac{1}{x}$ for x and obtain

$$sx^4 + rx^3 + qx^2 + qx^2 + px + 1 = 0.$$

   We find the condition for the roots of this equation in A.P.
   The required conditions are

$$r^3 = 4qrs - 8ps^2$$
$$\text{and} \quad (4qs + r^2)(36qs - 11r^2) = 1600s^3.$$

9. (a) i. 12.
   ii. 43.
   iii. 2.
   iv. −22.
   v. −5.

   (b) i. 1.
   ii. 1.
   iii. 7.

   (c) i. $2p^2$.
   ii. $3q$.
   iii. p^2.

10. (a) $p^2 - 2q$.
   (b) $-r/s$.
   (c) $3r - pq$.
   (d) $r^2 - 2qs$.
   (e) $pr - 4s$.

SECTION III

Transformations: Equation of Squared Differences of a Cubic: Cardan's Method of Solving a Cubic

3.9 Transformation of Equations

We begin with a few examples:

▶ Example 3.9.1 *Form an equation whose roots are three times those of the equation*

$$3x^3 - 5x^2 + 7 = 0.$$

Solution: We form an equation in y where $y = 3x$. Since $x = y/3$, we substitute this value of x in the given equation. We thus obtain

$$3\left(\frac{y}{3}\right)^3 - 5\left(\frac{y}{3}\right)^2 + 7 = 0,$$

$$\textbf{or,}\quad y^3 - 5y^2 + 63 = 0 \text{ (Required equation).}$$

Note that we could write the required equation in x; thus

$$x^3 - 5x^2 + 63 = 0.$$

▶ Example 3.9.2 *Form an equation whose roots are the negatives of the roots of the equation*

$$x^4 - 4x^3 + 6x^2 - x + 2 = 0.$$

Solution: We form an equation in y, where $y = -x$. Substituting $x = -y$ in the given equation we obtain the required equation in y as

$$(-y)^4 - 4(-y)^3 + 6(-y)^2 - (-y) + 2 = 0,$$

$$\textbf{or,}\quad y^4 + 4y^3 + 6y^2 + y + 2 = 0.$$

We could write the required equation as

$$x^4 + 4x^3 + 6x^2 + x + 2 = 0.$$

▶ Example 3.9.3 *Form an equation whose roots are less by 2 than the roots of the equation*

$$x^3 - 5x^2 + 3x - 5 = 0.$$

Solution: We form an equation in y, where $y = x - 2$ (or, $x = 2 + y$).
The required equation is

$$(y+2)^3 - 5(y+2)^2 + 3(y+2) - 5 = 0,$$

$$\textbf{or,}\quad y^3 + 3y^2\cdot 2 + 3y\cdot 2^2 + 2^3 - 5(y^2 + 4y + 4) + 3(y+2) - 5 = 0,$$

$$\textbf{or,}\quad y^3 + y^2 - 5y - 11 = 0.$$

We could write the equation as

$$x^3 + x^2 - 5x - 11 = 0.$$

Short-cut Rule: Replace x by x + 2 in the given equation.
We could also obtain by using synthetic method of division.

(Successive division by x − 2)

2	1	−5	3	−5
		2	−6	−6
2	1	−3	−3	**−11**
		2	−2	
2	1	−1	**−5**	
		2		
	1	**1**		

Less by 2
Multiplier: +2

The required equation is

$$x^3 + x^2 - 5x - 11 = 0.$$

▶ Example 3.9.4 *Form an equation whose roots are greater by 4 than the roots of the equation*

$$x^4 - 6x^3 - 38x^2 - 3x + 17 = 0$$

Solution: We form an equation in y, where $y = x + 4$. Substituting $x = y - 4$ in the given equation, we obtain the required equation as

$$\begin{aligned} y^4 - 22y^3 + 130y^2 - 243y + 61 &= 0 \\ \text{or, } x^4 - 22x^3 + 130x^2 - 243x + 61 &= 0 \end{aligned}$$

Short-cut Rule: Replace x by x − 4 in the given equation, or, by using synthetic method of division by −4.

(Successive division by x + 4)

−4	1	−6	−38	−3	17
		−4	40	−8	44
−4	1	−10	2	−11	**61**
		−4	56	−232	
−4	1	−14	58	**−243**	
		−4	72		
−4	1	−18	**130**		
		−4			
	1	**−22**			

Greater by 4
Multiplier: −4

The required equation is, therefore,

$$x^4 - 22x^3 + 130x^2 - 243x + 61 = 0.$$

► Example 3.9.5 *Form an equation whose roots are reciprocals of roots of the equation:*

$$ax^3 + bx^2 + cx + d = 0 \ (a \neq 0).$$

Solution: We form an equation in y, where

$$y = \frac{1}{x}, \left(\textbf{or, } x = \frac{1}{y}\right)$$

The required equation is

$$a\left(\frac{1}{y}\right)^3 + b\left(\frac{1}{y}\right)^2 + c\left(\frac{1}{y}\right) + d = 0,$$

$$\textbf{or, } \quad dy^3 + cy^2 + by + a = 0.$$

We could write the required equation as

$$dx^3 + cx^2 + bx + a = 0.$$

► Example 3.9.6 *By diminishing the roots of the equation*

$$x^3 - 6x^2 + 15x - 1 = 0$$

by a suitable number, form an equation in which the coefficients of the second degree term is zero.

Solution: Sum of the roots of the given equation is 6. Therefore if we diminish the roots by 6, then the sum of the roots of the transformed equation will be $6 - 3h$. We wish to get h such that $6 - 3h = 0$, i.e., we take $h = 2$. Dividing

$$x^3 - 6x^2 + 15x - 1$$

successively by $x - 2$, we get by the process of synthetic division.

$$\begin{array}{c|cccc}
2 & 1 & -6 & 15 & -1 \\
 & & 2 & -8 & 14 \\
\hline
2 & 1 & -4 & 7 & \mathbf{13} \\
 & & 2 & -4 & \\
\hline
2 & 1 & -2 & \mathbf{3} & \\
 & & 2 & & \\
\hline
 & \mathbf{1} & \mathbf{0} & &
\end{array}$$

Hence, the required equation is

$$y^3 + 3y + 13 = 0.$$

Otherwise. We take $y = x - h$ (or, $x = y + h$) and substitute in the given equation, thus

$$(y + h)^3 - 6(y + h)^2 + 15(y + h) - 1 = 0,$$

$$\textbf{or, } \quad y^3 + 3y^2(h - 2) + 3y(h^2 - 4h + 5) + h^3 - 6h^2 + 15h - 1 = 0.$$

Choose h so that coefficient of y^2 in the transformed equation is zero so that $h = 2$. Putting $h = 2$, the equation becomes

$$y^3 + 3y + 13 = 0.$$

For the most General Cubic

▶ Example 3.9.7 *By diminishing the roots of the general cubic equation*

$$a_0x^3 + 3a_1x^2 + 3a_2x + a_3 = 0$$

by a suitable number, form an equation in which the coefficients of the second degree term is zero.

Solution: Let us take $y = x - h$ so that $x = y + h$ and the given equation transforms into

$$\begin{aligned} &a_0(y+h)^3 + 3a_1(y+h)^2 + 3a_2(y+h) + a_3 = 0, \\ \text{or,}\quad &a_0y^3 + 3y^2(a_0h + a_1) + 3y(a_0h^2 + 2a_1h + a_2) \\ &+ a_0h^3 + 3a_1h^2 + 3a_2h + a_3 = 0. \end{aligned}$$

Choose h so that $a_0h + a_1 = 0$, i.e., take $h = -\frac{a_1}{a_0}$.
The reduced equation becomes

$$a_0y^3 + \frac{3(a_0a_2 - a_1^2)y}{a_0} + \frac{2a_1^3 - 3a_0a_1a_2 + a_0^2a_3}{a_0^2} = 0,$$

$$\text{or,}\quad y^3 + \frac{3(a_0a_2 - a_1^2)}{a_0^2}y + \frac{2a_1^3 - 3a_0a_1a_2 + a_0^2a_3}{a_0^3} = 0.$$

See that if the given cubic has three roots α, β, γ, then this transformed equation has roots

$$\alpha - h,\ \beta - h,\ \gamma - h,$$

$$\text{or,}\quad \alpha + \frac{a_1}{a_0},\ \beta + \frac{a_1}{a_0},\ \gamma + \frac{a_1}{a_0},$$

$$\text{or,}\quad \frac{a_0\alpha + a_1}{a_0},\ \frac{a_0\beta + a_1}{a_0},\ \frac{a_0\gamma + a_1}{a_0}.$$

Now sometimes it is necessary to make these roots in the form

$$a_0\alpha + a_1,\ a_0\beta + a_1,\ a_0\gamma + a_1.$$

For that we make a second transformation

$$z = a_0y\ \left(\text{or, } y = \frac{z}{a_0}\right).$$

The new transformed equation, then becomes

$$\left(\frac{z}{a_0}\right)^3 + \frac{3(a_0a_2 - a_1^2)}{a_0^2}\left(\frac{z}{a_0}\right) + \frac{2a_1^3 - 3a_0a_1a_2 + a_0^2a_3}{a_0^3} = 0,$$

$$\text{or,}\quad z^3 + 3Hz + G = 0,$$

where $H = a_0a_2 - a_1^2$ and $G = 2a_1^3 - 3a_0a_1a_2 + a_0^2a_3$.

The roots of this equation are $(z = a_0y)$.

$$a_0\alpha + a_1,\ a_0\beta + a_1,\ a_0\gamma + a_1.$$

The transformations which make

$$a_0x^3 + 3a_1x^2 + 3a_2x + a_3 = 0$$

a standard equation

$$z^3 + 3Hz + G = 0,$$

where H and G are, as defined above are very useful techniques and are to be remembered carefully.

▶ Example 3.9.8 *Find the equation whose roots are the squares of the roots of the equation*

$$x^3 + bx^2 + cx + d = 0.$$

[C.P. 1996, 2000]

Solution: Let $y = x^2$. Put $x = \sqrt{y}$ in the given equation and obtain

$$(\sqrt{y})^3 + b(\sqrt{y})^2 + c(\sqrt{y}) + d = 0,$$

$$\textbf{or,}\quad y^{3/2} + cy^{1/2} = -by - d.$$

Squaring, we get

$$y^3 + c^2y + 2cy^2 = b^2y^2 + d^2 + 2bdy,$$

$$\textbf{or,}\quad y^3 + (2c - b^2)y^2 + (c^2 - 2bd)y - d^2 = 0 \text{ (Required equation).}$$

▶ Example 3.9.9 *Find the equation whose roots are equal in magnitude but opposite in sign to the roots of the equation*

$$x^4 + 3x^3 - 7x^2 + 2x + 1 = 0.$$

[C.P. 1997]

Solution: Take $y = -x$ so that $x = -y$.

The required equation becomes

$$(-y)^4 + 3(-y)^3 - 7(-y)^2 + 2(-y) + 1 = 0,$$

$$\textbf{or,}\quad y^4 - 3y^3 - 7y^2 - 2y + 1 = 0$$

$$\textbf{or,}\quad x^4 - 3x^3 - 7x^2 - 2x + 1 = 0.$$

Equation of the Squared Differences of the Cubic

▶ Example 3.9.10 *Suppose*

$$x^3 - 12x + 65 = 0$$

is a cubic equation whose three roots are α, β, γ. *Form an equation whose roots are*

$$(\beta - \gamma)^2,\ (\gamma - \alpha)^2 \text{ and } (\alpha - \beta)^2.$$

Solution: Here

$$\begin{aligned}\alpha+\beta+\gamma &= 0,\\ \alpha\beta+\beta\gamma+\gamma\alpha &= -12,\\ \alpha\beta\gamma &= -65.\end{aligned}$$

To find an equation whose roots are

$$(\beta-\gamma)^2, (\gamma-\alpha)^2, (\alpha-\beta)^2.$$

Now,

$$\begin{aligned}(\beta-\gamma)^2 &= (\beta+\gamma)^2-4\beta\gamma\\ &= (-\alpha)^2-4\frac{\beta\gamma\alpha}{\alpha}\\ &= \alpha^2+\frac{4.65}{\alpha}.\end{aligned}$$

Similarly

$$\begin{aligned}(\gamma-\alpha)^2 &= \beta^2+\frac{4.65}{\beta}\\ \text{and } (\alpha-\beta)^2 &= \gamma^2+\frac{4.65}{\gamma}.\end{aligned}$$

We, therefore, make the transformation

$$y = x^2+\frac{4.65}{x}, \quad \textbf{or,} \quad xy = x^3+260,$$

$$\textbf{i.e.,} \quad x^3-xy+260 = 0.$$

But the given equation is

$$x^3-12x+65 = 0.$$

On subtraction

$$-xy+12x+195 = 0, \textbf{ or, } x = \frac{195}{y-12}.$$

Substituting this value of x in the given equation, we have

$$\left(\frac{195}{y-12}\right)^3-12\left(\frac{195}{y-12}\right)+65 = 0,$$

$$\begin{aligned}\textbf{or,}\quad & 65(y-12)^3-12\cdot 195(y-12)^2+(195)^3 = 0,\\ \textbf{or,}\quad & (y-12)^3-36(y-12)^2+(195)^2\cdot 3 = 0,\\ \textbf{or,}\quad & y^3-3y^2\cdot 12+3y\cdot 144-(12)^3-36(y^2-24y+144)+(195)^2\cdot 3 = 0,\\ \textbf{or,}\quad & y^3-72y^2+1296y+107163 = 0.\end{aligned}$$

This is the equation whose roots are $(\alpha-\beta)^2, (\beta-\gamma)^2, (\gamma-\alpha)^2$. This equation is called the equation of the **squared differences** of the cubic

$$x^3-12x+65 = 0.$$

▶ Example 3.9.11 *Find the equation of the squared differences of the cubic*

$$x^3 + px + q = 0.$$

Solution: Let the three roots of the given equation be α, β, γ. Then $\alpha + \beta + \gamma = 0$, $\alpha\beta + \beta\gamma + \gamma\alpha = p$, $\alpha\beta\gamma = -q$.

We require to find an equation whose roots are

$$(\alpha - \beta)^2, (\beta - \gamma)^2 \text{ and } (\gamma - \alpha)^2.$$

Now,

$$(\alpha - \beta)^2 = (\alpha + \beta)^2 - 4\alpha\beta = (-\gamma)^2 - \frac{4\alpha\beta\gamma}{\gamma} = \gamma^2 + \frac{4q}{\gamma}.$$

Similarly,

$$(\beta - \gamma)^2 = \alpha^2 + \frac{4q}{\alpha} \text{ and } (\gamma - \alpha)^2 = \beta^2 + \frac{4q}{\beta}.$$

We now make the transformation

$$y = x^2 + \frac{4q}{x}, \text{ or, } x^3 - xy + 4q = 0.$$

Given equation: $x^2 + px + q = 0$ so that, on subtraction,

$$-xy - px + 3q = 0, \text{ or, } x = \frac{3q}{p + y}.$$

Putting this value of x in the given equation we obtain

$$\left(\frac{3q}{p+y}\right)^3 + p\left(\frac{3q}{p+y}\right) + q = 0,$$

$$\text{or, } 27q^3 + 3pq(p+y)^2 + q(p+y)^3 = 0,$$

$$\text{or, } 27q^2 + 3p(p+y)^2 + (p+y)^3 = 0$$

(Taking out the common factor q)

$$\text{or, } y^3 + 6py^2 + 9p^2y + (4p^3 + 27q^2) = 0.$$

This is the required equation of the **squared differences** of the given cubic.

Note: In both the examples (3.9.10) and (3.9.11) the given cubic has no term of second degree. This fact has made the calculations easier. Now see the next example:

▶ Example 3.9.12 *Find the equation of the squared differences of the general cubic*

$$a_0x^3 + 3a_1x^2 + 3a_2x + a_3 = 0. \tag{5}$$

Solution: In this case, we shall first transform the given cubic in the form

$$z^3 + 3Hz + G = 0, \tag{6}$$

where

$$H = a_0a_2 - a_1^2,$$
$$G = a_0^2a_3 - 3a_0a_1a_2 + 2a_1^3.$$

[See worked out Ex. 3.9.7 above]

If the roots of the given cubic (1) be α, β, γ, then the roots of the transformed cubic (2) will be $a_0\alpha + a_1, a_0\beta + a_1, a_0\gamma + a_1$.

Now, by the method of **Ex. 3.9.11**, the equation of the, squared difference of the cubic (2) will be (put $p = 3H, q = G$)

$$t^3 + 18t^2H + 81tH + 27(G^2 + 4H^3) = 0$$

whose roots are

$$a_0^2(\alpha - \beta)^2, a_0^2(\beta - \gamma)^2, a_0^2(\gamma - \alpha)^2.$$

Remember. The following four cases will arise:

Case 1. $G^2 + 4H^3 < 0 \Rightarrow$ All the roots α, β, γ are real.

Case 2. $G^2 + 4H^3 > 0 \Rightarrow$ Two roots are imaginary and the third root is real.

Case 3. $G^2 + 4H^3 = 0 \Rightarrow$ Two roots are equal.

Case 4. $G = 0, H = 0 \Rightarrow$ All the three roots are equal.

► Example 3.9.13 *If α, β, γ be the roots of*

$$3x^3 - 11x^2 + 8x + 4 = 0$$

find the equation whose roots are

$$(\beta - \gamma)^2,\ (\gamma - \alpha)^2,\ (\alpha - \beta)^2.$$

Solution: Here, $a_0 = 3$; $3a_1 = -11$, i.e., $a_1 = -\frac{11}{3}$; $3a_2 = 8$, i.e., $a_2 = \frac{8}{3}$; $a_3 = 4$.

$$\therefore H = a_0a_2 - a_1^2 = 3\cdot\frac{8}{3} - \left(\frac{-11}{3}\right)^2 = 8 - \frac{121}{9} = -\frac{49}{9}.$$

$$G = a_0^2a_3 - 3a_0a_1a_2 + 2a_1^3 = 36 - 88 - \frac{2662}{27} = -\frac{4066}{27}.$$

Transformed equation is

$$z^3 + 3Hz + G = 0, \textbf{ or, } z^3 - \frac{49}{3}z - \frac{4066}{27} = 0.$$

Its roots are

$$3\alpha - \frac{11}{3},\ 3\beta - \frac{11}{3},\ 3\gamma - \frac{11}{3}.$$

Miscellaneous Solved Problems

(Using Transformations)

▶ Example 3.9.14 (a) *If α, β, γ be the roots of the equation $x^3 + px + q = 0$, find an equation whose roots are $\frac{1}{\beta+\gamma}, \frac{1}{\gamma+\alpha}, \frac{1}{\alpha+\beta}$ and hence find $\sum \frac{1}{\beta+\gamma}$.* [C.P. 2004]

Solution: Since $\alpha + \beta + \gamma = 0$, we write

$$\frac{1}{\beta+\gamma} = -\frac{1}{\alpha}; \quad \frac{1}{\gamma+\alpha} = -\frac{1}{\beta} \text{ and } \frac{1}{\alpha+\beta} = -\frac{1}{\gamma}.$$

So if we take the transformation $y = -\frac{1}{x}$, we can find an equation in y by putting $x = -\frac{1}{y}$ in the given equation. Thus $qy^3 - py^2 - 1 = 0$ is the **required equation** and its roots are

$$\frac{1}{\beta+\gamma}, \frac{1}{\gamma+\alpha}, \frac{1}{\alpha+\beta}. \quad \therefore \sum \frac{1}{\beta+\gamma} = \frac{p}{q}.$$

(b) *If α, β, γ be the roots of the cubic equation $x^3 + px + q = 0$ find an equation whose roots are*

$$\beta + \gamma - 2\alpha,$$
$$\gamma + \alpha - 2\beta$$
$$\text{and} \quad \alpha + \beta - 2\gamma.$$

Solution: In this case

$$\alpha + \beta + \gamma = 0.$$

Hence, we are to form an equation whose roots are

$$\beta + \gamma - 2\alpha = \beta + \gamma + \alpha - 3\alpha = -3\alpha;$$
$$\gamma + \alpha - 2\beta = \gamma + \alpha + \beta - 3\beta = -3\beta;$$
$$\alpha + \beta - 2\gamma = \alpha + \beta + \gamma - 3\gamma = -3\gamma.$$

That is, we are to form an equation whose roots are -3 times the roots of

$$x^3 + px + q = 0.$$

Take $y = -3x$, or, $x = -\frac{1}{3}y$. The required equation becomes

$$\left(-\frac{1}{3}y\right)^3 + p\left(-\frac{1}{3}y\right) + q = 0,$$
$$\text{or,} \quad y^3 + 9py - 27q = 0.$$

Notice that if it is required to find the value of the expression

$$(\alpha + \beta - 2\gamma)(\beta + \gamma - 2\alpha)(\gamma + \alpha - 2\beta)$$

we can form an equation

$$y^3 + 9py - 27q = 0$$

whose roots are

$$\alpha+\beta-2\gamma,$$
$$\beta+\gamma-2\alpha$$
$$\text{and}\quad \gamma+\alpha-2\beta.$$

Then the product of the roots $= (\alpha+\beta-2\gamma)(\beta+\gamma-2\alpha)(\gamma+\alpha-2\beta)$
$= 27q.$

▶ Example 3.9.15 *If α, β, γ be the roots of*

$$x^3+2x^2-5x-6=0,$$

find the value of

$$\alpha(\beta+\gamma)+\beta(\gamma+\alpha)+\gamma(\alpha+\beta).$$

[C.P. 1982]

Solution: We first form an equation whose roots are

$$\alpha\beta+\alpha\gamma,\ \beta\gamma+\alpha\beta,\ \gamma\alpha+\gamma\beta.$$

Here

$$\Sigma\alpha=-2,\ \Sigma\alpha\beta=-5,\ \alpha\beta\gamma=6.$$

But

$$\alpha(\beta+\gamma)=\alpha\beta+\alpha\gamma+\beta\gamma-\beta\gamma = -5-\frac{\alpha\beta\gamma}{\alpha}$$
$$= -5-\frac{6}{\alpha}.$$

Similarly,

$$\beta(\gamma+\alpha)=-5-\frac{6}{\beta}\ \text{ and }\ \gamma(\alpha+\beta)=-5-\frac{6}{\gamma}.$$

Hence, we take the transformation

$$y=-5-\frac{6}{x},\ \text{or,}\ \frac{6}{x}=-5-y,\ \text{i.e.,}\ x=-\frac{6}{y+5}.$$

The equation is transformed into

$$\left(\frac{-6}{y+5}\right)^3+2\left(\frac{-6}{y+5}\right)^2-5\left(\frac{-6}{y+5}\right)-6=0$$

which, on simplification, give

$$y^3+10y^2+13y-24=0.$$

This equation has roots

$$\alpha(\beta+\gamma),\ \beta(\gamma+\alpha),\ \gamma(\alpha+\beta).$$

∴ their sum $=-10$ **(Required result).**

▶ Example 3.9.16 *If α, β, γ be the roots of*

$$x^3 - 7x + 6 = 0,$$

find an equation whose roots are $(\alpha-\beta)^2, (\alpha-\gamma)^2, (\beta-\gamma)^2$. **(Equation of the squared differences of the cubic).**

Solution: Here

$$\begin{aligned}\alpha+\beta+\gamma &= 0,\\ \Sigma\alpha\beta &= -7,\\ \alpha\beta\gamma &= -6.\end{aligned}$$

$$(\alpha-\beta)^2 = (\alpha+\beta)^2 - 4\alpha\beta = (-\gamma)^2 - \frac{4\alpha\beta\gamma}{\gamma} = \gamma^2 + \frac{24}{\gamma}.$$

Similarly,

$$(\alpha-\gamma)^2 = \beta^2 + \frac{24}{\beta} \text{ and } (\beta-\gamma)^2 = \alpha^2 + \frac{24}{\alpha}.$$

The transformation, therefore, should be

$$y = x^2 + \frac{24}{x}, \text{ or, } x^3 - xy + 24 = 0.$$

But given equation

$$x^3 - 7x + 6 = 0.$$

∴ on subtraction, we get $x(7-y) + 18 = 0$. Hence, $x = \frac{18}{y-7}$.

The equation in y is, therefore, obtained by putting this value of x in the given equation thus

$$\left(\frac{18}{y-7}\right)^3 - 7\left(\frac{18}{y-7}\right) + 6 = 0,$$

$$\text{or, } 6(y-7)^3 - 126(y-7)^2 + (18)^3 = 0,$$

$$\text{or, } \mathbf{y^3 - 42y^2 + 441y - 400 = 0.}$$

▶ Example 3.9.17 *Transform the equation*

$$x^3 - 6x^2 + 4x - 7 = 0$$

into one which lacks the second term.

Solution: Let us diminish each root of the given equation by h.

Put $y = x - h$, **or,** $x = y + h$.

The transformed equation is

$$(y+h)^3 - 6(y+h)^2 + 4(y+h) - 7 = 0. \tag{1}$$

We choose h in such a manner that the coefficient of $y^2 = 3h - 6 = 0$; this gives $h = 2$.

∴ the equation which lacks the second term can be obtained by putting $h = 2$ in equation (1).

On simplification we shall obtain the required equation as

$$y^3 - 8y - 15 = 0.$$

▶ **Example 3.9.18** *If α, β, γ be the roots of*

$$x^3 + px^2 + qx + r = 0$$

form an equation whose roots are

$$\frac{\alpha}{\beta+\gamma}, \frac{\beta}{\gamma+\alpha}, \frac{\gamma}{\alpha+\beta}.$$

[C.P. 2001]

Solution: Here $\alpha + \beta + \gamma = -p$; hence, $\beta + \gamma = -p - \alpha$.
Then,

$$\frac{\alpha}{\beta+\gamma} = \frac{\alpha}{-p-\alpha};$$

similarly,

$$\frac{\beta}{\gamma+\alpha} = \frac{\beta}{-p-\beta}, \text{ etc.}$$

Take

$$y = \frac{x}{-p-x}, \text{ or, } x = \frac{-py}{1+y}.$$

Putting this value of x in the given equation and simplifying we get the transformed equation

$$(p^2 - pq - p^3 + r)y^3 + (p^2 - 2pq + 3r)y^2 + (3r - pq)y + r = 0.$$

▶ **Example 3.9.19** α, β, γ *are the roots of*

$$2x^3 + 3x^2 - x - 1 = 0.$$

Form the equation whose roots are

$$\frac{\alpha}{\beta+\gamma}, \frac{\beta}{\gamma+\alpha}, \frac{\gamma}{\alpha+\beta}.$$

[C.P. 1988]

Solution: Exactly similar to **Example 3.9.18** above.
Here

$$p = \frac{3}{2}, \quad q = -\frac{1}{2}, \quad r = -\frac{1}{2}.$$

▶ **Example 3.9.20** *If α, β, γ be the roots of the equation*

$$x^3 + qx + r = 0,$$

find the value of

$$\frac{1}{\beta+\gamma-\alpha} + \frac{1}{\gamma+\alpha-\beta} + \frac{1}{\alpha+\beta-\gamma}.$$

[C.P. 1988]

Solution: From the given equation we have $\Sigma\alpha = 0, \Sigma\alpha\beta = q$ and $\alpha\beta\gamma = -r$.

Now

$$\beta + \gamma - \alpha = \beta + \gamma + \alpha - 2\alpha = -2\alpha.$$

Similarly,

$$\begin{aligned} \gamma + \alpha - \beta &= -2\beta \\ \text{and } \alpha + \beta - \gamma &= -2\gamma. \end{aligned}$$

We form an equation whose roots are

$$\frac{1}{\beta + \gamma - \alpha}, \frac{1}{\gamma + \alpha - \beta} \text{ and } \frac{1}{\alpha + \beta - \gamma},$$

i.e., whose roots are $-\frac{1}{2\alpha}, -\frac{1}{2\beta}$ and $-\frac{1}{2\gamma}$.

We, therefore, use the transformation $y = -\frac{1}{2x}$, or, $x = -\frac{1}{2y}$.

The equation in y will be

$$\begin{aligned} \left(-\frac{1}{2y}\right)^3 + q\left(-\frac{1}{2y}\right) + r &= 0, \\ \text{or, } 8ry^3 - 4qy^2 - 1 &= 0. \end{aligned} \qquad (2)$$

We require to find the value of

$$\frac{1}{\beta + \gamma - \alpha} + \frac{1}{\gamma + \alpha - \beta} + \frac{1}{\alpha + \beta - \gamma},$$

i.e., the sum of the roots of the equation (1) $= \frac{4q}{8r} = \frac{1}{2}\frac{q}{r}$ **(Ans.)**

► Example 3.9.21 *If $\alpha, \beta, \gamma, \delta$ be the roots of*

$$x^4 - 3x^3 + 2x^2 + 7x - 5 = 0,$$

then find the equation whose roots are $\alpha - 2, \beta - 2, \gamma - 2, \delta - 2$.

Solution: Here the transformation is obviously $y = x - 2$, or, $x = y + 2$.

$\therefore$ the required equation is then

$$\begin{aligned} &(y + 2)^4 - 3(y + 2)^3 + 2(y + 2)^2 + 7(y + 2) - 5 = 0, \\ \text{or,} \quad &y^4 + 5y^3 + 8y^2 + 11y + 9 = 0. \end{aligned}$$

► Example 3.9.22 *If α, β, γ be the roots of*

$$x^3 + px^2 + qx + r = 0,$$

form the equation whose roots are $\alpha + \beta, \beta + \gamma, \gamma + \alpha$. Also find the value of $(\beta + \gamma)(\gamma + \alpha)(\alpha + \beta)$. [C.P. 1993, 2004]

Solution:

$$\alpha+\beta=\alpha+\beta+\gamma-\gamma=-p-\gamma$$
$$\beta+\gamma=\alpha+\beta+\gamma-\alpha=-p-\alpha$$
$$\gamma+\alpha=\alpha+\beta+\gamma-\beta=-p-\beta$$

From the relation between roots and coefficients

$$\Sigma\alpha=-p,\ \Sigma\alpha\beta=q,\ \alpha\beta\gamma=-r$$

$\therefore$ the transformation is $y=-p-x$, or, $x=-p-y$.
Putting the value of x in the given equation we get the equation in y

$$(-p-y)^3+p(-p-y)^2+q(-p-y)+r=0,$$

or, $[y^3+3y^2p+3yp^2+p^3]+p(y^2+2yp+p^2)-pq-qy+r=0,$

or, $-y^3-2y^2p-yp^2-pq-qy+r=0,$

or, $\mathbf{y^3+2y^2p+y(p^2+q)+pq-r=0}$ **(which is the required equation).**

Clearly $(\alpha+\beta)(\beta+\gamma)(\gamma+\alpha)=-(pq-r)=r-pq.$

▶ Example 3.9.23 *If α, β, γ be the roots of*

$$x^3+px-q=0,$$

find the equation whose roots are $\alpha^2+\beta^2$, $\beta^2+\gamma^2$, $\gamma^2+\alpha^2$.

Solution:

$$\alpha^2+\beta^2 = (\alpha+\beta)^2-2\alpha\beta=(-\gamma)^2-\frac{2\alpha\beta\gamma}{\gamma}\ (\because\ \alpha+\beta+\gamma=0)$$
$$= \gamma^2-\frac{2q}{\gamma}\ (\because\ \alpha\beta\gamma=q).$$

Similarly,

$$\beta^2+\gamma^2=\alpha^2-\frac{2q}{\alpha}\ \text{and}\ \gamma^2+\alpha^2=\beta^2-\frac{2q}{\beta}.$$

$\therefore$ transformation is

$$y=x^2-\frac{2q}{x},\ \text{or,}\ x^3-xy-2q=0.$$

Given equation is

$$x^3+px-q=0.$$

On subtraction,

$$x(p+y)+q=0,\ \text{or,}\ x=-q/(p+y).$$

Putting this value of x in $x^3+px-q=0$,

$$\left(-\frac{q}{p+y}\right)^3+p\left(-\frac{q}{p+y}\right)-q=0,$$

or, $(p+y)^3+p(p+y)^2+q^2=0$

or, $\mathbf{y^3+4py^2+5p^2y+2p^3+q^2=0.}$

► Example 3.9.24 *If α, β, γ be the roots of*

$$x^3 + px^2 + qx + r = 0,$$

then find an equation whose roots are $\beta\gamma - \alpha^2, \gamma\alpha - \beta^2, \alpha\beta - \gamma^2$. [C.P. 1999]

Solution: Here, $\Sigma\alpha = -p$, $\Sigma\alpha\beta = q$, $\alpha\beta\gamma = -r$.

$$\beta\gamma - \alpha^2 = \frac{\alpha\beta\gamma}{\alpha} - \alpha^2 = -\frac{r}{\alpha} - \alpha^2.$$

Similarly,

$$\gamma\alpha - \beta^2 = -\frac{r}{\beta} - \beta^2 \text{ and } \alpha\beta - \gamma^2 = -\frac{r}{\gamma} - \gamma^2.$$

We take the transformation as

$$y = -\frac{r}{x} - x^2, \text{ or, } x^3 + xy + r = 0.$$

Given

$$x^3 + px^2 + qx + r = 0.$$

Therefore, on subtraction

$$px^2 + qx - xy = 0, \quad \textbf{or,} \quad px + q - y = 0,$$

$$\textbf{or,} \quad x = \frac{y - q}{p}.$$

Putting $x = \dfrac{y-q}{p}$ in the given equation, we get the required equation:

$$\left(\frac{y-q}{p}\right)^3 + p\left(\frac{y-q}{p}\right)^2 + q\left(\frac{y-q}{p}\right) + r = 0.$$

On simplification, we get the required equation as

$$y^3 + (p^2 - 3q)y^2 + y(3q^2 - p^2q) + p^3r - q^3 = 0.$$

An Important Example on Use of transformations

► Example 3.9.25 *If α, β, γ are the roots of $x^3 + 3x + 1 = 0$ find the equation whose roots are $\frac{\alpha}{\beta} + \frac{\beta}{\alpha}, \frac{\beta}{\gamma} + \frac{\gamma}{\beta}, \frac{\gamma}{\alpha} + \frac{\alpha}{\gamma}$ and hence find the value of $\sum\left(\frac{\alpha}{\beta} + \frac{\beta}{\alpha}\right)$.* [C.P. 2005]

Solution:

$$\frac{\alpha}{\beta} + \frac{\beta}{\alpha} = \frac{\alpha^2 + \beta^2}{\alpha\beta} = \frac{(\alpha + \beta)^2 - 2\alpha\beta}{\alpha\beta} = \frac{(-\gamma)^2}{\alpha\beta} - 2$$

$$= \frac{\gamma^2}{\alpha\beta} - 2$$

[$\because$ α, β, γ are the roots of $x^3 + 3x + 1 = 0, \alpha + \beta + \gamma = 0,$ $\alpha\beta + \beta\gamma + \gamma\alpha = 3, \alpha\beta\gamma = -1$]

$$= \frac{\gamma^3}{\alpha\beta\gamma} - 2 = \frac{\gamma^3}{-1} - 2 = -\gamma^3 - 2.$$

Similarly,

$$\frac{\beta}{\gamma}+\frac{\gamma}{\beta}=-\alpha^2-2,\quad \frac{\gamma}{\alpha}+\frac{\alpha}{\beta}=-\beta^2-2.$$

We now make the transformation $y=-x^3-2$ or $x^3+y+2=0$.
Now using the given equation $x^3+3x+1=0$.
We get (by subtraction) $y-3x+1=0$, or $x=\frac{y+1}{3}$.
Putting $x=(y+1)/3$ in the given equation we obtain

$$\left(\frac{y+1}{3}\right)^3+3\frac{y+1}{3}+1=0,$$

$$\text{or,}\quad y^2+3y^2+30y+55=0 \text{ (Required equation)}$$

This equation has roots

$$\frac{\beta}{\gamma}+\frac{\gamma}{\beta},\ \frac{\gamma}{\alpha}+\frac{\alpha}{\gamma}\ \text{and}\ \frac{\alpha}{\beta}+\frac{\beta}{\gamma}.$$

Hence

$$\sum\left(\frac{\alpha}{\beta}+\frac{\beta}{\alpha}\right)=-3.$$

3.10 Cardan's Method of Solution of a Cubic Equation

This classical method of solving a cubic equation was originally invented by Ferro and Tartaglia, but was first published in 1545 A.D. by Cardan who had obtained it from Tartaglia after giving a solemn promise not to reveal it. In spite of these facts, it is generally called **Cardan's method.**

Any cubic equation can be put in the form

$$x^3+3Hx+G=0 \text{ (Standard form of a cubic).}$$

We solve by *Cardan's* method cubic equations of this form:

▶ Example 3.10.1 *Solve, by Cardan's method, the following cubic equation:*

$$x^3-15x-126=0.$$

[C.P. 1997]

Solution: Let $x=p+q$. Cubic both sides we obtain

$$x^3=p^3+q^3+3pq(p+q)=p^3+q^3+3pqx,$$

$$\text{or,}\quad x^3-3pqx-(p^3+q^3)=0.$$

Comparing it with the given cubic we get

$$-3pq=-15,\ \text{or, } pq=5\ \text{ (and hence } p^3q^3=125)$$

$$\text{and}\quad -(p^3+q^3)=-126,\ \text{or, } p^3+q^3=126.$$

Thus, p^3 and q^3 are the two roots of the quadratic equation

$$t^2 - 126t + 125 = 0, \quad \text{or,} \quad (t-125)(t-1) = 0,$$
$$\text{or,} \quad t = 125, \text{ or, } 1.$$

Take $p^3 = 1$ (say). Then $p = 1, w, w^2$, where w and w^2 are the imaginary cube roots of unity. If

$$w = -\frac{1}{2} + \frac{\sqrt{3}}{2}i, \text{ then } w^2 = -\frac{1}{2} - \frac{\sqrt{3}}{2}i.$$

Now since $pq = 5$, we see that

1. When $p = 1$, $q = 5$ so that one root of the given equation is $p + q = 6$.
2. When $p = w$, $q = \dfrac{5}{w} = \dfrac{5w^2}{w^3} = 5w^2$ so that another root of the given equation is
$$p + q = w + 5w^2.$$
3. When $p = w^2$, $q = \dfrac{5}{w^2} = \dfrac{5w}{w^3} = 5w$ and the third root of the given equation is
$$p + q = w^2 + 5w.$$

[*Ans.* The three roots are 6, $w + 5w^2$ and $w^2 + 5w$. Writing $w = -\frac{1}{2} + \frac{\sqrt{3}}{2}i$ and $w^2 = -\frac{1}{2} - \frac{\sqrt{3}}{2}i$ it can be easily verified that the three roots are $6, -3 - 2\sqrt{3}i, -3 + 2\sqrt{3}i$.

▶ **Example 3.10.2** *Solve, by Cardan's method, the cubic equation*

$$x^3 - 18x - 35 = 0.$$

Solution: Let $x = p + q$. Cubing both sides, we obtain

$$x^3 = p^3 + q^3 + 3pq(p+q) = p^3 + q^3 + 3pqx,$$
$$\text{or,} \quad x^3 - 3pqx - (p^3 + q^3) = 0.$$

Comparing it with the given cubic, we get

$$-3pq = -18, \quad \text{i.e.,} \quad pq = 6 \ \ (\text{and hence } p^3q^3 = 216),$$
$$-p^3 - q^3 = -35, \quad \text{i.e.,} \quad p^3 + q^3 = 35.$$

Thus p^3 and q^3 are the two roots of the quadratic equation.

$$t^2 - 35t + 216 = 0, \quad \text{or,} \quad (t-27)(t-8) = 0,$$
$$\text{or,} \quad t = 27, \text{ or, } 8.$$

Take $p^3 = 8$ (say). Then $p = 2, 2w, 2w^2$. Since $pq = 6$, $q = 3, 3w^2, 3w$
∴ the three roots of the given cube are $(p+q) = 5, 2w + 3w^2, 2w^2 + 3w$.
Writing

$$w = -\frac{1}{2} + \frac{\sqrt{3}}{2}i, \ w^2 = -\frac{1}{2} - \frac{\sqrt{3}}{2}i,$$

we may write the roots as

$$5, \ -\frac{5}{2} - \frac{\sqrt{3}}{2}i, \ -\frac{5}{2} + \frac{\sqrt{3}}{2}i.$$

▶ Example 3.10.3 *Solve, by Cardan's method, the cubic equation*

$$x^3 - 6x - 4 = 0.$$

Solution: We follow the steps of Example 3.10.1.

Let $x = p + q$. Then

$$x^3 - 3pqx - (p^3 + q^3) = 0.$$

Comparing it with the given equation

$$-3pq = -6, \quad \text{or,} \quad pq = 2 \ \ (\text{hence } p^3q^3 = 8)$$
$$-(p^3 + q^3) = -4, \quad \text{or,} \quad p^3 + q^3 = 4.$$

$\therefore$ p^3 and q^3 are roots of

$$t^2 - 4t + 8 = 0, \text{ or, } t = \frac{4 + \sqrt{-16}}{2} = 2 \pm 2i.$$

Let

$$p^3 = 2 + 2i = 2\sqrt{2}\left[\cos\left(2r\pi + \frac{\pi}{4}\right) + i\sin\left(2r\pi + \frac{\pi}{4}\right)\right].$$

[Writing $2 + 2i = r(\cos\theta + i\sin\theta)$. Find r and the general value of θ]

$$\therefore\ p = \sqrt{2}\left[\cos\frac{2r\pi + \frac{\pi}{4}}{3} + i\sin\frac{2r\pi + \frac{\pi}{4}}{3}\right], \text{ where } r = 0, 1, 2.$$

The three values of p are

$$\sqrt{2}\left(\cos\frac{\pi}{12} + i\sin\frac{\pi}{12}\right),$$
$$\sqrt{2}\left(\cos\frac{3\pi}{4} + i\sin\frac{3\pi}{4}\right),$$
$$\text{and}\quad \sqrt{2}\left(\cos\frac{17\pi}{12} + i\sin\frac{17\pi}{12}\right).$$

The corresponding values of q are (since $pq = 2$)

$$\frac{2}{\sqrt{2}\left(\cos\frac{\pi}{12} + i\sin\frac{\pi}{12}\right)} = \sqrt{2}\left\{\cos\left(\frac{-\pi}{12}\right) + i\sin\left(\frac{-\pi}{12}\right)\right\}$$
$$\sqrt{2}\left\{\cos\left(\frac{-3\pi}{4}\right) + i\sin\left(\frac{-3\pi}{4}\right)\right\}$$
$$\text{and}\quad \sqrt{2}\left\{\cos\left(\frac{-17\pi}{12}\right) + i\sin\left(\frac{-17\pi}{12}\right)\right\}$$

$\therefore$ the three roots of the given equation are $(p + q)$

$$2\sqrt{2}\cos\frac{\pi}{12},\ 2\sqrt{2}\cos\frac{3\pi}{4},\ 2\sqrt{2}\cos\frac{17\pi}{12},$$
$$\text{or,}\quad 2\sqrt{2}\cos\frac{\pi}{12},\ 2\sqrt{2}\left(\frac{-1}{\sqrt{2}}\right),\ 2\sqrt{2}\cos\left\{\frac{24\pi - 7\pi}{12}\right\}$$
$$\text{or,}\quad 2\sqrt{2}\cos\frac{\pi}{12},\ -2,\ 2\sqrt{2}\cos\frac{7\pi}{12}.$$

▶ Example 3.10.4 *Solve, by Cardan's method, the following cubic*
(a) $x^3 - 15x^2 - 33x + 847 = 0$; *(b)* $x^3 - 6x^2 - 6x - 7 = 0$. [C.P. 2005]

Solution: **There are the most general type of cubic equations. We shall transform them to the standard form and then use the steps outlined in above examples.**

(a) To put it in the standard form, let $x = y + h$. Then the equation becomes

$$(y+h)^3 - 15(y+h)^2 - 33(y+h) + 847 = 0,$$
$$\text{or,}\quad y^3 + 3(h-5)y^2 + 3(h^2 - 10h - 11)y + h^3 - 15h^2 - 33h + 847 = 0.$$

Choose h so that the coefficient of y^2 is zero, i.e., take $h = 5$. The equation thus reduces to

$$y^3 - 108y + 432 = 0 \text{ (standard form).}$$

We now solve this equation by Cardan's method. Knowing the three values of y, we may obtain the three values of x from $x = y + h$, where $h = 5$.

To solve: $y^3 - 108y + 432 = 0$ *by Cardan's method:*

Let $y = p + q$. Then $y^3 - 3pqy - (p^3 + q^3) = 0$.

Comparing,

$$-3pq = -108, \text{ or, } pq = 36 (p^3q^3 = 46656) \text{ and } p^3 + q^3 = -432.$$

$\therefore$ p^3 and q^3 are the roots of

$$t^2 + 432t + 46656 = 0, \quad \text{or,} \quad (t + 216)^2 = 0,$$
$$\text{i.e.,}\quad t = -216, -216.$$

Let $p^3 = -216 = (-6)^3$. Then, $p = -6, -6w, -6w^2$ and corresponding $q = -6$, $-6w, -6w$, respectively ($\because$ $pq = 36$).

$\therefore$ the three value of y are

$$-12,\ -6w - 6w^2,\ -6w^2 - 6w, \text{ or, } -12, 6, 6\ (\because\ w + w^2 = -1).$$

$\therefore$ the three values of x are $(x = y + 5) - 7, 11, 11$.

(b) To put the given equation in the standard form, let $x = y + h$. Then the given equation becomes $(y+h)^3 - 6(y+h)^2 - 7(y+h) - 7 = 0$. Choose h such that the coefficient of $y^2 = 0$. This will give $h = 2$, i.e., the standard form will be

$$y^3 - 18y - 35 = 0. \tag{1}$$

Now we apply Cardon's method to solve this equation in y. Put $y = u + v$. Cubing we get $y^3 - 3uvy - (u^3 + v^3) = 0$. Comparing with (1) we have $uv = 6$, $u^3 + v^3 = 35$, i.e., u^3 and v^3 are the roots of

$$t^2 - 35t + 216 = 0 \textbf{ or } t = 27 \text{ or } t = 8.$$

Let $u^3 = 27$, i.e., $u = 3$ then $v = 2$ ($\because$ $uv = 6$). One value of y is 5. Dividing $y^3 - 18y - 35$ by $y - 5$ we get $y^2 + 5y + 7$. So other two values of y are $\frac{-5 \pm \sqrt{3}i}{2}$. Since y has three values 5, $\frac{-5+\sqrt{3}i}{2}$, $\frac{-5-\sqrt{3}i}{2}$, three values of x are

$$5 + 2,\ \frac{-5+\sqrt{3}i}{2} + 2,\ \frac{-5-\sqrt{3}i}{2} + 2,$$
$$\text{i.e.,}\quad 7,\ -\frac{1}{2} + \frac{\sqrt{3}i}{2},\ -\frac{1}{2} - \frac{\sqrt{3}i}{2}.$$

General Discussions

Any cubic of the form

$$a_0x^3 + 3a_1x^2 + 3a_2x + a_3 = 0$$

can be reduced to

$$y^3 + 3Hy + G = 0$$

by the substitution $y = a_0x + a_1$, where

$$H = a_0a_2 - a_1^2, G = a_0^2a_3 - 3a_0a_1a_2 + 2a_1^3.$$

Cardan's method of solution of $y^3 + 3Hy + G = 0$.

Let $y = p + q$ so that

$$y^3 - 3pqy - (p^3 + q^3) = 0.$$

Comparing, $pq = -H$ and $p^3 + q^3 = -G$.
p^3 and q^3 are the roots of

$$t^2 + Gt - H^3 = 0, \text{ or, } t = -\frac{G \pm \sqrt{G^2 + 4H^3}}{2}.$$

Let

$$p^3 = -G + \frac{\sqrt{G^2 + 4H^3}}{2}.$$

Let the three cube roots be p, wp, w^2p.
The corresponding values of q are

$$\frac{-H}{p}, \frac{-H}{wp}, \frac{-H}{w^2p}, \text{ or, } -\frac{H}{p}, -w^2\frac{H}{p}, -w\frac{H}{p}.$$

$\therefore$ the three values of y are $(p + q)$,

$$p - \frac{H}{p}, wp - w^2\frac{H}{p}, w^2p - w\frac{H}{p}.$$

$\therefore$ the three roots of the general cubic are

$$\left(x = \frac{y - a_1}{a_0}\right) \cdot \frac{1}{a_0}\left\{p - \frac{H}{p} - a_1\right\},$$

$$\frac{1}{a_0}\left\{wp - w^2\frac{H}{p} - a_1\right\},$$

$$\frac{1}{a_0}\left\{w^2p - w\frac{H}{p} - a_1\right\}.$$

Observations on Nature of the Roots of the General Cubic

1. If $G^2 + 4H^3 > 0$, then two roots are imaginary and the third root is real.
2. If $G^2 + 4H^3 = 0$, then the **two roots are equal.**
   If $G = 0$, $H = 0$, then all the three roots are equal.
3. If $G^2 + 4H^3 < 0$, then all the three roots are **real.**

Exercises on Chapter 3(III)

1. If α, β, γ be the roots of
$$x^3 + 2x^2 - 2 = 0,$$
form the equation whose roots are

(a) $\frac{1}{\alpha}, \frac{1}{\beta}, \frac{1}{\gamma}$;
(b) $10\alpha, 10\beta, 10\gamma$;
(c) $\alpha - 1, \beta - 1, \gamma - 1$;
(d) $\alpha + 1, \beta + 1, \gamma + 1$;
(e) $\beta + \gamma, \gamma + \alpha, \alpha + \beta$;
(f) $\beta\gamma, \gamma\alpha, \alpha\beta$.

2. If $\alpha, \beta, \gamma, \delta$ be the roots of
$$x^4 - x^2 - 3x - 5 = 0,$$
form the equation whose roots are:

(a) $\frac{1}{\alpha}, \frac{1}{\beta}, \frac{1}{\gamma}, \frac{1}{\delta}$;
(b) $-2\alpha, -2\beta, -2\gamma, -2\delta$.

3. Use the transformation $y = x + 1$ to solve the equation
$$x^4 + 4x^3 - 7x^2 - 22x + 24 = 0.$$
[*Hints*: Reduced eqn: $y^4 - 13y^2 + 36 = 0$ or, $(y^2 - 9)(y^2 - 4)$]

4. (a) Diminish the roots of $x^3 - 9x^2 + 28x - 27 = 0$ by 3;
   (b) Diminish the roots of $2x^3 - 15x^2 + 31x - 12 = 0$ by 1. Hence, solve the equation, given that two of its roots differ by unity.

5. If α, β be the roots of $x^2 + px + q = 0$, form the equation whose roots are
$$(\alpha - 1)^2 \text{ and } (\beta - 1)^2.$$

6. The equation $x^3 + px + q = 0$ has three roots α, β, γ. Form the equation whose roots are

(a) $\dfrac{1-\alpha}{\alpha}, \dfrac{1-\beta}{\beta}, \dfrac{1-\gamma}{\gamma}$;

(b) $\alpha^2, \beta^2, \gamma^2$;

(c) $\dfrac{\beta\gamma}{\alpha}, \dfrac{\gamma\alpha}{\beta}, \dfrac{\alpha\beta}{\gamma}$;

(d) $\dfrac{\beta+\gamma}{\alpha^2}, \dfrac{\gamma+\alpha}{\beta^2}, \dfrac{\alpha+\beta}{\gamma^2}$;

(e) $\dfrac{\beta}{\gamma}+\dfrac{\gamma}{\beta}, \dfrac{\gamma}{\alpha}+\dfrac{\alpha}{\gamma}, \dfrac{\alpha}{\beta}+\dfrac{\beta}{\alpha}$.

7. If α, β, γ be the roots of

$$x^3 - px^2 + qx - r = 0,$$

form the equation whose roots are

(a) $\beta\gamma + \dfrac{1}{\alpha}, \gamma\alpha + \dfrac{1}{\beta}, \alpha\beta + \dfrac{1}{\gamma}$;

(b) $\dfrac{\alpha}{\beta+\gamma}, \dfrac{\beta}{\gamma+\alpha}, \dfrac{\gamma}{\alpha+\beta}$. [C.P. 1980; '88]

[**Hints.** (a) $y = \frac{r+1}{x}$, or, $x = \frac{r+1}{y}$. Transformed equation will be

$$ry^3 - q(1+r)y^2 + p(1+r)^2y - (1+r)^3 = 0.$$

(b) $y = \frac{x}{p-x}$, or, $x = \frac{py}{y+1}$. Transformed equation becomes

$$(pq-r)y^3 + (2pq - p^3 - 3r)y^2 + (pq - 3r)y - r = 0.]$$

8. (a) If $y = x + \dfrac{1}{x}$, express $x^2 + \dfrac{1}{x^2}$ in terms of y. Hence, solve

$$x^4 - 2x^3 - 6x^2 - 2x + 1 = 0$$

by dividing throughout by x^2.
[See the working method given under **Answers**]

(b) Use similar method as in (a) to solve the equation

$$x^4 - 5x^3 - 12x^2 - 5x + 1 = 0.$$

9. If α be a root of

$$x^4 + x^3 - 4x^2 - 4x + 1 = 0,$$

prove that $2 - \alpha^2$ is another root.

10. (a) Solve:

$$x^3 - 11x^2 + 37x - 35 = 0,$$

given that one root is $3 + \sqrt{2}$.

(b) Solve:

$$x^4 - 10x^3 + 26x^2 - 10x + 1 = 0,$$

given that one root is $2 - \sqrt{3}$.

11. If α, β, γ be the roots of

$$x^3 - 147x + 686 = 0,$$

find the equation whose roots are

$$(\alpha - \beta)^2, (\beta - \gamma)^2, (\gamma - \alpha)^2.$$

12. Find the equation of the squared differences of the cubic

(a) $x^3 + 6x^2 + 9x + 4 = 0$;

(b) $3x^3 - 11x^2 + 8x + 4 = 0$.

[*Hints*: (a) Remove the 2nd term by diminishing the roots by h, i.e., put $y = x - h$, or, $x = y + h$. Choose h so that coefficient of $y^2 = 0$. See that $h = -2$, and the transformed equation is $y^3 - 3y + 2 = 0$. Now find the equation of the squared differences of this cubic. Note that if α, β, γ be the roots the given equation, then $\alpha + 2$, $\beta + 2$, $\gamma + 2$ are the roots of the transformed cubic in y, so that the equation of the **squared differences** of the given cubic and that of the transformed cubic will be the same.]

Problems on Cardan's Method of solving a cubic.

Solve the following equations by Cardan's Method:

13. $x^3 - 9x + 28 = 0$.

14. $x^3 + 6x - 2 = 0$.

15. $x^3 - 6x - 9 = 0$. **[C.P. 1999]**

16. $x^3 - 18x + 35 = 0$.

17. $x^3 - 3x + 1 = 0$. **[C.P. 1992]**

18. $x^3 + x^2 - 9x + 12 = 0$.

19. $x^3 - 3x^2 + 3 = 0$.

20. $x^3 - 18x - 75 = 0$.

21. $x^3 + 12x - 12 = 0$. **[C.P. 2003]**

22. $x^3 - 12x + 65 = 0$. **[C.P. 2001]**

Answers

1. **(a)** $2x^3 = 2x + 1$;
(b) $x^3 + 20x^2 = 2000$;
(c) $x^3 + 5x^2 + 7x + 1 = 0$;
(d) $x^3 - x^2 - x - 1 = 0$;
(e) $x^3 + 4x^2 + 4x + 2 = 0$;
(f) $x^3 - 4x - 4 = 0$.

2. **(a)** $5x^4 + 3x^3 + x^2 - 1 = 0$;
(b) $x^4 - 4x^2 + 24x - 80 = 0$.

3. $y^4 - 13y^2 + 36 = 0; -4, -3, 1, 2$.

4. **(a)** $x^3 + x + 3 = 0$;
(b) $2x^3 - 9x^2 + 7x + 6 = 0; 4, 3, \frac{1}{2}$.

5. $x^2 - x(p^2 + 2p - 2q + 2) + (p + q + 1)^2 = 0$;

6. **(a)** $qx^3 + x^2(p + 3q) + x(2p + 3q) + p + q + 1 = 0$;
(b) $x^3 + 2px^2 + p^2x = q^2$;
(c) $qx^3 + p^2x^2 - 2pqx + q^2 = 0$;
(d) $qy^3 - py^2 - 1 = 0$;
(e) $q^3 + 3q^3y^2 + (3q^2 + p^3)qy + q(q^2 + 2p^3) = 0$.

8. (a) $x^2 + \frac{1}{x^2} = \left(x + \frac{1}{x}\right)^2 - 2 = y^2 - 2$. On dividing by x^2 we get

$$\left(x^2 + \frac{1}{x^2}\right) - 2\left(x + \frac{1}{x}\right) - 6 = 0, \quad \text{or, } y^2 - 2 - 2y - 6 = 0,$$

i.e., $y^2 - 2y - 8 = 0$. $\therefore$ $y = 4$, or, $y = -2$.

Now solve: $x + \frac{1}{x} = 4$ and also $x + \frac{1}{x} = -2$.

Required roots will be $-1, -1, 2 \pm \sqrt{3}$.

(b) $-1, -1, \frac{1}{2}(7 \pm 3\sqrt{5})$.

10. (a) $3 \pm \sqrt{2}, 5$; (b) $2 \pm \sqrt{3}, 3 \pm 2\sqrt{2}$.

11. $x^3 - 882x^2 + (147)^2 \cdot 9x = 0$.

12. (a) $x^3 - 18x^2 + 81x = 0$; (b) $x^3 - \frac{98}{9}x^2 + \frac{(49)^2}{81}x = 0$.

[In each case, two roots of the given equation are equal. Why?]

13. $-4, -w - 3w^2, -w^2 - 3w$.

14. $p - q, wp - w^2q, w^3p - wq$, where $p = \sqrt[3]{4}$ and $p = \sqrt[3]{2}$.

15. $3, w + 2w^2, w^2 + 2w$.

16. $-5, -2w - 3w^2, -2w^2 - 3w$.

17. $2\cos\frac{2\pi}{9}, 2\cos\frac{8\pi}{9}, 2\cos\frac{14\pi}{9}$.

18. $-4, -w - 2w^2, -w^2 - 2w$.

19. $1 + 2\cos\frac{2\pi r}{9}$ $(r = 1, 2, 4)$.

20. $p + q, wp + w^2q, w^2p + wp$, where $p = \sqrt[3]{3}, q = 2\sqrt[3]{9}$.

21. $p + q, wp + w^2p + w^2q, w^2p + w$, when $p = 2\sqrt[3]{2}, q = 2/\sqrt[3]{2}$.

SECTION IV

Location of Roots

[Descartes' Rule of Signs, Rolle's Theorem and their Simple Applications]

3.11 Descartes' Rule of Signs

In this section we shall examine the number and approximate location of real roots of a *polynomial whose coefficients are real.*

Preliminaries

1. When two consecutive coefficients of a polynomial $f(x)$ have the same sign, we say that there is a **continuation** of signs; but if they have opposite signs they present a **variation** of signs, e.g., in the polynomial

$$10x^{12} + 8x^{11} + 3x^{10} + 4x^9 - 6x^6 - 3x^4 + 2x^3 - 3x^2 + 8x - 9$$

there are **five variations** and **four continuations** of signs; for the signs of the successive coefficients are

$$+ + + + - - + - + -$$

2. If one or more of the coefficients be equal to zero we say that the polynomial is an **incomplete polynomial.**

3. In a complete polynomial $f(x)$, if x is replaced by $-x$, then a variation becomes a continuation and *vice versa*; see that

$$f(x) = x^3 - 3x^2 + 2x + 7; \quad f(-x) = -x^3 - 3x^2 - 2x + 7$$
$$+ \quad - \quad + \quad + \qquad\qquad - \quad - \quad - \quad +$$

I. Descartes' Rule of Signs

If $f(x)$ be a polynomial with real coefficients, the number of positive roots of $f(x) = 0$ is equal to the number of variations oj signs of its successive coefficients, or less than that number by an even number.

Two Immediate Corollaries

1. *The number of negative roots of $f(x) = 0$ is equal to the number of variations of signs of the coefficients of $f(-x)$ or less than that number by an even number.*

2. *Let $f(x)$ be a polynomial (with real coefficients) of degree n. Let μ be the number of variations in $f(x)$ and μ' be the number of variations in $f(-x)$. If $\mu + \mu' < n$, then the equation $f(x) = 0$ has imaginary roots [at least $n - (\mu + \mu')$ imaginary roots].*

Proof of Descartes' Rule is not included in the present text; rather we give more stress on its applications.

II. Applications

▶ Example 3.11.1 $x^4 + 3x^2 + 2x - 7 = 0$ *has one positive root, one negative root and two imaginary roots—verify by Descartes' Rule.*

Solution: We look to the sign-patterns of $f(x)$ and $f(-x)$:

$f(x)$: $+ + + -$ [one variation; hence, one positive root of $f(x) = 0$]

$f(-x)$: $+ + - -$ [one variation; hence, one negative root of $f(x) = 0$]

$\therefore$ the maximum number of real roots = 2.

Since the degree of $f(x)$ is four, $f(x) = 0$ has, in all, four roots. Hence, the number of imaginary roots is $4 - 2 = 2$.

N.B. The existence of zero-root is to be first investigated.

► Example 3.11.2 $x^7 - 3x^4 - 7x^2 - 8x = 0$.

Solution: Let

$$f(x) = x^7 - 3x^4 - 7x^2 - 8x = 0.$$

First, observe that the constant term is zero; hence one root of $f(x) = 0$ is zero. We then consider the roots of

$$\phi(x) = x^6 - 3x^3 - 7x - 8 = 0.$$

Sign-patterns of

$\phi(x)$: $+ - - -$ [one variation; hence, one positive root of $f(x) = 0$]
$\phi(-x)$: $- - - +$ [one variation; hence, one negative root of $f(x) = 0$]

$\phi(x) = 0$ has two real roots (non-zero).

The degree of $f(x) = 0$ is 7; there are, in all, seven roots. One root is zero, two non-zero real roots.

$\therefore$ it has 4 imaginary roots.

► Example 3.11.3 *Show that*

$$x^5 + x^3 - 2x^2 + x - 2 = 0$$

has at least two imaginary roots.

Solution: Verify that this equation has **no zero-root,** no **negative root** and either 3 positive roots or one positive root. Since the degree of the equation is 5, there will be either 2 imaginary roots or 4 imaginary roots, i.e., **at least two imaginary roots.**

A Few Short-Answer Questions

► Example 3.11.4 *If the coefficients of a polynomial* $f(x)$ *are all positive, then how many positive real roots can* $f(x) = 0$ *have?*

Solution: No variation of signs in $f(x)$, so there **cannot be any positive root** of $f(x) = 0$.

► Example 3.11.5 *If the polynomial* $f(x)$ *contains only even powers of* x *and all the coefficients are of the same sign, then how many real roots are there for* $f(x) = 0$*?*

Solution: No real root exists for $f(x) = 0$ [no variation of sign in $f(x)$ and also for $f(-x)$].

▶ Example 3.11.6 *If for the polynomial* $f(x)$ *the coefficients of even powers of* x *are all positive and the coefficients of odd powers of* x *are all negative, then how many negative roots are there for* $f(x) = 0$?

Solution: No negative root since $f(-x)$ has no variation of sign.

▶ Example 3.11.7 *Prove that* $x^4 + 16x^2 + 7x - 10 = 0$ *has exactly two imaginary roots.*

Solution: Sign pattern of $f(x) = x^4 + 16x^2 + 7x - 10$

$+ + + -$ (one variation of sign and hence one positive root)

Sign pattern of $f(-x) = x^4 + 16x^2 - 7x - 10$

$+ + - -$ (one variation of sign and hence one negative root)

$\therefore$ the remaining two roots must be imaginary.

III. A Few Important Results which help in Location of Real Roots

We shall take

$$f(x) = a_0x^n + a_1x^{n-1} + a_2x^{n-2} + \cdots + a_{n-1}x + a_n,$$

a polynomial in x of degree n ($a_0 \neq 0$ and all the coefficients are real numbers)

1. When x is sufficiently small ($x \to 0$) the sign of $f(x)$ is the same as that of the coefficient of the lowest power of x present.
   In particular if $a_n \neq 0$ (i.e., $f(x) = 0$ has no zero-root), then $f(x)$ will have the same sign as that of a_n (if $x \to 0$). We shall denote such a value of $f(x)$ by $f(0)$.

2. When x is sufficiently large ($x \to \infty$) the sign of $f(x)$ is the same as that of a_0 (if $a_0 = 0$ the sign of $f(x)$ will be the same as the coefficient of the highest power of x present). We denote such value of $f(x)$ by $f(\infty)$.
   Similar interpretation is to made for $f(-\infty)$.

Theorem 3.11.1 *If* a *and* b *are two distinct real numbers, where* $a < b$ *and if* $f(a)$ *and* $f(b)$ *are of opposite signs, then* $f(x) = 0$ *must have at least one real root, say* α, *where* $a < \alpha < b$.

e.g., There exists a root of $x^3 - 4x + 2 = 0$ lying between 1 and 2

$$\left.\begin{aligned} f(1) &= 1^3 - 4.1 + 2 = -1 \\ \text{and } f(2) &= 2^3 - 4.2 + 2 = 2, \end{aligned}\right\},$$

i.e., $f(1)$ and $f(2)$ are of opposite signs. **[C.P. 2001]**

More generally, *if* $f(a)$ *and* $f(b)$ *have opposite signs, then* $f(x) = 0$ *has an odd number of real roots between* a *and* b.

Theorem 3.11.2 *If* a *and* b *are two distinct real numbers, where* $a < b$ *and if* $f(a)$ *and* $f(b)$ *have the same sign, there is no real root on an even number of roots between* a *and* b.

We can illustrate the results of these theorems by the following graphs:

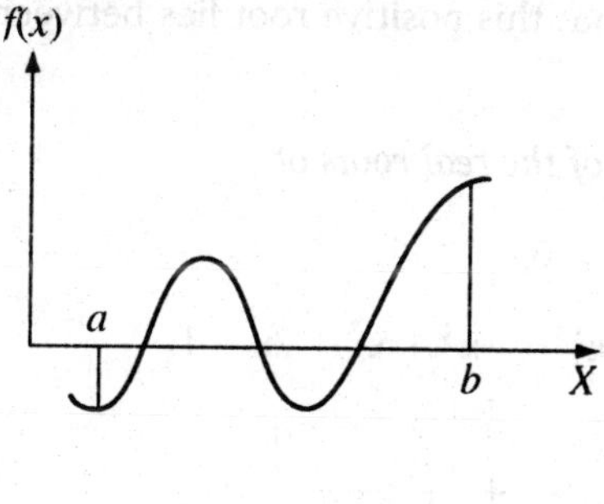

Three roots: f(a) <0, f(b) >0.

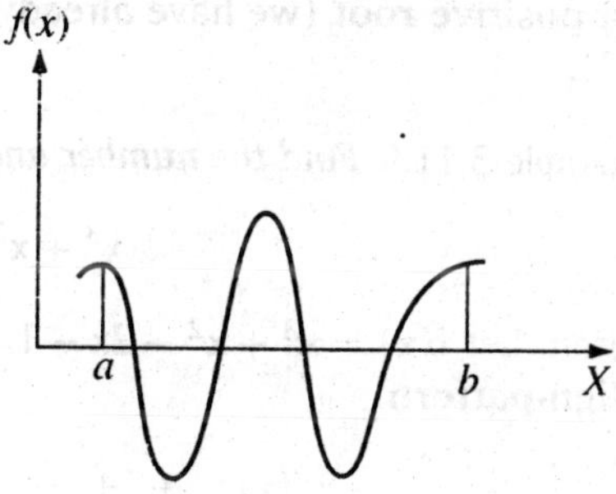

Four roots: f(a), f(b) >0.

Fig 3.1

Note that the **roots of $f(x) = 0$ are those values of x where the graph of $f(x)$ cuts the X-axis.** What happens if the graph is tangent to the X-axis? Such points correspond to multiple roots of $f(x) = 0$. The graphical approach is sufficient for the purpose of this text.

Corollary. (*a*) If the degree of $f(x)$ be odd, then $f(x) = 0$ has at least one real root whose sign is opposite to that of the constant term (provided, of course, the constant term $\neq 0$).

[$f(-\infty)$ and $f(\infty)$ have then opposite signs; hence, there exists at least one real root.

Let $f(\infty) = +$ve, $f(-\infty) = -$ve, $f(0) = a_n (\neq 0)$.

If $a_n > 0$, then the real roots lie between $-\infty$ and 0 (i.e., the root is $-$ve).

If $a_n < 0$, then the real root lies between 0 and ∞ (i.e., the root is $+$ve)]

Corollary. (*b*) If the degree of $f(x)$ be even and if the constant term is negative, then $f(x) = 0$ has at least one positive root and one negative root.

[Here $f(-\infty) = +$ve, $f(\infty) = +$ve, $f(0) = -$ve.]

Applications

▶ Example 3.11.8 $f(x) = x^3 - 3x - 3 = 0$ *has one positive root.*

Solution: Here $f(0) = -$ve and $f(\infty) = +$ve; hence, $f(x) = 0$ has one positive real root.

We may know the nature of this root more closely. See that

$$f(2) = 2^3 - 3.2 - 3 = -\text{ve and } f(3) = 3^3 - 3.3 - 3 = +\text{ve}.$$

$\therefore$ the positive root lies between 2 and 3.

Notice that:

$f(x): + - -$ [one variation and hence, by Descarte's Rule one positive root of $f(x) = 0$]

$f(-x): - + -$ [two negative roots of $f(x) = 0$ or no negative root].

$\therefore$ Descartes' Rule give that $f(x) = 0$ has either all three roots real (one positive, two negative) or else, one positive root and the other two imaginary. In any case, it **has a positive root** (we have already proved that this positive root lies between 2 and 3).

▶ Example 3.11.9 *Find the number and position of the real roots of*

$$x^3 + x^2 - 2x - 1 = 0.$$

Solution: Let $f(x) = x^3 + x^2 - 2x - 1$. Then $f(-x) = -x^3 + x^2 + 2x - 1$.

Sign-pattern.

$$f(x): \; + \; + \; - \; -; \; f(-x): \; - \; + \; + \; -.$$

By Descartes' Rule $f(x) = 0$ has *one* positive root and *either two or none* negative roots, i.e., $f(x) = 0$ has all its roots real or **else** one real positive root and two imaginary roots.

Again, we see that $f(0) = -1$, $f(1) = -1$ and $f(2) = 7$.

Since $f(1)$ and $f(2)$ are of opposite signs, the positive root lies between 1 and 2.

Moreover, we see that $f(0) = -1, f(-1) = 1, f(-2) = -1$, i.e., $f(x) = 0$ has two negative roots—one lying between -1 and 0 and another lying between -2 and -1. In other words, $f(x) = 0$ has **all its roots real (no complex roots).**

Rolle's Thoerem. *It is fundamental in real analysis that if a function* $f(x)$ *has a derivative at every point of the open interval* (a, b), *i.e.,* $a < x < b$ *and if it is continuous up to the ends of the interval (i.e., continuous in* $a \le x \le b$), *and if* $f(a) = f(b)$, *then* $f(x)$ *vanishes for* ***at least one value of*** x *in the open interval* (a, b).

In the algebraic form of this theorem (i.e., when $f(x)$ is a polynomial in x, which is always continuous and always derivable) we may state:

Theorem 3.11.3 **Rolle's Theorem.** *Between any two consecutive real roots of a polynomial equation* $f(x) = 0$, ***there lies at least one real root of*** $f'(x) = 0$.

See the graph (Fig. 3.2). If the graph of $f(x)$ cuts the X-axis at $x = a$ and at $x = b$, then clearly its tangent must be horizontal at some point α between a and b and therefore, $f'(\alpha) = 0$.

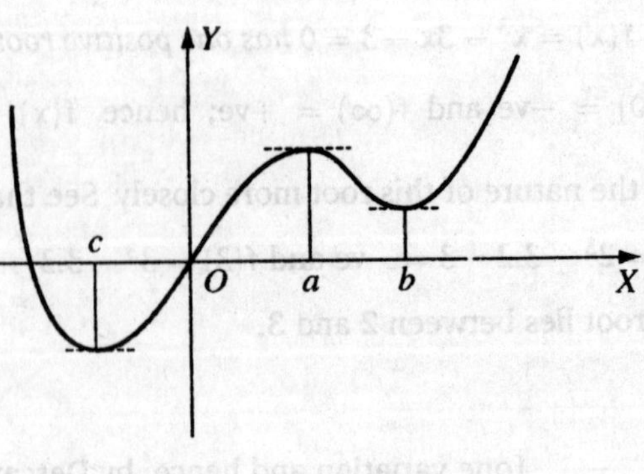

Fig 3.2

It follows as a corollary that

Between any two consecutive roots of $f'(x) = 0$, *either there exists just one root of* $f(x) = 0$ *or there may be no root of* $f(x) = 0$ (see Fig. 3.2).

[**Arguments for just one root.** If we assume that there are two roots α and p of $f(x) = 0$ lying between **two consecutive** real roots of $f'(x) = 0$, then by Rolle's Theorem, there must be a root of $f'(x) = 0$ between α and β—contrary to the hypothesis. Hence, etc.]

Use of Rolle's Theorem in Location of Real Roots

► Example 3.11.10 *Find the number of real roots and determine pairs of consecutive integers between which these roots lie*

$$x^4 - 8x^3 + 22x^2 - 24x + 5 = 0.$$

Solution: Let

$$f(x) = x^4 - 8x^3 + 22x^2 - 24x + 5.$$

- Then

$$\begin{aligned} f'(x) &= 4x^3 - 24x^2 + 44x - 24 = 4(x^3 - 6x^2 + 11x - 6) \\ &= 4(x-1)(x-2)(x-3). \end{aligned}$$

The real roots of $f(x) = 0$ must be separated by the roots of $f'(x) = 0$. The roots of $f'(x) = 0$ are 1, 2 and 3.

x :	$-\infty$	1	2	3	∞
$f(x)$:	+ve	−ve	−ve	−ve	+ve

i.e., $f(x)$ has one root less than 1 and one root greater than 3. Since $f(0) = +$ve and $f(1) = -$ve, the first root actually lies between 0 and 1. That there is no negative root is also borne by the fact that

$$f(-x) = x^4 + 8x^3 + 22x^2 + 24x + 5 \text{ (no variation)}.$$

► Example 3.11.11 *Find the range of values of* k *for which the equation* $x^4 - 26x^2 + 48x - k = 0$ *has four unequal roots.*

Solution: Let

$$f(x) = x^4 - 26x^2 + 48x - k.$$

$$\begin{aligned} \therefore\ f'(x) &= 4x^3 - 52x + 48 = 4(x^3 - 13x + 12) \\ &= 4(x-1)(x^2 + x - 12) = 4(x-1)(x+4)(x-3). \end{aligned}$$

The roots of $f(x) = 0$ must be separated by the roots of $f'(x) = 0$ The roots of $f'(x) = 0$ are clearly $-4, 1, 3$.

x :	$-\infty$	-4	1	3	∞
$f(x)$:	+ve	$-352 - k$	$23 - k$	$-9 - k$	+ve

In order that $f(x) = 0$ may have four unequal roots, the signs of $f(x)$ should be

+ve −ve +ve −ve +ve

Since $23 - k$ should be positive, hence, $k < 23$.
Since $-(9 + k)$ should be negative (i.e., $9 + k > 0$), hence, $k > -9$.
$\therefore$ k must lie between −9 and 23 (i.e., $-9 < k < 23$).

▶ Example 3.11.12 *Find of values of* k *if the equation*

$$f(x) = x^4 - 14x^2 + 24x - k$$

has four unequal real roots. Exactly similar to **Example 7.** [*Ans.* $8 < k < 11$.]

▶ Example 3.11.13 *Find the number and positions of the roots of*

$$f(x) = 2x^5 - 4x^4 - 9x - 2 = 0.$$

Solution:

$f(x)$: + − − −

(1 variation and hence one positive root)

$f(-x) = -2x^5 - 4x^4 + 9x - 2 = 0$

$f(-x)$: − − + −

(2 variations and hence $f(-x) = 0$ has at most 2 negative roots.)

x :	−2	−1	0	2	3
f(x) :	−112	1	−2	−20	133

$\therefore$ there are negative roots between $(-2, -1)$ and $(-1, 0)$ and one positive root between $(2, 3)$.

Exercises on Chapter 3(IV)

Apply Descartes' Rule to prove nos. 1–4.

1. $x^4 + 2x^2 - 7x - 5 = 0$ has two real roots and two complex roots. [C.P. 1990]

2. $x^4 + 2x^2 + 3x - a = 0$ $(a > 0)$ has one positive root, one negative root and two imaginary roots; same conclusion for $x^4 + x^2 + x - 1 = 0$. [C.P. 1991]

3. $x^5 - 3x^4 + 12x^2 + 7x - 4 = 0$ has at least two imaginary roots.

4. $x^{2n} - 1 = 0$ has only two real roots −1 and +1.

Find the number of real roots of the equations in nos. **5–8** and determine pairs of consecutive integers between which the roots lie:

5. $2x^5 - 4x^4 - 9x - 2 = 0$.

6. $x^3 + x^2 - 5x - 1 = 0$.

7. $x^5 - 6x + 2 = 0$.

8. $x^6 - x^5 - 10x + 7 = 0$.

9. Given that the roots of $x^4 - x^3 - 10x^2 - x + 1 = 0$ are all real, determine the number of positive roots and also the number of negative roots.

10. Show that $x^3 - qx + r = 0$ $(q, r > 0)$ has one negative root and two or no positive root.

11. Show that $ax^4 + bx^2 + cx - d$ $(a, b, c, d > 0)$ has one positive, one negative and two imaginary roots.

12. Show that $x^5 + 5x^4 - 20x^2 - 19x - 2 = 0$ has one root lying between 2 and 3 and also another root lying between -5 and -4.

13. Use Rolle's Theorem in locating the number and position (between two consecutive integers) of real roots of the following equations:

(a) $2x^5 - 4x^4 - 9x - 2 = 0$;

(b) $x^6 - x^5 - 10x + 7 = 0$;

(c) $27x^4 - 45x^2 - 27x + 4 = 0$;

(d) $x^3 - 7x + 2 = 0$;

(e) $x^4 + x^2 - 10x - 3 = 0$.

14. If $x^4 + 4x^3 - 8x^2 + k = 0$ has 4 unequal roots, prove that $0 < k < 3$.

Answers

5. One positive root in $(2, 3)$ and two negative roots: one in $(-1, 0)$ and the other in $(-2, -1)$.

6. One positive root in $(1, 2)$ and two negative roots: one in $(-1, 0)$ and the other in $(-3, -2)$.

7. 3 real roots: one in $(-2, -1)$, second in $(0, 1)$ and the third in $(1, 2)$.

8. 2 real roots: one in $(0, 1)$ and the other in $(1, 2)$.

9. Two positive and two negative.

13. (a) Two negative roots: one in $(-2, -1)$ and another in $(-1, 0)$ and one positive root in $(2, 3)$;

(b) two positive roots: one in $(0, 1)$ and another in $(1, 2)$;

(c) two positive roots: one in $(0, 1)$ and another in $(1, 2)$;

(d) three real roots: one in $(-3, -2)$, the second in $(0, 1)$ and the third in $(2, 3)$;

(e) two real roots: one in $(-1, 0)$ and the other in $(2, 3)$.

A Model Test Paper on Chapter 3

1. Use synthetic division to find the Quotient and the Remainder when
$$2x^4 - 6x^3 + 7x^2 - 5x + 1$$
is divided by $2x - 3$. **[Example 3.2.2]**

2. Show that
$$\frac{1}{x-\alpha} + \frac{1}{x-\beta} + \frac{1}{x-\gamma} = \frac{1}{x} \ (\alpha > \beta > \gamma > 0)$$
cannot have an imaginary root. **[Example 3.5.8]** **[C.P. 2003]**

3. Find the multiple roots, if any, of $x^3 + x^2 - 16x + 20 = 0$ and then solve it. **[Example 3.6.5]**

4. Find the condition that
$$ax^3 + 3bx^2 + 3cx + d = 0$$
may have two equal roots and if the equation has two equal roots, then prove that each equal root
$$= \frac{1}{2}\frac{bc - ad}{ac - b^2}.$$
[Example 3.6.6]

5. (a) Solve
$$x^6 - 4x^5 - 11x^4 + 40x^3 + 11x^2 - 4x - 1 = 0$$
if it is given that one root is $\sqrt{2} + \sqrt{3}$. **[Exercises of Section No. 14(a)]**

   (b) Find an equation of fourth degree with rational coefficients, one root being $\sqrt{2} + i\sqrt{3}$. **[Example 3.5.11]**

6. Find a cubic in x which vanishes when $x = 1$ and $x = -2$, and have the values 4 and 28 when $x = -11$ and $x = 2$ respectively. **[C.P. 1997]**
[*Hints*: Take $f(x) = (x - 1)(x + 2)(ax + b)$]

7. Solve $4x^3 - 24x^2 + 23x + 18 = 0$, given that its roots are in A.P. **[Example 3.8.8]**

8. If $x^3 + ax^2 + bx + c = 0$ has its roots in G.P., then prove that $b^3 = a^3c$. **[Example 3.8.10]**

9. Solve
$$x^4 + 4x^3 + 8x^2 - 120x + 900 = 0,$$
if two of its are of the form $a \pm ia$ and other two roots are of the form $b \pm ib$ $(a, b \text{ real})$. **[Example 3.8.13]**

10. If α, β, γ be the roots of $x^3 - 4x^2 + 2x - 1 = 0$, find:

(a) $\Sigma\alpha^2$; (b) $\Sigma\alpha^3$; (c) $\Sigma\alpha^2\beta$.

11. Find the equation of the squared differences of the cubic
$$x^3 + px + q = 0.$$
[Example 3.8.11]

12. If α, β, γ be the roots of
$$x^3 + px + q = 0,$$
find an equation whose roots are
$$\beta + \gamma - 2\alpha,\ \gamma + \alpha - 2\beta,\ \alpha + \beta - 2\gamma$$
[Example 3.7.14]

13. If α, β, γ be the roots of $x^3 + px^2 + qx + r = 0$ form the equation whose roots are

(a) $\alpha + \beta, \beta + \gamma, \gamma + \alpha$;
(b) $\alpha^2 + \beta^2, \beta^2 + \gamma^2, \gamma^2 + \alpha^2$.

14. Solve by Cardan's method, the cubic equation

(a) $x^3 - 12x + 65 = 0$. [C.P. 2000]
(b) $x^3 - 15x^2 - 33x + 847 = 0$. [Example 3.8.4]
(c) $x^3 + 12x - 12 = 0$. [C.P. 2003]

15. Solve the equation $4x^3 + 16x^2 - 9x - 36 = 0$, given that sum of two of its roots is zero. [Example 3.8.6]

16. Show that the equation
$$27x^4 - 48x^2 - 12x + 13 = 0$$
has two positive roots – one in $(0, 1)$ and the other in $(1, 2)$. [C.P. 2002]

17. Show that a root of $x^3 - 4x + 2 = 0$ lies between 1 and 2.

18. Form an equation whose roots are the squares of the roots of
$$x^3 + px^2 + qx + r = 0.$$

19. If $\alpha, \beta, \gamma, \delta$ be the roots of
$$x^4 - 3x^3 + 4x^2 - 5x + 6 = 0,$$
find the value of
$$(\alpha^2 + 3)(\beta^2 + 3)(\gamma^2 + 3)(\delta^2 + 3).$$

20. Show that the number of real roots of $x^n - 1 = 0$ is two or one according as n is even or odd.

21. State Descartes' Rule of signs and apply it to prove that the equation
$$x^7 - 2x^4 + 3x^3 - 1 = 0$$
has at least four imaginary roots.

22. If $\alpha, \beta, \gamma, \delta$ be the roots of

$$x^4 - 3x^3 + 4x^2 - 5x + 6 = 0,$$

find the value of

$$(\alpha^2 + 3)(\beta^2 + 3)(\gamma^2 + 3)(\delta^2 + 3).$$

23. If α, β, γ be the roots of $x^3 + x + 1 = 0$, then prove that

$$(\alpha^2 + 1)(\beta^2 + 1)(\gamma^2 + 1) = 1.$$ **[Example 3.8.17]**

24. If α, β, γ be the roots of

$$x^3 + px^2 + qx + r = 0$$

form the equation whose roots are

$$\frac{\alpha}{\beta + \gamma}, \frac{\beta}{\gamma + \alpha}, \frac{\gamma}{\alpha + \beta}.$$ **[Example 3.7.18]**

25. If α, β, γ be the roots of

$$x^3 + px^2 + qx + r = 0,$$

find an equation whose roots are

$$\beta\gamma - \alpha^2, \gamma\alpha - \beta^2, \alpha\beta - \gamma^2.$$ **[Example 3.7.24]**

Chapter 4

Determinants and its Applications

Determinants	
1.	Determinants up to the third order—Properties.
2.	Co-factors and minors.
3.	Product of two determinants.
4.	Adjoint—Symmetric and Skew-symmetric determinants.
5.	**Applications:** Solutions of Linear Equations with not more than three variables by **Cramer's Rule.**

4.1 Introduction

In the next chapter **(Chapter 5)** we shall introduce the concept of Matrix as an *arrangement* of numbers in rows and columns. In case the number of rows is equal to the number of columns, the matrix is called a **square matrix.** With every square matrix we associate a **number** obtained by some definite rule specified below. We call this number the **determinant of the square matrix.**

We shall restrict ourselves to determinants of **second and third orders only.**

In the second section of the present chapter we shall show the applications of determinants in solutions of linear equations by **Cramer's Rule.**

SECTION I

4.2 Determinants of Second Order

Let $a_{11}, a_{12}, a_{21}, a_{22}$ be **four numbers, real or complex.** The arrangement

$$A = \begin{pmatrix} a_{11} & a_{12} \\ a_{21} & a_{22} \end{pmatrix}$$

is a square matrix of order two. We have taken the double suffix notations: a_{11} occurs in the first row and first column; a_{21} occurs in 2nd row and 1st column, etc.

We associate a number $a_{11}a_{22}-a_{21}a_{12}$ (*Multiply the diagonal elements* a_{11} *and* a_{22} *and then multiply the elements of the other diagonal and then subtract*) with the matrix A. This number is called the **determinant of the matrix** A, denoted by

$$\det A, \text{ or, } |A|, \text{ or, } \begin{vmatrix} a_{11} & a_{12} \\ a_{21} & a_{22} \end{vmatrix}.$$

Hence, **our definition for a second order determinant** is

$$\begin{vmatrix} a_{11} & a_{12} \\ a_{21} & a_{22} \end{vmatrix} = a_{11}a_{22} - a_{21}a_{12}.$$

▶ **Example 4.2.1**

$$\begin{vmatrix} 3 & -5 \\ 2 & 1 \end{vmatrix} = 3\cdot 1 - 2\cdot(-5) = 3 + 10 = 13.$$

Properties of Second Order Determinants

Let

$$D = \begin{vmatrix} a_{11} & a_{12} \\ a_{21} & a_{22} \end{vmatrix} = a_{11}a_{22} - a_{21}a_{12}.$$

P1. The value of a determinant does not change if rows and columns are interchanged.

i.e., $\begin{vmatrix} a_{11} & a_{12} \\ a_{21} & a_{22} \end{vmatrix} = \begin{vmatrix} a_{11} & a_{21} \\ a_{12} & a_{22} \end{vmatrix}$ (both sides $= a_{11}a_{22} - a_{21}a_{12}$).

Since the interchange of rows and columns does not affect the value of D, Any property established about "rows" must also hold for "columns"

P2. If two rows of D are interchanged, the numerical value of D is unaltered, but its sign is changed.

$$\begin{vmatrix} a_{21} & a_{22} \\ a_{11} & a_{12} \end{vmatrix} = a_{21}a_{12} - a_{11}a_{22} = -(a_{11}a_{22} - a_{21}a_{12})$$
$$= -\begin{vmatrix} a_{11} & a_{12} \\ a_{21} & a_{22} \end{vmatrix}.$$

Similarly, for columns,

$$\begin{vmatrix} a_{12} & a_{11} \\ a_{22} & a_{21} \end{vmatrix} = -(a_{11}a_{22} - a_{21}a_{12})$$
$$= -\begin{vmatrix} a_{11} & a_{12} \\ a_{21} & a_{22} \end{vmatrix}.$$

P3. If two rows (or two columns) of D are identical, the value of D is zero.

$$\begin{vmatrix} a_{11} & a_{12} \\ a_{11} & a_{12} \end{vmatrix} = a_{11}a_{12} - a_{11}a_{12} = 0.$$

P4. If all the elements of one row (or of one column) of D are multiplied by a constant k, the new determinant equals kD.

$$\begin{vmatrix} a_{11} & a_{12} \\ ka_{21} & ka_{22} \end{vmatrix} = a_{11}ka_{22} - ka_{21}a_{12} = k(a_{11}a_{22} - a_{21}a_{12})$$

$$= k\begin{vmatrix} a_{11} & a_{12} \\ a_{21} & a_{22} \end{vmatrix}.$$

▶ Example 4.2.2

$$\begin{vmatrix} 63 & 27 \\ 35 & 16 \end{vmatrix} = 9\begin{vmatrix} 7 & 3 \\ 35 & 16 \end{vmatrix}$$

$$= 9 \times 7 \times \begin{vmatrix} 1 & 3 \\ 5 & 16 \end{vmatrix}$$

$$= 63(1 \times 16 - 5 \times 3) = 63.$$

P5. A determinant of the form $\begin{vmatrix} a_{11} + x_1 & a_{12} + y_1 \\ a_{21} & a_{22} \end{vmatrix}$ **is the sum of the two determinants**

$$\begin{vmatrix} a_{11} & a_{12} \\ a_{21} & a_{22} \end{vmatrix} + \begin{vmatrix} x_1 & y_1 \\ a_{21} & a_{22} \end{vmatrix}$$

$[\because (a_{11} + x_1)a_{22} - a_{21}(a_{12} + y_1) = (a_{11}a_{22} - a_{21}a_{12}) + (x_1a_{22} - a_{21}y_1)]$.

P6. Consequently,

$$\begin{vmatrix} a_{11} + ka_{21} & a_{12} + ka_{22} \\ a_{21} & a_{22} \end{vmatrix} = \begin{vmatrix} a_{11} & a_{12} \\ a_{21} & a_{22} \end{vmatrix} + k\begin{vmatrix} a_{21} & a_{22} \\ a_{21} & a_{22} \end{vmatrix}$$

$$= \begin{vmatrix} a_{11} & a_{12} \\ a_{21} & a_{22} \end{vmatrix} + k \cdot 0$$

$$= \begin{vmatrix} a_{11} & a_{12} \\ a_{21} & a_{22} \end{vmatrix}.$$

This establishes the following **most important property of determinants:**

The value of a determinant is not altered by adding to the elements of a row (or a column), the same (positive or negative) multiples of the corresponding elements of the other row (or column).

▶ Example 4.2.3

$$\begin{vmatrix} 56 & 17 \\ 43 & 13 \end{vmatrix} = \begin{vmatrix} 56 - 3 \times 17 & 17 \\ 43 - 3 \times 13 & 13 \end{vmatrix}$$

(*multiply* col_2 *by* -3 *and add with* col_1)

$$= \begin{vmatrix} 5 & 17 \\ 4 & 13 \end{vmatrix} = 65 - 68 = -3.$$

Exercises on Chapter 4(I)

(Second Order Determinants)

Evaluate the following determinants (using Properties 1–6 described above)

1. $\begin{vmatrix} 45 & 37 \\ 45 & 17 \end{vmatrix}$.

2. $\begin{vmatrix} 19 & 39 \\ 23 & 48 \end{vmatrix}$.

3. $\begin{vmatrix} x+2 & 2x+5 \\ 3x-1 & 2x-3 \end{vmatrix}$.

4. $\begin{vmatrix} x^2-y^2 & (x-y)^2 \\ x^2+xy & (xy-y^2) \end{vmatrix}$.

5. $\begin{vmatrix} px+qz & py+qz \\ rx+sz & ry+sz \end{vmatrix}$.

6. $\begin{vmatrix} 1931 & 1932 \\ 1933 & 1934 \end{vmatrix}$.

7. Find x for which $\begin{vmatrix} 2x-1 & 2x+1 \\ x+1 & 4x+2 \end{vmatrix} = 0$.

Answers

1. -900.
2. 15.
3. $-4x^2 - 12x - 1$.
4. $(x+y)(y-x)^3$.
5. $z(x-y)(ps-qr)$.
6. -2.
7. $x = 1$, or, $x = -\frac{1}{2}$.

4.3 Determinants of Third Order: Properties

Let

$$A = \begin{bmatrix} a_{11} & a_{12} & a_{13} \\ a_{21} & a_{22} & a_{23} \\ a_{31} & a_{32} & a_{33} \end{bmatrix}$$

be a square matrix of order three (i.e., having three rows and three columns). The first subscript indicates the row and the second subscript indicates the column in which the element is located, e.g., a_{32} occurs in the 3rd row and 2nd column.

The determinant of the matrix A.

$$= \det A = \begin{vmatrix} a_{11} & a_{12} & a_{13} \\ a_{21} & a_{22} & a_{23} \\ a_{31} & a_{32} & a_{33} \end{vmatrix}$$

$$= a_{11}\begin{vmatrix} a_{22} & a_{23} \\ a_{32} & a_{33} \end{vmatrix} - a_{12}\begin{vmatrix} a_{21} & a_{23} \\ a_{31} & a_{33} \end{vmatrix} + a_{13}\begin{vmatrix} a_{21} & a_{22} \\ a_{31} & a_{32} \end{vmatrix}$$

[Expansion in terms of the elements of first row]

$$\begin{aligned} = &+a_{11}(a_{22}a_{33} - a_{32}a_{23}) \\ &-a_{12}(a_{21}a_{33} - a_{31}a_{23}) \\ &+a_{13}(a_{21}a_{32} - a_{31}a_{22}). \end{aligned}$$

One can re-arrange this expansion, grouping it by elements of *first column:*

$$\det A = a_{11}(a_{22}a_{33} - a_{32}a_{23}) - a_{21}(a_{12}a_{33} - a_{32}a_{13}) + a_{31}(a_{12}a_{23} - a_{22}a_{13})$$
$$= a_{11}\begin{vmatrix} a_{22} & a_{23} \\ a_{32} & a_{33} \end{vmatrix} - a_{21}\begin{vmatrix} a_{12} & a_{13} \\ a_{32} & a_{33} \end{vmatrix} + a_{31}\begin{vmatrix} a_{12} & a_{13} \\ a_{22} & a_{23} \end{vmatrix}$$

[Expansion in terms of the elements of first column]

Rule. Write down the elements of the first row (or first column) with alternatively positive and negative signs. Multiply each element by a second order determinant obtained by deleting the row and the column in which that element occurs.

▶ Example 4.3.1

$$\begin{vmatrix} 3 & -2 & 2 \\ 6 & 1 & -1 \\ -1 & -2 & 4 \end{vmatrix} = +3\begin{vmatrix} 1 & -1 \\ -2 & 4 \end{vmatrix} - (-2)\begin{vmatrix} 6 & -1 \\ -1 & 4 \end{vmatrix} + 2\begin{vmatrix} 6 & 1 \\ -1 & -2 \end{vmatrix}$$

[Expansion in terms of first row]

$$= 3(4-2) + 2(24-1) + 2(-12+1)$$
$$= 3 \times 2 + 2 \times 23 + 2 \times (-11) = 6 + 46 - 22 = 30.$$

Properties

The properties stated before in the case of determinants of second order are also true for determinants of third order—in fact, **they are true for determinants of any order.**

P1. The value of a determinant of third order is unaltered if rows and columns are interchanged.

Let

$$D = \begin{vmatrix} a_{11} & a_{12} & a_{13} \\ a_{21} & a_{22} & a_{23} \\ a_{31} & a_{32} & a_{33} \end{vmatrix}$$ be a determinant of third order.

By definition,

$$D = a_{11}\begin{vmatrix} a_{22} & a_{23} \\ a_{32} & a_{33} \end{vmatrix} - a_{12}\begin{vmatrix} a_{21} & a_{23} \\ a_{31} & a_{33} \end{vmatrix} + a_{13}\begin{vmatrix} a_{21} & a_{22} \\ a_{31} & a_{32} \end{vmatrix}$$

or, in more simplified form:

$$D = a_{11}(a_{22}a_{33} - a_{32}a_{23}) - a_{12}(a_{21}a_{33} - a_{31}a_{23}) + a_{13}(a_{21}a_{32} - a_{31}a_{22}).$$

We can re-arrange and group it in terms of elements of the first column: thus,

$$D = a_{11}(a_{22}a_{33} - a_{32}a_{23}) - a_{21}(a_{12}a_{33} - a_{32}a_{13}) + a_{31}(a_{12}a_{23} - a_{22}a_{13}).$$

This shows that the **row expansion** of D and the **column expansion** of D are identical. In other words, *the value of* D *does not change if columns and rows are interchanged.*

The importance of this result is that:

Any property established for rows of a third order determinant must also hold for columns.

In the following discussions we shall establish the properties for **rows** and hence they can be assumed to be true for **columns** also.

P2. The interchange of two rows (or two columns) does not alter the numerical value of the determinant but only its sign is changed.

As before, let

$$D = \begin{vmatrix} a_{11} & a_{12} & a_{13} \\ a_{21} & a_{22} & a_{23} \\ a_{31} & a_{32} & a_{33} \end{vmatrix}$$

be a determinant of third order with elements arranged in double suffix notation.

By definition,

$$\begin{aligned} D &= a_{11}\begin{vmatrix} a_{22} & a_{23} \\ a_{32} & a_{33} \end{vmatrix} - a_{12}\begin{vmatrix} a_{21} & a_{23} \\ a_{31} & a_{33} \end{vmatrix} + a_{13}\begin{vmatrix} a_{21} & a_{22} \\ a_{31} & a_{32} \end{vmatrix} \\ &= a_{11}(a_{22}a_{33} - a_{32}a_{23}) - a_{12}(a_{21}a_{33} - a_{31}a_{23}) + a_{13}(a_{21}a_{32} - a_{31}a_{22}). \end{aligned} \tag{4.3.1}$$

Let us interchang two rows, say, first and second rows and write

$$D_1 = \begin{vmatrix} a_{21} & a_{22} & a_{23} \\ a_{11} & a_{12} & a_{13} \\ a_{31} & a_{32} & a_{33} \end{vmatrix}.$$

By definition,

$$\begin{aligned} D_1 &= a_{21}(a_{12}a_{33} - a_{32}a_{13}) - a_{22}(a_{11}a_{33} - a_{31}a_{13}) + a_{23}(a_{11}a_{32} - a_{31}a_{12}) \\ &= -[a_{11}(a_{22}a_{33} - a_{32}a_{23}) - a_{12}(a_{21}a_{33} - a_{31}a_{23}) \\ &\qquad + a_{13}(a_{21}a_{32} - a_{31}a_{22})] \\ &= -D \text{ [Compare 4.3.1]} \end{aligned}$$

Similar result is obtained by interchanging first and third rows or second and third rows. Similarly, we can verify the result by interchanging two columns.

Using this property, sometimes the evaluation becomes easier.

▶ Example 4.3.2

$$\begin{aligned} \begin{vmatrix} 7 & 8 & 9 \\ 0 & 0 & 1 \\ 6 & 9 & 2 \end{vmatrix} &= -\begin{vmatrix} 0 & 0 & 1 \\ 7 & 8 & 9 \\ 6 & 9 & 2 \end{vmatrix} \\ &= -1\begin{vmatrix} 7 & 8 \\ 6 & 9 \end{vmatrix} \\ &= -(63 - 48) = -15. \end{aligned}$$

Observations. We have observed that

$$\begin{aligned} D &= \begin{vmatrix} a_{11} & a_{12} & a_{13} \\ a_{21} & a_{22} & a_{23} \\ a_{31} & a_{32} & a_{33} \end{vmatrix} \\ &= a_{11}a_{22}a_{33} - a_{11}a_{32}a_{23} - a_{12}a_{21}a_{33} \\ &\quad + a_{12}a_{31}a_{23} + a_{13}a_{21}a_{32} - a_{13}a_{31}a_{22}. \end{aligned}$$

1. Each term in the expansion of D has 3 factors which belong to different rows (row 1, row 2 and row 3) and to different columns (Col. 1, Col. 2 and Col. 3).

2. Sets of such factors can be formed in $3 \times 2 \times 1 = 6$ ways, giving 6 terms in all.

3. The term $+a_{11}a_{22}a_{33}$ given by the diagonal (*from left-hand top corner to right-hand bottom corner*) is called the **leading term.**

4. By interchanging the 2 first suffixes (or 2 second suffixes) and changing the sign, the other terms can be obtained,

e.g., interchanging 2, 3 in the first suffixes, the term $+a_{11}a_{22}a_{33}$ leads to $-a_{11}a_{32}a_{23}$, and then interchanging 1, 3 we have $+a_{31}a_{12}a_{23}$ and so on.

These observations will help us to expand determinants of higher orders.

P3. If two rows (or two columns) of a determinant of third order are identical the value of the determinant is zero.

Let us consider a determinant D of third order, two of whose rows have identical elements.

If we interchange these two rows, the interchange makes no difference in D (since they are identical rows.)

But we know that whenever there is an interchange of two rows in a determinant, the numerical value remains unaltered but the sign is changed.

$$\therefore\ D = -D, \text{ or, } D = 0.$$

Since the property is true for rows it is also true for columns.

▶ Example 4.3.3 $\begin{vmatrix} 16 & 3 & 16 \\ 8 & 0 & 8 \\ 3 & 1 & 3 \end{vmatrix} = 0$ (*two columns are identical*).

P4. If all the elements of one row (or of one column) are multiplied by the same constant, the value of the determinant is multiplied by that constant.

We have seen that in the expansion of a determinant each term contains *one and only one element* from each row. Therefore, if it is given that each element in a row is multiplied by the same constant k, say, then each term in the expansion contains k as a factor and hence, k is a common factor of all the terms in the expansion;

hence the statement is true for rows and consequently true also for columns. See that

$$\begin{vmatrix} ka_{11} & ka_{12} & ka_{13} \\ a_{21} & a_{22} & a_{23} \\ a_{31} & a_{32} & a_{33} \end{vmatrix} = ka_{11}\begin{vmatrix} a_{22} & a_{23} \\ a_{32} & a_{33} \end{vmatrix} - ka_{12}\begin{vmatrix} a_{21} & a_{23} \\ a_{31} & a_{33} \end{vmatrix} + ka_{13}\begin{vmatrix} a_{21} & a_{22} \\ a_{31} & a_{32} \end{vmatrix}$$

$$= k\left[a_{11}\begin{vmatrix} a_{22} & a_{23} \\ a_{32} & a_{33} \end{vmatrix} - a_{12}\begin{vmatrix} a_{21} & a_{23} \\ a_{31} & a_{33} \end{vmatrix} + a_{13}\begin{vmatrix} a_{21} & a_{22} \\ a_{31} & a_{32} \end{vmatrix}\right]$$

$$= k\begin{vmatrix} a_{11} & a_{12} & a_{13} \\ a_{21} & a_{22} & a_{23} \\ a_{31} & a_{32} & a_{33} \end{vmatrix}.$$

▶ Example 4.3.4 $\begin{vmatrix} 21 & 9 & 0 \\ 9 & 15 & 18 \\ 6 & 7 & 5 \end{vmatrix} = 3\begin{vmatrix} 7 & 9 & 0 \\ 3 & 15 & 18 \\ 2 & 7 & 5 \end{vmatrix}.$

Note. If two rows or columns of a determinant are proportional, then its value is zero.

▶ Example 4.3.5 $\begin{vmatrix} 42 & 1 & 6 \\ 28 & 3 & 4 \\ 14 & 7 & 2 \end{vmatrix} = \begin{vmatrix} 7\times 6 & 1 & 6 \\ 7\times 4 & 3 & 4 \\ 7\times 2 & 7 & 2 \end{vmatrix} = 7\begin{vmatrix} 6 & 1 & 6 \\ 4 & 3 & 4 \\ 2 & 7 & 2 \end{vmatrix} = 7\cdot 0 = 0$

At this stage we wish to introduce two important terms:

Minors and Cofactors

Let

$$D = \begin{vmatrix} a_{11} & a_{12} & a_{13} \\ a_{21} & a_{22} & a_{23} \\ a_{31} & a_{32} & a_{33} \end{vmatrix}$$

be a determinant of third order. By **'Minor of an element of D'** we mean a second-order determinant obtained by deleting from D **the row and the column** in which that element occurs,

e.g., with reference to above D, minor of $a_{23} = \begin{vmatrix} a_{11} & a_{12} \\ a_{31} & a_{32} \end{vmatrix}$; the element a_{23} occurs in the *second row* and *the third column*. So we have deleted the second row and the third column; remains the second-order determinant $\begin{vmatrix} a_{11} & a_{12} \\ a_{31} & a_{32} \end{vmatrix} \rightarrow$ this is the minor of a_{23}.

When the minor of an element of a determinant is taken with a definite sign (+ or −), we call *such a signed minor as the* cofactor *of that element.* To determine the sign we use the following rule:

If a_{ij} be an element of D occurring in the ith row and jth column, then

$$\textbf{cofactor of } a_{ij} = (-1)^{i+j} \times \textbf{ minor of } a_{ij}.$$

We shall always denote the cofactor of an element by **the corresponding capital letters**, e.g., A_{ij} is the cofactor of a_{ij},
e.g., minor of

$$a_{23} = \begin{vmatrix} a_{11} & a_{12} \\ a_{31} & a_{32} \end{vmatrix}, \text{ but cofactor of}$$

$$a_{23} = (-1)^{2+3}\begin{vmatrix} a_{11} & a_{12} \\ a_{31} & a_{32} \end{vmatrix} = -\begin{vmatrix} a_{11} & a_{12} \\ a_{31} & a_{32} \end{vmatrix} = A_{23}.$$

The following scheme may be helpful for the readers to decide the signs of cofactors:

$$\begin{vmatrix} + & - & + \\ - & + & - \\ + & - & + \end{vmatrix}.$$

4.3.1 Evaluation of Determinants by Cofactors

We have defined

$$D = \begin{vmatrix} a_{11} & a_{12} & a_{13} \\ a_{21} & a_{22} & a_{23} \\ a_{31} & a_{32} & a_{33} \end{vmatrix}$$

$$= a_{11}\begin{vmatrix} a_{22} & a_{23} \\ a_{32} & a_{33} \end{vmatrix} - a_{12}\begin{vmatrix} a_{21} & a_{23} \\ a_{31} & a_{33} \end{vmatrix} + a_{13}\begin{vmatrix} a_{21} & a_{22} \\ a_{31} & a_{32} \end{vmatrix}$$

Note that $+\begin{vmatrix} a_{22} & a_{23} \\ a_{32} & a_{33} \end{vmatrix}$, $-\begin{vmatrix} a_{21} & a_{23} \\ a_{31} & a_{33} \end{vmatrix}$ and $+\begin{vmatrix} a_{21} & a_{22} \\ a_{31} & a_{32} \end{vmatrix}$ are, respectively, the cofactors of a_{11}, a_{12} and a_{13}; hence we may denote them by A_{11}, A_{12} and A_{13}, respectively. Consequently we may write

$$D = a_{11}A_{11} + a_{12}A_{12} + a_{13}A_{13}. \tag{4.3.2}$$

Interchanging the first two rows of D we get

$$-D = \begin{vmatrix} a_{21} & a_{22} & a_{23} \\ a_{11} & a_{12} & a_{13} \\ a_{31} & a_{32} & a_{33} \end{vmatrix}$$

$$= a_{21}\begin{vmatrix} a_{12} & a_{13} \\ a_{32} & a_{33} \end{vmatrix} - a_{22}\begin{vmatrix} a_{11} & a_{13} \\ a_{31} & a_{33} \end{vmatrix} + a_{23}\begin{vmatrix} a_{11} & a_{12} \\ a_{31} & a_{32} \end{vmatrix}.$$

$$\therefore D = -a_{21}\begin{vmatrix} a_{12} & a_{13} \\ a_{32} & a_{33} \end{vmatrix} + a_{22}\begin{vmatrix} a_{11} & a_{13} \\ a_{31} & a_{33} \end{vmatrix} - a_{23}\begin{vmatrix} a_{11} & a_{12} \\ a_{31} & a_{32} \end{vmatrix}.$$

Now notice that $-\begin{vmatrix} a_{12} & a_{13} \\ a_{32} & a_{33} \end{vmatrix}$, $+\begin{vmatrix} a_{11} & a_{13} \\ a_{31} & a_{33} \end{vmatrix}$ and $-\begin{vmatrix} a_{11} & a_{12} \\ a_{31} & a_{32} \end{vmatrix}$ are, respectively, the cofactors of a_{21}, a_{22} and a_{23}. Hence, we may also write

$$D = a_{21}A_{21} + a_{22}A_{22} + a_{23}A_{23}. \tag{4.3.3}$$

Similarly, the cofactors of a_{31}, a_{32} and a_{33} enter into

$$D = a_{31}\begin{vmatrix} a_{12} & a_{13} \\ a_{22} & a_{23} \end{vmatrix} - a_{32}\begin{vmatrix} a_{11} & a_{13} \\ a_{21} & a_{23} \end{vmatrix} + a_{33}\begin{vmatrix} a_{11} & a_{12} \\ a_{21} & a_{22} \end{vmatrix}$$

and we may write

$$D = a_{31}A_{31} + a_{32}A_{32} + a_{33}A_{33}. \tag{4.3.4}$$

Since, a determinant is unaltered when rows and columns are interchanged, we have also

$$\left.\begin{aligned} D &= +a_{11}A_{11} + a_{21}A_{21} + a_{31}A_{31} \\ &= +a_{12}A_{12} + a_{22}A_{22} + a_{32}A_{32} \\ &= +a_{13}A_{13} + a_{23}A_{23} + a_{33}A_{33} \end{aligned}\right\}. \tag{4.3.5}$$

The relations (4.3.2) to (4.3.5) lead to the conclusion:

P5. If we multiply the elements of any line (row or column) by their corresponding cofactors the sum of such products gives the value of the determinant.

We have the following important result in connection with cofactors:

P6. If the elements of any line (row or column) of a determinant of third order be multiplied in order by the cofactors of the corresponding elements of a parallel line, then the sum of such products is zero.

Let $D = \begin{vmatrix} a_{11} & a_{12} & a_{13} \\ a_{21} & a_{22} & a_{23} \\ a_{31} & a_{32} & a_{33} \end{vmatrix}$ be a determinant of third order.

We can expand D either as a row-expansion or as a column-expansion in six different ways, thus

$$\begin{aligned} D &= a_{11}A_{11} + a_{12}A_{12} + a_{13}A_{13} = a_{21}A_{21} + a_{22}A_{22} + a_{23}A_{23}, \text{ etc.} \\ \text{or,} &= a_{11}A_{11} + a_{21}A_{21} + a_{31}A_{31} = a_{12}A_{12} + a_{22}A_{22} + a_{32}A_{32}, \text{ etc.,} \end{aligned}$$

where A_{ij} is the cofactor of a_{ij}.

Now consider a sum of the form $a_{11}A_{21}+a_{12}A_{22}+a_{13}A_{23}$, where the elements of the first row are multiplied in order by the cofactors of corresponding elements of the second row. Then

$$\begin{aligned} & a_{11}A_{21} + a_{12}A_{22} + a_{13}A_{23} \\ &= -a_{11}\begin{vmatrix} a_{12} & a_{13} \\ a_{32} & a_{33} \end{vmatrix} + a_{12}\begin{vmatrix} a_{11} & a_{13} \\ a_{31} & a_{33} \end{vmatrix} - a_{13}\begin{vmatrix} a_{11} & a_{12} \\ a_{31} & a_{32} \end{vmatrix} \\ &= -\left[a_{11}\begin{vmatrix} a_{12} & a_{13} \\ a_{32} & a_{33} \end{vmatrix} - a_{12}\begin{vmatrix} a_{11} & a_{13} \\ a_{31} & a_{33} \end{vmatrix} + a_{13}\begin{vmatrix} a_{11} & a_{12} \\ a_{31} & a_{32} \end{vmatrix}\right] \\ &= -\begin{vmatrix} a_{11} & a_{12} & a_{13} \\ a_{11} & a_{12} & a_{13} \\ a_{31} & a_{32} & a_{33} \end{vmatrix} = 0 \ (\because \text{two rows become identical}). \end{aligned}$$

Other expressions with elements in any line multiplied by the cofactors of a parallel line will always lead to two lines identical and hence the value will be zero in all such cases.

P7. If each element of any row (or any column) is expressed as the sum of two numbers, the determinant can be expressed as the sum of two determinants whose remaining rows (or columns) are unaltered.

Consider the following determinant D of third order:

$$D = \begin{vmatrix} a_{11}+x_1 & a_{12}+y_1 & a_{13}+z_1 \\ a_{21} & a_{22} & a_{23} \\ a_{31} & a_{32} & a_{33} \end{vmatrix}.$$

Let us denote the cofactor of $a_{11}+x_1 = \begin{vmatrix} a_{22} & a_{23} \\ a_{32} & a_{33} \end{vmatrix}$ by A_{11}.

Similarly,

$$\text{the cofactor of } a_{12}+y_1 = -\begin{vmatrix} a_{21} & a_{23} \\ a_{31} & a_{33} \end{vmatrix} \text{ by } A_{12}$$

$$\text{and the cofactor of } a_{13}+z_1 = +\begin{vmatrix} a_{21} & a_{22} \\ a_{31} & a_{32} \end{vmatrix} \text{ by } A_{13}.$$

Then,

$$\begin{aligned} D &= (a_{11}+x_1)A_{11}+(a_{12}+y_1)A_{12}+(a_{13}+z_1)A_{13} \\ &= (a_{11}A_{11}+a_{12}A_{12}+a_{13}A_3)+(x_1A_{11}+y_1A_{12}+z_1A_{13}) \\ &= \begin{vmatrix} a_{11} & a_{12} & a_{13} \\ a_{21} & a_{22} & a_{23} \\ a_{31} & a_{32} & a_{33} \end{vmatrix} + \begin{vmatrix} x_1 & y_1 & z_1 \\ a_{21} & a_{22} & a_{23} \\ a_{31} & a_{32} & a_{33} \end{vmatrix}. \end{aligned}$$

Hence, the result. We observe here that this property also holds for columns.

Most Useful Property for Evaluation of Determinants

P8. The value of a determinant is not altered by adding to the elements of any row (or any column) the same multiples of the corresponding elements of any other rows (or columns).

Let $D = \begin{vmatrix} a_{11} & a_{12} & a_{13} \\ a_{21} & a_{22} & a_{23} \\ a_{31} & a_{32} & a_{33} \end{vmatrix}$ be a determinant of third order.

Suppose we add p times the elements of second row and q times the elements of third row with the corresponding elements of first row. We thus get a determinant

$$D_1 = \begin{vmatrix} a_{11}+pa_{21}+qa_{31} & a_{12}+pa_{22}+qa_{32} & a_{13}+pa_{23}+qa_{33} \\ a_{21} & a_{22} & a_{23} \\ a_{31} & a_{32} & a_{33} \end{vmatrix}.$$

We can easily verify $D_1 = D$. For,

$$D_1 = (a_{11}+pa_{21}+qa_{31})A_{11}+(a_{12}+pa_{22}+qa_{32})A_{12}+(a_{13}+pa_{23}+qa_{33})A_{13}$$

(where the cofactors of the elements of the first row of D_1 are A_{11}, A_{12}, A_{13}).

$$\begin{aligned} \therefore D_1 &= (a_{11}A_{11}+a_{12}A_{12}+a_{13}A_{13})+p(a_{21}A_{11}+a_{22}A_{12}+a_{23}A_{13}) \\ &\quad +q(a_{31}A_{11}+a_{32}A_{12}+a_{33}A_{13}) \\ &= D+0+0 \text{ (using P5 and P6)} \\ &= D. \end{aligned}$$

The reader may argue in the following manner also:

$$\begin{vmatrix} a_{11}+pa_{21}+qa_{31} & a_{12}+pa_{22}+qa_{32} & a_{13}+pa_{23}+qa_{33} \\ a_{21} & a_{22} & a_{23} \\ a_{31} & a_{32} & a_{33} \end{vmatrix}$$

$$= \begin{vmatrix} a_{11} & a_{12} & a_{13} \\ a_{21} & a_{22} & a_{23} \\ a_{31} & a_{32} & a_{33} \end{vmatrix} + \begin{vmatrix} pa_{21} & pa_{22} & pa_{23} \\ a_{21} & a_{22} & a_{23} \\ a_{31} & a_{32} & a_{33} \end{vmatrix} + \begin{vmatrix} qa_{31} & qa_{32} & qa_{33} \\ a_{21} & a_{22} & a_{23} \\ a_{31} & a_{32} & a_{33} \end{vmatrix}$$

$$= D + p\begin{vmatrix} a_{21} & a_{22} & a_{23} \\ a_{21} & a_{22} & a_{23} \\ a_{31} & a_{32} & a_{33} \end{vmatrix} + q\begin{vmatrix} a_{31} & a_{32} & a_{33} \\ a_{21} & a_{22} & a_{23} \\ a_{31} & a_{32} & a_{33} \end{vmatrix}$$

$$= D + p \times 0 + q \times 0 = D$$

($\because$ in each of the last two determinants two rows are identical).

► Example 4.3.6 (a) $\begin{vmatrix} 5 & 2 & 3 \\ 7 & 3 & 4 \\ 8 & 3 & 5 \end{vmatrix} = \begin{vmatrix} 5 & 5 & 3 \\ 7 & 7 & 4 \\ 8 & 8 & 5 \end{vmatrix}$ $(c_2' = c_2 + c_3) = 0$ ($\because$ *two cols. are identical*) **[C.P. 1995]**

(b) *Prove, without expanding, that* $\begin{vmatrix} 5 & 1 & 2 \\ 7 & 3 & 6 \\ 9 & 7 & 14 \end{vmatrix} = 0.$ **[C.P. 2004]**

See that the given determinant $= 2\begin{vmatrix} 5 & 1 & 1 \\ 7 & 3 & 3 \\ 9 & 7 & 7 \end{vmatrix} = 0$ ($\because$ *two columns are identical).*

► Example 4.3.7

$$\begin{vmatrix} 1 & 1 & 1 \\ a & b & c \\ a^2 & b^2 & c^2 \end{vmatrix} = \begin{vmatrix} 1-1 & 1 & 1 \\ a-b & b & c \\ a^2-b^2 & b^2 & c^2 \end{vmatrix} \quad [c_1' = c_1 + (-1)c_2]$$

$$= (a-b)\begin{vmatrix} 0 & 1 & 1 \\ 1 & b & c \\ a+b & b^2 & c^2 \end{vmatrix} \quad \begin{bmatrix}\text{Taking out the common} \\ \text{factor } (a-b) \text{ of the} \\ \text{first column}\end{bmatrix}$$

$$= (a-b)\begin{vmatrix} 0 & 1-1 & 1 \\ 1 & b-c & c \\ a+b & b^2-c^2 & c^2 \end{vmatrix} \quad (c_2' = c_2 - c_3)$$

$$= (a-b)(b-c)\begin{vmatrix} 0 & 0 & 1 \\ 1 & 1 & c \\ a+b & b+c & c^2 \end{vmatrix}$$

$$= (a-b)(b-c)\begin{vmatrix} 1 & 1 \\ a+b & b+c \end{vmatrix}$$

$$= (a-b)(b-c)(b+c-a-b) = (b-c)(c-a)(a-b).$$

Note. The students are advised to use this convenient notation. $r_3' = r_3 + \ell r_1 + m r_2$ will mean that starting from a determinant D we form a new determinant whose 3rd row (r_3') is the old 3rd row (r_3) plus ℓ times the first row (r_1) plus m times the second row (r_2).

Similar interpretation for notations like $c_3' = pc_1 + qc_2 + c_3$ for the new 3rd column.

Instead of transforming the first two columns step-by-step one can do the whole procedure in the same step:

See the following working:

$$\begin{vmatrix} 1 & 1 & 1 \\ a & b & c \\ a^2 & b^2 & c^2 \end{vmatrix} = \begin{vmatrix} 1 & 0 & 0 \\ a & b-a & c-a \\ a^2 & b^2-a^2 & c^2-a^2 \end{vmatrix} \left| \begin{array}{l} c_2' = c_2 - c_1 \\ c_3' = c_3 - c_1 \end{array} \right|$$

$$= \begin{vmatrix} b-a & c-a \\ b^2-a^2 & c^2-a^2 \end{vmatrix} = (b-a)(c-a) \begin{vmatrix} 1 & 1 \\ b+a & c+a \end{vmatrix}$$

$$= (b-a)(c-a)(c+a-b-a) = (b-c)(c-a)(a-b).$$

This determinant is known as an **Alternant.**

But never transform all the columns at the same time; at least one column should remain unaltered.

Similar caution for the transformation of rows.

4.4 Applications of Properties 1–8: Solved Problems

▶ Example 4.4.1 *Evaluate* $\begin{vmatrix} 1 & -2 & 3 \\ 2 & -1 & 4 \\ -2 & 3 & 1 \end{vmatrix}$.

Solution: It is often convenient to transform the given determinant in such a way that we may obtain at least two zeros in the first row.

$$\begin{vmatrix} 1 & -2 & 3 \\ 2 & -1 & 4 \\ -2 & 3 & 1 \end{vmatrix} = \begin{vmatrix} 1 & -2+2(1) & 3+(-3)1 \\ 2 & -1+2(2) & 4+(-3)2 \\ -2 & 3+2(-2) & 1+(-3)(-2) \end{vmatrix} \left[\begin{array}{l} c_2' = c_2 + 2c_1; \\ c_3' = c_3 + (-3)c_1; \\ c_1 \text{ remains unaltered} \end{array} \right]$$

$$= \begin{vmatrix} 1 & 0 & 0 \\ 2 & 3 & -2 \\ -2 & -1 & 7 \end{vmatrix} = 1 \begin{vmatrix} 3 & -2 \\ -1 & 7 \end{vmatrix} = 21 - 2 = \mathbf{19}.$$

▶ Example 4.4.2 *Express in factors the following determinant:*

$$D = \begin{vmatrix} x & y & 1 \\ x^2 & y^2 & 1 \\ x^3 & y^3 & 1 \end{vmatrix}$$

Solution:

$$D = xy\begin{vmatrix} 1 & 1 & 1 \\ x & y & 1 \\ x^2 & y^2 & 1 \end{vmatrix} = xy\begin{vmatrix} 1 & 1-1 & 1-1 \\ x & y-x & 1-x \\ x^2 & y^2-x^2 & 1-x^2 \end{vmatrix} \begin{bmatrix} c_2' = c_2 - c_1; \\ c_3' = c_3 - c_1 \end{bmatrix}$$

$$= xy\begin{vmatrix} 1 & 0 & 0 \\ x & y-x & 1-x \\ x^2 & (y-x)(y+x) & (1+x)(1-x) \end{vmatrix}$$

$$= xy(y-x)(1-x)\begin{vmatrix} 1 & 0 & 0 \\ x & 1 & 1 \\ x^2 & y+x & 1+x \end{vmatrix}$$

$$= xy(y-x)(1-x)\begin{vmatrix} 1 & 1 \\ y+x & 1+x \end{vmatrix}, \text{ (expanding in terms of first row)}$$

$$= xy(y-x)(1-x)\{(1+x)-(y+x)\}$$

$$\therefore D = xy(y-x)(x-1)(y-1).$$

▶ **Example 4.4.3** *Express* $\begin{vmatrix} 1 & 1 & 1 \\ \alpha & \beta & \gamma \\ \beta\gamma & \gamma\alpha & \alpha\beta \end{vmatrix}$ *as a product of three linear factors.*

Solution:

$$\begin{vmatrix} 1 & 1 & 1 \\ \alpha & \beta & \gamma \\ \beta\gamma & \gamma\alpha & \alpha\beta \end{vmatrix} = \begin{vmatrix} 1 & 1-1 & 1-1 \\ \alpha & \beta-\alpha & \gamma-\alpha \\ \beta\gamma & \gamma\alpha-\beta\gamma & \alpha\beta-\beta\gamma \end{vmatrix} \begin{bmatrix} c_2' = c_2 - c_1 \\ c_3' = c_3 - c_1 \end{bmatrix}$$

$$= \begin{vmatrix} 1 & 0 & 0 \\ \alpha & \beta-\alpha & \gamma-\alpha \\ \beta\gamma & -\gamma(\beta-\alpha) & -\beta(\gamma-\alpha) \end{vmatrix}$$

$$= (\beta-\alpha)(\gamma-\alpha)\begin{vmatrix} 1 & 0 & 0 \\ \alpha & 1 & 1 \\ \beta\gamma & -\gamma & -\beta \end{vmatrix}$$

$$= (\beta-\alpha)(\gamma-\alpha)\begin{vmatrix} 1 & 1 \\ -\gamma & -\beta \end{vmatrix} = (\beta-\alpha)(\gamma-\alpha)(-\beta+\gamma)$$

$$= (\alpha-\beta)(\beta-\gamma)(\gamma-\alpha).$$

▶ **Example 4.4.4** *Factorise* $\begin{vmatrix} a & b & c \\ a^2 & b^2 & c^2 \\ b+c & c+a & a+b \end{vmatrix}$. **[C.P. 1988]**

Solution:

$$\begin{vmatrix} a & b & c \\ a^2 & b^2 & c^2 \\ b+c & c+a & a+b \end{vmatrix}$$

$$= \begin{vmatrix} a+b+c & b+c+a & c+a+b \\ a^2 & b^2 & c^2 \\ b+c & c+a & a+b \end{vmatrix} \quad (r_1' = r_1 + r_3)$$

$$= (a+b+c)\begin{vmatrix} 1 & 1 & 1 \\ a^2 & b^2 & c^2 \\ b+c & c+a & a+b \end{vmatrix}$$

$$= (a+b+c)\begin{vmatrix} 1 & 1-1 & 1-1 \\ a^2 & b^2-a^2 & c^2-a^2 \\ b+c & (c+a)-(b+c) & (a+b)-(b+c) \end{vmatrix}$$

$$\begin{bmatrix} c_2' = c_2 - c_1 \\ c_3' = c_3 - c_1 \end{bmatrix}$$

$$= (a+b+c)\begin{vmatrix} 1 & 0 & 0 \\ a^2 & b^2-a^2 & c^2-a^2 \\ b+c & a-b & a-c \end{vmatrix}$$

$$= (a+b+c)(a-b)(c-a)\begin{vmatrix} 1 & 0 & 0 \\ a^2 & -(a+b) & c+a \\ b+c & 1 & -1 \end{vmatrix}$$

$$= (a+b+c)(a-b)(c-a)\begin{vmatrix} -(a+b) & c+a \\ 1 & -1 \end{vmatrix}$$

$$= (a+b+c)(a-b)(c-a) \times \{(a+b)-(c+a)\}$$

$$= (a+b+c)(b-c)(c-a)(a-b).$$

▶ **Example 4.4.5** $D = \begin{vmatrix} 5 & 2 & 3 \\ 7 & 3 & 4 \\ 9 & 4 & 5 \end{vmatrix} = 0$ (*without actual expansion*) **[C.P. 1989]**

Solution:

$$D = \begin{vmatrix} 5-(2+3) & 2 & 3 \\ 7-(3+4) & 3 & 4 \\ 9-(4+5) & 4 & 5 \end{vmatrix} \quad c_1' = c_1 - (c_2 + c_3)$$

$$= \begin{vmatrix} 0 & 2 & 3 \\ 0 & 3 & 4 \\ 0 & 4 & 5 \end{vmatrix} = 0.$$

▶ **Example 4.4.6** ***Prove without expansion that the determinant***

$$\begin{vmatrix} 1 & a & b+c \\ 1 & b & c+a \\ 1 & c & a+b \end{vmatrix} = 0.$$ **[C.P. 2003]**

Solution:

$$\begin{vmatrix} 1 & a+b+c & b+c \\ 1 & b+c+a & c+a \\ 1 & c+a+b & a+b \end{vmatrix} \quad (c_2' = c_2 + c_3)$$

$$= (a+b+c)\begin{vmatrix} 1 & 1 & b+c \\ 1 & 1 & c+a \\ 1 & 1 & a+b \end{vmatrix} = 0 \ (\because \text{two cols. are identical})$$

▶ **Example 4.4.7** *Evaluate (without expansion):* $\begin{vmatrix} 0 & a & b \\ -a & 0 & c \\ -b & -c & 0 \end{vmatrix}$. [C.P. 1990, 2005]

Solution:

$$D = \begin{vmatrix} 0 & a & b \\ -a & 0 & c \\ -b & -c & 0 \end{vmatrix} = \frac{1}{(-1)\cdot(-1)\cdot(-1)}\begin{vmatrix} 0 & -a & -b \\ a & 0 & -c \\ b & c & 0 \end{vmatrix}$$

(Multiply each of the rows by -1 and take the factor $\frac{1}{(-1)\cdot(-1)\cdot(-1)}$ outside the determinant)

$$= -\begin{vmatrix} 0 & -a & -b \\ a & 0 & -c \\ b & c & 0 \end{vmatrix} = -\begin{vmatrix} 0 & a & b \\ -a & 0 & c \\ -b & -c & 0 \end{vmatrix} \quad \text{(interchanging rows and columns)}$$

$$= -D$$

i.e., $D = -D$ or, $2D = 0$

i.e., $D = 0$.

(This is a third order skew symmetric determinant and its value is zero. **art. 4.5.6**)

▶ **Example 4.4.8** (a) *Prove that* $D = \begin{vmatrix} 1+a & 1 & 1 \\ 1 & 1+b & 1 \\ 1 & 1 & 1+c \end{vmatrix} = abc\left(1 + \frac{1}{a} + \frac{1}{b} + \frac{1}{c}\right)$

(b) *Show that* $\begin{vmatrix} 1+a_1 & a_2 & a_3 \\ a_1 & 1+a_2 & a_3 \\ a_1 & a_2 & 1+a_3 \end{vmatrix} = 1 + a_1 + a_2 + a_3.$ [C.P. 2001]

Solution: (a) Subtracting the top row from each of the others we get

$$\text{L.H.S. determinant } D = \begin{vmatrix} 1+a & 1 & 1 \\ 1-1-a & 1+b-1 & 1-1 \\ 1-1-a & 1-1 & 1+c-1 \end{vmatrix} = \begin{vmatrix} 1+a & 1 & 1 \\ -a & b & 0 \\ -a & 0 & c \end{vmatrix}.$$

Expanding with reference to the first row:

$$= (1+a)\begin{vmatrix} b & 0 \\ 0 & c \end{vmatrix} - 1\begin{vmatrix} -a & 0 \\ -a & c \end{vmatrix} + 1\begin{vmatrix} -a & b \\ -a & 0 \end{vmatrix}$$

$$= (1+a)bc + ac + ab = abc\left(1 + \frac{1}{a} + \frac{1}{b} + \frac{1}{c}\right).$$

Alternatively. Taking a, b, c common from the first, second and third columns respectively

$$D = abc\begin{vmatrix} 1+\frac{1}{a} & \frac{1}{a} & \frac{1}{a} \\ \frac{1}{b} & 1+\frac{1}{b} & \frac{1}{b} \\ \frac{1}{c} & \frac{1}{c} & 1+\frac{1}{c} \end{vmatrix}$$

$$= abc\begin{vmatrix} 1+\frac{1}{a}+\frac{1}{b}+\frac{1}{c} & \frac{1}{a}+1+\frac{1}{b}+\frac{1}{c} & \frac{1}{a}+\frac{1}{b}+1+\frac{1}{c} \\ \frac{1}{b} & 1+\frac{1}{b} & \frac{1}{b} \\ \frac{1}{c} & \frac{1}{c} & 1+\frac{1}{c} \end{vmatrix}$$

$$(r_1' = r_1 + r_2 + r_3)$$

$$= abc\left(1+\frac{1}{a}+\frac{1}{b}+\frac{1}{c}\right)\begin{vmatrix} 1 & 1 & 1 \\ \frac{1}{b} & 1+\frac{1}{b} & \frac{1}{b} \\ \frac{1}{c} & \frac{1}{c} & 1+\frac{1}{c} \end{vmatrix}$$

$$= abc\left(1+\frac{1}{a}+\frac{1}{b}+\frac{1}{c}\right)\begin{vmatrix} 1 & 0 & 0 \\ \frac{1}{b} & 1 & 0 \\ \frac{1}{c} & 0 & 1 \end{vmatrix} \quad (c_2' = c_2 - c_1 \text{ and } c_3' = c_3 - c_1)$$

$$= abc\left(1+\frac{1}{a}+\frac{1}{b}+\frac{1}{c}\right).$$

(b) Subtracting the top row from each of the others we get L.H.S. determinant

$$= \begin{vmatrix} 1+a_1 & a_2 & a_3 \\ -1 & 1 & 0 \\ -1 & 0 & 1 \end{vmatrix}.$$

Now expand in terms of first row

$$= (1+a_1)\begin{vmatrix} 1 & 0 \\ 0 & 1 \end{vmatrix} - a_2\begin{vmatrix} -1 & 0 \\ -1 & 1 \end{vmatrix} + a_3\begin{vmatrix} -1 & 1 \\ -1 & 0 \end{vmatrix}$$

$$= 1 + a_1 + a_2 + a_3 \quad \textbf{(Proved)}$$

▶ **Example 4.4.9** *Show that* $x = a + b + c$ *is a solution of the equation*

$$\begin{vmatrix} a-x & b & c \\ b & c-x & a \\ c & a & b-x \end{vmatrix} = 0$$ **[C.P. 1993]**

Solution: Putting $x = a+b+c$ in the left side determinant we get

$$\begin{vmatrix} -b-c & b & c \\ b & -a-b & a \\ c & a & -a-c \end{vmatrix} = \begin{vmatrix} -b-c+b+c & b & c \\ b-a-b+a & -a-b & a \\ c+a-a-c & a & -a-c \end{vmatrix} \quad (c_1' = c_1+c_2+c_3)$$

$$= \begin{vmatrix} 0 & b & c \\ 0 & -a-b & a \\ 0 & a & -a-c \end{vmatrix} = 0.$$

This proves that $x = a+b+c$ is a solution of the given equation.

▶ Example 4.4.10 *Prove that* $\begin{vmatrix} a-b-c & 2a & 2a \\ 2b & b-c-a & 2b \\ 2c & 2c & c-a-b \end{vmatrix} = (a+b+c)^3.$

[C.P. 1993]

Solution:

$$\text{Left side} = \begin{vmatrix} a-b-c+2b+2c & 2a+b-c-a+2c & 2a+2b+c-a-b \\ 2b & b-c-a & 2b \\ 2c & 2c & c-a-b \end{vmatrix} \quad (r_1' = r_1+r_2+r_3)$$

$$= (a+b+c)\begin{vmatrix} 1 & 1 & 1 \\ 2b & b-c-a & 2b \\ 2c & 2c & c-a-b \end{vmatrix}$$

$$= (a+b+c)\begin{vmatrix} 1 & 0 & 0 \\ 2b & b-c-a-2b & 0 \\ 2c & 0 & c-a-b-2c \end{vmatrix} \quad \begin{bmatrix} c_2' = c_2-c_1 \\ c_3' = c_3-c_1 \end{bmatrix}$$

$$= (a+b+c)\begin{vmatrix} -(a+b+c) & 0 \\ 0 & -(a+b+c) \end{vmatrix}$$

$$= (a+b+c)^3 = \text{Right side } \textbf{(proved)}.$$

4.4.1 Factors

It is often possible to factorise a determinant by using the Factor Theorem:

If $f(x)$ *be a polynomial* $a_0x^n + a_1x^{n-1} + \ldots + a_n$ (n *is a positive integer*) *where* $a_0, a_1, \ldots a_n$ *are real constants and if for* $x = \alpha$, $f(x)$ *vanishes, then* $(x-\alpha)$ *is a factor of* $f(x)$,

e.g., let $f(x) = x^2 - 5x + 6$; we see that when $x = 3$ is substituted,

$$f(3) = 3^2 - 5\cdot 3 + 6 = 0.$$

Hence, according to this theorem $(x-3)$ is a factor of $f(x)$.

4.4.2 Applications

▶ Example 4.4.11 *Factorise:* $D = \begin{vmatrix} 1 & 1 & 1 \\ a & b & c \\ a^3 & b^3 & c^3 \end{vmatrix}$.

Solution: On expanding D we shall get an expression in a, b, c; we may consider it a polynomial of degree 4 in a only.

If $a = b$, the first and second columns are identical and, therefore, $D = 0$. Hence, by the factor theorem, $(a - b)$ is a factor of D; similarly, we can prove that $(b - c)$ and $(c - a)$ are also factors of D.

But D is a homogeneous polynomial of 4th degree in a, b, c and is unaltered when a, b, c are replaced by b, c, a or by c, a, b; therefore, the remaining factor must be of the form $k(a + b + c)$ where k is a numerical constant (independent of a, b, c).

$$\therefore\ D = k(a - b)(b - c)(c - a)(a + b + c).$$

But the leading term is $+bc^3$, therefore, equating its coefficients we get $1 = k$.

$$\therefore\ D = (a - b)(b - c)(c - a)(a + b + c).$$

▶ Example 4.4.12 *If w be a cube root of unity* $(w \neq 1)$, *prove that* $(a + bw + cw^2)$ *is a factor of the determinant*

$$\begin{vmatrix} a & b & c \\ b & c & a \\ c & a & b \end{vmatrix}$$

and show that the determinant is equal to $-(a^3 + b^3 + c^3 - 3abc)$. [C.P. 2003]

Solution: Since w is a cube root of unity and $w \neq 1$, therefore,

$$1 + w + w^2 = 0,\ w^3 = 1,\ w^4 = w.$$

$$D = \begin{vmatrix} a & b & c \\ b & c & a \\ c & a & b \end{vmatrix} = \begin{vmatrix} a + wb + cw^2 & b & c \\ b + wc + aw^2 & c & a \\ c + wa + bw^2 & a & b \end{vmatrix} \quad (c_1' = c_1 + wc_2 + w^2c_3)$$

$$= \begin{vmatrix} a + bw + cw^2 & b & c \\ w^2(a + bw + cw^2) & c & a \\ w(a + bw + cw^2) & a & b \end{vmatrix} = (a + bw + cw^2) \begin{vmatrix} 1 & b & c \\ w & c & a \\ w^2 & a & b \end{vmatrix},$$

i.e., $a + bw + cw^2$ is a factor of D; in a similar manner, is also a factor of D.

Again,

$$D = \begin{vmatrix} a & b & c \\ b & c & a \\ c & a & b \end{vmatrix} = \begin{vmatrix} a + b + c & b & c \\ b + c + a & c & a \\ c + a + b & a & b \end{vmatrix} = (a + b + c) \begin{vmatrix} 1 & b & c \\ 1 & c & a \\ 1 & a & b \end{vmatrix}$$

i.e., $a + b + c$ is a factor of D.

But D is a homogeneous polynomial of 3rd degree in a, b, c.

$$\therefore\ D = k(a + b + c)(a + wb + w^2c)(a + w^2b + wc), \tag{4.4.1}$$

where k can at most be a numerical constant (independence of a, b, c).

But the second diagonal of D gives a term $-c^3$. Also the right side of (4.4.1) gives a term $kc \cdot w^2 c \cdot wc = kc^3$. Equating the two terms we get $k = -1$.

$$\therefore D = -(a+b+c)(a^2+b^2+c^2-ab-ac-bc)$$
$$\text{i.e., } D = -(a^3+b^3+c^3-3abc).$$

Another version of Example 4.4.12

▶ Example 4.4.13 *If α be any root of $x^3 = 1$, then prove that $a + b\alpha + c\alpha^2$ is always a factor of*

$$\begin{vmatrix} a & b & c \\ b & c & a \\ c & a & b \end{vmatrix}$$

and hence show that this determinant is equal to $-(a^3 + b^3 + c^3 = 3abc)$.

Solution: Since α is a root of $x^3 = 1$, it follows that $\alpha^3 = 1$, $\alpha^4 = \alpha^3 \cdot \alpha = \alpha$, $\alpha^5 = \alpha^3 \cdot \alpha^2 = \alpha^2$, etc.

Now,

$$D = \begin{vmatrix} a & b & c \\ b & c & a \\ c & a & b \end{vmatrix} = \begin{vmatrix} a+b\alpha+c\alpha^2 & b & c \\ b+c\alpha+a\alpha^2 & c & a \\ c+a\alpha+b\alpha^2 & a & b \end{vmatrix} (c_1'' = c_1 + \alpha c_2 + \alpha^2 c_3).$$

See that we can write

$$b + c\alpha + a\alpha^2 = \alpha^2(a + b\alpha + c\alpha^2)$$
$$\text{and } c + a\alpha + b\alpha^2 = \alpha(a + b\alpha + c\alpha^2) \ (\because \alpha^3 = 1, \alpha^4 = \alpha).$$

$$\therefore D = \begin{vmatrix} a+b\alpha^2+c\alpha^2 & b & c \\ \alpha^2(a+b\alpha^2+c\alpha^2) & c & a \\ \alpha(a+b\alpha^2+c\alpha^2) & a & b \end{vmatrix} = (a+b\alpha+c\alpha^2)\begin{vmatrix} 1 & b & c \\ \alpha^2 & c & a \\ \alpha & a & b \end{vmatrix}.$$

Thus, $(a + b\alpha + c\alpha)^2$ is a factor of D where α is any root of $x^3 = 1$.

But we know $x^3 = 1$ has three roots $1, \omega, \omega^2$ such that $1 + \omega + \omega^2 = 0, \omega^3 = 1$.

$\therefore$ putting $\alpha = 1, \omega, \omega^2$ successively we have three factors of D, namely $a + b + c$, $a + \omega b + \omega^2 c$ and $a + b\omega^2 + c\omega$.

$\therefore$ we can write $D = k(a+b+c)(a+\omega b+\omega^2 c)(a+\omega^2 b+\omega c)$, where k is a numerical constant, since D is a homogeneous polynomial of degree 3 in a, b, c.

Now, as in **Example 4.4.12** we obtain $k = -1$.

$$\therefore D = -(a+b+c)(a+\omega b+\omega^2 c)(a+\omega^2 b+\omega c)$$
$$= -(a^3+b^3+c^3-3abc).$$

▶ Example 4.4.14 *If x, y, z are distinct real numbers and*

$$D = \begin{vmatrix} x & x^2 & 1+x^3 \\ y & y^2 & 1+y^3 \\ z & z^2 & 1+z^3 \end{vmatrix} = 0,$$

then prove that $1 + xyz = 0$.

Solution: We have

$$D = \begin{vmatrix} x & x^2 & 1+x^3 \\ y & y^2 & 1+y^3 \\ z & z^2 & 1+z^3 \end{vmatrix} = \begin{vmatrix} x & x^2 & 1 \\ y & y^2 & 1 \\ z & z^2 & 1 \end{vmatrix} + \begin{vmatrix} x & x^2 & x^3 \\ y & y^2 & y^3 \\ z & z^2 & z^3 \end{vmatrix} \quad \text{(P.7)}$$

$$= \begin{vmatrix} x & x^2 & 1 \\ y & y^2 & 1 \\ z & z^2 & 1 \end{vmatrix} + xyz \begin{vmatrix} 1 & x & x^2 \\ 1 & y & y^2 \\ 1 & z & z^2 \end{vmatrix} \quad \left[\begin{array}{l}\text{Taking } x \text{ from 1st row,} \\ y \text{ from 2nd row, } z \\ \text{from 3rd row}\end{array}\right]$$

$$= (-1)^2 \begin{vmatrix} 1 & x & x^2 \\ 1 & y & y^2 \\ 1 & z & z^2 \end{vmatrix} + xyz \begin{vmatrix} 1 & x & x^2 \\ 1 & y & y^2 \\ 1 & z & z^2 \end{vmatrix}$$

$$= (1+xyz) \begin{vmatrix} 1 & x & x^2 \\ 1 & y & y^2 \\ 1 & z & z^2 \end{vmatrix} = (1+xyz) \begin{vmatrix} 1 & x & x^2 \\ 0 & y-x & y^2-x^2 \\ 0 & z-x & z^2-x^2 \end{vmatrix}$$

$$\left[\begin{array}{l} r_2' = r_2 - r_1; \\ r_3' = r_3 - r_1 \end{array}\right]$$

$$= (1+xyz)(y-x)(z-x) \begin{vmatrix} 1 & x & x^2 \\ 0 & 1 & y+x \\ 0 & 1 & z+x \end{vmatrix} \quad \left[\begin{array}{l}\text{Taking out } (y-x) \\ \text{from 2nd row and} \\ (z-x) \text{ from 3rd row}\end{array}\right]$$

$$= (1+xyz)(y-x)(z-x) \begin{vmatrix} 1 & y+x \\ 1 & z+x \end{vmatrix}$$

$$= (1+xyz)(y-x)(z-x)(z-y) = (1+xyz)(x-y)(y-z)(z-x).$$

$$\therefore D = 0 \Rightarrow 1 + xyz = 0$$

($\because x \neq y \neq z$; other factors cannot be zero; since z, y, z are distinct.)

4.5 Multiplication of Determinants

I. The product of two determinants of second order is also a determinant of second order.

4.5.1 Row-by-Row Multiplication Rule (Second Order)

Let $D_1 = \begin{vmatrix} a & b \\ c & d \end{vmatrix}$ and $D_2 = \begin{vmatrix} p & q \\ r & s \end{vmatrix}$ be two determinants of second order. Then

$$D_1 \cdot D_2 = \begin{vmatrix} ap+bq & ar+bs \\ cp+dq & cr+ds \end{vmatrix}$$

Rule. To obtain the **"ith row—jth column"** element of D_1D_2. Take the the ith row of D_1 and jth row of D_2—multiply the corresponding elements and add.
e.g., To determine the element in the 2nd row and 2nd column of D_1D_2 we take the 2nd row of D_1, namely (c, d) and 2nd row of D_2, namely (r, s)—multiply the corresponding elements and then add, i.e., the element is $cr + ds$.

Verification. Express $D_1 D_2$ as the sum of four determinants:

$$\begin{vmatrix} ap & ar \\ cp & cr \end{vmatrix} + \begin{vmatrix} ap & bs \\ cp & ds \end{vmatrix} + \begin{vmatrix} bq & ar \\ dq & cr \end{vmatrix} + \begin{vmatrix} bq & bs \\ dq & ds \end{vmatrix}$$

$$= pr\begin{vmatrix} a & a \\ c & c \end{vmatrix} + ps\begin{vmatrix} a & b \\ c & d \end{vmatrix} + qr\begin{vmatrix} b & a \\ d & c \end{vmatrix} + qs\begin{vmatrix} b & b \\ d & d \end{vmatrix}$$

$$= (ps - qr)\begin{vmatrix} a & b \\ c & d \end{vmatrix} = \begin{vmatrix} a & b \\ c & d \end{vmatrix}\begin{vmatrix} p & q \\ r & s \end{vmatrix} \quad \text{(other 2 det.=0)}$$

Since the value of the determinant does not change if rows and columns are interchanged, we may also obtain the product $D_1 D_2$ by **Row-by-Column Multiplication Rule:**

$$D_1 D_2 = \begin{vmatrix} ap + br & aq + bs \\ cp + dr & cq + ds \end{vmatrix}.$$

For element in the **ith row and jth col.** of $D_1 D_2$ we multiply the elements of ith row with the corresponding elements of jth column of D_2 and then add.

▶ **Example 4.5.1** *If* $D_1 = \begin{vmatrix} 2 & 3 \\ 4 & 5 \end{vmatrix}$, $D_1 = \begin{vmatrix} 3 & 8 \\ -1 & 0 \end{vmatrix}$. *Find* $D_1 \cdot D_2$.

Solution: $D_1 D_2$ (row-by-row rule)$= \begin{vmatrix} 2.3 + 3.8 & 2\cdot(-1) + 5.0 \\ 4.3 + 5.8 & 4\cdot(-1) + 5.0 \end{vmatrix} = \begin{vmatrix} 30 & -2 \\ 52 & -4 \end{vmatrix}.$

▶ **Example 4.5.2** *By forming the product*

$$\begin{vmatrix} a + ib & c + id \\ -(c - id) & a - ib \end{vmatrix} \cdot \begin{vmatrix} \alpha + i\beta & \gamma + i\delta \\ -(\gamma - i\delta) & \alpha - i\beta \end{vmatrix}$$

prove that $(a^2 + b^2 + c^2 + d^2)\cdot(\alpha^2 + \beta^2 + \gamma^2 + \delta^2) =$ **Sum of four squares**

$$= (a\alpha - b\beta + c\gamma - d\delta)^2 + (a\beta + b\alpha + c\delta + d\gamma)^2$$
$$+ (-a\gamma - b\delta + c\alpha + d\beta)^2 + (a\delta - b\gamma - c\beta + d\alpha)^2.$$

Solution:

$$\begin{vmatrix} a + ib & c + id \\ -(c - id) & a - ib \end{vmatrix} \cdot \begin{vmatrix} \alpha + i\beta & \gamma + i\delta \\ -(\gamma - i\delta) & \alpha - i\beta \end{vmatrix}$$

$$= \begin{vmatrix} (a+ib)(\alpha+i\beta) + (c+id)(\gamma+i\delta) & -(a+ib)(\gamma-i\delta) + (c+id)(\alpha-i\beta) \\ -(c-id)(\alpha+i\beta) + (a-ib)(\gamma+i\delta) & +(c-id)(\gamma-i\delta) + (a-ib)(\alpha-i\beta) \end{vmatrix}$$

(row-by-row multiplication)

$$= \begin{vmatrix} (a\alpha - b\beta + c\gamma - d\delta) + i(b\alpha + a\beta + d\gamma + c\delta) & (-a\gamma + c\alpha - b\delta + d\beta) + i(-b\gamma + a\delta + d\alpha - c\beta) \\ (-c\alpha + a\gamma - d\beta + b\delta) + i(d\alpha - c\beta - b\gamma + a\delta) & (c\gamma - d\delta + a\alpha - b\beta) + i(-d\gamma - c\delta - b\alpha - a\beta) \end{vmatrix}$$

$$= \begin{vmatrix} x + iy & x' + iy' \\ -(x' - iy') & x - iy \end{vmatrix},$$

where

$$x = a\alpha - b\beta + c\gamma - d\delta,\ y = a\beta + b\alpha + c\delta + d\gamma,$$
$$x' = -a\gamma - b\delta + c\alpha + d\beta,\ y' = a\delta - b\gamma - c\beta + d\alpha.$$

Hence,

$$\{(a+ib)(a-ib)+(c+id)(c-id)\}\{(\alpha+i\beta)(\alpha-i\beta)+(\gamma+i\delta)(\gamma-i\delta)\}$$
$$=\{(x+iy)(x-iy)+(x'+iy')(x'-iy')\}$$

or, $(a^2+b+c^2+d^2)(\alpha^2+\beta^2+\gamma^2+\delta^2)=x^2+y^2+x'^2+y'^2.$

What we have shown with the help of this problem is that: **"The product of two expressions, each being the sum of four squared quantities = sum of four squared expressions." (all the quantities are real numbers)**

II. The product of two determinants of third order is also a determinant of third order.

4.5.2 Row-by-Row Multiplication Rule (Third Order)

Let $D_1 = \begin{vmatrix} a_1 & b_1 & c_1 \\ a_2 & b_2 & c_2 \\ a_3 & b_3 & c_3 \end{vmatrix}$ and $D_2 = \begin{vmatrix} \alpha_1 & \beta_1 & \gamma_1 \\ \alpha_2 & \beta_2 & \gamma_2 \\ \alpha_3 & \beta_3 & \gamma_3 \end{vmatrix}$ be two determinants of third order. Then

$$D_1 \cdot D_2 = \begin{vmatrix} a_1\alpha_1+b_1\beta_1+c_1\gamma_1 & a_1\alpha_2+b_1\beta_2+c_1\gamma_2 & a_1\alpha_3+b_1\beta_3+c_1\gamma_3 \\ a_2\alpha_1+b_2\beta_1+c_2\gamma_1 & a_2\alpha_2+b_2\beta_2+c_2\gamma_2 & a_2\alpha_3+b_2\beta_3+c_2\gamma_3 \\ a_3\alpha_1+b_3\beta_1+c_3\gamma_1 & a_3\alpha_2+b_3\beta_2+c_3\gamma_2 & a_3\alpha_3+b_3\beta_3+c_3\gamma_3 \end{vmatrix}.$$

(Row-by-Row rule: The element in the ith row and jth column → Multiply the elements of ith row by the corresponding elements of the jth row and then add)

Verification. The determinant D_1D_2 (given above) can be expressed as the sum of $3 \times 3 \times 3 = 27$ determinants. One of these is

$$\begin{vmatrix} c_1\gamma_1 & a_1\alpha_2 & b_1\beta_3 \\ c_2\gamma_1 & a_2\alpha_2 & b_2\beta_3 \\ c_3\gamma_1 & a_3\alpha_2 & b_3\beta_3 \end{vmatrix} = \alpha_2\beta_3\gamma_1 \begin{vmatrix} c_1 & a_1 & b_1 \\ c_2 & a_2 & b_2 \\ c_3 & a_3 & b_3 \end{vmatrix}$$
$$= +\alpha_2\beta_3\gamma_1 \begin{vmatrix} a_1 & b_1 & c_1 \\ a_2 & b_2 & c_2 \\ a_3 & b_3 & c_3 \end{vmatrix}.$$

Another of these is

$$\begin{vmatrix} c_1\gamma_1 & a_1\alpha_2 & c_1\gamma_3 \\ c_2\gamma_1 & a_2\alpha_2 & c_2\gamma_3 \\ c_3\gamma_1 & a_3\alpha_2 & c_3\gamma_3 \end{vmatrix} = \alpha_2\gamma_1\gamma_3 \begin{vmatrix} c_1 & a_1 & c_1 \\ c_2 & a_2 & c_2 \\ c_3 & a_3 & c_3 \end{vmatrix} = 0.$$

Of the 27 determinants arranged in this way, all are zero except the 6 which contain all the letters a, b, c and each of these D_1 has as a factor.

∴ D_1 is a factor of the product (D_1D_2) and it is easy to check that the six non-zero determinants give

$$D_1(\alpha_1\beta_2\gamma_3 - \alpha_1\beta_3\gamma_2 - \alpha_2\beta_1\gamma_3 + \alpha_3\beta_1\gamma_2 + \alpha_2\beta_3\gamma_1 - \alpha_3\beta_2\gamma_1) = D_1D_2.$$

4.5.3 Row-by-Column Multiplication Rule (Third Order)

$$D_1D_2 = \begin{vmatrix} a_1\alpha_1+b_1\alpha_2+c_1\alpha_3 & a_1\beta_1+b_1\beta_2+c_1\beta_3 & a_1\gamma_1+b_1\gamma_2+c_1\gamma_3 \\ a_2\alpha_1+b_2\alpha_2+c_2\alpha_3 & a_2\beta_1+b_2\beta_2+c_2\beta_3 & a_2\gamma_1+b_2\gamma_2+c_2\gamma_3 \\ a_3\alpha_1+b_3\alpha_2+c_3\alpha_3 & a_3\beta_1+b_3\beta_2+c_3\beta_3 & a_3\gamma_1+b_3\gamma_2+c_3\gamma_3 \end{vmatrix}$$

[Row-by-Column Rule. Multiply the elements of the ith row with corresponding elements of jth column and then add → this will give the element in the ith row—jth column of D_1D_2]

► Example 4.5.3 *Let A_{ij} be the cofactor of a_{ij} in*

$$D = \begin{vmatrix} a_{11} & a_{12} & a_{13} \\ a_{21} & a_{22} & a_{23} \\ a_{31} & a_{32} & a_{33} \end{vmatrix} \neq 0 \ (say)$$

Prove that (i) $\begin{vmatrix} A_{22} & A_{23} \\ A_{32} & A_{33} \end{vmatrix} = a_{11} \times D$; (ii) $\begin{vmatrix} A_{11} & A_{12} & A_{13} \\ A_{21} & A_{22} & A_{23} \\ A_{31} & A_{32} & A_{33} \end{vmatrix} = D^2$. [C.P. 2002]

Solution: (i) The second order determinant $D_1 = \begin{vmatrix} A_{22} & A_{23} \\ A_{32} & A_{33} \end{vmatrix}$ is the same as the third order determinant $D_1 = \begin{vmatrix} 1 & 0 & 0 \\ A_{21} & A_{22} & A_{23} \\ A_{31} & A_{32} & A_{33} \end{vmatrix}$.

Now let us multiply (row-by-row) two-third order determinants D and D_1. Thus,

$$\begin{aligned} D \times D_1 &= \begin{vmatrix} a_{11} & a_{12} & a_{13} \\ a_{21} & a_{22} & a_{23} \\ a_{31} & a_{32} & a_{33} \end{vmatrix} \times \begin{vmatrix} 1 & 0 & 0 \\ A_{21} & A_{22} & A_{23} \\ A_{31} & A_{32} & A_{33} \end{vmatrix} \\ &= \begin{vmatrix} a_{11}+0+0 & a_{11}A_{21}+a_{12}A_{22}+a_{13}A_{23} & a_{11}A_{31}+a_{12}A_{32}+a_{13}A_{33} \\ a_{21}+0+0 & a_{21}A_{21}+a_{22}A_{22}+a_{23}A_{23} & a_{21}A_{31}+a_{22}A_{32}+a_{23}A_{33} \\ a_{31}+0+0 & a_{31}A_{21}+a_{32}A_{22}+a_{33}A_{23} & a_{31}A_{31}+a_{32}A_{32}+a_{33}A_{33} \end{vmatrix} \\ &= \begin{vmatrix} a_{11} & 0 & 0 \\ a_{21} & D & 0 \\ a_{31} & 0 & D \end{vmatrix} \quad \left[\text{since } \Sigma a_{ij}A_{ij} \begin{array}{l} = D, \text{ where } i = j \\ = 0, \text{ where } i \neq j \end{array}\right] \end{aligned}$$

$$\therefore D \times D_1 = a_{11}D^2.$$

Hence, $D_1 = a_{11}D$ $(\because D \neq 0)$.

This last equality $D_1 = a_{11}D$ holds even when $D = 0$. We can, in fact, obtain by direct substitution that $A_{22}A_{23} - A_{32}A_{23} = a_{11}D$.

$$\text{(ii)} \quad \begin{vmatrix} a_{11} & a_{12} & a_{13} \\ a_{21} & a_{22} & a_{23} \\ a_{31} & a_{32} & a_{33} \end{vmatrix} \times \begin{vmatrix} A_{11} & A_{12} & A_{13} \\ A_{21} & A_{22} & A_{23} \\ A_{31} & A_{32} & A_{33} \end{vmatrix}; \text{ [Use row-by-row rule]}$$

$$= \begin{vmatrix} a_{11}A_{11}+a_{12}A_{12}+a_{13}A_{13} & a_{11}A_{21}+a_{12}A_{22}+a_{13}A_{23} & a_{11}A_{31}+a_{12}A_{32}+a_{13}A_{33} \\ a_{21}A_{11}+a_{22}A_{12}+a_{23}A_{13} & a_{21}A_{21}+a_{22}A_{22}+a_{23}A_{23} & a_{21}A_{31}+a_{22}A_{32}+a_{23}A_{33} \\ a_{31}A_{11}+a_{32}A_{12}+a_{33}A_{13} & a_{31}A_{21}+a_{32}A_{22}+a_{33}A_{23} & a_{31}A_{31}+a_{32}A_{32}+a_{33}A_{33} \end{vmatrix}$$

$$= \begin{vmatrix} D & 0 & 0 \\ 0 & D & 0 \\ 0 & 0 & D \end{vmatrix} \quad \left(\because \Sigma a_{ij}A_{ij} \begin{array}{l} = D, \text{ if } i = j \\ = 0, \text{ if } i \neq j \end{array}\right)$$

$$\therefore D \times \begin{vmatrix} A_{11} & A_{12} & A_{13} \\ A_{21} & A_{22} & A_{23} \\ A_{31} & A_{32} & A_{33} \end{vmatrix} = D^3, \text{ or, } \begin{vmatrix} A_{11} & A_{12} & A_{13} \\ A_{21} & A_{22} & A_{23} \\ A_{31} & A_{32} & A_{33} \end{vmatrix} = D^2, \text{ if } D \neq 0.$$

Note. When $D = 0$, we can prove that the determinant

$$\begin{vmatrix} A_{11} & A_{12} & A_{13} \\ A_{21} & A_{22} & A_{23} \\ A_{31} & A_{32} & A_{33} \end{vmatrix} \text{ is also zero;}$$

for by (i) minors like $\begin{vmatrix} A_{22} & A_{23} \\ A_{32} & A_{33} \end{vmatrix} = a_{11} \times D$ and hence$=0$, if $D = 0$.

$\therefore$ the result $\begin{vmatrix} A_{11} & A_{12} & A_{13} \\ A_{21} & A_{22} & A_{23} \\ A_{31} & A_{32} & A_{33} \end{vmatrix} = D^2$ is valid (both sides being$=0$) even when $D = 0$.

4.5.4 Adjugate of a Determinant

If $D = \begin{vmatrix} a_{11} & a_{12} & a_{13} \\ a_{21} & a_{22} & a_{23} \\ a_{31} & a_{32} & a_{33} \end{vmatrix}$ then the determinant D' formed by the corresponding cofactors

$$\left[\text{i.e., } D' = \begin{vmatrix} A_{11} & A_{12} & A_{13} \\ A_{21} & A_{22} & A_{23} \\ A_{31} & A_{32} & a_{33} \end{vmatrix} \text{ or, in its transposed form } \begin{vmatrix} A_{11} & A_{21} & A_{31} \\ A_{12} & A_{22} & A_{32} \\ A_{13} & A_{23} & A_{33} \end{vmatrix}\right]$$

is called the **adjuagate of D** or **adjoint of D.**

4.5.5 Reciprocal Determinant

Some authors call D' *reciprocal determinant* of D. But the term cannot be fully justified because the word 'reciprocal' of a quantity is usually taken to be that quantity which, when multiplied by the given quantity, should give *unity*—this is not so in this case.

We give the following definition of **reciprocal determinant.**

If $D \neq 0$, the determinant formed by dividing every element of the adjugate of D by the value of the determinant D is equal to

$$\begin{aligned} D' \div D^3 &= D^2 \div D^3 \ (\because D' = D^2) \\ &= D^{-1}. \end{aligned}$$

This determinant is called the **reciprocal** (*or* **inverse**) *determinant of* D.

▶ Example 4.5.4 *If A_i, B_i, C_i are the cofactors of a_i, b_i, c_i $(i = 1, 2, 3)$ in*

$$\Delta = \begin{vmatrix} a_1 & b_1 & c_1 \\ a_2 & b_2 & c_2 \\ a_3 & b_3 & c_3 \end{vmatrix},$$

show that $\begin{vmatrix} B_1+C_1 & C_1+A_1 & A_1+B_1 \\ B_2+C_2 & C_2+A_2 & A_2+B_2 \\ B_3+C_3 & C_3+A_3 & A_3+B_3 \end{vmatrix} = 2\Delta^2.$ [C.P. 2004]

Solution: Given determinant can be expressed as the sum of 8 determinants, two of which are $\begin{vmatrix} B_1 & C_1 & A_1 \\ B_2 & C_2 & A_2 \\ B_3 & C_3 & A_3 \end{vmatrix}$ and $\begin{vmatrix} C_1 & A_1 & B_1 \\ C_2 & A_2 & B_2 \\ C_3 & A_3 & B_3 \end{vmatrix}$, others all vanish.

Hence the given determinant $= 2\begin{vmatrix} A_1 & B_1 & C_1 \\ A_2 & B_2 & C_2 \\ A_3 & B_3 & C_2 \end{vmatrix} = 2\Delta^2$ [using Ex. 4.5.3(ii)]

▶ Example 4.5.5 *If* $D = \begin{vmatrix} 1 & 1 & 1 \\ 1 & 2 & 3 \\ 3 & 1 & -5 \end{vmatrix} = -4\ (\neq 0)$. *What is D^{-1}?*

Solution: The adjugate determinant

$$D' = \begin{vmatrix} \begin{vmatrix} 2 & 3 \\ 1 & -5 \end{vmatrix} & -\begin{vmatrix} 1 & 3 \\ 3 & -5 \end{vmatrix} & \begin{vmatrix} 1 & 2 \\ 3 & 1 \end{vmatrix} \\ -\begin{vmatrix} 1 & 1 \\ 1 & -5 \end{vmatrix} & \begin{vmatrix} 1 & 1 \\ 3 & -5 \end{vmatrix} & -\begin{vmatrix} 1 & 1 \\ 3 & 1 \end{vmatrix} \\ \begin{vmatrix} 1 & 1 \\ 2 & 3 \end{vmatrix} & -\begin{vmatrix} 1 & 1 \\ 1 & 3 \end{vmatrix} & \begin{vmatrix} 1 & 1 \\ 1 & 2 \end{vmatrix} \end{vmatrix}$$

$$\text{or, } D' = \begin{vmatrix} -13 & 14 & -5 \\ 6 & -8 & 2 \\ 1 & -2 & 1 \end{vmatrix}$$

$$\therefore D^{-1} = \begin{vmatrix} -13/-4 & 14/-4 & -5/-4 \\ 6/-4 & -8/-4 & 2/-4 \\ 1/-4 & -2/-4 & 1/-4 \end{vmatrix} = \begin{vmatrix} 13/4 & -7/2 & 5/4 \\ -3/2 & 2 & -1/2 \\ -1/4 & 1/2 & -1/4 \end{vmatrix}.$$

Check. $D \cdot D^{-1} = \begin{vmatrix} 1 & 1 & 1 \\ 1 & 2 & 3 \\ 3 & 1 & -5 \end{vmatrix} \cdot \begin{vmatrix} 13/4 & -7/2 & 5/4 \\ -3/2 & 2 & -1/2 \\ -1/4 & 1/2 & -1/4 \end{vmatrix} = \begin{vmatrix} 1 & 0 & 0 \\ 0 & 1 & 0 \\ 0 & 0 & 1 \end{vmatrix} = 1.$

Hence, the name **reciprocal determinant** for D^{-1}.

4.5.6 Symmetric and Skew-symmetric Determinants

A determinant is **symmetric** (about the main diagonal) when its elements remain same after *transposition* (i.e., the determinant does not alter if we change rows into columns and columns into rows). A third order symmetric determinant has the form

$$\begin{vmatrix} a & h & g \\ h & b & f \\ g & f & c \end{vmatrix}$$

and its value is $abc + 2fgh - af^2 - bg^2 - ch^2$.

Note that in a symmetric determinant a pair of elements equidistant from the main diagonal are equal **both in magnitude and in sign.** But if the determinant be such that a pair of elements equidistant from the main diagonal are equal in magnitude but **opposite in sign,** we call such a determinant a **skew-symmetric determinant.** A third order skew-symmetric determinant has the form

$$D = \begin{vmatrix} 0 & h & g \\ -h & 0 & f \\ -g & -f & 0 \end{vmatrix}.$$

Points to Note

1. In a skew-symmetric determinant diagonal elements must all be zeros (Why?);
2. Multiply each row of skew-symmetric determinant by -1. The result is the transpose of the original determinant;

   e.g., $(-1)^3 D = \begin{vmatrix} 0 & -h & -g \\ h & 0 & -f \\ g & f & 0 \end{vmatrix}$ = transpose of D.

   We know that the value of a determinant and its transpose are equal (See **Property 1**). It, therefore, follows that if D is a skew-symmetric determinant of third order, then $(-1)^3 D = D$, i.e., $-D = D$; or, $2D = 0$; or, $D = 0$.

   [See **Example 4.4.7** (worked out).]

   Thus, a skew-symmetric determinant of third order is always zero. **[C.P. 2005]**

   We may remark here that the demonstration given above is true for any odd order skew-symmetric determinant. Hence, one may remember that **a skew-symmetric determinant of odd order is always zero.**
3. The general form of a symmetric and a skew-symmetric determinant of second order are as follows $\begin{vmatrix} a & f \\ f & b \end{vmatrix}$ and $\begin{vmatrix} 0 & -f \\ f & 0 \end{vmatrix}$.

   Notice that the skew-symmetric determinant of second order is a **perfect square.** In fact, **a skew-symmetric determinant of any even order is a perfect square** (the proof is beyond the scope of this text).
4. Note that the adjugate of a symmetric determinant

$$\begin{vmatrix} a & h & g \\ h & b & f \\ g & f & c \end{vmatrix}$$

is itself a symmetric determinant

$$\begin{vmatrix} A & H & G \\ H & B & F \\ G & F & C \end{vmatrix}.$$

It can be verified that the adjugate of the third-order skew-symmetric determinant

$$\begin{vmatrix} 0 & h & g \\ -h & 0 & f \\ -g & -f & 0 \end{vmatrix}$$

is symmetric but the adjugate of the second-order skew-symmetric determinant $\begin{vmatrix} 0 & f \\ -f & 0 \end{vmatrix}$ is skew-symmetric.

In fact, the adjugate of a skew-symmetric determinant is symmetric or skew-symmetric according as its order is odd or even.

5. If $D = \begin{vmatrix} a_1 & b_1 & c_1 \\ a_2 & b_2 & c_2 \\ a_3 & b_3 & c_3 \end{vmatrix}$, we can easily verify that $D^2 = D.D$ (*use row-by-row multiplication rule*) is a symmetric determinant. [C.P. 2001]

4.6 Solved Problems

▶ Example 4.6.1 *Prove that* $\begin{vmatrix} a & b & c \\ b & c & a \\ c & a & b \end{vmatrix} = -(a^3 + b^3 + c^3 - 3abc)$.

Solution:

$$\begin{aligned}
\begin{vmatrix} a & b & c \\ b & c & a \\ c & a & b \end{vmatrix} &= \begin{vmatrix} a+b+c & b & c \\ b+c+a & c & a \\ c+a+b & a & b \end{vmatrix} \quad [r_1' = r_1 + r_2 + r_3] \\
&= (a+b+c) \begin{vmatrix} 1 & b & c \\ 1 & c & a \\ 1 & a & b \end{vmatrix} \\
&= (a+b+c) \begin{vmatrix} 1 & b & c \\ 0 & c-b & a-c \\ 0 & a-b & b-c \end{vmatrix} \quad \begin{bmatrix} c_2' = c_2 - c_1 \\ c_3' = c_3 - c_1 \end{bmatrix} \\
&= (a+b+c)\{-(b-c)^2 - (a-b)(a-c)\} \\
&= (a+b+c)\{-a^2 - b^2 - c^2 + ab + ac + bc\} \\
&= -(a+b+c)(a^2 + b^2 + c^2 - ab - ac - bc) \\
&= -(a^3 + b^3 + c^3 - 3abc).
\end{aligned}$$

Note. Such determinants are called **Circulants.** See also another method of working out this Example (**Ex. 4.4.12**).

► Example 4.6.2 *Prove that the product of two expressions of the form* $a^3+b^3+c^3-3abc$ *is of the same form.* [C.H. 1982]

Solution:

$$\begin{vmatrix} a & b & c \\ b & c & a \\ c & a & b \end{vmatrix} \times \begin{vmatrix} x & y & z \\ y & z & x \\ z & x & y \end{vmatrix} = \begin{vmatrix} ax+by+cz & ay+bz+cx & az+bx+cy \\ bx+cy+az & by+cz+ax & bz+cx+ay \\ cx+ay+bz & cy+az+bx & cz+ax+by \end{vmatrix}$$

$$= \begin{vmatrix} \alpha & \gamma & \beta \\ \beta & \alpha & \gamma \\ \gamma & \beta & \alpha \end{vmatrix} = - \begin{vmatrix} \alpha & \beta & \gamma \\ \beta & \gamma & \alpha \\ \gamma & \alpha & \beta \end{vmatrix},$$

(interchanging 2nd and 3rd columns)

where $\alpha = ax+by+cz, \beta = bx+cy+az, \gamma = cx+ay+bz$.

i.e., $\{-(a^3+b^3+c^3-3abc\}\cdot\{-(x^3+y^3+z^3-3xyz)\}$
$= +(\alpha^3+\beta^3+\gamma^3-3\alpha\beta\gamma)$

► Example 4.6.3 *Prove that*

$$\begin{vmatrix} 2bc-a^2 & c^2 & b^2 \\ c^2 & 2ca-b^2 & a^2 \\ b^2 & a^2 & 2ab-c^2 \end{vmatrix} = (a^3+b^3+c^3-3abc)^2. \quad \text{[C.P. 2004]}$$

Solution: Since $\begin{vmatrix} a & b & c \\ b & c & a \\ c & a & b \end{vmatrix} = -(a^3+b^3+c^3-3abc)$,

therefore, $\begin{vmatrix} a & b & c \\ b & c & a \\ c & a & b \end{vmatrix} \cdot \begin{vmatrix} a & b & c \\ b & c & a \\ c & a & b \end{vmatrix} = (a^3+b^3+c^3-3abc)^2$

$$\begin{vmatrix} a & b & c \\ b & c & a \\ c & a & b \end{vmatrix} \times \begin{vmatrix} a & b & c \\ b & c & a \\ c & a & b \end{vmatrix} = \begin{vmatrix} a & b & c \\ b & c & a \\ c & a & b \end{vmatrix} \times \begin{vmatrix} -a & c & b \\ -b & a & c \\ -c & b & a \end{vmatrix}$$

(interchange col_2 and col_3; multiply each element of col_1 by -1)

$$= \begin{vmatrix} -a^2+bc+cb & -ab+ab+c^2 & -ac+b^2+ca \\ -ab+c^2+ab & -b^2+ac+ac & -bc+bc+a^2 \\ -ac+ac+b^2 & -bc+a^2+bc & -c^2+ab+ba \end{vmatrix} \quad \begin{bmatrix} \text{Multiply by using} \\ \text{row-by-row rule} \end{bmatrix}$$

$$= \begin{vmatrix} 2bc-a^2 & c^2 & b^2 \\ c^2 & 2ac-b^2 & a^2 \\ b^2 & a^2 & 2ab-c^2 \end{vmatrix}.$$

Hence, the result follows.

Some Useful Results to Remember

The following results are often found helpful while solving problems of products:

1. $\begin{vmatrix} 1 & 1 & 1 \\ \alpha & \beta & \gamma \\ \alpha^2 & \beta^2 & \gamma^2 \end{vmatrix} = (\alpha-\beta)(\beta-\gamma)(\gamma-\alpha)$; **[Alternants]**

2. $\begin{vmatrix} \alpha & \beta & \gamma \\ \beta & \gamma & \alpha \\ \gamma & \alpha & \beta \end{vmatrix} = -(\alpha^3+\beta^3+\gamma^3-3\alpha\beta\gamma)$ **[Circulants]**;

3. $\begin{vmatrix} \alpha & \beta & 0 \\ 0 & \alpha & \beta \\ \beta & 0 & \alpha \end{vmatrix} = \alpha^3+\beta^3$;

4. $\begin{vmatrix} A_{11} & A_{12} & A_{13} \\ A_{21} & A_{22} & A_{23} \\ A_{31} & A_{32} & A_{33} \end{vmatrix} = \begin{vmatrix} a_{11} & a_{12} & a_{13} \\ a_{21} & a_{22} & a_{23} \\ a_{31} & a_{32} & a_{33} \end{vmatrix}^2$; ($A_{ij}$ is the cofactor of a_{ij}.)

▶ **Example 4.6.4** *Find the value of the determinant* $\begin{vmatrix} \lambda & c & -b \\ -c & \lambda & a \\ b & -a & \lambda \end{vmatrix}$ *and hence obtain the value of*

$$\begin{vmatrix} a^2+\lambda^2 & ab+c\lambda & ca-b\lambda \\ ab-c\lambda & b^2+\lambda^2 & bc+a\lambda \\ ca+b\lambda & bc-a\lambda & c^2+\lambda^2 \end{vmatrix}.$$

Solution: It is better in this case to find by direct expansion the value of

$$D = \begin{vmatrix} \lambda & c & -b \\ -c & \lambda & a \\ b & -a & \lambda \end{vmatrix} = \lambda(\lambda^2+a^2+b^2+c^2).$$

Now the determinant

$$\begin{vmatrix} a^2+\lambda^2 & ab+c\lambda & ca-b\lambda \\ ab-c\lambda & b^2+\lambda^2 & bc+a\lambda \\ ca+b\lambda & bc-a\lambda & c^2+\lambda^2 \end{vmatrix}$$

is, in fact, the *adjugate* of D and hence its value $= D^2 = \lambda^2(\lambda^2+a^2+b^2+c^2)^2$. We leave the details for the students.

▶ **Example 4.6.5** *Prove that* $\begin{vmatrix} 1 & 1 & 1 \\ a & b & c \\ a^2 & b^2 & c^2 \end{vmatrix} = (a-b)(b-c)(c-a)$, *and hence prove that*

(i) $\begin{vmatrix} b^2c^2+a^2d^2 & bc+ad & 1 \\ c^2a^2+b^2d^2 & ca+bd & 1 \\ a^2b^2+c^2d^2 & ab+cd & 1 \end{vmatrix} = (b-c)(c-a)(a-b)\times(a-d)(b-d)(c-d),$

(ii) $\begin{vmatrix} 0 & (a-b)^2 & (a-c)^2 \\ (b-a)^2 & 0 & (b-c)^2 \\ (c-a)^2 & (c-b)^2 & 0 \end{vmatrix} = 2(a-b)^2(b-c)^2(c-a)^2.$

[C.P. 2002, 2005]

Solution: For the proof of the first part see **Example 4.3.7.**

(i) $$\begin{vmatrix} b^2c^2+a^2d^2 & bc+ad & 1 \\ c^2a^2+b^2d^2 & ca+bd & 1 \\ a^2b^2+c^2d^2 & ab+cd & 1 \end{vmatrix}$$

$$= \begin{vmatrix} (bc+ad)^2-2abcd & bc+ad & 1 \\ (ca+bd)^2-2abcd & ca+bd & 1 \\ (ab+cd)^2-2abcd & ab+cd & 1 \end{vmatrix}$$

$$= \begin{vmatrix} (bc+ad)^2 & bc+ad & 1 \\ (ca+bd)^2 & ca+bd & 1 \\ (ab+cd)^2 & ab+cd & 1 \end{vmatrix} - 2abcd \begin{vmatrix} 1 & bc+ad & 1 \\ 1 & ca+bd & 1 \\ 1 & ab+cd & 1 \end{vmatrix}$$

$$= \begin{vmatrix} \alpha^2 & \alpha & 1 \\ \beta^2 & \beta & 1 \\ \gamma^2 & \gamma & 1 \end{vmatrix}, \quad \left[\begin{array}{l}\text{where } \alpha = bc+ad,\ \beta = ca+bd, \\ \gamma = ab+cd \text{ and observe that the} \\ \text{second term}=0.\end{array}\right]$$

$$= -\begin{vmatrix} 1 & \alpha & \alpha^2 \\ 1 & \beta & \beta^2 \\ 1 & \gamma & \gamma^2 \end{vmatrix} \quad \text{(interchanging col. 1 and col. 3)}$$

$$= -\begin{vmatrix} 1 & 1 & 1 \\ \alpha & \beta & \gamma \\ \alpha^2 & \beta^2 & \gamma^2 \end{vmatrix} = -(\alpha-\beta)(\beta-\gamma)(\gamma-\alpha).$$

Now

$$\alpha - \beta = bc + ad - ac - bd = c(b-a) - d(b-a) = (b-a)(c-d).$$

Similarly,

$$\beta - \gamma = (b-c)(d-a) \text{ and } \gamma - \alpha = (a-c)(b-d).$$

$$\begin{aligned} \therefore \text{Given determinant} &= -(b-a)(c-d)(b-c)(d-a)(a-c)(b-d) \\ &= (b-c)(c-a)(a-b) \times (a-d)(b-d)(c-d). \end{aligned}$$

(ii) We can express the given determinant as the product of two determinants, thus

$$\begin{vmatrix} 0 & (a-b)^2 & (a-c)^2 \\ (b-a)^2 & 0 & (b-c)^2 \\ (c-a)^2 & (c-b)^2 & 0 \end{vmatrix}$$

$$= \begin{vmatrix} 1 & a & a^2 \\ 1 & b & b^2 \\ 1 & c & c^2 \end{vmatrix} \times \begin{vmatrix} a^2 & -2a & 1 \\ b^2 & -2b & 1 \\ c^2 & -2c & 1 \end{vmatrix} \quad \left[\begin{array}{l}\text{Use row-by-} \\ \text{row rule}\end{array}\right]$$

Now

$$\begin{vmatrix} 1 & a & a^2 \\ 1 & b & b^2 \\ 1 & c & c^2 \end{vmatrix} = \begin{vmatrix} 1 & 1 & 1 \\ a & b & c \\ a^2 & b^2 & c^2 \end{vmatrix} = (b-c)(c-a)(a-b)$$

and
$$\begin{vmatrix} a^2 & -2a & 1 \\ b^2 & -2b & 1 \\ c^2 & -2c & 1 \end{vmatrix} = -2\begin{vmatrix} a^2 & a & 1 \\ b^2 & b & 1 \\ c^2 & c & 1 \end{vmatrix}$$

$$= \ +2\begin{vmatrix} 1 & a & a^2 \\ 1 & b & b^2 \\ 1 & c & c^2 \end{vmatrix} = 2(b-c)(c-a)(a-b).$$

Hence, their product$=2(b-c)^2(c-a)^2(a-b)^2$.

▶ Example 4.6.6 *Prove that*

$$\begin{vmatrix} (b+c)^2 & a^2 & a^2 \\ b^2 & (c+a)^2 & b^2 \\ c^2 & c^2 & (a+b)^2 \end{vmatrix} = 2abc(a+b+c)^3.$$

Hence, show that

$$\begin{vmatrix} a^2 & (s-a)^2 & (s-a)^2 \\ (s-b)^2 & b^2 & (s-b)^2 \\ (s-c)^2 & (s-c)^2 & c^2 \end{vmatrix} = 2s^3(s-a)(s-b)(s-c),$$

where $2s = a+b+c$. **[C.P. 1995]**

Solution: Apply $c_1' = c_1 - c_3, c_2' = c_2 - c_3$ and obtain

$$\begin{vmatrix} (b+c)^2 - a^2 & 0 & a^2 \\ 0 & (c+a)^2 - b^2 & b^2 \\ c^2 - (a+b)^2 & c^2 - (a+b)^2 & (a+b)^2 \end{vmatrix}$$

$$= (a+b+c)^2\begin{vmatrix} b+c-a & 0 & a^2 \\ 0 & c+a-b & b^2 \\ c-a-b & c-a-b & (a+b)^2 \end{vmatrix}.$$

Now apply $r_3' = r_3 - (r_1 + r_2)$

$$= (a+b+c)^2\begin{vmatrix} b+c-a & 0 & a^2 \\ 0 & c+a-b & b^2 \\ -2b & -2a & 2ab \end{vmatrix}. \quad \text{Apply} \quad \begin{array}{l} c_1' = c_1 + \frac{1}{a}c_3 \\ c_2' = c_2 + \frac{1}{b}c_3 \end{array}$$

$$= (a+b+c)^2\begin{vmatrix} b+c & a^2/b & a^2 \\ b^2/a & c+a & b^2 \\ 0 & 0 & 2ab \end{vmatrix} = (a+b+c)^2 2ab\begin{vmatrix} b+c & a^2/b \\ b^2/a & c+a \end{vmatrix}$$

$$= 2ab(a+b+c)^2\{(b+c)(c+a) - ab\} = 2abc(a+b+c)^3.$$

In the second case, put $s-a = x, s-b = y, s-c = z$ so that

$$x+y+z = 3s - 2s = s.$$

The det. transforms into

$$\begin{vmatrix} (y+z)^2 & x^2 & x^2 \\ y^2 & (z+x)^2 & y^2 \\ z^2 & z^2 & (x+y)^2 \end{vmatrix}$$

which is $= 2xyz(x+y+z)^3$ by using the previous result.
$= 2(s-a)(s-b)(s-c)s^3$ **(Proved)**

▶ Example 4.6.7 *Prove that* $D = \begin{vmatrix} a^2 & bc & ac+c^2 \\ a^2+ab & b^2 & ac \\ ab & b^2+bc & c^2 \end{vmatrix} = 4a^2b^2c^2.$

Solution: Take a, b, c common from the first, second and third column, respectively.

$$D = abc\begin{vmatrix} a & c & a+c \\ a+b & b & a \\ b & b+c & c \end{vmatrix} = abc\begin{vmatrix} 0 & 2c & 2c \\ a+b & b & a \\ b & b+c & c \end{vmatrix}$$
$[r_1' = r_1 + r_3 - r_2]$

$$= abc\begin{vmatrix} 0 & 0 & 2c \\ a+b & b-a & a \\ b & b & c \end{vmatrix} \quad (c_2' = c_2 - c_3)$$

$$= abc\cdot 2c\begin{vmatrix} a+b & b-a \\ b & a \end{vmatrix} \quad \text{(Expanding in terms of first row)}$$

$$= 2abc^2\{ab+b^2-b^2+ab\}$$

$= 4a^2b^2c^2$ **(Proved).**

▶ Example 4.6.8 *Prove that* $\begin{vmatrix} b^2+c^2 & ab & ac \\ ab & c^2+a^2 & bc \\ ac & bc & a^2+b^2 \end{vmatrix} = 4a^2b^2c^2.$

Solution: We verify that

$$\begin{vmatrix} 0 & b & c \\ c & a & 0 \\ b & 0 & a \end{vmatrix} = -2abc \text{ (directly expanding in terms of first row)}$$

Now multiply the determinant $\begin{vmatrix} 0 & b & c \\ c & a & 0 \\ b & 0 & a \end{vmatrix}$ by itself.

▶ Example 4.6.9 *Prove that*

$$D = \begin{vmatrix} 1+a^2-b^2 & 2ab & -2b \\ 2ab & 1-a^2+b^2 & 2a \\ 2b & -2a & 1-a^2-b^2 \end{vmatrix} = (1+a^2+b^2)^3.$$

Solution: Apply $c_1' = c_1 - bc_3$

$$D = \begin{vmatrix} 1+a^2+b^2 & 2ab & -2b \\ 0 & 1-a^2+b^2 & 2a \\ b(1+a^2+b^2) & -2a & 1-a^2-b^2 \end{vmatrix}.$$

Apply $r_3' = r_3 - br_1$ and then expand in terms of first column.

▶ **Example 4.6.10** *Prove that*

$$D = \begin{vmatrix} a^2 & a^2-(b-c)^2 & bc \\ b^2 & b^2-(c-a)^2 & ca \\ c^2 & c^2-(a-b)^2 & ab \end{vmatrix}$$
$$= (b-c)(c-a)(a-b)(a+b+c)(a^2+b^2+c^2).$$

Solution: Apply $c_2' = c_2 - c_1 = -\begin{vmatrix} a^2 & (b-c)^2 & bc \\ b^2 & (c-a)^2 & ca \\ c^2 & (a-b)^2 & ab \end{vmatrix}$.

Apply $c_2' = c_2 + 2c_3 = -\begin{vmatrix} a^2 & b^2+c^2 & bc \\ b^2 & c^2+a^2 & ca \\ c^2 & a^2+b^2 & ab \end{vmatrix}$.

Apply $c_2' = c_2 + c_1$ and take $a^2+b^2+c^2$ common

$$= -(a^2+b^2+c^2)\cdot\begin{vmatrix} a^2 & 1 & bc \\ b^2 & 1 & ca \\ c^2 & 1 & ab \end{vmatrix} = (a^2+b^2+c^2)\begin{vmatrix} 1 & a^2 & bc \\ 1 & b^2 & ca \\ 1 & c^2 & ab \end{vmatrix}.$$

Make the first column (1 0 0) (i.e., $r_2' = r_2 - r_1$ and $r_3' = r_3 - r_1$).

▶ **Example 4.6.11** *Prove (without expansion) that*

$$D = \begin{vmatrix} x & y & z \\ x^2 & y^2 & z^2 \\ yz & zx & xy \end{vmatrix} = \begin{vmatrix} 1 & 1 & 1 \\ x^2 & y^2 & z^2 \\ x^3 & y^3 & z^3 \end{vmatrix}.$$

and hence deduce $D = (y-z)(z-x)(x-y)(yz+zx+xy)$.

Solution:

$$D = \begin{vmatrix} x & y & z \\ x^2 & y^2 & z^2 \\ yz & zx & xy \end{vmatrix}$$ Multiply col. 1 by x, col. 2 by y and col. 3 by z and take out $\frac{1}{xyz}$ as a factor.

$$= \frac{1}{xyz}\begin{vmatrix} x^2 & y^2 & z^2 \\ x^3 & y^3 & z^3 \\ xyz & zxy & xyz \end{vmatrix} = \begin{vmatrix} x^2 & y^2 & z^2 \\ x^3 & y^3 & z^3 \\ 1 & 1 & 1 \end{vmatrix} = -\begin{vmatrix} x^2 & y^2 & z^2 \\ 1 & 1 & 1 \\ x^3 & y^3 & z^3 \end{vmatrix}$$

$$= +\begin{vmatrix} 1 & 1 & 1 \\ x^2 & y^2 & z^2 \\ x^3 & y^3 & z^3 \end{vmatrix} = \begin{vmatrix} 1 & 0 & 0 \\ x^2 & y^2-x^2 & z^2-x^2 \\ x^3 & y^3-x^3 & z^3-x^3 \end{vmatrix} \quad \begin{matrix} c_2' = c_2 - c_1 \\ c_3' = c_3 - c_1 \end{matrix}$$

$$= (y-x)(z-x)\begin{vmatrix} y+x & z+x \\ y^2+xy+x^2 & z^2+xz+x^2 \end{vmatrix}$$

$$= (y-x)(z-x)\begin{vmatrix} x+y & z+x \\ y^2+xy+x^2-x(x+y) & z^2+xz+x^2-x(z+x) \end{vmatrix}$$
$$(r_2' = r_1 - x \cdot r_1)$$

$$= (y-x)(z-x)\begin{vmatrix} x+y & z+x \\ y^2 & z^2 \end{vmatrix}$$
$$= (y-x)(z-x)\{z^2x+yz^2-y^2z-x^2\}$$
$$= (y-x)(z-x)(z-y)(yz+zx+xy)$$
$$= (y-z)(z-x)(x-y)(xy+yz+zx).$$

▶ Example 4.6.12 *Prove that* $D = \begin{vmatrix} -a^2 & ab & ac \\ ab & -b^2 & bc \\ ac & bc & -c^2 \end{vmatrix} = 4a^2b^2c^2.$

Solution:

$$D = abc\begin{vmatrix} -a & b & c \\ a & -b & c \\ a & b & -c \end{vmatrix} = a^2b^2c^2\begin{vmatrix} -1 & 1 & 1 \\ 1 & -1 & 1 \\ 1 & 1 & -1 \end{vmatrix}$$

$$= a^2b^2c^2\begin{vmatrix} -1 & 1 & 1 \\ 0 & 0 & 2 \\ 0 & 2 & 0 \end{vmatrix}$$

$$= 4a^2b^2c^2 \ (r_2' = r_2 + r_1;\ r_3 = r_3 + r_1).$$

▶ Example 4.6.13 *Prove (without expanding) that*

$$\begin{vmatrix} b+c & c+a & a+b \\ q+r & r+p & p+q \\ y+z & z+x & x+y \end{vmatrix} = 2\begin{vmatrix} a & b & c \\ p & q & r \\ x & y & z \end{vmatrix}.$$ [C.P. 2002]

Solution: L.H.S. determinant can be expressed as the sum of eight determinants, namely

$$\begin{vmatrix} b & c & a \\ q & r & p \\ y & z & x \end{vmatrix};\ \begin{vmatrix} b & c & b \\ q & r & q \\ y & z & y \end{vmatrix};\ \begin{vmatrix} b & a & a \\ q & p & p \\ y & x & x \end{vmatrix};\ \begin{vmatrix} b & a & b \\ q & p & q \\ y & x & y \end{vmatrix};$$
$$\begin{vmatrix} c & c & a \\ r & r & p \\ z & z & x \end{vmatrix};\ \begin{vmatrix} c & c & b \\ r & r & q \\ z & z & y \end{vmatrix};\ \begin{vmatrix} c & a & q \\ r & p & p \\ z & x & x \end{vmatrix};\ \begin{vmatrix} c & a & b \\ r & p & q \\ z & x & y \end{vmatrix}.$$

Excepting the first and the last determinants all others vanish ($\because$ two columns are identical). Therefore, the L.H.S. determinant

$$= \begin{vmatrix} b & c & a \\ q & r & p \\ y & z & x \end{vmatrix} + \begin{vmatrix} c & a & b \\ r & p & q \\ z & x & y \end{vmatrix}$$

$$= -\begin{vmatrix} a & c & b \\ p & r & q \\ x & z & y \end{vmatrix} - \begin{vmatrix} a & c & b \\ p & r & q \\ x & y & z \end{vmatrix}$$

[interchanging col.1 and col.3] [interchanging col.1 and col.2]

$$= +\begin{vmatrix} a & b & c \\ p & q & r \\ x & y & z \end{vmatrix} + \begin{vmatrix} a & b & c \\ p & q & r \\ x & y & z \end{vmatrix}$$

[interchanging col.2 and col.3] [interchanging col.2 and col.3]

$$= 2\begin{vmatrix} a & b & c \\ p & q & r \\ x & y & z \end{vmatrix} = \text{R.H.S. determinant (Proved).}$$

▶ Example 4.6.14 *Prove that* $\begin{vmatrix} x^2+y^2+1 & x^2+2y^2+3 & x^2+3y^2+4 \\ y^2+2 & 2y^2+6 & 3y^2+8 \\ y^2+1 & 2y^2+3 & 3y^2+4 \end{vmatrix} = x^2y^2.$

[C.P. 2003]

Solution: L.H.S. determinant

$$= \begin{vmatrix} x^2 & x^2 & x^2 \\ 1 & 3 & 4 \\ y^2+1 & 2y^2+3 & 3y^2+4 \end{vmatrix} \quad \begin{bmatrix} r_1' = r_1 - r_3 \\ r_2' = r_2 - r_3 \end{bmatrix}$$

$$= x^2\begin{vmatrix} 1 & 1 & 1 \\ 1 & 3 & 4 \\ y^2+1 & 2y^2+3 & 3y^2+4 \end{vmatrix}$$

$$= x^2\begin{vmatrix} 1 & 0 & 0 \\ 1 & 2 & 3 \\ y^2+1 & y^2+2 & 2y^2+3 \end{vmatrix} \quad \begin{bmatrix} c_2' = c_2 - c_1 \\ c_3' = c_3 - c_1 \end{bmatrix}$$

$= x^2(4y^2+6-3y^2-6)$ (expanding in terms of first row)

$= x^2y^2 =$ R.H.S. **(Proved).**

▶ Example 4.6.15 *Define a symmetric determinant.*

If $D = \begin{vmatrix} a_1 & b_1 & c_1 \\ a_2 & b_2 & c_2 \\ a_3 & b_3 & c_3 \end{vmatrix}$ *be any third order determinant. Prove that its square* D^2 *is a symmetric determinant.*

[C.P. 2001]

Solution: A determinant D is said to be a **symmetric determinant** if any pair of elements equidistant from the main diagonal are equal both in magnitude and in sign, i.e., for all i and j, i.e., $a_{ij} = a_{ji}$. One property of the determinant is that its square is always a symmetric determinant.

Let $D = \begin{vmatrix} a_1 & b_1 & c_1 \\ a_2 & b_2 & c_2 \\ a_3 & b_3 & c_3 \end{vmatrix}$ be a determinant of third order.

Then

$$D^2 = \begin{vmatrix} a_1 & b_1 & c_1 \\ a_2 & b_2 & c_2 \\ a_3 & b_3 & c_3 \end{vmatrix} \begin{vmatrix} a_1 & b_1 & c_1 \\ a_2 & b_2 & c_2 \\ a_3 & b_3 & c_3 \end{vmatrix}$$

$$= \begin{vmatrix} a_1^2+b_1^2+c_1^2 & a_1a_2+b_1b_2+c_1c_2 & a_1a_3+b_1b_3+c_1c_3 \\ a_2a_1+b_2b_1+c_2c_1 & a_2^2+b_2^2+c_2^2 & a_2a_3+b_2b_3+c_2c_3 \\ a_3a_1+b_3b_1+c_3c_1 & a_3a_2+b_3b_2+c_3c_2 & a_3^2+b_3^2+c_3^2 \end{vmatrix} \quad \begin{bmatrix}\text{row-by-} \\ \text{row rule}\end{bmatrix}$$

See that any pair of elements equidistant from the main diagonal are equal and hence D^2 is a symmetric determinant.

▶ Example 4.6.16 *If the adjugate of a determinant*

$$D = \begin{vmatrix} a & h & g \\ h & b & f \\ g & f & c \end{vmatrix} \textit{ is } D' = \begin{vmatrix} A & H & G \\ H & B & F \\ G & F & C \end{vmatrix} \quad (D \neq 0),$$

then prove that

$$\frac{BC - F^2}{a} = \frac{CA - G^2}{b} = \frac{AB - H^2}{c} = D. \qquad \textbf{[C.P. 2005]}$$

Solution: To prove $BC - F^2 = aD$, we multiply D by $\begin{vmatrix} 1 & 0 & 0 \\ H & B & F \\ G & F & C \end{vmatrix}$.

$$\text{i.e.,} \begin{vmatrix} a & h & g \\ h & b & f \\ g & f & c \end{vmatrix} \times \begin{vmatrix} 1 & 0 & 0 \\ H & B & F \\ G & F & C \end{vmatrix} = \begin{vmatrix} a & aH+hB+gF & aG+hF+gC \\ h & hH+bB+fF & hG+bF+fC \\ g & gH+fB+cF & gF+fF+cC \end{vmatrix}$$

$$= \begin{vmatrix} a & 0 & 0 \\ h & D & 0 \\ g & 0 & D \end{vmatrix} = aD^2,$$

$$\text{i.e., } D\cdot(BC - F^2) = aD^2 \text{ or } BC - F^2 = aD \; (\because D \neq 0).$$

Similarly, prove $CA - G^2 = bD$ and $AB - H^2 = CD$.

▶ Example 4.6.17 *Show that* $\Delta = \begin{vmatrix} a^2 & 2ab & b^2 \\ b^2 & a^2 & 2ab \\ 2ab & b^2 & a^2 \end{vmatrix}$ *is a perfect square.* **[C.P. 2004]**

Solution: Δ = Product of two determinants

$$\begin{vmatrix} 0 & a & b \\ b & 0 & a \\ a & b & 0 \end{vmatrix} \cdot \begin{vmatrix} b & a & 0 \\ 0 & b & a \\ a & 0 & b \end{vmatrix} = \Delta_1 \cdot \Delta_2 \text{ (say)}.$$

But $\Delta_2 = \Delta_1$ (interchange col_3 and col_1 and then row_2 and row_3)
$\therefore\ \Delta = \Delta_1^2$ = a perfect square.

Exercises on Chapter 4(II)

(Third Order Determinants)

[A]

Evaluate the following determinants:

1. (a) $\begin{vmatrix} 1 & 1 & 0 \\ 0 & 1 & 1 \\ 1 & 0 & 1 \end{vmatrix}$; (c) $\begin{vmatrix} 1 & 1 & 2 \\ 2 & 1 & 1 \\ 1 & 2 & 1 \end{vmatrix}$; (e) $\begin{vmatrix} 1 & 1 & 1 \\ 4 & 6 & 9 \\ 5 & 6 & 7 \end{vmatrix}$;

(b) $\begin{vmatrix} 1 & 1 & 1 \\ -1 & 0 & 1 \\ -1 & -1 & 0 \end{vmatrix}$; (d) $\begin{vmatrix} 1 & 2 & 3 \\ 2 & 3 & 1 \\ 3 & 1 & 2 \end{vmatrix}$; (f) $\begin{vmatrix} -1 & 2 & 2 \\ 2 & -1 & 2 \\ 2 & 2 & -1 \end{vmatrix}$.

[*Ans.* (a) 2; (b) 1; (c) 4; (d) −18; (e) −1; (f) 27.]

2. Prove that

(a) $\begin{vmatrix} 7 & 11 & 13 \\ 17 & 19 & 23 \\ 29 & 31 & 37 \end{vmatrix} = 36$; (d) $\begin{vmatrix} 1 & 2 & 3 \\ 6 & 5 & 5 \\ 5 & 5 & 6 \end{vmatrix} = -2$;

(b) $\begin{vmatrix} 1^2 & 2^2 & 3^2 \\ 2^2 & 3^2 & 4^2 \\ 3^2 & 4^2 & 5^2 \end{vmatrix} = -8$; (e) $\begin{vmatrix} 46 & 38 & 16 \\ 27 & 21 & 9 \\ 18 & 17 & 7 \end{vmatrix} = -6$;

(c) $\begin{vmatrix} 6 & 5 & 5 \\ 5 & 6 & 5 \\ 5 & 5 & 6 \end{vmatrix} = 16$; (f) $\begin{vmatrix} 21 & 10 & 25 \\ 26 & 13 & 32 \\ 15 & 7 & 17 \end{vmatrix} = -8$.

3. Expand the following determinants:

(a) $\begin{vmatrix} a & 0 & c \\ a & b & 0 \\ 0 & b & c \end{vmatrix}$; (c) $\begin{vmatrix} a & h & g \\ h & b & f \\ g & f & c \end{vmatrix}$;

(b) $\begin{vmatrix} a & b & c \\ b & c & a \\ c & a & b \end{vmatrix}$; (d) $\begin{vmatrix} 1 & 1 & 1 \\ x & y & z \\ y+z & z+x & x+y \end{vmatrix}$;

(e) $\begin{vmatrix} 1 & a & -b \\ -a & 1 & c \\ b & -c & 1 \end{vmatrix}$; (f) $\begin{vmatrix} 1 & x & x^2 \\ x^2 & 1 & x \\ x & x^2 & 1 \end{vmatrix}$.

[*Ans.* (a) $2abc$; (b) $3abc - a^3 - b^3 - c^3$; (c) $abc + 2fgh - af^2 - bg^2 - ch^2$; (d) 0; (e) $1 + a^2 + b^2 + c^2$; (f) $(x^3 - 1)^2$.]

4. Express in factors the following determinants:

(a) $\begin{vmatrix} 1 & x & x^2 \\ 1 & y & y^2 \\ 1 & z & z^2 \end{vmatrix}$; (d) $\begin{vmatrix} 1 & 1 & 1 \\ x-y & y-z & z-x \\ (x-y)^2 & (y-z)^2 & (z-x)^2 \end{vmatrix}$;

(b) $\begin{vmatrix} 1 & 1 & 1 \\ a & b & c \\ a^3 & b^3 & c^3 \end{vmatrix}$; (e) $\begin{vmatrix} x & y & z \\ x^2 & y^2 & z^2 \\ yz & zx & xy \end{vmatrix}$;

(c) $\begin{vmatrix} 1 & 1 & 1 \\ a & b & c \\ bc & ca & ab \end{vmatrix}$; (f) $\begin{vmatrix} a & a & b+c \\ a+c & b & b \\ c & a+b & c \end{vmatrix}$.

[*Ans.* (a) $(x-y)(y-z)(z-x)$; (b) $(b-c)(c-a)(a-b)(a+b+c)$; (c) $(a-b)(b-c)(c-a)$; (d) $(x-2y+z)(y-2z+x)(z-2x+y)$; (e) $(x-y)(y-z)(z-x)(yz+zx+xy)$; (f) $4abc$.]

5. Prove that

(a) $\begin{vmatrix} x^2 & y^2 & z^2 \\ (y+z)^2 & (z+x)^2 & (x+y)^2 \\ yz & zx & xy \end{vmatrix} = -(y-z)(z-x)(x-y)(x+y+z)(x^2+y^2+z^2)$;

(b) $\begin{vmatrix} a & a-c & a-b \\ h & c+a & b-a \\ c & c-a & a+b \end{vmatrix} = 4abc$;

(c) $\begin{vmatrix} a-b & b-c & c-a \\ b-c & c-a & a-b \\ c-a & a-b & b-c \end{vmatrix} = 0 \quad (r_1' = r_1 + r_2 + r_3)$. **[C.P. 1981]**

(d) $\begin{vmatrix} b+c & a-c & a-b \\ b-c & c+a & b-a \\ c-b & c-a & a+b \end{vmatrix} = 8abc$.

6. Prove that $(a-1)$ is a factor of the determinant

$$D = \begin{vmatrix} a+1 & 2 & 3 \\ 3 & a+2 & 4 \\ 4 & 4 & a+3 \end{vmatrix}.$$

[Put $a = 1$, $D = 0$; hence $a - 1$ is a factor of D.] **[C.P. 1992]**

[B]

1. (a) Express the square of $\begin{vmatrix} a & b \\ c & d \end{vmatrix}$ as a second-order determinant.

(b) Express the square of $\begin{vmatrix} l_1 & m_1 & 0 \\ l_2 & m_2 & 0 \\ l_3 & m_3 & 0 \end{vmatrix}$ as a third-order determinant. What is the value of this product?

(c) Obtain the value of $\begin{vmatrix} a & b & 0 \\ c & 0 & a \\ 0 & c & b \end{vmatrix}$, express the square of this determinant as a third-order determinant.

2. If $A_1, B_1, C_1 \ldots$ be the cofactors of $a_1, b_1, c_1, \ldots$ in

$$\Delta = \begin{vmatrix} a_1 & b_1 & c_1 \\ a_2 & b_2 & c_2 \\ a_3 & b_3 & c_3 \end{vmatrix},$$

verify that $\begin{vmatrix} a_1 & b_1 & c_1 \\ a_2 & b_2 & c_2 \\ a_3 & b_3 & c_3 \end{vmatrix} \begin{vmatrix} 1 & A_2 & A_3 \\ 0 & B_2 & B_3 \\ 0 & C_2 & C_3 \end{vmatrix} = \begin{vmatrix} a_1 & 0 & 0 \\ a_2 & \Delta & 0 \\ a_3 & 0 & \Delta \end{vmatrix}$ and hence, deduce that, when

$$\Delta \neq 0 \begin{vmatrix} B_2 & C_2 \\ B_3 & C_3 \end{vmatrix} = a_1\Delta.$$

In a similar manner prove that, when $\Delta \neq 0$.

$$\begin{vmatrix} A_2 & B_2 \\ A_3 & B_3 \end{vmatrix} = c_1\Delta \text{ and } \begin{vmatrix} A_2 & C_2 \\ A_3 & C_3 \end{vmatrix} = -b_1\Delta.$$

3. Prove that

(a) $\begin{vmatrix} 2ab & a^2 & b^2 \\ a^2 & b^2 & 2ab \\ b^2 & 2ab & a^2 \end{vmatrix} = -(a^3 + b^3)^2;$

(b) $\begin{vmatrix} b^2 + c^2 & ab & ac \\ ab & c^2 + a^2 & bc \\ ca & cb & a^2 + b^2 \end{vmatrix} = 4a^2b^2c^2;$ $\left[D = \begin{vmatrix} 0 & c & b \\ c & 0 & a \\ b & a & 0 \end{vmatrix}. \text{ Find } D^2. \right]$

[C.H. 1980]

(c) $\begin{vmatrix} (a-x)^2 & (a-y)^2 & (a-z)^2 \\ (b-x)^2 & (b-y)^2 & (b-z)^2 \\ (c-x)^2 & (c-y)^2 & (c-z)^2 \end{vmatrix}$

$= 2(b-c)(c-a)(a-b)(y-z)(z-x)(x-y).$ [C.P. 1993]

[Express the given determinant $= \begin{vmatrix} a^2 - 2a & 1 \\ b^2 - 2b & 1 \\ c^2 - 2c & 1 \end{vmatrix} \times \begin{vmatrix} 1 & x & x^2 \\ 1 & y & y^2 \\ 1 & z & z^2 \end{vmatrix}$

(d) $\begin{vmatrix} a^2 & a^2 & (b+c)^2 \\ (a+c)^2 & b^2 & b^2 \\ c^2 & (a+b)^2 & c^2 \end{vmatrix} = 2abc(a+b+c)^3;$

(e) $\begin{vmatrix} (b+c)^2 & c^2 & b^2 \\ c^2 & (c+a)^2 & a^2 \\ b^2 & a^2 & (a+b)^2 \end{vmatrix} = 2(ab+bc+ca)^3.$

4. Verify that the skew-symmetric determinant $\begin{vmatrix} 0 & a & b \\ -a & 0 & c \\ -b & -c & 0 \end{vmatrix}$ is zero by changing the signs of all its elements. Also find its adjugate. Is it symmetric?

[C.P. 1987; 1993]

[C]

1. Without actually expanding (but using different properties) prove that

$$\begin{vmatrix} 1 & 1 & 1 \\ a & b & c \\ bc & ca & ab \end{vmatrix} = \begin{vmatrix} 1 & 1 & 1 \\ a & b & c \\ a^2 & b^2 & c^2 \end{vmatrix}.$$

2. Verify that $\begin{vmatrix} a+ib & c+id \\ -(c-id) & a-ib \end{vmatrix} = a^2+b^2+c^2+d^2 \ (i^2=-1).$

By forming the product of two determinants of this nature prove that the product of two quantities, each quantity being the sum of four squares, can be expressed as the sum of four squares.

Use this principle to express $(1^2+2^2+3^2+5^2)^2$ as sum of four squares.

[**Suggestions.** Multiply the two determinants

$$\begin{vmatrix} 1+2i & 3+5i \\ -(3-5i) & 1-2i \end{vmatrix} \text{ and } \begin{vmatrix} 5+2i & 3+i \\ -(3-i) & 5-2i \end{vmatrix}.]$$

3. Prove that

(a) $\begin{vmatrix} b^2c^2 & bc & b+c \\ c^2a^2 & ca & c+a \\ a^2b^2 & ab & a+b \end{vmatrix} = 0;$

(b) $\begin{vmatrix} -bc & bc+b^2 & bc+c^2 \\ ca+a^2 & -ca & ca+c^2 \\ ab+a^2 & ab+b^2 & -ab \end{vmatrix} = (bc+ca+ab)^2;$

(c) $\begin{vmatrix} x^2 & x^2-(y-z)^2 & yz \\ y^2 & y^2-(z-x)^2 & zx \\ z^2 & z^2-(x-y)^2 & xy \end{vmatrix} = (y-z)(z-x)(x-y)(x+y+z)(x^2+y^2+z^2).$

(d) $\begin{vmatrix} 1+a_1 & 1 & 1 \\ 1 & 1+a_2 & 1 \\ 1 & 1 & 1+a_3 \end{vmatrix} = a_1a_2a_3\left(1+\frac{1}{a_1}+\frac{1}{a_2}+\frac{1}{a_3}\right).$

[D]

1. Prove that

(a) $$\begin{vmatrix} -2a & a+b & c+a \\ b+a & -2b & b+c \\ c+a & c+b & -2c \end{vmatrix} = 4(a+b)(b+c)(c+a);$$

(b) $$\begin{vmatrix} a^2-bc & b^2-ca & c^2-ab \\ c^2-ab & a^2-bc & b^2-ca \\ b^2-ca & c^2-ab & a^2-bc \end{vmatrix} = (a^3+b^3+c^2-3abc)^2;$$

(c) $$\begin{vmatrix} ax-by-cz & ay+bx & cx+az \\ ay+bx & by-cz-ax & bz+cy \\ cx+az & bz+cy & cz-ax-by \end{vmatrix} = (a^2+b^2+c^2)(x^2+y^2+z^2)(ax+by+cz).$$

2. (a) If $a+b+c=0$ and

$$\begin{vmatrix} a-x & c & b \\ c & b-x & a \\ b & a & c-x \end{vmatrix} = 0,$$

prove that either x is zero or $x = \pm\sqrt{\frac{3}{2}(a^2+b^2+c^2)}$.

(b) Find x in terms of a, b, c if $\begin{vmatrix} x+a & b & c \\ c & x+b & a \\ a & b & x+c \end{vmatrix} = 0.$ [C.P. 1982]

[*Ans.* 0; $-(a+b+c)$]

(c) Prove that $\begin{vmatrix} a-b-c & 2a & 2a \\ 2b & b-c-a & 2b \\ 2c & 2c & c-a-b \end{vmatrix}$ is a perfect cube.

[Apply $r_1' = r_1 + r_2 + r_3$] [*Ans.* $2(a+b+c)^3$]

(d) If $2s = a+b+c$, prove that

$$\begin{vmatrix} a^2 & (s-a)^2 & (s-a)^2 \\ (s-b)^2 & b^2 & (s-b)^2 \\ (s-c)^2 & (s-c)^2 & c^2 \end{vmatrix} = 2s^3(s-a)(s-b)(s-c).$$

3. Prove that [Remember: $\text{cis}\,\theta = \cos\theta + i\sin\theta$; $\text{cis}(-\theta) = \cos\theta - i\sin\theta$].

(a) $$\begin{vmatrix} 1 & 1 & 1 \\ 1 & \omega & \omega^2 \\ 1 & \omega^2 & \omega \end{vmatrix} = -3\sqrt{3}i. \left(\omega = \text{cis}\,\frac{2\pi}{3}\right);$$

(b) $$\begin{vmatrix} 1 & \text{cis}\,\pi/3 & \text{cis}\,\pi/4 \\ \text{cis}(-\pi/3) & 1 & \text{cis}\,2\pi/3 \\ \text{cis}(-\pi/4) & \text{cis}(-2\pi/3) & 1 \end{vmatrix} = -2-\sqrt{2};$$

(c) $$\begin{vmatrix} \cos(\alpha-\beta) & \cos(\beta-\gamma) & \cos(\gamma-\alpha) \\ \cos(\alpha+\beta) & \cos(\beta+\gamma) & \cos(\gamma+\alpha) \\ \sin(\alpha+\beta) & \sin(\beta+\gamma) & \sin(\gamma+\alpha) \end{vmatrix} = \{\sin(\gamma-\alpha)\times\sin(\gamma-\beta)\times\sin(\alpha-\beta)\}.$$

4. Prove (Without actually expanding):

(a) $\begin{vmatrix} 1 & a & b+c \\ 1 & b & c+a \\ 1 & c & a+b \end{vmatrix} = 0.$

(b) $\begin{vmatrix} a & b & c \\ a^2 & b^2 & c^2 \\ b+c & c+a & a+b \end{vmatrix} = (a-b)(b-c)(c-a)(a+b+c);$ **[C.P. 1988]**

[*Hints*: Add row 1 with row 3 and obtain $(a+b+c)\begin{vmatrix} a & b & c \\ a^2 & b^2 & c^2 \\ 1 & 1 & 1 \end{vmatrix}$. Now proceed]

(c) $\begin{vmatrix} x & a & b \\ a & x & b \\ a & b & x \end{vmatrix} = (x-a)(x-b)(x+a+b);$ **[C.P. 1991]**

(d) $\begin{vmatrix} b+c & c+a & a+b \\ q+r & r+p & p+q \\ y+z & z+x & x+y \end{vmatrix} = 2\begin{vmatrix} a & b & c \\ p & q & r \\ x & y & z \end{vmatrix};$

(e) Prove that

$$\begin{vmatrix} x^2+y^2+1 & x^2+2y^2+3 & x^2+3y^2+4 \\ y^2+2 & 2y^2+6 & 3y^2+8 \\ y^2+1 & 2y^2+3 & 3y^2+4 \end{vmatrix} = x^2y^2.$$ **[C.P. 2003]**

(Express the left side as the product of two determinants) or see **Ex. 4.6.14.**

SECTION II

[Applications of Determinants: Solutions of Systems of Linear Equations]

4.7 Solutions of Linear Equations: Cramer's Rule

Let us begin with a system of two linear equations in two unknowns x and y:

$$\begin{cases} a_1x + b_1y = c_1 \quad (1) \\ a_2x + b_2y = c_2 \quad (2) \end{cases}$$

Multiply (1) by b_2 and (2) by b_1 and subtract, whence

$$x = \frac{c_1b_2 - c_2b_1}{a_1b_2 - a_2b_1}, \text{ provided } a_1b_2 - a_2b_1 \neq 0.$$

Similarly,

$$y = \frac{c_2a_1 - c_1a_2}{a_1b_2 - a_2b_1}, \ (a_1b_2 - a_2b_1 \neq 0).$$

We may write

$$x = \frac{\begin{vmatrix} c_1 & b_1 \\ c_2 & b_2 \end{vmatrix}}{\begin{vmatrix} a_1 & b_1 \\ a_2 & b_2 \end{vmatrix}}, \ y = \frac{\begin{vmatrix} a_1 & c_1 \\ a_2 & c_2 \end{vmatrix}}{\begin{vmatrix} a_1 & b_1 \\ a_2 & b_2 \end{vmatrix}}, \text{ provided } \begin{vmatrix} a_1 & b_1 \\ a_2 & b_2 \end{vmatrix} \neq 0.$$

From the last two results we observe the following:

1. The denominators for x and y are same, namely the determinant of the coefficients
$$\begin{vmatrix} a_1 & b_1 \\ a_2 & b_2 \end{vmatrix} = \Delta \text{ (say) and } \Delta \neq 0.$$
2. The numerator of x is the same as Δ **with the exception** that the column which contains the coefficients of x in (1) and (2) has been replaced by the column of the constants, namely by $\left.\begin{matrix} c_1 \\ c_2 \end{matrix}\right\}$.

Similar rule for the numerator of y.

▶ Example 4.7.1 *Solve:*
$$\left.\begin{aligned} 2x + 3y &= 8 \\ x - 2y &= -3 \end{aligned}\right\}.$$

Solution:
$$x = \frac{\begin{vmatrix} 8 & 3 \\ -3 & -2 \end{vmatrix}}{\begin{vmatrix} 2 & 3 \\ 1 & -2 \end{vmatrix}} = \frac{-16+9}{-4-3} = 1 \text{ and } y = \frac{\begin{vmatrix} 2 & 8 \\ 1 & -3 \end{vmatrix}}{\begin{vmatrix} 2 & 3 \\ 1 & -2 \end{vmatrix}} = \frac{-6-8}{-4-3} = 2.$$

The rule stated above is known as **Cramer's Rule.**

Cramer's Rule (For Solutions of a System of Three Linear Equations)

Statement. Let
$$\left.\begin{aligned} a_{11}x_1 + a_{12}x_2 + a_{13}x_3 &= c_1 \ (1) \\ a_{21}x_1 + a_{22}x_2 + a_{23}x_3 &= c_2 \ (2) \\ a_{31}x_1 + a_{32}x_2 + a_{33}x_3 &= c_3 \ (3) \end{aligned}\right\} \qquad (4.7.1)$$
be a system of three linear equations in three unknowns x_1, x_2, x_3.

Let the determinant of the coefficients $\Delta = \begin{vmatrix} a_{11} & a_{12} & a_{13} \\ a_{21} & a_{22} & a_{23} \\ a_{31} & a_{32} & a_{33} \end{vmatrix} \neq 0.$

Then the Cramer's Rule states:

The system of equations (4.7.1) has exactly one solution, namely, that given by
$$x_1 = \frac{\Delta_1}{\Delta}, \; x_2 = \frac{\Delta_2}{\Delta} \text{ and } x_3 = \frac{\Delta_3}{\Delta},$$
where Δ_i $(i = 1, 2, 3)$ *is the determinant obtained from* Δ *by replacing the elements of its ith column by the elements* c_1, c_2, c_3 *[i.e., the column of constants in (4.7.1)].*

Proof. Let A_{ij} be the cofactor of a_{ij} in Δ.

We multiply Eqn. (1) by A_{11}, Eqn. (2) by A_{21}, Eqn. (3) by A_{31} and then add the three relations: we then get
$$\begin{aligned} &x_1(a_{11}A_{11} + a_{21}A_{21} + a_{31}A_{31}) + x_2(a_{12}A_{11} + a_{22}A_{21} + a_{32}A_{31}) \\ &+ x_3(a_{13}A_{11} + a_{23}A_{21} + a_{33}A_{31}) = c_1A_{11} + c_2A_{21} + c_3A_{31}. \qquad (4.7.2) \end{aligned}$$

We know that if the elements of any line (row or column) be multiplied by their corresponding cofactors, the sum of such products = value of the determinant; but if they are multiplied by the cofactors of a parallel line, then the sum of such products is zero. Accordingly, from (4.6.2) we get

$$\Delta x_1 + x_2 \times 0 + x_3 \times 0 = c_1A_{11} + c_2A_{21} + c_3A_{31}$$

or, $$\Delta x_1 = \Delta_1 \left(\because c_1A_{11} + c_2A_{21} + c_3A_{31} = \begin{vmatrix} c_1 & a_{12} & a_{13} \\ c_2 & a_{22} & a_{23} \\ c_3 & a_{32} & a_{33} \end{vmatrix} \right).$$

Hence, $x_1 = \frac{\Delta_1}{\Delta}$ ($\because$ given $\Delta \neq 0$).

Similarly, multiplying Eqn. (1) by A_{12}, Eqn. (2) by A_{22}, Eqn. (3) by A_{32} and adding, we can prove that

$$x_2 = \frac{\Delta_2}{\Delta}.$$

Again, multiplying Eqn. (1) by A_{13}, Eqn. (2) by A_{23}, Eqn. (3) by A_{33} and adding, we can also prove that

$$x_3 = \frac{\Delta_3}{\Delta}.$$

$\therefore$ the system (4.7.1) has a solution $x_1 = \Delta_1/\Delta, x_2 = \Delta_2/\Delta$ and $x_3 = \Delta_3/\Delta$.

Note that what we have proved is that if the system (4.7.1) has a solution then

$$x_1 = \Delta_1/\Delta,\ x_2 = \Delta_2/\Delta \text{ and } x_3 = \Delta_3/\Delta.$$

Now, by direct substitution we can verify that these values of x_1, x_2, x_3 actually satisfy simultaneously the three equations of (4.7.1). Thus, we say that the system (4.7.1) has *one and only one solution*, namely $\{x_1 = \Delta_1/\Delta, x_2 = \Delta_2/\Delta$ and $x_3 = \Delta_3/\Delta\}$.

Important Observations

1. By a SOLUTION of the system (4.7.1) we mean that there exist definite values for x_1, x_2, x_3 which satisfy simultaneously the three equations of the system. When a solution exists for a system we say that the **system is consistent**; otherwise the system is **inconsistent.**

   What we have proved is that the system (4.7.1) **is consistent if $\Delta \neq 0$** and **that there is one and only one solution of the system.**

2. If $\Delta = 0$, the system (4.7.1) may or may not be consistent.
   If $\Delta = 0$ and at least one of $\Delta_1, \Delta_2, \Delta_3 \neq 0$, then the system is *certainly inconsistent.*

   But if $\Delta = 0$ and $\Delta_1 = \Delta_2 = \Delta_3 = 0$ then the system may or may not be consistent. We require the concept of rank of a matrix to decide whether a system of linear equation line (4.7.1) are consistant or not.

▶ Example 4.7.2 *Solve:*

$$\begin{aligned} x_1 - x_2 + x_3 &= 2, \\ 3x_1 - x_2 + 2x_3 &= -6, \\ 3x_1 + x_2 + x_3 &= -18. \end{aligned}$$

Here Cramer's Rule cannot be applied because

$$\Delta = \begin{vmatrix} 1 & -1 & 1 \\ 3 & -1 & 2 \\ 3 & 1 & 1 \end{vmatrix} = 0.$$

One can easily verify that here $\Delta_1 = \Delta_2 = \Delta_3 = 0$.

But it is possible to show that $x_1 = -\frac{1}{2}x_3 - 4$ and $x_2 = \frac{1}{2}x_3 - 6$.

$\therefore$ if $x_3 = 2$, then $x_1 = -5$ and $x_2 = -5$ and they form a solution of the system.

In fact, the system has **infinite number of solutions.** By assigning any value for x_3 we can get x_1 and x_2 and they give a solution.

3. If in the system (4.7.1) all the right-hand side constants are zero we call the system a set of **homogeneous linear equations.** Such a set is always satisfied by $x_1 = 0$, $x_2 = 0$, $x_3 = 0$ (called a **Trivial Solution**). Here obviously $\Delta_1 = 0$, $\Delta_2 = 0$, $\Delta_3 = 0$.

   Now solutions other than the trivial solution can exist if $\Delta = 0$, but we cannot definitely say this, unless we investigate further. One point, however, is clear that if a system of homogeneous equations has a solution other than the trivial one, than $\Delta = 0$, e.g., suppose we know, **in advance,** that the system

$$\left.\begin{aligned} ax + hy + gz &= 0 \\ hx + by + fz &= 0 \\ gx + fy + cz &= 0 \end{aligned}\right\} \tag{4.7.3}$$

   has a solution other then $(0, 0, 0)$ we can conclude

$$\begin{vmatrix} a & h & g \\ h & b & f \\ g & f & c \end{vmatrix} = 0. \tag{4.7.4}$$

   In this case, we also say that eliminating x, y, z in (4.7.3) we get (4.7.4).

4.7.1 Solved Problems

Use of Cramer's Rule.

▶ **Example 4.7.3** *Solve the following system of linear equations:*

$$\left.\begin{aligned} x_1 + 2x_2 - x_3 &= -3 \\ 3x_1 + x_2 + x_3 &= 4 \\ x_1 - x_2 + 2x_3 &= 6 \end{aligned}\right\}.$$

Solution: Here,

$$\Delta = \begin{vmatrix} 1 & 2 & -1 \\ 3 & 1 & 1 \\ 1 & -1 & 2 \end{vmatrix} = -3(\neq 0).$$

$\therefore$ we can apply Cramer's Rule for the solution of the system.

The unique solution of the system is given by

$$x_1 = \frac{\begin{vmatrix} -3 & 2 & -1 \\ 4 & 1 & 1 \\ 6 & -1 & 2 \end{vmatrix}}{\Delta} = \frac{-3}{-3} = 1;$$

$$x_2 = \frac{\begin{vmatrix} 1 & -3 & -1 \\ 3 & 4 & 1 \\ 1 & 6 & 2 \end{vmatrix}}{\Delta} = \frac{3}{-3} = -1;$$

$$x_3 = \frac{\begin{vmatrix} 1 & 2 & -3 \\ 3 & 1 & 4 \\ 1 & -1 & 6 \end{vmatrix}}{\Delta} = \frac{-6}{-3} = 2.$$

The required solution is $\{x_1 = 1, x_2 = -1, x_3 = 2\}$.

▶ **Example 4.7.4** *Solve the simultaneous equation:*

$$3x + 2y + 4z = 19,\ 2x - y + z = 3,\ 6x + 7y - z = 17.$$

Solution: Since

$$\Delta = \begin{vmatrix} 3 & 2 & 4 \\ 2 & -1 & 1 \\ 6 & 7 & -1 \end{vmatrix} = 78 (\neq 0),$$

we may apply Cramer's Rule. We can easily verify

$$\Delta_1 = 78,\ \Delta_2 = 156,\ \Delta_3 = 234.$$

∴ by Cramer's Rule

$$\left.\begin{aligned} x &= \Delta_1/\Delta = 78/78 = 1 \\ y &= \Delta_2/\Delta = 156/78 = 2 \\ z &= \Delta_3/\Delta = 234/78 = 3 \end{aligned}\right\}.$$

Exercises on Chapter 4(III)

(Solutions of Linear Equations: Cramer's Rule)

Solve the following system of equations (using Cramer's Rule):

1. $\left.\begin{aligned} x + y + z &= 2 \\ x + 2y + 3z &= 1 \\ 3x + y - 5z &= 4 \end{aligned}\right\}.$ **[C.P. 1989]**

2. $\left.\begin{aligned} 3x - 2y + z &= -1 \\ -x + y + 7z &= 1 \\ 4x - 3y - 2z &= -2 \end{aligned}\right\}.$ **[C.P. 2003]**

3. $\left.\begin{aligned} x + y + z &= 4 \\ x + 2y + 3z &= 9 \\ 3x + y + 4z &= 12 \end{aligned}\right\}.$

4. $\left.\begin{array}{rcl} x-y+2z &=& 1 \\ x+y+z &=& 2 \\ 2x-y+z &=& 5 \end{array}\right\}.$ **[C.P. 2001]**

5. $\left.\begin{array}{rcl} 3i_1-2i_2+4i_3 &=& 2 \\ i_1+3i_2-6i_3 &=& 8 \\ 2i_1-i_2-2i_3 &=& 0 \end{array}\right\}$, find i_3 only.

6. $\left.\begin{array}{rcl} x+y+z &=& 3 \\ 2x+3y+4z &=& 9 \\ x+2y-4z &=& -1 \end{array}\right\}.$ **[C.P. 1980]**

7. $\left.\begin{array}{rcl} \frac{1}{x}+\frac{2}{y}+\frac{1}{z} &=& \frac{1}{2} \\ \frac{4}{x}+\frac{2}{y}-\frac{3}{z} &=& \frac{2}{3} \\ \frac{3}{x}-\frac{4}{y}+\frac{4}{z} &=& \frac{1}{3} \end{array}\right\}$, find x only.

8. $\left.\begin{array}{rcl} x-3z &=& 1 \\ 2x-y-4z &=& 2 \\ y+z &=& 4 \end{array}\right\}.$ **[C.P. 1982, 1999]**

9. $\left.\begin{array}{rcl} x+2y+3z &=& 6 \\ 2x+4y+z &=& 7 \\ 3x+2y+9z &=& 14 \end{array}\right\}.$ **[C.P.1992]**

Answers

1. $x = 4, y = -3, z = 1.$

2. $x = 1, y = +2, z = 0.$

3. 1, 1, 2.

4. 3, 0, −1.

5. $i_3 = 1/2.$

6. $x = 1, y = 1, z = 1.$

7. $x = 6$. (**Check:** $y = 6, z = 12$)

8. $x = 5, y = 8/3, z = 4/3.$

9. $x = 1, y = 1, z = 1.$

10. $x = 5, y = 8/3, z = 4/3.$

Chapter 5

Introduction to Matrix Algebra

Matrices of Real Numbers	
1.	Equality of Matrices—Addition of Matrices—Multiplication of a Matrix by a Scalar.
2.	Multiplication of Matrices—Associated Properties
3.	Transpose of a Matrix—Its Properties
4.	Inverse of a Non-singular Square Matrix.
5.	Symmetric, Skew-symmetric Matrices, Scalar Matrices.
6.	Orthogonal Matrix: Elementary Operations of Matrices
7.	Rank of a Matrix—Determination of Rank either by considering minors or by sweep-out process
8.	Consistency and Solution of a System of Linear Equations with not more than three variables by Matrix method.

5.1 Introduction

There are many situations in both pure and applied branches of mathematics in which we have to deal with rectangular arrays of numbers, such as

$$\begin{bmatrix} a & b & c \\ d & e & f \end{bmatrix} \text{ or, } \begin{bmatrix} 1 & 4 \\ 7 & 9 \\ 8 & 3 \end{bmatrix}, \text{ etc.}$$

Such rectangular arrays are subjected to some rules of operations. These arrays are then called **Matrices**. We shall describe only the simplest aspects of the Matrix Theory.

5.2 Matrices of Real Numbers

Once for all, we shall mention that the objects (or elements) that we shall consider here are always **real numbers,** unless otherwise specified. Matrices of other type of objects will be left out of our discussions.

Definition. Given $m \times n$ real numbers, we arrange them in the form of a rectangular array of m rows and n columns; thus

$$A = \begin{bmatrix} a_{11} & a_{12} & a_{13} & \cdots\cdots & a_{1n} \\ a_{21} & a_{22} & a_{23} & \cdots\cdots & a_{2n} \\ a_{31} & a_{32} & a_{33} & \cdots\cdots & a_{3n} \\ \cdots & \cdots & \cdots & \cdots\cdots & \cdots \\ a_{m1} & a_{m2} & a_{m3} & \cdots\cdots & a_{mn} \end{bmatrix} \tag{5.2.1}$$

Such an array (subject to rules of operation defined below) is called a **real matrix of order** $m \times n$ (m by n).

We note that the element in the ith row and jth column is denoted by a_{ij} (e.g., a_{32} is the element in the third row and second column). Instead of square brackets [ ] a pair of parenthesis ( ) may be used to exhibit the elements of a matrix.

It is often convenient to abbreviate the symbol (5.2.1) to the form

$$[a_{ij}]_{m\times n} \text{ or, } (a_{ij})_{m\times n},$$

which means "the matrix of order m by n whose element in the ith row and jth column is a_{ij}."

$$(i = 1, 2, 3, \ldots m;\ j = 1, 2, 3, \ldots n).$$

Another convenient (and often used) procedure is to denote matrices by capital letters like A, B, C, X, Y, etc. when it is not needed to exhibit the elements explicitly.

Illustrations

1. $\begin{pmatrix} 2 & 5 & 3 \\ 1 & 3 & 2 \end{pmatrix}$ is a matrix which has two rows and three columns. So this matrix has the order 2×3 (**read:** 2 by 3)

2. $\begin{pmatrix} 2 & 1 \\ 5 & 3 \\ 3 & 2 \end{pmatrix}$ is a matrix of order 3×2; it has three rows and two columns.

3. $\begin{pmatrix} 1 & 3 & 2 \end{pmatrix}$ is a matrix of order 1×3; it has *only one row.* Such matrices **(having only one row)** are called **row vectors.**

4. $\begin{pmatrix} 1 \\ 3 \\ 2 \end{pmatrix}$ is a matrix of order 3×1; it has *only one column.* Such matrices are called **column vectors.**

5. If in a matrix the number of rows be equal to the number of columns we call it a *square matrix.* Thus, if in the matrix $(a_{ij})_{m\times n}$ we take $m = n$, then we call it a square matrix of order n; thus

(a) $\begin{bmatrix} 2 & 5 \\ 1 & -2 \end{bmatrix}$ is a square matrix of order two.

(b) $\begin{bmatrix} 1 & 3 & 2 \\ 2 & 5 & 0 \\ 1 & 0 & 1 \end{bmatrix}$ is a square matrix of order three.

6. With the set of two linear equations

$$\begin{aligned} 3x + 8y + 1 &= 0 \\ 4x - y - 7 &= 0. \end{aligned}$$

We usually associate two matrices—one, the **Coefficient Matrix**

$$\begin{bmatrix} 3 & 8 \\ 4 & -1 \end{bmatrix} \text{ (Coeff. of } x \text{ and } y)$$

and the other, the **Augmented matrix.**

$$\begin{bmatrix} 3 & 8 & 1 \\ 4 & -1 & -7 \end{bmatrix} \left(\begin{array}{l} \text{Containing coeff. of } x \text{ and } y \text{ and} \\ \text{also a column of constant terms.} \end{array} \right)$$

7. Let a_{ij} = 1, if there is some relationship between i and j.
= 0, if there is no relation between i and j.

Suppose that there are three girls A, B and C. A and B are friends; B and C are also friends but A and C are not. We represent this relationship with the help of a matrix

$$\begin{array}{c} \\ A \\ B \\ C \end{array} \begin{array}{c} \begin{array}{ccc} A & B & C \end{array} \\ \begin{pmatrix} 0 & 1 & 0 \\ 1 & 0 & 1 \\ 0 & 1 & 0 \end{pmatrix} \end{array}$$

The binary relations thus give rise to matrices, where the entries are either 0 or 1. These are called **adjacency matrices** in social science subjects.

Diagonals of a Square Matrix

Let A be a square matrix of order n, given by

$$A = \begin{bmatrix} a_{11} & a_{12} & a_{13} & \dots & a_{1n} \\ a_{21} & a_{22} & a_{23} & \dots & a_{2n} \\ a_{31} & a_{32} & a_{33} & \dots & a_{3n} \\ \dots & \dots & \dots & \dots & \dots \\ a_{n1} & a_{n2} & a_{n3} & \dots & a_{nn} \end{bmatrix}.$$

The elements $\{a_{11}, a_{22}, a_{33}, \dots, a_{nn}\}$ along the diagonal from left-hand top corner to the right-hand bottom corner is called the **main diagonal** of the matrix A and the other diagonal with elements $\{a_{n1}, a_{n-1,2}, \dots, a_{1n}\}$ is called the **second diagonal.**

5.3 Operations on Matrices

We shall first define

Equality of two matrices. *Two matrices A and B are said to be equal if, and only if, they have the same order and their corresponding elements are equal; we then write* $A = B$.

If $A = (a_{ij})_{m\times n}$ and $B = (b_{ij})_{m\times n}$ then $A = B$ if and only if $a_{ij} = b_{ij}$ for every pair of subscripts i and j.

Illustrations

1. $\begin{pmatrix} a & b \\ c & d \end{pmatrix} = \begin{pmatrix} p & q \\ r & s \end{pmatrix}$, if and only if $a = p$, $b = q$, $c = r$ and $d = s$.

2. Note that $\begin{bmatrix} 1 \\ 2 \end{bmatrix} \neq \begin{bmatrix} 1 & 0 \\ 2 & 0 \end{bmatrix}$. Why?

3. Suppose we require to find x and y, where it is given that

$$\begin{bmatrix} 3 & x+1 \\ y & 5 \end{bmatrix} = \begin{bmatrix} x-1 & 5 \\ x-3y & 5 \end{bmatrix}.$$

Solution: By definition of equality, $3 = x - 1$, $x + 1 = 5$, $y = x - 3y$ and $5 = 5$; whence $x = 4$ and $y = 1$.

4. We can easily verify that

(a) For any matrix A, $A = A$ (**Reflexive property**);

(b) If $A = B$, then $B = A$ (**Symmetric property**);

(c) If $A = B$ and $B = C$, then $A = C$, (**Transitive property**).

We know that any relation having reflexive, symmetric and transitive property is called an **equivalence relation.**

Thus, *equality of matrices* is an **equivalence relation.**

Matrix Operations

We consider, one by one, four operations:

I. **Addition of matrices.**

II. **Subtraction of matrices.**

III. **Multiplication of matrix by a scalar.**

IV. **Multiplication of a matrix by a matrix.**

I. Addition of two matrices

Two matrices are said to be conformable for addition only when they are of the same order; the sum is also a matrix of the same order obtained by adding the corresponding elements.

If $A = (a_{ij})_{m\times n}$ and $B = (b_{ij})_{m\times n}$ then

$$A + B = (c_{ij})_{m\times n}, \text{ where } c_{ij} = a_{ij} + b_{ij}.$$

For example,

1. $$\begin{bmatrix} 1 & 0 & -1 \\ 2 & -1 & 3 \end{bmatrix} + \begin{bmatrix} -2 & \frac{1}{2} & 5 \\ -1 & 2 & 0 \end{bmatrix} = \begin{bmatrix} -1 & \frac{1}{2} & 4 \\ 1 & 1 & 3 \end{bmatrix}.$$

2. We cannot add the two matrices $\begin{bmatrix} 2 & 1 \\ 3 & -5 \end{bmatrix}$ and $\begin{bmatrix} 6 \\ 4 \end{bmatrix}$; for, they are of different orders and as such *not conformable for addition.*

Theorem 5.3.1 *The addition of matrices is both commutative and associative, that is, if* A, B, C *are conformable for addition, then*

(i) $A + B = B + A$, *and*

(ii) $A + (B + C) = (A + B) + C$.

Proof. Let A, B, C be arbitrary matrices, each of order m by n (elements are all real numbers).

Suppose $A = (a_{ij})_{m\times n}$, $B = (b_{ij})_{m\times n}$, $C = (c_{ij})_{m\times n}$
Then, by definition

$$\begin{aligned} A + B &= (d_{ij})_{m\times n}, \text{ where } d_{ij} = a_{ij} + b_{ij} \\ B + A &= (e_{ij})_{m\times n}, \text{ where } e_{ij} = b_{ij} + a_{ij}. \end{aligned}$$

Now, $d_{ij} = e_{ij}$, because the elements are real numbers $a_{ij} + b_{ij} = b_{ij} + a_{ij}$ and so

$$A + B = B + A.$$

This proves (i).
Again,

$$A + (B + C) = (a_{ij})_{m\times n} + (f_{ij})_{m\times n} = (k_{ij})_{m\times n},$$

where

$$\begin{aligned} & f_{ij} = b_{ij} + c_{ij} \\ \text{and} \quad & k_{ij} = a_{ij} + f_{ij} = a_{ij} + (b_{ij} + c_{ij}) = a_{ij} + b_{ij} + c_{ij} \\ \text{and} \quad & (A + B) + C = (l_{ij})_{m\times n} + (c_{ij})_{m\times n} = (t_{ij})_{m\times n}, \end{aligned}$$

where

$$\begin{aligned} & l_{ij} = a_{ij} + b_{ij} \\ \text{and} \quad & t_{ij} = l_{ij} + c_{ij} = (a_{ij} + b_{ij}) + c_{ij} = a_{ij} + b_{ij} + c_{ij}. \end{aligned}$$

Since $k_{ij} = t_{ij}$, we conclude that

$$A + (B + C) = (A + B) + C.$$

This proves (ii).

Null Matrix

A matrix (of any order) is called a **null matrix** (or a **zero matrix**) if all its elements are zero. We use a capital letter O to denote a null matrix.

$$\begin{pmatrix} 0 & 0 \\ 0 & 0 \end{pmatrix}, \begin{pmatrix} 0 & 0 & 0 \\ 0 & 0 & 0 \end{pmatrix}, \begin{pmatrix} 0 & 0 \\ 0 & 0 \\ 0 & 0 \end{pmatrix}, \begin{pmatrix} 0 & 0 & 0 \\ 0 & 0 & 0 \\ 0 & 0 & 0 \end{pmatrix}$$

are zero matrices of order 2 × 2, 2 × 3, 3 × 2, 3 × 3, respectively. We may use the notation like $O_{(m\times n)}$ to denote a null matrix of order m by n.

If $A = (a_{ij})_{m\times n}$ then it is clear that

$$A + O_{m\times n} = A; \text{ also } O_{m\times n} + A = A.$$

II. Subtraction of two Matrices

If A *and* B *are two matrices of same order, then the difference* A − B *is a matrix of same order whose elements are obtained by subtracting the elements of* B *from the corresponding elements of* A. For example,

$$\begin{bmatrix} 2 & 1 & -3 \\ -4 & 0 & 1 \end{bmatrix} - \begin{bmatrix} 1 & -1 & 2 \\ 0 & 1 & 0 \end{bmatrix} = \begin{bmatrix} 2-1 & 1+1 & -3-2 \\ -4-0 & 0-1 & 1-0 \end{bmatrix}$$
$$= \begin{bmatrix} 1 & 2 & -5 \\ -4 & -1 & 1 \end{bmatrix}.$$

More generally, if

$$A = (a_{ij})_{m\times n}, B = (b_{ij})_{m\times n}, \text{ then } A - B = (c_{ij})_{m\times n},$$

where $c_{ij} = a_{ij} - b_{ij}$.

III. Multiplication of a Matrix by a Scalar

For our purpose by a 'scalar' we shall mean 'a definite real number' (complex number, if specified).

To multiply a matrix A *by a scalar* α *is just to multiply every element of* A *by* α. For example, multiplying the matrix

$$\begin{bmatrix} 1 & 0 & -2 & 3 \\ 5 & -3 & 7 & 4 \end{bmatrix}$$

by 4, we get the matrix

$$\begin{bmatrix} 4 & 0 & -8 & 12 \\ 20 & -12 & 28 & 16 \end{bmatrix}.$$

If $A = (a_{ij})_{m\times n}$ and α is a scalar, we get $\alpha A = (c_{ij})_{m\times n}$, where $c_{ij} = \alpha a_{ij}$.

▶ Example 5.3.1 *If* $A = (a_{ij})_{m\times n}$, *then* $(-1)A = (-a_{ij})_{m\times n}$.

We write $(-1)A = -A$ and call it the negative of the matrix A.

The students should now justify that $A - B = A + (-B)$, i.e., to subtract B from A is the same as to add negative of B with A.

► Example 5.3.2 $0 \times A = (0 \cdot a_{ij})_{m \times n} = O_{(m \times n)}$, *i.e., if we multiply any matrix A by the scalar 0 then we obtain a null matrix.*

► Example 5.3.3 *If X, A, B are three matrices of same order, then the matrix equation* $X + A = B$ *is same as to state* $X = B - A$.

If $4\begin{bmatrix} 2 & 1 & 3 \\ -1 & 0 & 1 \end{bmatrix} + X = 6\begin{bmatrix} -1 & 3 & 4 \\ 5 & 1 & 0 \end{bmatrix}$ *then find the matrix* X.

Solution: Given condition is

$$\begin{bmatrix} 8 & 4 & 12 \\ -4 & 0 & 4 \end{bmatrix} + X = \begin{bmatrix} -6 & 18 & 24 \\ 30 & 6 & 0 \end{bmatrix}.$$

$$\therefore X = \begin{bmatrix} -6 & 18 & 24 \\ 30 & 6 & 0 \end{bmatrix} - \begin{bmatrix} 8 & 4 & 12 \\ -4 & 0 & 4 \end{bmatrix} = \begin{bmatrix} -14 & 14 & 12 \\ 34 & 6 & -4 \end{bmatrix}.$$

The operation of multiplying a matrix by a scalar has these basic properties:

(i) $(\alpha + \beta)A = \alpha A + \beta A$ (*Distributive property*)

(ii) $\alpha(A + B) = \alpha A + \alpha B$ (*Distributive property*)

(iii) $\alpha(\beta A) = (\alpha\beta)A$ (*Associative property*),

where $A = (a_{ij})_{m \times n}$ and α, β are arbitrary scalars.

All three properties can be readily proved by appealing to the definition, e.g., in the case of the first, we have

$$\begin{aligned} (\alpha + \beta)A &= [(\alpha + \beta)a_{ij}]_{m \times n} = [\alpha a_{ij} + \beta a_{ij}]_{m \times n} \\ &= [\alpha a_{ij}]_{m \times n} + [\beta a_{ij}]_{m \times n} = \alpha A + \beta A. \end{aligned}$$

The students are advised to prove the other two properties.

IV. Multiplication of two matrices

Preliminaries. We take a row vector, say $\begin{bmatrix} 2 & 1 & -2 & -3 \end{bmatrix}$ and also a column vector

$$\begin{bmatrix} 4 \\ 0 \\ 5 \\ -2 \end{bmatrix}.$$

We notice that both matrices have same number of elements (namely, 4). Let us multiply the corresponding elements and add; thus, we get

$$2 \times 4 + 1 \times 0 + (-2) \times 5 + (-3) \times (-2) = 4.$$

Such a product will be called **inner product** of two **single line matrices** (one is a row vector and the other a column vector) having the same number of elements. In a more general way, let us consider a particular row-vector $[a_{i1}\ a_{i2}\ \ldots\ a_{in}]$ with n elements (we may call it ith row-vector) and also consider a particular column vector with n elements:

$$\begin{bmatrix} b_{1j} \\ b_{2j} \\ b_{3j} \\ \cdots \\ b_{nj} \end{bmatrix}.$$

We may call it the jth column-vector. The inner product of these two single-line matrices

$$= a_{i1}b_{1j} + a_{i2}b_{2j} + a_{i3}l_{3j} + \ldots + a_{in}b_{nj}$$

$$= \sum_{r=1}^{n} a_{ir}b_{rj}.$$

Conformability condition for multiplication of two matrices.

In order to find the product AB of two matrices A and B we first impose one condition: **the number of columns in A must be equal to the number of rows in B. If this condition is not satisfied, then the product AB is said to be not defined.**

But if this condition is satisfied, then we use the following rule for multiplication:

The element in the ith row and jth column of AB = the inner product of the ith row of A and jth column of B.

We first give an example:

$$\begin{bmatrix} 2 & 3 \\ 3 & -4 \\ -5 & 6 \end{bmatrix}_{3\times 2} \times \begin{bmatrix} x & y \\ z & t \end{bmatrix}_{2\times 2} = \begin{bmatrix} 2x+3z & 2y+3t \\ 3x-4z & 3y-4t \\ -5x+6z & -5y+6t \end{bmatrix}_{3\times 2}$$

$$A \quad \times \quad B \quad = \quad AB.$$

Notice that A has 2 columns and B has two rows; hence AB is defined. The element in the first row and first column is obtained by taking the inner product of the first row of A, i.e., [2 3] and the first column of B $\begin{bmatrix} x \\ z \end{bmatrix}$. Similarly, other elements of the product AB is obtained.

We can now give a formal definition:

Definition.

1. *Two matrices A and B are said to be conformable for the product AB if the number of columns in A is equal to the number of rows in B.*

2. *Let A be a matrix of order $m \times n$ and B be a matrix of order $n \times p$. Then they are conformable for the product AB.*

By definition, AB will be a matrix of order $m \times p$ and its elements in ith row and kth column will be = the inner product of the ith row of A and kth column of B.

[Inner product, as already mentioned, means the product of the corresponding elements of the ith row of A and kth column of B and the adding the results].

Symbolically, if

$$A = (a_{ij})_{m\times n} \text{ and } B = (b_{jk})_{n\times p} \text{ then } AB = (c_{ik})_{m\times p},$$

where $c_{ik} = a_{i1}b_{1k} + a_{i2}b_{2k} + \ldots + a_{in}b_{nk} = \sum_{j=1}^{n} a_{ij}b_{jk}.$

Illustrations

▶ Example 5.3.4 $\begin{bmatrix} 1 & -1 & 3 \\ 2 & 0 & 1 \end{bmatrix}_{2\times3} \times \begin{bmatrix} 1 & 2 & 0 \\ 0 & -1 & 1 \\ 1 & 3 & -1 \end{bmatrix}_{3\times3}$

$$= \begin{pmatrix} 1\times1+-1\times0+3\times1 & 1\times2+(-1)\times(-1)+3\times3 & 1\times0+(-1)\times1+3\times-1 \\ 2\times1+0\times0+1\times1 & 2\times2+0\times-1+1\times3 & 2\times0+0\times1+1\times-1 \end{pmatrix}$$

$$= \begin{bmatrix} 4 & 12 & -4 \\ 3 & 7 & -1 \end{bmatrix}_{2\times3}.$$

▶ Example 5.3.5 $[a\ b]_{1\times2} \times \begin{bmatrix} 1 & 2 & 4 \\ -1 & 3 & -2 \end{bmatrix}_{2\times3} = \begin{bmatrix} a-b & 2a+3b & 4a-2b \end{bmatrix}_{1\times3}.$

▶ Example 5.3.6 $[a\ b\ c]_{1\times3} \cdot \begin{bmatrix} d \\ e \\ f \end{bmatrix}_{3\times1} = \begin{bmatrix} ad+be+cf \end{bmatrix}_{1\times1}.$

Observations. In the product AB we say that B is **premultiplied** by A and that A is **post multiplied** by B. Oridinarily, AB may not be equal to BA.

Suppose A has order $m \times n$ and B has order $n \times p$, then AB is defined and has order $m \times p$ but BA is not defined unless we know that $p = m$, i.e., the fact that A is conformable to B for the product AB *does not necessarily imply* that B is conformable to A for the product BA.

Now even if $m = p$, we need not have AB = BA; because the construction of the elements in the two cases are different; e.g., the element in the ith row and jth column of AB = the inner product of the ith row of A and jth column of B but the element in the ith row and jth column of BA = the inner product of ith row of B and jth column of A. **The two results, therefore, need not be the same.**

Conclusion: Matrix Multiplication is not, in general, Commutative.

See the examples given below:

▶ Example 5.3.7

$$\begin{bmatrix} 0 & 1 & 2 & 3 \\ 3 & 2 & 1 & 0 \end{bmatrix}_{2\times4} \times \begin{bmatrix} 0 & 3 \\ 1 & 2 \\ 2 & 1 \\ 3 & 0 \end{bmatrix}_{4\times2} = \begin{bmatrix} 14 & 4 \\ 4 & 14 \end{bmatrix}_{2\times2}$$

but $$\begin{bmatrix} 0 & 3 \\ 1 & 2 \\ 2 & 1 \\ 3 & 0 \end{bmatrix}_{4\times2} \times \begin{bmatrix} 0 & 1 & 2 & 3 \\ 3 & 2 & 1 & 0 \end{bmatrix}_{2\times4} = \begin{bmatrix} 9 & 6 & 3 & 0 \\ 6 & 5 & 4 & 3 \\ 3 & 4 & 5 & 6 \\ 0 & 3 & 6 & 9 \end{bmatrix}_{4\times4}.$$

The two right-hand side matrices are of different orders and hence they cannot be equal.

▶ Example 5.3.8

$$\begin{bmatrix} 2 & -1 \\ -1 & 2 \end{bmatrix} \times \begin{bmatrix} 1 & 4 \\ -1 & 1 \end{bmatrix} = \begin{bmatrix} 3 & 7 \\ -3 & -2 \end{bmatrix}$$

$$\textit{but} \begin{bmatrix} 1 & 4 \\ -1 & 1 \end{bmatrix} \times \begin{bmatrix} 2 & -1 \\ -1 & 2 \end{bmatrix} = \begin{bmatrix} -2 & 7 \\ -3 & 3 \end{bmatrix}.$$

This example shows that matrix multiplication may not be commutative even in the case of square matrices of same order, i.e., even if the conformability conditions are satisfied for AB as well as BA.

Note. The students should not have the impression that we never have AB = BA. There are cases where AB may be equal to BA. **What we wish to stress upon is that AB ≠ BA, in general.**

The following example is interesting:

▶ Example 5.3.9 $\begin{bmatrix} 1 & 2 & 0 \\ 1 & 1 & 0 \\ -1 & 4 & 0 \end{bmatrix} \times \begin{bmatrix} 0 & 0 & 0 \\ 0 & 0 & 0 \\ 2 & 3 & 9 \end{bmatrix} = \begin{bmatrix} 0 & 0 & 0 \\ 0 & 0 & 0 \\ 0 & 0 & 0 \end{bmatrix}.$

Here neither factor on the left side is a zero matrix but the product is a zero matrix. This illustrates the fact that

AB may be equal to 0 without having either A or B being 0.

Lastly we take an example to illustrate the fact

AB = AC without having B = C, even though A ≠ 0.

▶ Example 5.3.10 $A = \begin{bmatrix} 1 & 2 & 0 \\ 1 & 1 & 0 \\ -1 & 4 & 0 \end{bmatrix}, B = \begin{bmatrix} 1 & 2 & 3 \\ 1 & 1 & -1 \\ 2 & 2 & 2 \end{bmatrix}, C = \begin{bmatrix} 1 & 2 & 3 \\ 1 & 1 & -1 \\ 1 & 1 & 1 \end{bmatrix}.$

Then

$$AB = \begin{bmatrix} 3 & 4 & 1 \\ 2 & 3 & 2 \\ 3 & 2 & -7 \end{bmatrix} = AC,$$

but it is easily observed that B ≠ C (even though A ≠ 0).

Summary. In matrix multiplication

1. The rule AB = BA does not hold good in every case.

2. From AB = 0 we should not always conclude that either A = 0 or B = 0.

3. From AB = AC, or BA = CA we cannot, in general, conclude that B = C, even if A ≠ 0.

V. Two Important Theorems on Multiplication of Matrices

We shall, however, prove that the associative and distributive laws do hold in matrix multiplication:

Theorem 5.3.2 *The multiplication of matrices is associative, i.e., let* $A = (a_{ij})_{m\times n}$ *and* $B = (b_{jk})_{n\times p}$ *and let* $C = (c_{kr})_{p\times q}$. *Then this theorem states* $(AB)C = A(BC)$.

[Note here that i ranges from 1 to m, j ranges from 1 to n, k ranges from 1 to p and r ranges from 1 to q.]

Proof. By definition,

$$AB = (d_{ik})_{m\times p}, \text{ where } d_{ik} = \sum_{j=1}^{n} a_{ij}b_{jk}.$$

Again, $(AB)_{m\times p}$ and $(C)_{p\times q}$ are conformable for the product $(AB)C$ and has order $m \times q$.

$$(AB)C = (d_{ik})_{m\times p}\cdot(c_{kr})_{p\times q} = (e_{ir})_{m\times q},$$

where

$$\begin{aligned} e_{ir} &= \sum_{k=1}^{p} d_{ik}c_{kr} = \sum_{k=1}^{p}\left(\sum_{j=1}^{n} a_{ij}b_{jk}\right)c_{kr} \\ &= \sum_{k=1}^{p}\left(\sum_{j=1}^{n} a_{ij}b_{jk}c_{kr}\right) \quad \left[\text{Multiplying each term of } \sum_{j=1}^{n} a_{ij}b_{jk} \text{ by } c_{kr}\right]. \end{aligned}$$

In the same way,

$$BC = (b_{jk})_{n\times p}\cdot(c_{kr})_{p\times q} = (f_{jr})_{n\times q},$$

where

$$f_{jr} = \sum_{k=1}^{p} b_{jk}c_{kr}$$

$$\text{and } A(BC) = (a_{ij})_{m\times n}(f_{jr})_{n\times q} = (g_{ir})_{m\times q},$$

where

$$\begin{aligned} g_{ir} &= \sum_{j=1}^{n} a_{ij}f_{jr} = \sum_{j=1}^{n} a_{ij}\left(\sum_{k=1}^{p} b_{jk}c_{kr}\right) \\ &= \sum_{j=1}^{n}\left(\sum_{k=1}^{p} a_{ij}b_{jk}c_{kr}\right) \quad \left[\text{Multiplying each term of } \sum_{k=1}^{p} b_{jk}c_{kr} \text{ by } a_{ij}\right] \end{aligned}$$

In a finite sum the order of summation does not affect the total sum, so we have

$$e_{ir} = g_{ir} = \sum_{k=1}^{p}\sum_{k=1}^{n}(a_{ij}b_{jk}c_{kr}).$$

Thus, $(AB)C$ and $A(BC)$ have the same order $m \times q$ and the corresponding elements e_{ir} and g_{ir} equal; hence $(AB)C = A(BC)$.

Theorem 5.3.3 *Matrix multiplication is distributive with respect to addition.*

Let

$$A = (a_{ij})_{m\times n},\ B = (b_{jk})_{n\times p}\ \textit{and}\ C = (c_{jk})_{n\times p}.$$

Then the theorem states that $A(B+C) = AB + AC$.

Proof. $B + C = (b_{jk})_{n\times p} + (c_{jk})_{n\times p} = (d_{jk})_{n\times p}$, where $d_{jk} = b_{jk} + c_{jk}$.

Now,

$$A\cdot(B+C) = (a_{ij})_{m\times n}\cdot(d_{jk})_{n\times p} = (e_{ik})_{m\times p},$$

where

$$\begin{aligned} e_{ik} &= \sum_{j=1}^{n} a_{ij}d_{jk} = \sum_{j=1}^{n} a_{ij}(b_{jk}+c_{jk}) \\ &= \sum_{j=1}^{n}(a_{ij}b_{jk}+a_{ij}c_{jk}) = \sum_{j=1}^{n} a_{ij}b_{jk} + \sum_{j=1}^{n} a_{ij}c_{jk}. \end{aligned}$$

Again

$$AB = (a_{ij})_{m\times n}\cdot(b_{jk})_{n\times p} = (f_{ik})_{m\times p},$$

where

$$f_{ik} = \sum_{j=1}^{n} a_{ij}b_{jk}.$$

$$AC = (a_{ij})_{m\times n}\cdot(c_{jk})_{n\times p} = (g_{ik})_{m\times p},$$

where

$$g_{ik} = \sum_{j=1}^{n} a_{ij}c_{jk}.$$

$$\therefore\ AB + AC = (f_{ik})_{m\times p} + (g_{ik})_{m\times p} = (h_{ik})_{m\times p},$$

where

$$h_{ik} = f_{ik} + g_{ik} = \sum_{j=1}^{n} a_{ij}b_{jk} + \sum_{j=1}^{n} a_{ij}c_{jk}.$$

Thus, $A(B+C)$ and $AB+AC$ have the same order $m\times p$ and their corresponding elements e_{ik} and h_{ik} are equal.

$\therefore\ A(B+C) = AB + AC$ **(Left distributive law).**

One may similarly prove that, (assuming conformability),

$(D+E)F = DF + EF$ **(Right distributive law).**

▶ Example 5.3.11 *Assuming conformability, wherever necessary, prove that*

$$(A+B)\times(C+D) = AC + AD + BC + BD$$

Solution: Use the first distributive law and then the second, we thus get

$$\begin{aligned} \text{Left side} &= (A+B)\times C + (A+B)\times D \ \text{(Left distributive law)} \\ &= A\times C + B\times C + A\times D + B\times D \ \text{(Right distributive law)} \\ &= AC + AD + BC + BD. \end{aligned}$$

One can also use the second distributive law and then the first, and arrive at the same result.

5.4 The Transpose of a Matrix

Let

$$A = \begin{bmatrix} 2 & -1 & 0 \\ 0 & 2 & 4 \end{bmatrix}.$$

We construct another matrix

$$\begin{bmatrix} 2 & 0 \\ -1 & 2 \\ 0 & 4 \end{bmatrix}$$

by interchanging rows and columns of A.

This new matrix is denoted by A^T or A' and is called the **transpose of A.**

Definition. The matrix A^T (or A') of order n by m obtained by interchanging rows and columns in a matrix A of order m by n is called the transpose of A.

Observe that the element a_{ij} in the ith row and jth column of A stands in the jth row and ith column of A^T.

5.4.1 Properties of Transposes

Theorem 5.4.1 *If A^T and B^T are the transposes of the matrices A and B respectively, and if k is a scalar, then the following relations hold good:*

(i) $(A^T)^T = A$; **(ii)** $(A+B)^T = A^T + B^T$;
(iii) $(kA)^T = kA^T$; **(iv)** $(AB)^T = B^T \times A^T$.

One can easily prove (or verify) **(i)** and **(iii)**.
We, therefore, give the proofs of **(ii)** and **(iv)**.

To prove (ii): $(A+B)^T = A^T + B^T$.

Proof. Let $A = (a_{ij})_{m\times n}$ and $B = (b_{ij})_{m\times n}$.

Step 1. $A+B$ *is a matrix of order* $m \times n$.

$\therefore$ $(A+B)^T$ is a matrix of order $n \times m$.
Again, A^T is a matrix of order $n \times m$ and B^T is a matrix of order $n \times m$.
$\therefore$ $A^T + B^T$ is a matrix of order $n \times m$.
$\therefore$ we see that $(A+B)^T$ and $A^T + B^T$ are both of same order.

Step 2. *We need only to check that their corresponding elements are equal.*

The element in the ith row and jth column of A^T and B^T are respectively a_{ji} and b_{ji} ($\because$ they are elements in the jth row and ith column of A and B, respectively) and hence the element in the ith row and jth column of $A^T + B^T$ is $a_{ji} + b_{ji}$.

Now $A+B = (c_{ij})_{m\times n}$, where $c_{ij} = a_{ij} + b_{ij}$.

$\therefore$ the element in the ith row and jth column of $(A+B)^T$ is $c_{ji} = a_{ji} + b_{ji}$.
($\because$ it is the element in the jth row and ith column of $A+B$).

Hence, we see that the elements in the ith row and jth column of $A^T + B^T$ and $(A+B)^T$ are same.

$\therefore$ $(A+B)^T = A^T + B^T$. [From Steps 1 and 2]

An Important Law

To prove (iv): $(AB)^T = B^T \times A^T$: [C.P. 1980]

Proof. Let $A = (a_{ij})_{m\times n}$ and $B = (b_{jk})_{n\times p}$.

$\therefore$ $AB = (c_{ik})_{m\times p}$, where $c_{ik} = \sum_{j=1}^{n} a_{ij}b_{jk}$.

Step 1. $(AB)^T$ *is a matrix of order* p *by* m *since* AB *is a matrix of order* $m \times p$.

Since B^T is a matrix of order $p \times n$ and A^T is a matrix of order $n \times m$, we find that they are conformable for the product $B^T \cdot A^T$ and this product is a matrix of order p by m.

$\therefore$ $(AB)^T$ and $B^T \cdot A^T$ are matrices of same order p by m.

Step 2. *We need now to check that the corresponding elements of* $(AB)^T$ *and* $B^T A^T$ *are same.*

The element in the kth row and ith column of $B^T A^T$

$$= \text{the inner product of the kth row of } B^T \text{ and ith column of } A^T$$

$$= \text{the inner product of the kth column of } B \text{ and ith row of } A$$

$$= a_{i1}b_{1k} + a_{i2}b_{2k} + \ldots + a_{in}b_{nk} = \sum_{j=1}^{n} a_{ij}b_{jk}$$

[Note that the elements of ith row of A are $a_{i1}, a_{i2}, \ldots, a_{in}$ and the elements of kth column of B are $b_{1k}, b_{2k}, \ldots, b_{nk}$]

Again, the element in the kth row and ith column of $(AB)^T$

$$= \text{the element in the kth column and ith row of } AB$$

$$= \text{the inner product of the ith row of } A \text{ and kth column of } B$$

$$= \sum_{j=1}^{n} a_{ij}b_{jk} \text{ (as before).}$$

$\therefore$ the corresponding elements of $(AB)^T$ and $B^T A^T$ are same.

$\therefore$ $(AB)^T = B^T A^T$ ($\because$ they are of same order p by m and their corresponding elements are equal).

▶ Example 5.4.1 (a) *Let* $A = \begin{bmatrix} 1 & -1 & 2 \\ 2 & 1 & 0 \end{bmatrix}$ *and* $B = \begin{bmatrix} 1 & 3 \\ 2 & 0 \\ -1 & 1 \end{bmatrix}$.

$\therefore$ $AB = \begin{bmatrix} -3 & 5 \\ 4 & 6 \end{bmatrix}$ and hence $(AB)^T = \begin{bmatrix} -3 & 4 \\ 5 & 6 \end{bmatrix}$.

But $B^T = \begin{bmatrix} 1 & 2 & -1 \\ 3 & 0 & 1 \end{bmatrix}$ and $A^T = \begin{bmatrix} 1 & 2 \\ -1 & 1 \\ 2 & 0 \end{bmatrix}$

$\therefore$ $B^T A^T = \begin{bmatrix} -3 & 4 \\ 5 & 6 \end{bmatrix}$

$\therefore$ it is verified that $B^T A^T = (AB)^T$ in this case

(b) *If* $A = \begin{bmatrix} 1 & 1 & 2 \\ 2 & -1 & 3 \end{bmatrix}$ *and* $B = \begin{bmatrix} 1 & -2 \\ -2 & 0 \\ 1 & 1 \end{bmatrix}$, *verify that* $B^TA^T = (AB)^T$.

[C.P. 2004]

Exactly similar method.

▶ Example 5.4.2 *Prove* $(ABC)^T = C^TB^TA^T$.

Proof. Let us write $ABC = (AB)C$.

$$\begin{aligned} \therefore (ABC)^T &= \{(AB)C\}^T = C^T(AB)^T \\ &= C^T(B^TA^T) = C^TB^TA^T \text{ [Using **Property (iv)** of transposes]} \end{aligned}$$

5.5 Special Types of Matrices

I. Diagonal Matrix. Scalar Matrix. Identity (or Unit) Matrix.

(i) If in a square matrix A all the elements are zero *excepting the elements of the main diagonal* then we call it a **diagonal matrix.** Such a matrix is of the form

$$A = \begin{bmatrix} a_{11} & 0 & 0 & 0 & \dots & 0 \\ 0 & a_{22} & 0 & 0 & \dots & 0 \\ 0 & 0 & a_{23} & 0 & \dots & 0 \\ \dots & \dots & \dots & \dots & \dots & \dots \\ 0 & 0 & 0 & 0 & \dots & a_{nn} \end{bmatrix}_{n\times n}.$$

(ii) A square matrix in which each element of the main diagonal equals a scalar α (say) and all other elements are zero, is called a *scalar matrix.*

e.g., $\begin{bmatrix} \alpha & 0 & 0 & 0 & \dots & 0 \\ 0 & \alpha & 0 & 0 & \dots & 0 \\ \dots & \dots & \dots & \dots & \dots & \dots \\ 0 & 0 & 0 & 0 & \dots & \alpha \end{bmatrix}_{n\times n}$ is a **scalar matrix.**

Thus, a scalar matrix is a diagonal matrix whose (main) diagonal elements are all equal, α, say.

In matrix multiplication, a scalar matrix behaves just like a scalar; for example, see that

$$\begin{bmatrix} \alpha & 0 & 0 \\ 0 & \alpha & 0 \\ 0 & 0 & \alpha \end{bmatrix} \begin{bmatrix} a_{11} & a_{12} & a_{13} \\ a_{21} & a_{22} & a_{23} \\ a_{31} & a_{32} & a_{33} \end{bmatrix} = \begin{bmatrix} \alpha a_{11} & \alpha a_{12} & \alpha a_{13} \\ \alpha a_{21} & \alpha a_{22} & \alpha a_{23} \\ \alpha a_{31} & \alpha a_{32} & \alpha a_{33} \end{bmatrix}$$

$$= \alpha \begin{bmatrix} a_{11} & a_{12} & a_{13} \\ a_{21} & a_{22} & a_{23} \\ a_{31} & a_{32} & a_{33} \end{bmatrix}.$$

(iii) The scalar matrix of order n corresponding to the scalar 1 (unity) is called an **Identity Matrix** or **Unit Matrix** of order n, denoted by I_n. We may simply write I instead of I_n when the order need not be emphasised. Thus,

$$I_2 = \begin{bmatrix} 1 & 0 \\ 0 & 1 \end{bmatrix},\ I_3 = \begin{bmatrix} 1 & 0 & 0 \\ 0 & 1 & 0 \\ 0 & 0 & 1 \end{bmatrix}, \text{ etc.}$$

If A is a square matrix of order n and I is the unit matrix of order n, then

$$AI = IA = A.$$

The student can easily verify

$$\begin{bmatrix} \alpha & 0 & 0 & \dots & 0 \\ 0 & \alpha & 0 & \dots & 0 \\ \dots & \dots & \dots & \dots & \dots \\ 0 & 0 & 0 & \dots & \alpha \end{bmatrix}_{n\times n} = \alpha I_n.$$

Thus, every scalar matrix is related to the unit matrix by such a relation.

II. Given a square matrix A of order n, we define

$$A^2 = A\cdot A,\ A^3 = A^2\cdot A,\ A^4 = A^3\cdot A \text{ and so on.}$$

We may write, for convenience, $A^\circ = I$.

If $a_\circ, a_1, a_2, \dots, a_n$ are scalars we may form the matrix expression

$$a_\circ I + a_1 A + a_2 A^2 + \dots + a_n A^n;\ (a_n \neq 0).$$

We call this expression **a polynomial in the square matrix A of degree n.** Just like we can write $f(x) = x^2 - 5x + 6$, we may have $f(A) = A^2 - 5A + 6I$. Again, since $f(x) = (x-2)(x-3)$ we may write here $f(A) = (A-2I)(A-3I)$.

▶ Example 5.5.1 $x^2+1=0$ *has two solutions* i *and* $-i$. $A^2 + I_2 = 0$ *has two solutions* iI_2 *and* $-iI_2$,

$$\textit{i.e.,} \begin{bmatrix} i & 0 \\ 0 & i \end{bmatrix} \textit{and} \begin{bmatrix} -i & 0 \\ 0 & -i \end{bmatrix}.$$

Solution: But it may be observed that besides these two solutions, the matrix equation may have other solutions; for example $A^2 + I_2 = 0$ has a solution $\begin{bmatrix} 1 & 2 \\ -1 & -1 \end{bmatrix}$.

One can easily check.

We just remark here that:

The fact that a second degree equation in x has two and only two roots does not hold true in Matrix Algebra, where the number of roots is at least two.

▶ Example 5.5.2 *If* $A = \begin{bmatrix} 1 & 2 \\ -3 & 0 \end{bmatrix}$, *what is the matrix corresponding to the polynomial* $A^2 + 3A + 5I$.

Solution:

$$A^2 = A\cdot A = \begin{bmatrix} 1 & 2 \\ -3 & 0 \end{bmatrix}\begin{bmatrix} 1 & 2 \\ -3 & 0 \end{bmatrix}$$

$$= \begin{bmatrix} -5 & 2 \\ -3 & -6 \end{bmatrix}.$$

$$\therefore\ A^2 + 3A + 5I = \begin{bmatrix} -5 & 2 \\ -3 & -6 \end{bmatrix} + \begin{bmatrix} 3 & 6 \\ -9 & 0 \end{bmatrix} + \begin{bmatrix} 5 & 0 \\ 0 & 5 \end{bmatrix}$$

$$= \begin{bmatrix} 3 & 8 \\ -12 & -1 \end{bmatrix}.$$

III. If $AB = BA$, we say that A and B **commute.**
If $AB = -BA$, we say that A and B **anticommute.**

▶ Example 5.5.3 *Let* $\sigma_1 = \begin{bmatrix} 0 & 1 \\ 1 & 0 \end{bmatrix}$, $\sigma_2 = \begin{bmatrix} 0 & -i \\ i & 0 \end{bmatrix}$, $\sigma_3 = \begin{bmatrix} 1 & 0 \\ 0 & -1 \end{bmatrix}$, *where* $i^2 = -1$.

Verify that each of the matrices anticommutes with the others. (*These are called* **Pauli Spin Matrices** *used in the study of electron spin in Quantum Mechanics*).

IV. A square matrix A such that $A^p = 0$, for some positive integer p is called **Nilpotent.** If p is the least positive integer for which $A^p = 0$, then A is said to be **nilpotent of index p.**

▶ Example 5.5.4 $A = \begin{bmatrix} 1 & 1 & 3 \\ 5 & 2 & 6 \\ -2 & -1 & -3 \end{bmatrix}$ *is nilpotent of index* 3.

Verify that $A^2 \neq 0$, $A^3 = 0$; hence A is a nilpotent matrix of index 3.

V. A square matrix A is called **idempotent matrix** if $A^2 = A$.

▶ Example 5.5.5 $A = \begin{bmatrix} 2 & -2 & -4 \\ -1 & 3 & 4 \\ 1 & -2 & -3 \end{bmatrix}$ *is idempotent, since* $A^2 = A$. (*Verify*).

VI. Symmetric and Skew-symmetric Matrices

Let us take a square matrix $A = (a_{ij})_{n \times n}$.
A square matrix A is called **symmetric** if $A^T = A$;
i.e., if $a_{ij} = a_{ji}$ for all pairs of subscripts i and j.
On the other hand A is called **skew-symmetric** if $A^T = -A$, i.e., if $a_{ij} = -a_{ji}$ for all i, j. (Often instead of writing A^T we may use the notion A' for the transpose of A)

(i) A symmetric matrix is of the form

$$\begin{bmatrix} 5 & 2 \\ 2 & 3 \end{bmatrix}, \begin{bmatrix} 0 & 1 & 2 \\ 1 & 2 & 8 \\ 2 & 8 & 4 \end{bmatrix}, \begin{bmatrix} a & h & g \\ h & b & f \\ g & f & c \end{bmatrix} \text{ etc.}$$

that is, they are *symmetrical about the main diagonal.*

(ii) *In a skew-symmetric matrix the elements of the main diagonal are all zero* ($\because a_{ij} = -a_{ji}$ and so $a_{ii} = -a_{ii}$; or $a_{ii} = 0$), e.g.,

$$\begin{bmatrix} 0 & a \\ -a & 0 \end{bmatrix}, \begin{bmatrix} 0 & 1 & -2 \\ -1 & 0 & 5 \\ 2 & -5 & 0 \end{bmatrix}, \begin{bmatrix} 0 & a & b & c \\ -a & 0 & d & e \\ -b & -d & 0 & f \\ -c & -e & -f & 0 \end{bmatrix}$$

(iii) *If A is a skew-symmetric matrix, then* A^2 *is symmetric.*

$$[\because A^2 = A \cdot A = -A^T \cdot -A^T = (A^T)^2 = (A^2)^T].$$

▶ **Example 5.5.6** *Show that if A is any matrix, then the matrices A′A and AA′ are both symmetric.*

Solution: Let A be a matrix of order $m \times n$.

Then A' is a matrix of order $n \times m$.

$\therefore$ AA' is a square matrix of order $m \times m$ and $A'A$ is also a square matrix of order $n \times n$.

Now $(AA')' = (A')' \cdot A' = A \cdot A'$, so that AA' is symmetric and

$$(A'A)' = A' \cdot (A')' = A' \cdot A,$$

so that $A'A$ is also symmetric.

An Important Theorem

Theorem 5.5.1 *A square matrix A can be expressed in a unique way as a sum of a symmetric matrix and a skew-symmetric matrix.* [C.P. 2002]

Proof. Let A be any square matrix. We can write

$$A = \frac{1}{2}(A + A') + \frac{1}{2}(A - A') = P + Q \text{ (say)},$$

where

$$P = \frac{1}{2}(A + A') \text{ and } Q = \frac{1}{2}(A - A').$$

See that

$$P' = \frac{1}{2}(A + A')' = \frac{1}{2}[A' + (A')'] = \frac{1}{2}(A' + A) = P;$$

$$Q' = \frac{1}{2}(A - A')' = \frac{1}{2}[A' - (A')'] = \frac{1}{2}(A' - A) = -\frac{1}{2}(A - A') = -Q$$

so that P is a symmetric matrix and Q is a skew-symmetric matrix. Thus we can express any square matrix A on the sum of a symmetric and a skew-symmetric matrix.

To prove that the representation is **unique**, let $A = R + S$ where R is symmetric and S is skew-symmetric. Then we are to prove that $R = P$ and $S' = Q$.

We have

$$A' = (R + S)' = R' + S' = R - S \ (\because\ R' = R \text{ and } S' = S).$$

$$\therefore \quad A + A' = 2R \text{ and } A - A' = 2S,$$

$$\text{i.e.,} \quad R = \frac{1}{2}(A + A') = P, \ \ S = \frac{1}{2}(A - A') = Q.$$

Hence, the representation is unique.

▶ **Example 5.5.7** *Let* $A = \begin{bmatrix} 1 & 0 & 0 \\ -1 & -2 & -1 \\ -2 & 3 & -2 \end{bmatrix}$. *Verify* $A + A'$ *is symmetric and* $A - A'$ *is skew-symmetric and express A as the sum of a symmetric and a skew-symmetric matrix.* [C.P. 1999]

VII. Inverse of a Square Matrix

Definition. If A and B be two square matrices of order n such that $AB = BA = I_n$ (I_n = Identity matrix of order n) then B is called an inverse of A and we write $B = A^{-1}$. One may also say that A is an inverse of B and write $A = B^{-1}$.

Thus,

$$AA^{-1} = A^{-1}A = I_n.\ B^{-1}B = BB^{-1} = I_n.$$

▶ Example 5.5.8 *Let* $A = \begin{bmatrix} 1 & 2 & 3 \\ 1 & 3 & 3 \\ 1 & 2 & 4 \end{bmatrix}$ *and* $B = \begin{bmatrix} 6 & -2 & -3 \\ -1 & 1 & 0 \\ -1 & 0 & 1 \end{bmatrix}$.

Verify that $AB = BA = \begin{bmatrix} 1 & 0 & 0 \\ 0 & 1 & 0 \\ 0 & 0 & 1 \end{bmatrix} = I_3$.

Then we may conclude, $A = B^{-1}$ or $B = A^{-1}$.

▶ Example 5.5.9 *If* $A = \begin{bmatrix} 3 & 1 \\ -1 & 2 \end{bmatrix}$, *show that A satisfies* $A^2 - 5A + 7I_2 = 0$. *Hence, deduce the inverse of A.*

Since, $A = \begin{bmatrix} 3 & 1 \\ -1 & 2 \end{bmatrix}$;

$$A^2 = A \cdot A = \begin{bmatrix} 3 & 1 \\ -1 & 2 \end{bmatrix}\begin{bmatrix} 3 & 1 \\ -1 & 2 \end{bmatrix} = \begin{bmatrix} 8 & 5 \\ -5 & 3 \end{bmatrix}.$$

$$\therefore\ A^2 - 5A + 7I_2 = \begin{bmatrix} 8 & 5 \\ -5 & 3 \end{bmatrix} - \begin{bmatrix} 15 & 5 \\ -5 & 10 \end{bmatrix} + \begin{bmatrix} 7 & 0 \\ 0 & 7 \end{bmatrix} = \begin{bmatrix} 0 & 0 \\ 0 & 0 \end{bmatrix} = 0.$$

Now,

$$A^2 - 5A = -7I_2;\ \text{i.e.,}\ -\frac{1}{7}A(A - 5I_2) = I_2.$$

Hence, the inverse of A is

$$\begin{aligned} -\frac{1}{7}(A - 5I_2) &= -\frac{1}{7}\left\{\begin{bmatrix} 3 & 1 \\ -1 & 2 \end{bmatrix} - \begin{bmatrix} 5 & 0 \\ 0 & 5 \end{bmatrix}\right\} \\ &= -\frac{1}{7}\begin{bmatrix} -2 & 1 \\ -1 & -3 \end{bmatrix} = \begin{bmatrix} 2/7 & -1/7 \\ 1/7 & 3/7 \end{bmatrix}. \end{aligned}$$

Check.

$$\begin{aligned} \begin{bmatrix} 3 & 1 \\ -1 & 2 \end{bmatrix}\begin{bmatrix} 2/7 & -1/7 \\ 1/7 & 3/7 \end{bmatrix} &= \begin{bmatrix} 2/7 & -1/7 \\ 1/7 & 3/7 \end{bmatrix}\begin{bmatrix} 3 & 1 \\ -1 & 2 \end{bmatrix} \\ &= \begin{bmatrix} 1 & 0 \\ 0 & 1 \end{bmatrix} = I_2. \end{aligned}$$

We shall prove that **not every square matrix has an inverse. Only non singular square matrices have inverses,** (i.e., only those square matrices whose determinants $\neq 0$ have inverses)

Moreover, **if an inverse of A exists, then this is unique, i.e., we shall never get more than one inverse of a given square matrix** A.

Hence, we use the term **the inverse** of A, instead of **an inverse** of A. A matrix with an inverse is said to be **invertible.**

Theorem 5.5.2

(i) *If* A^{-1} *exists, then* $(A^{-1})^{-1} = A$;

(ii) *Also* $(A')^{-1}$ *exists and* $(A')^{-1} = (A^{-1})'$,

where A *is any square matrix of order* n *and* A' *is the transpose of* A.

Proof.

(i) Since A^{-1} exists, by definition $AA^{-1} = A^{-1}A = I$.

This shows that A^{-1} has A as *an inverse* and since inverse, when exists, is unique, we may write $(A^{-1})^{-1} = A$.

(ii) Now $(AA^{-1})' = (A^{-1}A)' = I' = I$ and so $(A^{-1})'A' = A'(A^{-1})' = I$

$\therefore$ A' has $(A^{-1})'$ as an inverse and since inverse when exists, is unique, therefore,

$$(A')^{-1} = (A^{-1})'.$$

Theorem 5.5.3 *If* A^{-1} *and* B^{-1} *exist, where* A *and* B *are both square matrices of order* $n \times n$, *then* $(AB)^{-1} = B^{-1} \cdot A^{-1}$.

Proof.

$$\begin{aligned}
(AB)(B^{-1}A^{-1}) &= A(BB^{-1})A^{1} \text{ (Associative Law)} \\
&= A(I)A^{-1} \ (\because\ BB^{-1} = I) \\
&= (AI)A^{-1} \text{ (Associative Law)} \\
&= AA^{-1} \ (\because\ AI = A) \\
&= I \\
\text{and } (B^{-1}A^{-1})(AB) &= B^{-1}(A^{-1}A)B = B^{-1}(IB) \\
&= B^{-1}B = I.
\end{aligned}$$

$\therefore$ AB has $B^{-1}A^{-1}$ as an inverse and by uniqueness property we write

$$(AB)^{-1} = B^{-1}A^{-1}.$$

Extension of the property described in Theorem 5.5.3.

For three non-singular square matrices A, B, C we have

$$(A \cdot B \cdot C)^{-1} = C^{-1} \cdot B^{-1} \cdot A^{-1}$$

Proof. Since A, B, C are all non-singular square matrices, therefore, their inverses A^{-1}, B^{-1}, C^{-1} all exist so that

$$A \cdot A^{-1} = A^{-1} \cdot A = I,\ B \cdot B^{-1} = B^{-1}B = I,\ C \cdot C^{-1} = C^{-1} \cdot C = I.$$

$$\begin{aligned}\therefore \quad & (C^{-1}\cdot B^{-1}\cdot A^{-1})\cdot(A\cdot B\cdot C) \\ & = (C^{-1}B^{-1})\cdot(A^{-1}A)\cdot(BC) \text{ [Associative Property]} \\ & = C^{-1}B^{-1}\cdot I\cdot(BC) = C^{-1}B^{-1}\cdot(BC) \\ & = C^{-1}(B^{-1}B)C = C^{-1}\, I\, C \\ & = C^{-1}C = I.\end{aligned}$$

Similarly, $(ABC)\cdot(C^{-1}B^{-1}A^{-1}) = I$, i.e., $C^{-1}B^{-1}A^{-1}$ is the inverse of ABC,

$$\text{or, } \mathbf{(ABC)^{-1} = C^{-1}B^{-1}A^{-1}}.$$

Characteristic Properties of Inverses

1. *The inverse of a square matrix, whenever exists, is unique.* **(Uniqueness Theorem)**

2. *Only those square matrices whose determinants $\neq 0$ possess inverses.* **(Existence Theorem)**

Definition. A square matrix A is called ***non-singular*** if the determinant of the matrix $A \neq 0$. But if $\det A = 0$, then the matrix is a ***singular matrix.***

Theorem 5.5.4 **Uniqueness Theorem.** *Let A be a non-singular square matrix of order n. Assume that there exists an inverse A^{-1} of the matrix A so that*

$$A^{-1}\cdot A = A\cdot A^{-1} = I_n. \text{ (Identity matrix of order } n) \tag{1}$$

Suppose there exists another matrix X which has the property

$$A\cdot X = X\cdot A = I_n. \tag{2}$$

Then we shall prove $X = A^{-1}$. In other words, if the inverse of A exists, then it is unique.

Proof. Using Distributive Law and the two results (1) and (2) given above, we obtain

$$\begin{aligned} & A\cdot(X - A^{-1}) = A\,X - A\cdot A^{-1} = I_n - I_n = 0 \\ \therefore \quad & A^{-1}\{A(X - A^{-1})\} = A^{-1}\cdot 0 = 0 \\ \text{or,} \quad & (A^{-1}A)(X - A^{-1}) = 0 \text{ (using Associative Law)} \\ \text{or,} \quad & I_n(X - A^{-1}) = 0 \text{ [using (i)]} \\ \therefore \quad & X - A^{-1} = 0 \text{ or, } \mathbf{X = A^{-1}} \textbf{ (Proved).}\end{aligned}$$

Theorem 5.5.5 **Existence Theorem.** *If A is a non-singular square matrix, there exists an inverse of A and conversely, if inverse of A exists then A must be non-singular. In other words:*

A square matrix A has an inverse if and only if $\det A \neq 0$.

Step 1. *In order to show the existence of an inverse of A, we first introduce the concept of **adjoint of A** (written as **adj A**):*

Let $A = (a_{ij})_{n\times n}$ be a square matrix of order n.

We denote the cofactor of a_{ij} in $\det A$ by A_{ij}. We construct a **matrix of the cofactors of the elements of A** and then **transpose** it. Such a matrix is called the **Adjugate or Adjoint of A**. That is,

adj A = transposed matrix of the cofactors of A.

Step 2. *We now establish the following important result:*

$$\mathbf{A}\cdot \mathit{adj}\,\mathbf{A} = \mathit{adj}\,\mathbf{A}\cdot\mathbf{A} = \det\mathbf{A}\cdot\mathbf{I_n},$$

where I_n *is the unit matrix of order* n *and* $\det A$ *is the determinant of* A.

Before we establish the result in **Step 2**, we give an illustration.

Let $A = \begin{pmatrix} 1 & 2 \\ 3 & 5 \end{pmatrix}$. See that $\text{adj}\,A = \begin{pmatrix} 5 & -2 \\ -3 & 1 \end{pmatrix}$ and $|A| = -1$.

Now,

$$A\cdot\frac{\text{adj}\,A}{|A|} = \begin{pmatrix} 1 & 2 \\ 3 & 5 \end{pmatrix}\cdot\frac{1}{-1}\begin{pmatrix} 5 & -2 \\ -3 & 1 \end{pmatrix}$$

$$= \begin{pmatrix} 1 & 2 \\ 3 & 5 \end{pmatrix}\cdot\begin{pmatrix} -5 & 2 \\ 3 & -1 \end{pmatrix} = \begin{bmatrix} 1 & 0 \\ 0 & 1 \end{bmatrix} = I_2.$$

Similar $\dfrac{\text{adj}\,A}{|A|}\cdot A = I_2$.

Hence we observe that for the given matrix A,

$$A\cdot\frac{\text{adj}\,A}{|A|} = \frac{\text{adj}\,A}{|A|}\cdot A = I_2. \qquad \text{[C.P. 2001]}$$

Proof of Step 2. We see that **A· adj A =**

$$\begin{bmatrix} a_{11} & a_{12} & a_{13} & \cdots & a_{1n} \\ a_{21} & a_{22} & a_{23} & \cdots & a_{2n} \\ \cdots & \cdots & \cdots & \cdots & \cdots \\ a_{n1} & a_{n2} & a_{n3} & \cdots & a_{nn} \end{bmatrix}\begin{bmatrix} A_{11} & A_{21} & A_{31} & \cdots & A_{n1} \\ A_{12} & A_{22} & A_{32} & \cdots & A_{n2} \\ \cdots & \cdots & \cdots & \cdots & \cdots \\ A_{1n} & A_{2n} & A_{3n} & \cdots & A_{nn} \end{bmatrix}$$

$$= \begin{bmatrix} a_{11}A_{11} + a_{12}A_{12} + \ldots + a_{1n}A_{1n} & \cdots & a_{11}A_{n1} + a_{12}A_{n2} + \ldots + a_{1n}A_{nn} \\ a_{21}A_{11} + a_{22}A_{12} + \ldots + a_{2n}A_{2n} & \cdots & a_{21}A_{n1} + a_{22}A_{n2} + \ldots + a_{2n}A_{nn} \\ \ldots\ \ldots\ \ldots\ \ldots\ \ldots & \cdots & \ldots\ \ldots\ \ldots\ \ldots\ \ldots \\ a_{n1}A_{11} + a_{n2}A_{12} + \ldots + a_{nn}A_{1n} & \cdots & a_{n1}A_{n1} + a_{n2}A_{n2} + \ldots + a_{nn}A_{nn} \end{bmatrix}$$

$$= \begin{bmatrix} \det A & 0 & 0 & \cdots & 0 \\ 0 & \det A & 0 & \cdots & 0 \\ \cdots & \cdots & \cdots & \cdots & \cdots \\ 0 & 0 & 0 & \cdots & \det A \end{bmatrix} \left(\because \Sigma a_{ij}A_{ij} \begin{array}{l} = D \text{ if } i = j \\ = 0 \text{ if } i \neq j \end{array} \right) \qquad (3)$$

$$= \det A \begin{bmatrix} 1 & 0 & 0 & \cdots & 0 \\ 0 & 1 & 0 & \cdots & 0 \\ \cdots & \cdots & \cdots & \cdots & \cdots \\ 0 & 0 & 0 & \cdots & 1 \end{bmatrix} = \mathbf{\det A\cdot I_n}.$$

We may also prove that $(\text{adj}\,A)\cdot A = \det A\cdot I_n$.

Consequently

$$\mathbf{A\cdot(adj\,A) = (adj\,A)\cdot A = \det A\cdot I_n}. \qquad (4)$$

Step 3. *With this knowledge in mind we can prove the* Existence Theorem *which states:*

A Square matrix A has an inverse if and only if det A $\neq 0$.

Sufficient Condition: Let A be a non-singular matrix so that det A $\neq 0$ and let us define a matrix B as follows:

$$B = \frac{\operatorname{adj} A}{\det A} = \frac{1}{\det A}\begin{bmatrix} A_{11} & A_{21} & A_{31} & \dots & A_{n1} \\ A_{12} & A_{22} & A_{32} & \dots & A_{n2} \\ \dots & \dots & \dots & \dots & \dots \\ A_{1n} & A_{2n} & A_{3n} & \dots & A_{nn} \end{bmatrix}. \tag{5}$$

Then from (4) (dividing both sides by det A) we get

$$A\cdot\frac{\operatorname{adj} A}{\det A} = \frac{\operatorname{adj} A}{\det A}\cdot A = I_n, \text{ or, } A\cdot B = B\cdot A = I_n.$$

Hence, B [as defined by (5)] is an inverse of A.

Conversely, if B is any square matrix of order n satisfying

$$A\cdot B = B\cdot A = I_n$$

$\therefore$ $\det(AB) = \det(BA) = \det I_n$

or, $\det A\cdot \det B = 1$ ($\because$ $\det(AB) = \det A\cdot\det B$).

$\therefore$ det A cannot be zero,

i.e., if an inverse of A exists, then A is non-singular. (Thus det A $\neq 0$ is the necessary condition for the existence of A^{-1}).

Note. By Uniqueness Theorem, B is the only inverse of A.

Thus, for a non-singular square matrix A of order n there exists a unique square matrix B of order n such that AB = BA = I_n. B is then the inverse of A and we write B = A^{-1}.

▶ Example 5.5.10 (a) *Find the inverse of the matrix* $A = \begin{bmatrix} 5 & 3 \\ -2 & 2 \end{bmatrix}$. **[C.P. 2005]**

Solution:

Step 1. $\det A = \begin{vmatrix} 5 & 3 \\ -2 & 2 \end{vmatrix} = 5.2 - (-2)(3) = 16(\neq 0)$. *Since* det A $\neq 0$, *the inverse of A exists.*

Step 2. *Matrix of the cofactors of* $A = \begin{bmatrix} 2 & +2 \\ -3 & 5 \end{bmatrix}$.

adj A = *transpose of the matrix of cofactors of* $A = \begin{bmatrix} 2 & -3 \\ 2 & 5 \end{bmatrix}$.

$\therefore$ inverse of $A = A^{-1} = \frac{1}{\det A}\operatorname{adj} A = \frac{1}{16}\begin{bmatrix} 2 & -3 \\ 2 & 5 \end{bmatrix}$ **(Reqd. Answer)**

(b) *To find the inverse of the matrix*

$$A = \begin{bmatrix} 8 & 4 & 2 \\ 2 & 8 & 4 \\ 1 & 2 & 8 \end{bmatrix}.$$

Step 1.

$$\det A = \begin{vmatrix} 8 & 4 & 2 \\ 2 & 8 & 4 \\ 1 & 2 & 8 \end{vmatrix} = 8\begin{vmatrix} 8 & 4 \\ 2 & 8 \end{vmatrix} - 4\begin{vmatrix} 2 & 4 \\ 1 & 8 \end{vmatrix} + 2\begin{vmatrix} 2 & 8 \\ 1 & 2 \end{vmatrix}$$

$$= 8(64-8) - 4(16-4) + 2(4-8) = 392.$$

Then, $\det A \neq 0$; hence, inverse of A exists.

Step 2. *Matrix of the cofactors of A*

$$= \begin{bmatrix} +\begin{vmatrix} 8 & 4 \\ 2 & 8 \end{vmatrix} & -\begin{vmatrix} 2 & 4 \\ 1 & 8 \end{vmatrix} & +\begin{vmatrix} 2 & 8 \\ 1 & 2 \end{vmatrix} \\ -\begin{vmatrix} 4 & 2 \\ 2 & 8 \end{vmatrix} & +\begin{vmatrix} 8 & 2 \\ 1 & 8 \end{vmatrix} & -\begin{vmatrix} 8 & 4 \\ 1 & 2 \end{vmatrix} \\ +\begin{vmatrix} 4 & 2 \\ 8 & 4 \end{vmatrix} & -\begin{vmatrix} 8 & 2 \\ 2 & 4 \end{vmatrix} & +\begin{vmatrix} 8 & 4 \\ 2 & 8 \end{vmatrix} \end{bmatrix} = \begin{bmatrix} 56 & -12 & -4 \\ -28 & 62 & -12 \\ 0 & -28 & 56 \end{bmatrix}.$$

$$\therefore \text{ adj } A = \text{transpose of matrix of cofactors of } A = \begin{bmatrix} 56 & -28 & 0 \\ -12 & 62 & -28 \\ -4 & -12 & 56 \end{bmatrix}.$$

$$\therefore \text{ inverse of } A = A^{-1} = \frac{1}{\det A} \text{ adj } A$$

$$= \frac{1}{392}\begin{bmatrix} 56 & -28 & 0 \\ -12 & 62 & -28 \\ -4 & -12 & 56 \end{bmatrix} = \begin{bmatrix} \frac{1}{7} & -\frac{1}{14} & 0 \\ -\frac{3}{98} & \frac{31}{196} & -\frac{1}{14} \\ -\frac{1}{98} & -\frac{3}{98} & \frac{1}{7} \end{bmatrix}.$$

Check. $A \cdot A^{-1} = \begin{bmatrix} 8 & 4 & 2 \\ 2 & 8 & 4 \\ 1 & 2 & 8 \end{bmatrix}\begin{bmatrix} \frac{1}{7} & -\frac{1}{14} & 0 \\ -\frac{3}{98} & \frac{31}{196} & -\frac{1}{14} \\ -\frac{1}{98} & -\frac{3}{98} & \frac{1}{7} \end{bmatrix} = \begin{bmatrix} 1 & 0 & 0 \\ 0 & 1 & 0 \\ 0 & 0 & 1 \end{bmatrix}.$

▶ Example 5.5.11 *To find the inverse of*

$$A = \begin{bmatrix} 1 & 4 & 0 \\ -1 & 2 & 2 \\ 0 & 0 & 2 \end{bmatrix}.$$

Step 3. $\det A = 12$. *(verify), so that the determinant is not zero and hence, A^{-1} exists.*

Step 4. *adj* $A = \begin{bmatrix} 4 & -8 & 8 \\ 2 & 2 & -2 \\ 0 & 0 & 6 \end{bmatrix}$; *verify.*

$$\therefore A^{-1} = \begin{bmatrix} \frac{1}{3} & -\frac{2}{3} & \frac{2}{3} \\ \frac{1}{6} & \frac{1}{6} & -\frac{1}{6} \\ 0 & 0 & \frac{1}{2} \end{bmatrix}.$$

Check. $AA^{-1} = A^{-1}A = I$.

VIII. Orthogonal Matrix

A square matrix A of order n is called an **orthogonal matrix** if

$$A \cdot A^T = A^T \cdot A = I.$$

By the property of Uniqueness of Inverse we now conclude $A^T = A^{-1}$.

▶ Example 5.5.12 *Verify that*

$$A_1 = \begin{bmatrix} 1 & 6 \\ 0 & -1 \end{bmatrix} \text{ and } A_2 = \frac{1}{3}\begin{bmatrix} -1 & 2 & -2 \\ -2 & 1 & 2 \\ 2 & 2 & 1 \end{bmatrix}$$

are both orthogonal matrices.

Solution: We need only to check that $A_1 \cdot A_1^T = I_2$ and $A_2 \cdot A_2^T = I_3$. See that

$$A_1 \cdot A_1^T = \begin{bmatrix} 1 & 0 \\ 0 & -1 \end{bmatrix}\begin{bmatrix} 1 & 0 \\ 0 & -1 \end{bmatrix} = \begin{bmatrix} 1 & 0 \\ 0 & 1 \end{bmatrix} = I_2.$$

Again

$$A_2 \cdot A_2^T = \frac{1}{3}\begin{bmatrix} -1 & 2 & -2 \\ -2 & 1 & 2 \\ 2 & 2 & 1 \end{bmatrix} \cdot \frac{1}{3}\begin{bmatrix} -1 & -2 & 2 \\ 2 & 1 & 2 \\ -2 & 2 & 1 \end{bmatrix} = \begin{bmatrix} 1 & 0 & 0 \\ 0 & 1 & 0 \\ 0 & 0 & 1 \end{bmatrix} = I_3.$$

IX. Solutions of a set of Linear Equations by inverting the Coefficient Matrix.

We consider the problem of solving simultaneously a set of n linear equations in n unknowns:

$$\begin{aligned} a_{11} \; x_1 + a_{12} \; x_2 + a_{13} \; x_3 + \ldots + a_{1n} \; x_n &= b_1 \\ a_{21} \; x_1 + a_{22} \; x_2 + a_{23} \; x_3 + \ldots + a_{2n} \; x_n &= b_2 \\ \ldots \quad \ldots\ldots \quad \ldots\ldots \quad \ldots\ldots \quad &\ldots \\ a_{n1} \; x_1 + a_{n2} \; x_2 + a_{n3} \; x_3 + \ldots + a_{nn} \; x_n &= b_n. \end{aligned} \tag{S}$$

[$x_1, x_2, \ldots, x_n$ are n unknowns; a_{ij}, b_j are all constants.]

The problem is to find all possible sets of values of $x_1, x_2, \ldots, x_n$ which will simultaneously satisfy all these n equations. Such a set of values is called a **solution of the system (S)**.

The coefficients of this system define what we call the **coefficient matrix** A, thus

$$A = \begin{bmatrix} a_{11} & a_{12} & \ldots & a_{1n} \\ a_{21} & a_{22} & \ldots & a_{2n} \\ \ldots & \ldots & \ldots & \ldots \\ a_{n1} & a_{n2} & \ldots & a_{nn} \end{bmatrix}$$

We shall write two column matrices X and B defined by

$$X = \begin{bmatrix} x_1 \\ x_2 \\ x_3 \\ \vdots \\ x_n \end{bmatrix} \text{ and } B = \begin{bmatrix} b_1 \\ b_2 \\ b_3 \\ \vdots \\ b_n \end{bmatrix}.$$

The system (S) can then be written in the compact form: **AX = B.**

If we assume that $\det A \neq 0$, then the system (S) has exactly one set of solution; for, in that case, A^{-1} exists and hence

$$\begin{aligned} & A^{-1}\cdot(AX) = A^{-1}\cdot B \\ \text{or,}\quad & (A^{-1}A)X = A^{-1}B \\ \text{or,}\quad & IX = A^{-1}B. \\ \therefore\quad & X = A^{-1}B. \end{aligned}$$

Again, substituting $X = A^{-1}B$ in $AX = B$, we see that X satisfies $AX = B$.

This gives our desired unique set of solution. We illustrate by an example, where 3 linear equations with 3 unknowns are given:

▶ Example 5.5.13 *Solve the following system of linear equations*

$$\left.\begin{aligned} x + 2y + z &= 4 \\ x - y + z &= 5 \\ 2x + 3y - z &= 1 \end{aligned}\right\}.$$

Solution: The given system of equations can be written as $AX = B$, where

$$A = \begin{bmatrix} 1 & 2 & 1 \\ 1 & -1 & 1 \\ 2 & 3 & -1 \end{bmatrix},\ X = \begin{bmatrix} x \\ y \\ z \end{bmatrix} \text{ and } B = \begin{bmatrix} 4 \\ 5 \\ 1 \end{bmatrix}.$$

The solution is given by $X = A^{-1}B$, provided A^{-1} exists.

We, therefore, obtain A^{-1}.

Step 1. $\det A = 9(\neq 0)$; *hence* A^{-1} *exists.*

Step 2. *Matrix of cofactors of* $A = \begin{bmatrix} -2 & 3 & 5 \\ 5 & -3 & 1 \\ 3 & 0 & -3 \end{bmatrix}$.

$$\therefore\ adj\,A = \begin{bmatrix} -2 & 5 & 3 \\ 3 & -3 & 0 \\ 5 & 1 & -3 \end{bmatrix}.$$

$$\therefore\ A^{-1} = \begin{bmatrix} -\frac{2}{9} & \frac{5}{9} & \frac{1}{3} \\ \frac{1}{3} & -\frac{1}{3} & 0 \\ \frac{5}{9} & \frac{1}{9} & -\frac{1}{3} \end{bmatrix}.$$

Now the solution set is given by

$$X = A^{-1}B = \begin{bmatrix} -\frac{2}{9} & \frac{5}{9} & \frac{1}{3} \\ \frac{1}{3} & -\frac{1}{3} & 0 \\ \frac{5}{9} & \frac{1}{9} & -\frac{1}{3} \end{bmatrix}\begin{bmatrix} 4 \\ 5 \\ 1 \end{bmatrix} = \begin{bmatrix} \frac{20}{9} \\ -\frac{1}{3} \\ \frac{22}{9} \end{bmatrix}, \text{ or, } \begin{bmatrix} x \\ y \\ z \end{bmatrix} = \begin{bmatrix} \frac{20}{9} \\ -\frac{1}{3} \\ \frac{22}{9} \end{bmatrix}.$$

$\therefore$ **by definition of equatility of two matrices**

$$x = \frac{20}{9},\ y = -\frac{1}{3},\ z = \frac{22}{9}.$$

Check. It is always advisable to see that the set satisfies the given system of equations.

Observations. The students should note that the solution $X = A^{-1}B$ is just the matrix form of the solution that we derived by Cramer's Rule. In fact, we may state *Cramer's Rule* thus:

If $\det A \neq 0$, *the system* $AX = B$ *has exactly one solution, namely, that given by*

$$x_j = \frac{\det A_j}{\det A}\ (j = 1, 2, 3, \ldots, n),$$

where A_j *is the matrix obtained from* A, *by replacing the* j*th column of* A *by the column of* b*'s.*

Proof. We have already established $X = A^{-1}B$.

Thus, if the matrix equation $AX = B$ has a solution, it must be given by $X = A^{-1}B$. However, by direct substitution we see that

$$A(A^{-1}B) = (A^{-1}A)B = IB = B,$$

i.e., $A^{-1}B$ is indeed a solution of $AX = B$ and hence is the *one and only solution thereof.*

To express the solution in the scalar form, we note that

$$X = \frac{1}{\det A}\begin{bmatrix} A_{11} & A_{21} & \ldots & A_{n1} \\ A_{12} & A_{22} & \ldots & A_{n2} \\ \ldots & \ldots & \ldots & \ldots \\ A_{1n} & A_{2n} & \ldots & A_{nn} \end{bmatrix}\begin{bmatrix} b_1 \\ b_2 \\ \vdots \\ b_n \end{bmatrix}$$

or,
$$\begin{bmatrix} x_1 \\ x_2 \\ x_3 \\ \vdots \\ x_n \end{bmatrix} = \frac{1}{\det A}\begin{bmatrix} A_{11}b_1 + A_{21}b_2 + \ldots + A_{n1}b_n \\ A_{12}b_1 + A_{22}b_2 + \ldots + A_{n2}b_n \\ A_{13}b_1 + A_{23}b_2 + \ldots + A_{n3}b_n \\ \vdots \\ A_{1n}b_1 + A_{2n}b_2 + \ldots + A_{nn}b_n \end{bmatrix}$$

$$\begin{cases} x_1 = \dfrac{1}{\det A}(A_{11}b_1 + A_{21}b_2 + \ldots + A_{n1}b_n) = \dfrac{\det A_1}{\det A} \\ x_2 = \dfrac{1}{\det A}(A_{12}b_1 + A_{22}b_2 + \ldots + A_{n2}b_n) = \dfrac{\det A_2}{\det A} \\ \ldots \ldots \ldots \ldots \ldots \ldots \ldots \ldots \ldots \ldots \\ x_n = \dfrac{1}{\det A}(A_{1n}b_1 + A_{2n}b_2 + \ldots + A_{nn}b_n) = \dfrac{\det A_n}{\det A} \end{cases}$$

This is, therefore, the proof of Cramer's Rule by the use of inversion of matrices.

5.6 Solved Problems

▶ Example 5.6.1 *An example showing* AB ≠ BA (*One is defined and the other is not*).

Let $A = \begin{bmatrix} 1 & 2 \\ 3 & 0 \\ 4 & 1 \end{bmatrix}_{3\times 2}$ *and* $B = \begin{bmatrix} 0 & 1 & 0 \\ 0 & 2 & 1 \\ 2 & 3 & 0 \end{bmatrix}_{3\times 3}$. *Find* BA. *Can you find* AB?

Solution: B has three columns and A has three rows; hence BA exists.

$$BA = \begin{bmatrix} 0 & 1 & 0 \\ 0 & 2 & 1 \\ 2 & 3 & 0 \end{bmatrix}_{3\times ③} \begin{bmatrix} 1 & 2 \\ 3 & 0 \\ 4 & 1 \end{bmatrix}_{③\times 2} = \begin{bmatrix} 3 & 0 \\ 10 & 1 \\ 11 & 4 \end{bmatrix}.$$

We cannot find AB in this case, since the number of columns of A is 2 and the number of rows of B is 3, i.e., they are not equal.

∴ since we cannot define AB, we cannot write BA = AB.

▶ Example 5.6.2 *An example showing* AB ≠ BA *even if both are defined.*

Let $A = \begin{bmatrix} 1 & 0 \\ 0 & -1 \end{bmatrix}_{2\times 2}$, $B = \begin{bmatrix} 0 & 1 \\ 1 & 0 \end{bmatrix}_{2\times 2}$. *Prove that* AB ≠ BA.

Solution: Here both AB and BA exist.

$$AB = \begin{bmatrix} 1 & 0 \\ 0 & -1 \end{bmatrix} \begin{bmatrix} 0 & 1 \\ 1 & 0 \end{bmatrix} = \begin{bmatrix} 0 & 1 \\ -1 & 0 \end{bmatrix}$$

$$BA = \begin{bmatrix} 0 & 1 \\ 1 & 0 \end{bmatrix} \begin{bmatrix} 1 & 0 \\ 0 & -1 \end{bmatrix} = \begin{bmatrix} 0 & -1 \\ 1 & 0 \end{bmatrix}.$$

Clearly AB ≠ BA.

▶ Example 5.6.3 *Give an example where* AB = BA.

Let $A = \begin{bmatrix} 1 & 2 & 1 \\ 3 & 4 & 2 \\ 1 & 3 & 2 \end{bmatrix}$, $B = \begin{bmatrix} 10 & -4 & -1 \\ -11 & 5 & 0 \\ 9 & -5 & 1 \end{bmatrix}$. *Prove that* AB = BA.

Solution:

$$AB = \begin{bmatrix} 1 & 2 & 1 \\ 3 & 4 & 2 \\ 1 & 3 & 2 \end{bmatrix}_{3\times 3} \cdot \begin{bmatrix} 10 & -4 & -1 \\ -11 & 5 & 0 \\ 9 & -5 & 1 \end{bmatrix} = \begin{bmatrix} -3 & 1 & 0 \\ 4 & -2 & -1 \\ -5 & 1 & 1 \end{bmatrix}$$

$$BA = \begin{bmatrix} 10 & -4 & -1 \\ -11 & 5 & 0 \\ 9 & -5 & 1 \end{bmatrix} \begin{bmatrix} 1 & 2 & 1 \\ 3 & 4 & 2 \\ 1 & 3 & 2 \end{bmatrix} = \begin{bmatrix} -3 & 1 & 0 \\ 4 & -2 & -1 \\ -5 & 1 & 1 \end{bmatrix}$$

Hence, AB = BA.

▶ Example 5.6.4 *Give an example where* AB = 0, *but neither* A = 0 *nor* B = 0.

Solution: $A = \begin{bmatrix} 0 & 1 \\ 0 & 0 \end{bmatrix}, B = \begin{bmatrix} 1 & 0 \\ 0 & 0 \end{bmatrix}$.

Here $AB = \begin{bmatrix} 0 & 0 \\ 0 & 0 \end{bmatrix} = 0$ but neither $A = 0$, nor $B = 0$.

▶ **Example 5.6.5** *If* AB = 0, *does it necessarily imply* BA = 0?

Solution: Let $A = \begin{bmatrix} 0 & 1 \\ 0 & 0 \end{bmatrix}, B = \begin{bmatrix} 1 & 0 \\ 0 & 0 \end{bmatrix}$. Here AB = 0 **(Ex. 5.6.4).**

But $BA = \begin{bmatrix} 1 & 0 \\ 0 & 0 \end{bmatrix}\begin{bmatrix} 0 & 1 \\ 0 & 0 \end{bmatrix} = \begin{bmatrix} 0 & 1 \\ 0 & 0 \end{bmatrix} \neq 0$.

▶ **Example 5.6.6** *Find the product of*

$$A = \begin{bmatrix} 0 & c & -b \\ -c & 0 & a \\ -b & -a & 0 \end{bmatrix}, B = \begin{bmatrix} a^2 & ab & ac \\ ab & b^2 & bc \\ ac & bc & c^2 \end{bmatrix}.$$

Solution: Here $AB = \begin{bmatrix} 0 & 0 & 0 \\ 0 & 0 & 0 \\ 0 & 0 & 0 \end{bmatrix}$ but neither $A = 0$ nor $B = 0$.

▶ **Example 5.6.7** *If* $A = [x\ y\ z]$, $B = \begin{bmatrix} a & h & g \\ h & b & f \\ g & f & c \end{bmatrix}$ *and* $C = \begin{bmatrix} x \\ y \\ z \end{bmatrix}$, *find* ABC.

Solution:

$$\begin{aligned}
ABC &= (AB)C \text{ or, } A(BC) \text{ [Associative Law]} \\
AB &= [x\ y\ z]_{1\times③} \begin{bmatrix} a & h & g \\ h & b & f \\ g & f & c \end{bmatrix}_{③\times 3} \\
&= [ax + hy + gz \ \ hx + by + fz \ \ gx + fy + cz]_{1\times 3} \\
(AB)C &= [ax + hy + gz \ \ hx + by + fz \ \ gx + fy + cz]_{1\times 3} \cdot \begin{bmatrix} x \\ y \\ z \end{bmatrix}_{3\times 1} \\
&= [(ax + hy + gz)x + (hx + by + fz)y + (gx + fy + cz)z]_{1\times 1} \\
&= [ax^2 + by^2 + cz^2 + 2hxy + 2gzx + 2fyz]_{1\times 1}.
\end{aligned}$$

▶ **Example 5.6.8** *Under what condition the matrix equation*

$$A^2 - B^2 = (A + B)(A - B) \textit{ true?}$$

Solution:

$$\begin{aligned}
A^2 - B^2 = (A + B)(A - b) &\implies A^2 - B^2 = A^2 - AB + BA - B^2 \\
&\qquad\qquad \text{(Associative Law)} \\
&\implies A^2 - B^2 - (A^2 - AB + BA - B^2) = 0 \\
&\implies A^2 - B^2 - A^2 + AB - BA + B^2 = 0 \\
&\implies AB - BA = 0 \\
&\implies AB = BA
\end{aligned}$$

i.e., the given matrix equation is true only if AB = BA.

▶ **Example 5.6.9** *If* $A = \begin{bmatrix} 1 & 2 & 2 \\ 2 & 1 & 2 \\ 2 & 2 & 1 \end{bmatrix}$, *show that* $A^2 - 4A - 5I = 0$; *hence obtain* A^{-1}. [C.P. 2002]

Solution:

$$A^2 = A \cdot A = \begin{bmatrix} 1 & 2 & 2 \\ 2 & 1 & 2 \\ 2 & 2 & 1 \end{bmatrix}\begin{bmatrix} 1 & 2 & 2 \\ 2 & 1 & 2 \\ 2 & 2 & 1 \end{bmatrix} = \begin{bmatrix} 9 & 8 & 8 \\ 8 & 9 & 8 \\ 8 & 8 & 9 \end{bmatrix}$$

$$-4A = -4\begin{bmatrix} 1 & 2 & 2 \\ 2 & 1 & 2 \\ 2 & 2 & 1 \end{bmatrix} = \begin{bmatrix} -4 & -8 & -8 \\ -8 & -4 & -8 \\ -8 & -8 & -4 \end{bmatrix}$$

$$-5I = -5\begin{bmatrix} 1 & 0 & 0 \\ 0 & 1 & 0 \\ 0 & 0 & 1 \end{bmatrix} = \begin{bmatrix} -5 & 0 & 0 \\ 0 & -5 & 0 \\ 0 & 0 & -5 \end{bmatrix}.$$

$$\therefore A^2 - 4A - 5I = \begin{bmatrix} 9 & 8 & 8 \\ 8 & 9 & 8 \\ 8 & 8 & 9 \end{bmatrix} + \begin{bmatrix} -4 & -8 & -8 \\ -8 & -4 & -8 \\ -8 & -8 & -4 \end{bmatrix} + \begin{bmatrix} -5 & 0 & 0 \\ 0 & -5 & 0 \\ 0 & 0 & -5 \end{bmatrix}$$

$$= \begin{bmatrix} 5 & 0 & 0 \\ 0 & 5 & 0 \\ 0 & 0 & 5 \end{bmatrix} + \begin{bmatrix} -5 & 0 & 0 \\ 0 & -5 & 0 \\ 0 & 0 & -5 \end{bmatrix} = \begin{bmatrix} 0 & 0 & 0 \\ 0 & 0 & 0 \\ 0 & 0 & 0 \end{bmatrix} = 0.$$

Now

$$A \cdot (A - 4I) = 5I \quad \therefore \quad A \cdot \frac{1}{5}(A - 4I) = I$$

$$\text{i.e.,} \quad \frac{1}{5}(A - 4I) = A^{-1} \text{ i.e.,} \quad A^{-1} = \frac{1}{5}\left\{\begin{bmatrix} 1 & 2 & 2 \\ 2 & 1 & 2 \\ 2 & 2 & 1 \end{bmatrix} - \begin{bmatrix} 4 & 0 & 0 \\ 0 & 4 & 0 \\ 0 & 0 & 4 \end{bmatrix}\right\}$$

$$\text{i.e.,} \quad A^{-1} = \frac{1}{5}\begin{bmatrix} -3 & 2 & 2 \\ 2 & -3 & 2 \\ 2 & 2 & -3 \end{bmatrix}.$$

Check. $AA^{-1} = I$, i.e., $\begin{bmatrix} 1 & 2 & 2 \\ 2 & 1 & 2 \\ 2 & 2 & 1 \end{bmatrix} \cdot \frac{1}{5}\begin{bmatrix} -3 & 2 & 2 \\ 2 & -3 & 2 \\ 2 & 2 & 3 \end{bmatrix}$

$$= \frac{1}{5}\begin{bmatrix} 5 & 0 & 0 \\ 0 & 5 & 0 \\ 0 & 0 & 5 \end{bmatrix} = \begin{bmatrix} 1 & 0 & 0 \\ 0 & 1 & 0 \\ 0 & 0 & 1 \end{bmatrix} = I.$$

▶ **Example 5.6.10** *Show that the matrix* $\begin{bmatrix} 2 & -2 & -4 \\ -1 & 3 & 4 \\ 1 & -2 & 3 \end{bmatrix}$ *is idempotent.*

Solution: **[A square matrix A is called idempotent if $A^2 = A$]**

Let

$$A = \begin{bmatrix} 2 & -2 & -4 \\ -1 & 3 & 4 \\ 1 & -2 & 3 \end{bmatrix}.$$

Then

$$A^2 = \begin{bmatrix} 2 & -2 & -4 \\ -1 & 3 & 4 \\ 1 & -2 & 3 \end{bmatrix} \begin{bmatrix} 2 & -2 & -4 \\ -1 & 3 & 4 \\ 1 & -2 & 3 \end{bmatrix} = \begin{bmatrix} 2 & -2 & -4 \\ -1 & 3 & 4 \\ 1 & -2 & 3 \end{bmatrix} = A.$$

Since $A^2 = A$, therefore, the matrix A is an idempotent matrix.

▶ Example 5.6.11 *If B is an idempotent matrix, show that $A = I - B$ is also idempotent and that $AB = BA = 0$.*

Solution: Since B is an idempotent matrix, $B^2 = B$.

Now

$$\begin{aligned} A^2 = (I-B)^2 &= (I-B)\cdot(I-B) \\ &= I - I\cdot B - B\cdot I + B\cdot B \\ &= I - I\cdot B - BI + B \\ &= I - B - B + B \ (\because\ IB = B = BI) \\ &= I - B = A. \end{aligned}$$

Since $A^2 = A$, therefore $A = I - B$ is idempotent.

Again,

$$AB = (I-B)\cdot B = IB - B^2 = B - B = 0.$$

Similarly

$$BA = B(I-B) = BI - B^2 = B - B = 0.$$

$$\therefore\ AB = BA = 0.$$

▶ Example 5.6.12 (a) *A square matrix A such that $A^2 = I$ is called **involutary**. Show that A is involutory if and only if $(I+A)(I-A) = 0$.*

Solution: Suppose that A is involutory matrix of order n. Then $A^2 = I$,

$$\begin{aligned} \text{i.e.,}\quad & I - A^2 = 0,\ \text{or},\ I^2 - A^2 = 0, \\ \text{or,}\quad & (I+A)(I-A) = 0\ (\because\ IA = AI = A). \end{aligned}$$

Conversely suppose $(I+A)(I-A) = 0$,

$$\begin{aligned} \text{i.e.,}\quad & I^2 + AI - IA - A^2 = 0 \\ \text{i.e.,}\quad & I^2 - A^2 = 0\ (\because\ AI = IA) \\ \text{or,}\quad & A^2 = I\ (\because\ I^2 = I). \end{aligned}$$

(b) *If* $A + I = \begin{bmatrix} 1 & 3 & 4 \\ -1 & 1 & 3 \\ -2 & -3 & 1 \end{bmatrix}$, *evaluate* $(A + I)\cdot(A - I)$, *where* I *is the* 3×3 *identity matrix.* [C.P. 2005]

Solution:

$$A+I = \begin{bmatrix} 1 & 3 & 4 \\ -1 & 1 & 3 \\ -2 & -3 & 1 \end{bmatrix} \Rightarrow A = \begin{bmatrix} 1 & 3 & 4 \\ -1 & 1 & 3 \\ -2 & -3 & 1 \end{bmatrix} - \begin{bmatrix} 1 & 0 & 0 \\ 0 & 1 & 0 \\ 0 & 0 & 1 \end{bmatrix}$$

$$= \begin{bmatrix} 0 & 3 & 4 \\ 1 & 0 & 3 \\ -2 & -3 & 0 \end{bmatrix}.$$

$$(A+I)(A-I) = A^2 - I^2 \text{ (see above)}$$

$$= \begin{bmatrix} 0 & 3 & 4 \\ -1 & 0 & 3 \\ -2 & -3 & 0 \end{bmatrix}^2 - \begin{bmatrix} 1 & 0 & 0 \\ 0 & 1 & 0 \\ 0 & 0 & 1 \end{bmatrix} \quad (\because I^2 = I)$$

$$= \begin{bmatrix} -11 & -12 & 9 \\ -6 & -12 & -4 \\ 3 & -6 & -17 \end{bmatrix} - \begin{bmatrix} 1 & 0 & 0 \\ 0 & 1 & 0 \\ 0 & 0 & 1 \end{bmatrix}$$

$$= \begin{bmatrix} -12 & -12 & 9 \\ -6 & -13 & -4 \\ 3 & -6 & -18 \end{bmatrix} \quad \text{(Reqd. Ans)}$$

▶ Example 5.6.13 *A square matrix* A *is called a* **nilpotent matrix** *if there exists a positive integer* m *such that* $A^m = 0$ *(null matrix). If* m *is the least positive integer such that* $A^m = 0$, *then* m *is called the index of the nilpotent matrix* A. *Show that the matrix.*

$$A = \begin{bmatrix} ab & +b^2 \\ -a^2 & -ab \end{bmatrix}$$

is nilpotent of index 2.

Solution:

$$A^2 = A\cdot A = \begin{bmatrix} ab & b^2 \\ -a^2 & -ab \end{bmatrix} \cdot \begin{bmatrix} ab & b^2 \\ -a^2 & -ab \end{bmatrix}$$

$$= \begin{bmatrix} a^2b^2 - b^2a^2 & ab^3 - ab^3 \\ -a^3b + a^3b & -a^2b + a^2b^2 \end{bmatrix} = \begin{bmatrix} 0 & 0 \\ 0 & 0 \end{bmatrix} = 0.$$

Hence, A is nilpotent of index two.

▶ Example 5.6.14 *If* A, B *are symmetric matrices, then prove that* $A + B$ *is also symmetric.*

Solution: Since A and B are symmetric, $A' = A$ and $B' = B$.

Now $(A+B)' = A'+B' = A+B$, i.e., $(A+B)' = A+B$, i.e., $A+B$ is a symmetric matrix.

▶ Example 5.6.15 ***Remember** that any square matrix A can be expressed uniquely as*

$$A = \tfrac{1}{2}(A + A') + \tfrac{1}{2}(A - A') = P + Q, \quad (\textit{See Theorem } 5.5.1)$$

where P is a symmetric matrix and Q is a skew-symmetric matrix. **[C.P. 2001]**

Now express $A = \begin{bmatrix} 3 & 2 & -6 \\ 0 & -1 & 4 \\ 5 & -2 & 0 \end{bmatrix}$ *as the sum of a symmetric and a skew-symmetric matrix.*

Solution:

$$A + A' = \begin{bmatrix} 3 & 2 & -6 \\ 0 & -1 & 4 \\ 5 & -2 & 0 \end{bmatrix} + \begin{bmatrix} 3 & 0 & 5 \\ 2 & -1 & -2 \\ -6 & 4 & 0 \end{bmatrix}$$

$$= \begin{bmatrix} 6 & 2 & -1 \\ 2 & -2 & 2 \\ -1 & 2 & 0 \end{bmatrix} \text{ (a symmetric matrix)}$$

$$A - A' = \begin{bmatrix} 3 & 2 & -6 \\ 0 & -1 & 4 \\ 5 & -2 & 0 \end{bmatrix} - \begin{bmatrix} 3 & 0 & 5 \\ 2 & -1 & -2 \\ -6 & 4 & 0 \end{bmatrix}$$

$$= \begin{bmatrix} 0 & 2 & -11 \\ -2 & 0 & 6 \\ 11 & -6 & 0 \end{bmatrix} \text{ (a skew-symmetric matrix).}$$

$$\therefore A = \frac{1}{2}\begin{bmatrix} 6 & 2 & -1 \\ 2 & -2 & 2 \\ -1 & -2 & 0 \end{bmatrix} + \frac{1}{2}\begin{bmatrix} 0 & 2 & -11 \\ -2 & 0 & 6 \\ 11 & -6 & 0 \end{bmatrix}$$

= sum of a symmetric matrix and a skew-symmetric matrix.

▶ Example 5.6.16 *If*

$$A = \begin{bmatrix} 2 & -1 \\ 1 & 3 \end{bmatrix} \textit{ and } A^2 - 5A + 7I_2 = 0,$$

then prove that $A^{-1} = \frac{1}{7}\begin{bmatrix} 3 & 1 \\ -1 & 2 \end{bmatrix}$. **[C.P. 2000, 2003]**

Solution: Since $A^2 - 5A + 7I_2 = 0$, we write

$$A(A - 5I_2) = -7I_2$$

$$\text{or,}\quad A \cdot \frac{1}{7}(5I_2 - A) = I_2 \implies A^{-1} = \frac{1}{7}(5I_2 - A)$$

$$\text{i.e.,}\quad A^{-1} = \frac{1}{7}\left\{5\begin{bmatrix} 1 & 0 \\ 0 & 1 \end{bmatrix} - \begin{bmatrix} 2 & -1 \\ 1 & 3 \end{bmatrix}\right\}$$

$$= \frac{1}{7}\left\{\begin{bmatrix} 5 & 0 \\ 0 & 5 \end{bmatrix} - \begin{bmatrix} 2 & -1 \\ 1 & 3 \end{bmatrix}\right\} = \frac{1}{7}\begin{bmatrix} 3 & 1 \\ -1 & 2 \end{bmatrix}.$$

Note. In this problem, we can actually prove that if $A = \begin{bmatrix} 2 & -1 \\ 1 & 3 \end{bmatrix}$, then $A^2 - 5A + 7I_2 = 0$ is true.

$$A^2 = \begin{bmatrix} 2 & -1 \\ 1 & 3 \end{bmatrix}\begin{bmatrix} 2 & -1 \\ 1 & 3 \end{bmatrix} = \begin{bmatrix} 3 & -5 \\ 5 & 8 \end{bmatrix};$$

$$\begin{aligned} A^2 - 5A + 7I_2 &= \begin{bmatrix} 3 & -5 \\ 5 & 8 \end{bmatrix} - 5\begin{bmatrix} 2 & -1 \\ 1 & 3 \end{bmatrix} + 7\begin{bmatrix} 1 & 0 \\ 0 & 1 \end{bmatrix} \\ &= \begin{bmatrix} 3 & -5 \\ 5 & 8 \end{bmatrix} - \begin{bmatrix} 10 & -5 \\ 5 & 15 \end{bmatrix} + \begin{bmatrix} 7 & 0 \\ 0 & 0 \end{bmatrix} \\ &= \begin{bmatrix} 3-10+7 & -5+5+0 \\ 5-5+0 & 8-15+7 \end{bmatrix} = \begin{bmatrix} 0 & 0 \\ 0 & 0 \end{bmatrix}. \end{aligned}$$

▶ **Example 5.6.17** *If A is a symmetric matrix, then prove that adj A is also symmetric.*

Solution: Since A is symmetric, $A' = A$. Now $(\text{adj}\,A)' = \text{adj}\,A' = \text{adj}\,A$ ($\because A' = A$) i.e., $(\text{adj}\,A)' = \text{adj}\,A$, therefore, adj A is a symmetric matrix.

▶ **Example 5.6.18** *Find the inverse of the matrix* $A = \begin{bmatrix} 0 & 1 & 2 \\ 1 & 2 & 3 \\ 3 & 1 & 1 \end{bmatrix}$.

Solution: We have

$$\begin{aligned} |A| &= \begin{bmatrix} 0 & 1 & 2 \\ 1 & 2 & 3 \\ 3 & 1 & 1 \end{bmatrix} = -1\begin{bmatrix} 1 & 3 \\ 3 & 1 \end{bmatrix} + 2\begin{bmatrix} 1 & 2 \\ 3 & 1 \end{bmatrix} \\ &= 8 + 2\cdot(-5) = -2. \end{aligned}$$

Since $\det A \neq 0$, A^{-1} exists.
We now obtain the matrix of the cofactors

$$= \begin{bmatrix} \begin{bmatrix} 2 & 3 \\ 1 & 1 \end{bmatrix} & -\begin{bmatrix} 1 & 3 \\ 3 & 1 \end{bmatrix} & \begin{bmatrix} 1 & 2 \\ 3 & 1 \end{bmatrix} \\ -\begin{bmatrix} 1 & 2 \\ 1 & 1 \end{bmatrix} & \begin{bmatrix} 0 & 2 \\ 3 & 1 \end{bmatrix} & -\begin{bmatrix} 0 & 1 \\ 3 & 1 \end{bmatrix} \\ \begin{bmatrix} 1 & 2 \\ 2 & 3 \end{bmatrix} & -\begin{bmatrix} 0 & 2 \\ 1 & 3 \end{bmatrix} & \begin{bmatrix} 0 & 2 \\ 1 & 2 \end{bmatrix} \end{bmatrix} = \begin{bmatrix} -1 & 8 & -5 \\ 1 & -6 & 3 \\ -1 & 2 & -1 \end{bmatrix}.$$

$$\text{Adjoint of } A = \begin{bmatrix} -1 & 1 & -1 \\ 8 & -6 & 2 \\ -5 & 3 & -1 \end{bmatrix}$$

$$\begin{aligned} \therefore A^{-1} &= \frac{1}{\det A}\begin{bmatrix} -1 & 1 & -1 \\ 8 & -6 & 2 \\ -5 & 3 & -1 \end{bmatrix} = -\frac{1}{2}\begin{bmatrix} -1 & 1 & -1 \\ 8 & -6 & 2 \\ -5 & 3 & -1 \end{bmatrix} \\ &= \begin{bmatrix} \frac{1}{2} & -\frac{1}{2} & \frac{1}{2} \\ -4 & 3 & -1 \\ \frac{5}{2} & -\frac{3}{2} & \frac{1}{2} \end{bmatrix}. \end{aligned}$$

$$\textbf{Check: } A\cdot A^{-1} = \begin{bmatrix} 0 & 1 & 2 \\ 1 & 2 & 3 \\ 3 & 1 & 1 \end{bmatrix} \cdot \begin{bmatrix} \frac{1}{2} & -\frac{1}{2} & \frac{1}{2} \\ -4 & 3 & -1 \\ \frac{5}{2} & -\frac{3}{2} & \frac{1}{2} \end{bmatrix} = \begin{bmatrix} 1 & 0 & 0 \\ 0 & 1 & 0 \\ 0 & 0 & 1 \end{bmatrix} = I_3.$$

▶ Example 5.6.19 *Show that the matrix*

$$\frac{1}{3}\begin{bmatrix} 1 & 2 & 2 \\ 2 & 1 & -2 \\ -2 & 2 & -1 \end{bmatrix}$$

is orthogonal. Hence, find A^{-1}.

Solution: It is orthogonal, then $A\cdot A^T = I$.

Here let $A = \dfrac{1}{3}\begin{bmatrix} 1 & 2 & 2 \\ 2 & 1 & -2 \\ -2 & 2 & -1 \end{bmatrix}$, $A^T = \frac{1}{3}\begin{bmatrix} 1 & 2 & -2 \\ 2 & 1 & 2 \\ 2 & -2 & -1 \end{bmatrix}$.

$$\therefore\ A\cdot A^T = \frac{1}{9}\begin{bmatrix} 1 & 2 & 2 \\ 2 & 1 & -2 \\ -2 & 2 & -1 \end{bmatrix}\begin{bmatrix} 1 & 2 & -2 \\ 2 & 1 & 2 \\ 2 & -2 & -1 \end{bmatrix}$$

$$= \frac{1}{9}\begin{bmatrix} 9 & 0 & 0 \\ 0 & 9 & 0 \\ 0 & 0 & 9 \end{bmatrix} = \begin{bmatrix} 1 & 0 & 0 \\ 0 & 1 & 0 \\ 0 & 0 & 1 \end{bmatrix} = I_3.$$

$\therefore$ A is an orthogonal matrix.
Now $A\cdot A^T = I_3$ and inverse is unique, so

$$A^T = A^{-1},\ \text{i.e., } A^{-1} = \frac{1}{3}\begin{bmatrix} 1 & 2 & -2 \\ 2 & 1 & 2 \\ 2 & -2 & -1 \end{bmatrix}.$$

▶ Example 5.6.20 *Find the inverse of the matrix*

$$\begin{bmatrix} 1 & 1 & 2 \\ 2 & -1 & 3 \\ 3 & -1 & -1 \end{bmatrix}$$

and using it solve the following system of linear equations:

$$\left.\begin{aligned} x + y + 2z &= 4 \\ 2x - y + 3z &= 9 \\ 3x - y - z &= 2 \end{aligned}\right\}.$$ **[C.P. 1998]**

Solution: Coefficient matrix $A = \begin{bmatrix} 1 & 1 & 2 \\ 2 & -1 & 3 \\ 3 & -1 & -1 \end{bmatrix}$.

Let $X = \begin{bmatrix} x \\ y \\ z \end{bmatrix}$ and let $B = \begin{bmatrix} 4 \\ 9 \\ 2 \end{bmatrix}$.

$\therefore$ the given system of equations can be written as

$$\begin{aligned} & AX = B. \\ \therefore\quad & A^{-1}(AX) = A^{-1}B \text{ (assuming } A^{-1} \text{ exists)} \\ \textbf{or,}\quad & (A^{-1}A)X = A^{-1}B, \\ \textbf{or,}\quad & IX = A^{-1}B, \text{ i.e., } X = A^{-1}B. \end{aligned}$$

We then find A^{-1}. Here

$$\begin{aligned} \det A &= \begin{bmatrix} 1 & 1 & 2 \\ 2 & -1 & 3 \\ 3 & -1 & -1 \end{bmatrix} = 1\cdot(1+3) - 1(-2-9) + 2\cdot(-2+3) \\ &= 17 \neq 0, \text{ hence } A^{-1} \text{ exists.} \\ A^{-1} &= \frac{1}{17}\operatorname{adj} A = \frac{1}{17}\begin{bmatrix} 4 & 11 & 1 \\ -1 & -7 & 4 \\ 5 & 1 & -3 \end{bmatrix} = \frac{1}{17}\begin{bmatrix} 4 & -1 & 5 \\ 11 & -7 & 1 \\ 1 & 4 & -3 \end{bmatrix}. \end{aligned}$$

$$\begin{aligned} \therefore X &= A^{-1}B \Longrightarrow \begin{bmatrix} x \\ y \\ z \end{bmatrix} \\ &= \frac{1}{17}\begin{bmatrix} 4 & -1 & 5 \\ 11 & -7 & 1 \\ 1 & 4 & -3 \end{bmatrix}\begin{bmatrix} 4 \\ 9 \\ 2 \end{bmatrix} = \frac{1}{17}\begin{bmatrix} 17 \\ -17 \\ 34 \end{bmatrix} \\ &= \begin{bmatrix} 1 \\ -1 \\ 2 \end{bmatrix} \Longrightarrow x = 1,\ y = -1,\ z = 2 \textbf{ (Reqd. Soln.)} \end{aligned}$$

▶ Example 5.6.21 *If*

$$A = \begin{bmatrix} 1 & -1 & 0 \\ 2 & 3 & 4 \\ 0 & 1 & 2 \end{bmatrix},\ C = \begin{bmatrix} 2 & 2 & -4 \\ -4 & 2 & -4 \\ 2 & -1 & 5 \end{bmatrix},\ \textit{verify } AC = CA = 6I_3.$$

Use this fact to solve the system of equations:

$$\left.\begin{aligned} x - y &= 3 \\ 2x + 3y + 43 &= 17 \\ y + 2z &= 7 \end{aligned}\right\}$$

[C.P. 1997]

Solution: It is easy to check $AC = CA = 6I_3$.

$$\Rightarrow\quad A\cdot\frac{1}{6}C = \frac{1}{6}C\cdot A = I_3.$$

$$\therefore\quad A^{-1} = \frac{1}{6}C = \frac{1}{6}\begin{bmatrix} 2 & 2 & -4 \\ -4 & 2 & -4 \\ 2 & -1 & 5 \end{bmatrix}.$$

The given system of equations can be written as $AX = B$, where

$$A = \begin{bmatrix} 1 & -1 & 0 \\ 2 & 3 & 4 \\ 0 & 1 & 2 \end{bmatrix},\ X = \begin{bmatrix} x \\ y \\ z \end{bmatrix},\ B = \begin{bmatrix} 3 \\ 17 \\ 7 \end{bmatrix}.$$

Now, $AX = B \Rightarrow X = A^{-1}B$, i.e.,

$$\begin{bmatrix} x \\ y \\ z \end{bmatrix} = \frac{1}{6}\begin{bmatrix} 2 & 2 & -4 \\ -4 & 2 & -4 \\ 2 & -1 & 5 \end{bmatrix}\begin{bmatrix} 3 \\ 17 \\ 7 \end{bmatrix} = \frac{1}{6}\begin{bmatrix} 12 \\ -6 \\ 24 \end{bmatrix} = \begin{bmatrix} 2 \\ -1 \\ 4 \end{bmatrix}$$

which gives $x = 2$, $y = -1$, $z = 4$. **(Reqd. Solutions)**

5.7 Rank of a Matrix: Its Determination

Definition 5.7.1 *From a matrix A if we* **retain** *any r rows and any r columns and delete all other rows or columns, then we get a* **square submatrix** *of order r. Its determinant is called a* **minor of order r.**

A matrix A is said to be of rank r [we write $r(A) = r$] if, and only if, A has at least one minor of order $r \neq 0$, while all minors of order more than $r = 0$.

5.7.1 Determination of Rank by Considering Minors

▶ Example 5.7.1 *What is the rank of* $A \begin{bmatrix} 3 & 2 & -2 \\ 0 & 3 & -3 \end{bmatrix}$?

Solution: There cannot be a **minor of order three** or of **order more than three.** But there exists at least one minor of order $2 \neq 0$, e.g., $\begin{vmatrix} 3 & 2 \\ 0 & 3 \end{vmatrix} = 9(\neq 0)$. Hence, $r(A) = 2$.

▶ Example 5.7.2 *Let* $A = \begin{bmatrix} 2 & 1 & -1 \\ 0 & 3 & -2 \\ 2 & 4 & -3 \end{bmatrix}$.

Solution: There can be only one submatrix of order 3, namely A itself. Its determinant

$$\begin{vmatrix} 2 & 1 & -1 \\ 0 & 3 & -2 \\ 2 & 4 & -3 \end{vmatrix} = 0.$$

Hence, we can say, all minors of order 3 or more than 3, vanish.

But there exists a submatrix of order 2, say $\begin{bmatrix} 2 & 1 \\ 0 & 3 \end{bmatrix}$. Its determinant $= \begin{vmatrix} 2 & 1 \\ 0 & 3 \end{vmatrix} = 6(\neq 0)$, i.e., there exists at least one minor of order $2 \neq 0$. **Therefore rank of A = 2.**

▶ Example 5.7.3 *Exactly as in* **Example 5.7.2.** *We can prove that the rank of the matrix*

$$A = \begin{bmatrix} 1 & 2 & 3 \\ 2 & 3 & 4 \\ 3 & 5 & 7 \end{bmatrix}$$

is two, since $\begin{vmatrix} 1 & 2 \\ 2 & 3 \end{vmatrix} = 3 - 4 = -1\ (\neq 0)$ *while* $|A| = 0$.

Note. A square matrix A of order n is **non-singular** if its rank = n, that is if $|A| \neq 0$. Otherwise A is called a **singular matrix.** The matrix of Example 5.7.3 is **singular.**

Now verify the following assertions:

▶ Example 5.7.4 (a) *The matrix* $A = \begin{bmatrix} 0 & 1 & 2 & 3 \\ 0 & 0 & 1 & -1 \\ 0 & 0 & 0 & 0 \end{bmatrix}$ *has the* **rank two.**

(b) *The matrix* $A = \begin{bmatrix} 4 & -1 & 2 \\ 3 & 4 & 0 \\ 1 & 0 & 0 \end{bmatrix}$ *has the* **rank 3.**

It is a non-singular matrix.

(c) *The matrix* $A = \begin{bmatrix} 1 & a & b & 0 \\ 0 & c & d & 1 \\ 1 & a & c & 0 \\ 0 & c & d & 1 \end{bmatrix}$ *is singular since* $\det A = 0$. *Here all minors of order three are zero since in every square submatrix of order 3, two rows will be identical. At least one minor of order two is* $\begin{vmatrix} 1 & 0 \\ 0 & 1 \end{vmatrix} \neq 0$. *Hence rank of this matrix is* **two.**

(d) *The matrix* $\begin{bmatrix} 2 & 0 & 7 \\ 3 & 3 & 6 \\ 2 & 2 & 4 \end{bmatrix}$ *has rank* **two** *(Why?)*

(e) *The matrix* $\begin{bmatrix} 3 & -2 \\ -6 & 4 \end{bmatrix}$ *has rank* **one** **[C.P. 1991]**

Since only minor of order two, namely $\begin{bmatrix} 3 & -2 \\ -6 & 4 \end{bmatrix} = 0$, *and elements are not all zero, i.e., at least one minor of order one, say* $|3|$ *is not zero.*

(f) *The rank of unit matrix of order 3, namely*

$$A = \begin{bmatrix} 1 & 0 & 0 \\ 0 & 1 & 0 \\ 0 & 0 & 1 \end{bmatrix} \text{ is three.}$$ **[C.P. 2004]**

We have $|A| = 1$. *Therefore A is a non-singular matrix.*

(g) *Let* $A = \begin{bmatrix} 3 & 1 & 2 \\ 6 & 2 & 4 \\ 3 & 1 & 2 \end{bmatrix}$. *Here* $|A| = 0$, *since the first two columns are proportional. Also each two-rowed minor of A is equal to zero. But A is not a null matrix. Hence its* **rank is one.**

(h) $A = \begin{bmatrix} 2 & 4 & 3 & 2 \\ 3 & 5 & 1 & 4 \end{bmatrix}$. *Its rank* = 2 *(Why?)*

(i) *The rank of the matrix* $\begin{pmatrix} 1 & 1 & 2 \\ 3 & 5 & 2 \\ 4 & 8 & 0 \end{pmatrix}$ *is two.* **[C.P. 2003]**

(j) *The rank of the matrix* $\begin{pmatrix} 1 & 2 & 3 \\ 2 & 1 & 0 \\ 0 & 1 & 2 \end{pmatrix}$ *is two.* [C.P. 2001]

For (i) *and* (j): *Give arguments as in **Example 5.7.2** above.*

▶ Example 5.7.5 *Prove that in a plane, three points whose rectangular coordinates are* $(x_1, y_1), (x_2, y_2)$ *and* (x_3, y_3) *are collinear, if and only if the rank of the matrix*

$$\begin{bmatrix} x_1 & y_1 & 1 \\ x_2 & y_2 & 1 \\ x_3 & y_3 & 1 \end{bmatrix} \text{ is less than three.}$$

Solution: First suppose that the three points are collinear and suppose that they lie on the line whose equation is $ax+by+c=0$. Since the three points $(x_1, y_1), (x_2, y_2)$ and (x_3, y_3) has on the line, we write

$$\left.\begin{aligned} ax_1 + by_1 + c &= 0 \\ ax_2 + by_2 + c &= 0 \\ ax_3 + by_3 + c &= 0 \end{aligned}\right\}.$$

Eliminating a, b, c we obtain $\begin{vmatrix} x_1 & y_1 & 1 \\ x_2 & y_2 & 1 \\ x_3 & y_3 & 1 \end{vmatrix} = 0$.

$\therefore$ The rank of the matrix $A = \begin{vmatrix} x_1 & y_1 & 1 \\ x_2 & y_2 & 1 \\ x_3 & y_3 & 1 \end{vmatrix}$ is less than three.

Conversely, if the rank of the matrix is less than three, then $\begin{vmatrix} x_1 & y_1 & 1 \\ x_2 & y_2 & 1 \\ x_3 & y_3 & 1 \end{vmatrix} = 0$.

$\therefore$ The area of the triangle whose vertices are $(x_1, y_1), (x_2, y_2)$ and (x_3, y_3) is equal to zero. Hence the three coplanar points lie on a line.

5.7.2 Determination of Rank by Elementary Transformations

The method of determining the rank of a matrix by considering minors, though simple, are laborious if the matrix is of order more than three. So we introduce some easier method—*Method of Elementary operations on the matrix*, also known as *Sweep-out Process*.

Preliminaries

1. Elementary Transformations (E-Transformations)

The following operations, called **Elementary Transformations,** on a matrix **do no change its order or its rank.**

(α) The interchange of the ith and jth rows, denoted by R_{ij}.

The interchange of the ith and jth columns, denoted by C_{ij}.

(β) The multiplication of every element of the ith row by a non-zero scalar k, denoted by $R_i(k)$.

The multiplication of every element of the jth column by a non-zero scalar k, denoted by $C_j(k)$.

(γ) The addition to the elements of the ith row of k times the corresponding elements of the jth row (k is a non-zero scalar), denoted by $R_{ij}(k)$. A similar meaning of the symbol $C_{ij}(k)$.

The R-transformations are called **elementary row transformation.**

The C-transformations are called **elementary column transformation.**

2. **Equivalent Matrices:** Two matrices A and B are said to be **equivalent** $(A \sim B)$ if one can be obtained from the other by a sequence of elementary transformations.

Equivalent matrices have the *same order* and the *same rank.*

▶ Example 5.7.6 *We apply a sequence of elementary transformations* $R_{21}(-2)$, $R_{31}(1)$, $H_{32}(-1)$ *on the matrix* A *and obtain.*

$$A = \begin{bmatrix} 1 & 2 & -1 & 4 \\ 2 & 4 & 3 & 5 \\ -1 & -2 & 6 & -7 \end{bmatrix} \sim \begin{bmatrix} 1 & 2 & -1 & 4 \\ 0 & 0 & 5 & -3 \\ -1 & -2 & 6 & -7 \end{bmatrix}$$

$$\sim \begin{bmatrix} 1 & 2 & -1 & 4 \\ 0 & 0 & 5 & -3 \\ 0 & 0 & 5 & -3 \end{bmatrix}$$

$$\sim \begin{bmatrix} 1 & 2 & -1 & 4 \\ 0 & 0 & 5 & -3 \\ 0 & 0 & 0 & 0 \end{bmatrix} = B.$$

We can then write $A \sim B$. See that the order of B = order of $A = 3 \times 4$. The rank of B is two (why?). Hence the rank of A is **also two.** (A and B are equivalent and they have the same order and same rank).

3. **The Normal Form of a Matrix** (also called **Canonical Form**)

By means of elementary transformations any matrix A of rank $r(> 0)$ can be brought in any one of the forms

$$I_r, \quad \begin{pmatrix} I_r & 0 \\ 0 & 0 \end{pmatrix}, \quad \begin{pmatrix} I_r & 0 \end{pmatrix}, \quad \begin{bmatrix} I_r \\ 0 \end{bmatrix}$$

called the **normal form of A.**

[Both row and column transformations may be used here. The element 1 should come in the first row and first column. Then both the first row and first column can be cleared of other non-zero elements. Then try to bring 1 in the second row and second column and clear of other non-zero elements and so on.]

▶ Example 5.7.7 *Let* $A = \begin{bmatrix} 1 & 2 & -1 & 4 \\ 2 & 4 & 3 & 5 \\ -1 & -2 & 6 & -7 \end{bmatrix}$.

Apply the following sequence of elementary transformations: $R_{21}(-2)$, $R_{31}(1)$, $C_{21}(-2)$, $C_{31}(1)$, C_{23}, $C_2\left(\frac{1}{5}\right)$, $R_{32}(-1)$, $C_{42}(3)$.

$$A = \begin{bmatrix} 1 & 2 & -1 & 4 \\ 2 & 4 & 3 & 5 \\ -1 & -2 & 6 & -7 \end{bmatrix} \sim \underset{R_{21}(-2);R_{31}(1)}{\begin{bmatrix} 1 & 2 & -1 & 4 \\ 0 & 0 & 5 & -3 \\ 0 & 0 & 5 & -3 \end{bmatrix}} \sim \underset{C_{31}(-2)}{\begin{bmatrix} 1 & 0 & -1 & 4 \\ 0 & 0 & 5 & -3 \\ 0 & 0 & 5 & -3 \end{bmatrix}}$$

$$\sim \underset{C_{31}(1)}{\begin{bmatrix} 1 & 0 & 0 & 4 \\ 0 & 0 & 5 & -3 \\ 0 & 0 & 5 & -3 \end{bmatrix}} \sim \underset{C_{23}}{\begin{bmatrix} 1 & 0 & 0 & 4 \\ 0 & 5 & 0 & -3 \\ 0 & 5 & 0 & -3 \end{bmatrix}} \sim \underset{C_2\left(\frac{1}{5}\right)}{\begin{bmatrix} 1 & 0 & 0 & 4 \\ 0 & 1 & 0 & -3 \\ 0 & 1 & 0 & -3 \end{bmatrix}}$$

$$\sim \underset{R_{32}(-1)}{\begin{bmatrix} 1 & 0 & 0 & 4 \\ 0 & 1 & 0 & -3 \\ 0 & 0 & 0 & 0 \end{bmatrix}} \sim \underset{C_{42}(3)}{\begin{bmatrix} 1 & 0 & 0 & 4 \\ 0 & 1 & 0 & 0 \\ 0 & 0 & 0 & 0 \end{bmatrix}} \sim \underset{C_{41}(-4)}{\begin{bmatrix} 1 & 0 & 0 & 0 \\ 0 & 1 & 0 & 0 \\ 0 & 0 & 0 & 0 \end{bmatrix}}$$

$$= \begin{bmatrix} I_2 & 0 \\ 0 & 0 \end{bmatrix} \text{ (Normal Form).}$$

Important Result: *The rank of the matrix, when reduced to Normal form, is the order of the unit matrix,* e.g., in the above example, rank $A = 2$.

4. Echelon Form of a matrix: Definition:

A matrix A is said to be in **Echelon form** if

(i) Every row of A which has all its entries 0 (zero) occurs below every row which has a non-zero entry.

(ii) The first non-zero entry in each non-zero row is equal to 1 (unity). (Some authors do not insist that first non-zero element must be 1)

(iii) The number of zeros before the first non-zero element in a row is less than the number of such zeros in the next row.

Important Result: *The rank of a matrix in* ***Echelon form*** *is equal to the number of NON-ZERO rows of the matrix.*

▶ Example 5.7.8 *See that the matrix A given below is in Echelon form.*

$$A = \begin{bmatrix} 0 & 1 & 2 & 3 \\ 0 & 0 & 1 & -1 \\ 0 & 0 & 0 & 0 \end{bmatrix}.$$

since the number of non-zero rows in this echelon form is two we conclude that rank $(A) = 2$.

*You could reduce it to normal form and would get the same conclusion—**rank is two.***

▶ **Example 5.7.9** *Find the rank of the matrix*

$$A = \begin{bmatrix} 1 & 2 & -1 & 3 \\ 2 & 4 & -4 & 7 \\ -1 & -2 & -1 & -2 \end{bmatrix}$$

(i) *by reducing* A *to* ***Echelon form;***

(ii) *by reducing* A *to* ***Normal form.***

Solution: Apply the sequence of elementary transformations on A $R_{21}(-2)$; $R_{31}(1)$; $C_{21}(-2)$; $C_{31}(1)$; $C_{41}(-3)$; C_{24}; $R_{32}(-1)$.

$$\text{(i) } A \sim \underset{R_{21}(-2)}{\begin{bmatrix} 1 & 2 & -1 & 3 \\ 0 & 0 & -2 & 1 \\ -1 & -2 & -1 & -2 \end{bmatrix}} \sim \underset{R_{31}(1)}{\begin{bmatrix} 1 & 2 & -1 & 3 \\ 0 & 0 & -2 & 1 \\ 0 & 0 & -2 & 1 \end{bmatrix}} \sim \underset{C_{21}(-2)}{\begin{bmatrix} 1 & 0 & -1 & 3 \\ 0 & 0 & -2 & 1 \\ 0 & 0 & -2 & -1 \end{bmatrix}}$$

$$\sim \underset{C_{31}(1)}{\begin{bmatrix} 1 & 0 & 0 & 3 \\ 0 & 0 & -2 & 1 \\ 0 & 0 & -2 & 1 \end{bmatrix}} \sim \underset{C_{41}(-3)}{\begin{bmatrix} 1 & 0 & 0 & 0 \\ 0 & 0 & -2 & 1 \\ 0 & 0 & -2 & 1 \end{bmatrix}} \sim \underset{C_{24}}{\begin{bmatrix} 1 & 0 & 0 & 0 \\ 0 & 1 & -2 & 1 \\ 0 & 1 & -2 & 0 \end{bmatrix}}$$

$$\sim \underset{R_{32}(-1)}{\begin{bmatrix} 1 & 0 & 0 & 0 \\ 0 & 1 & -2 & 0 \\ 0 & 0 & 0 & 0 \end{bmatrix}}$$

(Reduced Echelon matrix with two non-zero rows and hence rank = 2)

(ii) We can proceed further to reduce it to normal form: Apply the elementary transformation $C_{32}(2)$:

$$\sim \underset{C_{32}(2)}{\begin{bmatrix} 1 & 0 & 0 & 0 \\ 0 & 1 & 0 & 0 \\ 0 & 0 & 0 & 0 \end{bmatrix}} \sim \begin{bmatrix} I_2 & 0 \\ 0 & 0 \end{bmatrix} \quad \textbf{(Normal form)}$$

Since the unit matrix to which the normal form is reduced is I_2; hence the rank of A = 2.

▶ **Example 5.7.10** *Reduce the matrix*

$$A = \begin{bmatrix} 1 & -1 & 2 & -3 \\ 4 & 1 & 0 & 2 \\ 0 & 3 & 0 & 4 \\ 0 & 1 & 0 & 2 \end{bmatrix}$$

to the normal form and hence determine its rank.

Solution:

$$A = \begin{bmatrix} 1 & -1 & 2 & -3 \\ 4 & 1 & 0 & 2 \\ 0 & 3 & 0 & 4 \\ 0 & 1 & 0 & 2 \end{bmatrix} \xrightarrow[\substack{C_{31}(-2)\\ C_{41}(3)}]{C_{21}(1)} \begin{bmatrix} 1 & 0 & 0 & 0 \\ 4 & 5 & -8 & 14 \\ 0 & 3 & 0 & 4 \\ 0 & 1 & 0 & 2 \end{bmatrix}$$

$$\xrightarrow{R_{21}(-4)} \begin{bmatrix} 1 & 0 & 0 & 0 \\ 0 & 5 & -8 & 14 \\ 0 & 3 & 0 & 4 \\ 0 & 1 & 0 & 2 \end{bmatrix} \xrightarrow{R_{24}} \begin{bmatrix} 1 & 0 & 0 & 0 \\ 0 & 1 & 0 & 2 \\ 0 & 3 & 0 & 4 \\ 0 & 5 & -8 & 14 \end{bmatrix}$$

$$\xrightarrow{C_{42}(-2)} \begin{bmatrix} 1 & 0 & 0 & 0 \\ 0 & 1 & 0 & 0 \\ 0 & 3 & 0 & -2 \\ 0 & 5 & -8 & 4 \end{bmatrix} \xrightarrow[R_{42}(-5)]{R_{32}(-3)} \begin{bmatrix} 1 & 0 & 0 & 0 \\ 0 & 1 & 0 & 0 \\ 0 & 0 & 0 & -2 \\ 0 & 0 & -8 & 4 \end{bmatrix}$$

$$\xrightarrow{C_{34}} \begin{bmatrix} 1 & 0 & 0 & 0 \\ 0 & 1 & 0 & 0 \\ 0 & 0 & -2 & 0 \\ 0 & 0 & 4 & -8 \end{bmatrix} \xrightarrow{C_3(-\frac{1}{2})} \begin{bmatrix} 1 & 0 & 0 & 0 \\ 0 & 1 & 0 & 0 \\ 0 & 0 & 1 & 0 \\ 0 & 0 & -2 & -8 \end{bmatrix}$$

$$\xrightarrow{R_{43}(2)} \begin{bmatrix} 1 & 0 & 0 & 0 \\ 0 & 1 & 0 & 0 \\ 0 & 0 & 1 & 0 \\ 0 & 0 & 0 & -8 \end{bmatrix} \xrightarrow{C_4(-\frac{1}{8})} \begin{bmatrix} 1 & 0 & 0 & 0 \\ 0 & 1 & 0 & 0 \\ 0 & 0 & 1 & 0 \\ 0 & 0 & 0 & 1 \end{bmatrix}$$

$= [I_4]$ **Normal form with unit matrix I_4 and hence rank $A = 4$**

5.8 Consistency and Solution of System of Linear Equations with not more than three variables

1. Consider a system of m linear equations in n unknowns $x_1, x_2, \cdots, x_n$.

$$\left.\begin{aligned} a_{11}x_1 + a_{12}x_2 + \cdots + a_{1n}x_n &= b_1 \\ a_{21}x_1 + a_{22}x_2 + \cdots + a_{2n}x_n &= b_2 \\ \cdots\cdots\cdots\cdots\cdots\cdots\cdots\cdots &\cdots\cdots \\ a_{m1}x_1 + a_{m2}x_2 + \cdots + a_{mn}x_n &= b_m \end{aligned}\right\}. \qquad \text{(T)}$$

If the right side constants (b's) are **all zero** then the system is called **Homogeneous system;** otherwise the system of equations is **Non-homogeneous.**

We shall consider those cases where the coefficients (a's) and the constant terms (b's) are all real numbers (not all zero).

But a **solution of the system** (S) we mean a set of values of $x_1, x_2, \cdots, x_n$ which satisfy simultaneously the m equations. When the system has **a solution,** we say that the system of equations (S) is **consistent;** otherwise the system is said to be in-consistent. A consistent system of equations may have **just one solution** or may have infinitely many solutions.

Remember: A system of equations have

either No solution
or Just one Solution (Unique solution)
or infinitely many solutions.

We usually take

$$A = \begin{bmatrix} a_{11} & a_{12} & \cdots & a_{1n} \\ a_{21} & a_{22} & \cdots & a_{2n} \\ \cdots & \cdots & \cdots & \cdots \\ a_{m1} & a_{m2} & \cdots & a_{mn} \end{bmatrix}$$

and call it the **coefficient matrix** of the system **(S)**.

Also we write

$$X = \begin{bmatrix} x_1 \\ x_2 \\ \vdots \\ x_n \end{bmatrix} \text{ and } B = \begin{bmatrix} b_1 \\ b_2 \\ \vdots \\ b_m \end{bmatrix}$$

so that the system (S) can be written in the form of a single matrix equation $AX = B$.

The matrix

$$[A \quad B] = \begin{bmatrix} a_{11} & a_{12} & \cdots & a_{1n} & b_1 \\ a_{21} & a_{22} & \cdots & a_{2n} & b_2 \\ \cdots & \cdots & \cdots & \cdots & \cdots \\ a_{m1} & a_{m2} & \cdots & a_{mn} & b_n \end{bmatrix}$$

is known as the **Augmented Matrix** of the system of equations (S).

2. Condition of Consistency: A very important theorem.

Theorem 5.8.1 *The system of equations* $AX = B$ *is consistent (i.e., possesses at least one solution) if and only if*

$$\begin{aligned} &\textit{Rank of the coefficient matrix } A \\ = \ &\textit{Rank of the Augmented matrix } [A \quad B] \end{aligned}$$

The proof is beyond the scope of the present text. But the students would assume the truth of the theorem and apply it whenever necessary. We shall further odd:

Theorem 5.8.2 *If* A *is an* n*-rowed non-singular matrix (i.e.,* A *is a square matrix of order* n *and* $\det A \neq 0$*) then the system has just one solution (i.e., unique solution).*

3. Illustrative Examples

▶ Example 5.8.1 (a) *Suppose we have 2 equations in 2 unknowns* x *and* y :

$$\left.\begin{aligned} 3x + 4y &= 7 \\ 6x + 8y &= 13 \end{aligned}\right\}.$$

Solution: Here the coefficient matrix $\begin{bmatrix} 3 & 4 \\ 6 & 8 \end{bmatrix}$ has the **rank one** but the Augmented matrix $\begin{bmatrix} 3 & 4 & 7 \\ 6 & 8 & 13 \end{bmatrix}$ has the **rank two** (at least one minor of order two, say $\begin{vmatrix} 3 & 7 \\ 6 & 13 \end{vmatrix} \neq 0$).

$\therefore$ By Theorem 5.8.1 the system has no solution (i.e., **Inconsistent**).

Check: Multiply the first equation by 2 and obtain $6x + 8y = 14$. Comparing with the second equation we see that $14 = 13$ which is certainly not true.

(b) *Suppose we have 3 equations in 4 unknown:*

$$\left.\begin{aligned} x_1 - 2x_2 + x_3 - x_4 &= -1 \\ 3x_1 - 2x_3 + 3x_4 &= -4 \\ 5x_1 - 4x_2 + x_4 &= -3 \end{aligned}\right\}.$$

Solution: Here the coefficient matrix

$$A = \begin{bmatrix} 1 & -2 & 1 & -1 \\ 3 & 0 & -2 & 3 \\ 5 & -4 & 0 & 1 \end{bmatrix}_{3\times 4}$$

has rank 2 but the Augmented matrix

$$[A \quad B] = \begin{bmatrix} 1 & -2 & 1 & -1 & -1 \\ 3 & 0 & -2 & 3 & -4 \\ 5 & -4 & 0 & 1 & -3 \end{bmatrix}_{3\times 5} \text{ has rank 3.}$$

[One can find the ranks by reducing the matrix into *Echelon form* or *Normal form*:]

By **Theorem 5.8.1** the system of equations has no solution (or **Inconsistent**)

Check: Multiply the first equation by 3 and add with second equation. We then obtain

$$6x_1 - 6x_2 + x_3 = -7.$$

Again add the first and third equations; we obtain

$$6x_1 - 6x_2 + x_3 = -4.$$

We arrive at the conclusion $-7 = -4$ which is certainly not true, i.e., the system is *inconsistent.*

(c) *Again it can be checked that for the system of equations*

$$\left.\begin{aligned} x_1 - 2x_2 + x_3 - x_4 &= -1 \\ 3x_1 - 2x_3 + 3x_4 &= 4 \\ 5x_1 - 4x_2 + x_4 &= 2 \end{aligned}\right\}$$

the coefficient matrix $\begin{bmatrix} 1 & -2 & 1 & -1 \\ 3 & 0 & -2 & 3 \\ 5 & -4 & 0 & 1 \end{bmatrix}$

and the augmented matrix $\begin{bmatrix} 1 & -2 & 1 & -1 & -1 \\ 3 & 0 & -2 & 3 & 4 \\ 5 & -4 & 0 & 1 & 2 \end{bmatrix}$

have the same rank 2.

Solution: Hence the system has **solution** or **solutions**. In other words, the system is **consistent**.

How to solve? Here number of unknowns = 4
and rank of coefficient matrix = 2
Their difference = 4 − 2 = 2

We shall be able to solve for two of the unknowns in terms of the other two. Solve the first two equations for x_1 and x_2 in terms of x_3 and x_4.

We easily obtain

$$\left.\begin{aligned} x_1 &= \tfrac{2}{3}x_3 - x_4 + \tfrac{4}{3} \\ x_2 &= \tfrac{5}{3}x_3 - x_4 + \tfrac{7}{6} \end{aligned}\right\}.$$

If we put these values of x_1 and x_2 in the third equation, we shall see that it is identically satisfied. Thus given x_3 and x_4 arbitrary values we can uniquely determine x_1 and x_2 and thereby we get a solution for x_1, x_2, x_3, x_4; say, for example, we take $x_3 = 0$, $x_4 = 0$. then clearly $x_1 = \frac{4}{3}$, $x_2 = \frac{7}{6}$ and we get one solution $\left(\frac{4}{3}, \frac{7}{6}, 0, 0\right)$. Since x_3 and x_4 can be given arbitrary values, we get here infinitely many solutions for this system of equations.

Note that we could as well express x_3 and x_4 in terms of x_1 and x_2. Only point to see is that the determinant of the coefficients of the two variables (which we express in terms of the other two) must not be zero.

(d) *Consider the following system of 2 equations in three unknowns:*

$$\left.\begin{aligned} x_1 + 2x_2 + x_3 &= 4 \\ 2x_1 + x_2 + 5x_3 &= 5 \end{aligned}\right\}$$

Solution: The given system can be written as $AX = B$ where

$$A = \text{coefficient matrix} = \begin{bmatrix} 1 & 2 & 1 \\ 2 & 1 & 5 \end{bmatrix};\ X = \begin{bmatrix} x_1 \\ x_2 \\ x_3 \end{bmatrix};\ B = \begin{bmatrix} 4 \\ 5 \end{bmatrix}$$

The system is consistent since ranks of A and (A, B) are equal and equals to 2.

Number of unknowns = 3, rank = 2; difference = 3 − 2 = 1

We can express two of the unknowns in terms of the third unknown.

We are to choose those two unknowns whose coefficient determinant is not zero.

(a) x_1, x_2 : coeff. det $= \begin{vmatrix} 1 & 2 \\ 2 & 1 \end{vmatrix} = -3\ (\neq 0)$

So we can express x_1, x_2 in terms of x_3:

$$\left.\begin{aligned} x_1 + 2x_2 &= 4 - x_3 \\ 2x_1 + x_2 &= 5 - 5x_3 \end{aligned}\right\} \Rightarrow x_1 = 2 - 3x_3,\ x_2 = 1 + x_3$$

choosing $x_3 = 0$, we get **One solution** (2,1,0) other solutions can be obtained by putting x_3 = any other value.

(b) x_1, x_3 : coeff. det $= \begin{vmatrix} 1 & 1 \\ 2 & 5 \end{vmatrix} = 3\ (\neq 0)$.

So we can express x_1, x_3 in terms of x_2:

$$\left.\begin{array}{l} x_1 + x_3 = 4 - 2x_2 \\ 2x_1 + 5x_3 = 5 - x_2 \end{array}\right\} \Rightarrow x_1 = 5 - 3x_2;\ x_3 = -1 + x_2$$

Choosing $x_2 = 0$, we get **One solution** $(5, 0, -1)$.

Other solutions can be obtained by putting x_2 = any other value.

(c) x_2, x_3 : coeff. det $= \begin{vmatrix} 2 & 1 \\ 1 & 5 \end{vmatrix} = 9\ (\neq 0)$.

So we can express x_2, x_3 in terms of x_1.

Here one solution will be $(0, \frac{5}{3}, \frac{7}{3})$ if we put $x_1 = 0$.

Other solutions can be obtained by putting x_1 = any other value.

We have thus proved that the given system is consistent and it has infinitely many solutions.

4. Homogeneous Equations

We have already mentioned that if all the constants on the right side of the linear system of equations (S) are zero, then the system is called a **system of homogeneous linear equations.** So in matrix form, we write the system (S) as $AX = 0$. Now rank (A) is the same as rank $[A\ \ 0]$ and as such the set of homogeneous equations is **always consistent**. In fact $(x_1 = 0, x_2 = 0, \ldots, x_n = 0)$ is always a solution of the homogeneous system. This solution is called **Trivial solution.** We are, therefore, in this case, interested to know whether or not there are solutions other than the Trivial Solution.

Theorem 5.8.3 *A necessary and sufficient condition that a system of m homogeneous linear equations in n unknowns $(m \leq n)$ of the matrix equation $AX = 0$ have non-trivial solutions is that the coefficient matrix A has a rank less than the number of unknowns. In case $m > n$, the coefficient matrix A has a rank $< n$ and the system has non-trivial solutions.*

In case $m = n$, then the condition for existence of non-trivial solutions is that the coefficient matrix must be singular ($\det A = 0$ *or rank of* A *is less than* n*).*

Note 1. *In the system of homogeneous linear equations in n unknowns $AX = 0$ $(m \leq n)$ if the rank of A is r, there r of the unknowns (suitably chosen) can be expressed as linear functions of the remaining $n - r$ unknowns. Giving arbitrary values to these $n - r$ unknowns we get an infinite number of solutions.*

5. Illustrative Examples

▶ Example 5.8.2 (a) *The system of equations*

$$\left.\begin{aligned} x - y + z &= 0 \\ x + 2y - z &= 0 \\ 2x + y + 3z &= 0 \end{aligned}\right\}$$

is a set of 3 linear homogeneous equations in 3 unknowns.

Solution: Here the coefficient matrix

$$A = \begin{bmatrix} 1 & -1 & 1 \\ 1 & 2 & -1 \\ 2 & 1 & 3 \end{bmatrix}$$

has rank 3 ($\because$ det $A \neq 0$). Hence the system **has only the trivial solution** $x = 0, y = 0, z = 0$ and no other solution.

(b) *Consider the homogeneous system of 3 linear equations:*

$$\left.\begin{aligned} x - 2y + z &= 0 \\ x - 2y - z &= 0 \\ 2x - 4y - 5z &= 0 \end{aligned}\right\}$$

Solution: This system has the coefficient matrix $A = \begin{bmatrix} 1 & -2 & 1 \\ 1 & -2 & -1 \\ 2 & -4 & -5 \end{bmatrix}$ which has rank = 2 (less than the number of unknowns)

Hence if possesses **non-trivial solutions:**

From the first two equations $z = 0$ and $x = 2y$. The third equation satisfied is identically. Since y is arbitrary, we write the solution as: $(2y, y, 0)$ **or** $y(2, 1, 0)$.

Otherwise: The given system is equivalent to $AX = 0$,

$$\text{or} \quad \begin{bmatrix} 1 & -2 & 1 \\ 1 & -2 & -1 \\ 2 & -4 & -5 \end{bmatrix} \begin{bmatrix} x \\ y \\ z \end{bmatrix} = 0.$$

By E.T. $R_{21}(-1), R_{31}(-2)$ we have

$$\begin{bmatrix} 1 & -2 & 1 \\ 0 & 0 & -2 \\ 0 & 0 & -7 \end{bmatrix} \begin{bmatrix} x \\ y \\ z \end{bmatrix} = 0. \qquad \text{(I)}$$

The coefficient matrix has rank 2; Number of unknown = 3, Difference = 1.

$\therefore$ Two variables can be can be expressed in terms of one variable.

From the reduced (I) we find the gives system is equivalent to

$$x - 2y + z = 0; \quad z = 0.$$

Thus $z = 0$, $x = 2y$. Solution is $y = t$, $x = 2t$, $z = 0$ (t *arbitrary*).

(c) *Solve completely the system of equations:*

$$\left.\begin{array}{r} x+\quad y+\ z=0 \\ 2x-\quad y-3z=0 \\ 3x-\ 5y+4z=0 \\ x+17y+4z=0 \end{array}\right\}.$$

Solution: The given system of equations is equivalent to $AX = 0$,

$$\text{where } A = \begin{bmatrix} 1 & 1 & 1 \\ 2 & -1 & -3 \\ 3 & -5 & 4 \\ 1 & 17 & 4 \end{bmatrix}.$$

We first find the rank of A by reducing to echelon form:

$$A \sim \begin{bmatrix} 1 & 1 & 1 \\ 0 & -3 & -5 \\ 0 & -8 & 1 \\ 0 & 16 & 3 \end{bmatrix} \sim \begin{bmatrix} 1 & 1 & 1 \\ 0 & -3 & -5 \\ 0 & -24 & 3 \\ 0 & 48 & 9 \end{bmatrix}$$

[Apply: $R_{21}(-2), R_{31}(-3), R_{41}(-1)$] [Apply: $R_3(3), R_4(3)$]

$$\sim \begin{bmatrix} 1 & 1 & 1 \\ 0 & -3 & -5 \\ 0 & 0 & 43 \\ 0 & 0 & -71 \end{bmatrix} \sim \begin{bmatrix} 1 & 1 & 1 \\ 0 & -3 & -5 \\ 0 & 0 & 43 \\ 0 & 0 & 0 \end{bmatrix}$$

[Apply: $R_{32}(-8), R_{42}(16)$] [Apply: $R_{43}\left(\frac{71}{43}\right)$]

$\therefore$ rank of $A = 3 =$ the number of unknowns (B)

$\therefore$ the gives system has no non-zero solution

$\therefore$ here the only solution is $x = 0$, $y = 0$, $z = 0$.

Exercises on Chapter 5

[A]

[Addition, Scalar Multiplication and Multiplication of Matrices]

1. (a) If $A = \begin{bmatrix} 1 & 2 \\ 3 & 4 \end{bmatrix}$ and $B = \begin{bmatrix} 2 & 1 \\ 4 & 3 \end{bmatrix}$, find

$$A + B, A - B, (A - B)\cdot(A + B) \text{ and } A^2 - B^2.$$

(Note that the last two results are not equal. Why?)

(b) If $A = \begin{bmatrix} 1 & 2 \\ 2 & 4 \end{bmatrix}$, $B = \begin{bmatrix} 4 & -2 \\ -2 & 1 \end{bmatrix}$ and $C = \begin{bmatrix} 1 & 2 \\ 3 & 4 \end{bmatrix}$.

Verify that $(A-B)(A+B) = A^2-B^2$ but see that $(A-C)(A+C) \neq A^2-C^2$.

[C.P. 2004]

(c) If $A = \begin{bmatrix} 0 & 1 \\ 1 & 1 \end{bmatrix}$, $B = \begin{bmatrix} 0 & -1 \\ 1 & 0 \end{bmatrix}$ be two matrices, then show that $(A+B)\cdot(A-B) \neq A^2 - B^2$. [C.P. 2003]

2. If $A = \begin{bmatrix} 2 & 1 & 2 \\ 3 & 5 & 7 \\ 1 & 0 & 1 \end{bmatrix}$ and $B = \begin{bmatrix} -3 & 1 & 0 \\ 6 & 2 & 1 \\ 1 & -1 & 2 \end{bmatrix}$, evaluate $A+B$, $A-B$ and AB, BA.

[Here is an example where $AB \neq BA$ though both are defined]

3. If $A = \begin{bmatrix} 2 & 3 \\ 4 & -1 \end{bmatrix}$, $B = \begin{bmatrix} 3 & -2 \\ 2 & 1 \end{bmatrix}$, $C = \begin{bmatrix} 1 & 2 \\ 3 & 4 \end{bmatrix}$.

Verify: $A\cdot(BC) = (AB)\cdot C$ and $A\cdot(B+C) = AB + AC$.
Also $(A+B)\cdot C = AC + BC$.

4. If $A = \begin{bmatrix} 2 & 3 & 1 \\ 0 & 1 & -1 \\ -1 & 0 & 2 \end{bmatrix}$, $U = \begin{bmatrix} 1 \\ 1 \\ 0 \end{bmatrix}$ and $V = \begin{bmatrix} 1 \\ 1 \\ -2 \end{bmatrix}$ evaluate

$$A\cdot U, A^2U, AV \text{ and } A^2V \quad (A^2 = A\cdot A).$$

5. Find the products AB and BA, where

$$A = \begin{bmatrix} 1 & 2 & 0 \\ 3 & 0 & 4 \\ 0 & 5 & 1 \end{bmatrix} \text{ and } B = \begin{bmatrix} -1 & 0 & 3 \\ 2 & 0 & 4 \\ -3 & 1 & 0 \end{bmatrix}.$$

Are the above products equal? [C.P. 1987]

6. Find x, y, z and t, where the matrices

$$\begin{bmatrix} x+y & y-z \\ 5-t & 7+x \end{bmatrix} = \begin{bmatrix} t-x & z-t \\ z-y & x+z+t \end{bmatrix}$$

[C.P. 1987]

[See that $x+y = t-x$; $y-z = z-t$; $5-t = z-y$; $7+x = x+z+t$. Solve and obtain $x = 1$, $y = 2$, $z = 3$, $t = 4$.]

7. If $A = \begin{bmatrix} 4 & 2 \\ -1 & 1 \end{bmatrix}$ and $I = \begin{bmatrix} 1 & 0 \\ 0 & 1 \end{bmatrix}$, prove that $(A-2I)\cdot(A-3I) = \begin{bmatrix} 0 & 0 \\ 0 & 0 \end{bmatrix}$.

[C.P. 1992, 1988]

[See that left side

$$= \left\{\begin{bmatrix} 4 & 2 \\ -1 & 1 \end{bmatrix} - \begin{bmatrix} 2 & 0 \\ 0 & 2 \end{bmatrix}\right\}\cdot\left\{\begin{bmatrix} 4 & 2 \\ -1 & 1 \end{bmatrix} - \begin{bmatrix} 3 & 0 \\ 0 & 3 \end{bmatrix}\right\}$$

$$= \begin{bmatrix} 2 & 2 \\ -1 & -1 \end{bmatrix}\cdot\begin{bmatrix} 1 & 2 \\ -1 & -2 \end{bmatrix} = \begin{bmatrix} 0 & 0 \\ 0 & 0 \end{bmatrix}.$$

Here is an example which shows that the product of two non-zero matrices may be a zero-matrix.]

8. Find the matrix $A^2 - 3A - 13I$, where $A = \begin{bmatrix} 2 & 5 \\ 3 & 1 \end{bmatrix}$ and $I = \begin{bmatrix} 1 & 0 \\ 0 & 1 \end{bmatrix}$.

[C.P. 1991]

9. Given $A = \begin{bmatrix} 1 & 3 & 2 \\ 2 & 4 & 9 \\ 1 & 0 & 0 \end{bmatrix}$, $B = \begin{bmatrix} 3 & -1 & 2 \\ 4 & 2 & 5 \\ 2 & 0 & 3 \end{bmatrix}$ and $C = \begin{bmatrix} 4 & 1 & 2 \\ 0 & 3 & 2 \\ 1 & -2 & 3 \end{bmatrix}$,

   (a) Compute: $A + B, B + C, C + A, A - C, -2A, B - C$.

   (b) Verify: $A + (B - C) = (A + B) - C$.

   (c) Find a matrix X such that $A + X = B$.

10. Complete the Multiplication Table:

•	α	β	γ	δ	ω	ϵ	τ	θ
α								
β								
γ								
δ								
ω								
ϵ								
τ								
θ								

where

$$\alpha = \begin{bmatrix} 1 & 0 \\ 0 & 1 \end{bmatrix}, \beta = \begin{bmatrix} 0 & 1 \\ -1 & 0 \end{bmatrix}, \gamma = \begin{bmatrix} 0 & -1 \\ 1 & 0 \end{bmatrix}, \delta = \begin{bmatrix} -1 & 0 \\ 0 & -1 \end{bmatrix}$$

$$\omega = \begin{bmatrix} i & 0 \\ 0 & -i \end{bmatrix}, \epsilon = \begin{bmatrix} -i & 0 \\ 0 & i \end{bmatrix}, \tau = \begin{bmatrix} 0 & -i \\ -i & 0 \end{bmatrix}, \theta = \begin{bmatrix} 0 & i \\ i & 0 \end{bmatrix} \; [i^2 = -1].$$

Verify that the product of any two of the above eight matrices is also a matrix of the same set.

11. Given

$$A = \begin{bmatrix} 1 & -3 & 2 \\ 2 & 1 & -3 \\ 4 & -3 & -1 \end{bmatrix}, B = \begin{bmatrix} 1 & 4 & 1 & 0 \\ 2 & 1 & 1 & 1 \\ 1 & -2 & 1 & 2 \end{bmatrix}, C = \begin{bmatrix} 2 & 1 & -1 & -2 \\ 3 & -2 & -1 & -1 \\ 2 & -5 & -1 & 0 \end{bmatrix}.$$

Verify that $AB = AC$.

[Here is an example which shows that in Matrix multiplication AB may be equal to AC, without being $B = C$, even when $A \neq 0$.]

12. Given $A = \begin{bmatrix} 1 & -1 & 1 \\ 2 & 0 & 1 \end{bmatrix}$, $B = \begin{bmatrix} 1 & -1 & 0 \\ 0 & 1 & -1 \\ 1 & 1 & 1 \end{bmatrix}$, $C = \begin{bmatrix} 1 & 0 \\ 0 & 1 \\ 1 & 1 \end{bmatrix}$.

Verify:

(a) $(AB)C = A(BC)$ (Associative Law)

(b) $A(B+C) = AB + AC$ (Left Distributive Law)

(c) $(A+B)C = AC + BC$ (Right Distributive Law).

13. If $A = \begin{bmatrix} 2 & -3 & -5 \\ -1 & 4 & 5 \\ 1 & -3 & -4 \end{bmatrix}$, $B = \begin{bmatrix} -1 & 3 & 5 \\ 1 & -3 & -5 \\ -1 & 3 & 5 \end{bmatrix}$ and $C = \begin{bmatrix} 2 & -2 & -4 \\ -1 & 3 & 4 \\ 1 & -2 & -3 \end{bmatrix}$

(a) Verify: $AB = BA = 0$ (Null Matrix)

(b) Verify: $AC = A$, $CA = C$.

(c) Verify: $A^2 - B^2 = (A-B)\cdot(A+B)$; $(A \pm B)^2 = A^2 + B^2$.

14. (a) Prove that $[1 \ \ 2 \ \ 3 \ \ 4]\cdot \begin{bmatrix} 1 \\ 2 \\ 3 \\ 4 \end{bmatrix} = [30]$.

(b) Verify: $[2 \ \ 1 \ \ -10]\cdot \begin{bmatrix} 4 & -1 & 2 \\ -1 & 0 & 1 \\ 2 & 1 & 0 \end{bmatrix}\cdot[1 \ \ 2 \ \ 3 \ \ 4] = [-46]$.

(c) Find the product $\begin{bmatrix} 1 \\ 2 \\ 3 \\ 4 \end{bmatrix}\cdot[1 \ \ 2 \ \ 3 \ \ 4]$.

15. Let

$$A = \begin{bmatrix} 2 & 0 \\ 3 & -1 \\ -5 & 3 \\ 1 & 4 \end{bmatrix}, \ B = \begin{bmatrix} -1 & 1 \\ 0 & -3 \\ 3 & 0 \\ -2 & 1 \end{bmatrix}, \ C = \begin{bmatrix} 0 & 1 \\ -1 & 0 \end{bmatrix},$$

$$D = \begin{bmatrix} 0 & 1 & 2 \\ -1 & 0 & 3 \end{bmatrix}, \ E = \begin{bmatrix} 0 & 1 & -1 & 1 \\ 0 & 0 & 1 & -1 \\ 0 & 0 & 0 & 1 \\ 0 & 0 & 0 & 0 \end{bmatrix}.$$

Determine the following sums and products, when they exist:

$$A - 2B; \quad 2A + 3B; \quad A + C; \quad AC;$$
$$AD; \quad (A+B)D; \quad DE; \quad A^T(BC); \quad D^TB^T.$$

(T stands for Transpose)

16. (a) If $A = \begin{bmatrix} 3 & 2 \\ -2 & -1 \end{bmatrix}$, $I = \begin{bmatrix} 1 & 0 \\ 0 & 1 \end{bmatrix}$. Verify that $(A - I)^2 = 0$.

(b) If $I = \begin{bmatrix} 1 & 0 \\ 0 & 1 \end{bmatrix}$ and $J = \begin{bmatrix} 0 & 1 \\ -1 & 0 \end{bmatrix}$. Show that $J^2 = -I$.

17. If $P = \begin{bmatrix} \cos\theta & \sin\theta \\ -\sin\theta & \cos\theta \end{bmatrix}$ and $= Q = \begin{bmatrix} \cos\phi & \sin\phi \\ -\sin\phi & \cos\phi \end{bmatrix}$.
Show that PQ can be in the same form as that of P or Q.

18. (a) Show that the matrix $A = \begin{bmatrix} 1 & 2 & 3 \\ 3 & -2 & 1 \\ 4 & 2 & 1 \end{bmatrix}$ satisfies the equation

$$A^2 - 23A - 40I = 0.$$

Hence, obtain A^{-1}.

(b) If $A = \begin{bmatrix} 2 & -1 \\ -1 & 2 \end{bmatrix}$, verify that $A^2 - 4A + 3I = 0$ and hence, find A^{-1}.

[B]

(Different Types of Matrices)

1. Verify that the following two matrices are orthogonal matrices:

(a) $\frac{1}{3}\begin{bmatrix} 1 & 2 & 2 \\ 2 & 1 & -2 \\ -2 & 2 & -1 \end{bmatrix}$; (b) $\frac{1}{3}\begin{bmatrix} 1 & -2 & 2 \\ -2 & 1 & 2 \\ -2 & -2 & -1 \end{bmatrix}$. [C.P. 2002]

2. If $A = \begin{bmatrix} 0 & 0 & 1 \\ 0 & 1 & 0 \\ 1 & 0 & 0 \end{bmatrix}$, verify $A^{-1} = A$ (such matrices are called **self-reciprocal**).

3. Show that $S = \begin{bmatrix} \frac{1}{\sqrt{2}} & -\frac{1}{\sqrt{2}} \\ \frac{1}{\sqrt{2}} & \frac{1}{\sqrt{2}} \end{bmatrix}$ is orthogonal.

If $P = \begin{bmatrix} 1 & 3 \\ 3 & 1 \end{bmatrix}$, show that SPS^T is a **diagonal matrix.**

4. The non-zero matrix A is **nilpotent** of order $k \geq 2$ if $A^k = 0$ and $A^{k-1} \neq 0$. Hence,

$$(I - A)^{-1} = I + A + A^2 + \cdots + A^{k-1}].$$

Show that the matrix

$$A = \begin{bmatrix} 1 & 1 & 3 \\ 5 & 2 & 6 \\ -2 & -1 & -3 \end{bmatrix}$$

is *nilpotent* of order 3 and, hence, find the inverse of

$$\begin{bmatrix} 0 & 1 & -1 \\ -5 & -1 & -6 \\ 2 & 1 & 4 \end{bmatrix}.$$

5. Show that the two matrices

$$A = \begin{bmatrix} 0 & 1 & -1 \\ 4 & -3 & 4 \\ 3 & -3 & 4 \end{bmatrix} \text{ and } B = \begin{bmatrix} 4 & 3 & 3 \\ -1 & 0 & -1 \\ -4 & -4 & -3 \end{bmatrix}$$

are **involutory** (i.e., *square of the matrix = Identity Matrix*)

6. If $A = \begin{bmatrix} 1 & -1 & 1 \\ 2 & -1 & 0 \\ 1 & 0 & 0 \end{bmatrix}$, verify: $A^3 = A \cdot A^2 = A^2 \cdot A = I$, and hence, obtain A^{-1}.

7. (a) Verify that $A = \begin{bmatrix} 2 & -2 & -4 \\ -1 & 3 & 4 \\ 1 & -2 & -3 \end{bmatrix}$ is **idempotent** (i.e., the square of the matrix is the matrix itself, i.e., $A^2 = A$).

(b) Show that if $AB = A$ and $BA = B$, then A and B are idempotent.

8. (a) If A and B are square matrices of order n, then prove that AB is **symmetric** if, and only if, A and B commute.

(b) If A is a square matrix, verify that $A + A^T$ is **symmetric** and $A - A^T$ is skew-symmetric.

(c) Let A be a square matrix of order m and P be a matrix of order $m \times n$. If A be **skew-symmetric**, then prove that P^TAP is **skew-symmetric.** What can you say about P^TAP if A is **symmetric**?

9. A square matrix A is **idempotent** i.e., $A^2 = A$ and $(A - A^T)^2 = 0$. Prove that

$$A \cdot A^T + A^T A = A + A^T$$

and hence deduce that $(A \cdot A^T)^2 = AA^T$, i.e., $A \cdot A^T$ is **idempotent.**

[C]

(Invertible Matrices)

1. Determine which of the following matrices have inverses and obtain the inverse when it exists:

(a) $\begin{bmatrix} 2 & -1 \\ -4 & 3 \end{bmatrix}$; (b) $\begin{bmatrix} 1 & -1 \\ -1 & 1 \end{bmatrix}$; (c) $\begin{bmatrix} \cos\theta & \sin\theta \\ -\sin\theta & \cos\theta \end{bmatrix}$.

2. Prove that $\begin{bmatrix} a & b \\ c & d \end{bmatrix}^{-1} = \dfrac{1}{ad - bc}\begin{bmatrix} d & -b \\ -c & a \end{bmatrix}$, given $ad - bc \neq 0$.

3. (a) Find the inverse of the matrix $A = \begin{bmatrix} 2 & 1 & 4 \\ 1 & 0 & 2 \\ 2 & 3 & 1 \end{bmatrix}$ and hence, solve the set of equations

$$\left.\begin{aligned} 2x + y + 4z &= 2 \\ x + 2z &= 3 \\ 2x + 3y + z &= -6 \end{aligned}\right\}.$$

(b) Find the inverse of the matrix $\begin{bmatrix} 2 & 1 & 1 \\ 1 & -1 & 0 \\ 2 & 1 & -1 \end{bmatrix}$ and using it, solve the following system of equations:

$$\left.\begin{aligned} 2x + y + z &= 5 \\ x - y &= 0 \\ 2x + y - z &= 1 \end{aligned}\right\}.$$ **[C.P. 2005]**

4. Show that if $A^2 - A + I = 0$, then A has an inverse and find this inverse.

5. If $S = \begin{bmatrix} 0 & 1 & 0 \\ 0 & 0 & 1 \\ 0 & 0 & 0 \end{bmatrix}$, evaluate S^2 and S^3.

If $A = aI + bS + cS^2$ and $B = a^2I - abS + (b^2 - ac)S^2$, where a, b, c are constants and I is the third order unit matrix, evaluate the product AB. Deduce that if $a \neq 0$, A has an inverse and find this inverse.

6. If $A = \begin{bmatrix} 1 & 0 & 0 \\ -1 & -2 & -1 \\ 2 & 3 & 2 \end{bmatrix}$, show that $A^3 - A = A^2 - I$ and hence find A^{-1}.

7. Solve the following system of equations (*using inverse of matrices*):

(a) $$\left.\begin{aligned} 2x + y + z &= 1 \\ x - y + 2z &= -1 \\ 3x + 2y - z &= 4 \end{aligned}\right\};$$

(b) $$\left.\begin{aligned} x - y + z &= 1 \\ x + y + 2z &= 0 \\ 2x - y + 3z &= 2 \end{aligned}\right\};$$

(c) $$\left.\begin{aligned} 2x + 5y + 3z &= 9 \\ 3x + y + 2z &= 3 \\ x + 2y - z &= 6 \end{aligned}\right\};$$ **[C.P. 1993]**

(d) $$\left.\begin{aligned} x + 2y + 3z &= 6 \\ 3x - 2y + z &= 2 \\ 4x + 2y + z &= 7 \end{aligned}\right\};$$ **[C.P. 1991]**

(e) $$\left.\begin{aligned} x - 3z &= 1 \\ 2x - y - 4z &= 2 \\ y + 2z &= 4 \end{aligned}\right\};$$

(f) $$\left.\begin{aligned} 5x + 6y &= 29 \\ -7x + 8y &= 11 \end{aligned}\right\}.$$ **[C.P. 1991]**

(g) $$\left.\begin{aligned} x + y + z &= 4 \\ 2x - y + 3z &= 1 \\ 3x + 2y - z &= 1 \end{aligned}\right\}.$$ **[C.P. 2002, 2004]**

8. If $A = \begin{pmatrix} 0 & 0 & 1 \\ 0 & 1 & 0 \\ 1 & 0 & 0 \end{pmatrix}$, show that $A^{-1} = A$. (Such matrices are called *self-reciprocal.*)

9. If $\alpha + i\beta = \begin{pmatrix} \alpha & \beta \\ -\beta & \alpha \end{pmatrix}$, verify that $(\alpha + i\beta)^{-1} = \begin{pmatrix} \alpha & \beta \\ -\beta & \alpha \end{pmatrix}^{-1}$.

10. If $A = \begin{bmatrix} 1 & 0 & -1 \\ 1 & 0 & 0 \\ 0 & 1 & 0 \end{bmatrix}$ and $I = \begin{bmatrix} 1 & 0 & 0 \\ 0 & 1 & 0 \\ 0 & 0 & 1 \end{bmatrix}$, verify that $(A^{-1})^2 + A = I$.

11. If $A = \begin{bmatrix} 0 & -\cot\theta/2 \\ \cot\theta/2 & 0 \end{bmatrix}$ and $I = \begin{bmatrix} 1 & 0 \\ 0 & 1 \end{bmatrix}$, obtain $(I+A)(I-A)^{-1}$.

12. If $A = \begin{bmatrix} 2 & -1 \\ -1 & 2 \end{bmatrix}$, verify that $A^2 - 4A + 3I = 0$ and hence, obtain A^{-1}. Using this value of A^{-1} solve:

$$\left.\begin{array}{rcl} 2x - y & = & 3 \\ -x + 2y & = & -3 \end{array}\right\}.$$

13. Show that the matrix $A = \begin{pmatrix} 1 & 2 & 2 \\ 2 & 1 & 2 \\ 2 & 2 & 1 \end{pmatrix}$ satisfies the equation

$$A^2 - 4A - 5I = 0,$$

where I is the 3×3 unit matrix; hence, obtain A^{-1}. **[C.P. 1962, 2002]**

14. Find the matrix X from the equation $AX = B$, where

$$A = \begin{pmatrix} 1 & -1 & 0 \\ 0 & 1 & -1 \\ 1 & 1 & 1 \end{pmatrix} \text{ and } B = \begin{pmatrix} 2 \\ 1 \\ 7 \end{pmatrix}$$

[C.P. 1961]

15. Prove by taking a third order matrix, that the inverse of a non-singular symmetric matrix is symmetric and that of a non-singular skew symmetric matrix is skew-symmetric.

16. Solve, by matrix inversion, the equation

$$\begin{array}{rcl} x + 2y + 3z & = & 14 \\ 2x - y + 5z & = & 15 \\ 2y + 4z - 3x & = & 13. \end{array}$$

[C.P. 1988]

[The given system of equations can be written as $AX = B$,

$$\text{where } A = \begin{bmatrix} 1 & 2 & 3 \\ 2 & -1 & 5 \\ 2 & 4 & -3 \end{bmatrix}, \; X = \begin{bmatrix} x \\ y \\ z \end{bmatrix} \text{ and } B = \begin{bmatrix} 14 \\ 15 \\ 13 \end{bmatrix}$$

$$\text{then } X = A^{-1}B = \frac{1}{57}\begin{bmatrix} 14 & 2 & -13 \\ 23 & -13 & -1 \\ -1 & -8 & 5 \end{bmatrix}\begin{bmatrix} 14 \\ 15 \\ 13 \end{bmatrix} = \begin{bmatrix} 1 \\ 2 \\ 3 \end{bmatrix}.$$

17. Find the inverse of $\begin{bmatrix} 1 & 1 & 1 \\ 1 & 2 & 3 \\ 3 & 1 & -5 \end{bmatrix}$ and hence, solve. [C.P. 1990]

$$\left.\begin{aligned} x+y+z &= 2 \\ x+2y+3z &= 1 \\ 3x+y-5y &= 12 \end{aligned}\right\}.$$

[D]

(Determination of Rank of a Matrix)

1. The rank of a matrix is the order of highest order non-vanishing minor of the matrix: Use this definition to find the ranks of the following matrices:

(a) $\begin{bmatrix} 2 & -4 & 3 & 1 \\ 1 & -2 & 1 & -4 \end{bmatrix}$;

(b) $\begin{bmatrix} 3 & 1 & 2 \\ 6 & 2 & 4 \\ 3 & 1 & 2 \end{bmatrix}$;

(c) $\begin{bmatrix} 2 & 4 & 3 & 2 \\ 3 & 5 & 1 & 4 \end{bmatrix}$;

(d) $\begin{bmatrix} 1 & 2 & 3 \\ 2 & 3 & 4 \\ 0 & 2 & 2 \end{bmatrix}$;

(e) $\begin{bmatrix} 1 & 2 & 3 \\ 3 & 4 & 5 \\ 4 & 5 & 6 \end{bmatrix}$;

(f) $\begin{bmatrix} 1 & 0 & 0 \\ 0 & 1 & 0 \\ 0 & 0 & 1 \end{bmatrix}$. [C.P. 2004]

2. What is the rank of a square matrix of order three; if every element of the matrix is unity?

3. Prove that if at least one of the elements a, b, c, d, e, f is not zero, the skew symmetric matrix

$$\begin{bmatrix} 0 & a & b & c \\ -a & 0 & d & e \\ -b & -d & 0 & f \\ -c & -e & -f & 0 \end{bmatrix}$$

has the rank ≥ 2.

4. Verify that the matrix.

$$\begin{bmatrix} 0 & 2 & 3 & 8 \\ 0 & 0 & 1 & -1 \\ 0 & 0 & 0 & 0 \end{bmatrix}$$

is in Echelon form. What is the rank of A?

5. If $A = \begin{bmatrix} 0 & 1 & 0 & 0 \\ 0 & 0 & 1 & 0 \\ 0 & 0 & 0 & 1 \\ 0 & 0 & 0 & 0 \end{bmatrix}$, find A^2 and then show that both A and A^2 are in Echelon form. What are their ranks?

6. Reduce the matrix A to B by E.T. (Elementary Transformer) where

$$A = \begin{bmatrix} 1 & 2 & 3 & 0 \\ 2 & 4 & 3 & 2 \\ 3 & 2 & 1 & 3 \\ 0 & 8 & 7 & 5 \end{bmatrix}, \quad B = \begin{bmatrix} 1 & 2 & 3 & 0 \\ 0 & 0 & -3 & 2 \\ 0 & -4 & -8 & 3 \\ 0 & 0 & 0 & 0 \end{bmatrix}.$$

What is the rank of A?

7. Find the rank of the following matrices by reducing to Echelon form, where

(a) $A = \begin{bmatrix} 5 & 3 & 14 & 4 \\ 0 & 1 & 2 & 1 \\ 1 & -1 & 2 & 0 \end{bmatrix}$; (b) $B = \begin{bmatrix} 8 & 1 & 3 & 6 \\ 0 & 3 & 2 & 2 \\ -8 & -1 & -3 & 4 \end{bmatrix}$.

8. By reducing each of the following matrices to Normal forms find their ranks:

(a) $\begin{bmatrix} 1 & 1 & -1 \\ 2 & -3 & 4 \\ 3 & -2 & 3 \end{bmatrix}$;

(b) $\begin{bmatrix} -1 & -2 & -1 \\ 6 & 12 & 6 \\ 5 & 10 & 5 \end{bmatrix}$;

(c) $\begin{bmatrix} 0 & -1 & -2 \\ 8 & 9 & 10 \\ 8 & 8 & 8 \end{bmatrix}$;

(d) $\begin{bmatrix} -8 & 7 & -10 \\ 48 & -42 & 60 \\ 40 & -35 & 50 \end{bmatrix}$;

(e) $\begin{bmatrix} 6 & 1 & 3 & 8 \\ 4 & 2 & 6 & -1 \\ 10 & 3 & 9 & 7 \\ 16 & 4 & 12 & 15 \end{bmatrix}$;

(f) $\begin{bmatrix} 2 & 3 & -1 & -1 \\ 1 & -1 & -2 & -4 \\ 3 & 1 & 3 & -2 \\ 6 & 3 & 0 & -7 \end{bmatrix}$;

(g) $\begin{bmatrix} 1 & 2 & 1 & 2 \\ 1 & 3 & 2 & 2 \\ 2 & 4 & 3 & 4 \\ 3 & 7 & 4 & 6 \end{bmatrix}$;

(h) $\begin{bmatrix} 3 & -2 & 0 & -1 \\ 0 & 2 & 2 & 1 \\ 1 & -2 & -3 & 2 \\ 0 & 1 & 2 & 1 \end{bmatrix}$;

(i) $\begin{bmatrix} 2 & -2 & 0 & 0 \\ 4 & 2 & 0 & 2 \\ 1 & -1 & 0 & 3 \\ 1 & -2 & 1 & 2 \end{bmatrix}$;

(j) $\begin{bmatrix} 6 & 1 & 3 & 8 \\ 4 & 2 & 6 & -1 \\ 10 & 3 & 9 & 7 \\ 16 & 4 & 12 & 15 \end{bmatrix}$;

9. Choose λ so that

$$\text{rank of } \begin{bmatrix} \lambda & 2 & -2 \\ 4 & 2\lambda & -1 \\ 6 & 6 & \lambda \end{bmatrix} = \text{rank of } \begin{bmatrix} \lambda & 2 & -2 & 1 \\ 1 & 2\lambda & -1 & 3 \\ 6 & 6 & \lambda & -1 \end{bmatrix} = 3$$

(E)

(Consistency and Solutions of Linear equations)

1. Calculate the ranks of the coefficient matrix and the augmented matrix of the following systems of linear equations and then examine whether the system is solvable or not:

(a) $\left.\begin{array}{r}x-5y+3z=-1\\2x-y-z=5\\5x-7y+z=2\end{array}\right\};$ (b) $\left.\begin{array}{r}x_1-x_2+x_3+x_4=-2\\x_1+x_2-x_3+2x_4=1\\3x_1-x_2+x_3+2x_4=2\end{array}\right\}.$

2. Show that the system $3x+4y=7$, $2.25x+3y=1$ has no solution (Inconsistent).

3. Find the solutions of (if they exist):

(a) $\left.\begin{array}{r}x+2y+z=4\\2x+y+5z=5\end{array}\right\};$ (b) $\left.\begin{array}{r}x+2y+3z+4\omega=7\\2x+y+z+2\omega=3\end{array}\right\}.$

4. Find non-trivial solutions, if they exist:

(a) $\left.\begin{array}{r}2x_1+5x_2+6x_3=0\\x_1-3x_2-8x_3=0\\3x_1+x_2-4x_3=0\end{array}\right\};$ (b) $\left.\begin{array}{r}4x-y+6z=0\\2x+7y+12z=0\\x-4y+3z=0\\5x-5y+3z=0\end{array}\right\}.$

5. Show that the equations are not consistent

$$\begin{aligned}x+y+z&=-3\\3x+y-2z&=-2\\2x+4y+7z&=7.\end{aligned}$$

6. Show that equations

$$\begin{aligned}x+y+z&=6\\x+2y+3z&=14\\x+4y+7z&=30\end{aligned}$$

are consistent and solve them.

7. Investigate for what values of λ, μ the simultaneous equations

$$\begin{aligned}x+y+z&=6\\x+2y+3z&=10\\x+2y+\lambda z&=\mu.\end{aligned}$$

have (a) No Solution; (b) Unique solution; (c) an infinite number of solutions.

[The Augmented matrix $[A \quad B] = \left[\begin{array}{ccc|c} 1 & 1 & 1 & 6 \\ 1 & 2 & 3 & 10 \\ 1 & 2 & \lambda & \mu \end{array}\right]$.

Apply E.T. $R_{21}(-1)$, $R_{31}(-1)$, $R_{32}(-1)$

$$[A, B] \sim \left[\begin{array}{ccc|c} 1 & 1 & 1 & 6 \\ 0 & 1 & 2 & 4 \\ 0 & 0 & \lambda - 3 & \mu - 10 \end{array}\right].$$

If $\lambda \neq 3$ we have rank $[A \quad B] = 3 =$ rank A.

In this case the system is consistent. rank $A = 3 =$ the number of unknowns, therefore the given system possesses an **Unique solution.**

If $\lambda = 3$, $\mu \neq 10$, we have rank $[A \quad B] = 3$ but rank $A = 2$. In this case the system is in consistent (No solution)

If $\lambda = 3$, $\mu = 10$, we have rank $[A \quad B] =$ rank $A = 2$. The given system is again consistent. Since the rank A < the number of unknowns therefore in this case the system has an infinite number of solutions]

8. Investigate for what values of a, b the equations

$$\begin{aligned} x + 2y + 3z &= 4 \\ x + 3y + 4z &= 5 \\ x + 3y + az &= b \end{aligned}$$

have (a) no solution; (b) Unique solution; (c) an infinite number of solutions.

Answers

[A]

1. $\begin{bmatrix} 3 & 3 \\ 7 & 7 \end{bmatrix}, \begin{bmatrix} -1 & 1 \\ -1 & 1 \end{bmatrix}, \begin{bmatrix} 4 & 4 \\ 4 & 4 \end{bmatrix}, \begin{bmatrix} -1 & 5 \\ -5 & 9 \end{bmatrix}$.

2. $\begin{bmatrix} -1 & 2 & 2 \\ 9 & 7 & 8 \\ 2 & -1 & 3 \end{bmatrix}, \begin{bmatrix} 5 & 0 & 2 \\ -3 & 3 & 6 \\ 0 & 1 & -1 \end{bmatrix}, \begin{bmatrix} 2 & 2 & 5 \\ 28 & 6 & 19 \\ -2 & 0 & 2 \end{bmatrix}, \begin{bmatrix} -3 & 2 & 1 \\ 19 & 16 & 27 \\ 1 & -4 & -3 \end{bmatrix}$.

4. $\begin{bmatrix} 5 \\ 1 \\ 1 \end{bmatrix}, \begin{bmatrix} 12 \\ 2 \\ -7 \end{bmatrix}, \begin{bmatrix} 3 \\ 3 \\ -5 \end{bmatrix}, \begin{bmatrix} 10 \\ 8 \\ -13 \end{bmatrix}$.

5. $AB = \begin{bmatrix} 3 & 0 & 11 \\ -15 & 4 & 9 \\ 7 & 1 & 20 \end{bmatrix}, BA = \begin{bmatrix} -1 & 13 & 3 \\ 2 & 24 & 4 \\ 0 & -6 & 4 \end{bmatrix}, AB \neq BA.$

8. Null matrix $\begin{bmatrix} 0 & 0 \\ 0 & 0 \end{bmatrix}$.

9. **(a)** $\begin{bmatrix} 4 & 2 & 4 \\ 6 & 6 & 14 \\ 3 & 0 & 3 \end{bmatrix}, \begin{bmatrix} 7 & 0 & 4 \\ 4 & 5 & 7 \\ 3 & -2 & 6 \end{bmatrix}, \begin{bmatrix} 5 & 4 & 4 \\ 2 & 7 & 11 \\ 2 & -2 & 3 \end{bmatrix}, \begin{bmatrix} -3 & 2 & 0 \\ 2 & 1 & 7 \\ 0 & 2 & -3 \end{bmatrix},$

$\begin{bmatrix} -2 & -6 & -4 \\ -4 & -8 & -18 \\ -2 & 0 & 0 \end{bmatrix}, \begin{bmatrix} -1 & -2 & 0 \\ 4 & -1 & -3 \\ 1 & 2 & 0 \end{bmatrix}.$

(b) $X = \begin{bmatrix} 2 & -4 & 0 \\ 2 & -2 & -4 \\ 1 & 0 & 3 \end{bmatrix}.$

14. **(c)** $\begin{bmatrix} 1 & 2 & 3 & 4 \\ 2 & 4 & 6 & 8 \\ 3 & 6 & 9 & 12 \\ 4 & 8 & 12 & 16 \end{bmatrix}.$

15. $\begin{bmatrix} 4 & -2 \\ 3 & 5 \\ -11 & 3 \\ 5 & 2 \end{bmatrix}; \begin{bmatrix} 1 & 3 \\ 6 & -11 \\ -1 & 6 \\ -4 & 11 \end{bmatrix}$; Not defined; $\begin{bmatrix} 0 & 2 \\ 1 & 3 \\ -3 & -5 \\ -4 & 1 \end{bmatrix}; \begin{bmatrix} 0 & 2 & 4 \\ 1 & 3 & 3 \\ -3 & 5 & 1 \\ -4 & 1 & 14 \end{bmatrix}.$

18. **(a)** $\frac{1}{40}(A^2 - 23I) = \frac{1}{40}\begin{bmatrix} -4 & 4 & 8 \\ 1 & -11 & 8 \\ 14 & 6 & -6 \end{bmatrix}.$

(b) $\frac{1}{3}\begin{pmatrix} 2 & 1 \\ 1 & 2 \end{pmatrix}.$

[B]

4. $\begin{bmatrix} 2 & 1 & 3 \\ 8 & 6 & 15 \\ -3 & -2 & -5 \end{bmatrix}$. **6.** $A^{-1} = A^2 = \begin{bmatrix} 0 & 0 & 1 \\ 0 & -1 & 2 \\ 1 & -1 & 1 \end{bmatrix}.$

[C]

1. **(a)** $\frac{1}{2}\begin{bmatrix} 3 & 1 \\ 4 & 2 \end{bmatrix}$; **(b)** no inverse exists; **(c)** $\begin{bmatrix} \cos\theta & -\sin\theta \\ \sin\theta & \cos\theta \end{bmatrix}.$

3. **(a)** $x = 3, y = -4, z = 0.$
(b) $x = 1, y = 1, z = 2.$

4. $I - A.$

5. $S^2 = \begin{bmatrix} 0 & 0 & 1 \\ 0 & 0 & 0 \\ 0 & 0 & 0 \end{bmatrix}$, $S^3 = 0$, $AB = a^3 I$, $A^{-1} = a^{-3}B.$

7. **(a)** $x = 1, y = 0, z = -1;$
(b) $x = -1, y = -1, z = 1;$
(c) $x = 1, y = 2, z = -1;$
(d) $x = y = z = 1;$
(e) $x = 4, y = 2, z = 1;$
(f) $x = 3, y = 4.$
(g) $x = 1, y = 3, z = 2.$

11. $\begin{bmatrix} \operatorname{cosec}^2 \theta/2 & 0 \\ 0 & \operatorname{cosec}^2 \theta/2 \end{bmatrix}$.

12. $x = 1, y = -1$.

13. $\frac{1}{5}(A - 4I) = \frac{1}{5}\begin{bmatrix} -3 & 2 & 2 \\ 2 & -3 & 2 \\ 2 & 2 & -3 \end{bmatrix}$.

14. $X = \begin{bmatrix} x \\ y \\ z \end{bmatrix} = \begin{bmatrix} 4 \\ 2 \\ 1 \end{bmatrix}$.

16. $x = 1, y = 2, z = 3$.

17. $x = 2, y = 1, z = -1$.

[D]

1. **(a)** 2; **(b)** 1; **(c)** 2; **(d)** 3; **(e)** 2; **(f)** 3.

2. 1.

4. 2.

5. 3, 2.

6. 3.

7. (a) 3; (b) 3.

8. **(a)** 2; **(b)** 1; **(c)** 2; **(d)** 1; **(e)** 2; **(f)** 3; **(g)** 3; **(h)** 4; **(i)** 3; **(j)** 2.

9. $\lambda = 1$.

[E]

3. **(a)** $x = 2 - 3t, y = 1 + t, z = t$;

(b) $x = \frac{z}{3} - \frac{1}{3}, y = \frac{11}{3} - \frac{5}{3}z - 2w$ (z.w arbitrary).

4. (a) $x_1 = k, x_2 = -k, x_3 = \frac{1}{2}k$ (k = arbitrary).

6. rank A = 2, rank [A B] = 2. Hence consistent. $x = t - 2, y = 8 - 2t, z = t$ (t arbitrary) one solution is $(-2, 8, 0)$ by putting $t = 0$.

8. **(a)** No solution when $a = 4, b \neq 5$.

(b) Unique solution for all values of b provided $a \neq 4$.

(c) An infinite number of solutions when $a = 4, b = 5$.

Part II

Modern Algebra

Chapter 6

Basic Concepts: Sets, Mappings and Binary Operations

Basic Concepts: Sets, Mappings and Binary operations	
1.	**Basic Concepts:** Sets, subsets, equality of sets. Operations on sets: Union, intersection and complement. Verification of the laws of Algebra of Sets and De Morgan's laws. Cartesian product of sets.
2.	**Mappings:** One-one and onto Mappings. Composition of mappings—Concepts only. Identity and Inverse mappings.
3.	**Binary operations** in a set. Identity element: Inverse element.

Notations often used:

$\Longrightarrow$ Stands for '*implies*'

$\Longleftrightarrow$ Stands for "*implies and implied by*"

$\forall$ Stands for "*For all*"

6.1 Introduction

In Modern Algebra we build up mathematical structures like Groups, Rings, Fields, Vector spaces, etc. But before we advance, we shall explain the more basic concepts: **concept of a set** and the **concept of mapping**. We should also know the notion of **binary operation** on a set.

6.2 What is a set?

1. **Definition:** A set is any **well-defined** (or **well-determined**) collection of objects.

The objects in a given set are called its **elements** or **members**. The set is thus **a single entity** composed of objects (or members) of any sort.

2. We use capital letters A, B, C, X, Y, etc. to denote sets and we use small letters a, b, c, x, y, etc. to denote members of the sets.

3. If a is a member of a set S, then we write $a \in S$ (**read as** 'a belongs to S'). We often write $a \notin S$ to mean 'a does not belong to S'.

▶ Example 6.2.1 *Let* **N** *be the set of all positive integers (or natural numbers). Then* $3 \in \mathbf{N}$ *but* -3 *or* $\frac{1}{3} \notin \mathbf{N}$.

(**Read:** 3 *belongs to* **N** *but* -3 *or* $\frac{1}{3}$ *does not belong to* **N**.)

4. Often we require to specify a set by a **defining property** P. We then write:

$$S = \{x : x \text{ satisfies } P\}$$

(**Read:** S consists of all those objects x which obeys a property P.)

▶ Example 6.2.2 *The set* **N** *of natural numbers can be specified by*

$$\mathbf{N} = \{n : n \text{ is a positive integer}\}.$$

(**Read:** N *contains members like* n, *where* n *is a positive integer*)

5. We usually write the elements of a set within two curly brackets { }.

▶ Example 6.2.3 $\mathbf{N}$ = *set of positive integers* $= \{1, 2, 3, 4, 5, 6, \ldots\}$.

▶ Example 6.2.4 *We have examples of sets like*

Set of vowels in English alphabet $= \{a, e, i, o, u\}$.
Set of days in a week = {Mon., Tues., ..., Sat., Sun}
Set of officially recognised languages in a country like India = {Hindi, Bengali, Tamil, Telegu, etc.}
Set of books on a shelf.

Whenever possible, we shall display the members; otherwise we shall specify the elements by some characteristic property.

6. **Finite and Infinite sets.** A set S is said to be **finite** if it has a definite number of elements in it; otherwise it is called an **infinite set**. If, in particular, a set has only one member, then it is called a **singleton set**.

▶ Example 6.2.5 $\{1, 2, 3, 4, 5, 6, 7, 8, 9, 10\}$ *is a* **finite set** *containing the first ten natural numbers.*

{President of India} is a **singleton set**.

The set of all integers $\{0, \pm 1, \pm 2, \pm 3, \ldots\}$ *is an example of an* **infinite** *set.*

Standard notations for some sets	
N:	the set of all natural numbers (or positive integers)
Z:	the set of all integers (positive, negative and zero)
$\mathbf{Z}^+$:	the set of all positive integers
Q:	the set of all rational numbers
R:	the set of all real numbers
C:	the set of all complex numbers

7. **Empty set.** A set defined by some well-defined rule may sometimes does not yield any element. Such a set is called an **Empty set** or **Null set** or **Void set**.

▶ Example 6.2.6 $\{n : n \in \mathbf{N} \text{ and } 4 < n < 5\}$
$\{x : x^2 - 2 = 0 \text{ and } x \in \mathbf{Q}\}$
$\{x : x \in \mathbf{R} \text{ and } x^2 < 0\}$

are all examples of **empty sets**. *Empty set is a unique set. We shall denote an* **empty set** *by the symbol* ϕ *(phi) or by* $\{\ \}$.

8. **Subsets.** If every element $a \in A$ is also an element of the set B, then we say that 'A is a subset of B'. We shall then write $A \subseteq B$

(Read as: 'A is a subset of B' or 'A is contained in B'). If $A \subseteq B$, then we may also write $B \supseteq A$ and **read:** 'B contains A'.

▶ Example 6.2.7

- $\{1, 2, 3\}$ *is a subset of* $\{1, 2, 3, 4, 5\}$. *We may write*
$$\{1,2,3\} \subseteq \{1,2,3,4,5\} \text{ or, } \{1,2,3,4,5\} \supseteq \{1,2,3\}.$$
- *The set of all equilateral triangles in a plane is a subset of the set of all isosceles triangles in the plane.*

9. **Equality of sets.** Two sets A and B are said to be equal if they consist of the same elements. We then write $A = B$.

Note that $A = B \implies A \subseteq B$ and $B \subseteq A$,

$$\text{i.e., } A = B \implies \begin{cases} \text{every member of } A \text{ is in } B, \text{ i.e., } A \subseteq B \\ \text{and} \\ \text{every member of } B \text{ is in } A, \text{ i.e., } B \subseteq A. \end{cases}$$

▶ Example 6.2.8 $\{1, 3\} = \{1, 1, 1, 3, 3\}$.

▶ Example 6.2.9 $\{3, 4\} = \{x : x^2 - 7x + 12 = 0\}$.

10. **Proper subset.** If $A \subseteq B$ and $A \neq B$, then we often write $A \subset B$.

(Read as: A is a **proper subset** of B or A is **strictly contained** in B).

Note that: $A \subset B$, if every member of A is in B but not every member of B is in A, i.e., $\exists$ at least one member in B which is not in A.

11. (a) For any set A, $A \subseteq A$ (**every member of A is certainly in A**).
    (b) Also, the nullset $\phi \subseteq$ any set A.
    [**Logic** is that no element of ϕ exists such that it is *not a member of* A. In fact, there is no element in ϕ; so the logic is correct]

12. In view of above, while collecting all possible subsets of any arbitrary set A, we shall **include** the set A itself and the null set ϕ.

▶ Example 6.2.10 *Let* $A = \{a, b, c\}$. *What are the possible subsets of* A?

Solution:

(a) ϕ is a subset of A (*with number of members* $= 0$)
(b) $\{a\}, \{b\}, \{c\}$ are subsets of A (*with number of members = one in each*)
(c) $\{a, b\}, \{a, c\}, \{b, c\}$ are also subsets of A (*with two members each*)
(d) $\{a, b, c\}$, i.e., A itself a subset (*with three members*)
$\therefore$ the number of possible subsets of $A = \mathbf{8}$.

▶ Example 6.2.11 *Can you prove: "A finite set with* n *elements has, in all,* 2^n *subsets"?* [*Ans.* ${}^nC_0 + {}^nC_1 + {}^nC_2 + \cdots + {}^nC_n (= 2^n)$.]

Power set of A: The set of all subsets of a set A is often called the **power set** of A, denoted by $\mathbf{P}(A)$ or by 2^A.

▶ Example 6.2.12 *Are the following statements meaningful?*

$$a \in \{a\};\ \{a\} \subseteq \{a\};\ a \in \{a, b\};\ \{a\} \subseteq \{a, b\}.$$

What can you say about the following statements?

$$a = \{a\},\ a \subseteq \{a, b\},\ \{a\} \in \{a, b\}$$

[*Ans.* The first set of statements are all meaningful, but in the second set none has a meaning.]

13. **Universal set.** In any logical discussion it is customary to take a **fixed set** S (or U) such that all sets under discussion are subsets of that fixed set S (or U). We call that fixed set S **Universe** or **Universal set.**

▶ Example 6.2.13 *Let us throw a dice. For this experiment*

$$S \text{ or } Universe = \{1, 2, 3, 4, 5, 6\}.$$

6.3 Operations on Sets

Union — Intersection — Complement — Symmetric Difference

What is an Operation on Sets? It is a **rule** by which with one, two or more given sets we can associate a **unique set**. We shall, in the following discussions, introduce some important operations on sets, namely, **Union, Intersection, Complementation and Symmetric Difference.**

I. Union

Let A and B be any two sets (supposed to be subsets of the fixed universal set S). Then

> A union B (written as $A \cup B$) is a **set** whose elements are members of A or B or of both.
> i.e., $A \cup B$ is a set consisting of all objects that are members of at least one of the sets A, B.

$A \cup B = \{x : x \in A \text{ or } x \in B \text{ or } x \in \text{ both } A \text{ and } B\}$

▶ Example 6.3.1 $\{1,2\} \cup \{2,3,4\} = \{1,2,3,4\}$.

Extension to three or more sets. $A \cup B \cup C$ is the set consisting of all objects that are members of at least one of the sets A, B, C. Clearly the order in which the sets appear does not matter:

We can write $A \cup B \cup C = B \cup A \cup C$, etc.
Also $(A \cup B) \cup C = A \cup (B \cup C)$ **(Associative Property of union of sets)**
Similarly the union of more than three sets can be defined. We often write

$$\bigcup_{i=1}^{n} A_i = A_1 \cup A_2 \cup A_3 \cup \cdots \cup A_n.$$

▶ Example 6.3.2 *Let* $A = \{1,3\}, B = \{3,5\}$ *and* $C = \{5,6,7\}$. *Suppose we require to write down the set* $A \cup B \cup C$. *It is usual that we first obtain* $A \cup B = \{1,3,5\}$ *and then obtain* $(A \cup B) \cup C = \{1,3,5\} \cup \{5,6,7\} = \{1,3,5,6,7\}$. *There is no harm if we first obtain* $B \cup C = \{3,5,6,7)$ *and then obtain*

$$A \cup (B \cup C) = \{1,3\} \cup \{3,5,6,7\} = \{1,3,5,6,7\}.$$

In either way we obtain $A \cup B \cup C = \{1,3,5,6,7\}$.

II. Intersection

Let A and B be two sets (supposed to be subsets of the universal set S). Then

> A intersection B (written as $A \cup B$) is a set consisting of objects that are common to A and B,
> i.e., $A \cap B = \{x/x \in A \text{ and } x \in B\}$.

▶ Example 6.3.3 (*i*) $\{1,2,4\} \cap \{2,3,5\} = \{2\}$;

(*ii*) *Let* $A = \{x : x$ *is an integer and* $1 \leq x \leq 10\}$ *and* $B = \{x : x$ *is a prime integer less than* $10\}$. *Verify that* $A \cap B = B$. [C.P. 2004]

Solution: (ii) $A = \{1,2,3,4,5,6,7,8,9,10\}$, $B = \{2,3,5,7\}$.

$\therefore$ $A \cap B = \{2,3,5,7\} = B$. Hence the required verification.

The following statements are easily found to be true

$$A \cap A = A, A \cap \phi = \phi, \phi \cap \phi = \phi,$$
$$A \cap B = B \cap A \text{ (\textbf{commutative property of intersection})}$$

Extension to three or more sets. $A \cap B \cap C$ is the set of objects that are common to A, B and C. The order in which the sets appear does not matter. Also we could write $A \cap B \cap C$ as:

$$(A \cap B) \cap C \text{ or } A \cap (B \cap C)$$

(called the **Associative property of intersection**).

For n sets $A_1, A_2, A_3, \ldots, A_n$, we write

$$\bigcap_{i=1}^{n} A_i = A_1 \cap A_2 \cap A_3 \cap \cdots \cap A_n.$$

Disjoint sets. When $A \cap B = \phi$, i.e., when A, B have no common elements, we say that A, B are **disjoint** or **non intersecting.**

III. Complements

Let A and B be two sets (subsets of the universal set S). Then the **difference** $A - B$ is the set of those elements in A which do not belong to B.

$$A - B = \{x/x \in A \text{ but } x \notin B\}$$

▶ Example 6.3.4 $A = \{1, 2, 3\}, B = \{2\}$. *Then* $A - B = \{1, 3\}$. *What is* $B - A$ *in this case?* [*Ans.* $B - A = \phi$]

▶ Example 6.3.5 $A = \{1, 2, 3\}, B = \{2, 3, 5\}$. *Then* $A - B = \{1\}, B - A = \{5\}$.

We use the term **complement** in case $A \subseteq S$ (S is the universe) and $S - A$ is then called '**complement of A**'.

▶ Example 6.3.6 *Let* $S = \{1, 2, 3, 4\}; A = \{1, 4\}$. *Then* $S - A = \{2, 3\}$ *is the complement of* A, *denoted by* A^c *or* A'.

(**Read as:** *A complement or A slashed or A prime*)

Note 1. $S' = \phi$, *and* $\phi' = S$. *Also if* $A \subseteq S$, *then* $A \cup A' = S, A \cap A' = \phi, (A')' = A$.

IV. Symmetric difference

Symmetric difference of two sets A and B, denoted by $A \triangle B$.

This is defined to be the set $(A - B) \cup (B - A)$.

▶ Example 6.3.7 $A = \{1, 2, 3, 4\}, B = \{3\}$. *Then* $A - B = \{1, 2, 4\}$ *and* $B - A = \{\ \} = \phi$.

$\therefore (A - B) \cup (B - A) = \{1, 2, 4\} \cup \phi = \{1, 2, 4\}$, or, $A \triangle B = \{1, 2, 4\}$.

▶ Example 6.3.8

$A = \{1, 2, 3\}, \quad B = \{3\}, \quad C = \{2, 3, 6\}$

$A - B = \{1, 2\}, \quad B - A = \{\ \}, \quad (A - B) \cup (B - A) = \{1 - 2\} = A \triangle B$

$A - C = \{1\}, \quad C - A = \{6\}, \quad (A - C) \cup (C - A) = \{1, 6\} = A \triangle C.$

V. Venn Diagrams [named after British logician John Venn (1834-1923)]

Venn diagrams provide a useful method for displaying relations between subsets of a fixed set S **(Universal set)**. S is represented by the interior of a rectangle and subsets of S are represented by regions, e.g., the interiors of circles, inside the rectangle. Points on the boundaries of all regions involved are ignored.

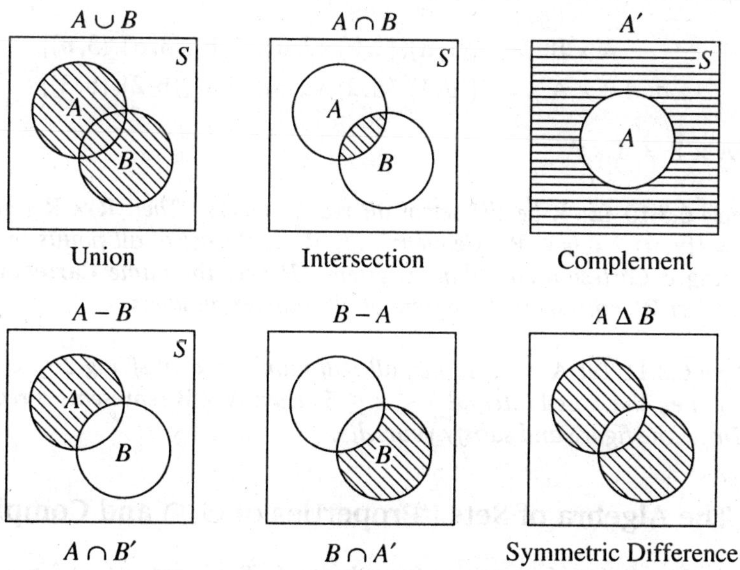

Fig. 6.1 Venn Diagrams of operations on Sets

Fig. 6.1 shows Venn diagrams representing operations on sets: $A \cup B, A \cap B, A'$, $A - B$, $B - A$ and $A \triangle B$ respectively. It is clearly from the definition of **difference** that, when A, B are subsets of S,

$$\begin{aligned} A - B &= A \cap B' \text{ and } B - A = B \cap A' \\ \therefore\ A \triangle B &= (A \cap B') \cup (B \cap A'). \end{aligned}$$

VI. Cartesian Product of two non-empty sets A, B

Ordered pair of elements

Given two elements a, b. When we write (a, b) where a is the first element and b the second, we call (a, b) **an ordered pair.** But (b, a) is also an ordered pair where the first element is b and the second is a. Clearly the pair $(a, b) \neq (b, a)$ unless $a = b$.

Cartesian Product of the Sets A, B

We denote by $A \times B$ *the set of all ordered pairs* (a, b) where $a \in A$ and $b \in B$, i.e.,

$$A \times B = \{(a, b) : a \in A \text{ and } b \in B\}$$

Such a set $A \times B$ is called the **Cartesian product** of the sets A, B in the order A, B.

$\therefore$ $B \times A$ is the **Cartesian product** of the sets A, B in the order B, A, i.e.,

$$B \times A = \{(b, a) : b \in B \text{ and } a \in A\}$$

Clearly $A \times B$ is not necessarily equal to $B \times A$, unless $A = B$.

▶ Example 6.3.9 *Let* $A = \{1, 2, 3\}$ *and* $B = \{a, b\}$. *Then*

$$\begin{aligned} A \times B &= \{(1,a),(1,b),(2,a),(2,b),(3,a),(3,b)\} \\ \textit{and } B \times A &= \{(a,1),(a,2),(a,3),(b,1),(b,2),(b,3)\}. \end{aligned}$$

Clearly $A \times B \neq B \times A$.

▶ Example 6.3.10 *Let* **R** *be the set of all real numbers. Then* $\mathbf{R} \times \mathbf{R}$ *(often written as* $\mathbf{R}^2$*)* $= \{(x, y) : x, y \in \mathbf{R}\}$. *Geometrically* $\mathbf{R}^2$ *is the set of all points in a plane by constructing a Cartesian co-ordinate system. Hence, the name* **Cartesian Product** *comes. In fact* $\mathbf{R}^2$ *represents the system of all complex numbers.*

▶ Example 6.3.11 *Let* $A = [2, 3]$, *i.e., all real numbers* x *satisfying* $2 \leq x \leq 3$ *and let* $B = [4, 5]$, *i.e., the closed interval* $4 \leq x \leq 5$. *Then* $A \times B$ *represents a rectangle in a plane (Draw the figure and satisfy yourself).*

6.4 The Algebra of Sets ('Properties of ∪, ∩ and Complement')

We give below a list of identities for subsets A, B, C, etc. of an Universal Set S. They can be proved by using definitions of different operations:

Laws of Algebra of Sets	
Associative Laws	
1. (a) $A \cup (B \cup C) = (A \cup B) \cup C$	**1.** (b) $A \cap (B \cap C) = (A \cap B) \cap C$
Commutative Laws	
2. (a) $A \cup B = B \cup A$	**2.** (b) $A \cap B = B \cap A$
Identity Laws	
3. (a) $A \cup \phi = A$	**3.** (b) $A \cap S = A$
4. (a) $A \cup S = S$	**4.** (b) $A \cap \phi = \phi$
Idempotent Laws	
5. (a) $A \cup A = A$	**5.** (b) $A \cap A = A$
Distributive Laws	
6. (a) $A \cap (B \cup C) = (A \cap B) \cup (A \cap C)$	**6.** (b) $A \cup (B \cap C) = (A \cup B) \cap (A \cup C)$
Complement Laws	
7. (a) $A \cup A' = S$	**7.** (b) $A \cap A' = \phi$
8. (a) $S' = \phi$	**8.** (b) $\phi' = S$
9. (a) $(A')' = A$	**9.** (b) $(A')' = A$
De Morgan's Laws	
10. (a) $(A \cup B)' = A' \cap B'$	**10.** (b) $(A \cap B)' = A' \cup B'$

In each line we have a pair of identities, called **dual identities.** Right side identity is obtained formally from the left side identity by replacing $\cup$ by $\cap$, $\cap$ by $\cup$; $'$ by $'$; ϕ by S; S by ϕ.

The proofs of these identities, particularly, **Associative Laws (1.a and 1.b), Distributive Laws (6.a and 6.b)** and **De Morgan's Laws (10.a and 10.b)** are to be learnt carefully. In proving equality of two sides, prove

L.H.S. $\subseteq$ R.H.S. and R.H.S. $\subseteq$ L.H.S.

Hence, it will follow that R.H.S.=L.H.S.

We illustrate by proving **De Morgan's Laws 10.a:** $(\mathbf{A} \cup \mathbf{B})' = \mathbf{A}' \cap \mathbf{B}'$.

Proof. Let $X = (A \cup B)'$, $Y = A' \cap B'$. To prove $X = Y$.

We first prove: $X \subseteq Y$.

Let $x \in X$, i.e., let $x \in (A \cup B)' \implies x \notin A \cup B$ (but $x \in S$)

$\implies x \notin A$ and $x \notin B$

$\implies x \in A'$ and $x \in B'$

$\implies x \in A' \cap B'$,

i.e., $x \in X \implies x \in Y$ **or,** $X \subseteq Y$

Next we prove: $Y \subseteq X$

Let $y \in Y$, i.e., $y \in A' \cap B' \implies y \in A'$ and $y \in B'$

$\implies y \notin A$ **or,** $y \notin B$

$\implies y \notin A \cup B$

$\implies y \in (A \cup B)'$, i.e. $y \in X$,

i.e., $y \in Y \implies y \in X$ **or,** $Y \subseteq X$

Hence, it follows $\mathbf{X} = \mathbf{Y}$.

Next we show how to prove **10.(b)**, assuming the truth of **10.(a)**. Assuming that **10.(a)** holds for any pair of subsets of S, we use it for the pair of subsets A' and B' of S, i.e.,

$$(A' \cup B')' = (A')' \cap (B')' = A \cap B$$

Hence, **taking complements** of both sides

$$\{(A' \cup B')'\}' = (A \cap B)'$$

or, $\quad A' \cup B' = (A \cap B)'$; this proves **10.(b)**.

But the students can prove **10.(b)** by following arguments of the proof of **10.(a)** as given above.

We ask our students to write similar proof for **1.(a)**, **1.(b)** and **6.(a)** and **6.(b)**.

To prove 6.(a): $\mathbf{A} \cap (\mathbf{B} \cup \mathbf{C}) = (\mathbf{A} \cap \mathbf{B}) \cup (\mathbf{A} \cap \mathbf{C})$ **(Distributive law)**

Proof. Let $X = A \cap (B \cup C)$ and $Y = (A \cap B) \cup (A \cap C)$.

To prove $X = Y$ we use two steps (a) $X \subseteq Y$ and (b) $Y \subseteq X$.

(a) Let x be any element of X, i.e., $x \in A \cap (B \cup C)$

$\implies x \in A$ and $x \in B \cup C$

$\implies x \in A$ and ($x \in B$ or, $x \in C$)

$\implies$ ($x \in A$ and $x \in B$) **or,** ($x \in A$ and $x \in C$)

$\implies x \in A \cap B$ **or,** $x \in A \cap C$

$\implies x \in (A \cap B) \cup (A \cap C)$

$\implies x \in Y$,

i.e., any arbitrary element x of X is also a member of Y, i.e., $X \subseteq Y$.

(b) Similarly taking an arbitrary element $y \in Y$ and proceeding as above it can be proved $Y \subseteq X$.

$\therefore$ we conclude $X = Y$.

The other distributive law, i.e., **6.b** $A \cup (B \cap C) = (A \cup B) \cap (A \cup C)$ can be proved exactly in a similar way.

To prove the Associative Law: 1.(a) $A \cup (B \cup C) = (A \cup B) \cup C$.

Proof. Let $x \in A \cup (B \cup C)$

$$\begin{aligned} &\implies x \in A \text{ or, } x \in (B \cup C) \\ &\implies x \in A \text{ or, } (x \in B \text{ or, } x \in C) \\ &\implies (x \in A \text{ or, } x \in B), \text{ otherwise } x \in C \\ &\implies x \in A \cup B \text{ or, } x \in C \\ &\implies x \in \{(A \cup B) \cup C\}, \end{aligned}$$

i.e., $A \cup (B \cup C) \subseteq (A \cup B) \cup C$.

Similarly we can prove $(A \cup B) \cup C \subseteq A \cup (B \cup C)$.

Hence, **1.**(a) follows.

6.5 Enumerative Results for Finite Sets

If A is a finite set, we shall denote by $n(A)$, the number of elements in A. If A and B are disjoint, so that $A \cap B = \phi$, then clearly

$$n(A \cup B) = n(A) + n(B). \text{ (See Fig. 6.2)}$$

But for any pair of sets A, B, the following rule holds:

$$n(A \cup B) = n(A) + n(B) - n(A \cap B). \text{ (See Fig. 6.3)}$$

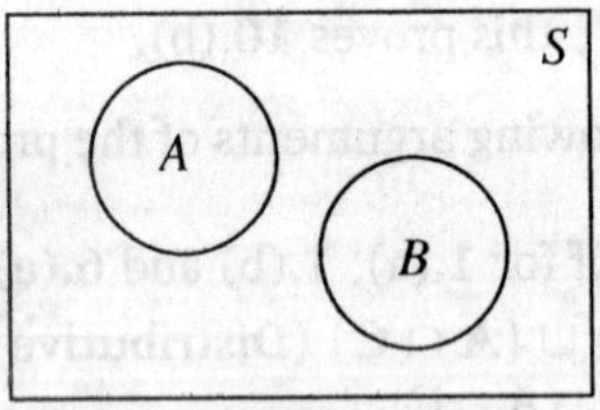

Fig 6.2

For Disjoint Sets:
$n(A \cup B) = n(A) + n(B)$

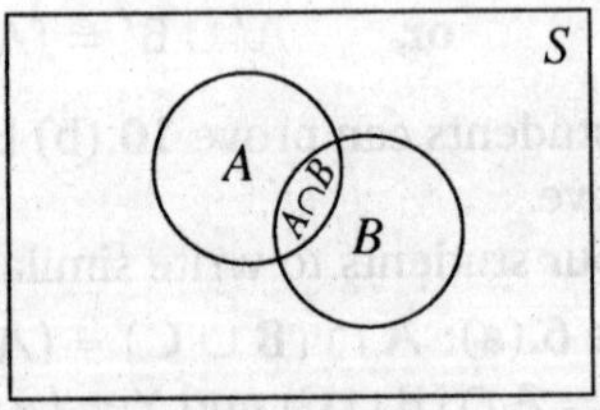

Fig 6.3

In general:
$n(A \cup B) = n(A) + n(B) - n(A \cap B)$

[See Fig. **6.3**. $A = (A - B) \cup (A \cap B)$ and so

$$n(A) = n(A - B) + n(A \cap B),$$

since $A - B$ and $A \cap B$ are disjoint.

Also from Fig. **6.3**, $A \cup B = B \cup (A - B)$ and so $n(A \cup B) = n(B) + n(A - B)$, since B and $A - B$ are disjoint.

Hence, by subtraction,

$$\begin{aligned} n(A) - n(A \cup B) &= n(A - B) + n(A \cap B) - n(B) - n(A - B) \\ &= n(A \cap B) - n(B) \\ \text{or, } n(A \cup B) &= n(A) + n(B) - n(A \cap B).] \end{aligned}$$

We can extend the result to three sets A, B, C when we have

$$\begin{aligned} n(A \cup B \cup C) &= n(A) + n(B) + n(C) - n(A \cap B) \\ &\quad - n(B \cap C) - n(C \cap A) + n(A \cap B \cap C). \end{aligned}$$

▶ Example 6.5.1 *Each student in a class of 40 studied at least one of the subjects— English, Bengali and Mathematics. 16 studied English, 22 Mathematics and 26 Bengali. 5 studied English and Mathematics, 14 Bengali and Mathematics and 2 all three (i.e., English, Bengali and Mathetics). Determine how many students studied English and Bengali but not Mathematics.* [See Fig. 6.4]

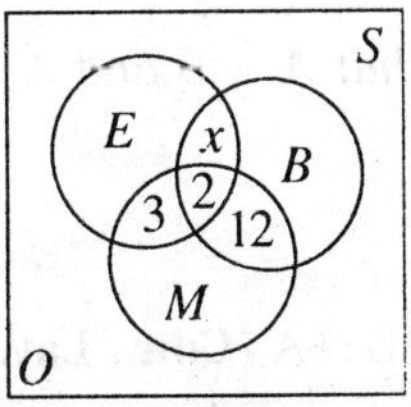

Fig 6.4

Solution: Let E, B, M denote the sets of students studying English, Bengali and Mathematics respectively. We refer to the Venn diagram Fig. 6.4 where we have partitioned the sets into disjoint sets. Let

$$n(E \cap B \cap M) = 2.$$

Then

$$\begin{aligned} n(E \cap M) &= 3 + 2 = 5, \\ n(B \cap M) &= 12 + 2 = 14. \end{aligned}$$

Let $n(E \cap B) = x + 2$, where x is the number of students studying E and B but not mathematics [i.e., $n(E \cap B \cap M')$]. Then

$$\begin{aligned} n(E \cup B \cup M) &= n(E) + n(B) + n(M) - n(E \cap B) - n(B \cap M) \\ &\quad - n(E \cap M) + n(E \cap B \cap M), \\ \text{i.e., } 40 &= 16 + 26 + 22 - (x + 2) - (14) - 5 + 2 \end{aligned}$$

whence $x = 5$ (Reqd. number)

6.6 Solved Problems

▶ Example 6.6.1 *Let* $A = \{2, 4, 6, 8\}$ *and* $B = \{-1, 0, 2\}$. *Find* $A \cup B$ *and* $A \cap B$. *What are the elements of the set* $(A \cup B) - (A \cap B)$? **[C.P. 1995]**

Solution:

$$A \cup B = \{-1, 0, 2, 4, 6, 8\} \text{ and } A \cap B = \{2\}$$

$$\therefore \quad (A \cup B) - (A \cap B) = \{-1, 0, 2, 4, 6, 8\} - \{2\} = \{-1, 0, 4, 6, 8\}.$$

► Example 6.6.2 *If* $A = \{2, 16, 32\}$, $B = \{1, 4, 8, 16\}$ *and if the universe* $S = \{1, 2, 4, 8, 16, 32\}$, *obtain* $(A \cap B)'$. *Then verify that* $(A \cap B)' = A' \cup B'$. (**dash** *denotes* **complement**) **[C.P. 1996]**

Solution: $A \cap B = \{16\}$. Therefore, w.r.t the universal set S, we have

$$\begin{aligned}(A \cap B)' &= S - (A \cap B)\\ &= \{1, 2, 4, 8, 16, 32\} - \{16\}\\ &= \{1, 2, 4, 8, 32\}.\end{aligned}$$

Now

$$\begin{aligned}A' &= S - A = \{1, 2, 4, 8, 16, 32\} - \{2, 16, 32\} = \{1, 4, 8\}\\ B' &= S - B = \{1, 2, 4, 8, 16, 32\} - \{1, 4, 8, 16\} = \{2, 32\}\\ \therefore A' \cup B' &= \{1, 4, 8\} \cup \{2, 32\} = \{1, 2, 4, 8, 32\}.\end{aligned}$$

See that $(A \cap B)' = \{1, 2, 4, 8, 32\} = A' \cup B'$.

► Example 6.6.3 *If* A *and* B *are any two sets, prove that* $A - B$ *and* $A \cap B$ *are two disjoint sets.* **[C.P. 1995]**

Solution:

$$\begin{aligned}A - B &= A \cap B' \text{ and } A \cap B = B \cap A \textbf{ (Com. Law)}\\ \therefore (A - B) \cap (A \cap B) &= (A \cap B') \cap (B \cap A) = A \cap (B' \cap B) \cap A\\ &= A \cap \phi \cap A = A \cap A = \phi \textbf{ (Idempotent Law)}\end{aligned}$$

$\therefore$ $A - B$ and $A \cap B$ are two disjoint sets.

► Example 6.6.4 (a) *For the sets* $A = \{a, b, c\}$ *and* $B = \{1, 2\}$, *verify the Cartesian product* $A \times B \neq B \times A$. **[C.P. 1996]**

(b) *For the sets* $A = \{1, 2, 3\}$ *and* $B = \{4, 5\}$. *Verify* $A \times B \neq B \times A$. **[C.P. 2004]**

Solution: (a)

$$\begin{aligned}A \times B &= \{(a, 1), (b, 1), (c, 1), (a, 2), (b, 2), (c, 2)\}\\ B \times A &= \{(1, a), (1, b), (1, c), (2, a), (2, b), (2, c)\}.\end{aligned}$$

We see that $A \times B \neq B \times A$

{see that an element $(a, 1)$ is different from $(1, a)$}.

(b) Exactly similar. Try yourself.

► Example 6.6.5 *Write down all the subsets of the set* $\{x, y, z\}$. *What is the power set of* $\{x, y, z\}$? **[C.P. 1997, '99]**

Solution: The subsets of $\{x, y, z\}$ are

$$\phi, \{x, y, z\}, \{x, y\}, \{x, z\}, \{y, z\}, \{x\}, \{y\}, \{z\}.$$

The collection of all possible subsets of $\{x, y, z\}$ will be called its **power set, i.e., the power set** of $\{x, y, z\}$ is

$$\{\phi, \{x, y, z\}, \{x, y\}, \{x, z\}, \{y, z\}, \{x\}, \{y\}, \{z\}\}.$$

▶ Example 6.6.6 *If* $A = \{5,6,7,8,9\}, B = \{2,4,6,8,10,12\}$ *and* $C = \{3,6,9,12\}$; *then verify that* $A - (B \cup C) = (A - B) \cap (A - C)$. [C.P. 1997]

Solution:

$$B \cup C = \{2,4,6,8,10,12\} \cup \{3,6,9,12\} = \{2,3,4,6,8,9,10,12\}$$
$$A - (B \cup C) = \{5,6,7,8,9\} - \{2,3,4,6,8,9,10,12\} = \{5,7\}$$
$$A - B = \{5,6,7,8,9\} - \{2,4,6,8,10,12\} = \{5,7,9\}$$
$$A - C = \{5,6,7,8,9\} - \{3,6,9,12\} = \{5,7,8\}$$
$$\therefore \quad (A - B) \cap (A - C) = \{5,7,9\} \cap \{5,7,8\} = \{5,7\}.$$

Clearly, $A - (B \cup C) = \{5,7\} = (A - B) \cap (A - C)$.

▶ Example 6.6.7 *Find the simplest form of* $(A \cap B)' \cup (A' \cap B)'$ *where* A *and* B *are any two subsets of* S *and the notation* $(')$ *stands for compliment of a set w.r.t.* S. [C.P. 1998]

Solution:

$$\begin{aligned}
(A \cap B)' &= A' \cup B' \text{ (De Morgan's Law)} \\
(A' \cap B)' &= (A')' \cup B' \text{ (De Morgan]s Law)} \\
&= A \cup B' \\
\therefore\ (A \cap B)' \cup (A' \cap B)' &= (A' \cup B') \cup (A \cup B') = (A' \cup B') \cup (B' \cup A) \\
&= A' \cup (B' \cup B') \cup A = A' \cup B' \cup A = S.
\end{aligned}$$

▶ Example 6.6.8 *If* $A = \{a,b,c,d\}$ *and* $B = \{a,b,c\}$ *be two sets, determine*

$$(A \times B) \cap (B \times A) \text{ and } (A \times B) - (B \times A).$$

[C.P. 1998]

Solution:

$$\begin{aligned}
A \times B &= \{(a,a),(a,b),(a,c),(b,a),(b,b),(b,c),(c,a),(c,b), \\
&\qquad (c,c)(d,a),(d,b),(d,c)\} \\
B \times A &= \{(a,a),(a,b),(a,c),(a,d),(b,a),(b,b),(b,c),(b,d), \\
&\qquad (c,a),(c,b),(c,c),(c,d)\} \\
\therefore\ (A \times B) \cap (B \times A) &= \{(a,a),(b,b),(c,c),(a,b),(a,c),(b,a),(b,c),(c,a),(c,b)\} \\
\text{and } (A \times B) - (B \times A) &= \{(d,a),(d,b),(d,c)\}.
\end{aligned}$$

▶ Example 6.6.9 *For the three sets* $A = \{p,q,r\}$, $B = \{s,t,u\}$ *and* $C = \{s,u\}$, *verify* $A \times (B - C) = (A \times B) - (A \times C)$. [C.P. 1999]

Solution:

$$\begin{aligned}
B - C &= \{s,t,u\} - \{s,u\} = \{t\} \\
A \times (B - C) &= \{p,q,r\} \times \{t\} = \{(p,t),(q,t),(r,t)\}. \qquad \text{(i)}
\end{aligned}$$

$$\begin{aligned} A \times B &= \{p, q, r\} \times \{s, t, u\} \\ &= \{(p,s),(p,t),(p,u),(q,s),(q,t),(q,u),(r,s),(r,t),(r,u)\}. \\ A \times C &= \{p, q, r\} \times \{s, u\} \\ &= \{(p,s),(p,u),(q,s),(q,u),(r,s),(r,u)\}. \end{aligned}$$

$$\therefore\ (A \times B) - (A \times C) = \{(p,t),(q,t),(r,t)\}. \qquad \text{(ii)}$$

From (i) and (ii) we see that **A × (B − C) = A × B − A × C.**

▶ Example 6.6.10 *Verify that* $A - (B \cup C) = (A - B) \cap (A - C)$, *where*
(a) $A = \{1,2,5\}$, $B = \{2,4,6\}$, $C = \{5,7\}$. [C.P. 2000]
(b) $A = \{5,6,7,8,9\}$, $B = \{2,4,6,8,10,12\}$ *and* $C = \{3,6,9,12\}$. [C.P. 2004]

Solution: (a)

$$B \cup C = \{2,4,6\} \cup \{5,7\} = \{2,4,5,6,7\}$$
$$A - (B \cup C) = \{1,2,5\} - \{2,4,5,6,7\} = \{1\} \qquad \text{(i)}$$
$$A - B = \{1,2,5\} - \{2,4,6\} = \{1,5\}$$
$$A - C = \{1,2,5\} - \{5,7\} = \{1,2\}$$
$$\therefore\ (A - B) \cap (A - C) = \{1,5\} \cap \{1,2\} = \{1\}. \qquad \text{(ii)}$$

From (i) and (ii) we see that $A - (B \cup C) = (A - B) \cap (A - C)$.
(b) Exactly similar. Try yourself.

▶ Example 6.6.11 *Vefify that* $(A \cap B)' = A' \cup B'$, *where* $A = \{2,3,4\}$, $B = \{3,4,5\}$, $B = \{3,4,5\}$ *and the universal set* $U = \{1,2,3,4,5,6,7,30\}$ *where* A' *and* B' *are respectively complements of* A *and* B *in* U. *Also verify that* $A \times B \neq B \times A$. [C.P. 2000]

Solution:

$$A \cap B = \{2,3,4\} \cap \{3,4,5\} = \{3,4\}$$
$$(A \cap B)' = U - (A \cap B) = \{1,2,3,4,5,6,7,30\} - \{3,4\} = \{1,2,5,6,7,30\}$$
$$A' = U - A = \{1,2,3,4,5,6,7,30\} - \{2,3,4\} = \{1,5,6,7,30\}$$
$$B' = U - B = \{1,2,3,4,5,6,30\} - \{3,4,5\} = \{1,2,6,7,30\}$$
$$\therefore\ A' \cup B' = \{1,2,5,6,7,30\} = (A \cap B)' \ \textbf{(Proved)}$$

Again

$$\begin{aligned} A \times B &= \{2,3,4\} \times \{3,4,5\} \\ &= \{(2,3),(2,4),(2,5),(3,3),(3,4),(3,5),(4,3),(4,4),(4,5)\} \\ B \times A &= \{3,4,5\} \times \{2,3,4\} \\ &= \{(3,2),(3,3),(3,4),(4,2)(4,3),(4,4),(5,2),(5,3),(5,4)\}. \end{aligned}$$

Clearly $A \times B \neq B \times A$.

▶ Example 6.6.12 *Let* S (*the universal set*) $= \{1,2,3,4,5,6,7\}$ *and* A, B *are its two subsets given by* $A = \{1,5,6\}$, $B = \{5,6,7\}$, *verify that* $(A \cap B)' = A' \cup B'$, *where dash* ′ *denotes complement with respect to* S. [C.P. 2001]

Solution: **See Ex. 6.6.2 above. Similar Problem.**

▶ Example 6.6.13 *Write down all the subsets of the set* $A = \{a, b, c\}$. **[C.P. 2001]**

Solution: **See Ex. 6.6.5. Similar problem.**

▶ Example 6.6.14 *If* $A = \{1, 2\}, B = \{1, 2, 3\}$, *find* $(A \times B) \cap (B \times A)$. **[C.P. 2001]**

Solution:

$$A \times B = \{(1,1), (1,2), (1,3), (2,1), (2,2), (2,3)\}$$
$$B \times A = \{(1,1), (1,2), (2,1), (2,2), (3,1), (3,2)\}$$
$$\therefore \quad (A \times B) \cap (B \times A)\} = \{(1,1), (1,2), (2,1), (2,2)\}.$$

▶ Example 6.6.15 *If* $A = \{1, 2\}, B = \{2, 3\}$ *and* $C = \{3, 4\}$ *are three given sets then find:* (i) $(A \times B) \cup (A \times C)$ *and* (ii) $(A \times B) \cap (A \times C)$.

Solution:

$$A \times B = \{(1,2), (1,3), (2,2), (2,3)\}$$
$$A \times C = \{(1,3), (1,4), (2,3), (2,4)\}$$
$$\therefore \quad (A \times B) \cup (A \times C) = (1,2), (1,3), (2,2), (2,3), (1,4), (2,4)\}$$
$$\text{and} \quad (A \times B) \cap (A \times C) = \{(1,3), (2,3)\}.$$

▶ Example 6.6.16 *If* $A = \{5, 6, 7, 8, 9\}, B = \{2, 4, 6, 8, 10, 12\}, C = \{3, 6, 9, 12\}$, *then verify that* $A \cup (B \cap C) = (A \cup B) \cap (A \cup C)$. **[C.P. 2003]**

Solution:

$$B \cap C = \{2, 4, 6, 8, 10, 12\} \cap \{3, 6, 9, 12\} = \{6, 12\}$$
$$A \cup (B \cap C) = \{5, 6, 7, 8, 9\} \cup \{6, 12\} = \{5, 6, 7, 8, 9, 12\} \qquad \text{(i)}$$
$$A \cup B = \{5, 6, 7, 8, 9\} \cup \{2, 4, 6, 8, 10, 12\} = \{2, 4, 5, 6, 7, 8, 9, 10, 12\}$$
$$A \cup C = \{5, 6, 7, 8, 9\} \cup \{3, 6, 9, 12\} = \{3, 5, 6, 7, 8, 9, 12\}$$
$$\therefore \quad (A \cup B) \cap (A \cup C) = \{2, 4, 5, 6, 7, 8, 9, 10, 12\} \cap \{3, 5, 6, 7, 8, 9, 12\}$$
$$= \{5, 6, 7, 8, 9, 12\}. \qquad \text{(ii)}$$

From (i) and (ii) we see the required equality follows.

▶ Example 6.6.17 *Write down the subsets of the set* $\{2, 3\}$. **[C.P. 2003]**

Solution: $\{2, 3\}, \phi, \{2\}, \{3\}$.

▶ Example 6.6.18 *If* $A = \{3, 5, 6, 8\}, B = \{-3, 0, 6\}$, *find* $(A \cup B) - (A \cap B)$. **[C.P. 2002]**

Solution:

$$A \cup B = \{-3, 0, 3, 5, 6, 8\}$$
$$A \cap B = \{6\}$$
$$\therefore \quad (A \cup B) - (A \cap B) = \{-3, 0, 3, 5, 6, 8\} - \{6\} = \{-3, 0, 3, 5, 8\}.$$

► Example 6.6.19 *In an examination of* 100 *students,* 70 *passed in Mathematics,* 65 *passed in Physics and* 55 *passed in Chemistry.*

Of these students, 50 *passed in Mathematics and Physics,* 45 *passed in Mathematics and Chemistry,* 40 *passed in Physics and Chemistry.* 35 *students passed in all the three subjects.*

(i) *How many students passed in exactly two of the three subjects?*

(ii) *How many students passed in exactly one of the three subjects?*

(iii) *How many students failed in all the three subjects?* [C.P. 2002]

Solution: **See Fig. 6.5.**

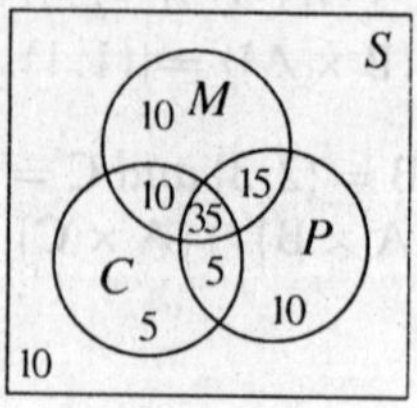

Fig 6.5

Given

$$n(M) = 70,\ n(P) = 65,\ n(C) = 55,\ n(M \cap P) = 50,$$
$$n(M \cap C) = 45,\ n(P \cap C) = 40,\ n(M \cap P \cap C) = 35$$
$$n(M \cap C \cap P') = 45 - 35 = 10 \text{ (Passed in M \& C)}$$
$$n(M \cap P \cap C') = 50 - 35 = 15 \text{ (Passed in M \& P)}$$
$$n(P \cap C \cap M') = 40 - 35 = 5 \text{ (Passed in P \& C)}$$

$\therefore$ number of students passed exactly two of the three subjects $= 10+15+5 = 30$.

Clearly the number of students passed in

Mathematics only $= 70 - (15 + 35 + 10) = 10$
Chemistry only $= 55 - (10 + 35 + 5) = 5$
Physics only $= 65 - (15 + 35 + 5) = 10$.

Hence, the number of students who passed exactly in one of the three subjects = 25.

Total number of students who passed in at least one of the three subjects

$$= 10 + 15 + 35 + 10 + 5 + 5 + 10 = 90.$$

$\therefore$ number of students who failed in all the three subjects $= 100 - 90 = 10$.

Now Try yourself

► Example 6.6.20 *In a survey of* 1000 *families in Calcutta, TV sets, Radiograms and Tape Recorders were counted. Each family had at least one of these appliances.* 400 *had no Tape Recorder,* 380 *no Radiogram,* 542 *no TV sets,* 294 *has both a Radiogram and a TV set,* 277 *both a Tape Recorder and a Radiogram,* 190 *both a Tape Recorder and a TV set.*

How many families had all the three appliances?
How many had only a Radiogram, no Tape Recorder, no TV set? [*Ans.* 83; 132]

6.7 Mappings

I. Definition

Let A and B be two given non-empty sets. Suppose that there exists a rule of which associates to **each member of** A a **unique** (one and only one) member of B. Then f is called a **function** or a **mapping** from A to B which we denote by

$$f : A \to B. \text{ (Read as: f maps A into B)}$$

A is called the **domain** and B, the **co-domain** of the mapping f.

If $x \in A$, we denote by $f(x)$, the **unique member** of B which f associates with x. We call $f(x)$, the **image** of x under f or **the value of** f at x. The element x is often called the **pre-image**.

Each element of A has a **unique image** but each element of B need not become the image of only one element of A. There can be more than one element of A which have the **same image** in B. Thus our mapping may be **many** to **one** but NOT **one** to **many**.

We define the **range** of f to consist of those elements of B which appear as the image of an element in A. Thus *range of a function* is the image of its domain and we may write the range as $f(A)$, i.e.,

$$f(A) = \{f(x) : x \in A\} \text{ clearly } f(A) \subseteq B.$$

For a diagrammatic representation of a mapping and its range $f(A)$: see **Fig. 6.6** and **Fig. 6.7**.

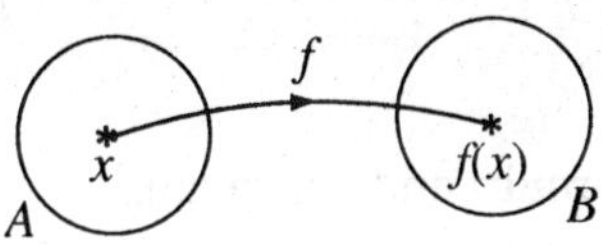

Fig. 6.6 Image of *x* under *f*

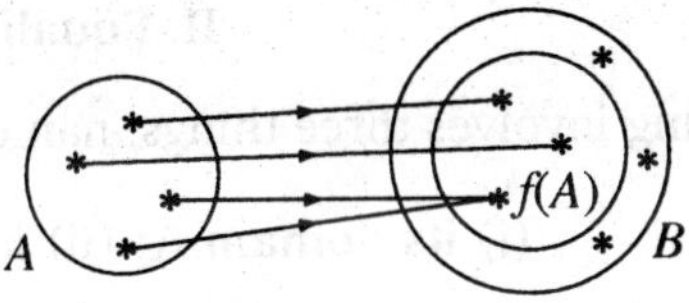

Fig. 6.7 Image of *x* under *f*

▶ Example 6.7.1 *Let* $A = \{1, 2, 3\}$, $B = \{1, 2, 3, 4, 5, 6, 7\}$. *Let* f *assign to each member in* A *its square. Then* f *is* **not a mapping from** A *into* B. *Since the element* $3 \in A$ *has no image in* B. *But our definition states that every member of the domain* A *must have a unique image in* B.

▶ Example 6.7.2 *Let* $A = \{a, b, c\}$ *and* $B = \{x, y, z\}$. *Suppose the rule* f *is such that*

$$f(a) = x, \; f(b) = y, \; f(a) = z, \; f(c) = z.$$

Then f *is not a function from* A *to* B. *Because the element* $a \in A$ *has two images* x *and* $z \in B$ *(i.e., not unique image for* a*). But our definition states that for each element of the domain there must exist an unique image in the co-domain.*

▶ Example 6.7.3 *Let* $A = \{1, 2, 3, 4, 5\}$ *and* $B = \{1, 2, 3, 4, 5, 6, 7\}$. *Suppose* A *is the domain and* B *the co-domain of a mapping* f *given by*

$$1 \to 2, \; 2 \to 3, \; 3 \to 5, \; 4 \to 7, \; 5 \to 7$$

f is a mapping of A into B, because each element of A is associated with a unique member of B. Note that not every element of B is an image of some element of A (e.g., 1,4,6 are not images). See that the element $7 \in B$ is **an image of more than one element of A (namely 4 and 5).** *What is the image set f(A) or range of f?*

[*Ans.* $f(A) = \{2, 3, 5, 7\}$. See that here $f(A) \subset B$.]

▶ Example 6.7.4 *A mapping* $f : \mathbf{R} \to \mathbf{R}$ *is defined by (where* **R** *is the set of all real numbers)*

$$f(x) = x^2 + 2, \textit{ for all } x \in \mathbf{R}.$$

The image set $= \{x^2 + 2/x \in \mathbf{R}\}$ *or* $\{y \in \mathbf{R}/y = x^2 + 2 \textit{ for } x \in \mathbf{R}\}$.

▶ Example 6.7.5 *A mapping* $g : \mathbf{R} - \{1\} \to \mathbf{R}$ *is defined by the formula*

$$g(x) = \frac{1}{x-1}, \ \forall x \in \mathbf{R} - \{1\}.$$

The image set of g is

$$\{y : y \in \mathbf{R} \textit{ and } y = \frac{1}{x-1} \textit{ for } x \in \mathbf{R} - \{1\}.$$

The domain of g is $\mathbf{R} - \{1\}$.

II. Equality of Mappings

A mapping involves three things, namely

(i) **its domain** A; (ii) **its co-domain** and (iii) **a rule.**

Consequently: Two mappings $f : A \to B$ and $g : A_1 \to B_1$ are said to be equal and we write $f = g$, if (i) $A = A_1$; (ii) $B = B_1$, (iii) $f(x) = g(x) \forall x \in A$.

Note 1. $\left.\begin{array}{l} f : \mathbf{R} \to \mathbf{R} \textit{ defined by } f(x) = x^2, \forall x \in \mathbf{R} \\ \textit{and } \ g : \mathbf{R} \to \mathbf{R}^+ \textit{ defined by } g(x) = x^2, \forall x \in \mathbf{R} \end{array}\right\}$
are not equal *because they have different co-domains even though they have the same image.*

III. Different Types of Mappings

1. Injective Mapping (or Injection)

In general, mappings are **many-to-one** (i.e., more than one element in the domain may have the same image element in the range).

We shall now define a mapping which is **one-to-one.**

Definition: *Suppose f maps A into B (i.e.,* $f : A \to B$*). The mapping f is said to be* **one-to-one** *(or* **injective***), if distinct elements of A have distinct images in B, i.e.,*

$$\text{if } f(x) = f(y) \implies x = y \ (x, y \in A).$$

In one-to-one mapping an element in B **either** *has no pre-image in* A **or** *has only one pre-image in* A.

How to prove that a given function f is one-to-one or injective?

If we are given a function f and we are to prove that it is one-one, we can do so by showing that if $f(x_1) = f(x_2)$, then $x_1 = x_2$, where x_1 and x_2 are arbitrary element of the domain of f. This means, of course, $x_1 \neq x_2 \implies f(x_1) \neq f(x_2)$.

▶ Example 6.7.6 *Let* $f : R \to R$ *and* $f(x) = 3x - 2$, $\forall x \in R$. *Prove that this mapping is one-to-one.*

Solution: Let $x, y \in$ domain R of f. Then $f(x) = f(y)$ gives

$$3x - 2 = 3y - 2 \implies x = y,$$

i.e., f is **one-to-one** or **injective.**

2. Surjective Mapping or Onto-mapping

Definition: *In the mapping* $f : A \to B$, *in general* $f(A) \subseteq B$. *But if* $f(A) = B$, *then the mapping is* **surjective** *or* **onto**, *i.e.,* f *is an onto mapping if every element of the co-domain* B *is image of* **some element** *in* A. *We then say "*f *maps* A *onto* B*" and we write*

$$f : A \xrightarrow{\text{onto}} B.$$

How to prove that a given function $f : A \to B$ is an onto function?

To prove that f is onto, we show that if $y \in B$, then there exists $x \in A$ such that $y = f(x)$. Then

$$y \in B \Rightarrow y \in f(A).$$

Since y is an arbitrary element of B, we get, every element of B is an element of $f(A)$, i.e., $B \subseteq f(A)$. But we know that $f(A) \subseteq B$.

Hence, when $f(A) = B$, the function is an onto function. So take any element of B and see whether it belongs to $f(A)$ or not. If that element $\in f(A)$, then the function is onto.

▶ Example 6.7.7 *The mapping* $f : R \to R$, *when*

$$\forall x \in R, f(x) = 2x + 3.$$

Prove that it is an onto mapping.

Solution: Here both the domain and co-domain consist of all real numbers. Let y be any arbitrary real number.

If $y = f(x) = 2x + 3$, we have $x = \frac{y-3}{2} \in R$. Thus $f\left(\frac{y-3}{2}\right) = y$,

i.e., any arbitrary real number y is the f-image of the element $\frac{y-3}{2}$ which is also a real number. Hence, f is onto.

► Example 6.7.8 *The mapping* $f : R \to R$, *where*

$$f(x) = \begin{cases} x^2, & x \geq 0 \\ x, & x < 0 \end{cases}$$

is an onto mapping. *If y is any member of the co-domain, i.e., any real number in this case, then there is a member x of the domain, i.e., also a real number in this case, such that* $y = f(x)$. *If* $y \geq 0$ *then* $x = \sqrt{y}$ *and if* $y < 0$ *then* $x = y$. *In this case every member of the codomain is the image of some member of the domain.*

The mapping $g : \mathbf{R} \to \mathbf{R}$ *defined by*

$$g(x) = \begin{cases} x+1, & x \geq 0 \\ x, & x < 0 \end{cases}$$

is not an onto mappings.

Here there are real numbers (for instance $\frac{1}{2}$*) which are not images of any member of the domain which is also* R.

3. Bijective Mapping

A mapping $f : A \to B$ is called bijective if it is both one-to-one and onto.

► Example 6.7.9 (a) *The mapping* $f : R \to R$ *defined by*

$$f(x) = 3x - 2 \ \forall x \in R$$

is both one-to-one and onto.

One-to-one, *because*

$$f(x) = f(y) \textbf{ or, } 3x - 2 = 3y - 2 \Rightarrow x = y$$

onto, *because if y is any real number that*

$$f\left(\frac{y+2}{3}\right) = y, \ \frac{y+2}{3} \in R.$$

(b) *Show that the function* $f : R \to R$, *given by*

$$f(x) = |x|, \quad x \in R$$

is neither injective nor surjective. **[C.P. 2004]**

Solution: Two distinct real numbers (say 2 and −2) have the same image 2 in R.

∴ the mapping is **not injective.**

The mapping is **not surjective** for there exist real number like $-\frac{1}{2}, -\frac{3}{4}$, etc. for which no pre-image exists.

4. Identity mapping

Let $f : A \to A$ be defined by $f(x) = x\ \forall x \in A$, i.e., each element of A be mapped on itself. Then f is called the **identity function** or **identity mapping.** We write I_A to denote the identity mapping of a set A:

$$I_A(x) = x\ \forall x \in A.$$

▶ Example 6.7.10 *Let* $A = \{1, 2, 3, 4, 5\}$. *If* $\exists$ *a mapping* f *such that*

$$f(1) = 1,\ f(2) = 2,\ f(3) = 3,\ f(4) = 4,\ f(5) = 5.$$

Then f *is the identity mapping* I_A. *See that identity mapping is always one-one and onto.*

5. Constant function

A function $f : A \to B$ is called a **constant function** if the same element $b \in B$ is assigned to every element in A.

i.e., $f : A \to B$ is a constant function if the **range** of f consists of only one element.

▶ Example 6.7.11 *Let* $f : R \to R$ *be defined by* $f(x) = 10\ \forall x \in R$. *Then* f *is a constant function since* 10 *is assigned to every element* $\in R$.

6. Bijective mapping: Inverse mapping

A mapping which is both **injective** and **surjective** is called **bijective:** a bijective mapping is called a **bijection.**

Bijective mappings are particularly important:

If $f : A \to B$ is a bijective mapping then for each $y \in B$ there exists a unique $x \in A$ such that $f(x) = y$.

[**Existence of** x follows from the fact that f is **surjective;**
Uniqueness of x follows from the fact that f is **injective.**]

Hence, we can define another mapping **from B into A** (denoted by $f^{-1} : B \to A$) (**read:** inverse of f maps B into A) by the rule:

if $y \in B$ then $f^{-1}(y) = x$, where x is a unique element of A such that $f(x) = y$.

Notice that f^{-1} is defined only when f is bijective. It is easy to see that f^{-1} is itself bijective and that its inverse is f again.

$$\left[\left(f^{-1}\right)^{-1} = f\right]$$

7. Involutory mapping

A bijective mapping from A onto A is **involutory** if $f = f^{-1}$ and **properly involutory** if $f = f^{-1}$ but $f \neq I_A$ (identity mapping).

8. Geometrical representations of different kinds of mapping

I. Three mappings are shown in Fig. 6.8; (The elements of A are shown by dots and those of B by crosses). Now observe: (i) is neither injective nor surjective; (ii) is injective but not surjective; (iii) is surjective but not injective.

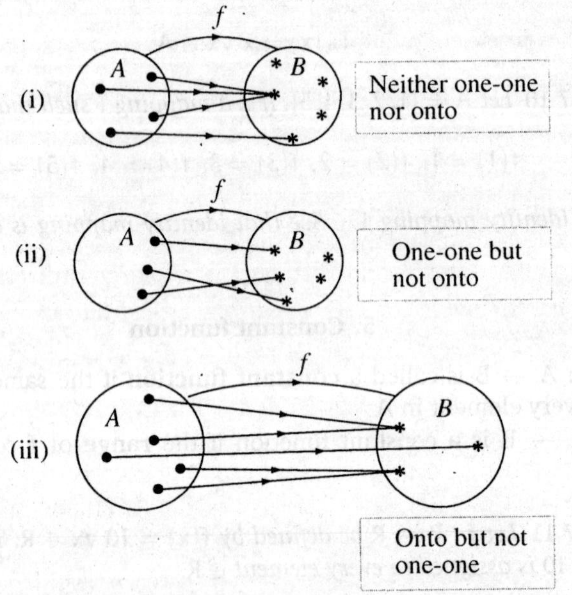

Fig. 6.8

Fig. 6.9(i) represents bijective mapping $f : A \to B$.

This mapping generates an inverse mapping $f^{-1} : B \to A$, [see Fig. 6.9(ii)]

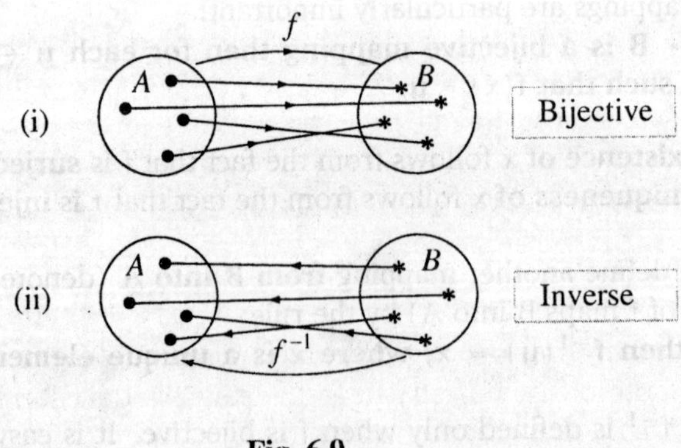

Fig. 6.9

II. The mapping $f : R \to R$ by the rule: if $x \in R$, then $f(x) = e^x \in R$ is injective, ($\because$ if $e^x = e^{x'}$, then $x = x'$).

But it is not surjective because if y is any negative real number or zero, there exists no real x such that $e^x = y$.

On the other hand, the mapping $\phi : R \to R^+$ (R^+=set of positive real numbers) defined by: if $x \in R$, then $\phi(x) = e^x$ is both injective and surjective, i.e.,

ϕ is bijective. Hence there exists an inverse mapping $\phi^{-1} : R^+ \to R$ and it is defined by: if $y \in R^+$, then $\phi^{-1}(y) = \log y$.

III. Let R be the set of all real numbers and f is a mapping given by $x \to Kx$ (where K is any **given** real number).

If $K \neq 0$, f is one-to-one and onto and its inverse f^{-1} is given by $x \to K^{-1}x$.

This bijective mapping is *properly involutory* if $K \neq 1$.

If $K = 0$, f is not one-to-one.

IV. Let A and B be two lines (regarded as set of points) in a plane. Let x, y denote points of A, B respectively; let ϕ be the mapping defined by taking y as the orthogonal projection of x on B. Then ϕ is one-to-one if and only if A and B are not at right angles; in this exceptional case, ϕ maps A into B and $\phi(A)$ is a single point (singleton set).

V. If S is the set of points in a plane and $\phi(x)$ is the reflection of the point x in a fixed line then ϕ is **involutory.**

VI. Let S denote the set of all points in the Cartesian plane, where any individual point is represented by pairs (x, y) of real numbers. The mapping, $f : S \to S$ defined by

$$f(x, y) = (x + 1, y + 2).$$

This mapping is an example of mapping referred to as a **translation.** If the image of the point (x, y), be taken as (x', y'), then the mapping can be represented by

$$x' = x + 1,\ y' = y + 2.$$

In set notation we express the mapping as

$$\begin{aligned} f &= \{((x, y), (x + 1, y + 2)) \,|\, (x, y) \in S\} \\ \text{or } f &= \left\{\left((x, y), (x', y')\right) \middle| (x, y) \in S \text{ and } x' = x + 1, y' = y + 2\right\}. \end{aligned}$$

We often think of it as a change in the location of coordinate axes—the points remain fixed but the axes is shifted causing a change in the coordinates of the points (*Translation of origin, axes remain parallel*).

Thus, in Fig. 6.10(i) observe how the points $P(2, 0)$, $Q(1, 3)$, $R(-1, 3)$ and $S(-2, -3)$ change their coordinates to $(3, 2)$, $(2, 5)$, $(0, 5)$ and $(-1, -1)$, respectively; the axes has shifted

Otherwise, we may visualise the mapping equally correctly as a shifting of points in a plane where the coordinate axes remain fixed. See Fig. 6.10(ii): the image points of $P(2, 0), Q(1, 3), R(-1, 3)$ and $S(-2, -3)$ are, respectively, indicated by P', Q', R' and S'. The arrow is drawn to picture the **translation** which results from the mapping.

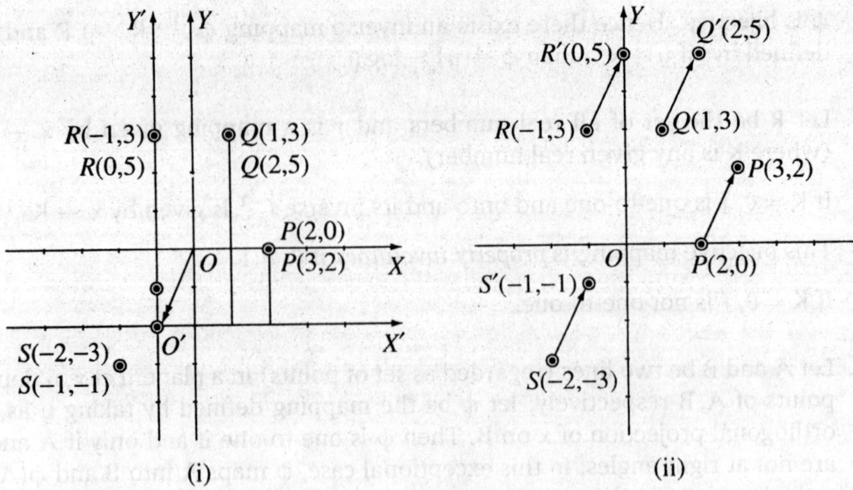

Fig. 6.10

6.8 Multiplication of Mappings (Compositions)

Let A, B, C be sets (not necessarily distinct) and suppose that

$$f \text{ maps A into B } (f : A \to B)$$

$$\text{and} \quad g \text{ maps B into C } (g : B \to C).$$

Then, f maps each element of A to its image $f(x)$, which is an element of B.

Since f maps B into C, the element $f(x) \in B$ will be mapped into $g[f(x)]$ which is an element of C.

Thus, the result of applying first f and then g is a new mapping from A into C. Such a mapping is called the **product of f by g** denoted by $g \circ f$. We also call such a mapping a *Composite mapping*. We have thus, $g \circ f$ maps A into C.

Fig. 6.11 illustrates the definition of product of two mappings.

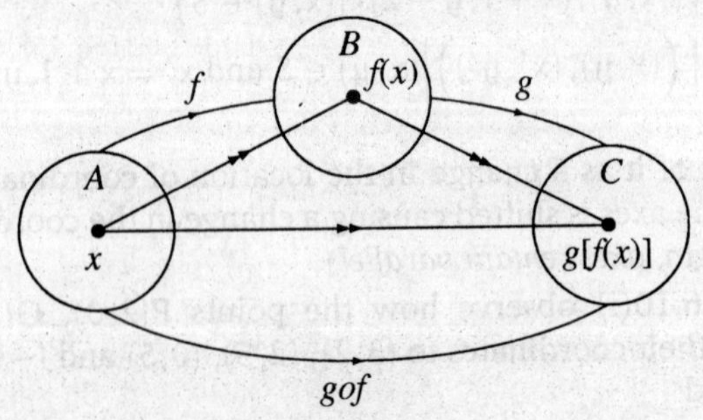

Fig. 6.11

It is evident from the definition that $g \circ f$ is defined only if for each $x \in A$ $f(x)$ is element of the domain of g so that we can take its g-image. Hence for the composition $g \circ f$ to exist, **the range of f must be a subset of the domain of g.**

► Example 6.8.1 *Two mappings* $f : \mathbf{R} \to \mathbf{R}$ *and* $g : \mathbf{R} \to \mathbf{R}$ *are defined by* $f(x) = x^2 + 1$ *and* $g(x) = x$. *Determine the composite mappings* $f \circ g$ *and* $g \circ f$. [C.P. 2001]

Solution: Clearly, range of f is a subset of domain of g and range of g is a subset of Dom. (f).

Now $(g \circ f)(x) = g(f(x)) = g(x^2+1) = x^2+1$
and $(f \circ g)(x) = f(g(x)) = f(x) = x^2+1$. Here $fg = gf$.

► Example 6.8.2 *Let* $A = \{1,2,3\}$, $B = \{a,b\}$, $C = \{l,m,n\}$.

Let f take 1 to a, 2 to b, 3 to a. We describe this situation by writing

$$f = \begin{pmatrix} 1 & 2 & 3 \\ a & b & a \end{pmatrix}.$$

Let g take a to n and b to m. Then $g = \begin{pmatrix} a & b \\ n & m \end{pmatrix}$.

Then the product $g(f) = \begin{pmatrix} 1 & 2 & 3 \\ n & m & n \end{pmatrix}$.

($\because$ for example, f takes 1 to a and then g takes a to n, so gf takes 1 to n and so on).

► Example 6.8.3 *Let* $A = B = C = \{1,2,3\}$. *Then let us consider two mappings* α *and* β *such that*

$$\alpha = \begin{pmatrix} 1 & 2 & 3 \\ 2 & 3 & 1 \end{pmatrix} \text{ and } \beta = \begin{pmatrix} 1 & 2 & 3 \\ 1 & 3 & 2 \end{pmatrix}.$$

Then

$$\alpha\beta = \begin{pmatrix} 1 & 2 & 3 \\ 2 & 1 & 3 \end{pmatrix} = \textit{Product of } \beta \textit{ by } \alpha$$

$$\text{and } \beta\alpha = \begin{pmatrix} 1 & 2 & 3 \\ 3 & 2 & 1 \end{pmatrix} = \textit{Product of } \alpha \textit{ by } \beta.$$

We ask the students to note:

The **multiplication of mappings** is different from the ordinary multiplication of real or complex numbers. In the first place the **product gf exists only** if the image set of f is the same or subset of the domain of g. See that in Ex. 6.8.2 above the product gf exists but fg does not exist. Again, even if both the products gf and fg exist, they need not be equal (see Ex. 6.8.3 above).

Definition: *Two mappings* f *and* g *are said to* **commute** *if* $fg = gf$.

► Example 6.8.4 *Let S be the set of all integers and consider the three mappings from S into S:*

$$f : x \to x^2; \; g : x \to |x|; \; h : x \to x+1.$$

h *is the only one of these mappings which is one-to-one, and its inverse is* $x \to x-1$. hf *is* $x \to x^2+1$ *and* fh *is* $x \to (x+1)^2$; *hence* f, h *do not* commute *even though both exist.*

Theorem 6.8.1 **The product of two one-to-one mappings is one-to-one; and the product of two bijective mappings is bijective.**

Proof. Follows from our definitions.

Let $f : A \to B$ and $g : B \to C$ be two injective (one-to-one) mappings. Suppose that x and x' are any two distinct elements of A. Then $f(x)$ and $f(x')$ are distinct elements of B. Hence, $g[f(x)]$ and $g[f(x')]$ are also distinct elements of C ($\because$ g is injective). Thus, $g[f(x)], g[f(x')]$ are distinct when x, x' are distinct, i.e., gf is injective.

Again, let f and g be both surjective and let z be any element of C. Because g is surjective there is some y in B such that $g(y) = z$ and because f is also surjective there is some x in A such that $f(x) = y$. Therefore, $gf(x) = g\{f(x)\} = g(y) = z$.

This, proves that for every z in C there is an x in A such that

$$gf(x) = z, \text{ i.e., } gf \text{ is surjective.}$$

Putting together the two statements we get:
If f, g are both bijective, then gf is also bijective.

Theorem 6.8.2 **Multiplication of mappings is associative.**

Proof. Let $f : A \to B$, $g : B \to C$, $h : C \to D$ be mappings.
For every x in A.

$$h(gf)x = h[g\{f(x)\}] = (hg)f(x)$$

Hence, $h(gf) = (hg)f$.

6.9 Binary Operations

We shall make a distinction between the terms: BINARY OPERATION IN A SET S and BINARY OPERATION ON A SET S.

A **binary operation in a set** S is a rule or a description by which **certain ordered pairs** (a, b) of elements a and b of the set S can be combined to form a unique element c of the set S.

On the other hand if the rule assigns to **every ordered pair** of elements of S a **unique element of S** then we shall say that the rule is a **binary operation on S.**

Addition is a binary **operation on the set N** of natural numbers because **every ordered pair** (a, b) of natural numbers gives a unique sum $a + b \in N$. [N is closed under addition]

But **subtraction** is a **binary operation in N** because **some ordered** pairs (a, b) of natural numbers give a unique element $a - b \in$ **N**. [**N** is not closed under subtraction]

Instead of the using signs like +, −, etc. we may use any arbitrary symbol, say '∘', for a binary operation.

A Formal Definition. *Let* A *be a non-empty set.* A binary operation '∘' **on** A *is a mapping of* $A \times A$ *(or,* A^2*) into* A *which maps* **every ordered pair** $(a, b) \in A^2$ *into a unique element* $c \in A$.

We denote this unique element by $a \circ b$ (**read,** a circle b or a composition b). We say in this case that "A *is closed under the binary operation* "∘".

[Other notations for an *arbitrary* binary operation are $*$ $(a * b)$ or $\wedge$ $(a \wedge b)$, etc.]

One may use familiar notations + (**additive** operation) or (**multiplicative** operation). Even when the elements are not numbers.

Thus, one may write $a + b$ or $a \cdot b$ (or simply ab) where $a, b \in$ an arbitrary set A. What is important to note is that:

An operation on A is a mapping of $A \times A$ into A, where the Domain of the mapping is $A \times A$ and the Range is A.

6.10 Solved Examples

▶ Example 6.10.1 *Let the set be P (the set of all positive integers); and the operation is ordinary subtraction.*

Can we map $P \times P$ into P? [See that $2 - 3 \notin P$]

Answer is **no.** This operation is **not** an operation **on** P, or, *P is not closed under subtraction.*

We can at most say that subtraction is **an operation in** P (but certainly **not** on P), since for some pairs $(a, b) \in P^2$, $a - b$ may belong to P.

We give six more examples: They are all examples of operations **on** different sets:

▶ Example 6.10.2 *Let the set be* $A = \{a, b, c\}$. *We define an operation on A. We shall use the symbol '∘' for this operation. The results can be read from the following table (called Multiplication Table or Composition Table):*

$a \circ a = b$, $a \circ b = c$,
$a \circ c = b$, $b \circ a = b$, *etc.*

∘	a	b	c
a	b	c	b
b	b	a	c
c	a	c	c

Multiplication Table; operation is '∘'

▶ Example 6.10.3 *Let the set be P(X) where P(X) is the set consisting of all subsets of a given set X (including* ϕ *and X), i.e.,* **power set of X.** *Then*

(i) *"Union" is a binary operation* **on** *P(X) since P(X) is closed under 'Union'.*

(ii) *"Intersection" is also a binary operation* **on** *P(X).*

(iii) *"Difference of two sets" is again a binary operation* **on** *P(X).*

▶ Example 6.10.4 *Let S be the set of all* 2×2 *real matrices.*

S is then closed under 'matrix addition' and also under "matrix multiplication". But an operation of the form $A^{-1}B$ (where A and B are two matrices $\in S$) is **not** an operation **on** S (Why?) [The pair $(A, B) \in S \times S$ if and only if A is non singular—S is not closed under this operation].

▶ Example 6.10.5 *M is the set of all mappings:* $X \to X$ *(X is a given set). Then*

(i) *Composition of mappings in which* $f \circ g$ *corresponds to* $(f, g) \in M \times M$ *is an* **operation on** *M. (M is closed under this operation).*

(ii) *Forming* $f^{-1} \circ g$ *is an operation* **in** *(but* **not on***) M, since the rule assigns an element of M to pair* $(f, g) \in M^2$ *if and only if f is a bijection. (M is not closed under this operation.)*

▶ Example 6.10.6 *Let V be the set of all vectors in 3-dimensional space. Then,*

(i) *Addition of vectors is an operation* **on** V.

$$(\vec{a}+\vec{b} \in V, \text{ where } \vec{a}, \vec{b} \in V^2)$$

V is closed under addition of vectors.

(ii) *Forming vector product is also an operation* **on** V.

$$(\vec{a}\times\vec{b} \in V, \text{ where } \vec{a}, \vec{b} \in V^2)$$

V is closed under vector product. But

(iii) *Forming scalar product is* **not** *an operation* **on** V.

(It is not even an operation in V, $\vec{a}\cdot\vec{b} \notin V$ **for any pair of vectors** $\vec{a}, \vec{b}$**).**

▶ Example 6.10.7 *Let* Z_3 *be the set of residues* $\{0, 1, 2\}$ *modulo* 3.

Then (i) addition modulo 3, and (ii) multiplication modulo 3 are both **operations on** Z_3. The following two tables will assert the truth of this statement:

+	0	1	2
0	0	1	2
1	1	2	0
2	2	0	1

Addition Modulo 3

•	0	1	2
0	0	0	0
1	0	1	2
2	0	2	1

Multiplication modulo 3

Properties for a binary operation

An operation on a set by itself does not lead to useful results unless it obeys some special characteristic properties. The properties with which we are mostly interested are the following:

(i) **Closure property;** (ii) **Commutative property;** (iii) **Associative property;** (iv) **Existence of identity;** (v) **Existence of Inverse.**

(i) If $\circ$ is a binary operation on a set A and if x, y are elements of A, then $x \circ y \in A$. Our definition of binary operation on a set implies closure property.

Closure property (**Cl.**)

(ii) If $\circ$ is a binary operation on a set A and if x, y are elements of A, then we say that x and y *commute* (with respect to $\circ$) if $x \circ y = y \circ x$

Commutative property (**Comm.**)

(iii) A binary operation $\circ$ on a set A is *associative* if it satisfies

$$(x \circ y) \circ z = x \circ (y \circ z), \text{ for all } x, y, z \in A$$

Associative property (**Ass.**)

(iv) Let A be any set and $\circ$ be a binary operation on A.

If there exists an element $e \in A$ which satisfies $e \circ x = x$ for any $x \in A$, then we say that e is a *left-identity element;* when it satisfies $x \circ e = x$ for any $x \in A$, then we say that e is a *right-identity element.*

If e is both a left-identity and a right-identity, then e is called an identity element (with respect to $\circ$). *Existence of Identity* (**Id.**)

(v) Let A be a set and $\circ$ be a binary operation on A.

If for any element $x \in A$, there exists an element $x^{-1} \in A$ such that $x^{-1} \circ x = e$ (left-identity element), then x^{-1} is the *left-inverse* of x; if $x \circ x^{-1} = e$ (right-identity), then x^{-1} is the *right-inverse* or x.

If x^{-1} is both a left-inverse and a right-inverse of x, then we say that x^{-1} is the *inverse of* x. *Existence of Inverse* (**In.**)

▶ Example 6.10.8 *Let R be the set of all real numbers and let the operation be ordinary* ADDITION (+). *Then*

(i) *For each pair* $a, b \in R$, $a + b \in R$. **Cl.**

(ii) *For any three real numbers, it is known,* $a + (b + c) = (a + b) + c$. **Ass.**

(iii) *The real number* 0 *obeys* $x + 0 = 0 + x = x$ *for any* $x \in R$. **Id.**
0 is here both right identity and left identity.

(iv) *For every* $x \in R$ *there exists* $-x \in R$ *such that* $x + (-x) = (-x) + x = 0$. **In.**
Here $(-x)$ *is both right inverse and left inverse.*

Since the symbol for the operation is + (*Addition*) *we call* 0 *(zero) the* **additive identity** *and* $-x$, **the additive inverse** *of* x.

(v) *For any pair* x, y *of real numbers* $x + y = y + x$ **Comm.**

Note 1. *If a set on which an operation* $\circ$ *is defined obeys (i)—(iv), then we say that the set forms a* **group** *under* $\circ$*; if, moreover, (v) holds the group is* **commutative** *(or* **Abelian***).*

We shall discuss the concept of group in the next chapter.
Conclusion. The set R of real numbers is a commutative group under addition.

▶ Example 6.10.9 *Let* $R - \{0\}$ *be the set of non-zero real numbers and let the operation be ordinary* **multiplication** (•). *Then*

(i) $R - \{0\}$ *is* **closed** *under* (•) : $a \cdot b$ (*or simply* ab) $\in R - \{0\}$, *for any pair*

$$(a, b) \in R - \{0\}.$$

(ii) **Associative Law** *holds:* $a(bc) = (ab)c$ *for any* $a, b, c \in R - \{0\}$.

(iii) *There exists the real number* 1 *such that* $1 \cdot x = x \cdot 1 = x$ *for any* $x \in R - \{0\}$. (1 *is called the Multiplicative Identity.*)

(iv) *For any* $x \in R-\{0\}$ *there exists* $1/x$ *or* x^{-1} *such that*

$$\frac{1}{x}\cdot x = x\cdot\frac{1}{x} = 1.$$

($\frac{1}{x}$ *or,* x^{-1} *is called Multiplicative Inverse of* x.)

(v) *For any two elements* $x, y \in R-\{0\}$, *we have* $x\cdot y = y\cdot x$. *Thus, the properties* **Cl., Ass., Id.** *and* **In.** *hold and, moreover, the property* **Comm.** *also holds.*

Conclusion. The set $R-\{0\}$ of non-zero real numbers is a commutative group under multiplication.

▶ Example 6.10.10 S = *the set of all* 2×2 *real matrices and the operation "matrix addition".*

Check the properties: **Cl., Ass., Id., In., Comm.**
Additive identity is $\begin{bmatrix} 0 & 0 \\ 0 & 0 \end{bmatrix}$. If $A \in S$, then $-A$ is the additive inverse of A.

Conclusion. The set S forms a commutative group under matrix addition.

▶ Example 6.10.11 V, *the set of all vectors in three-dimensional space under vector addition.*

This has properties **Cl., Ass., Id., In.** and **Comm.** (Check).
Additive identity is the null vector $\vec{0}$.
If $\vec{a} \in V$, then the additive inverse of $\vec{a}$ is $-\vec{a}$.
Conclusion. *V* forms a commutative group under vector addition.

▶ Example 6.10.12 S_1 = *the set of all non-singular* 2×2 *real matrices under the operation 'Matrix Multiplication'.*

Properties satisfied are **Cl., Ass., Id.,** $\left\{\text{the unit matrix} \begin{bmatrix} 1 & 0 \\ 0 & 1 \end{bmatrix}\right\}$ and **In.** (A^{-1} corresponds to the matrix $A \in S_1$).
But in this case commutative property (**Comm.**) does not hold good.

e.g.,

$$\begin{bmatrix} 1 & -1 \\ 0 & 1 \end{bmatrix}\begin{bmatrix} 0 & 1 \\ -1 & 1 \end{bmatrix} = \begin{bmatrix} 1 & 0 \\ -1 & 1 \end{bmatrix}$$

$$\begin{bmatrix} 0 & 1 \\ -1 & 1 \end{bmatrix}\begin{bmatrix} 1 & -1 \\ 0 & 1 \end{bmatrix} = \begin{bmatrix} 0 & 1 \\ -1 & 2 \end{bmatrix}.$$

They are not equal.
Conclusion. S_1 forms a group under matrix multiplication but the group is non-commutative.

▶ Example 6.10.13 *Let* $Z = \{0, \pm 1, \pm 2, \ldots\}$. *The binary operation* $\circ$ *on Z is defined by* $a \circ b = a - b \; (a, b \in z)$

(i) *Find* $x \in Z$ *such that* $x \circ 3 = 3$.

(ii) *Show that the operation is neither commutative nor associative.* **[C.P. 1991]**

Solution:

(i) $x \circ 3 = x - 3$. Therefore, $x \circ 3 = 3$ leads to $x - 3 = 3$ and hence $x = 6 \in Z$ satisfies this relation.

(ii) The composition $a \circ b = a - b$ is not always equal to $b \circ a = b - a$,

e.g., $3 \circ 7 = 3 - 7 = -4$ but $7 \circ 3 = 7 - 3 = 4$

$\therefore$ the operation 0 is *not commutative.*

It is not also associative, e.g.,

$$\begin{aligned} 3 \circ (7 \circ 1) &= 3 \circ (7 - 1) = 3 \circ 6 = 3 - 6 = -3 \\ \text{but } (3 \circ 7) \circ 1 &= (3 - 7) \circ 1 = -4 \circ 1 = -4 - 1 = -5. \end{aligned}$$

Thus, in general, it is not true that $a \circ (b \circ c) = (a \circ b) \circ c (a, b, c \in Z)$.

Exercises on Chapter 6

[**Notations used:** N or **P** = the set of all positive integers; Z = the set of all integers, including zero; Q = the set of all rational numbers; R = the set of all real numbers; C = the set of all complex numbers.

$$\begin{aligned} [a, b] &= \text{the set of all real numbers between } a \text{ and } b, \text{ including } a \text{ and } b; \\ &= \text{the closed interval from } a \text{ to } b; \\ (a, b) &= \text{the set of all real numbers between } a \text{ and } b \text{ but } \textbf{not including } a \text{ or } b \\ &= \text{the open interval from } a \text{ to } b. \\ (a, \infty) &= \text{the set of all real numbers greater than } a \\ (-\infty, \infty) &= \text{the set of all real numbers.]} \end{aligned}$$

A
[Basic Concepts]

1. Are the following statements true?

(a) $2 \in \{2, 3\}$; **(d)** $\frac{1}{4} \in P$; **(g)** $\sqrt{2} \in R$;

(b) $3 \in \{4, 1, 5\}$; **(e)** $\frac{1}{2} \in Q$; **(h)** $(\sqrt{-1})^3 \in C$;

(c) $5 \in P$; **(f)** $\sqrt{2} \in Q$; **(i)** $(\sqrt{-1})^2 \notin P$.

2. Check the truth of the following statements:

(a) $\{3, 4\}$ is a subset of $\{3, 4\}$; **(d)** $\{3, a, b, c\} = \{3, 3, 3, a, a, c, b, c\}$;

(b) $P \subseteq Q$; **(e)** $R \subseteq C$;

(c) $Z \subseteq Q$; **(f)** $\{2, 3\} = \{3, 4\}$.

3. True or False?

(a) $Z = \{x/x \text{ is real and } x^2 \in P\}$.

(b) $Q = \{x/x = a/b, \text{ where } b \neq 0, a, b \in Z\}$.

4. 'If $a = 3$ and $A = \{x/3x = 9\}$, then $a = A$—True? Give reasons.

5. Let $A = \{1, 2, 3\}$. Check which of the following statements are correct:

(a) $2 \in A$; (b) $2 \subset A$; (c) $\{2\} \in A$; (d) $\{2\} \subset A$.

6. True or False?

(a) $\{0\} = \phi$; (b) $\{\phi\} = \phi$; (c) $\{\phi\} = \{0\}$.

7. List the elements of the following sets:

(a) the set of all vowels in the English alphabet;
(b) the set of months in a year; the set of days in a week (English Calendar);
(c) the set of months in a year which begin with M (English Calendar);
(d) the set of all positive integers lying between $2\frac{1}{2}$ and $15\frac{3}{4}$;
(e) the set of all factors of 15;
(f) $\{n/n \in P \text{ and } n^2 - 5n + 6 = 0\}$;
(g) $\{x/x \in P \text{ and } 3 < x < 4\}$.

8. True or False? (A = any arbitrary set).

(a) $\phi \in A$; (c) $A \subseteq A$; (e) $A \in \{A\}$.
(b) $\phi \subseteq A$; (d) $A \in A$;

9. Remember: The collection of all subsets of

$$A = \{1, 2\} \text{ is } P(A) = [\phi, \{1, 2\}, \{1\}, \{2\}].$$

(Known as **power set** of A). Find the power set of $\{1, 2, 3\}$.

Find the number of elements in the power set of $\{1, 2, 3, 4, \ldots, n\}$.

10. (a) $A = \{x/x \text{ is a figure in the number } 2, 4, 2, 3, 2\}$,
$B = \{2, 3, 4\}$. Is $A \subseteq B$? Is $B \subseteq A$?

(b) $A = \{1, 2\}$, $B = \{2, 4, 6\}$, $C = \{1, 3, 4, 6\}$, $D = \{1, 2, 3, 4, 6\}$.
Write down a set whose subsets are A, B, C, D.

B
[Operations on sets]

1. Let $A = \{1, 2, 3\}$ and $B = \{1, 2, 4\}$. Find $A \cup B$, $A \cap B$, $A \times B$, $B \times A$.
Is $A \times B = B \times A$ in the present problem?

2. P is the set of all positive integers. Three subsets of P are

$$\begin{aligned} F &= \{a/a \in P \text{ and } a < 10\} \\ G &= \{a/a \in P \text{ and } a > 5\}, \\ \text{and } H &= \{a/a \in P \text{ and } a \text{ divisible by } 3\}. \end{aligned}$$

Obtain the following sets:

$$F \cap G,\ F \cap H,\ G \cap H,\ F \cup G,\ F \cup H,\ G \cup H.$$

3. If S (the Universal set)$=\{1,2,3,4,5,6,7,8\}$, and A, B, C are its three subsets given by
$A = \{1,5,6\}, B = \{2,3,5,7,8\}$ and $C = \{1,3,6,8\}$, obtain the following sets:

$$A \cap B' \cap C \text{ and } (A \cup C) \cap (B \cup C').$$

4. Let $A = \{1,2,3,\{4,5,6\}\}$ and $B = \{1,2,\{4,6\}\}$. Find $A \cap B$ and $A \cup B$.

5. If $A - \{x/x \in R \text{ and } 0 < x \leq 3\}$ and $B = \{x/x \subset R, \text{ and } 3 \leq x \leq 5\}$, find

$$A \cap B,\ A \cap B' \text{ and } A \cup B'.$$

6. If $A = (1,3]$, $B = [3,6)$, and $C = (4,\infty)$, find $A \cup B$, $A \cap C$, $B \cap C'$ and $(A' \cup B) \cap C'$.

7. Check the truth of the following results:

   (a) $\{1,2,3,4\} - \{1,2,3,4\} = \phi$;

   (b) If $a, b \in R$ and $a < b$, then $[a,b] - (a,b) = \{a,b\}$.

8. Let $A = \{1,2,3\}$ and $B = \{1,5\}$. Are the following assertions on Cartesian products valid for the set A, B? Justify.

   (a) $A^2 \neq B^2$;

   (b) $A \times B \neq B \times A$;

   (c) $(A \times B) \times A \neq A \times (B \times A)$.

[Applications of Laws of Algebra of Sets]

9. The *difference* between two sets A and B, denoted by $A - B$, is the set of all elements in A but not in B.

   Verify (either using this definition or using a Venn diagram) that $A - B = A \cap B'$.

   Now prove the following: (using Laws of Algebra of Sets)

   (a) $A - B = A - (A \cap B) = (A \cup B) - B$,

   e.g., $A - (A \cap B) = A \cap (A \cap B)' = A \cap (A' \cup B')$ [De Morgan's Law]
   $= (A \cap A') \cup (A \cap B')$ [Distributive Law]
   $= \phi \cup (A \cap B') = A \cap B' = A - B$;

(b) $A \cap (B - C) = (A \cap B) - (A \cap C)$;

(c) $A - (B \cup C) = (A - B) \cap (A - C)$;

(d) $(A \cup B) - C = (A - C) \cup (B - C)$;

(e) $A - (B - C) = (A - B) \cup (A \cap C)$;

(f) $(A - B) - C = A - (B \cup C)$;

(g) $(A - C) \cap (B - C) = (A \cap B) - C$.

10. If S be the Universe = $\{a, b, c, d, e, f\}$ and $A = \{a, c, d, f\}$ $B \cap C = \{a, b, f\}$, A, B, C being subsets of S, find:

(a) $(A \cup B) \cap (A \cup C)$; (b) $B' \cup C'$.

[*Hints*: (a) Use Distributive Law. (b) Use De Morgan's Law.]

11. Notice that we may proceed as

$$(A' \cup B') \cup (A \cap B \cap C')$$
$$= (A \cap B)' \cup \{(A \cap B) \cap C'\} \text{ [De Morgan's Law]}$$
$$= \{(A \cap B)' \cup (A \cap B)\} \cap \{(A \cap B)' \cup C')\} \text{ [Distributive Law]}$$
$$= S \cap \{(A' \cup B') \cup C') = A' \cup B' \cup C'.$$

Thus, we have proved: $(A' \cup B') \cup (A \cap B \cap C') = A' \cup B' \cup C'$.

Now prove in a similar manner that $(A' \cap B' \cap C) \cup (B \cap C) \cup (A \cap C) = C$.

12. We define the *symmetric difference* of two sets A and B (denoted by $A \triangle B$) by

$$A \triangle B = (A - B) \cup (B - A).$$

Verify that

(a) $A \triangle B = (A \cup B) - (A \cap B)$;

(b) $A \triangle (B \triangle C) = (A \triangle B) \triangle C$;

(c) $A \triangle \phi = A$; $A \triangle A = \phi$;

(d) $A \triangle B = B \triangle A$;

(e) $A \cap (B \triangle C) = (A \cap B) \triangle (A \cap C)$.

Enumerative results for finite sets

13. Each student in a class of 40 students studied at least one of the subjects—English, French and Bengali.

16 studied English, 22 Bengali and 26 French.

5 studied English and Bengali, 14 French and Bengali and 2 studied all the three languages.

Determine how many students studied English and French but not Bengali.

14. It is known that in a group of people, each of whom speaks at least one of the languages English, German and Russian, 31 speak English, 36 speak German and 27 speak Russian, 10 speak both English and German, 9 both English and Russian, 11 both German and Russian. Prove that the group contains **at least** 64 people and **at most** 73 people.

C
[On Mappings]

1. **(a)** For the mapping $f : R \to R$ (f maps the set of real numbers into itself), defined by $f(x) = x^2 - 1$ for any $x \in R$ obtain $f(0), f(2), f(-3)$. State the image set (or Range) $f(R)$.

   (b) For the mapping $g : [-3, 2] \to R$ defined by $g(x) = 3x + 4$, for any $x \in [-3, 2]$, state the image set of g.

   (c) For the mapping $h : [0, \pi] \to R$ defined by $h(x) = \cos x$, for any $x \in [0, \pi]$, state the image set of h.

   (d) Mapping $f : R \to R$ defined by $f(x) = \frac{1}{2}(x + |x|)$; find $f(3), f(-3), f(0)$ and state the image set of f.

2. For each of the following mappings state whether it is (i) injective, (ii) surjective. If it is not surjective, state its image; if it is not injective, specify an element of the co-domain that is the image of more than one element of the domain.

   (a) $f : R^+ \to R^+$ defined by $f(x) = x^2 + 1$, for any $x \in R^+$. (R^+ is the set of all **positive** real numbers).

   (b) $g : R \to R$ defined by $g(x) = \begin{cases} 2x + 1, & \text{when } x \geq 0, \\ x - 1, & \text{when } x < 0. \end{cases}$

   (c) $h : R \to R$ defined by $h(x) = \begin{cases} x^2, & \text{when } x \geq 0, \\ 2x + 3, & \text{when } x < 0. \end{cases}$

   (d) $k : R \to R^+$ defined by $k(x) = \begin{cases} x^2 + 1, & \text{when } x \geq 0, \\ 1 - x, & \text{when } x < 0. \end{cases}$

3. For each of the following mappings determine whether it is (i) injective, (ii) surjective. Find the inverse mapping of the mapping which is bijective.

   (a) $f : R \to R$, defined by $f(x) = x + 2$.

   (b) $g : R \to R^+$, defined by $g(x) = x^2$.

   (c) $h : R \to R$, defined by $h(x) = \begin{cases} x^2 - 1, & \text{when } x \geq 0, \\ -x^2 - 1 & \text{when } x < 0. \end{cases}$

   (d) $k : R \to R$, defined by $k(x) = \begin{cases} 2x, & \text{when } x \text{ is an integer}, \\ x, & \text{when } x \text{ is not an integer}. \end{cases}$

   (e) $u : Z \to Z$, defined by $u(n) = n^3$.

   (f) $v : Z \to Z$, defined by $v(n) = \begin{cases} n + 2, & \text{when } n \text{ is even}, \\ n, & \text{when } n \text{ is odd}. \end{cases}$

4. Let $A = \{1, 3, 5, 7, 9, \ldots\}$ be the set of odd positive integers and $B = \{2, 4, 6, 8, \ldots\}$ be the set of even positive integers. Give examples of mappings from A to B such that
   (a) f_1 is bijective;
   (b) f_2 is neither injective nor surjective;
   (c) f_3 is injective but not surjective;
   (d) f_4 is surjective but not injective.

5. Mappings $f : R \to R$ and $g : R \to R$ are defined by

$$f(x) = x^2,\ g(x) = \begin{cases} x-1, & \text{when } x \geq 0, \\ -x, & \text{when } x < 0. \end{cases}$$

Find the formulae for the composite mappings $f \circ f$, $f \circ g$, $g \circ f$ and $g \circ g$.

6. $f : R \to [-1, 1]$ and $g : [-1, 1] \to R$ are two mappings defined by $f(x) = \sin x$ and $g(x) = \sqrt{(1 - x^2)}$. Find the formulae for the composite mappings $g \circ f$ and $f \circ g$ specifying in each case the domain and co-domain. Find also the image sets of $g \circ f$ and $f \circ g$.

7. (a) Let f map A into B and I_A and I_B are the identity mappings on A and B, respectively. Show that the composition $f \circ I_A = f$ and the composition $I_B \circ f = f$.
   (b) Let $f : A \to B$, $g : B \to A$ be two mappings such that

$$g \circ f = I_A \text{ and } f \circ g = I_B.$$

   Show that f is bijective and that $g = f^{-1}$.
   (c) Let $f : A \to B$ and $g : B \to C$ be two bijective mappings; prove that $g \circ f$ is also a bijection and that $(g \circ f)^{-1} = f^{-1} \circ g^{-1}$.

8. Let $A = \{1, 2, 3, \ldots, n\}$. Consider the mapping $f : A \to A$. Prove that f is an ONTO mapping if, and only if, it is one-to-one.

9. If $A = R - \{0, 1\}$ (all real numbers except 0 and 1), mappings $f_i : A \to A$ $(i = 1, 2, 3, 4, 5, 6)$ are defined by

$$f_1(x) = x,\ f_2(x) = 1 - x,\ f_3(x) = \frac{x-1}{x},$$
$$f_4(x) = \frac{1}{x},\ f_5(x) = \frac{1}{1-x} \text{ and } f_6(x) = \frac{x}{x-1}.$$

Show that the set $\{f_1, f_2, f_3, f_4, f_5, f_6\}$ is closed under the composition of mappings.

D
[On Binary Operations]

1. With suitable illustrations explain the difference between 'A binary operation defined **in a set**' and 'A binary operation defined **on a set**' (i.e., closed under the operation).

2. Take the set N of natural numbers; determine which of the following define **operations in** or **operations on N**:

(a) addition;
(b) multiplication;
(c) subtraction;
(d) raising a to b-th power (i.e., a^b);
(e) forming $\sqrt{ab}$;
(f) forming $a - 2b$;
(g) a/b;
(h) $|a - b|$;
(i) forming the set $\{a, b\}$;
(j) forming the g.c.d. of a, b. [a, b are any two elements of **N**]. Remember $0 \notin \mathbf{N}$.

3. Let S be a set defined by

$$S = \left\{ \begin{bmatrix} 1 & a \\ 0 & 1 \end{bmatrix}, \ a \in R \right\} \text{ (R=set of all real numbers)..}$$

Is S closed under matrix addition? Under matrix multiplication? If in defining S, we replace R by Z (the set of all integers) what would your answers be?

4. If $S = \{1, 2, 3, 4, 5\}$, determine whether or not S is closed under addition (mod 6) and under multiplication (mod 6). What would your answers be if (mod 6) were replaced by (mod 5)?

5. For the set of real numbers, we define two binary operations $*$ and $\wedge$ (in terms of ordinary arithmetical operations) as follows: $a*b = 2ab$ and $a \wedge b = a + 2b$.

(a) Is $*$ Commutative? If not, give a counter example;
(b) Is $\wedge$ Commutative? If not, give a counter example;
(c) Prove or disprove that $*$ is associative;
(d) Prove or disprove that $\wedge$ is associative;
(e) Prove or disprove that $*$ is left-distributive over $\wedge$;
(f) Prove or disprove that $\wedge$ is left-distributive over $*$.

Remember. We define

Left-distributive Law for $*$ over $\wedge$: $a * (b \wedge c) = (a * b) \wedge (a * c)$
Right-distributive Law for $*$ over $\wedge$: $(a \wedge b) * c = (a * c) \wedge (b * c)$].

6. We define an operation $*$ on R (set of real numbers) in the following way:

$$\forall x, y \in R, \ x * y = xy + 1.$$

Show that the operation is commutative but not associative.

7. Let R be the set of all real numbers and R^* be the set of real numbers excluding 0 and 1 (i.e., $R^* = R - \{0, 1\}$).

We now consider that set F of mappings $\{f_1, f_2, f_3, f_4, f_5, f_6\}$ of R^* into R defined for each $x \in R^*$ by:

$$f_1 : x \to x; \qquad f_4 : x \to \frac{1}{x};$$

$$f_2 : x \to \frac{1}{1-x}; \qquad f_5 : x \to \frac{x}{x-1};$$

$$f_3 : x \to \frac{x-1}{x}; \qquad f_6 : x \to 1 - x.$$

Show that the composition of mappings (e.g., f_3, f_2, first apply f_2 and then f_3; thus, $f_3 f_2 : x \to \frac{1}{1-(x-1)/x} = x = f_1$) is a binary operation on F.

[Write a multiplication table and then check.]

8. Let $G = \{a, b, c, d\}$ be a set with four elements. We define an operation "•" on G by the Table given below

•	a	b	c	d
a	a	b	c	d
b	b	c	c	d
c	c	d	d	a
d	d	b	b	c

Investigate whether this operation is:

(a) Associative; **(b)** Commutative; **(c)** has a unit element; **(d)** has an inverse for each element.

9. Give an example of a binary operation ($\wedge$) on a finite set which is

(a) Commutative but not associative;

(b) associative but not commutative;

(c) neither commutative nor associative.

[*Hints*: (b) $G = \{1, 2\}$. Define

$$1 \wedge 1 = 1,\ 1 \wedge 2 = 2,\ 2 \wedge 1 = 1,\ 2 \wedge 2 = 2.$$

Not commutative but associative.

(c) Define $1 \wedge 1 = 1,\ 1 \wedge 2 = 1,\ 2 \wedge 1 = 2,\ 2 \wedge 2 = 1$; neither commutative nor associative.]

10. Define a binary operation "$*$" on the set S of non-negative integers

$$a * b = a^2 + b^2 \cdot \forall a,\ b \in S.$$

Does the operation have (i) a unit element? (ii) an inverse element?

Hints and Answers

A
[Basic Concepts]

1. (a), (c), (e), (g), (h) are true: others not true.

2. (a), (b), (c), (d), (e) are all true but (f) is not true.

3. (a), (b) both true.

4. No.

5. Only (a) and (d) are correct.

6. None correct.

7. (a) $\{a, e, i, o, u\}$; (b) {Jan., Feb., ..., Dec.}, {Sun, Mon., ..., Sat.}; (c) {March, May}; (d) {3,4,5, ..., 15}; (e) {1,3,5,15}; (f) {2,3}; (g) ϕ.

8. (b), (c), (e) only are true.

9. $P\{1,2,3\} = [\phi, \{1,2,3\}, \{1,2\}, \{1,3\}, \{2,3\}, \{1\}, \{2\}, \{3\}]$;

$$^nC_0 + {}^nC_1 + {}^nC_2 + \cdots + {}^nC_n = 2^n.$$

10. (a) Both true, i.e., here $A = B$; (b) $\{1, 2, 3, 4, 6\}$.

B

[Operations on sets]

1. $A \cup B = \{1,2,3,4\}$, $A \cap B = \{1,2\}$, $A \times B = \{(1,1), (1,2), (1,4), (2,1), (2,2), (2,4), (3,1), (3,2), (3,4)\}$, $B \times A = \{(1,1), (1,2), (1,3), (2,1), (2,2), (2,3), (4,1), (4,2), (4,3)\}$. Note that $A \times B \neq B \times A$.

2. $F \cap G = \{6, 7, 8, 9\}$; $F \cap H = \{3, 6, 9\}$; $G \cap H = \{6, 9, 12, 15, \ldots\}$; $F \cup G = \{1, 2, 3, 4, \ldots\}$; $F \cup H = \{1, 2, 3, 4, 5, 6, 7, 8, 9, 12, 15, 18, \ldots\}$. $G \cup H = \{3, 6, 7, 8, 9, 10, 11, 12, 13, 14, 15, \ldots\}$.

3. $\{1, 6\}$ and $\{3, 5, 8\}$.

4. $\{1, 2\}, \{1, 2, 3, \{4, 5\}, \{4, 5, 6\}\}$.

5. $\{3\}, (0, 3), (-\infty, 3]$ and $(5, \infty)$.

6. $(1, 6), \phi, [3, 4], (-\infty, 1] \cup [3, 4]$.

7. Both (a) and (b) are true.

8. The statements can be verified to be true.

9. (b) Right side

$$\begin{aligned} &= (A \cap B) \cap ((A \cap C)' = (A \cap B) \cap (A' \cup C') \\ &= (A \cap B \cap A') \cup (A \cap B \cap C') = \phi \cup A \cap (B \cap C') \\ &= A \cap (B - C) = \text{Left side.} \end{aligned}$$

(g)

$$\begin{aligned} (A \cap C') \cap (B \cap C') &= (A \cap C') \cap (C' \cap B) \\ &= A \cap (C' \cap C') \cap B = A \cap C' \cap B \\ &= (A \cap B) \cap C' \text{ (using Associative Law)} \\ &= (A \cap B) - C. \end{aligned}$$

10. (a) $\{a, b, c, d, f\}$; (b) $\{c, d, e\}$.

11. (i)
$$\begin{aligned} A\Delta B &= (A-B)\cup(B-A) = (A\cap B')\cup(B\cap A') \\ &= (A\cap B')\cup B\cap(A\cap B')\cup A' \text{ (Distributive)} \\ &= \{(A\cup B)\cap(B'\cup B)\}\cap\{(A\cup A')\cap(B'\cup A')\} \\ &= \{(A\cup B)\cap S\}\cap\{S\cap(B'\cup A')\} = (A\cup B)\cap(B'\cup A') \\ &= (A\cup B)\cap(A\cap B)' = (A\cup B)-(A\cap B). \end{aligned}$$

12. **5 students** studied English and French but not Bengali.

13. Let x be the number of people who speak all the three languages. Now proceed as in examples worked out in page **272** or **277**. Note that the least number when $x = 0$ and greatest number when $x = 9$ (Why? Can you explain?).

C
[On mappings]

1. **(a)** $-1, 3, 8$; $[-1, \infty) = f(R)$, image set; **(b)** $[-5, 10]$; **(c)** $[-1, 1]$; **(d)** $3, 0, 0$; $[0, \infty) = R^+$.

2. **(a)** Injective, not Surjective, image set $[1, \infty)$.
(b) Injective, not Surjective, image set $(-\infty, -1) \cup [1, \infty)$.
(c) Surjective, not Injective, $h(1) = h(-1) = 1$.
(d) Neither Injective nor surjective $k(1) = k(-1) = 2$, Image set $[1, \infty)$.

3. **(a)** (i), (ii); $f^{-1}(y) = y - 2$.
(b) (ii).
(c) (i), (ii); $h^{-1}(y) = \begin{cases} \sqrt{1+y}, & \text{when } y \geq -1 \\ -\sqrt{-1-y}, & \text{when } y < -1 \end{cases}$
(d) (i),
(e) (i),
(f) (i), (ii); $v^{-1}(m) = \begin{cases} m-2 & (m \text{ even}) \\ m & (m \text{ odd}). \end{cases}$

4. Possible answers:

(a) $f_1(s) = s + 1$; **(b)** $f_2(s) = 2$ for all s; **(c)** $f_3(s) = 2s$; **(d)** $f_4(s) = |s - 2| + 1$.

5. $f\circ f(x) = x^4$; $f\circ g(x) = \begin{cases} (x-1)^2, & \text{when } x \geq 0 \\ x^2, & \text{when } x < 0 \end{cases}$

$g\circ f(x) = x^2 - 1$; $g\circ g(x) = \begin{cases} x-2, & x \geq 1 \\ 1-x, & 0 \leq x < 1 \\ -x-1, & x < 0 \end{cases}$

6. $g\circ f(x) = |\cos x| : R \to R$, image $[0, 1]$.
$f\circ g(x) = \sin[\sqrt{(1-x^2)}] : [-1, 1] \to [-1, 1]$, image $[0, \sin 1]$.

D
[On Binary Operations]

2. **In** $\rightarrow$ (c), (e), (f), (g), (h).

On $\rightarrow$ all the rest excepting (i) which is neither **In** nor **On.**

3. No; Yes; No (for addition), Yes (for multiplication)

4. Not closed; closed.

5. **(a)** $*$ is commutative.

(b) $\wedge$ is not commutative, e.g., $3 \wedge 4 \neq 4 \wedge 3$.

8. **(a)** Not associative, e.g., $c \cdot (d \cdot b) = c \cdot b = d$ but $(c \cdot d) \cdot b = a \cdot b = b$.

(b) Not commutative, e.g., $b \cdot c = c$ but $c \cdot b = d$.

(c) Unit element is a.

(d) No, e.g., b has no inverse.

An Appendix to Chapter Six

Binary Relations: Equivalence Relations[1]

Relations occur frequently in mathematics and in everyday life. The symbol $<$ (is less than) describes a relation on the set of real numbers. The expression "*is the mother of*" describes a relation on a set of people. Similar connective expressions which yield relations are '*is equal to*', '*is the wife of*', '*is a multiple of*', '*is included in*', etc. Thus, some connective expression connecting two elements (hence the word BINARY) of a set A yields a relation R on A. If a pair of elements a, b of A satisfies the connective expression, we write $(a, b) \in R$ or $a \mathfrak{R} b$ (**read:** a is related to b). In practice we collect all ordered pairs (a, b) satisfying the connective expression and form a SET OF SUCH ORDERED PAIRS. This set determines the relation R on the set A. Thus, in the Cartesian product $A \times A$ we leave out pairs (x, y) which do not satisfy the connective expression:

Definition: *A binary relation* R *on a set* A *is a subset of the Cartesian product* $A \times A$. *The subset consists of pairs* $(a, b), a, b \in A$ *where* a *and* b *are connected by* **certain stated condition.**

If $(a, b) \in R$, we write $a \mathfrak{R} b$ [**read:** a is related to b by the relation $\mathfrak{R}$].

If $(a, b) \notin R$, we say that 'a is **not related** to b'.

▶ Example 1. *Let* $A = \{1, 2, 3, 4\}$. *We shall write* $a \mathfrak{R} b$ *if, and only if,* $b = a + 1$. *Then,* R *is given by* $R = \{(1, 2), (2, 3), (3, 4)\}$.

More generally, if A and B are two sets then a binary relation R from A to B assigns to each pair $(a, b) \in A \times B$ **exactly one** of the statements:

[1] Not included in the C.U. Three year General Degree Course.

a is related to b or a is not related to b.

i.e., either $a\,\mathfrak{R}\,b$ or 'a is **not related** to b' ($a \not\mathfrak{R}\, b$)

The relation R is a set of all such ordered pairs which obey $a\,\mathfrak{R}\,b$. Some authors prefer to write $a \sim b$ instead of $a\,\mathfrak{R}\,b$.

▶ Example 2. A = *Set of rooms numbered* 1, 2, 3, 4;
B = *Set of boarders* X, Y, Z.

Mr. X is assigned room 1, Mr. Y room 3 and Mr. Z room 4.

Then $\{(1, X), (3, Y), (4, Z)\} \subseteq A \times B$ determines a **relation.**

In any relation R the set formed by the first elements of the pairs is called the **Domain** of the relation [D(R)] and the set formed by the second elements of the pairs is called the **Range** of the relation.

[e.g., in Ex. 2 above $D(R) = \{1, 3, 4\}$ and the range=$\{X, Y, Z\}$]

It is natural **to graph a relation** in the ways that equations are graphed in elementary algebra and analytic geometry. See Figs. 6.12 and 6.13 (given below):

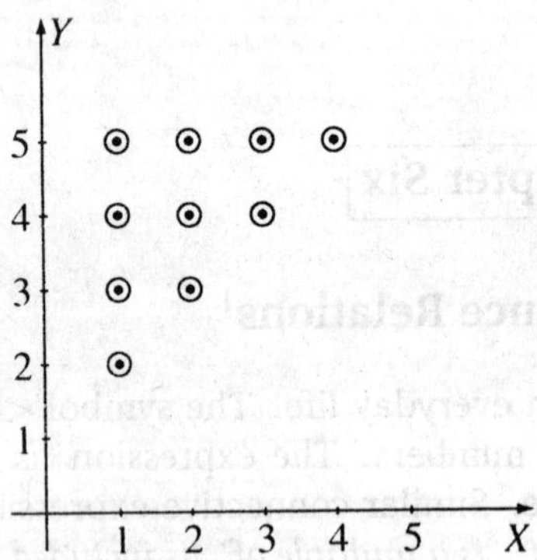

Fig. 6.12: Graph of the relation $\{(x, y)/x < y\}$ on the set $A = \{1, 2, 3, 4, 5\}$.

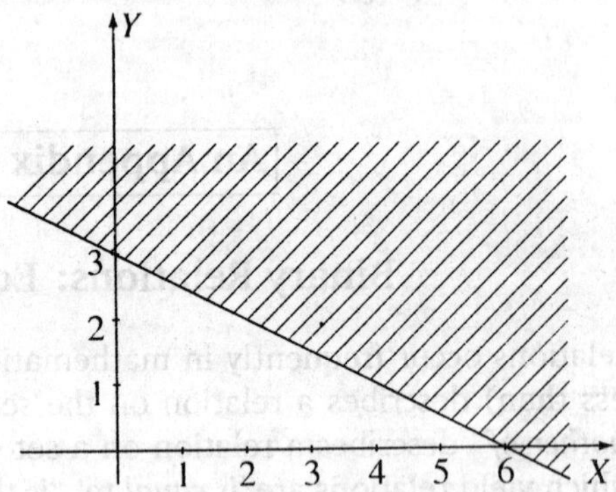

Fig. 6.13: Graph of the relation $\{(x, y)/x + 2y \geq 6\}$ on the set of real numbers.

Equivalence Relations

We shall be mostly concerned with relations which have some particular properties as stated in the following definition:

Definition: *Let A be a non-empty set and R be a subset of the Cartesian product A × A. Then R is called an* Equivalence Relation *in A if the following properties hold:*

P.(i) **Reflexive property:** *For all* $a \in A$, $a\,\mathfrak{R}\,a$ (*a is related to a itself*) *i.e.*, $(a, a) \in R$ *for all* $a \in A$.

P.(ii) **Symmetric property:** *If for some* a, b *in* A, $a\,\mathfrak{R}\,b$ *holds, then* $b\,\mathfrak{R}\,a$ *also holds. i.e.,* $a\,\mathfrak{R}\,b \Rightarrow b\,\mathfrak{R}\,a$ *for* $a, b \in A$. *In other words, if* $(a, b) \in R$ *then* $(b, a) \in R$.

P.(iii) **Transitive property:** *Whenever* $a\,\mathfrak{R}\,b$ *and* $b\,\mathfrak{R}\,c$ *hold, then* $a\,\mathfrak{R}\,c$ *also holds. i.e.,* $a\,\mathfrak{R}\,b$ *and* $b\,\mathfrak{R}\,c \Rightarrow a\,\mathfrak{R}\,c$. *In other words, if* $(a, b) \in R$ *and* $(b, c) \in R$, *then* $(a, c) \in R$.

Illustrations

1. The relation '**is equal to**' is the simplest Equivalence Relation. Suppose we take A=set of all real numbers. If $a, b, c \in A$, then clearly.

$$a = a; b = b; c = c; \; a = b \Rightarrow b = a; \; a = b \text{ and } b = c \Rightarrow a = c.$$

Clearly **P.(i)**, **P.(ii)** and **P.(iii)** are all satisfied, i.e., equality relation on real numbers is an equivalence relation.

2. The relation R determined by $x \,\mathfrak{R}\, y$, if and only if, x and y are of the same age $(x, y$ belonging to a set P of people). This is an equivalence relation.

[Check: P.(i), P.(ii) and P.(iii)].

3. Define R by "is the mother of" on the set P of people. **Check:** This relation is **neither reflexive, nor symmetric** nor **transitive.** Therefore, it is not an equivalence relation on P.

4. The relation $<$ on a set A of real numbers is *Transitive* $[a < b, b < c \Rightarrow a < c]$ but it is not **reflexive,** not also **symmetric.** So $<$ is not an equivalence relation.

5. Let $A = \{1, 3, 5, 7\}$.

One subset of $A \times A$ is $\{(1,1), (3,3), (5,5), (7,7)\}$ = relation R, say.

Check: This relation R on A is an equivalence relation.

6. Let T be the set of triangles in a plane. Define R by the expression "**is similar to**". This is an equivalence relation T (Check the three properties).

7.

Relations	on the set	Properties satisfied
(i) **is acquainted with**	P (of all people)	Reflexive and Symmetric but not transitive (Not an equivalence Relation).
(ii) **is a multiple of**	N (all positive integers)	Reflexive and Transitive but not Symmetric.
(iii) **Likes**	P (of all people)	Not Transitive and not Symmetric but *perhaps* reflexive.
(iv) **is perpendicular to**	L (Set of all lines in a plane)	Not an Equivalence relation
(v) **Congruent to**	T (Set of all triangles in a plane)	Equivalence relation. (Why?)
(vi) **is parallel to**	L (Set of all lines)	Equivalence relation.

8. **An Important Example: Congruence Relation.** Let Z be the set of all integers (positive, negative, including 0). We define: $a \,\mathfrak{R}\, b$ (some authors write $a \sim b$) if and only if $a - b$ is divisible by **a fixed positive integer** m (say)

i.e., $a - b = mq$ where $a, b, q \in Z$.

We call such a relation: *Congruence Relation Modulo* m on the set Z.

In case $a - b = mq$, we write

$$a \equiv b(\textit{mod } m) \text{ (read: } a \text{ is congruent to } b \text{ modulo } m)$$

Such a relation is an Equivalence Relation, because it is

(a) *Reflexive:* For any $a \in Z, a - a = 0. m[a \equiv a(\bmod m)]$.

(b) *Symmetric:* $a - b = mq \Rightarrow b - a = m(-q)$,

i.e., $a \equiv b(\bmod m) \Rightarrow b \equiv a(\bmod m)$, and

(c) *Transitive:* $a - b = mq, b - c = mk$

$$\Rightarrow a - c = m(q + k) \; (\because q, k \in Z, q + k \in Z),$$

i.e., $a \equiv b(\bmod m)$ and $b \equiv c(\bmod m) \Rightarrow a \equiv c(\bmod m)$.

Equivalence Class

If R is an equivalence relation on a given set A, then the set of all elements in A equivalent to a fixed element $x \in A$ is called an *Equivalent class modulo* R *in* A. The equivalence set is denoted by $[x]$:

$$[x] = \{y \in A / x \,\mathfrak{R}\, y\}.$$

We may use $x \sim y$ (**read:** x is equivalent to y modulo R) in place of $x \,\mathfrak{R}\, y$.

▶ Example 3. *Consider the equivalence relation given by* Congruence modulo 3 *on the set Z of all integers. This equivalence relation generates three Equivalence classes* $[0], [1], [2]$.

$$\begin{aligned}[0] &= \{\dots \dots, -12, -9, -6, -3, 0, 3, 6, 9, 12, \dots \dots\}\\ [1] &= \{\dots \dots, -11, -8, -5, -2, 1, 4, 7, 10, 13, \dots \dots\}\\ [2] &= \{\dots \dots, -10, -7, -4, -1, 2, 5, 8, 11, 14, \dots \dots\}\end{aligned}$$

These classes are also known as **Congruence classes** or **Residue classes modulo 3. We observe** that

(i) Any two elements of the same equivalence class are equivalent (e.g., $0 \sim 3, 3 \sim 6, -12 \sim 9$, etc. in $[0]$).

This implies that any element of the class can serve as a representative for the class (e.g., $[0] = [3] = [-9]$, etc.)

(ii) No element of Z belongs to more than one equivalence class.

(iii) Any element of Z belongs to one, and only one, of the three equivalence classes $[0], [1]$ and $[2]$.

What we have done is **"partitioning the set Z into three Equivalence sets".** [By *Partition* of a set $S \neq \phi$ we mean breaking the set S into **pairwise disjoint subsets** such that **every element** of S belongs to **one and only one** of these subsets, i.e., a partition of a set $S(\neq \phi)$ is a set P of non-empty subsets of S, s.t. each $x \in S$ belongs to one and only one member of P.

Theorem: An equivalence relation on a non-empty set S induces a partition of the set S into equivalent classes.

Proof. Suppose that in S we have an equivalence relation denoted by $\sim$. For any element $x \in S$ we form an equivalence class $[x]$ which is the set of all elements in S equivalent to x.

(i) $x \in [x]$ because $x \sim x$.

(ii) Suppose y and z are any two elements of $[x]$. Then $x \sim y$ and $x \sim z$, by definition. Using the symmetric property, we know that $y \sim x$. Now, $y \sim x$ and $x \sim z \Rightarrow y \sim z$ (Transitive property).

Hence, any two elements of the same equivalence class are equivalent. In particular, if $y \sim x$, then $[x] = [y]$; i.e., any element of an equivalence class will serve as a representative for that class.

(iii) No element of S can belong to more than one equivalence class. Suppose that $x \in [a]$ and $x \in [b]$. Then by (ii) $[x] = [a] = [b]$.

(iv) For **any element** λ in S we may form $[\lambda]$, so we have shown that each element in S belongs to one and only one of the equivalence classes of A, formed with respect to the equivalence relation $\sim$.

These arguments thus prove that any equivalence relation on S provides a partition of S into equivalence classes.

The converse of the theorem is also true.

Converse Theorem. If the sets $X_1, X_2, \ldots, X_n$ form a partition of a set S, then they serve to define an equivalence relation $\sim$ on S, where $x \sim y$ if and only if x and y belong to the same set X_i for some i.

That this is an equivalence relation, it is easy to verify.

(i) Any element x is in the same set with itself, so that $x \sim x$ (*Reflexive*).

(ii) Suppose $x \sim y$ (i.e., x and y belong to the same set X_i), then y and x are in the same set X_i, so that $y \sim x$ (*Symmetric*).

(iii) If $x \sim y$ and $y \sim z$, this says that x and y are in the same set and y and z are in the same set. Since in a partition (by definition) no element is in more than one set, x and z must be in the same set, so that $x \sim z$ (*Transitive*).

$\therefore$ the relation $\sim$ is an equivalence relation.

Binary Relations Problems for Practice

► **Example 1.** (i) *Let us define a relation* $\sim$ *on the set W on all people by taking* $x \sim y$ *when 'x and y' have the same age'* $(x, y \in W)$.

Is it an equivalence relation? Justify your answer.

(ii) *Next suppose* $x \sim y$ *only when 'x is the mother of y'. Show that this relation is neither Reflexive nor Symmetric nor Transitive.*

► **Example 2.** *Show that the relation* $>$ *(greater than) on the set of real numbers is Transitive but not reflexive or symmetric.*

► **Example 3.** *Give an example of a relation on a set*

(i) *which is Symmetric and Transitive but not Reflexive;*

[*Ans.* $x \sim y$ $(x, y \in R$ *only when* $x - y \neq 0)$]

(ii) *which is Reflexive and Symmetric but not Transitive;*

[*Ans.* $x \sim y$ $(x, y \in R)$ *only when* $x - y = 0, 3$ *or* -3]

(iii) *which is Symmetric but neither Reflexive nor Transitive.*

[*Ans.* $x \sim y (x, y \in R)$ *only when* $x \neq y$.]

► **Example 4.** *Which of the following relations on the set Z of all integers are equivalence relations?*

(i) $x \,\mathfrak{R}\, y$ *when* $|x - y|$ *is of the form 5k for some integer k.*

(ii) $x \,\mathfrak{R}\, y$ *when* $x + 2y$ *is divisible by 3.*

(iii) $x \,\mathfrak{R}\, y$ *when* $x \leq y$. [*Ans.* (i) and (ii) are equivalence relations but (iii) is not.]

► **Example 5.** *Let Z be the set of all integers. If* $a, b, \in Z$, *we define*

$$a \equiv b (mod\ 5),$$

if $a - b$ *has 5 a factor. Prove that '*$\equiv$*' is an equivalence relation on Z. Exhibit the equivalence (or residue) classes.*

► **Example 6.** *Let V be the set of all points* $P(x, y)$ *in the xy-plane. Define a relation* $\mathfrak{R}$ *by writing* $(x, y) \,\mathfrak{R}\, (z, t)$ *if and only if* $x^2 + y^2 = z^2 + t^2$.

Show that $\mathfrak{R}$ *is an equivalence relation on V. Describe geometrically the partition of V into equivalence classes. What is the equivalence class determined by the point* $A(1, 1)$? [*Ans.* $\{0\}$ *and circles with centre O, where O is* $(0, 0)$; $\{(x, y) : x^2 + y^2 = 2\}$.]

Chapter 7

Introduction of Group Theory

Introduction of Group Theory	
1.	**Group**—Definitions and Examples taken from various branches (examples from number system, roots of unity, 2 × 2 real matrices, non-singular real matrices of a fixed order). Elementary properties using the definition of Group.
2.	**Definition and Examples of a subgroup.** Statement of the necessary and sufficient condition—its applications.

7.1 Introduction

Under the heading *Modern Algebra* we shall study the following *algebraic structures* or **mathematical systems:**

Groups, Rings, Fields and Vector spaces.

By an algebraic structure (or mathematical system) we mean a **non-empty set** on which one or more binary operations are defined and a set of axioms which are imposed on these operations.

The simplest algebraic structure that we shall study in this chapter is a **Group.** We shall define a group and also a subgroup with various examples.

Some special types of groups like *Cyclic groups* or *Permutation groups* will be introduced in the later part of this chapter as APPENDIX. They are meant for more curious students.

7.2 Group: Definition

Definition: *A mathematical system* $(G, \circ)$ *consisting of a* **non-empty set** G *and a binary operation* $\circ$ *defined on* G, *is called a Group under the operation* $\circ$ *provided it satisfies the following four axioms:*

G1. Closure property: G is closed under $\circ$.

This means: for any two elements a, b of G, $a \circ b \in G$.

G2. Associative property: The operation $\circ$ obeys associative law in G.

i.e., for any three elements $a, b, c \in G$,

$$a \circ (b \circ c) = (a \circ b) \circ c.$$

G3. Existence of Identity: There exists an element $e \in G$, such that

$$a \circ e = e \circ a = a, \ \forall a \in G.$$

(e is then called an *Identity element.*)

G4. Existence of Inverse: For every element $a \in G$ there exists an element $a' \in G$ such that

$$a \circ a' = a' \circ a = e.$$

Note 1. *We shall prove that if an identity element e exists in the group* $(G, \circ)$, *then it is* unique *(one and only one). Inverse of an element, when exists, is also unique.*

Besides the four axioms **G1–G4**, if the system obeys the following commutative law, then the group is called an **Abelian group** or a **Commutative group.**

G5. Commutative Property: For ANY two elements

$$a, b \in G, \ a \circ b = b \circ a.$$

A Few Points to Remember

1. When the operation $\circ$ need not be specifically mentioned (i.e., understood); we shall denote the group $(G, \circ)$ simply by G.
2. Usually we shall write '+' for additive operations and '•' for multiplicative operations. Addition or Multiplication may not be the usual type as we do in number-systems.

   (a) When additive operation $(+)$ is used, we take

   the identity element $e = 0$ (zero)
   and the inverse of $a = -a$ (negative of a).

   (b) When multiplicative operation $(\bullet)$ issued, we take

   the identity element $e = 1$ (unity)
   and the inverse of $a = a^{-1}$.

   In multiplicative operation the symbol '•' is often omitted; e.g., for $a \bullet b$ we may often write ab.

3. We shall use the following symbols:

Z = the set of all integers $\{0, \pm 1, \pm 2, \ldots\}$
Q = the set of all rational numbers
R = the set of all real numbers
C = the set of all complex numbers.

Order of a Group

If a group $(G, \circ)$ contains a *finite number of elements*, say n elements, we call it a finite group of **order n**.

An *infinite group*, on the other hand, does not contain an infinite number of elements. We shall soon prove that the set $\{1, -1, i, -i\}$ is a group under the operation "multiplication".

This is an example of a **finite group of order 4**.

We shall also prove that the set Z of all integers forms a group under addition. This is an example of an **infinite group**.

For the sake of illustrations we give below examples of two groups. Further important examples are given under **Art. 7.6.**

▶ Example 7.2.1 *The set Z of all integers $\{0, \pm 1, \pm 2, \ldots\}$ is an additive group. This group is an infinite group and it is also abelian.*

Solution: We have the infinite set $Z = \{0, \pm 1, \pm 2, \ldots\}$

and the operation $\circ$ is the ordinary addition $(+)$.

1. **Closure Property:** We know that the sum of two integers is also an integer. Hence, $\{Z, +\}$ obeys closure property.

2. **Associative Property:** Addition of integers obeys associative law.

$$\text{e.g., } 2 + (3 + 8) = (2 + 3) + 8.$$

3. **Existence of Identity:** There exists an element 0 (zero) $\in$ Z such that $a + o = o + a = a$, for every integer a. (o is the *additive identity*.)

4. **Existence of Inverse:** Given any integer a, we can obtain another integer $-a$ such that

$$a + (-a) = (-a) + a = 0,$$

i.e., $(-a)$ is the additive inverse of a.

e.g., the inverse of -7 is $-(-7) = +7$.

In view of the fact that the operation + obeys all the four group axioms, we conclude that $\{Z, +\}$ is an additive group. Moreover,

5. **Commutative property** also holds here:

$$a + b = b + a, \text{ for any two integers } a, b.$$

$\therefore$ $\{\mathbf{Z}, +\}$ *is an infinite abelian additive group.*

Exactly in a similar way we can verify that

(a) The set Q of all rational numbers $\{\mathbf{Q}, +\}$ is an infinite abelian additive group under the operation $+$.

(b) The set **R** of all real numbers $\{\mathbf{R}, +\}$ and the set **C** of all complex number $\{\mathbf{C}, +\}$ are **both infinite abelian groups under addition.**

Groupoid: Semigroup: Monoid: Group: Abelian Group

A mathematical system $(G, \circ)$ consisting of a non-empty set G and a binary operation $\circ$, associated with G, is called

(i) A GROUPOID, if the closure property holds,

$$\text{i.e., } \forall a, b \in G, a \circ b \in G. \ \textbf{(G1)}$$

(ii) A SEMIGROUP, if both the closure property and associative property hold,

$$\begin{aligned} &\text{i.e.,} && \forall a, b \in G, \ a \circ b \in G \\ &\textbf{and} && \forall a, b, c \in G, \ (a \circ b) \circ c = a \circ (b \circ c). \ \textbf{(G1 and 2)} \end{aligned}$$

(iii) A MONOID, if the closure property, associative property and identity element e of G exist. **(G1, 2, and 3)**

(iv) A GROUP, if the closure property, associative property, existence of identity and inverse of every element of G exists. **(G1, 2, 3, 4)**

(v) A GROUP is an ABELIAN GROUP if the commutative property holds for every pair of elements. **(G1, G2, G3, G4 and G5).**

▶ Example 7.2.2 *In each case verify that*

(i) $Q - \{0\}$, *set of non-zero rationals.*
(ii) $R - \{0\}$, *set of non-zero reals.*
(iii) $C - \{0 + i0\}$, *set of non-zero complex numbers.*

} form *an infinite abelian, multiplicative group.*

Solution: They are all infinite, abelian multiplicative groups. We discuss (iii) and leave others as exercises for the students:

1. $C - \{0 + i0\}$ is the set of all complex numbers excluding the zero $0 + i0$ of the complex number system. We shall denote this set by C^*.

2. We define the binary operation of multiplication of two **non-zero complex** numbers:

$$(a + ib) \bullet (c + id) = (ac - bd) + i(ad + bc).$$

This multiplication gives a member of C^*, since

$$(ac - bd) + i(ad + bc) \in C^*$$

(Note that both $ac - bd$ and $ad + bc$ cannot be simultaneously zero. Why?)

Thus, the set C^* is closed under the operation of multiplication, i.e., $(C^*, \bullet)$ is a *Groupoid.*

3. Let $a + ib, c + id, e + if \in C^*$. Check

$$\begin{aligned}&\{(a+ib)\bullet(c+id)\}\bullet(e+if)\\ &=\{(ac-bd)+i(bc+ad)\}\bullet(e+if)\\ &=\{(ac-bd)e-(bc+ad)f\}+i(bc+ad)e+(ac-bd)f\}. \qquad \text{(i)}\end{aligned}$$

Also,

$$\begin{aligned}&(a+ib)\bullet\{(c+id)\bullet(e+if)\}\\ &=(a+ib)\bullet\{(ce-df)+i(de+cf)\}\\ &=\{a(ce-df)-b(de+cf)\}+i\{b(ce-df)+a(de+cf)\}. \qquad \text{(ii)}\end{aligned}$$

The right-hand sides of (i) and (ii) give the same complex number. Thus, the Associative law holds in C^*.

i.e., $(C^*, \bullet)$ is not only a *groupoid* but it is also a *semigroup.*

4. In the semi-group $(C^*, \bullet)$ the element $1 + i0 \in C^*$ is the identity element, for

$$(a+ib)\bullet(1+i0)=(1+i0)\bullet(a+ib)=a+ib,$$

i.e., $(C^*, \bullet)$ is not only a *semigroup* but also a *Monoid.*

5. Let $a + ib \in C^*$, a, b are real numbers but $ab \neq 0$. If possible, let there exist $x + iy$ $(x, y$ real but not both zero) such that

$$\begin{aligned}&(x+iy)\bullet(a+ib)=1+i0\\ \Rightarrow\quad &ax-by=1,\ ay+bx=0\\ \Rightarrow\quad &x=\frac{a}{a^2+b^2},\ y=-\frac{b}{a^2+b^2},\end{aligned}$$

i.e., $\dfrac{a}{a^2+b^2}+i\left(-\dfrac{b}{a^2+b^2}\right)$ is a **left-inverse** of $a + ib$.

Again,

$$\frac{a}{a^2+b^2}-\frac{bi}{a^2+b^2}$$

is also the right inverse of $a + ib$. [i.e., $(a+ib)\cdot\left(\frac{a}{a^2+b^2}-\frac{bi}{a^2+b^2}\right)=1+i0$]
i.e., for every non-zero complex number an inverse exists.

$\therefore$ $(C^*, \bullet)$ is a **group** under the operation Multiplication (.) (Multiplicative group).

6. Check $(a+ib)\bullet(c+id)=(c+id)\bullet(a+ib)$.

Thus, $(C^*, \bullet)$ is an **Abelian group** under multiplication.

7. Since C^* contains an infinite number of elements, $(C^*, \bullet)$ is an *infinite, Abelian, Multiplicative group.*

We give below an illustration of a Finite Group:

▶ Example 7.2.3 *Cube roots of unity:* $\{1, \omega, \omega^2\}$ *under multiplication forms a group.*

	1	ω	ω^2
1	1	ω	ω^2
ω	ω	ω^2	1
ω^2	ω^2	1	ω

[C.P. 1995]

Solution: Usually for a finite set of elements we draw up a *multiplication table* or, *an addition table* (as the case may be).

We know that $x^3 = 1$ has three values $1, \omega, \omega^2$ such that

$$1 + \omega + \omega^2 = 0,\ \omega^3 = 1 \text{ and hence } \omega^4 = \omega,\ \omega^5 = \omega^2, \text{ etc.}$$

With these facts in mind we construct a multiplication table.

1. Clearly, from the body of the table we at once conclude that under multiplication the set $S = \{1, \omega, \omega^2\}$ is **closed.**

2. *Associative Law* holds for the set of all complex numbers.

   The given set $S = \left\{1 + i\cdot 0, -\frac{1}{2} + \frac{\sqrt{3}}{2}i, -\frac{1}{2} - \frac{\sqrt{3}}{2}i\right\}$ is a subset of the complex number system. Hence, the associative law of multiplication also holds in S.

3. From the first row and first column we see that 1 is the *multiplicative Identity.*

4. In every row we find one element is 1. Thus, every element has multiplicative inverse, e.g., Inverse of 1 is 1, Inverse of ω is ω^2, Inverse of ω^2 is ω.

5. Commutative property holds

   ∴ The set $S = \{1, \omega, \omega^2\}$ is finite **Abelian multiplicative** group or order three.

▶ Example 7.2.4 *Fourth Roots of Unity* $\{1, i, -1, -i\}$ *forms an Abelian multiplicative group of order four:*

For details see **Art. 7.6. Example 7.6.8.**

7.3 Elementary Properties of Groups Using Definition

Property I. **The identity element e of a group $(G, \circ)$ is unique.** [C.P. 2003]

Proof. We have defined an identity element $e \in G$ of a group $(G, \circ)$ by the following property

$$\forall a \in G,\ a \circ e = e \circ a = a. \tag{i}$$

If possible, let $e_1 \in G$ be any other identity element. Then

$$\forall a \in G,\ a \circ e_1 = e_1 \circ a = a. \tag{ii}$$

Taking $a = e_1$ in (i) we get

$$e_1 \circ e = e \circ e_1 = e_1. \tag{iii}$$

Taking $a = e$ in (ii) we get

$$e \circ e_1 = e_1 \circ e = e. \tag{iv}$$

Hence, from (iii) and (iv), $e_1 = e$.
Thus, the identity element, when exists, is unique.

Property II. **For each $a \in (G, \circ)$ the inverse element a^{-1} is unique.**

Proof. By our definition of inverse.
For each $a \in G$, $\exists$ an element $a^{-1} \in G$ such that

$$a \circ a^{-1} = a^{-1} \circ a = e \text{ (identity element)}.$$

If possible, let a_1 be another inverse of a. Then

$$a \circ a_1 = a_1 \circ a = e.$$

$$\begin{aligned} \text{Hence,}\quad a^{-1} &= a^{-1} \circ e = a^{-1} \circ (a \circ a_1) \\ &= (a^{-1} \circ a) \circ a_1 \quad \text{(associative law)} \\ &= e \circ a_1 = a_1. \end{aligned}$$

Thus a^{-1} is unique.

Property III. **The equations $a \circ x = b$ and $y \circ a = b$ have unique solutions in $(G, \circ)$ for x and y, for each pair of elements a, b in G.** [C.P. 1992]

Proof. We show in corresponding columns the steps involved for the two equations. In each case, we start by **assuming** that a solution exists and deduce that it must be a certain element of the group. Then, we check that this element is, in fact, a solution.

$a \circ x = b$ (i)	$y \circ a = b$ (ii)
$\Rightarrow a^{-1} \circ (a \circ x) = a^{-1} \circ b$	$\Rightarrow (y \circ a) \circ a^{-1} = b \circ a^{-1}$
$\Rightarrow (a^{-1} \circ a) \circ x = a^{-1} \circ b$ (associative law)	$\Rightarrow y \circ (a \circ a^{-1}) = b \circ a^{-1}$ (associative law)
$\Rightarrow e \circ x = a^{-1} \circ b$	$\Rightarrow y \circ e = b \circ a^{-1}$
$\Rightarrow x = a^{-1} \circ b$ $[a^{-1} \circ b \in G]$	$\Rightarrow y = b \circ a^{-1}$ $[b \circ a^{-1} \in G]$
Also $x = a^{-1} \circ b$ does satisfy (i) for	Also $y = b \circ a^{-1}$ does satisty (ii) for,
$a \circ (a^{-1} \circ b) = (a \circ a^{-1}) \circ b = e \circ b = b$	$(b \circ a^{-1}) \circ a = b \circ (a^{-1} \circ a) = b \circ e = b$
Thus (i) has unique solution $x = a^{-1} \circ b$.	Thus (ii) has unique solution $y = b \circ a^{-1}$.

Note 1. *If $a \circ x = e$, then x is called a* **right inverse** *of a. Then* **Property II** *shows that in a group, a right inverse is* **unique** *and is, in fact, $a^{-1} \circ e$, i.e., a^{-1}, the group inverse itself.*

Similarly if $y \circ a = e$, then y is the **left inverse** *of a. Again* **Property II** *shows that in a group, a left inverse is unique and is the group inverse a^{-1} itself.*

Property IV. The Cancelation Laws

$x \circ y = x \circ z \Rightarrow y = z$ **(Left cancellation law)**

$y \circ x = z \circ x \Rightarrow y = z$ **(Right cancellation law)**

Proof.

$$\begin{aligned} & x \circ y = x \circ z \\ \Rightarrow\ & x^{-1} \circ (x \circ y) = x^{-1} \circ (x \circ z) \\ \Rightarrow\ & (x^{-1} \circ x) \circ y = (x^{-1} \circ x) \circ z \\ \Rightarrow\ & e \circ y = e \circ z \text{ (assoc. law)} \\ \Rightarrow\ & y = z. \end{aligned}$$

[C.P. 1991]

$$\begin{aligned} & y \circ x = z \circ x \\ \Rightarrow\ & (y \circ x) \circ x^{-1} = (z \circ x) \circ x^{-1} \\ \Rightarrow\ & y \circ (x \circ x^{-1}) = z \circ (x \circ x^{-1}) \\ \Rightarrow\ & y \circ e = z \circ e \text{ (assoc. law)} \\ \Rightarrow\ & y = z. \end{aligned}$$

Property V. $(a^{-1})^{-1} = a$, **i.e., inverse of a^{-1} is a.** [C.P. 1990]

Proof. $a \circ a^{-1} = e = a^{-1} \circ a$

These two equations imply that a is an inverse of a^{-1}. But an inverse is unique; hence $(a^{-1})^{-1} = a$.

In additive notation this property can be written as

$$-(-a) = a.$$

Property VI. $(a \circ b)^{-1} = b^{-1} \circ a^{-1}$, **for any two elements $a, b \in G$.** [C.P. 1995]

Proof. $(a \circ b) \circ (b^{-1} \circ a^{-1}) = a \circ (b \circ b^{-1}) \circ a^{-1} = (a \circ e) \circ a^{-1} = a \circ a^{-1} = e.$

Also $(b^{-1} \circ a^{-1}) \circ (a \circ b) = b^{-1} \circ (a^{-1} \circ a) \circ b = b^{-1} \circ (e \circ b) = b^{-1} \circ b = e.$

Thus, $(a \circ b) \circ (b^{-1} \circ a^{-1}) = (b^{-1} \circ a^{-1}) \circ (a \circ b) = e.$

$\therefore$ an Inverse of $a \circ b$ is $b^{-1} \circ a^{-1}$. But an Inverse is unique.

Hence, $(a \circ b)^{-1} = b^{-1} \circ a^{-1}$.

In multiplicative notation we write $(ab)^{-1} = b^{-1}a^{-1}$.

In additive notation we write $-(a + b) = (-b) + (-a)$.

Extension (a) $(a \circ b \circ c)^{-1} = c^{-1} \circ b^{-1} \circ a^{-1}$ for any three elements of a group $(G, \circ)$.

Proof. Associative Law gives $(a \circ b) \circ c = a \circ (b \circ c)\ \forall a, b, c \in G$

So while writing either side we may simply write $a \circ b \circ c$.

Thus, $a \circ b \circ c = a \circ (b \circ c)$

$$\begin{aligned} \therefore (a \circ b \circ c)^{-1} &= \{a \circ (b \circ c)\}^{-1} = (b \circ c)^{-1} \circ a^{-1} \text{ (Prop. VI.)} \\ &= (c^{-1} \circ b^{-1}) \circ a^{-1} \text{ (using Prop. VI again)} \\ &= c^{-1} \circ b^{-1} \circ a^{-1}. \end{aligned}$$

Using the Method of Induction one may obtain

$$(a_1 \circ a_2 \circ a_3 \circ \cdots \cdots \circ a_n)^{-1} = a_n^{-1} \circ a_{n-1}^{-1} \circ \cdots \cdots \circ a_1^{-1}$$

for all positive integers $n \geq 2$.

We shall write: $a \circ a \circ \cdots \circ a = a^n$

$$\therefore\ (a^n)^{-1} = (a \circ a \circ \cdots \circ a)^{-1} = a^{-1} \circ a^{-1} \circ \cdots \circ a^{-1} \text{ (n terms)}$$
$$= (a^{-1})^n$$

For either side we write a^{-n}.
Thus, $(a^n)^{-1} = (a^{-1})^n = a^{-n}$.

Note 2. ***For convenience of writing we shall use multiplicative notations. Thus, instead of*** $a \circ b$ ***we may write*** $a \bullet b$ ***or simply*** ab.

Property VII. **We introduce the terms {left identity and left inverse} and {right identity and right inverse}.**

Definition of a group based upon Left Axioms:

Let G be a non-empty set associated with a binary operation $\bullet$ (multiplicative notation). Then the algebraic structure $(G, \bullet)$ is a group if the operation $\bullet$ satisfies the following axioms:

G1. Closure property, i.e., $a \cdot b \in G, \forall a, b \in G$.

G2. Associative property, i.e., $a \cdot (b \cdot c) = (a \cdot b) \cdot c, \forall a, b, c \in G$.
(Each side is written as $a \cdot b \cdot c$ or simply, abc)

G3. Existence of left identity: There exists an element e_1 (called **left identity**) in G such that

$$e_1 \cdot a = a \ \forall a \in G.$$

G4. Existence of left inverse: Every element of G possesses left inverse. That is, for every element $a \in G$, there exists an element $a_1 \in G$ such that

$$a_1 \bullet a = e_1.$$

Then this a_1 is called *left inverse* of a.

This definition is equivalent to the definition given in **Art. 7.2.** Obviously if the axioms of **Art. 7.2** hold good then the axioms given in this definition also hold good.

The reader can easily understand that in **G3** we may introduce *right identity* e_2 if $\exists e_2 \in G$ such that $\forall a, a \bullet e_2 = a$ and then in the **G4** we shall introduce *right inverse* a_2 for a if $\exists a_2 \in G$ such that

$$a \cdot a_2 = e_2 \, \forall a \in G.$$

Starting with left axioms stated above if we can prove that the left identity e_1 is also the right identity e_2 and that the left inverse of an element is also its right inverse, then the definition based on left axioms will be equivalent to the definition given in **Art. 7.2** because then $e_1 = e_2 = e$ and $a_1 = a_2 = a$.

We first prove the *left cancellation law* and use it to prove the other results.

I. Left Cancellation Law: If $a, b, c \in G$, then $a \bullet b = a \bullet c \Rightarrow b = c$.

Proof. Since $a \in G$, therefore, $a^{-1} \in G$ such that

$$a^{-1} \bullet a = e_1 \text{ (}e_1\text{ is the left identity and } a^{-1} \text{ is the left inverse)}$$

Now,

$$
\begin{aligned}
& a \bullet b = a \circ c \\
\Rightarrow \quad & a^{-1} \bullet (a \bullet b) = a^{-1} \bullet (a \circ c) \\
\Rightarrow \quad & (a^{-1} \bullet a) \bullet b = (a^{-1} \bullet a) \bullet c \quad \text{(Associative Property)} \\
\Rightarrow \quad & e_1 \bullet b = e_1 \bullet c \quad (\because\ a^{-1} \text{ is the left inverse of } a) \\
\Rightarrow \quad & b = c \quad (\because\ e_1 \text{ is the left identity})
\end{aligned}
$$

II. The left identity e_1 is also the right identity e_2, i.e., if e_1 is the left identity, then $a \cdot e_1 = a\ \forall a \in G$. (i.e., e_1 will behave like the right identity e_2). [C.P. 1990]

Proof. Let $a \in G$ and e_1 the left identity (i.e., $e_1 \cdot a = a$).

Since 'a' possesses left inverse also, therefore, $\exists a^{-1} \in G$ such that $a^{-1} \cdot a = e_1$.

Now we have
$$
\begin{aligned}
a^{-1} \bullet (a \bullet e_1) &= (a^{-1} \bullet a) \bullet e_1 \quad \text{(Associativity)} \\
&= e_1 \bullet e_1 \quad (\because\ a^{-1} \bullet a = e_1) \\
&= e_1 \quad (\because\ e_1 \text{ is left identity}) \\
&= a^{-1} \bullet a. \quad (\text{using again } a^{-1} \bullet a = e_1)
\end{aligned}
$$

Thus, $a^{-1} \bullet (a \bullet e_1) = a^{-1} \bullet a$

$$\Rightarrow a \bullet e_1 = a, \text{ by left cancellation law.}$$

Thus, e_1 is also the right identity. In fact, e_1 is the identity e of **Art. 7.2**, since

$$e_1 \bullet a = a = a \bullet e_1 \bullet a = a = a \bullet e_1$$

(Compare $a \bullet e = e \bullet a = a$ in **Art. 7.2.**).

III. The left inverse of an element is also its right inverse, i.e., if a^{-1} is the left inverse of a, then $a \bullet a^{-1} = e_1 = e_2 = e$. [C.P. 1990]

Proof. Let $a \in G$ and e be the identity element ($e = e_1 = e_2$).

Let a^{-1} be the left inverse of a, i.e., $a^{-1} \bullet a = e$.

To prove $a \cdot a^{-1} = e$

We have
$$
\begin{aligned}
a^{-1} \bullet (a \bullet a^{-1}) &= (a^{-1} \bullet a) \bullet a^{-1} = e \bullet a^{-1} \\
&= a^{-1} \quad \text{(taking } e \text{ as the left identity)} \\
&= a^{-1} \bullet e \quad \text{(taking } e \text{ also as the right identity)}
\end{aligned}
$$

$\therefore$ by left cancellation law, we get

$$a \bullet a^{-1} = e,$$

i.e., a^{-1} is the right inverse of a. In fact a^{-1} **is the inverse** of a, **i.e.,**

$$a^{-1} \bullet a = e = a \bullet a^{-1} \text{ (compare Art. 7.2.)}$$

Note 3. *We can also define a group with the help of right axioms only.*

However, **we cannot assume**, in defining a group, the existence of *left identity* and the existence of *right inverse*,

or, **we cannot assume**, in defining a group, the existence of *right identity* and the existence of *left inverse*.

7.4 Integral Powers of an Element of a Group

I. Composition denoted multiplicatively

Definition: *Suppose* G *is a group under an operation which we shall denote by multiplicative notation* ($a \bullet b$ *or simply* ab)

If a is an element of a group G under multiplication, then we define

1. $a^{\circ} = e$ (where e is the identity element of G.)
2. $a^{1} = a$ and $a^{n+1} = a^{n} \bullet a$ for every natural number n.
3. $a^{-n} = (a^{-1})^{n} = (a^{n})^{-1}$

(Note that $a^{n} = a \bullet a \bullet a \cdots a$ (n such factors).

II. Composition denoted additively

Definition 7.4.1 *Suppose* G *is a group under an operation which we shall denote by additive notation* $(a + b)$. *If* a *is any element of the group* G *under addition* $(+)$, *then we define*

1. $0a = e$ where e is the identity element of G.
2. If n is a positive integer.

$$\underbrace{a + a + \cdots + a}_{n \text{ such terms}} \text{ is denoted by } na$$

and $(-a) + (-a) + \cdots + (-a) = n(-a) = -(na)$ where $(-na)$ denotes the additive inverse of na.

Using the definitions given above, we can establish:

III. Laws of Exponents (Multiplicative Notation)

1. $a^{m} \bullet a^{n} = a^{m+n}$, where m and n are integers (positive or negative or zero).
2. $(a^{m})^{n} = a^{mn}$.
3. $e^{n} = e$.

IV. Laws of Exponents (Additive Notation)

1′. $ma + na = (m + n)a, a \in G, m, n$ are **integers**

2′. $n(ma) = nma, a \in G, n, m$ are **integers**

3′. $m(a + b) = ma + mb, a, b \in G, m$ is any integer (if *the Group* G *is abelian*)

The proofs are given in **Art. 7.6** as illustrative examples. (See **Example 7.6.12**).

V. Order of an element: Definition

Suppose G is a group, the operation being denoted by multiplicative notation

$$(a \bullet b \text{ or } ab).$$

By the **order of an element** $a \in G$, denoted by $O(a)$, we mean the least positive integer n, if one such integer exists, such that

$$a^n = e \quad (e \text{ is the identity element of G}).$$

If there no positive integer n such that $a^n = e$, then the element a is of *infinite order* (or of order zero).

In additive notation, we write $na = e$ in place of $a^n = e$.

▶ Example 7.4.1 *Suppose ∃ a positive integer* m *such that* $a^m = e$, *then the order of* a *is finite and* $o(a) \leq m$, i.e., *when* $a^m = e$, *the order of* a *cannot be more than* m. *At the most* $O(a) = m$. *If* m *is the least positive integer such that* $a^m = e$, *then* $O(a) = m$.

▶ Example 7.4.2 *In any group the identity element* e *is always of* order one. *We have*

$$e^1 = e \Rightarrow O(e) = 1.$$

Also if $O(a) = 1$, *then* $a^1 = a = e$.

7.5 Sub-Groups

Definition: *A non-empty subset* H *of a group* G *under the operation* ∘ *is called a subgroup of* G *if under the same operation* ∘, H *itself forms a group.* [C.P. 1993]

Points to Remember

1. **Every set is a subset of itself. Therefore, if G is a group then G itself is a subgroup of G.**

   If e is the identity element of G, then {e} is a subset of G and this subset is a group under the same composition as that of G. Therefore, {e} is a subgroup of G. Thus G and {e} are always subgroups of any group G.

   They are called TRIVIAL or IMPROPER subgroups.

   A subgroup of G other than the trivial ones is a PROPER SUBGROUP of G.

2. **The identity element of any subgroup of a group G is the same as that of the parent group, i.e., identity element of G.)**

   Proof. Let H be a subgroup of G (operation in multiplicative notation, say)

   Let e be the identity of G and e′ be the identity of H. To prove $e' = e$.

   Let $h \in H$, then obviously $h \in G$ also.

   We have $e' \circ h = h$ as well as $e \circ h = h$,

   $$\text{i.e., } e' \circ h - e \circ h \Rightarrow e' = e \text{ (right cancellation law of G).}$$

3. *The inverse of any element of a subgroup* H *of a group* G *is the same as the inverse of this element regarded as the element of* G.

Proof. Let e be the identity element of G and hence it is also the identity element of H.

Let $h \in H$. Suppose h' is the inverse of h in H, and h'' is the inverse of h in G. Then we have

$$h' \bullet e = e \text{ and } h'' \bullet e = e.$$

$\therefore$ in G we have $h' \cdot e = h'' \cdot e \Rightarrow h' = h''$ (Right Cancellation Law)

4. It is easy to see that the order of any element of a subgroup H of a group G is the same as the order of the element regarded as a member of the parent group G.

5. Examples of Subgroups

▶ Example 7.5.1 $G = \{1, -1, i, -i\}$ *is a multiplicative group of order four.* $H = \{1, -1\}$ *is a subset of* G *and is itself also a multiplicative group.*

$\therefore$ H is a subgroup of G.

▶ Example 7.5.2 *The additive group of all integers* $\{0, \pm 1, \pm 2, \ldots\}$ *is a subgroup of the additive group of rational numbers.*

▶ Example 7.5.3 *The multiplicative group of* ***positive*** *rational numbers* Q^+ *is a subgroup of the multiplicative group of all* ***non-zero*** *rational numbers* Q.

In the following discussions we shall use **multiplicative notation** of a group G.

An Important Characteristic Property of a Subgroup

Theorem 7.5.1 **A necessary and sufficient condition for a non-empty subset H of a group G to be a subgroup is that $a \in H$, $b \in H \implies ab^{-1} \in H$, where b^{-1} is the inverse of b in G.**

We may state this condition in the *following language:*

Theorem 7.5.2 **If H is a non-empty subset of a group G such that $a^{-1}b$ is in H for every pair of elements, a, b (not necessarily distinct) in H, then H is a subgroup of G and conversely.**

Sufficiency Condition

Proof. Assume (1) $H \neq \varphi$ and
(2) $a^{-1}b \in H$, for every pair $a, b \in H$.

I. Existence of Identity: We have

$$a \in H,\ a^{-1} \in H \Rightarrow aa^{-1} \in H \text{ [by cond. (2)]}$$

i.e., $e \in H$.

Thus the identity e is an element of h.

II. **Existence of Inverse:** Since $e \in H$, we have for any arbitrary element $y \in H$ (and so $y^{-1} \in H$)

$$y^{-1}e \in H \text{ by cond. (2)}$$

But $y^{-1}e = y^{-1}$, i.e., every element $y \in H$ has the inverse $y^{-1} \in H$.
The inverse must be unique and two-sided.

III. **Closure Property:** Next consider two elements $a, b \in H$. By (II), $a^{-1} \in H$.

$$\therefore \text{ using (2), } (a^{-1})^{-1} \cdot b \in H, \text{ i.e., } ab \in H.$$

That is, H is closed under the multiplicative operation.

IV. **Associativity holds in H**—since this property holds for elements of G and the elements of H are also elements in G.

These verify that all the group axioms for H hold, and H is a group under the same multiplicative operation as in G.

This proves the *sufficiency* of the condition.

Necessary Condition

Necessity of the Condition. Let H be a subgroup of a multiplicative group G. Hence, H contains at least the identity element of G and so $H \neq \phi$, i.e., H is non-empty.

If $a, b \in H$, then a^{-1} and b also $\in$ H.

$$\therefore a^{-1}b \in H \text{ (since H is closed under multiplication).}$$

Thus, the condition necessarily follows.
This completely establishes that the condition is both necessary and sufficient.

Note 1. *In the additive notation the theorem can be stated as:*
If G is an additive group and H is non-empty subset of G, then H is a subgroup under the same operation—addition—if and only if $a - b \in H$, for every pair of elements $a, b \in H$.

Another test for recognising a subgroup

Theorem 7.5.3 **If G is a Group (with multiplicative notations) and $H \subseteq G$, then H will be a subgroup of G if and only if**

1. **for any two elements $a, b \in H$, $ab \in H$.**
2. **for any element $a \in H$, a^{-1} also $\in$ H.**

Hints for the proof.
First, let H be a subgroup of G. Then by group axioms (1) and (2) follow (Necessary condition).

Next, let (1) and (2) hold for the subset H of G. Since $a \in H$, $a^{-1} \in H$; clearly $a^{-1}a \in H$, i.e., $e \in H$ where e is the identity element of G. Existence of inverse is clear from (2).

Associative law holds in H ($\because$ it holds in the super set G).
$\therefore$ all the group axioms hold in H, i.e., H is **a subgroup of** G.

Intersection of two subgroups

Theorem 7.5.4 **Let H and K be two subgroups of a group G. Then their intersection $H \cap k$ is also a subgroup of G.**

Proof. We shall use multiplicative notations.

The intersection $H \cap K$ is not empty (since, at least, e the identity of G belongs to all its subgroups).

Let $a, b \in H \cap K$. Then $a, b \in H$ and also $a, b \in K$.

Since H is a subgroup of G, $a^{-1}b \in H$ and also for similar reason $a^{-1}b \in K$.

$$\therefore\ a^{-1}b \in H \cap K.$$

That is, by **Theorem 7.5.1 or 7.5.2** above, $H \cap K$, is a subgroup of G.

Note 2. *With similar arguments one can prove:*

The intersection of a finite number of subgroups of a Group G, *is also a subgroup of* G.

Note 3. ***The union of the two subgroups may or may not be a subgroup.***

Let G be the additive group of all integers $\{0, \pm1, \pm2, \pm3, \ldots\}$

Then take the subsets

$$\begin{aligned} H_1 &= \{\cdots, -6, -4, -2, 0, 2, 4, 6, \ldots\} \\ H_2 &= \{\cdots, -12, -9, -6, -3, 0, 3, 6, 9, 12, \ldots\} \end{aligned}$$

H_1 and H_2 are both subgroups of the group G (*check*).

Now, $H_1 \cup H_2 = \{\cdots, -4, -3, 2, 0, 2, 3, 4, \ldots\}$

Obviously $H_1 \cup H_2$ is not closed under addition.

[Take, for example, two elements $2 \in H_1$ and $3 \in H_2$ but their sum

$$2 + 3 = 5 \notin H_1 \cup H_2].$$

$\therefore\ H_1 \cup H_2$ cannot be a subgroup of G.

Now take another subgroup H_3 of the Group G, where

$$H_3 = \{\cdots, -8, -4, 0, 4, 8, \ldots\}$$

Here the union $H_1 \cup H_3 = \{\cdots, -8, -6, -4, -2, 0, 2, 4, 6, 8, \ldots\}$ is a subgroup of G.

We have thus proved: *The union of two subgroups is not necessarily a subgroup.*

However, we can prove that

The union of two subgroups is a subgroup if and only if one is contained in the other.

Proof. Suppose H_1 and H_2 are two subgroups of a group G. Without loss of generality, let $H_1 \subseteq H_2$. Then $H_1 \cup H_2 = H_2$. But H_2 is a subgroup of G. Therefore, $H_1 \cup H_2$ is also a subgroup of G.

Conversely, suppose $H_1 \cup H_2$ is a subgroup. To prove:

$$\text{either } H_1 \subseteq H_2 \text{ or } H_2 \subseteq H_1.$$

If possible, let us assume that H_1 be not a subset of H_2 and H_2 is also not a subset of H_1. H_1 is not a subset of $H_2 \Rightarrow \exists$ at least one element

$$a \in H_1 \text{ but } a \notin H_2. \tag{1}$$

Similarly, if H_2 is not a subset of $H_1, \exists$ some element

$$b \in H_2 \text{ but } b \notin H_1. \tag{2}$$

Then, from (1) and (2), a, b both belong to $H_1 \cup H_2$.

Now, $H_1 \cup H_2$ is a subgroup of G (by our supposition), therefore,

$ab = c$ (say) $\in H_1 \cup H_2$, i.e., $ab \in H_1$ or $ab \in H_2$.

Suppose $ab = c \in H_1$. Then $b = a^{-1}c \in H_1$.

[$\because$ H_1 is a subgroup, therefore, $a \in H_1 \Rightarrow a^{-1} \in H_1$]

But from (2), we have $b \notin H_1$. So we get a contradiction.

Again, if $ab = c \in H_2$, then $a = cb^{-1} \in H_2$, i.e., $a \in H_2$, which is again a contradiction.

Hence, we must have to assume either $H_1 \subseteq H_2$ or $H_2 \subseteq H_1$.

Note 4.

(i) *An abelian group* **cannot** *have a non-abelian subgroup (operation on H is commutative because the elements of H are elements of G and in G they are commutative).*

(ii) *A non-abelian group* **can have** *an abelian subgroup.*

(iii) *A non-abelian group* **can have** *a non-abelian subgroup.*

▶ Example 7.5.4 *Let G be the additive group of all integers. Prove that the set of all multiples of integers by a fixed integer m is a subgroup of G.*

Solution: We have, $G = \{\cdots \cdots, -3, -2, -1, 0, 1, 2, 3, \ldots \cdots\}$ is the additive group of all integers. Let m be a fixed integer.

Then $H = \{\cdots \cdots, -3m, -2m, -m, 0, m, 2m, 3m, \ldots \cdots\}$ is clearly a **subset** of G.

To prove that H is a subgroup of G.

Let $a = rm$, $b = sm$ be two elements of H, where r and s are some integers.

The additive inverse of b is $-b$ or $-sm = (-s)m$.

We have $a + (-b) = rm + (-s)m = (r - s)m \in H$ ($\because$ $r - s$ is some integer)

Thus $a, b \in H \Rightarrow a + (-b) \in H$. Hence, H is a subgroup of G.

▶ Example 7.5.5 *Let a be an element of a group $(G, \bullet)$. The set $H = \{a^n : n$ is any integer$\}$ of all integral powers of a is a subgroup of G.*

Solution: We have $a \in G$. To prove that

$$H = \{\cdots \cdots, a^{-2}, a^{-1}, a^0, a^1, a^2, a^3, \ldots \cdots\}$$

is a subgroup of G.

Let a^r, a^s be any two elements of H, where r and s are some integers. The inverse of a^s in $(G, \bullet)$ is a^{-s}.

Now $a_r \cdot (a^s)^{-1} = a^r \cdot a^{-s} = a^{r-s} \in H$ ($\because$ $r - s$ is also some integer)

Therefore, H is a subgroup of G.

Note 5. *In such a case* H *is called a subgroup of* G *generated by* a, *e.g.,*

$$H = \{\cdots\ \cdots, 3^{-3}, 3^{-2}, 3^{-1}, 0, 3^1, 3^2, 3^3, \ldots\ \cdots\}$$

is a subgroup of the multiplicative group G *of non-zero rational numbers generated by* 3.

▶ Example 7.5.6 *Show that the set of all* 2×2 *real matrices* $\begin{pmatrix} a & b \\ c & d \end{pmatrix}$ *with* $ad-bc=1$, *is a subgroup of the group of all* 2×2 *non-singular real matrices w.r.t. matrix multiplication.*

Solution: See **Example 7.6.9** where we have proved that the set M of all non-singular 2×2 real matrices forms a non-abelian group under matrix multiplication.

To prove that

$$H = \left\{\begin{pmatrix} a & b \\ c & d \end{pmatrix} : ad-bc = 1, a, b, c, d \text{ are all reals}\right\}$$

is a subgroup of M.

Take two elements $\alpha, \beta \in H$, where

$$\alpha = \begin{pmatrix} a & b \\ c & d \end{pmatrix}, ad-bc = 1 \text{ and } \beta = \begin{pmatrix} x & y \\ z & w \end{pmatrix}, xw-yz = 1.$$

Then

$$\beta^{-1} = \frac{1}{xw-yz}\begin{pmatrix} w & -y \\ -z & x \end{pmatrix} = \begin{pmatrix} w & -y \\ -z & x \end{pmatrix} \quad (\because xw-yz=1)$$

Then

$$\alpha\cdot\beta^{-1} = \begin{pmatrix} a & b \\ c & d \end{pmatrix}\begin{pmatrix} w & -y \\ -z & x \end{pmatrix} = \begin{pmatrix} aw-bz & -ay+bx \\ cw-dz & -ey+dx \end{pmatrix}$$

where $(aw-bz)(-cy+dx)-(cw-dz)(-ay+bx) = (ad-bc)(xw-yz) = 1\cdot 1 = 1$, i.e., $\alpha\cdot\beta^{-1} \in H$.

Hence, $\alpha, \beta \in H \Rightarrow \alpha\cdot\beta^{-1} \in H$.

$\therefore$ H is a subgroup of M.

7.6 Examples of Groups

I. Groups from number systems

▶ Example 7.6.1 **Additive Groups:**

(i) *The following sets form Abelian groups under ordinary addition:* **Z** (*the set of all integers*), **Q** (*the set of all rational numbers*), **R** (*the set of all real numbers*), **C** (*the set of all complex numbers*) :

See **Example 7.2.1.**

(ii) *Now it is easy to verify that* **Z**, **Q** *and* **R** *are subgroups of the additive group of* **C**.

(iii) *The set* **N** *of all natural numbers* $\{1, 2, 3, 4, \ldots\}$ **does not form** *a group under addition though* it is closed under addition *and* the operation obeys associative property *but there exists* **no identity element** *and as such inverse cannot be defined.*

(iv) *The set W of all non-negative integers* $\{0, 1, 2, 3, \ldots\}$ **does not form** *a group under addition though the closure property, associative property and existence of identity all hold but the inverse of non-zero elements do not exist.*

(v) *The set S of all odd integers does not form a group under addition: Closure property and associative property hold but* no identity element *exists.* [C.P. 1987]

▶ Example 7.6.2 (a) **Multiplicative Groups**

(i) *The set* $\{\mathbf{C} - \{0 + i0\}\}$ *forms an abelian group under multiplication.* (*See* **Example 7.2.2.)**

(ii) $\mathbf{Q} - \{0\}$ *and* $\mathbf{R} - \{0\}$ *can similarly be shown to be abelian groups under multiplication.*

They are both **subgroups** *of the multiplicative group of* $\mathbf{C} - \{0\}$.

(iii) *The set* **Q does not form a group** *under multiplication since* ***0*** *has no multiplicative inverse, i.e., every member of* **Q** *does not posses a multiplicative inverse. Similar remarks for the set* **R** *(set of all real numbers) and* **C** *(the set of all complex numbers).*

(iv) *The set* **N** *of natural numbers of the set* $\mathbf{W} - \{0\}$ *do not form groups under multiplication since the multiplicative inverses do not exist.*

▶ Example 7.6.2 (b) **Subtraction**

The set **W** *of all* **non-zero** *integers* $\{\pm 1, \pm 2, \pm 3, \ldots\}$ *does not form a group under subtraction.* [C.P. 1988]

Solution: Under the operation "Subtraction", both closure property and associative property hold in **W** but there exists no identity element e such that

$$a - e = a \, \forall a \in W \text{ (Remember, 0 is not a member of } W)$$

$\therefore$ under subtraction W is a **semigroup** but **not a group.**

▶ Example 7.6.3 *Let the binary operation 0 be defined on Z, the set of all integers, by* $a \circ b = a + b + 1, \forall a, b \in Z$. *Does the identity element exist? If so, find the identity element. What is the inverse of* $a \in Z$? [C.P. 1997]

Solution: If e is the identity element under the operation 0, then for any element $a \in Z$, we have

$$\begin{aligned} & a \circ e = e \circ a = a \\ \text{i.e.,}\quad & a + e + 1 = e + a + 1 = a \\ \Longrightarrow\quad & e = -1 \in Z. \end{aligned}$$

If b is the inverse of a under $\circ$ then

$$
\begin{aligned}
& a \circ b = b \circ a = -1 \\
\text{i.e.,}\quad & a + b + 1 = b + a + 1 = -1 \\
\Longrightarrow\quad & b = -a - 2 \\
\text{i.e.,}\quad & -2 - a \text{ is the inverse of } a.
\end{aligned}
$$

▶ Example 7.6.4 *Let Q be the set of all rational numbers. An operation $*$ is defined by*

$$a * b = a + b + ab, \forall a, b \in Q.$$

Is $(Q, *)$ a group? Justify your assertion. **[C.P. 1998]**

Solution: We assert that $(Q, *)$ is a group. We now justify the assertion by verifying each and every group axiom.

1. The sum and product of two rational numbers is a rational number. So, if $a, b \in Q$, then $a + b + ab \in Q$.

   Hence, $a * b \in Q$. **Closure property holds for $*$.**

2. Let a, b, c be three elements of Q. Then

$$
\begin{aligned}
(a * b) * c &= (a + b + ab) * c \\
&= a + b + ab + c + (a + b + ab)c \\
&= a + b + c + ab + ac + bc + abc
\end{aligned}
$$

Again

$$
\begin{aligned}
a * (b * c) &= a * (b + c + bc) \\
&= a + b + c + bc + a(b + c + bc) \\
&= a + b + c + bc + ab + ac + abc \\
&= a + b + c + ab + ac + bc + abc.
\end{aligned}
$$

   $\therefore\ (a * b) * c = a * (b * c)$, i.e., **Associative property holds.**

3. 0 is the **identity element**, since for any element $a \in Q$

$$a * 0 = 0 * a = a + 0 + a \cdot 0 = a$$

4. Given $a \in Q$, we can find $b \in Q$ such that $a * b = b * a = 0$,

   i.e., $a + b + ab = b + a + ba = 0$, i.e., $b = \frac{-a}{1+a} \in Q$.

   Then b is the inverse of a under the operation $*$.

   i.e., **Inverse of any element of Q exists under $*$.**

   Hence, $(Q, *)$ is a group.

▶ Example 7.6.5 *Let Z be the set of all integers. Define an operation $*$ by*

$$a * b = a + b - 2\ \forall a, b \in Z.$$

Show that $(Z, *)$ is a group. **[C.P. 1999]**

Solution: Give similar justifications as in the previous Example. See that $e = 2$ and $4 - a$ is the inverse of $a \in Z$.

▶ Example 7.6.6 *Show that the set of all complex numbers z with $|z| = 1$, forms a group with respect to multiplication of complex numbers.*

Solution: Let $S = \{z : z \in \mathbf{C} \text{ and } |z| = 1\}$, **C** being the set of all complex numbers.

1. Let $z_1, z_2 \in S$. Then, by construction of S, $|z_1| = 1$ and $|z_2| = 1$. Now the product of two complex numbers is a complex number. See that $|z_1 \cdot z_2| = |z_1||z_2| = 1$. Thus $z_1, z_2 \in S \Rightarrow z_1 \cdot z_2 \in S$. (**Closure property**)

2. For any three complex numbers a, b, c

$$a \cdot (b \cdot c) = (a \cdot b) \cdot c.$$

So, in S, **Associative property** holds.

3. $1 + i0 \in S$ ($\because |1 + i0| = 1$).

$1 + i \cdot 0 = 1$ is the **identity element** of S, since for any $a \in S$, $a \cdot (1 + 10) = (1 + i0) \cdot a = a$.

4. Given $z \in S$, we get $z \cdot \frac{1}{z} = \frac{1}{z} \cdot z = 1$ $\left(\text{Also } \left|\frac{1}{z}\right| = \frac{1}{|z|} = \frac{1}{1}\right)$

So every element of S has an inverse.

Note that $0 + i0 \notin S$ since $|0 + i0| \neq 1$.

Thus S forms a group under multiplication.

▶ Example 7.6.7 *Verify that*

(i) $(S, \bullet)$ *is an infinite Abelian Multiplicative Group,*

where $S = \{x/x$ *is a real number of the form* $a + b\sqrt{2}$, *where* a *and* b *are rational numbers, not both zero*$\}$.

(ii) $(T, \bullet)$ *is an infinite Abelian Multiplicative group, where* $T = \{z/z \in C^*$ *and* z *is of the form* $a + ib\sqrt{5}$, a *and* b *are rational numbers but not both are zero*$\}$.

Solution: We discuss (i) and leave (ii) as an Exercise.

We first observe that S has infinite number of elements.

1. Take any two elements $a + b\sqrt{2}, c + d\sqrt{2} \in S$ (a, b, c, d are rational numbers, a, b not both zero and c, d not both zero).

See that $\left(a + b\sqrt{2}\right)\left(c + d\sqrt{2}\right) = (ac+2bd) + (bc+ad)\sqrt{2}$. $ac + 2bd$, $bc + ad$ are both rational numbers but not both zero and hence

$$(ac + 2bd) + (bc + ad)\sqrt{2} \in S.$$

Thus, the closure property **G1** holds and S is a **Groupoid** under multiplication.

2. **Multiplication of Real numbers satisfies associative property and it is easy to check:**

$$\left(a+b\sqrt{2}\right)\bullet\left\{\left(c+d\sqrt{2}\right)\bullet\left(e+f\sqrt{2}\right)\right\}$$
$$=\left\{\left(a+b\sqrt{2}\right)\bullet\left(c+d\sqrt{2}\right)\right\}\bullet\left(e+f\sqrt{2}\right).$$

Thus, under multiplication S is associative (i.e., G2 holds).

i.e., S is a semigroup under multiplication.

3. **There exists an element $e = 1 + 0\sqrt{2} \in S$ such that**

$$\left(1+0\sqrt{2}\right)\bullet\left(a+b\sqrt{2}\right)=\left(a+b\sqrt{2}\right)\cdot\left(1+0\sqrt{2}\right)=a+b\sqrt{2}$$

for any element $a + b\sqrt{2} \in S$.

i.e., Axiom G3 holds and under multiplication S is a monoid.

4. **Given an element $a+b\sqrt{2} \in S$ we try to obtain $x+y\sqrt{2}(x, y$ rationals and not both zero) such that**

$$\left(a+b\sqrt{2}\right)\left(x+y\sqrt{2}\right)=\left(x+y\sqrt{2}\right)\left(a+b\sqrt{2}\right)=1+0\sqrt{2}.$$

In that case, we require to have

$$ax + 2by = 1 \text{ and } bx + ay = 0,$$

from which if follows $x = \frac{a}{a^2-b^2}$ and $y = -\frac{b}{a^2-2b^2}$.

Thus, x and y are rational numbers but x and y are not both zero. Hence Inverse of $a + b\sqrt{2}$ exists, namely

$$\left(\frac{a}{a^2-b^2}-\frac{b}{a^2-2b^2}\sqrt{2}\right)$$

Thus, the axiom G4 also holds. Hence, $(S, \bullet)$ is a multiplicative group.

5. **Clearly,** $\left(a+b\sqrt{2}\right)\bullet\left(c+d\sqrt{2}\right) = ac + 2bd + (bc+ad)\sqrt{2}$
$= ca + 2db + (cb+da)\sqrt{2}$
($\because$ in Q, multiplication is commutative)
$= \left(c+d\sqrt{2}\right)\bullet\left(a+b\sqrt{2}\right),$

i.e., Commutative property G5 holds.

Accordingly $(S, \bullet)$ is an *infinite, Abelian, multiplicative group.*

II. Sets whose elements are Matrices

▶ **Example 7.6.8** *The set of all 2 × 2 real matrices forms an infinite additive Abelian group—Verify.*

Solution: Let S be the set of all 2×2 real matrices and the operation be "addition (+) of two square matrices of order two."

We verify that $(S, +)$ is an abelian group.

1. Let

$$\alpha = \begin{bmatrix} a_1 & b_1 \\ a_2 & b_2 \end{bmatrix} \text{ and } \beta = \begin{bmatrix} c_1 & d_1 \\ c_2 & d_2 \end{bmatrix}$$

where the elements a_1, b_1, a_2, b_2 and c_1, d_1, c_2, d_2 are real numbers.

Then

$$\alpha + \beta = \begin{bmatrix} a_1 + c_1 & b_1 + d_1 \\ a_2 + b_2 & c_2 + d_2 \end{bmatrix} \in S \left(\because \begin{array}{l} \text{sum of two real numbers} \\ \text{is a real number.} \end{array} \right)$$

Thus, under matrix addition the set S of second order real matrices is closed, i.e., the group axiom **G1** holds.

2. It is easy to check:

$$\alpha + (\beta + \gamma) = (\alpha + \beta) + \gamma,$$

where α, β, γ are any three real matrices of second order. Here also we use the fact that addition of real numbers obey associative law. Thus, the group axiom **G2** holds.

3. There exists an element $\mathbf{0} = \begin{bmatrix} 0 & 0 \\ 0 & 0 \end{bmatrix} \in S$ such that

$$\alpha + \mathbf{0} = \mathbf{0} + \alpha = \alpha, \text{ for any element } \alpha = \begin{bmatrix} a_1 & b_1 \\ a_2 & b_2 \end{bmatrix} \in S.$$

See that

$$\begin{bmatrix} a_1 & b_1 \\ a_2 & b_2 \end{bmatrix} + \begin{bmatrix} 0 & 0 \\ 0 & 0 \end{bmatrix} = \begin{bmatrix} a_1 + 0 & b_1 + 0 \\ a_2 + 0 & b_2 + 0 \end{bmatrix} = \begin{bmatrix} a_1 & b_1 \\ a_2 & b_2 \end{bmatrix}.$$

Also,

$$\begin{bmatrix} 0 & 0 \\ 0 & 0 \end{bmatrix} + \begin{bmatrix} a_1 & b_1 \\ a_2 & b_2 \end{bmatrix} = \begin{bmatrix} 0 + a_1 & 0 + b_1 \\ 0 + a_2 & 0 + b_2 \end{bmatrix} = \begin{bmatrix} a_1 & b_1 \\ a_2 & b_2 \end{bmatrix}.$$

Here $\mathbf{0}$ is the identity element of S, i.e., the group axiom **G3** is satisfied.

4. For any element $\begin{bmatrix} a_1 & b_1 \\ a_2 & b_2 \end{bmatrix} \in S$ there exists

another element $\begin{bmatrix} -a_1 & -b_1 \\ -a_2 & -b_2 \end{bmatrix} \in S$ such that

$$\begin{bmatrix} a_1 & b_1 \\ a_2 & b_2 \end{bmatrix} + \begin{bmatrix} -a_1 & -b_1 \\ -a_2 & -b_2 \end{bmatrix} = \begin{bmatrix} 0 & 0 \\ 0 & 0 \end{bmatrix} = \begin{bmatrix} -a_1 & -b_1 \\ -a_2 & -b_2 \end{bmatrix} + \begin{bmatrix} a_1 & b_1 \\ a_2 & b_2 \end{bmatrix},$$

i.e., the Group axiom **G4** holds.

This shows that {S, +} is an additive group.

Moreover, we see that

5. $$\begin{bmatrix} a_1 & b_1 \\ a_2 & b_2 \end{bmatrix} + \begin{bmatrix} c_1 & d_1 \\ c_2 & d_2 \end{bmatrix} = \begin{bmatrix} a_1 + c_1 & b_1 + d_1 \\ a_2 + c_2 & b_2 + d_2 \end{bmatrix}$$

$$= \begin{bmatrix} c_1 + a_1 & d_1 + b_1 \\ c_2 + a_2 & d_2 + b_2 \end{bmatrix} = \begin{bmatrix} c_1 & d_1 \\ c_2 & d_2 \end{bmatrix} + \begin{bmatrix} a_1 & b_1 \\ a_2 & b_2 \end{bmatrix},$$

i.e., the commutative property holds under matrix addition, i.e., the group axiom **G5** holds.

Hence, finally we conclude (since S contains infinite number of elements): (S, +) is an infinite, abelian additive group.

▶ Example 7.6.9 *The set M of all non-singular 2 × 2 real matrices forms an infinite non-abelian group under matrix multiplication.*

Solution: Here the set is

$$M = \left\{ \begin{bmatrix} a & b \\ c & d \end{bmatrix} : a, b, c, d \text{ are real and } ad - bc \neq 0 \right\}$$

1. **Operation is Matrix Multiplication (•)**

$$\begin{bmatrix} a & b \\ c & d \end{bmatrix} \bullet \begin{bmatrix} p & q \\ r & s \end{bmatrix} = \begin{bmatrix} ap + br & aq + bs \\ cp + dr & cq + ds \end{bmatrix}$$

[a, b, c, d; p, q, r, s are real numbers. $ad - bc \neq 0$, $ps - qr \neq 0$]

We observe that $ap + br$, $aq + bs$, $cp + dr$, $cq + ds$ are all real numbers and

$$(ap + br)(cq + ds) - (cp + dr)(aq + bs) = (ad - bc)(ps - qr) \neq 0$$

(since neither $ad - bc$ nor $ps - qr$ is zero).

This proves that under the operation Matrix multiplication the set M is closed, i.e., **G1** holds in this case.

2. Matrix multiplication obeys associative law. From this mother property we can conclude {M, •} obeys Associative property, i.e., **G2** holds.

3. **Identity element** $I = \begin{bmatrix} 1 & 0 \\ 0 & 1 \end{bmatrix} \in M$.

One can easily verify $A \cdot I = I \cdot A = A$ for any $A \in M$, i.e., the group axiom **G3** also holds.

4. **Inverse element:**

Given an element $A = \begin{bmatrix} a & b \\ c & d \end{bmatrix}$, $ad - bc \neq 0 \in M$.

We see whether an element $B \in M$ exists such that

$$A \cdot B = B \cdot A = I.$$

Let $B = \begin{bmatrix} x & y \\ z & w \end{bmatrix}$. Then we must have

$$\begin{bmatrix} a & b \\ c & d \end{bmatrix}\begin{bmatrix} x & y \\ z & w \end{bmatrix} = \begin{bmatrix} ax+bz & ay+bw \\ cx+dz & cy+dw \end{bmatrix} = \begin{bmatrix} 1 & 0 \\ 0 & 1 \end{bmatrix}$$

or, $$\left.\begin{matrix} ax+bz=1, & ay+bw=0 \\ cx+dz=0, & cy+dw=1 \end{matrix}\right\}.$$

We can solve for x, z, y, w provided $ad - bc \neq 0$.

Then

$$x = \frac{d}{ad-bc},\ z = -\frac{c}{ad-bc}$$

$$y = \frac{b}{ad-bc},\ w = \frac{a}{ad-bc},$$

i.e., $$\frac{1}{ad-bc}\begin{bmatrix} d & -b \\ -c & a \end{bmatrix} \text{ is the inverse of } \begin{bmatrix} a & b \\ c & d \end{bmatrix}.$$

In other words, the group axiom **G4** holds here.

Hence, we conclude ($\because$ **G1–G4** hold) the set $\{M, \bullet\}$ forms a group under matrix multiplication.

5. The group is *Non-Abelian,* since, for instance

$$\begin{bmatrix} 1 & -1 \\ 0 & 1 \end{bmatrix} \cdot \begin{bmatrix} 0 & 1 \\ -1 & 1 \end{bmatrix} \neq \begin{bmatrix} 0 & 1 \\ -1 & 1 \end{bmatrix} \cdot \begin{bmatrix} 1 & -1 \\ 0 & 1 \end{bmatrix}.$$

The left side $= \begin{bmatrix} 1 & 0 \\ -1 & 1 \end{bmatrix} \neq$ the right side $= \begin{bmatrix} 0 & 1 \\ -1 & 2 \end{bmatrix}$.

▶ Example 7.6.10 *The set* M_1 *of all* 2×2 *real matrices of determinant* $+1$ *forms a non-Abelian infinite group under matrix multiplication.*

1. Take two matrices $A, B \in M_1$;

   By the given condition $\det A = \det B = +1$.

   Then by the rule of matrix multiplication $A \cdot B$ is also a real 2×2 matrix whose determinant is $+1$.

   $$[\because\ \det(AB) = \det A \bullet \det B = 1 \cdot 1 = 1].$$

   Thus, M_1 is closed under matrix multiplication. **(G1 holds)**

2. The associative law holds, since the law holds for all matrix multiplications. **(G2 holds)**

3. **Identity element is** $I = \begin{bmatrix} 1 & 0 \\ 0 & 1 \end{bmatrix}$.

   (Since $A \cdot I = I \cdot A = A$, for any $A \in M_1$) **(G3 holds)**

4. If $A \in M_1$, then $\det A = +1$, so that A^{-1} exists, also A^{-1} is a real 2×2 real matrix and $A^{-1} \cdot A = A \cdot A^{-1} = I$; moreover,

$$\det A^{-1} = \frac{1}{\det A} = 1 \; [\det(AA^{-1}) = \det I, \; \therefore \; \det A \cdot \det A^{-1} = 1.$$

$\therefore$ M_1 is a group under matrix multiplication.

This group is **Non-Abelian**, since, for instance,

$$\begin{bmatrix} 1 & -1 \\ 0 & 1 \end{bmatrix} \cdot \begin{bmatrix} 0 & 1 \\ -1 & 1 \end{bmatrix} \neq \begin{bmatrix} 0 & 1 \\ -1 & 1 \end{bmatrix} \cdot \begin{bmatrix} 1 & -1 \\ 0 & 1 \end{bmatrix}.$$

► Example 7.6.11 *The set M_2 of all matrices of the form*

$$\begin{bmatrix} a & b \\ -b & a \end{bmatrix} \quad (a, b \textit{ are real numbers, not both zero})$$

forms an Abelian group under matrix multiplication.

Verify all the group axioms as in the previous two examples.

III. Finite Groups

An important example of a finite group of order 4

► Example 7.6.12 *The set $S = \{1, i, -1, -i\}$ forms an Abelian group under multiplication of complex numbers. These elements are the fourth roots of unity, i.e., solutions of* $x^4 = 1$.

When the number of elements is finite it is convenient to draw up a multiplication table as follows:

•	1	i	−1	−i
1	1	i	−1	−i
i	i	−1	−i	1
−1	−1	−i	+1	i
−i	−i	1	i	−1

1. If $a, b \in S$, then clearly, from the table, $a.b \in S$ and so S is closed under multiplication. (**G1**)

2. For any three complex numbers a, b, c we know that

$$a \cdot (b \cdot c) = (a \cdot b) \cdot c \; \textbf{(G2)} \qquad \textit{(Associative law holds)}$$

3. $1 + i \cdot 0 = 1 \in S$ and 1 is an identity, since for every $a \in S$, we find from the table (1st line) $1 \cdot a = a \cdot 1 = 1$. **G3** *(Existence of Identity)*

4. If $a \in S$, there exists $a^{-1} \in S$ such that $a \cdot a^{-1} = a^{-1} \cdot a = 1$. Check from the table: $1 \cdot 1 = 1$, $i \cdot (-i) = 1$, $(-1) \cdot (-1) = 1$, etc., so that $(1)^{-1} = 1$, $(i)^{-1} = -i$, $(-i)^{-1} = i \cdot (-1)^{-1} = -1$. **G4** *(Existence of Inverse)*

Thus, S forms a finite (since S contains only four elements) group under multiplication of order 4.

Since $a \cdot b = b \cdot a$ for any complex numbers a, b it holds for elements of S also and hence it follows that the group $(S, \bullet)$ is *Abelian*.

Note 1. We draw the attention of the readers that the set S can be written as:

$$S = \{1, i, i^2, i^3\} \qquad (\because i^2 = -1, i^3 = -i \text{ and } i^4 = 1)$$

▶ Example 7.6.13 *The set*

$$M_4 = \left\{ \begin{bmatrix} 1 & 0 \\ 0 & 1 \end{bmatrix}, \begin{bmatrix} 0 & 1 \\ -1 & 0 \end{bmatrix}, \begin{bmatrix} -1 & 0 \\ 0 & -1 \end{bmatrix}, \begin{bmatrix} 0 & -1 \\ 1 & 0 \end{bmatrix} \right\}$$

forms an Abelian group under matrix multiplication. [C.H. 2003]

This is also a finite group of order 4.

We may systematically check the group axioms and verify that M_4 *is a finite Abelian group of order four under the usual matrix multiplication.*

Let us call

$$\alpha = \begin{bmatrix} 1 & 0 \\ 0 & 1 \end{bmatrix}, \beta = \begin{bmatrix} 0 & 1 \\ -1 & 0 \end{bmatrix}, \gamma = \begin{bmatrix} -1 & 0 \\ 0 & -1 \end{bmatrix}, \delta = \begin{bmatrix} 0 & -1 \\ 1 & 0 \end{bmatrix}.$$

We construct a multiplication table:

$\circ$	α	β	γ	δ
α	α	β	γ	δ
β	β	γ	δ	α
γ	γ	δ	α	β
δ	δ	α	β	γ

1. If $a, b \in M$, then clearly from the body of the Table we see that $a \cdot b \in M$, i.e., M **is closed under matrix multiplication.**

2. Matrix multiplication is, in general, **Associative**.

3. From the first row and first column it is clear that α behaves like an **identity element.**

4. Observe that every row has the element α. So we can check that every element of M has an inverse:

   e.g., $\beta \cdot \delta = \delta \cdot \beta = \alpha$, i.e., δ is the inverse of β or β is the inverse of δ. Likewise we can find inverse of other elements of M.

$\therefore$ M forms a group under matrix multiplication.

Moreover, from the table we can easily check that

$$\alpha, \beta \in M \implies \alpha \cdot \beta = \beta \cdot \alpha$$

Thus M is an abelian group under Matrix Multiplication.
Or, we proceed as follows:
If $A = \begin{bmatrix} 0 & 1 \\ -1 & 0 \end{bmatrix}$, then

$$A^2 = \begin{bmatrix} 0 & 1 \\ -1 & 0 \end{bmatrix}\begin{bmatrix} 0 & 1 \\ -1 & 0 \end{bmatrix} = \begin{bmatrix} -1 & 0 \\ 0 & -1 \end{bmatrix},$$

$$A^3 = A^2 \cdot A = \begin{bmatrix} -1 & 0 \\ 0 & -1 \end{bmatrix}\begin{bmatrix} 0 & 1 \\ -1 & 0 \end{bmatrix} = \begin{bmatrix} 0 & -1 \\ 1 & 0 \end{bmatrix},$$

$$A^4 = A^3 \cdot A = \begin{bmatrix} 0 & -1 \\ 1 & 0 \end{bmatrix}\begin{bmatrix} 0 & 1 \\ -1 & 0 \end{bmatrix} = \begin{bmatrix} 1 & 0 \\ 0 & 1 \end{bmatrix} = I.$$

$\therefore$ the set $M_3 = \{I, A, A^2, A^3\}$, where $I = A^4$.
We now construct the Multiplication Table (putting $I = A^4$, where necessary)

•	I	A	A^2	A^3
I	I	A	A^2	A^3
A	A	A^2	A^3	I
A^2	A^2	A^3	I	A
A^3	A^3	I	A	A^2

The group axioms can now be easily verified.
That this group is Abelian can be proved from the fact that

$$A^m \cdot A^n = A^{m+n} = A^{n+m} = A^n \cdot A^m$$

for each non-negative integers $m, n (A^\circ = I)$.

Observations. We wish to draw the attention of our readers that the basic structures of **Example 7.6.12** and **Example 7.6.13** are the same. We can rewrite the set $\{1, i, -1, -i\}$ as $\{1, i, i^2, i^3\}$, where $i^4 = 1$ and then draw up the following multiplication table:

•	1	i	i^2	i^3
1	1	i	i^2	i^3
i	i	i^2	i^3	1
i^2	i^2	i^3	1	i
i^3	i^3	1	i	i^2

The two groups (of **Example 7.6.12** and **Example 7.6.13**) are identical in structure in the following sense:

Define a mapping from $\{1, i, i^2, i^3\}$ to $\{I, A, A^2, A^3\}$ by the rule:

$$i^n \to A^n (n = 0, 1, 2, 3).$$

This mapping is a **bijection** and, further, **any product** of elements in $\{1, i, i^2, i^3\}$ is mapped to the **corresponding product in** $\{I, A, A^2, A^3\}$. Two such groups are said to be **isomorphic to each other.**

In general, remember:

Definition 7.6.1 *Two groups* $\{G, \circ\}$ *and* $\{G', *\}$, *where* $\circ$ *denotes the operation on* G *and* $*$ *the operation on* G′, *are said to be* **isomorphic** *to each other if there exists a bijection* $f : G \to G'$ *such that*

$$\left.\begin{array}{rl} \textit{if} & f \textit{ maps } a \textit{ to } a' \\ \textit{and} & f \textit{ maps } b \textit{ to } b' \end{array}\right\}, \textit{ then } f \textit{ maps } a \circ b \textit{ to } a' * b',$$

i.e., for every pair a, b *of elements of* G, $f(a \circ b) = f(a) * f(b)$.

Two isomorphic groups have the same structure although they may differ in notation and in the nature of their elements. We add two more examples which are isomorphic to the groups of **Example 7.6.12, Example 7.6.13.**

▶ Example 7.6.14 *Check that the set* $\{1, 2, 3, 4,\}$ *forms an Abelian group under multiplication* (*modulo* 5).[1] *See the Table given below:*

•	1	2	3	4
1	1	2	3	4
2	2	4	1	3
3	3	1	4	2
4	4	3	2	1

Multiplication Table (mod 5)

Since

$$3 \equiv 2^3 (\text{mod } 5) \text{ and } 2^4 \equiv 1 (\text{mod } 5)$$

The group is, in fact, formed by

$$\{1, 2, 2^2\ 2^3\}$$

under the multiplication (mod 5) and $2^4 \equiv 1$ (mod 5).

Note the rearrangement in the listing of the elements 3, 4. We draw up the multiplication table again according to this rearrangement. From this table isomorphism will be clear.

	1	2	2^2	2^3
1	1	2	2^2	2^3
2	2	2^2	2^3	1
2^2	2^2	2^3	1	2
2^3	2^3	1	2	2^2

▶ Example 7.6.15 *Check that the set of matrices.*

$$S = \left\{ \begin{bmatrix} 1 & 1 \\ 0 & 0 \end{bmatrix}, \begin{bmatrix} i & i \\ 0 & 0 \end{bmatrix}, \begin{bmatrix} -1 & -1 \\ 0 & 0 \end{bmatrix}, \begin{bmatrix} -i & -i \\ 0 & 0 \end{bmatrix} \right\}$$

forms an Abelian group under matrix multiplication.

[1] For the concept of 'Congruence Modulo m'. See Appendix to Chapter 6.

If we write $I = \begin{bmatrix} 1 & 1 \\ 0 & 0 \end{bmatrix}$ and $A = \begin{bmatrix} i & i \\ 0 & 0 \end{bmatrix}$, then

$$S = \{I, A, A^2, A^3\} \text{ where } A^4 = I.$$

This group is also **isomorphic** to the groups shown in **Example 7.6.12** or **Example 7.6.13.**

IV. Integral Exponents (See Art. 7.4)

If a is an element of a multiplicative group G, then we define

1. $a^\circ = e$, where e is the identity element of G.
2. $a^1 = a$, and $a^{n+1} = a^n.a$ (or, simply $a^n\ a$) for every natural number n.
3. $a^{-n} = (a^{-1})^n = (a^n)^{-1}$, for every natural number n.

With these definitions in mind we shall prove the following result:

► Example 7.6.16 *If a is an element of a group G, with identity element e, and if m and n are integers (positive, negative or zero), then the following laws of exponents hold:*

(i) $e^n = e$. (ii) $a^{m+n} = a^m \cdot a^n$. (iii) $(a^m)^n = a^{mn}$.

Proof. (i) By definition,

$$\begin{aligned} e^2 &= e^1 \cdot e = e \cdot e = e, \\ e^3 &= e^2 \cdot e = e \cdot e = e, \text{ etc.} \\ e^{-2} &= (e^{-1})^2 = e^2 = e \text{ (inverse of } e = e; e \cdot e = e\text{), etc.} \end{aligned}$$

[By finite induction the law can be proved to be true for any natural number n.]

Proof. (ii)

Case 1. $m > 0, n > 0$.

We prove $a^{m+n} = a^m \cdot a^n$ by induction.

For $n = 1$, the result is true by definition $(\because\ a^{m+1} = a^m \cdot a^1)$

For any fixed $n = k$ (say), assume $a^{m+k} = a^m \cdot a^k$.

$$\begin{aligned} \text{Then } a^{m+k+1} &= a^{m+k} \cdot a^1 && \text{(by definition)} \\ &= (a^m \cdot a^k) \cdot a^1 && \text{(assumed condition)} \\ &= a^m \cdot (a^k \cdot a^1) && \text{(Associativity)} \\ &= a^m \cdot a^{k+1} && \text{(by definition)} \end{aligned}$$

Thus, the law is true for $n = k + 1$ if it is assumed to be true for $n = k$. But it is true for $n = 1$; hence for $n = 2$ and so for $n = 3$ and so on for all natural numbers n.

Case 2. $m < 0, n < 0$.

Let $m = -p, n = -q\ (p, q > 0)$

$$\begin{aligned} \text{Then } a^m \cdot a^n &= a^{-p} \cdot a^{-q} = (a^{-1})^p \cdot (a^{-1})^q \\ &= (a^{-1})^{p+q} = a^{-(p+q)} \\ &= a^{-p+(-q)} = a^{m+n}. \end{aligned}$$

Case 3. $m > 0, n < 0$ **but** $m + n > 0$.

Let $n = -q (q > 0)$.
Since $m + n > 0$, it also follows that $m > q$.

$$\begin{aligned} a^m \cdot a^n &= a^m \cdot a^{-q} = a^m \cdot (a^{-1})^q \\ &= (a^{m-q} \cdot a^q) \cdot (a^{-1})^q \text{ (using previous case)} \\ &= a^{m-q} \cdot \{a^q \cdot (a^{-1})^q\} \text{ (Associativity)} \\ &= a^{m-q} \cdot \{a^q \cdot (a^q)^{-1}\} \\ &= a^{m-q} \cdot e = a^{m-q} = a^{m+n}. \end{aligned}$$

Case 4. $m > 0, n < 0$ but $m + n < 0$.
Let $n = -q (q > 0)$. In this case, $m < q$.

$$\begin{aligned} a^m \cdot a^n &= a^m \cdot a^{-q} = a^m \cdot (a^{-1})^q \\ &= a^m \cdot \{(a^{-1})^m \cdot (a^{-1})^{q-m}\} \text{ (using previous case)} \\ &= \{a^m \cdot (a^{-1})^m\} \cdot (a^{-1})^{q-m} \text{ (Associativity)} \\ &= \{a^m \cdot (a^m)^{-1}\} \cdot (a^{-1})^{q-m} \\ &= e \cdot (a^{-1})^{q-m} = (a^{-1})^{q-m} \\ &= a^{m-q} = a^{m+n}. \end{aligned}$$

Thus, in all cases the law (ii) is true.

Proof. (iii) We shall prove $(a^m)^n = a^{mn}$, by induction, if $n > 0$. Clearly, it is true for $n = 1$.

Assume $(a^m)^k = a^{mk}$ for any fixed positive integer k.

Now see that $(a^m)^{k+1} = (a^m)^1 = a^{mk} \cdot a^m = a^{mk+m} = a^{m(k+1)}$, i.e., the law is true for $n = k + 1$, if it is assumed to be true for $n = k$. But since it is true for $n = 1$, it is true for $n = 2$ and consequently for $n = 3$ and so on. Hence, it is true for all natural numbers n.

Next if $n < 0$,

$$\begin{aligned} (a^m)^n &= (a^m)^{-q}, \text{ taking } n = -q (q > 0) \\ &= \{(a^m)^q\}^{-1} = (a^{mq})^{-1} = a^{-mq} = a^{m(-q)} = a^{mn}. \end{aligned}$$

Note 2. *If the multiplicative group* G *is commutative, then* $(ab)^n = a^n b^n$, *for any integer* n. *If the group is non-commutative this rule is not true in general.*

▶ Example 7.6.17 *If* G *is a multiplicative group, and* $a, b \in G$, *then we have already proved as* **Property VI, Art. 7.3**

$$(a \cdot b)^{-1} = b^{-1} \cdot a^{-1}.$$

Using this property, prove that

$$\{(a \cdot b) \cdot b\} \cdot \{(a \cdot b)^{-1} \cdot a\} = a \cdot b, \ (a, b \in G)$$

Solution: L.H.S.

$$\begin{aligned} &= \{(a \cdot b) \cdot b\} \cdot \{(a \cdot b)^{-1} \cdot a\} \\ &= \{(a \cdot b) \cdot b\} \cdot \{(b^{-1} \cdot a^{-1}) \cdot a\} \\ &= (a \cdot b) \cdot b \cdot (b^{-1} \cdot a^{-1}) \ a \\ &= (a \cdot b) \cdot (b \cdot b^{-1}) \cdot (a^{-1} \cdot a) \text{ (Associative property)} \\ &= (a \cdot b) \cdot 1 \cdot 1 \ (\because b \cdot b^{-1} = b^{-1} \cdot b = 1, \ a^{-1} \cdot a = a \cdot a^{-1} = 1) \\ &= a \cdot b = \text{R.H.S. } (\textbf{Proved}) \end{aligned}$$

▶ Example 7.6.18 *For any two elements* a *and* b *of a multiplicative group* G, *prove that*

$$(a\cdot b)^2 = a^2\cdot b^2 \text{ if and only if } a\cdot b = b\cdot a \text{ in } G.$$

Solution: First suppose that G is an abelian group. Then

$$\begin{aligned}(a\cdot b)^2 &= (a\cdot b)\cdot(a\cdot b) = a\cdot(b\cdot a)\cdot b \text{ (Associative Property)}\\ &= a\cdot(a\cdot b)\cdot b \ (\because \text{ we have assumed G is abelian})\\ &= (a\cdot b)\cdot(b\cdot b) \text{ (Associative Property)}\\ &= a^2\cdot b^2.\end{aligned}$$

Next we prove the converse. We assume $(a.b)^2 = a^2.b^2 \forall a, b \in G$.

$$\begin{aligned}(a\cdot b)^2 = a^2\cdot b^2 &\Rightarrow (a\cdot b)\cdot(a\cdot b) = (a\cdot a)\cdot(b\cdot b)\\ &\Rightarrow a\cdot(b\cdot a)\cdot b = a\cdot(a\cdot b)\cdot b\\ &\Rightarrow (b\cdot a)\cdot b = (a\cdot b)\cdot b \text{ (left cancellation law)}\\ &\Rightarrow b\cdot a = a\cdot b \text{ (right cancellation law)}\end{aligned}$$

Thus $(a\cdot b)^2 = a^2\cdot b^2 \Rightarrow b\cdot a = a\cdot b \ \forall a, b \in G$,

i.e., G is an abelian group.

▶ Example 7.6.19 *If* $a^2 = e \ \forall a \in G$, *prove that* G *is an abelian group.* [C.P. 2000]

Solution: Let a, b be any two elements of G.
Then $a\cdot b \in G$ and hence by the given condition $(a\cdot b)^2 = e$.
Now, $(a\cdot b)^2 = e \Rightarrow (a\cdot b)\cdot(a\cdot b) = e$, i.e., $(ab)^{-1} = a\cdot b$

$$\text{or, } b^{-1}\cdot a^{-1} = a\cdot b. \qquad (1)$$

But $a^2 = e$, **i.e.,** $a\cdot a = e$, **i.e.,** $a^{-1} = a$. Similarly $b^{-1} = b$.
Therefore, (1) gives $b\cdot a = a\cdot b \ \forall a, b \in G$, i.e., G is an abelian group.

▶ Example 7.6.20 *Let* $S = \{-3, -2, -1, 0, +1, +2, +3\}$. *Does the mathematical system* $(S, +)$ *form a group.*

Solution: Clearly $(S, +)$ is **not** a group. The very first Group Axiom—**Closure property** does not hold. See that $(-3) + (-2) = -5 \notin S$.

Exercises on Chapter 7(I)

A
[On Group Axioms]

1. Prove that each of the following sets of numbers forms an infinite Abelian group under the composition of ordinary **addition:**

 (a) The set Z of all integers $\{\cdots, -3, -2, -1, 0, 1, 2, 3, \ldots\}$;

 (b) The set Q of all rational numbers (*Zero including*);

(c) The set R of all real numbers (*Zero including*);

(d) The set C of all complex numbers (*Zero of the complex number system, namely* $0 + i0$ *including*).

2. Prove that each of the following sets forms an infinite Abelian group under ordinary **multiplication:**

(a) The set Q^* of all rational numbers excluding zero, i.e., $Q - \{0\}$;

(b) The set R^* of all real numbers excluding zero, i.e., $R - \{0\}$;

(c) The set C^* of all complex numbers excluding $0 + i0$;

(d) The set of real numbers of the form $a + b\sqrt{5}$, where a, b are rational numbers, but not simultaneously zero.

3. **(a)** Does the set of all odd integers form a group under addition? What can you say about the set of even integers (*including zero*) under the operation of addition?

(b) Prove or disprove: The set of all integers (*excluding zero*) which are multiples of 2 is an infinite multiplicative group.

4. **(a)** The set of cube roots of unity;

(b) The set of nth roots of unity (cis $2\pi/n$).

Prove that each set forms a **finite Abelian group** under multiplication.

5. Prove that the set

$$\left\{\begin{bmatrix} 1 & 0 \\ 0 & 1 \end{bmatrix}, \begin{bmatrix} -1 & 0 \\ 0 & 1 \end{bmatrix}, \begin{bmatrix} 1 & 0 \\ 0 & -1 \end{bmatrix}, \begin{bmatrix} -1 & 0 \\ 0 & -1 \end{bmatrix}\right\}$$

of four matrices forms a finite multiplicative Abelian group. **[C.P. 1981]**

6. Below are a few sets; corresponding to each set certain binary operation has been defined. Determine whether or not each of these sets forms a group under the given operation. In case the set forms a group, verify whether it is Abelian or not:

	Description of the set	**Given operation**
(i)	The set $G_1 = R - \{1\}$ (R = set of all real numbers)	Operation $*$ defined by $a * b = a + b - ab$ $(a, b \in G_1)$.
(ii)	The set $G_2 = Z - \{1\}$ (Z = set of all integers)	Operation $*$ similar to (i).
(iii)	The set G_3=all real numbers of the form $a + b\sqrt{2}$, a, b are integers (Zero inclusive)	Operation is *ordinary addition*, (+).
(iv)	The set $G_4 = \{x, y\}$	Operation (o) defined by table $\begin{array}{c\|cc} \circ & x & y \\ \hline x & x & y \\ y & y & x \end{array}$

	Description of the set	Given operation
(v)	The set $G_5 = \{1,2,3,4,5,6\}$	Operation ($\bullet$) is *Multiplication modulo* 7.
	The set $G_6 = \{1,5,7,11\}$	Operation ($\bullet$) is *Multiplication modulo* 12
	The set $G_7 = \{0,1,2,3,4\}$	Operation ($+$) is *Addition modulo* 5.
(vi)	The set $G_8 = \{0,1,2,3,\dots m-1\}$, where m is a fixed positive integer.	Operation ($\wedge$) defined by $a \wedge b = a+b$ if $a+b<m$; $= r$ if $a+b = m+r$, $(0 \le r < m)$.
(vii)	The set G_9=set of all the complex numbers z such that $\lvert z\rvert = 1$.	Operation ($\bullet$) is *multiplication of Complex numbers.*
(viii)	The set G_{10} = set of all $m \times n$ matrices whose elements are real numbers.	Operation is ($+$) *matrix addition.*
(ix)	The set G_{11} = set of all non-singular square matrices of order n, the elements being real numbers.	Operation ($\bullet$) is *matrix multiplication.*

7. Prove that the following sets do not form groups under the stated operations:

(a) Set of all odd integers under *ordinary multiplication.*

(b) Set of all rational numbers (including zero) under *ordinary multiplication.*

(c) Set of all vectors under the operation of *vector product of two vectors.*

(d) Set of all square non-singular matrices of order n whose elements are integers under the operation of *matrix multiplication.*

(e) The set $\{-1,0,1\}$ under ordinary *multiplication.*

(f) The set $\{0,1,2,3,4,5\}$ under *multiplication modulo* 6.

(g) The set $\{1,2,3,4,5,6\}$ under *addition modulo* 7.

8. Let $S = \{2,4,6,8\}$. Draw up a multiplication table *multiplication* ($\bullet$) *modulo* 10. Prove that S is a group under the operation *multiplication modulo* 10. Find the solution of $8 \cdot x = 2$, $y \cdot 4 = 2$. Find $8^5, 4^\circ, 2^{-3}$.

[*Hints:* If a is any element of a group, $a^\circ = e$ (identity element)
$a^{-n} = (a^{-1})^n$ and $a^{n+1} = a^n \cdot a$. Draw up the table. See that 6 is the identity element.]

9. If in a multiplicative group G, $a^2 = e$ (the identity element) $\forall a \in G$, then prove that G is necessarily Abelian. **[C.P. 1980]**

10. Prove that a group $(G,*)$ is Abelian if for every element $a \in G$, $a = a^{-1}$.

11. Prove that a multiplicative group G is Abelian if and only if

(a) $(ab)^{-1} = a^{-1}b^{-1}$, for every pair $a, b \in G$.

(b) $(ab)^2 = a^2b^2$, for every pair $a, b \in G$.

12. On a set G with at least two elements an operation $*$ is defined: $a * b = b$, for every pair $a, b \in G$. Prove that $(G, *)$ is a semigroup, but the operation $*$ is not commutative and that G has no identity element.

B
[On Subgroups]

1. Prove that the set of even integers (including 0) is a subgroup of the additive group of all integers.
2. Prove that the set of all integers is a subgroup of the additive group of all rational numbers.
3. Prove that the set Q^* of non-zero rationales is a subgroup of the multiplicative group of real numbers of the form $a + b\sqrt{2}$, a, b are rational numbers, not simultaneously zero.
4. Q^* is a subgroup of the multiplicative group of non-zero complex numbers—justify.
5. Consider the set $\{0, 1, 2, 3, 4, 5\}$. Under 'addition modulo 6' prove that the set forms a group. With the help of a counter example (using this group) show that the **union** of two of its subgroups is not necessarily a subgroup. What can you say about the intersection of two of its subgroups?

Hints and Answers

A

3. (a) The set of odd integers does not form a group under addition but the set of even integers forms a group under addition.

6. (a) Closure, Identity (0) can be easily checked. Associativity also can be verified.

$$\begin{aligned}(a * b) * c &= (a + b - ab) * c \\ &= a + b - ab + c - (a + b - ab)c \\ &= a + b + c - ab - ac - bc + abc \\ &= a * (b * c) \ (\textbf{Check}).\end{aligned}$$

But 0 and 2 are the **only two** elements which have inverses. Hence, G_2 does not form a group under the operation $*$.

All others are Abelian groups; only (IX) is a non-abelian group.

9. Let $a, b \in G$, then $ab \in G$. So $(ab)^2 = e$, i.e., $(ab)(ab) = e$. Now $ba = e(ba)e = a^2(ba)b^2 = a(ab\ ab)b = a\ e\ b = ab$.

10. $a * b = (a * b)^{-1} = b^{-1} * a^{-1} = b * a$.

11. (a) $(ab)^{-1} = b^{-1}a^{-1}$ (by elementary property of Groups)

Also here it is given $(ab)^{-1} = a^{-1}b^{-1}$.

$\therefore b^{-1}a^{-1} = a^{-1}b^{-1}$ and hence $(a^{-1}b^{-1})^{-1} = (b^{-1}a^{-1})^{-1}$, i.e., $ba = ab$.

Conversely, if G is commutative $(ab)^{-1} = b^{-1}a^{-1} = a^{-1}b^{-1}$.

A Few More Important Concepts Connected With Group[2]

I. Cyclic Groups.

II. Groups of Permutations.

III. Lagrange's Theorem.

I. Cyclic Groups

A special class of subgroups is of particular interest in the study of group theory. Such subgroups are sets of group elements such that each member of the set can be expressed as an integral power of a single element of the set. We first prove the following:

Theorem 1. **If a is element of a Group G, then the set**

$$H = \{a^k / k \text{ is any integer}\}$$

is a subgroup of G (multiplicative notations will be used).

Proof. First, H is not empty, since $a = a^1$ is an element of H. Next, suppose a^m and a^n are any two elements of H. Then

$$(a^n)^{-1} \cdot a^m = a^{-n} \cdot a^m = a^{-n+m}.$$

Since H contains all integral (positive or negative or zero) powers of a.

$$a^{-n+m} \in H.$$

What we have proved is that H is a non-empty set with this property.
(Inverse of an element of H). (that element of H) $\in$ H.
Hence, by sufficiency condition of Theorem 1 of Art. 7.5 H is a subgroup of G.

Remarks. We have seen that if $a \in G$, then a set H containing all integral powers of a is a subgroup. It can be clearly seen that of all the subgroups of G containing a, H is the smallest. Because if K be any subgroup of G containing a, then K must have all integral powers of a **plus** *perhaps other elements* of G, i.e., $K \supset H$. **H is, therefore, the smallest subgroup of G containing a.**

Definition: *A group H (or a subgroup H of a group G) is called* **cyclic** *if and only if there exists an element a in H such that the entire set H is the set of all integral powers of a, i.e., $H = \{a^k / k$ is any integer$\}$.*

The element a is called the **generator** of the group H. We say that the group H has been generated by a.

A **cyclic group** may be finite or infinite. While there are infinitely many integers k such that $a^k \in H$, it often happens that only a finite number of elements a^k are distinct; in this case we say that the cyclic group is **finite.** Otherwise the cyclic group is **infinite.**

See Example 7.6.12. The group $S = \{1, -1, i, -i\}$, rewritten as $S = \{1, i, i^2, i^3\}$ where $i^4 = 1$, is a **finite cyclic group** with operation *multiplication;* either i or $-i$ can be taken as a generator of the group S. Note that $i^1 = i, i^2 = -1, i^3 = -i, i^4 = 1$, etc. All other integral powers of i are equal to one or the other of the four quantities $1, -1, i, -i$. Every finite cyclic group has properties similar to this.

[2]These concepts are very important, but they are not included in the **C.U. B.Sc. General Course** Syllabus.

Order of an element $a \in G$

If a is an element of a group G, we may define the **order** of the element a as the smallest positive integer n such that $a^n = e$, where e is the identity element of G. If no such n exists, then the order of a is **infinity.**

We shall now prove the following:

Theorem 2. **Let a cyclic group G be generated by the element a. Let n be the order of the element a. Then the n elements of the set**

$$\{a, a^2, a^3, \ldots, a^{n-1}, a^n = e\} \tag{1}$$

must be all distinct and, in fact, this set constitutes the group G. (e **is the identity element of G.**)

Proof. **Step 1.** First we prove that the elements of the set (1) are all distinct.

If possible, let $a^i = a^j$, where $0 < i < j \le n$. Then clearly

$$0 < j - i < n$$

and $a^j \cdot a^{-i} = a^j \cdot (a^i)^{-1} = a^j \cdot (a^j)^{-1}$ $[\because\ a^j = a^i$, (assumed)]

$$\because \text{i.e., } a^{j-i} = e \ [\because\ a^j \cdot (a^j)^{-1} = e]$$

But then this contradicts the fact that n is the order of a.

$\therefore$ we conclude that the elements of (1) are all distinct.

Step 2. Next we prove that any integral power of a is some member of the set (1).

Let k be any integer. There exist integers q and r such that

$$k = nq + r \text{ with } 0 \le r < n.$$

Then $a^k = a^{nq+r} = a^{nq} \cdot a^r = (a^n)^q \cdot a^r = e^q \cdot a^r = a^r$.

Hence, a^k is some member of the set (1).

Step 3. See that $(a^i)^{-1} \cdot (a^j) = a^{-i} \cdot a^j = a^{j-i}$ which is an element of (1).

$\therefore$ by **Theorem 1, Art. 7.5** the set (1) constitutes of group.

Note 3. *The students will do well to prove by themselves this theorem by taking additive notations na in place of a^n.*

Examples of cyclic groups

▶ Example 21. *The cube roots of unity $\{1, \omega, \omega^2\}$ is a cyclic group under multiplication of order 3; ω is a generator $\rightarrow$ what is the order of ω?* [C.P. 1988]

[*Ans.* Three, which is the same as the number of elements of the cyclic group.]

▶ Example 22. *Consider the additive group $\{0, 1, 2, 3, 4, 5\}$ with operation of addition modulo 6. See the Table:*

This group is cyclic with generator 1 or 5.

In this group, 1 has the order 6,
2 has order 3,
3 has order 2,
4 has order 3,
5 has order 6,
0 has order 1.

+	0	1	2	3	4	5
0	0	1	2	3	4	5
1	1	2	3	4	5	0
2	2	3	4	5	0	1
3	3	4	5	0	1	2
4	4	5	0	1	2	3
5	5	0	1	2	3	4

Addition modulo 6.

▶ Example 23. *The group*

$$G = \left\{ \cdots\cdots, \frac{1}{16}, \frac{1}{8}, \frac{1}{4}, \frac{1}{2}, 1, 2, 4, 8, 16, \cdots\cdots \right\}$$

is cyclic for the operation multiplication *with generator* $\frac{1}{2}$.

A Few elementary properties of cyclic groups

Theorem 3. *If* G *is a cyclic group, then* G *must be abelian.*

Proof. Let G be a cyclic group with generator a.
Then a^m and $a^n \in G$ for some integers m, n.
Hence, $a^m \cdot a^n = a^{m+n} = a^{n+m} = a^n \cdot a^m$, i.e., G must be abelian.

Note 4. *If* G *is non-abelian, then* G *is not cyclic.*

Theorem 4. *Let* G *be a cyclic group and let* H *be a subgroup of* G. *Then* H *must be cyclic, i.e., every subgroup of a cyclic group is cyclic.*

Proof. Let G be cyclic and generated by a (G may be finite or infinite).
One subgroup is $\{e\}$ and is cyclic.
Any other subgroup H contains at least one element $a^s \neq e$ and hence also $(a^s)^{-1} = a^{-1}$. Thus, there is at least one positive exponent h such that a^h is in H. Let m be the least among all such positive exponents; then, by its group structure, H contains all powers of a^m, i.e., $\{a^m\}$, *the set containing all powers of* a^m, is a subset of H, i.e., $\{a^m\} \subseteq H$.
Next, if $a^j \in H$, we write

$$j = mq + r, \text{ where } q, r \text{ are integers and } 0 \leq r < m.$$

As H contains a^j and a^m, it contains

$$a^j (a^m)^{-q} = a^{j-mq} = a^r.$$

Since m is the least positive integer such that $a^m \in H$; this will be consistent with the definition of m if and only if $r = 0$; hence $a^j = a^{mq} = (a^m)^q$.
i.e., any element $a^j \in H$ is also a member of $\{a^m\}$.

$$\therefore H \subseteq \{a^m\}$$
$$\therefore H = \text{cyclic group } \{a^m\}.$$

II. Groups of Permutations

Preliminaries

Suppose we have a set of n distinct objects labelled, for convenience,

$$1, 2, 3, 4, \ldots \cdots, n.$$

The operation of replacing 1 by a_1, 2 by $a_2, \ldots, n$ by a_n to give some arrangement $a_1, a_2, a_3, \ldots, a_n$ of the n objects is called a permutation P and is denoted by the symbol

$$P = \begin{pmatrix} 1 & 2 & 3 & \cdots & n \\ a_1 & a_2 & a_3 & \cdots & a_n \end{pmatrix} \quad (1)$$

indicating that each element in the first row is to be replaced by the element directly below it in the second row.

For n objects there are $n!$ arrangements or permutations, each of which may be written in the form like (I). The order in which the columns of the permutation symbol (I) are placed is irrelevant. We may just as well write

$$P = \begin{pmatrix} 2 & 1 & n & \cdots & 3 \\ a_2 & a_1 & a_n & \cdots & a_3 \end{pmatrix} \text{ or, } \begin{pmatrix} 1 & 3 & 2 & \cdots & n \\ a_1 & a_3 & a_2 & \cdots & a_n \end{pmatrix}$$

and so on.

We show below the set of six $(= 3!)$ permutations of three objects which we label 1, 2, 3.

$$\text{(II)} \quad \begin{cases} P_1 = \begin{pmatrix} 1 & 2 & 3 \\ 1 & 2 & 3 \end{pmatrix}, \quad P_2 = \begin{pmatrix} 1 & 2 & 3 \\ 3 & 1 & 2 \end{pmatrix}, \quad P_3 = \begin{pmatrix} 1 & 2 & 3 \\ 2 & 3 & 1 \end{pmatrix}, \\ P_4 = \begin{pmatrix} 1 & 2 & 3 \\ 2 & 1 & 3 \end{pmatrix}, \quad P_5 = \begin{pmatrix} 1 & 2 & 3 \\ 3 & 2 & 1 \end{pmatrix}, \quad P_6 = \begin{pmatrix} 1 & 2 & 3 \\ 1 & 3 & 2 \end{pmatrix}. \end{cases}$$

[3]**The Product of two such permutations:** P_iP_j $(i, j = 1, 2, 3, \ldots, 6)$ is defined as the permutation obtained by first performing P_j and then P_i.

For example, P_6, P_2 means first perform P_2 and then P_6:

To evaluate P_6P_2 we see that by P_2, 1 is replaced by 3, and by P_6, 3 is replaced by 2. Hence, by P_6P_2, **1 is replaced by 2.** Similarly by P_2, 2 is replaced by 1 and by P_6, 1 is replaced 1. Hence, by P_6P_2, **2 is replaced by 1.** Now check that by P_6P_2 **3 is replaced by 3.**

We now write

$$P_6P_2 = \begin{pmatrix} 1 & 2 & 3 \\ 1 & 3 & 2 \end{pmatrix}\begin{pmatrix} 1 & 2 & 3 \\ 3 & 1 & 2 \end{pmatrix} = \begin{pmatrix} 1 & 2 & 3 \\ 2 & 1 & 3 \end{pmatrix} = P_4.$$

A more convenient way is the following: $P_6\,P_2 \to$ first write P_2, then P_6

$$\underset{P_2}{\begin{pmatrix} 1 & 2 & 3 \\ 3 & 1 & 2 \end{pmatrix}} \quad \underset{P_6}{\begin{pmatrix} 1 & 2 & 3 \\ 1 & 3 & 2 \end{pmatrix}} = \underset{P_2}{\begin{pmatrix} 1 & 2 & 3 \\ 3 & 1 & 2 \end{pmatrix}} \quad \underset{P_6}{\begin{pmatrix} 3 & 1 & 2 \\ 2 & 1 & 3 \end{pmatrix}}$$

$$= \underset{\substack{\text{as if middle} \\ \text{row } 3\ 1\ 2 \\ \text{cancels}}}{\begin{pmatrix} 1 & 2 & 3 \\ 2 & 1 & 3 \end{pmatrix}} = P_4.$$

---

[3]Some authors prefer to define $P_iP_j \to$ First perform P_i and then P_j. In conformity with our def. of composition of two functions we have defined $P_iP_j \to$ first perform P_j, then P_i.

Other products may be obtained in the same way.

Identity permutations: Included in the set of six permutations there is one which leaves the original arrangement unaltered, namely P_1. This permutation is called the *Identity permutation.*

Inverse permutation: To every permutation P_i there exists another permutation P_i^{-1}, called *the inverse of* P_i, which *undoes the work* of P_i. For example, the inverse of

$$P_2 = \begin{pmatrix} 1 & 2 & 3 \\ 3 & 1 & 2 \end{pmatrix}$$

$$\text{is } P_2^{-1} = \begin{pmatrix} 3 & 1 & 2 \\ 1 & 2 & 3 \end{pmatrix} = \begin{pmatrix} 1 & 2 & 3 \\ 2 & 3 & 1 \end{pmatrix} = P_3$$

$$\text{since } P_2P_2^{-1} = \begin{pmatrix} 1 & 2 & 3 \\ 3 & 1 & 2 \end{pmatrix}\begin{pmatrix} 1 & 2 & 3 \\ 2 & 3 & 1 \end{pmatrix} = \begin{pmatrix} 1 & 2 & 3 \\ 1 & 2 & 3 \end{pmatrix} = P_1 \text{ (the identity).}$$

(First perform P_2^{-1} and then P_2)

In fact, it may be verified from the multiplication table given below that all products and inverses are elements of the set of permutations $\{P_1, P_2, \ldots, P_6\}$ and we can verify the group axioms. We arrive at the following conclusion:

The set $\{P_1, P_2, \ldots, P_6\}$ forms a non-abelian group of order 6. This group is what we call a **symmetric group.**

	P_1	P_2	P_3	P_4	P_5	P_6
P_1	P_1	P_2	P_3	P_4	P_5	P_6
P_2	P_2	P_3	P_1	P_6	P_4	P_5
P_3	P_3	P_1	P_2	P_5	P_6	P_4
P_4	P_4	P_5	P_6	P_1	P_2	P_3
P_5	P_5	P_6	P_4	P_3	P_1	P_2
P_6	P_6	P_4	P_5	P_2	P_3	P_1

More formal definition

A permutation is a one-to-one mapping of a set onto itself. When the set is finite (having n elements, say), we speak of a permutation of n symbols. When the set is infinite, we speak of a transformation.

When we consider a **permutation of n symbols**, we use integers $1, 2, 3, \ldots, n$ to represent the n objects of the set. We usually write the integers in their natural order on one line and the image elements on the line below, so that the image of each integer appears directly below that integer. For any arrangement $a_1, a_2, \ldots, a_n$ of the integers $1, 2, 3, \ldots, n$ the permutation P which maps 1 into a_1, 2 into $a_2, \ldots, n$ into a_n is indicated by

$$P = \begin{pmatrix} 1 & 2 & 3 & \cdots & n \\ a_1 & a_2 & a_3 & \cdots & a_n \end{pmatrix}.$$

We may write: image of 1 by $P(1) = a_1$, image of 3 by $P(3) = a_3$, etc.

Instead of P, the notation θ, α, β, etc. may be used for the mapping. If α and β be two permutations, we define the **product of α and β** by $\alpha\beta(i) = \alpha\{\beta(i)\}$ for each $i = 1, 2, 3, \ldots, n$.

i.e., first performs the mapping β and then α.

Equality of two permutations means that every element of the set $\{1, 2, \ldots, n\}$ has the same image element under each of the two permutations.

We shall denote the set of all permutations on the integers $1, 2, 3, \ldots, n$ by the symbol S_n. The **identity permutation** for S_n is denoted by ϵ :

$$\epsilon = \begin{bmatrix} 1 & 2 & 3 & \cdots & n \\ 1 & 2 & 3 & \cdots & n \end{bmatrix}$$

If a permutation α in S_n maps the integer k onto a_k, then the **inverse permutation** α^{-1} will map a_k onto k.

i.e., if $\alpha(k) = a_k$, then $\alpha^{-1}(a_k) = k$.

$$\therefore \ \alpha^{-1}\alpha(k) = \alpha^{-1}\{\alpha(k)\} = \alpha^{-1}(a_k) = k, \text{i.e., } \alpha^{-1}\alpha = \epsilon.$$

Thus, $\alpha^{-1}\alpha$ is the identity mapping. It can be checked that $\alpha\,\alpha^{-1} = \epsilon$ also. Hence, α^{-1} is the multiplicative inverse of α.

Theorem 5. **S_n denotes the set of all permutations of n symbols denoted by $1, 2, 3, \ldots, n$. Then S_n forms a non-abelian (multiplicative) group of order $n!(n \geq 3)$.**

[The binary operation O used in Group definition is here the product of two permutations.]

Proof.

1. Every product of two permutations is, by definition, another permutations.

   $\therefore$ closure property holds.

2. Multiplication is associative. In fact, if α carries k into a_k, β carries a_k into b_k, and γ carries b_k into c_k, then $\beta\alpha$ carries k into b_k and $\gamma\beta$ carries a_k into c_k.

   $\therefore$ $\gamma(\beta\alpha)$ carries k into c_k and $(\gamma\beta)\alpha$ also carries k into c_k.

   Hence, $\gamma(\beta\alpha) = (\gamma\beta)\alpha$.

3. There is a left-identity ϵ, since $\epsilon\alpha = \alpha$ for all $\alpha \in S_n$.

4. Each permutation has a left inverse;

   For $\begin{bmatrix} 1 & 2 & \cdots & n \\ a_1 & a_2 & \cdots & a_n \end{bmatrix}$, the inverse is $\begin{bmatrix} a_1 & a_2 & \cdots & a_n \\ 1 & 2 & \cdots & n \end{bmatrix}$.

5. The set S_n has $n!$ elements, since there are $n!$ arrangements of n integers $1, 2, 3, \ldots, n$.

   Hence, the set of permutations S_n forms a multiplicative **group of order** $n!$.

   [This group is known as the **symmetric group of degree** n—the term "degree" refers to the number of integers (*Not the number of elements of the group*) involved in the permutation.]

6. This group is non-commutative as can be verified below:

$$\alpha = \begin{bmatrix} 2 & 3 & 1 & 4 & \cdots \\ 1 & 3 & 2 & 4 & \cdots \end{bmatrix} \text{ and } \beta = \begin{bmatrix} 1 & 2 & 3 & 4 & \cdots \\ 2 & 3 & 1 & 4 & \cdots \end{bmatrix}$$

See that

$$\alpha\beta = \begin{bmatrix} 1 & 2 & 3 & 4 & \cdots \\ 1 & 3 & 2 & 4 & \cdots \end{bmatrix} \text{ and } \beta\alpha = \begin{bmatrix} 1 & 2 & 3 & 4 & \cdots \\ 3 & 2 & 1 & 4 & \cdots \end{bmatrix},$$

so that $\alpha\beta \neq \beta\alpha$.

Observations

What is the **symmetric group** S_3 for the set of 3 integers $\{1, 2, 3\}$?

S_3 contains 3! elements:

$$P_\circ = \begin{bmatrix} 1 & 2 & 3 \\ 1 & 2 & 3 \end{bmatrix} = \epsilon; \quad P_1 = \begin{bmatrix} 1 & 2 & 3 \\ 1 & 3 & 2 \end{bmatrix}; \quad P_2 = \begin{bmatrix} 1 & 2 & 3 \\ 2 & 1 & 3 \end{bmatrix}$$

$$P_3 = \begin{bmatrix} 1 & 2 & 3 \\ 2 & 3 & 1 \end{bmatrix}; \quad P_4 = \begin{bmatrix} 1 & 2 & 3 \\ 3 & 1 & 2 \end{bmatrix}; \quad P_5 = \begin{bmatrix} 1 & 2 & 3 \\ 3 & 2 & 1 \end{bmatrix}.$$

We have just proved that the set $S_3 = \{P_\circ, P_1, P_2, \ldots, P_5\}$ forms a group of order $3! = 6$ under the operation "product of permutations". The degree of S_3 is 3 because only three integers are involved in the permutation.

Remember. A group whose elements are permutations is a **permutation group.** Symmetric group is a permutation group but any permutation group need not be a symmetric group, i.e., a permutation group need not contain all the permutations of a set.

See that $\{P_\circ, P_3, P_4\}$ will form a permutation group, a subgroup of S_3 but it is not a symmetric group.

Transpositions and Cycles

Consider now the elements of S_3. It has two types of elements:

Type 1. Those which keep one symbol fixed.

e.g., P_1 or P_2 or P_5.

Type 2. Those which change the symbol in a symbols in a cylic order like

$1 \longrightarrow 2,\ 2 \longrightarrow 3,\ 3 \longrightarrow 1$ (e.g., P_3) [Clockwise]
$1 \longrightarrow 3,\ 3 \longrightarrow 2,\ 2 \longrightarrow 1$ (e.g., P_4) [Anticlockwise]

Elements of the first type are called **transpositions,** and elements of the second type are called **cycles.**

We now give formal definitions:

Definition 1. *Let A be an arbitrary non-empty set with n elements which may be denoted by $1, 2, 3, \ldots, n$. The set S_n of all permutations of A forms a group with respect to the multiplication of permutations. We call this group* **symmetric group of n symbols.** *The degree of S_n is n and the order of S_n is $n!$.*

Definition 2. *A group S is called a* **permutation group** *if its elements are permutations. All symmetric groups are permutation groups but there may exist permutation groups which are not symmetric groups.*

Definition 3. *Let A be a non-empty finite set. A permutation P of the finite set A is a* **cycle of length** k *if and only if there exist distinct elements* $a_1, a_2, \ldots, a_k$ *such that* $P(a_1) = a_2, P(a_2) = a_3, \ldots, P(a_{k-1}) = a_k, P(a_k) = a_1$ *and for every a in A other than* $a_1, a_2, \ldots, a_k, P(a) = a$.

Such a permutation is called a k-cycle, denoted by

$$(a_1\ a_2\ a_3,\ \ldots\ a_{k-1}\ a_k),\ \text{e.g.,}\ \begin{bmatrix} 1 & 2 & 3 & 4 & 5 & 6 \\ 1 & 4 & 2 & 6 & 5 & 3 \end{bmatrix}$$

is a 4-cycle and can be denoted by (2 4 6 3). One may represent this cycle as (4 6 3 2) or (6 3 2 4), etc.

Remember. In cyclic notation the *identity permutation* ϵ is usually denoted by (1). In all cases, an integer not appearing in the cycle is presumed to be mapped onto itself by the permutation.

Definition 4. *Two cycles* P_1 *and* P_2 *are said to be disjoint if no integer appears in both* P_1 *and* P_2.

e.g., the cycles (1 3 5) and (2 4 6) are *disjoint* but the cycles (1 3 5) and (2 3 4) are *not disjoint.*

Remember. Every permutation can be written as a product of disjoint cycles (**See Theorem 7.7.6** below).

Definition 5. *A cycle of length two is called a* **Transposition.**

e.g., in $P = \begin{bmatrix} 1 & 2 & 3 & 4 & 5 & 6 & 7 \\ 2 & 1 & 7 & 6 & 5 & 4 & 3 \end{bmatrix}$, (1, 2), (3, 7), (4, 6) are *transpositions.*

Remember. Every permutation can be expressed as a product of transposition. (See **Theorem 7.7.7** below), but not in a unique way.

$$\begin{aligned} \text{e.g.,}\ (2\,4\,6\,8) &= (6\,8)\,(4\,8)\,(2\,8) \\ &= (2\,8)\,(2\,6)\,(2\,4) \\ &= (4\,8)\,(2\,4)\,(6\,8)\,(4\,8)\,(2\,6). \end{aligned}$$

It is interesting, however, that a given permutation can never be expressed as both a product of an odd number of transpositions and also as a product of an even number of transpositions.

Theorem 6. **Every permutation of a finite set is a cycle or it can be expressed as a product of two or more disjoint cycles.**

Proof. Let α be any permutation of a finite non-empty set A.

Select any element $a_1 \in A$ and determine $\alpha(a_1)$; let us suppose that $\alpha(a_1) = a_2$. If $a_1 = a_2$, then (a_1) is the first cycle in the factorisation of α.

First Cycle. Next suppose $a_1 \neq a_2$, let $\alpha(a_2) = a_3$ and so on we proceed. Having found a_i as the image of the preceding element at any given step, we then determine $\alpha(a_i)$.

Since α is one-to-one, $\alpha(a_i)$ cannot be any of the previous elements $a_2, a_3, \ldots, a_{i-1}$. If $\alpha(a_i) = a_1$, *we close the cycle and* $(a_1\ a_2\ a_3\ \cdots\ a_i)$ *is the first factor of* α.

But if $\alpha(a_i) \neq a_1$, we write $\alpha(a_i) = a_i + 1$ and continue. Since A is a finite set, this process must continue only up to a finite number of steps and then terminates.

Next Cycles. After the first factor is found, we select any **remaining element** of A and proceed as before to determine a second factor. Continuing in this way, we ultimately exhaust all the elements of A and determine the factorisation of α.

Note 5. *For a more formal proof one should use mathematical induction.*

▶ Example 24. $\alpha = \begin{pmatrix} 1 & 2 & 3 & 4 & 5 & 6 & 7 & 8 \\ 2 & 3 & 5 & 8 & 7 & 6 & 1 & 4 \end{pmatrix}$.

We begin with any integer, say $\alpha(1) = 2$; so we must determine $\alpha(2) = 3$. Continuing we see that $\alpha(3) = 5$, $\alpha(5) = 7$ and $\alpha(7) = 1$.

Since 1 is the integer we started with, we write down the cycle

$$\alpha_1 = (1\ 2\ 3\ 5\ 7).$$

Next consider any integer not affected by α_1, say 4.

See that $\alpha(4) = 8$ and $\alpha(8) = 4$, i.e., α has a second factor

$$\alpha_2 = (4\ 8).$$

Since there are no further integers mapped by α, we conclude

$$\alpha = \alpha_1\ \alpha_2.$$

Theorem 7. **Every permutation of a finite set A containing more than one element is a transposition or it can be expressed as a product of transpositions.**

Proof.

1. The identity element ϵ in the set of all permutations of A can be factored into $(1\ 2)\ (2\ 1)$.

[This means product of two permutations

$$\begin{bmatrix} 1 & 2 & 3 & \cdots & n \\ 2 & 1 & 3 & \cdots & n \end{bmatrix}\begin{bmatrix} 1 & 2 & 3 & \cdots & n \\ 2 & 1 & 3 & \cdots & n \end{bmatrix} = \begin{bmatrix} 1 & 2 & 3 & \cdots & n \\ 1 & 2 & 3 & \cdots & n \end{bmatrix}$$

Proceed from **right to left:** $1 \to 2$, $2 \to 1$ and hence $1 \to 1$
$2 \to 1$, $1 \to 2$ and hence $2 \to 2$; others not changed.]

Note that this requires that A must possess at least two elements.

2. To prove the theorem we need only to show that every cycle $(a_1\ a_2\ \cdots\ a_k)$ for $k > 1$ can be expressed as a product of transpositions, since we have already proved (Theorem 6) that every permutation equals a product of cycles. One can directly verify that

$$\begin{aligned}(a_1\ a_2\ a_3) &= (a_1\ a_3)\ (a_1\ a_2)\\ &= \begin{bmatrix} a_1 & a_2 & a_3 & a_4 & \cdots & a_n \\ a_3 & a_2 & a_1 & a_4 & \cdots & a_n \end{bmatrix} \begin{bmatrix} a_1 & a_2 & a_3 & \cdots & a_n \\ a_2 & a_1 & a_3 & \cdots & a_n \end{bmatrix}\\ &= \begin{bmatrix} a_1 & a_2 & a_3 & a_4 & \cdots & a_n \\ a_2 & a_3 & a_1 & a_4 & \cdots & a_n \end{bmatrix}.\end{aligned}$$

In fact, we can verify that

$$(a_1\ a_2\ \cdots\ a_k)(k > 1) = (a_1\ a_k)\ (a_1 a_{k-1})\ (a_1 a_{k-2}) \cdots (a_1 a_3)(a_1 a_2)$$

Use Mathematical induction for more formal proof.
i.e., assuming the result for $k = m$

$$(a_1\ a_2 \cdots\ a_m) = (a_1\ a_m)(a_1 a_{m-1}) \cdots (a_1\ a_3)(a_1\ a_2)$$

we get, $(a_1\ a_{m+1})(a_1\ a_m)(a_1 a_{m-1}) \cdots (a_1 a_3)(a_1\ a_2)$

$$\begin{aligned}&= (a_1\ a_{m+1})(a_1\ a_2 \cdots a_m).\ \text{[assumed]}\\ &= \begin{bmatrix} a_1 & a_2 & a_3 & \cdots & a_{m-1} & a_m & a_{m+1} \\ a_{m+1} & a_2 & a_3 & \cdots & a_{m-1} & a_m & a_1 \end{bmatrix} \begin{bmatrix} a_1 & a_2 & \cdots & a_m & a_{m+1} \\ a_2 & a_3 & \cdots & a_1 & a_{m+1} \end{bmatrix}\\ &= \begin{bmatrix} a_1 & a_2 & \cdots & a_m & a_{m+1} \\ a_2 & a_3 & \cdots & a_{m+1} & a_1 \end{bmatrix} = (a_1\ a_2\ \cdots\ a_m\ a_{m+1})\end{aligned}$$

= the result for $k = m + 1$; **hence, etc.**

Even and Odd Permutations

Definition 6. *A permutation is called an* **even permutation** *if it can be expressed as the product of an even number of transpositions.*

A permutation is called an **odd permutation** *if it can be expressed as the product of an odd number of transpositions.*

▶ Example 25. *The identity permutation ϵ is even, since ϵ can be factored thus:*

$$\epsilon = (1\ 2)\ (2\ 1)$$

▶ Example 26. $\begin{pmatrix} 1 & 2 & 3 & 4 & 5 & 6 \\ 2 & 1 & 4 & 3 & 6 & 5 \end{pmatrix} = (1\ 2)(3\ 4)(5\ 6)$ = *an* **odd** *permutation.*

▶ Example 27. $\begin{pmatrix} 1 & 2 & 3 & 4 & 5 & 6 \\ 3 & 4 & 5 & 6 & 1 & 2 \end{pmatrix} = (1\ 3\ 5)\ (2\ 4\ 6) = (1\ 5)\ (1\ 3)\ (2\ 6)\ (2\ 4)$ = *an* **even** *permutation.*

▶ Example 28. *Show that* $\begin{pmatrix} 1 & 2 & 3 & 4 & 5 & 6 \\ 4 & 3 & 1 & 2 & 6 & 5 \end{pmatrix}$ *is an even permutation and*

$$\begin{pmatrix} 1 & 2 & 3 & 4 & 5 & 6 & 7 \\ 4 & 3 & 1 & 2 & 6 & 7 & 5 \end{pmatrix}$$

is an odd permutation. **[C.P. 2003]**

Alternating group

We have already observed that the set of all permutations of n symbols is a multiplicative group of *order* $n!$, called the **symmetric group S_n of degree n.**

Consider the set of even permutations only; let us denote it by A_n. Clearly $A_n \subset S_n$. Now we see that

(i) the product of two even permutations is also even;

(ii) multiplication of all, and therefore, of even permutations is associative;

(iii) the identity permutation is even;

(iv) the inverse of every even permutation is also even.

$\therefore$ we can now say that the set A_n of even permutations forms a group of order $\frac{1}{2}(n!)$; this group A_n is a group under the same operation (viz., *multiplication of permutations*) as that of S_n. Hence, A_n is a **subgroup** of S_n.

This group of even permutations is called the **alternating group** A_n (*of degree* n).

The students can easily check.

(i) the product of two even permutations is even,

(ii) the product of two odd permutations is even,

(iii) the product of an odd permutation and an even permutation is odd,

(iv) the product of an even permutation and an odd permutation is also odd.

Accordingly, if P is even, then P^{-1} is also even

$$(\because \; PP^{-1} = \epsilon = \text{even}).$$

▶ Example 29. *If* $A_n = S_n$, *then* $n = 1$. (*Why*).

▶ Example 30. *Consider the subset*

$$A_3 = \left\{ \begin{pmatrix} 1 & 2 & 3 \\ 1 & 2 & 3 \end{pmatrix}, \begin{pmatrix} 1 & 2 & 3 \\ 2 & 3 & 1 \end{pmatrix}, \begin{pmatrix} 1 & 2 & 3 \\ 3 & 1 & 2 \end{pmatrix} \right\}$$
$$= \{P_1, P_2, P_3\} \textit{ (say)}.$$

Here $P_1 = \epsilon = \text{even}$; $P_2 = (1\ 2\ 3) = (1\ 3)\ (1\ 2)$ and $P_3 = (1\ 3\ 2) = (1\ 2)(1\ 3)$ are all even. The multiplication table is given below:

•	P_1	P_2	P_3
P_1	P_1	P_2	P_3
P_2	P_2	P_3	P_1
P_3	P_3	P_1	P_2

Clearly that set A_3 is a group under multiplication of permutations. Thus, A_3 is a set of even permutations forming a subgroup of S_3. In other words, A_3 is an *alternating group*.

▶ **Example 31.** $\begin{bmatrix} 1 & 2 & 3 & 4 & 5 & 6 & 7 \\ 2 & 5 & 6 & 4 & 7 & 3 & 1 \end{bmatrix} = (1\,2\,5\,7)\,(3\,6)$. *Check.*

▶ **Example 32.** $(1\,2\,3\,4) = (1\,4)\,(1\,3)\,(1\,2)$. *Check.*

▶ **Example 33.** $(1\,2\,3)^2 = (1\,2)\,(1\,3)$.

▶ **Example 34.** $(1\,3)\,(1\,4) = (1\,4\,3) = (1\,2\,3)\,(1\,4\,2) = (1\,2\,3)\,(1\,2\,4)^2$.

▶ **Example 35.** *The permutations in* S_3 *are* ϵ, $(1\,2\,3), (1\,3\,2), (1\,2), (1\,3), (2\,3)$.

▶ **Example 36.** $\begin{pmatrix} 1 & 2 & 3 & 4 & 5 & 6 & 7 & 8 \\ 3 & 6 & 7 & 1 & 8 & 2 & 5 & 7 \end{pmatrix} = (1\,3\,4)\,(2\,6)\,(5\,8\,7)$

$= (1\,4)\,(1\,3)\,(2\,6)\,(5\,7)\,(5\,8) =$ *an odd permutation.*

▶ **Example 37.** $\begin{pmatrix} 1 & 2 & 3 & 4 & 5 & 6 & 7 & 8 \\ 8 & 2 & 6 & 3 & 7 & 4 & 5 & 1 \end{pmatrix}$

$= (1\,8)\,(3\,6\,4)\,(5\,7) =$ *an even permutation.* (*Why?*)

▶ **Example 38.** *Given that*

$$\alpha = \begin{bmatrix} 1 & 2 & 3 & 4 & 5 \\ 2 & 1 & 4 & 3 & 5 \end{bmatrix} = (1\,2)\,(3\,4) \text{ and } \beta = (\,1 \quad 3 \quad 2 \quad 4 \quad 5\,)$$

obtain: $\alpha^2; \beta^2; \alpha\beta; \beta\alpha$.

III. Lagrange's Theorem

Theorem 8. **The order of each subgroup of a finite group is a divisor of the order of the group. That is:**

If H is a subgroup of a finite group G, then the order of H (i.e., the number of elements of H) is a factor of the order of G.

Preliminaries. Suppose G is a group and H is any subgroup of G. Let $a \in G$. Then the set

$$Ha = \{ha : h \in H\}$$

is called the **right coset** of H in G, generated by a. Similarly the set

$$aH = \{ah : h \in H\}$$

is called a **left coset** of H in G, generated by a. Obviously Ha and aH are both subsets of G.

(i) H *itself is a right as well as a left coset.*

$$(\because\ He = H = eH, \text{ } e \text{ being the identity element of G})$$

(ii) *No right coset of left coset can be empty.*

Since H is a subgroup of G, the identity element e of G belongs to H. So if Ha is a right coset of H in G generated by a, then $ea = a \in Ha$.

Consequently no right coset can be empty.

Similarly, $ae = a \in aH$ and hence no left coset can be empty.

(iii) If G is an abelian group then we have $ah = ha, \forall h \in H$. Therefore, $Ha = aH$.

However, if G is not abelian, then we **may** have $aH = Ha$ or $aH \neq Ha$.

Proof of Lagrange's Theorem:
Let G be a finite group of order n, i.e., $O(G) = n$.
Let H be a subgroup of G and let $O(H) = m$.
i.e., suppose there are m distinct members of H; namely

$$h_1, h_2, h_3, \ldots, h_m.$$

Let $a \in G$. Then Ha is the right coset of H in G generated by a and we have

$$Ha = \{h_1a, h_2a, h_3a, \ldots, h_ma\}$$

$\therefore$ Ha has m distinct members, since

$$h_ia = h_ja \text{ would imply } h_i = h_j.$$

Therefore, each right coset of H in G has m distinct members.

Again, any two distinct right cosets of H in G are disjoint, i.e., they have no element in common. Since G is a finite group, the number of distinct right cosets of H in G will be finite $= k$ (say). The union of these k distinct right cosets of H in G is equal to G. Thus, if $Ha_1, Ha_2, \ldots Ha_k$ are the k distinct right cosets of H in G, then $G = Ha_1 \cup Ha_2 \cup Ha_3 \cup \cdots \cup Ha_k$.

$\therefore$ number of elements in G=number of elements in Ha_1 + number of elements in $Ha_2 + \cdots +$ number of elements in Ha_k.

$$\text{i.e.,} \quad O(G) = m + m + \cdots + m \text{ (k terms)} = km$$

$$\therefore \quad k = \frac{O(G)}{m} \text{ or, } O(H)k = O(G)$$

$$\text{or,} \quad O(H) = O(G)/k = \frac{n}{k}, \text{ i.e., } O(H) \text{ is a divisor of } O(G).$$

Observations.

1. If G is a group of order 6, say, then there can be no subgroup of G of order 5 or 4.

2. If m is a divisor of n, then it is not necessary that the finite group G of order n must have a subgroup of order m, e.g., the alternating group A_4 of degree 4 is of order 12. It can be seen that there is no subgroup of A_4 of order 6, though 6 is a divisor of 12. In other words, the converse of Lagrange's theorem is not true.

3. *If the order of a group* G *is a prime number* p, *then* G *must be cyclic and every element of* G *other than the identity element* e *is a generator of* G.

Proof. If $a \neq e$ and $a \in G$, then a generates a cyclic subgroup

$$H = \{a^k / k \text{ is an integer}\}.$$

The order of H divides p (a prime number) and hence the order must be p, i.e., $G = H$ and a is a generator of G.

4. *If the order of a group* G *is the positive integer* n, *then* $\forall a \in G, a^n = e$, *where* e *is the identity element of* G.

Proof. Let $a \in G$. Then $H = \{a^k/k \text{ is an integer}\}$ is a subgroup of G. Since G is finite, the order of H is also finite. Let $O(H) = m$.

Then by our results in cyclic groups we know that $a^m = e$. By Lagrange's Theorem, m is a divisor of n.

That is $n = md$ for some integer d.

$$\therefore\ a^n = a^{md} = (a^m)^d = e^d = e.$$

▶ Example 39. *Find all the permutations* $P_o, P_1, P_2, \ldots, P_5$ *of 3 elements*

$$G = \{1, 2, 3\}.$$

Prove that under the rule of "product of permutations", the set $G = \{P_o, P_1, P_2, \cdots, P_5\}$ forms a group. Prove that $\{P_o, P_1\}, \{P_o, P_2\}$ and $\{P_o, P_5\}, \{P_o, P_3, P_4\}$ are subgroups of G. Is there any other subgroup? Verify the truth of Lagrange's Theorem.

Exercises on Chapter 7(II)

Cyclic groups, Symmetric groups, Alternating groups and Miscellaneous

1. Prove that

   (a) $\{1, i, -1, -i\}$ is a cyclic group with generators i or $-i$ under the operation of *product of two complex numbers.*

   (b) $\{1, \omega, \omega^2\}$, where ω is an imaginary cube root of unity is a *multiplicative cyclic group* with generators ω or ω^2.

   (c) The set whose elements are the n roots of $x^n = 1$ forms a cyclic group of order n under the multiplication of complex numbers.

2. Suppose that n is the smallest positive integer such that $x^n = e$ for any element x in a group G. Show that for $a \in G$ $a^r = a^s$ if and only if $r = s \pmod n$.

3. Prove that the set $\{1, a, a^2, \ldots, a^{n-1}\}$, where $a = \exp\cdot(2\pi i/n)$ forms a cyclic group of order n under multiplication.

4. Show that the four functions

$$f_1(x) = x,\ f_2(x) = -x,\ f_3(x) = 1/x,\ f_4(x) = -1/x$$

form a group under '*substitution of one function into another*'. Verify that this group is *isomorphic* with the matrix group whose corresponding elements are

$$\begin{pmatrix} 1 & 0 \\ 0 & 1 \end{pmatrix}, \begin{pmatrix} 1 & 0 \\ 0 & -1 \end{pmatrix}, \begin{pmatrix} -1 & 0 \\ 0 & 1 \end{pmatrix}, \begin{pmatrix} -1 & 0 \\ 0 & -1 \end{pmatrix}.$$

5. **(a)** Prove that every group of order 2 is cyclic.

(b) Prove: Every group of order 3 is also cyclic.

6. Consider a rectangular coordinate system in a plane and the set of four transformations I, X, Y, Z with the following representations:

$$I : (x, y) \to (x, y) \qquad X : (x, y) \to (x, -y);$$
$$Y : (x, y) \to (-x, y); \qquad Z : (x, y) \to (-x, -y).$$

I is the identity transformation. X replaces every point P by its reflection in the x-axis, Y is the reflection in the y-axis, Z is the reflection in the origin. These four transformations have the following multiplication table:

0	I	X	Y	Z
I	I	X	Y	Z
X	X	I	Z	Y
Y	Y	Z	I	X
Z	Z	Y	X	I

Verify that the four transformations form a commutative group.

Remember. Every group of four elements whose elements multiply according to the above table is called a **Klein Four-Group.**

Now prove that if any group of order 4 is not **cyclic,** then it is of **Klein's** type.

7. Verify:

(a) The permutation $\begin{pmatrix} 1 & 2 & 3 & 4 & 5 & 6 & 7 \\ 2 & 5 & 6 & 4 & 7 & 3 & 1 \end{pmatrix}$ is the product of the disjoint cycles $(1\ 2\ 5\ 7)\ (3\ 6)$.

(b) $(1\ 4)\ (1\ 3)\ (1\ 2) = (1\ 2\ 3\ 4)$.

(c) $(1\ 2\ 3)^2 = (1\ 3\ 2) = (1\ 2)\ (1\ 3)$.

8. Calculate $P_1P_2, P_2P_1, P_1^{-1}, (P_1P_2)^{-1}, (P_2P_1)^{-1}$ if

$$P_1 = \begin{pmatrix} 1 & 2 & 3 & 4 & 5 & 6 \\ 2 & 3 & 6 & 5 & 4 & 1 \end{pmatrix} \text{ and } P_2 = \begin{pmatrix} 1 & 2 & 3 & 4 & 5 & 6 \\ 1 & 3 & 5 & 6 & 2 & 4 \end{pmatrix}.$$

9. Write down the elements of S_3 (symmetric group of degree 3, i.e., three elements). Make a multiplication table for S_3.

10. Are the following permutations even or odd?

$$\begin{pmatrix} 1 & 2 & 3 & 4 \\ 2 & 1 & 4 & 3 \end{pmatrix}; \begin{pmatrix} 1 & 2 & 3 & 4 & 5 \\ 5 & 3 & 2 & 4 & 1 \end{pmatrix}; \begin{pmatrix} 1 & 2 & 3 & 4 & 5 & 6 \\ 3 & 5 & 2 & 1 & 6 & 4 \end{pmatrix};$$
$$\begin{pmatrix} 1 & 2 & 3 & 4 & 5 & 6 & 7 \\ 4 & 3 & 1 & 2 & 6 & 7 & 5 \end{pmatrix}; \begin{pmatrix} 1 & 2 & 3 & 4 & 5 & 6 \\ 3 & 1 & 4 & 6 & 2 & 5 \end{pmatrix}. \begin{pmatrix} 1 & 2 & 3 & 4 & 5 & 6 \\ 4 & 1 & 6 & 3 & 2 & 5 \end{pmatrix}$$

11. What is a symmetric group S_n of order n? How many elements are there in S_n? Find all elements of S_1, S_2, S_3 and exhibit multiplication tables for these groups. Is such a group Abelian? Justify. [C.P. 1982]

What is an alternating group A_n of degree n? Write down the elements of A_1, A_2, A_3. Draw the multiplication table for A_3.

[The number of elements in A_1, A_2, A_3 is respectively $1, 1, 3$.]

12. Prove that every group of prime order is cyclic.

13. "Every subgroup of a cyclic group is also cyclic"—prove this statement.

14. Give an example of a finite abelian group which is not cyclic.

15. Give an example of a group of order 4 which is not cyclic.

Hints and Answers

5. (b) Let e, a, b be the elements, e the identity. Then $ae = a$, a^2, ab are also three elements. Hence, either (i) $a^2 = e$, $ab = b$; or (ii) $ab = e$, $a^2 = b$: case (i) is not possible since $ab = b$ implies $a = e$.

Hence, $b = a^2$, and the elements of the group are $a^\circ = e$, $a^1 = a$, $a^2 = b$.

i.e., a group of order 3 is cyclic.

6. Last part. Suppose that the group has elements e, a, b, c and is not cyclic. The elements are, in some order, the same as a, a^2, ab, ac. If $a^2 = b$, either (i) $ab = e$, $ac = c$, or (ii) $ab = c$, $ac = e$.

(a) is impossible since, $ac = c$ implies $a = e$.

(b) is impossible since $a^2 = b$, $ab = c$ implies $a^2 = c$ which would make the group cyclic. Hence, $a^2 \neq b$. Similarly $a^2 \neq c$.

Since $a^2 \neq a$, it follows that $a^2 = e$ and either (i) $ab = b$, $ac = c$ or (ii) $ab = c$, $ac = b$.

(i) is impossible because it implies $a = e$.

Hence, $a^2 = e$, $ab = c$, $ac = b$.

Similarly

$$ba = c, \quad b^2 = c, \quad bc = a$$
$$ca = b, \quad cb = a, \quad c^2 = e.$$

Hence, the group is of **Klien's type.**

8. $\begin{pmatrix} 1 & 2 & 3 & 4 & 5 & 6 \\ 3 & 5 & 4 & 2 & 6 & 1 \end{pmatrix}, \begin{pmatrix} 1 & 2 & 3 & 4 & 5 & 6 \\ 2 & 6 & 4 & 1 & 3 & 5 \end{pmatrix},$
$\begin{pmatrix} 1 & 2 & 3 & 4 & 5 & 6 \\ 6 & 1 & 2 & 5 & 4 & 3 \end{pmatrix}, \begin{pmatrix} 1 & 2 & 3 & 4 & 5 & 6 \\ 1 & 5 & 2 & 6 & 3 & 4 \end{pmatrix},$
$\begin{pmatrix} 1 & 2 & 3 & 4 & 5 & 6 \\ 6 & 4 & 1 & 3 & 2 & 5 \end{pmatrix}, \begin{pmatrix} 1 & 2 & 3 & 4 & 5 & 6 \\ 4 & 1 & 5 & 3 & 6 & 2 \end{pmatrix}.$

10. even; even; odd; odd; even.

12. Since G is a group of prime order p, therefore, G must contain at least 2 elements (why?). So $\exists$ an element $a \in G$ s.t. $a \neq$ the identity element e. Since $a \neq e$, therefore, $O(a) \geq 2$. Let $O(a) = m$. If H is the cyclic subgroup of G generated by a, then $O(H) = O(a) = m$. By Lagrange's theorem m must be a divisor of p. But p is prime and $m \geq 2$. Hence, $m = p$, i.e., $H = G$. Since H is cyclic, therefore, G is cyclic and a is a generator of G.

14. $G = \left\{ \begin{pmatrix} 1 & 0 \\ 0 & 1 \end{pmatrix}, \begin{pmatrix} -1 & 0 \\ 0 & 1 \end{pmatrix}, \begin{pmatrix} 1 & 0 \\ 0 & -1 \end{pmatrix}, \begin{pmatrix} -1 & 0 \\ 0 & -1 \end{pmatrix} \right\}$

G is an abelian multiplication group of order 4. Except the identity element $\begin{pmatrix} 1 & 0 \\ 0 & 1 \end{pmatrix}$, each element has order two. Now G is a group of order 4 and G contains no element of order 4. Therefore, G is not a cyclic group. Hence, G is a finite abelian group which is not cyclic.

Chapter 8

Rings, Integral Domains and Fields

Rings, Integral Domains and Fields	
1.	**Definitions** and **Examples** of—(*i*) Ring; (*ii*) Field, (*iii*) Sub-ring, (*iv*) Sub-field.
2.	Simple consequences of the definitions.

8.1 Introduction

In the previous chapter we studied sets with a single binary operation satisfying some properties. In the present chapter we shall study sets with two binary operations satisfying some specific properties. We shall define three such structures, namely, **Rings, Integral domains** and **Fields**. We shall give the definitions of these algebraic structures along with suitable examples.

8.2 Definition of a Ring

Definition 1. *A* ***ring*** *is an algebraic structure* $\{R, +, \bullet\}$ *consisting of a* non-empty *set* R *on which are defined two binary operations* $+$ *and* $\bullet$ *(called* addition *and* multiplication *respectively) satisfying the following postulates:*

(i) R *forms an Abelian group under addition* $(+)$,

(ii) R *is closed under multiplication* $(\bullet)$,
(iii) *Multiplication* $(\bullet)$ *is associative,*
} *i.e.,* R *is a semigroup under* $\bullet$

(iv) *The left and right distributive laws hold in* R.

Writing in detail the implications of (i), (ii), (iii), (iv) we obtain the following list of conditions for a ring $\{R, +, \bullet\}$.

Addition properties	Multiplication properties
(i) R forms an Abelian group under addition (+), that is,	
A_1. R is closed under (+) (Cl.)	M_1. R is closed under (•) (Cl.)
A_2. $\forall a, b, c \in R$, $a + (b + c) = (a + b) + c$. (Ass.)	M_2. Multiplication is associative, that is, $\forall a, b, c \in R$. $a \cdot (b \cdot c) = (a \cdot b) \cdot c$ (Ass.)
A_3. $\exists 0 \in R$, such that $a + 0 = 0 + a = a$, (Id.)	
A_4. For each $a \in R, \exists -a \in R$, such that $a + (-a) = (-a) + a = 0$. (In.)	
A_5. $\forall a, b \in R, a+b = b+a$ (Comm.)	

AND Distributive properties

D_1. Left distributive law: $\forall a, b, c \in R, a \cdot (b + c) = a \cdot b + a \cdot c$

D_2. Right distributive law: $\forall a, b, c \in R, (b + c) \cdot a = b \cdot a + c \cdot a$

Notes.

1. We have used two symbols + and • for the two binary operations; we could as well use any other two symbols, e.g., $*$ and $\circ$, and would still call them *addition* and *multiplication*.

2. We shall, for convenience, often write ab in place of $a \cdot b$, i.e., we may suppress the *dot sign*.

3. We shall write $b - c$ in place of $b + (-c)$; in this notation we can write the distributive laws as:

$$a \cdot (b - c) = a \cdot b - a \cdot c$$
$$\text{and} \quad (a - b) \cdot c = a \cdot c - b \cdot c.$$

4. Since $\{R, +\}$ is a group, the identity element 0 and the inverse of an element $a \in R$ are **both unique.**

5. The notation $\{R, +, \bullet\}$ stands for. "R is a ring under + and •". Often, when the operations + and • are supposed to be known we shall simply say "R is a ring".

8.3 Special Types of Rings: Integral Domains: Field

I. A Ring with Unity: A ring $\{R, +, \bullet\}$ is called a **ring with unity** if, in addition to the aforesaid axioms of a ring, it satisfies also the following axiom:

M_3. **(Multiplicative Identity)** $\exists$ an element $e \in R$, such that for any element

$$a \in R, a \bullet e = e \bullet a = a \ (e \neq 0) \qquad \textbf{(Id.)}$$

[This element, when exists, is often denoted by 1 and called **one.**]

Thus, a *ring with unity* is a set R with two operations + and • satisfying conditions

$$\mathbf{A_1—A_5,\ M_1,\ M_2,\ D_1,\ D_2} \text{ and } \mathbf{M_3}.$$

II. **A Commutative ring:** A ring $(R, +, \bullet]$ is called a *commutative ring* if, in addition to the *axioms of a ring,* it satisfies also the following axiom:

$\mathbf{M_5}$. For any two elements $a, b \in R$, $a \bullet b = b \bullet a$ **(Comm.)** Thus a *commutative ring* is a set R with two operations + and • satisfying the condition

$$\mathbf{A_1—A_5,\ M_1,\ M_2,\ D_1} \text{ (or, } \mathbf{D_2}\text{) and } \mathbf{M_5}.$$

Notes.

1. We have used $\mathbf{M_5}$ (in parity with $\mathbf{A_5}$) for the multiplicative commutative property.
2. For a commutative ring $(b + c) \bullet a = a \bullet (b + c)$ and hence we need not list both $\mathbf{D_1}$ and $\mathbf{D_2}$, since one can be deduced from the other. We shall usually list $\mathbf{D_1}$.

III. **A Commutative ring with unity:** A non-empty set R with two binary operations + and • satisfying the axioms:

$$\mathbf{A_1—A_5,\ M_1,\ M_2,\ M_3,\ M_5,\ D_1}$$

is called a *commutative ring with unity.*

Before we proceed further, we shall first explain the concept of

Divisors of Zero. Let 0 be the additive identity of a ring R (i.e., 0 is the identity element with respect to addition (+)). Any **non-zero element** $a \in R (a \neq 0)$ is called a zero divisor if there exists $b \in R (b \neq 0)$ such that

$$a \cdot b = 0.$$

In this case, b is also a zero-divisor. In fact, we say that a is the **left zero-divisor** and b, the **right zero-divisor.**

e.g., consider $\begin{bmatrix} 1 & 0 \\ 0 & 0 \end{bmatrix} \cdot \begin{bmatrix} 0 & 0 \\ 1 & -1 \end{bmatrix} = \begin{bmatrix} 0 & 0 \\ 0 & 0 \end{bmatrix}$, **each matrix on the left is a zero divisor. See that,** by our definition, 0 is not a zero divisor.

IV. **A Ring with divisors of zero.** A ring $\{R, +, \bullet\}$ is called a *ring with zero-divisors,* if for two elements $a, b \in R$, with $a \neq 0, b \neq 0$, we have

$$a \cdot b = 0 \ \textbf{or} \ b \cdot a = 0.$$

V. **A Ring without divisors of zero.** A ring $\{R, +, \bullet\}$ is called a *ring without zero-divisors,* if for two elements $a, b \in R$, the relation $a.b = 0$ always implies **either** $a = 0$ or, $b = 0$, if not both, i.e.,

$$a \neq 0,\ b \neq 0 \in R \Rightarrow ab \neq 0.$$

VI. Integral domain. An *integral domain* is a commutative ring R with unity in which there are no zero-divisors. Thus, an integral domain is a set R with at least two elements and on which two operations, + and • are defined satisfying the conditions

$$\mathbf{A_1—A_5,\ M_1,\ M_2,\ M_3,\ M_5,\ D_1}$$

and also the condition:

I. If $a, b \in R$ and $a \cdot b = 0$, then either $a = 0$ or $b = 0$, if not both,

$$\text{i.e., } a \neq 0,\ b \neq 0 \Rightarrow ab \neq 0.$$

In defining an Integral domain some authors do not take the axiom $\mathbf{M_3}$. According to them, **A commutative ring is an integral domain if it has no zero divisors.**

Note. An integral domain has *at least two elements:* this ensures the fact that the additive identity (zero) and unity element (one) are distinct.

We can prove that the conditions "*without having zero divisors*" and "*cancellation laws hold*" are equivalent conditions:

Cancellation Laws:

With $a \neq 0,\ a \cdot b = a \cdot c \Rightarrow b = c$ (Left cancellation law)

With $a \neq 0,\ b \cdot a = c \cdot a \Rightarrow b = c$ (Right cancellation law)

In fact, a ring R has no zero-divisor if it satisfies the cancellation laws.

Proof. Suppose R has no zero divisors.

Let $a \cdot b = a \cdot c\ (a \neq 0)$

Then $a \cdot (b - c) = 0 \Rightarrow b - c = 0$ or $b = c$.

$$\left.\begin{array}{l}\text{Thus } a \cdot b = a \cdot c \Rightarrow b = c \\ \text{Similarly } b \cdot a = c \cdot a \Rightarrow b = c\end{array}\right\} \text{Cancellation Laws.}$$

Conversely, let R satisfy the cancellation laws for multiplication.

Suppose $a \cdot b = 0\ (a \neq 0)$. Then $a \cdot b = a \cdot 0 \Rightarrow b = 0$, by Cancellation Laws.

Similarly $a \cdot b = 0\ (b \neq 0)$. Then $a \cdot b = 0 \cdot b \Rightarrow a = 0$.

Consequently R has no **zero divisor.**

Hence, an integral domain *may also be defined* **as a commutative ring R with unity in which the cancellation laws hold.**

VII. Division Ring or Skew Field. A ring (with at least two elements) is called a *division ring* or a *skew field* if the ring is a ring with unity element and is such that every non-zero element has a *two-sided multiplicative inverse.*

Thus, a division ring satisfies

$$\mathbf{A_1—A_5,\ M_1,\ M_2,\ M_3,\ D_1,\ D_2}$$

and also the condition.

$\mathbf{M_4}$. (*Multiplicative inverse*). Every non-zero element of R has two-sided multiplicative inverse in R, i.e., for each $a \neq 0 \in R, \exists a^{-1} \in R$ such that

$$a \cdot a^{-1} = a^{-1} \cdot a = 1.$$

[A division ring is an additive Abelian group, a non-commutative multiplicative group of non-zero elements, the two operations being connected by the distributive laws.]

Finally, we define a field:

VIII. Field. A commutative division ring is a field.

For the convenience of the students we state below a complete list of properties for a field F:

Definition of a field

A field F is a set with at least two elements in which two binary operations, addition (+) and multiplication (•) are defined; these operations obey the following properties:

Additive properties			Multiplicative properties		
A_1.	F is closed under +	(Cl.)	M_1.	F is closed under	(Cl.)
A_2.	$\forall a, b, c, \in F$			$\forall a, b, c \in F,$	
	$a + (b + c) = (a + b) + c$	(Ass.)	M_2.	$a \cdot (b \cdot c) = (a \cdot b) \cdot c$	(Ass.)
A_3.	$\exists 0 \in F$ such that $\forall a \in F$		M_3.	$\exists 1 \in F$ such that $\forall a \in F$	
	$a + 0 = 0 + a = a$	(Id.)		$a \cdot 1 = 1 \cdot a = a$	(Id.)
A_4.	For each $a \in F, \exists - a \in F$		M_4.	For each $a \in F - \{0\}$	
	such that			$\exists a^{-1} \in F - \{0\}$ such that	
	$a + (-a) = (-a) + a = 0$	(In.)		$a \cdot a^{-1} = a^{-1} \cdot a = 1$	(In.)
A_5.	$\forall a, b \in F, a+b = b+a$	(Comm.)	M_5.	$\forall a, b \in F, a \cdot b = b \cdot a$	(Comm.)

AND Distributive property

D_1. $\forall a, b, c \in F, a \cdot (b + c) = a \cdot b + a \cdot c.$

[Because of Commutative property D_2 is not needed to specify]

Subrings and Subfields

1. **Subrings:** Let S be a non-empty subset of R and let R be a ring with operations + and •. If we find that S is also a ring under the same two operations, then S is called a **subring** of R.

   e.g., by checking the ring axioms one can easily see that the set Z of all integers and its subset S of even integers (including zero) are both rings under ordinary addition and multiplication; **hence S is a subring of Z.**

2. **Subfields:** A subset S (with at least two elements) of a field F is called a *subfield* of F if S is also a field defined under the same two operations under which F is defined.

   (e.g., the field Q of rational numbers is **a subfield of the field R** of real numbers under ordinary addition and multiplication.)

   The following conditions are both necessary as well as sufficient (**assumed without proof**):

3. **Necessary and Sufficient conditions for a subset S of a ring R to become a subring:**

   Let R be a ring. A non-empty subset S of the set R is a sub-ring of R if

   (a) $a \in S, b \in S \Rightarrow a - b \in S$

   (b) $a \in S, b \in S \Rightarrow a \cdot b \in S$.

4. **Necessary and sufficient conditions for a subset k of a field F to become a subfield:**

   Let F be a field. Any non-empty subset K of the set F will be a subfield of F if

   (a) $a \in K, b \in K \Rightarrow a - b \in K$

   (b) $a \in K, b \in K (b \neq 0) \Rightarrow ab^{-1} \in K$.

We shall give some more examples of **subrings/subfields** in the next article.

8.4 Solved Problems: Examples of Rings, Fields, etc.

▶ Example 8.4.1 *Let Z be the set of all integers, i.e.,*

$$Z = \{0, \ \pm 1, \ \pm 2, \ \pm 3, \ \cdots \ \cdots\}$$

Prove that under the operations—ordinary addition $(+)$ *and the ordinary multiplication* $(\bullet)$*—Z forms a* **Ring.**

Moreover verify that this ring is a **commutative ring with unity** *and that it is an* **integral domain** *but* **not a field.** [C.P. 1989]

Solution: It is easy to check that under addition Z *is an abelian group.*

A$_1$. *Closure property:* Sum of two integers is itself an integer.

A$_2$. *Associative property:* For any three integers a, b, c we have

$$(a+b)+c = a+(b+c)$$

A$_3$. $\exists$ an element $0 \in Z$ such that $\forall a \in Z$

$$a + 0 = 0 + a = a.$$

i.e., 0 is the *additive identity element.*

A$_4$. For each integer a $\exists$ an other integer $-a$ such that $a + (-a) = (-a) + a = 0$.

A$_5$. *Addition is commutative:* $a + b = b + a$ for any two integer a, b.

Next we check that under multiplication $(\bullet)$, Z is a *semigroup* (i.e., closure property and associative property hold).

M$_1$. For any two integers a, b, the product $a \cdot b$ is also an integer and hence $\in Z$. i.e., Z is *closed under multiplication.*

M_2. *Multiplication is associative:* For any three integers a, b, c we have always

$$a \cdot (b \cdot c) = (a \cdot b) \cdot c.$$

The two distributive laws also hold, i.e., for any three integers a, b, c we have

D_1. $a \cdot (b + c) = a \cdot b + a \cdot c.$

D_2. $(b + c) \cdot a = b \cdot a + c \cdot a.$

$\therefore$ Z is clearly a **Ring.**

Moreover, we see that multiplication is commutative and multiplicative identity (namely, 1) exists.

M_5. *Commutative:* For any two integers a, b, we always have $a \cdot b = b \cdot a$.

M_3. *Existence of Multiplicative Identity:* $\exists$ an element $1 \in Z$ such that for any integer a, $a \bullet 1 = 1 \bullet a = a$.

$\therefore$ Z is a **Commutative Ring with Unity.**

We next observe that Z has no divisor of zero, i.e., whenever the product of two integers is zero, at least one of them must be zero itself.

i.e., $a \cdot b = 0 \Rightarrow$ either $a = 0$ or $b = 0$.

$\therefore$ Z is an **Integral domain.**

Lastly, we check that Z is not a *field* because for any non-zero integer a, **there exists no integer** b such that $a \cdot b = 1$.

In other words, $\{Z, +, \bullet\}$ **is an integral domain** but **not a field.**

▶ Example 8.4.2 *Prove that the set* E *of all even integers, i.e., the set*

$$E = \{\cdots -4, -2, 0, 2, 4, \cdots\}$$

forms a Commutative Ring but it is not a ring with unity and it is certainly not a field. **[C.P. 1987, '95]**

Solution: First we check, as in **Example 8.4.1** above, that under ordinary addition E is an Abelian group:

A_1. *Closure property:* Sum of two even integers is also an even integer, i.e., $\forall a, b \in E$, $a + b \in E$.

A_2. *Associative property holds:* $a + (b + c) = (a + b) + c$ for any three even integers $a, b, c \in E$.

A_3. $\exists$ an element $0 \in E$ which forms the additive identity, i.e., for any even integer a we have $a + 0 = 0 + a = a$.

A_4. For any even integer a, we have an even integer $-a$ such that $a + (-a) = (-a) + a = 0$, i.e., $-a$ is the additive identity of a.

A_5. $a + b = b + a$, for any two members of E. (*commutative property*)

$\therefore$ E is an **Abelian group** under addition. Next we consider the operation – multiplication. Obviously

M_1. *Closure property holds:* Multiplication of two even integers is an even integer.

M_2. *Associative property also holds:* $a \cdot (b \cdot c) = (a \cdot b) \cdot c$ for any three even integers $a, b, c \in E$.

M_5. *Commutative property holds:* $a \cdot b = b \cdot a, \forall a, b \in E$.

Moreover, **D_1**. *Distributive Law holds,* i.e., $a \cdot (b + c) = a \cdot b + a \cdot c$ and by commutative property the other distributive law also holds.

$\therefore$ E is a **Commutative Ring.**

But we observe that $\exists$ no **multiplicative identity here.**

No even integer p exists such that $a \cdot p = a, \forall a \in E$.

i.e., the commutative ring of even integers is not a ring with unity and hence no multiplicative inverse can arise, i.e., **it is not a field.**

▶ Example 8.4.3 *The set of all odd integers* $\{\cdots -5, -3, -1, 1, 3, 5, \ldots\}$ *does not form a ring with ordinary addition and multiplication.*

[Even the closure property does not hold. There are elements, say 3 and 5, whose sum is 8 (not an odd integer), i.e., the sum may not always be a number of the set.]

▶ Example 8.4.4 *Prove that the set* S *of all real-valued continuous functions defined on the closed interval* $[0, 1]$ *is a Ring. Check that it is a Ring with zero divisors.*

Solution: Let f and g be two real-valued functions continuous on $[0, 1]$. We define addition $f + g$ and multiplication $f \cdot g$ in the following manner:

$$\begin{aligned} (f+g)(x) &= f(x) + g(x) \\ \text{and } (f \cdot g)(x) &= f(x) \cdot g(x) \end{aligned}$$

Moreover, $(-f)(x) = -f(x)$.

A_1. We know that the sum of two real-valued continuous functions is a real-valued continuous function in the same interval.

Therefore, the *closure property for addition* holds.

A_2. Clearly,

$$\left.\begin{aligned} [(f+g)+h](x) &= (f+g)(x) + h(x) \\ &= [f(x) + g(x)] + h(x) \\ &= f(x) + [g(x) + h(x)] \\ &= [f + (g+h)](x) \end{aligned}\right\} \text{f, g, h are any three functions} \in S$$

i.e., Associative property for addition holds.

A_3. $\exists$ a function, say $\theta(x) \in S$, defined by $\theta(x) = 0, \forall x \in [0, 1]$.

Then for any function $f(x) \in S$ we have

$$(f + \theta)(x) = f(x) + \theta(x) = f(x) \forall x \in [0, 1]$$

Similarly, $(\theta + f)(x) = f(x) \forall x \in [0, 1]$

$\therefore$ $\theta(x)$ is *the additive identity* of S.

A₄. $[(-f) + f](x) = (-f)(x) + f(x) = -f(x) + f(x) = \theta(x)$

Similarly, $[f + (-f)]x = \theta(x)$

$\therefore$ every element of S has an *additive inverse.*

A₅. $(f + g)(x) = f(x) + g(x) = g(x) + f(x) = (g + f)(x)$.

$\therefore$ Commutative property for addition holds.

i.e., S forms an Abelian group with respect to Addition.

M₁. The product of two real-valued continuous functions is also a real-valued continuous function. Hence, under multiplication S *is closed.*

M₂. *Associative property* holds for multiplications for three function f, g, h we have

$$\begin{aligned}[(f\cdot g)\cdot h](x) &= (f\cdot g)(x)\cdot h(x) = \{f(x)\cdot g(x)\}\cdot h(x)\\ &= f(x)\cdot\{g(x)\cdot h(x)\} = [f\cdot(g\cdot h)](x).\end{aligned}$$

D₁. Left distributive Law holds:

$$\begin{aligned}[f\cdot(g+h)](x) &= f(x)\cdot(g+h)(x) = f(x)\cdot[g(x)+h(x)]\\ &= f(x)\cdot g(x) + f(x)\cdot h(x)\\ &= (f\cdot g)(x) + (f\cdot h)(x) = [f\cdot g + f\cdot h](x)\\ \text{i.e., } f\cdot(g+h) &= f\cdot g + f\cdot h.\end{aligned}$$

D₂. Similarly Right distributive law also holds, i.e.,

$$(g + h)\cdot f = g\cdot f + h\cdot f.$$

These axioms lead to the conclusion that S is a **Ring.** *To show that S is a ring with zero divisor,* we take the *following example:*

Let

$$\begin{aligned}f(x) &= \tfrac{1}{2} - x, \text{ when } 0 \le x \le \tfrac{1}{2}\\ &= 0, \quad \text{when } \tfrac{1}{2} \le x \le 1\\ \text{and } g(x) &= 0, \quad \text{when } 0 \le x \le \tfrac{1}{2}\\ &= x - \tfrac{1}{2}, \text{ when } \tfrac{1}{2} \le x \le 1.\end{aligned}$$

Both the functions f and g are continuous in [0, 1].

Check that they are continuous even at $x = \frac{1}{2}$.

Here neither $f(x) = \theta(x)$ nor $g(x) = \theta(x)$, where $\theta(x)$ is the additive identity.

But observe that

$$\begin{aligned}(f\cdot g)(x) &= f(x)\cdot g(x)\\ &= \begin{cases} f(x)\cdot 0, & \text{when } 0 \le x \le \tfrac{1}{2}\\ 0\cdot g(x), & \text{when } \tfrac{1}{2} \le x \le 1\end{cases}\\ &= 0 \ \forall x \in [0, 1]\\ &= \theta(x).\end{aligned}$$

This shows that $\exists$ functions $f, g \in S$ such that $f \cdot g = \theta$ but neither $f = \theta$ nor $g = \theta$.

$\therefore$ the ring S has *zero divisor.*

Note 1. *If* $e(x) \in S$, *where* $e(x) = 1 \, \forall x \in [0, 1]$, *then it is easy to see that* e *behaves like the Multiplicative identity of* S.

$\therefore$ S is a *ring with unity.*
Moreover

$$\begin{aligned}(fg)(x) &= f(x)\cdot g(x) = g(x)\cdot f(x)\\ &= (g\cdot f)(x) \text{ [Commutative Property holds]}\end{aligned}$$

Hence, S is a *commutative ring with unity and with zero divisor.*
See that $\{S, +, \bullet\}$ is NOT an *Integral Domain* since an integral domain has no zero divisor.

▶ Example 8.4.5 *Let* $S = \{0, 1, 2, 3\}$ *and the operations are*

Addition $(+)$ *modulo* 4
and *Multiplication* $(\bullet)$ *modulo* 4.

Prove that the system $\{S, +, \bullet\}$ *is a commutative ring with unity but it is not an integral domain.*

Solution: Use the two Tables given below and check that S is an additive Abelian group and with respect to multiplication, closure property, Associativity and Distributive Laws hold so that S becomes a **Ring.** Then verify that $\{S, +, \bullet\}$ obeys Commutative Law with respect to multiplication and also it has a multiplicative identity (namely, 1). Thus S is a commutative ring with unity but it has zero divisors (see that 2.2 = 0) so that it is NOT an *Integral Domain.*

+	0	1	2	3
0	0	1	2	3
1	1	2	3	0
2	2	3	0	1
3	3	0	1	2

•	0	1	2	3
0	0	0	0	0
1	0	1	2	3
2	0	2	0	2
3	0	3	2	1

▶ Example 8.4.6 (*An example of a non-commutative ring*). *If* M *is the set of all* 2×2 *real matrices, then* $\{M, +, \bullet\}$, *where the operations* + *and* • *are respectively matrix addition and matrix multiplication, is a ring; it has a unity element* $\begin{bmatrix} 1 & 0 \\ 0 & 1 \end{bmatrix}$ (*Identity matrix*) *but it is* **non-commutative:** **[C.P. 1981]**

Check the ring axioms: $\mathbf{A_1}$—$\mathbf{A_5}, \mathbf{M_1}, \mathbf{M_2}, \mathbf{D_1}, \mathbf{D_2}$.
Now see that

$$\begin{pmatrix} 1 & 1 \\ 0 & 0 \end{pmatrix}\begin{pmatrix} 0 & 0 \\ 1 & 1 \end{pmatrix} = \begin{pmatrix} 1 & 1 \\ 0 & 0 \end{pmatrix}, \text{ but}$$

$$\begin{pmatrix} 0 & 0 \\ 1 & 1 \end{pmatrix}\begin{pmatrix} 1 & 1 \\ 0 & 0 \end{pmatrix} = \begin{pmatrix} 0 & 0 \\ 1 & 1 \end{pmatrix};$$

the two products do not give the same matrix, i.e., Commutative Law does not hold.

▶ Example 8.4.7 *Let* $S = \left\{ \begin{bmatrix} x & y \\ 0 & z \end{bmatrix}, x, y, z \text{ are even integers} \right\}$. *Then* $\{S, +, \bullet\}$ *where* + *and* • *are respectively matrix addition and matrix multiplication, is a* **non-commutative ring with no unity element:**

We exhibit the checking in detail:

A₁. $A = \begin{bmatrix} x_1 & y_1 \\ 0 & z_1 \end{bmatrix}$ and $B = \begin{bmatrix} x_2 & y_2 \\ 0 & z_2 \end{bmatrix}$ are elements of S.

$A + B = \begin{bmatrix} x_1 + x_2 & y_1 + y_2 \\ 0 & z_1 + z_2 \end{bmatrix}$; since $x_1 + x_2, y_1 + y_2$ and $z_1 + z_2$ are even integers, it follows that $A + B \in S$, i.e., S is closed under matrix addition.

A₂. Addition is associative (in fact, matrix addition always obeys the associative law).

A₃. Let $0 = \begin{bmatrix} 0 & 0 \\ 0 & 0 \end{bmatrix}$. Then the null matrix 0 is the additive identity for S,

$$A + 0 = 0 + A = A, \; A \in S.$$

A₄. If $A = \begin{bmatrix} x & y \\ 0 & z \end{bmatrix} \in S$, then $\begin{bmatrix} -x & -y \\ 0 & -z \end{bmatrix}$ is also in S and is such that their sum is 0, i.e., this matrix is the additive inverse of A.

A₅. From **A₁** (above) it is clear, $A + B = B + A$.

Hence, $\{S, +\}$ is an Abelian group under addition.

M₁. Taking A, B is in **A₁** we have

$$A \cdot B = \begin{bmatrix} x_1 & y_1 \\ 0 & z_1 \end{bmatrix} \cdot \begin{bmatrix} x_2 & y_2 \\ 0 & z_2 \end{bmatrix} = \begin{bmatrix} x_1x_2 & x_1y_2 + y_1z_2 \\ 0 & z_1z_2 \end{bmatrix}$$

and $A \cdot B \in S$, since the entries are all even integers.

Hence, S is closed under matrix multiplication (•).

M₂. Multiplication here is associative since matrix multiplication always obeys the associative law.

Finally, the distribution laws **D₁** and **D₂** both hold in S since they hold, in general, for all 2 × 2 matrices.

Conclusion: $\{S, +, \bullet\}$ is a ring.

This ring is **non-commutative:** Using A, B as in **A₁** we have

$$A \cdot B = \begin{bmatrix} x_1x_2 & x_1y_2 + y_1z_2 \\ 0 & z_1z_2 \end{bmatrix} \text{ and } B \cdot A = \begin{bmatrix} x_2x_1 & x_2y_1 + y_2z_1 \\ 0 & z_2z_1 \end{bmatrix}.$$

We can easily choose matrices $A, B \in S$ for which $x_1y_2 + y_1z_2 \neq x_2y_1 + y_2z_1$, and so, in general, $AB \neq BA$; take, for instance,

$$A = \begin{bmatrix} 2 & 0 \\ 0 & 0 \end{bmatrix} \text{ and } B = \begin{bmatrix} 2 & 2 \\ 0 & 2 \end{bmatrix}.$$

Here

$$AB = \begin{bmatrix} 4 & 4 \\ 0 & 0 \end{bmatrix} \text{ and } BA = \begin{bmatrix} 4 & 0 \\ 0 & 0 \end{bmatrix}; \; AB \neq BA.$$

This non-commutative ring has **no unity element** also.

If $E = \begin{bmatrix} a & b \\ 0 & c \end{bmatrix} \in S$ (a, b, c are even integers is a unity element for S, then we should have for each

$$A = \begin{bmatrix} x & y \\ 0 & z \end{bmatrix} \in S, \; AE = A$$

and so $$\begin{bmatrix} xa & xb + yc \\ 0 & zc \end{bmatrix} = \begin{bmatrix} x & y \\ 0 & z \end{bmatrix}.$$

Consequently $xa = x, zc = z$ and $xb + yc = y$ for all even integers x, y, z. These conditions imply and $a = 1, c = 1, b = 0$. Since a and c must be even, it follows that the *ring has no unity element.*

▶ Example 8.4.8 *If V is the set of all vectors in 3-dimensional space R^3, then it is easy to check that* $\{V, +, \times\}$ **is not a ring,** *where + is the addition of vectors and × is the* **vector product** (*or* **cross product**) *of two vectors.*

Check: $\{V, +\}$ is an additive Abelian group [additive identity is $\vec{0}$; additive inverse of $\vec{a}$ is $-\vec{a}$.]

Now V is closed under the cross-product of two vectors, but the associative law

$$\vec{a} \times (\vec{b} \times \vec{c}) = (\vec{a} \times \vec{b}) \times \vec{c}$$

is not, in general, true.

Then $\{V, \times\}$ is not a *semigroup*.

Hence, $\{V, +, \times\}$ does not form a ring.

▶ Example 8.4.9 *Let* $R = \{w, x, y, z\}$ *and let addition* (+) *and multiplication* (•) *be defined by the following tables. Prove that the set* $\{R, +, \bullet\}$ *is a ring but* **not a commutative ring.**

Check the ring axioms; checking associativity and distributivity would require checking each possible combinations of elements—64 cases for each law. One may check several cases for each law. The ring is not commutative, e.g., $y \cdot z = w$ but $z \cdot y = y$ and hence $y \cdot z \neq z \cdot y$.

+	w	x	y	z
w	w	x	y	z
x	x	w	z	y
y	y	z	w	x
z	z	y	x	w

•	w	x	y	z
w	w	w	w	w
x	w	x	y	z
y	w	w	w	w
z	w	x	y	w

Examples of Fields:

▶ Example 8.4.10 *The set* $S = \{0, 1, 2, 3, 4, 5, 6\}$ *form a field under addition (mod 7) and multiplication (mod 7).*

Set up two tables—One for addition (modulo 7) and the other multiplication (modulo 7). Now verify the axioms

$$A_1—A_5, M_1—M_5, D_1.$$

For checking A_2, M_2, and D_1 one can simply say that addition and multiplication of the set of *all integers* obey associative laws and they are connected by distributive laws; as a particular case the results will be true here.

▶ Example 8.4.11 *The set S of 2×2 real matrices of the form* $\begin{bmatrix} x & y \\ -y & x \end{bmatrix}$, $x, y \in R$, *forms a field under matrix addition and matrix multiplication.* [C.P. 1982, 1993]

We show the checking in detail:
First check that $\{S, +\}$ is an Abelian group.

A_1. $A = \begin{bmatrix} x_1 & y_1 \\ -y_1 & x_1 \end{bmatrix}$, $B = \begin{bmatrix} x_2 & y_2 \\ -y_2 & x_2 \end{bmatrix}$ are in S, then

$$A + B = \begin{bmatrix} x_1 + x_2 & y_1 + y_2 \\ -(y_1 + y_2) & x_1 + x_2 \end{bmatrix}$$, and this is clearly in S.

Hence, S is closed under matrix addition.

A_2. The associative property of addition holds since it holds for all 2×2 matrices.

A_3. The matrix $O = \begin{bmatrix} 0 & 0 \\ 0 & 0 \end{bmatrix}$ is in S, and so is an additive identity for S.

A_4. If $A = \begin{bmatrix} x & y \\ -y & x \end{bmatrix}$ is in S, then $-A = \begin{bmatrix} -x & (-y) \\ -(-y) & -x \end{bmatrix}$ is in S and is an additive inverse of A in S.

A_5. The commutative property of addition holds in S since it holds for all 2×2 real matrices.

Hence, $\{S, +\}$ is an Abelian group.

Multiplication Properties:

M_1. In the notation used for A_1 above,

$$A \cdot B = \begin{bmatrix} x_1 & y_1 \\ -y_1 & x_1 \end{bmatrix} \cdot \begin{bmatrix} x_2 & y_2 \\ -y_2 & x_2 \end{bmatrix} = \begin{bmatrix} x_1x_2 - y_1y_2 & x_1y_2 + x_2y_1 \\ -(x_1y_2 + x_2y_1) & x_1x_2 - y_1y_2 \end{bmatrix}$$

and clearly $A \cdot B \in S$.

Hence, S is closed under matrix multiplication.

M_2. The associative property of multiplication holds in S since it holds for all 2×2 matrices.

M₃. The matrices $I = \begin{bmatrix} 1 & 0 \\ 0 & 1 \end{bmatrix}$ is in S and is a multiplicative identity for S.

M₄. If $A = \begin{bmatrix} x & y \\ -y & x \end{bmatrix}$ is in S and $a \neq 0$, then x and y are not both zero. It follows, therefore, that $\det A = x^2 + y^2 \neq 0$ so that the matrix is non-singular and consequently A^{-1} exists:

$$A^{-1} = \frac{1}{x^2 + y^2}\begin{bmatrix} x & -y \\ y & x \end{bmatrix},$$ Using the usual method of finding the inverse of a non-singular matrix.

and A^{-1} is clearly is S and is such that $A \cdot A^{-1} = A^{-1} \cdot A = I$.

M₅. By symmetry of the product formed in **M₁**, $AB = BA$, so that multiplication is commutative.

Finally, we have the distributive property:

D₁. The distributive property holds S since it holds for all 2×2 matrices.

Hence, S *forms a field under matrix addition and multiplication.*

▶ Example 8.4.12 Q $=$ *set of all rational numbers:* $\{Q, +, \bullet\}$, *where* $+$ *and* $\bullet$ *are ordinary addition and multiplication of rational numbers, is a* **ring.**

We can further show that it is a commutative ring with unity element (namely, 1). This ring has no divisors of zero and hence Q is an **integral domains.** Further if a be a non-zero rational number, then a^{-1} (or, $1/a$) is its inverse and $a^{-1} \in Q$. So finally we can prove that Q is a field.

▶ Example 8.4.13 *The set C of complex numbers also forms a field. A typical element of C is* $\alpha + i\beta(\alpha, \beta$ *real*$)$.

Check A₁—A₅, M₁—M₅ and **D₁** all hold, additive identity is $0 + i0$ denoted by O and multiplicative identity is $1 + i0 = e$ (say).

The inverse of

$$\alpha + i\beta = \frac{1}{\alpha + i\beta} = \frac{\alpha - i\beta}{\alpha^2 + \beta^2} = \frac{\alpha}{\alpha^2 + \beta^2} + i\left(\frac{-\beta}{\alpha^2 + \beta^2}\right) \in C.$$

(The non-zero element $\alpha + i\beta \Rightarrow \alpha \neq 0, \beta \neq 0 \Rightarrow \alpha^2 + \beta^2 \neq 0]$ (simultaneously)

▶ Example 8.4.14 *The set* T *of all real numbers of the form* $a + b\sqrt{2}$ *where* a, b *are rational numbers, forms a field under the operations of addition* $(+)$ *and multiplication* $(\bullet)$ *of real numbers.* **[C.P. 1982, 1987, 1990, 2004]**

We first check Additive axioms.

A₁. $\left(a + b\sqrt{2}\right) + \left(c + d\sqrt{2}\right) = (a + c) + \sqrt{2}(b + d) \in T$ $\quad(\because\ a + c, b + d$ are rationals)

A_2. Associative property for addition:

$$\left(a+b\sqrt{2}\right)+\left\{\left(c+d\sqrt{2}\right)+\left(e+f\sqrt{2}\right)\right\}$$
$$=\left\{\left(a+b\sqrt{2}\right)+\left(c+d\sqrt{2}\right)\right\}+\left(e+f\sqrt{2}\right).$$

A_3. $0+\sqrt{2}\cdot 0$ is the additive identity:

Check: $\left(a+b\sqrt{2}\right)+\left(0+\sqrt{2}\cdot 0\right)=a+b\sqrt{2}$, etc.

A_4. $-\left(a+b\sqrt{2}\right)$ is the additive inverse of $\left(a+b\sqrt{2}\right)$.

A_5. Commutative property for addition is obvious.

Hence, $\{T,+\}$ is an Abelian group.

Next we consider Multiplicative axioms:

M_1. $\left(a+b\sqrt{2}\right)\cdot\left(c+d\sqrt{2}\right)=(ac+2bd)+\sqrt{2}(bc+ad)\in T,$

since $ac+2bd$, $bc+ad$ are rationals, a, b, c, d being all rationals.

M_2. Multiplication is associative—

$$\left(a+b\sqrt{2}\right)\cdot\left\{\left(c+d\sqrt{2}\right)\cdot\left(e+f\sqrt{2}\right)\right\}=\left\{\left(a+b\sqrt{2}\right)\cdot\left(c+d\sqrt{2}\right)\right\}\cdot\left(e+f\sqrt{2}\right)$$

M_3. $1+0\sqrt{2}$ is the multiplicative identity:

$$\left(1+0\sqrt{2}\right)\cdot\left(a+b\sqrt{2}\right)=a+b\sqrt{2}, \text{ etc.}$$

M_4. Let $a+b\sqrt{2}\in T$ where a, b (not simultaneously zero) are rationals.

The inverse $=\dfrac{1}{a+b\sqrt{2}}=\dfrac{a}{a^2-2b^2}+\dfrac{-b}{a^2-2b^2}\sqrt{2}\in T.$

($a^2-2b^2\neq 0$ since this will imply $a/b=\sqrt{2}$, i.e., ratio of two rationals=an irrational—not possible).

M_5. Multiplication is commutative

D_1. Distributive laws hold.

(can be easily checked)

$\therefore$ **T forms a field.**

{ These two laws hold for **all real numbers** and hence in T which is a subset of R, these two laws also hold.

Try a similar problem:

Assuming that the set F of all real numbers of the form $a+b\sqrt{3}$ where a, b are rational number forms a ring. With respect to the usual addition and multiplication, show that F is a field. **[C.P. 1992]**

▶ Example 8.4.15 *Let S be the set of all square matrices of the form* $\begin{pmatrix} 0 & 0 \\ a & b \end{pmatrix}$, *where* a, b *are integers. Determine whether S forms a ring under matrix addition and matrix multiplication. Does it form a field?* **[C.P 1987]**

Solution: **Additive axioms**

A_1. Let $\alpha = \begin{pmatrix} 0 & 0 \\ a & b \end{pmatrix}$, $\beta = \begin{pmatrix} 0 & 0 \\ c & d \end{pmatrix}$ where a, b, c, d are all integers.

Then $\alpha + \beta = \begin{pmatrix} 0 & 0 \\ a+c & b+d \end{pmatrix} \in S$, since $a+c, b+d$ are also integers, i.e., *closure property* holds.

A_2. If $\gamma = \begin{pmatrix} 0 & 0 \\ e & f \end{pmatrix}$, e, f are integers, then it is easy to check:

$$(\alpha + \beta) + \gamma = \alpha + (\beta + \gamma) \text{ (Associativity).}$$

A_3. $\exists$ a matrix $\begin{pmatrix} 0 & 0 \\ 0 & 0 \end{pmatrix} \in S$ which is the *additive identity:* thus,

$$\begin{pmatrix} 0 & 0 \\ a & b \end{pmatrix} + \begin{pmatrix} 0 & 0 \\ 0 & 0 \end{pmatrix} = \begin{pmatrix} 0 & 0 \\ 0 & 0 \end{pmatrix} + \begin{pmatrix} 0 & 0 \\ a & b \end{pmatrix} = \begin{pmatrix} 0 & 0 \\ a & b \end{pmatrix}.$$

A_4. If $A = \begin{pmatrix} 0 & 0 \\ a & b \end{pmatrix}$, a, b are integers, then take a matrix

$$B = \begin{pmatrix} 0 & 0 \\ -a & -b \end{pmatrix} \in S,$$

since $-a, -b$ also are integers such that

$$A + B = B + A = \begin{pmatrix} 0 & 0 \\ 0 & 0 \end{pmatrix},$$

i.e., B is *the additive inverse* of A.

A_5. *Commutative Law holds:*

$$\begin{aligned}\begin{pmatrix} 0 & 0 \\ a & b \end{pmatrix} + \begin{pmatrix} 0 & 0 \\ c & d \end{pmatrix} &= \begin{pmatrix} 0 & 0 \\ a+c & b+d \end{pmatrix} = \begin{pmatrix} 0 & 0 \\ c+a & d+b \end{pmatrix} \\ &= \begin{pmatrix} 0 & 0 \\ c & d \end{pmatrix} + \begin{pmatrix} 0 & 0 \\ a & b \end{pmatrix},\end{aligned}$$

i.e., under matrix addition S is an abelian group.

Multiplicative axioms

M_1. *Closure property:*

$$\begin{pmatrix} 0 & 0 \\ a & b \end{pmatrix} \cdot \begin{pmatrix} 0 & 0 \\ c & d \end{pmatrix} = \begin{pmatrix} 0 & 0 \\ bc & bd \end{pmatrix} \in S,$$

since bc, bd also are integers.

M_2. Matrix Multiplication is always associative.

D. Matrix Multiplication is distributive (both *right* and *left*).

So, we conclude that S forms a ring *w.r.t.* matrix addition and multiplication. But this ring is neither commutative nor it has a multiplicative identity

$$\begin{pmatrix} 0 & 0 \\ a & b \end{pmatrix} \cdot \begin{pmatrix} 0 & 0 \\ c & d \end{pmatrix} = \begin{pmatrix} 0 & 0 \\ bc & bd \end{pmatrix}$$

$$\text{but} \begin{pmatrix} 0 & 0 \\ c & d \end{pmatrix} \cdot \begin{pmatrix} 0 & 0 \\ a & b \end{pmatrix} = \begin{pmatrix} 0 & 0 \\ da & db \end{pmatrix}$$

and the two products are not equal.

If, at all, $\exists$ exists a multiplication identity $\begin{pmatrix} 0 & 0 \\ x & y \end{pmatrix}$, then

$$\begin{pmatrix} 0 & 0 \\ a & b \end{pmatrix} \cdot \begin{pmatrix} 0 & 0 \\ x & y \end{pmatrix} = \begin{pmatrix} 0 & 0 \\ a & b \end{pmatrix}$$

$\Rightarrow$ $bx = a$ and $by = b$. See that x need not be an integer. Thus $\exists$ no multiplicative identity in this case.

$\therefore$ S cannot, in any case, form a Field.

Try the following problem yourself.

Q. *Examine whether the set of all matrices of the form* $\begin{pmatrix} a & b \\ 0 & 0 \end{pmatrix}$ (a, b *real numbers*) *form a subring of the ring of all* 2×2 *real matrices under usual matrix addition and matrix multiplication.* **[C.P. 1992, 2004]**

The set of matrices of the form $\begin{pmatrix} a & b \\ 0 & 0 \end{pmatrix}$, where $a, b \in \mathbf{R}$, form a ring under matrix addition and multiplication (**check**) and hence a subring of the ring of all 2×2 matrices under matrix addition and multiplication.

► Example 8.4.16 *Let* $(G, +)$ *be a commutative group with at least two elements with* 0 *as its identity element.*

A binary operation $\times$ (*cross*) *is defined as* G *by the following rule:*

$$\forall a, b \in G, \ a \times b = 0.$$

Show that the structure $\{G, +, \times\}$ *forms a Ring.*

Solution: With respect to addition G is **given** to be a commutative group and G has 0 as its identity (additive) and at least one more element. Thus, in our notation, the axiom A_1—A_5 all hold. We now consider the operation multiplication $(\times)$.

M_1. *Closure property holds:* Since for all $a, b \in G$

$$a \times b = 0, \text{ where } 0 \in G.$$

M_2. *Associative property also holds:* Let a, b, c be elements of G. Then

$$a \times (b \times c) = a \times 0 = 0 \; (\because \; b \times c = 0)$$

Also $(a \times b) \times c = 0 \times c = 0$. Therefore, $a \times (b \times c) = (a \times b) \times c$.

D. Distributive Laws: For any three elements $a, b, c \in G$

$$\begin{aligned} & a \times (b \times c) = 0 \text{ since } b + c \in G \\ & a \times b = 0 \text{ and } a \times c = 0; \text{ therefore, } a \times b + a \times c = 0 + 0 = 0. \\ \therefore \quad & a \times (b + c) = a \times b + a \times c \text{ (Left Distributive Law)} \end{aligned}$$

Similarly, Right Distributive Law also holds.

Hence, we conclude that $(G, +, \times)$ form a **Ring**.

▶ Example 8.4.17 *Assuming that the set* F *of all* 2 × 2 *real matrices of the form* $\begin{bmatrix} a & b \\ -b & a \end{bmatrix}$ *forms a Ring w.r.t. matrix addition* (+) *and matrix multiplication* (•)), *show that* F *is a field.* [C.P. 1993]

Solution: Given: $\{F, +, \bullet\}$ is a ring (A_1—A_5, M_1, M_2, D_1, D_2 hold)

In order to prove that F is a field, we require to verify that

(i) multiplicative identity exists (M_3).

(iv) commutative property w.r.t. multiplication also holds (M_5).

(v) for every non-zero element of F there exists the multiplicative inverse (M_4').

Note that $\underset{\sim}{0} = \begin{bmatrix} 0 & 0 \\ 0 & 0 \end{bmatrix}$ is the additive identity of F.

M_3. Let $\alpha = \begin{bmatrix} 1 & 0 \\ 0 & 1 \end{bmatrix}$ be a member of F.

Then for any element $\beta = \begin{bmatrix} a & b \\ -b & a \end{bmatrix}$ (a, b are real) $\in$ F.

We have by using the rule of matrix multiplication

$$\alpha \cdot \beta = \begin{bmatrix} 1 & 0 \\ 0 & 1 \end{bmatrix} \begin{bmatrix} a & b \\ -b & a \end{bmatrix} = \begin{bmatrix} a & b \\ -b & a \end{bmatrix} = \beta.$$

Similarly,

$$\beta \cdot \alpha = \begin{bmatrix} a & b \\ -b & a \end{bmatrix} \begin{bmatrix} 1 & 0 \\ 0 & 1 \end{bmatrix} = \begin{bmatrix} a & b \\ -b & a \end{bmatrix} = \beta.$$

i.e., $\alpha \cdot \beta = \beta \cdot \alpha = \beta$ for all $\beta \in F$.

i.e., α is the multiplicative identity.

M_5.

$$\begin{bmatrix} a & b \\ -b & a \end{bmatrix}\begin{bmatrix} c & d \\ -d & c \end{bmatrix} = \begin{bmatrix} ac-bd & ad+bc \\ -bc-ad & -bd+ac \end{bmatrix}$$
$$= \begin{bmatrix} ac-bd & ad+bc \\ -(ad+bc) & ac-bd \end{bmatrix}$$

and
$$\begin{bmatrix} c & d \\ -d & c \end{bmatrix}\begin{bmatrix} a & b \\ -b & a \end{bmatrix} = \begin{bmatrix} ca-db & cb+da \\ -da-cb & -db+ca \end{bmatrix}$$
$$= \begin{bmatrix} ac-bd & ad+bc \\ -(ad+bc) & ac-bd \end{bmatrix}.$$

[Note. a, b, c, d are all real numbers. So $ac = ca$, etc.]

$\therefore$ Commutative Law of multiplication holds.

M_4. Given $\alpha = \begin{bmatrix} a & b \\ -b & a \end{bmatrix} \in F$. Its determinant $a^2 + b^2 = 0$ iff $a = 0, b = 0$.

We take $\alpha \neq 0 = \begin{bmatrix} 0 & 0 \\ 0 & 0 \end{bmatrix}$. We investigate the existence of α^{-1}

$$\text{s.t. } \alpha \cdot \alpha^{-1} = \alpha^{-1} \cdot \alpha = \begin{bmatrix} 1 & 0 \\ 0 & 1 \end{bmatrix}.$$

Our previous knowledge of determinants of inverse gives

$$\alpha^{-1} = \frac{1}{\det \alpha} \operatorname{adj} \alpha = \frac{1}{a^2 + b^2}\begin{bmatrix} a & -b \\ b & a \end{bmatrix}.$$

Thus, every non-zero element of F possesses an inverse.

Hence, our conclusion is that the Ring $\{F, +, \bullet\}$ obeys all the field axioms, i.e., F is a *field.*

8.5 Elementary Theorems: Simple Consequences of the Definitions

A ring may, or may not, possess a unity element, but when one such element exists it is unique.

Theorem 8.5.1. *If a ring* $\{R, +, \bullet\}$ *does contain an unity element* e (*with respect to multiplication*) *then* e *is unique.* [C.P. 1982]

Proof. By definition an element e is *unity element* only if

$$e \cdot a = a \cdot e = a \text{ for every } a \in R.$$

If possible, let e' be also a unity of R. Then e and e' both belong to R. Taking e as unity and e', any element of R, we have

$$e \cdot e' = e' \cdot e = e'.$$

Taking e' as unity and $e \in R$ we have

$$e' \cdot e = e \cdot e' = e.$$

Hence, $e = e'$. In other words, the unity element e is unique.

Inverses (multiplicative) of elements in a ring do not necessarily belong to the ring, as in the case of integers. However, if x, an element of a ring R has an inverse x^{-1} and $x^{-1} \in R$, then this inverse is unique.

Theorem 8.5.2. *If in a ring* $\{R, +, \bullet\}$, *an element* $a \in R$ *has a multiplicative inverse* $a^{-1} \in R$, *then it is unique.*

Proof. By definition the inverse x^{-1} of x must satisfy the condition

$$x \cdot x^{-1} = x^{-1} \cdot x = e \text{ (e is the multiplicative identity)}$$

Suppose that x' is also an inverse for x. Then

$$\begin{aligned} x^{-1} &= x^{-1} \cdot e && (\because \text{ e is the unity in R}) \\ &= x^{-1} \cdot (x \cdot x') && (\because x' \text{ is an inverse of } x, x \cdot x' = e) \\ &= (x^{-1} \cdot x) \cdot x' && (\text{Associative law}) \\ &= e \cdot x' && (\because x^{-1} \text{ is an inverse of } x, x^{-1} \cdot x = e) \\ &= x' && (\because \text{ e is the unity in R}) \end{aligned}$$

Thus, $x^{-1} = x'$ and the inverse is unique.

Theorem 8.5.3. *If* R *is a ring, then for all* $a, b, c \in \mathbf{R}$.
(i) $a \cdot 0 = 0 \cdot a = 0$; (ii) $a \cdot (-b) = -(a \cdot b) = (-a) \cdot b$;
(iii) $(-a) \cdot (-b) = a \cdot b$; (iv) $a \cdot (b \cdot c) = a \cdot b - a \cdot c$;
(v) $(b - c) \cdot a = b \cdot a - c \cdot a$.

[C.P. 2005]

Proof.

(i) We have

$$\begin{aligned} a \cdot 0 &= a \cdot (0 + 0) \; [\because 0 + 0 = 0] \\ &= a \cdot 0 + a \cdot 0 \; [\text{by left distributive law}] \\ \therefore 0 + a \cdot 0 &- a \cdot 0 + a \cdot 0 \; [\because a \cdot 0 \in R \text{ and } 0 + a \cdot 0 = a \cdot 0] \end{aligned}$$

Now R is a group with respect to addition, therefore applying right cancellation law for addition in R, we get $0 = a \cdot 0$.

Similarly we have

$$\begin{aligned} 0 \cdot a &= (0 + 0) \cdot a \\ &= 0 \cdot a + 0 \cdot a \; (\text{by right distributive law}) \\ \therefore 0 + 0 \cdot a &= 0 \cdot a + 0 \cdot a \; (\because 0 + 0 \cdot a = 0 \cdot a). \end{aligned}$$

Applying right cancellation law for addition in R, we get $0 = 0 \cdot a$.

$\therefore$ finally we have $a \cdot 0 = 0 = 0 \cdot a$ **(Proved)**

(ii) We have

$$a\cdot[-b+b]=a\cdot 0\ (\because\ -b+b=0)$$

$$\Rightarrow\quad a\cdot(-b)+(a\cdot b)=a\cdot 0=0\ \left[\begin{array}{l}\text{using left distributive}\\ \text{law and the result (i)}\end{array}\right]$$

$$\Rightarrow\quad a\cdot(-b)=-(a\cdot b)\ [\text{since in a ring } a+b=0\Rightarrow a=-b]$$

Similarly we have

$$(-a+a)\cdot b=0\cdot b$$

$$\Rightarrow\quad (-a)\cdot b+a\cdot b=0\cdot b=0$$

$$\Rightarrow\quad (-a)\cdot b=-(a\cdot b)\ [\text{since in a ring } a+b=0\Rightarrow a=-b]$$

$\therefore$ finally we have $a\cdot(-b)=-(a\cdot b)=(-a)\cdot b$.

(iii) We have

$$\begin{aligned}(-a)\cdot(-b) &= -[(-a)\cdot b]\ [\text{since } a\cdot b=-(a\cdot b)]\\ &= -[-(a\cdot b)]\ [\text{since } (-a)\cdot b=-(a\cdot b)]\\ &= a\cdot b\ \left[\begin{array}{l}\text{since R is a group with respect to addition}\\ \text{and in a group, we have } -(-a)=a.\end{array}\right]\end{aligned}$$

[An Important Particular Case of (ii) and (iii)

If 1 is the identity element in a ring (R, +, •), then exactly in a similar way we can prove

(α) $(-1)\cdot a=-a=a\cdot(-1)$ for all $a\in R$

(β) $(-1)\cdot(-1)=1$. [C.P. 2003]

For (α) see that $(-1+1)\cdot a=0\cdot a=0$

$\therefore$ by right distributive law

$$-1\cdot a+1\cdot a=0$$

$$\text{or,}\quad (-1)\cdot a+a=0\ (\because\ 1\cdot a=a)$$

$$\therefore\quad (-1)a=-a\ (\because\ a+b=0\Rightarrow a=-b).$$

Similarly $a\cdot(-1+1)=a\cdot 0=0$

$$\text{or,}\quad a\cdot(-1)+a\cdot 1=0\ [\text{left distributive law}]$$

$$\text{or,}\quad a\cdot(-1)+a=0\ (\because\ a\cdot 1=a)$$

$$\text{or,}\quad a\cdot(-1)=-a\ (\because a+b=0\Rightarrow a=-b)$$

$$\therefore\quad (-1)\cdot a=-a=a\cdot(-1).$$

For (β) write

$$\begin{aligned}(-1)\cdot(-1) &= -[(-1)\cdot 1]\ [\text{since } a\cdot(-1)=-(a\cdot 1)]\\ &= -[-1]\ [\text{since } (-1)\cdot a=-a]\\ &= 1\ (\text{since R is a group w.r.t. addition and in a group } -(-a)=a)]\end{aligned}$$

(iv) **To prove:** $a \cdot (b - c) = a \cdot b - a \cdot c \ \forall a, b, c \in R$.

We have

$$\begin{aligned} a \cdot (b - c) &= a \cdot [b + (-c)] \\ &= a \cdot b + a \cdot (-c) \text{ (left distributive law)} \\ &= a \cdot b + [-(a \cdot c)] \ (\because \ a \cdot (-c) = -(a \cdot c)] \\ &= a \cdot b - a \cdot c. \end{aligned}$$

(v) **To prove:** $(b - c) \cdot a = b \cdot a - c \cdot a \ \forall a, b, c \in R$

We have

$$\begin{aligned} (b - c) \cdot a &= [b + (-c)] \cdot a \\ &= b \cdot a + (-c) \cdot a \text{ (right distributive law)} \\ &= b \cdot a + [-(c \cdot a)] \\ &= b \cdot a - c \cdot a. \end{aligned}$$

Equivalent Definitions of an Integral domain

Theorem 8.5.4. *Let R be a ring with respect to addition (+) and multiplication (•). The two statements*

(i) *R has no divisors of zero (either right or left); and*

(ii) *Cancellation laws for multiplication hold in R*

$$[\text{i.e., } ab = ac \Rightarrow b = c; \ ba = ca \Rightarrow b = c]$$

are equivalent statements, i.e., one can be replaced by the other.

Proof. Suppose that R has no divisors of zero, i.e., for every pair $a, b \in R$ for which $ab = 0$, either $a = 0$ or $b = 0$.

Take three elements $a, b, c \in R$, where $a \neq 0$ in such a way that

$$ab = ac.$$

Then $ab - ac = a \cdot (b - c) = 0$.
But since $a \neq 0$ and R has no divisors of zero, it follows

$$b - c = 0, \text{ i.e., } b = c.$$

Also if $ac = bc \ (c \neq 0)$, then $ac - bc = 0$ and $(a - b) \cdot c = 0$ which implies

$$a - b = 0 \ (\because \ c \neq 0), \text{ i.e., } a = b.$$

Thus, the cancellation laws hold, i.e., (ii) follows from (i).
Next, suppose that both the cancellation laws holds in R.
Then, if $a \cdot b = 0$ with $a \neq 0$, we have

$$\begin{aligned} a \cdot b &= a \cdot 0 \ (\because \ a \cdot 0 = 0) \\ \therefore \ b &= 0 \ (\because \text{ left cancellation law holds.}) \end{aligned}$$

Similarly, $a \cdot b = 0$ with $b \neq 0$ will imply $a = 0$.
Hence, R has no divisors of zero, i.e., (i) follows from (ii).

Note 1. ***This justifies our definition of an integral domain is that it is a commutative ring with unity with the following additional condition:***

either, *without having any divisor of zero,*
or, *cancellation laws for multiplication hold.*

We have below an example of a *commutative ring with unity* which is not an *integral domain.*

▶ Example 8.5.1 S = {0, 1, 2, 3, 4, 5}; *operations are addition* (*mod* 6) (+) *and multiplication* (*mod* 6) (•). *We construct two tables—one for addition* (*mod* 6) *and the other for multiplication* (*mod* 6).

Check: {S, + (mod 6), • (mod 6)} is a commutative ring with unity.

[Axioms: $\mathbf{A_1}$—$\mathbf{A_5}$, $\mathbf{M_1}$, $\mathbf{M_2}$, $\mathbf{M_3}$, $\mathbf{M_5}$, $\mathbf{D_1}$ are satisfied.]

Now see that ∃ elements, say, for examples 2 and 3, such that

$$2 \cdot 3 = 6 \equiv 0 \pmod 6$$

but neither $2 = 0 \pmod 6$ nor $3 = 0 \pmod 6$. That is S contains divisors of zero. Hence, S is not an *integral domain* though it is a *commutative ring with unity.*

Theorem 8.5.5. *A division ring has no zero-divisor.*

Proof. Let R be a division ring. Let a, b be two elements of R. Suppose $a \neq 0$, but $a, b = 0$. Our theorem will be proved if we can show that in this case b must be zero.
Since R is a division ring, for the non-zero element a, $a^{-1} \in R$, such that

$$a^{-1} \cdot a = a \cdot a^{-1} = 1.$$

Now,

$$\begin{aligned} b = 1 \cdot b = (a^{-1} \cdot a) \cdot b &= a^{-1} \cdot (a \cdot b) \text{ [Associative law]} \\ &= a^{-1} \cdot 0 = 0. \end{aligned}$$

Thus, for $a \neq 0$, if $a \cdot b = 0$, then $b = 0$, i.e., R has no zero-divisor.

Theorem 8.5.6. *A ring* R *is a division ring if and only if the relations* $a \cdot x = b$ *and* $y \cdot a = b$ *have unique solutions in* R, *where* a *is a non-zero element of* R *and* $b \in R$.

Proof. Let R be a division ring. Then non-zero elements of R have *unique inverses.*
Given $a \neq 0$. *Clearly* $a^{-1} \in R$, $a^{-1} \cdot b$ or $b \cdot a^{-1} \in R$.
Suppose $b \neq 0$ (also).
It is easy to verify that $x = a^{-1} \cdot b$ is a solution of $a \cdot x = b$.
See that $a \cdot x = a \cdot (a^{-1} \cdot b) = (a \cdot a^{-1})b = 1 \cdot b = b$.
This solution is unique. For if $x = d$ be another solution, then

$$\begin{aligned} a \cdot d = b &\Rightarrow a^{-1} \cdot (a \cdot d) = a^{-1} \cdot b \Rightarrow (a^{-1} \cdot a) \cdot d = a^{-1} \cdot b \\ &\Rightarrow 1 \cdot d = a^{-1} \cdot b \Rightarrow d = a^{-1} \cdot b. \end{aligned}$$

Similarly $y\cdot a = b$ have unique solution $y = b\cdot a^{-1}$.

Next even if $b = 0$, then $x = 0, y = 0$ are unique solutions.

Thus, the existence of unique solutions of $a\cdot x = b$ and $y\cdot a = b$ follows necessarily from the fact that R is a division ring.

Conversely, let R be a ring and let the equations $a\cdot x = b$ and $y\cdot a = b$ have unique solutions in R.

In particular, taking $b = a$ there are unique solutions in R, of $a\cdot x = a$ and $y\cdot a = a$.

By definition this unique solution is unity 1. Thus, $1 \in R$.

Also existence of unique solutions of $a\cdot x = 1$ and $y\cdot a = 1$ $(a \neq 0)$ shows that $a^{-1} \in R$. Hence, R is a *division ring.*

Our next question is:

How are fields and integral domains related, apart from the fact that both are commutative rings with unity element?

The answer to this question will be given by the following theorem:

Theorem 8.5.7. *Every field is an integral domain but not every integral domain is a field.*

Proof. By definition, F is a field if it is a commutative ring with unity and is such that *every non-zero element of* F *has a multiplicative inverse in* F. We shall prove that F is an integral domain also, i.e., we shall prove that F *has no divisors of zero.*

Let there be two elements $a, b \in F$, so that $a\cdot b = 0$.

We shall prove that either $a = 0$ or $b = 0$.

If $a \neq 0$, then a has a multiplicative inverse $a^{-1} \in F$.

Then

$$\begin{aligned} a\cdot b = 0 \;&\Rightarrow\; a^{-1}\cdot(a\cdot b) = a^{-1}\cdot 0 \\ &\Rightarrow\; (a^{-1}\cdot a)\cdot b = 0 \quad [\because\ a^{-1}\cdot 0 = 0 \text{ and by associative law } a^{-1}\cdot(a\cdot b) = (a^{-1}\cdot a)\cdot b] \\ &\Rightarrow\; e\cdot b = 0 \ [e \text{ is the multiplicative identity of F.}] \\ &\Rightarrow\; b = 0 \ (\because\ e\cdot b = b) \end{aligned}$$

Thus, if $a \neq 0$, then $a\cdot b = 0 \Rightarrow b = 0$.

Similarly, if $b \neq 0$, then $a\cdot b = 0 \Rightarrow a = 0$.

Hence, F is an integral domain.

That the converse is not true is obvious, since, e.g., Z, the set of all integers is an integral domain but is not a field under the usual addition and multiplication. Every non-zero integer has not a multiplicative inverse in Z.

On the other hand it is true that

Theorem 8.5.8. *Every finite integral domain is a field.*

Proof. Let R be a finite integral domain and let

$$R - \{0\} = \{a_1, a_2, a_3, \ldots, a_n\}$$

be the n non-zero elements of R. They are all distinct.

Let a be any element of $R - \{0\}$.

Since $a \neq 0$ and $a_j \neq 0$ $(j = 1, 2, 3, \ldots, n)$ it follows that

$$a \cdot a_j \neq 0 \ (j = 1, 2, 3, \ldots, n).$$

Hence, if $S = \{a \cdot a_1, a \cdot a_2, a \cdot a_3, \ldots \cdots, a \cdot a_n\}$, then $S \subseteq R - \{0\}$,
But since $a \neq 0$,

$$a \cdot a_j = a \cdot a_k \Rightarrow a_j = a_k, \text{ using cancellation.}$$

Hence, the n displayed members of S are distinct and so

$$S = R - \{0\}.$$

Since the unity element e of R belongs to $R - \{0\}$, it follows that $e \in S$. Thus, there is an integer i such that $a \cdot a_i = e$ and so, by commutativity, $a \cdot a_i = a_i \cdot a = e$. Therefore, a_i is an inverse for a in R, i.e., $a \neq 0 \in R$ is invertible and so R is a field.

Note 2. *If we define an integral domain as a commutative ring with no zero divisor, then after proving that the elements of S are distinct we should justify that $\exists$ an element $e \in S$ s.t. $a \cdot e = a, \forall a \in S$.*

8.6 Further Solved Examples

▶ Example 8.6.1 *If Z be the set of all integers, show that $\{2Z, +, \bullet\}$ is a commutative ring with respect to usual addition and multiplication. Is it a ring with unity?*

[C.P. 1987, 1995, 1997, 2003]

[See Ex. 8.4.2: Worked out]

It has no multiplicative identity and as such it is not a ring with unity.

▶ Example 8.6.2 *Assuming that the set T of all real numbers of the form $a + b\sqrt{2}$ (a, b rationals), forms a ring with respect to the usual addition and multiplication, show that T is a field.*

[C.P. 1990, 2004]

[See Ex. 8.4.14: Worked out]

▶ Example 8.6.3 *Show that the set M of all 2×2 real matrices of the form $\begin{pmatrix} a & b \\ -b & a \end{pmatrix}$ is a ring under matrix addition and multiplication. Prove that M is a field.*

[C.P. 1982, 1993, 1996]

[See Ex. 8.4.11: Worked out]

▶ Example 8.6.4 *Prove that the set of all 2 × 2 real matrices of the form $\begin{pmatrix} a & b \\ 0 & c \end{pmatrix}$ is a subring of all 2 × 2 real matrices under matrix addition and multiplication.*

[C.P. 1998]

Solution: Let M be a set of all 2 × 2 real matrices. Let A and B be two square real matrices of order two. Then $A, B \in M$. We know that $A + B$ and AB both $\in M$. i.e., M is closed **under matrix addition and multiplication.** Both addition and multiplication obey **Associative Laws:**

$$A + (B + C) = (A + B) + C, \forall A, B, C \in M$$

$$\text{and} \quad A \cdot (B \cdot C) = (A \cdot B) \cdot C, \ \forall A, B, C \in M.$$

Addition of Matrices is **Commutative,**

$$\text{i.e.,}\quad A+B=B+A,\ \forall A,B\in M$$
$$\exists\, 0=\begin{pmatrix}0 & 0\\ 0 & 0\end{pmatrix}\in M$$

such that $0+A=A$, $\forall A\in M$, i.e., 0 is the additive identities.

Further multiplication of matrices is distributive with respect to addition.

$$\text{i.e.}\ \ \forall A,B,C \text{ we have } \begin{cases} A\cdot(B+C) = A\cdot B+A\cdot C \text{ (left distributive law)}\\ (B+C)\cdot A = B\cdot A+C\cdot A \text{ (right distributive law)}\end{cases}$$

$\therefore$ **M** is a **ring** with the given operations—addition of matrices and multiplication of matrices.

The square matrix $\begin{pmatrix}a & b\\ 0 & c\end{pmatrix}$ are special cases of the elements of M and the set N of such matrices also form a ring $N\subseteq M$ and hence **N** forms a subring of the ring M.

Question. Is M a commutative ring with unity element? Find the multiplicative identity element, if it exists. Does this ring possess zero divisors?

Answer. Multiplication of matrices is not, in general, a commutative composition.

$$\text{e.g., if } A = \begin{bmatrix}2 & 4\\ 3 & 5\end{bmatrix},\ B=\begin{bmatrix}1 & 2\\ 0 & 1\end{bmatrix},$$
$$\text{then } AB = \begin{bmatrix}2 & 8\\ 3 & 1\end{bmatrix} \text{ and } BA=\begin{bmatrix}8 & 14\\ 3 & 5\end{bmatrix},$$

Then $AB\neq BA$, so the ring of matrices does not form a commutative ring.

If I be the unit matrix of order 2×2, i.e., if $I=\begin{pmatrix}1 & 0\\ 0 & 1\end{pmatrix}$, then $I\in M$ and we have $AI=IA=A$ for all matrices; $A\in M$, i.e., I is the multiplicative identity.

The ring possesses zero divisor. For example if

$$A=\begin{bmatrix}0 & 1\\ 0 & 1\end{bmatrix},\ B=\begin{bmatrix}1 & 1\\ 0 & 0\end{bmatrix},\ \text{then } AB=\begin{bmatrix}0 & 0\\ 0 & 0\end{bmatrix}=0.$$

Then the product of two *non-zero elements* of the ring is equal the zero element of the ring.

▶ Example 8.6.5 *Prove that the commutative group of all real numbers of the form* $a+b\sqrt{2}$ $(a, b$ *rational) under usual addition is a field under usual multiplication.*

[C.P. 1998]

[See Ex. 8.4.14: Worked out]

▶ Example 8.6.6 *Prove that in a ring* $(\mathbf{R}, +, \bullet)$ *with unity element* 1,

$$(-1)\cdot a=-a,\ \forall a\in R.$$

Solution: See Theorem 8.5.3(ii). This is an important particular case.

▶ Example 8.6.7 *Considering the set of all matrices*

$$M = \left\{ \begin{pmatrix} a & b \\ 2b & a \end{pmatrix},\ a, b \textit{ rational} \right\}$$

as a ring under matrix addition and matrix multiplication show that M forms a field. **[C.P. 1999, 2005]**

Solution: Here {M, +, •} is given to be a ring under matrix addition (+) and matrix multiplication (•).

It is easy to see that

$$A = \begin{pmatrix} a & b \\ 2b & a \end{pmatrix},\ B = \begin{pmatrix} c & d \\ 2d & c \end{pmatrix};\ AB = BA$$

i.e., the ring is a commutation ring and also we see that $\begin{pmatrix} 1 & 0 \\ 0 & 1 \end{pmatrix}$ is the unity element.

Let $A = \begin{pmatrix} a & b \\ 2b & a \end{pmatrix}$ be a non-zero element of M. Then $(a, b) \neq (0, 0)$.

Det $A = a^2 - 2b^2 \neq 0$, since $(a, b) \neq (0, 0)$ and a, b are rational. Hence, A^{-1} exists,

$$A^{-1} = \frac{1}{a^2 - 2b^2} \begin{pmatrix} a & -b \\ -2b & a \end{pmatrix}.$$

∴ each non-zero element of M has an inverse so that {M, +, •} is a field.

Note. If a, b **are real numbers,** then the ring of matrices $\begin{pmatrix} a & b \\ 2b & a \end{pmatrix}$ **contains divisor of zero,**

$$\text{e.g.,} \begin{pmatrix} \sqrt{2} & 1 \\ 2 & \sqrt{2} \end{pmatrix} \begin{pmatrix} -\sqrt{2} & 1 \\ 2 & -\sqrt{2} \end{pmatrix} = \begin{pmatrix} 0 & 0 \\ 0 & 0 \end{pmatrix}.$$

Hence the set M of matrices $\begin{pmatrix} a & b \\ 2b & a \end{pmatrix}$ **where a, b are real numbers** cannot be an integral domain and hence cannot be a field (because every field is an integral domain). **[C.P. 2003]**

▶ Example 8.6.8 (a) *Give an example of a ring which is not a field.* **[C.P. 2000]**

(b) *Define a ring with unit element. Give an example of a ring without unity.* **[C.P. 1999]**

Solution: (a) Ring of integers (Z, +, •) forms a ring but not a field, since non-zero elements like 2, 3 have no inverses.

The ring of matrices

$$\left\{ \begin{pmatrix} a & b \\ 2b & a \end{pmatrix},\ a, b \in R \right\}$$

does not form a field. The non-zero matrix $\begin{pmatrix} \sqrt{2} & 1 \\ 2 & \sqrt{2} \end{pmatrix}$ has no inverse (det = 0) and as such this ring of matrices is not a field.

(b) The set {2Z, +, •} forms a commutative ring but it has no multiplicative identity and so it is not a ring with unity.

► Example 8.6.9 *Given that the set of all* 2×2 *matrices of the form* $\begin{pmatrix} x & y \\ 0 & z \end{pmatrix}$ *where* x, y, z *are integers forms a ring with respect to matrix addition and matrix multiplication. Is it a commutative ring? Justify your answer.* **[C.P. 2000]**

Answer. No.

► Example 8.6.10 *Show that the set* $\{a + b\sqrt{2}, a, b$ *are integers*$\}$ *forms a commutative ring with unity, with respect to ordinary addition and multiplication* (**Compare Ex. 8.6.2**) **[C.P. 2001]**

► Example 8.6.11 *Show that the ring of all end integers is not a field.* (*It has no unity element w.r.t. multiplication*)

► Example 8.6.12 (**Compare Ex. 8.6.7 above**) *Show that the ring of matrices*

$$S = \left\{ \begin{pmatrix} a & b \\ 2b & a \end{pmatrix}, \ a, b \in R \right\}$$

does not form a field. **[C.P. 2003]**

Solution: **Given:** S is a ring under matrix addition and matrix multiplication. It is also a ring with unity, $\begin{pmatrix} 1 & 0 \\ 0 & 1 \end{pmatrix}$ being the unity element.

Let $A = \begin{pmatrix} a & b \\ 2b & a \end{pmatrix}$ be a non-zero element of S.

Then

$$(a, b) \neq (0, 0) \cdot \det A = \begin{vmatrix} a & b \\ 2b & a \end{vmatrix} = a^2 - 2b^2.$$

Since a, b are real numbers there exist non-zero elements such then $a^2 - 2b^2 = 0$,

$$\text{e.g., } a = \sqrt{2}, b = 1 \Rightarrow a^2 - 2b^2 = 0$$

$\therefore$ the non-zero matrix $\begin{pmatrix} \sqrt{2} & 1 \\ 2 & \sqrt{2} \end{pmatrix}$ has no inverse.

$\therefore$ $\{S, +, \bullet\}$ is NOT a field.

Exercises on Chapter 8

Remember. (*if necessary, check*) : The set Q of all rational numbers, the set R of all real numbers and the set of all complex numbers form *fields*—called **number fields.**

A

1. Verify the truth of the following statements (All the relevant axioms are to be checked):

   (a) The set of all even integers (including 0) forms a commutative ring but it is not a ring with unity (operations are ordinary *addition* and *multiplication*). Can it form a field? **[C.P. 1981]**

(b) The set of all integers forms a commutative ring with unity (usual addition and multiplication).

(c) The set of all real numbers of the form $a + b\sqrt{2}$ where a and b are integers forms a commutative ring with unity; same assertion even if a and b are rational numbers (ordinary addition and multiplication).

[C.P. 1980]

2. Determine which of the following subsets of the set R of real numbers form rings under the usual addition and multiplication. If a subset does form a ring, determine whether it is an integral domain or a field:

(a) $S_1 = \{3m : m \in Z\}$;

(b) $S_2 = \{4k + 1 : k \in Z\}$;

(c) $S_3 = \left\{x + y\sqrt{2} : x, y \in Z\right\}$;

(d) $S_4 = \left\{x + y\sqrt{2} : x, y \in Q\right\}$;

(e) $S_5 = \left\{a + \sqrt[3]{2}\, b : a, b \in Z\right\}$

[Z = set of all integers, Q = set of all rational numbers]

3. Let S be a subset of Q (set of all rational numbers) defined by

$$S = \{a/7; a \in Z\}, \text{ Z being the set of all integers.}$$

Under the operations

(a) usual addition (+), and

(b) multiplication (•) defined by $\frac{a}{7} \cdot \frac{b}{7} = \frac{ab}{7}, \forall a, b \in Z$.

Prove that S forms a ring.

4. The set $\{0, 2, 4, 6, 8\}$ forms a field under addition and multiplication modulo 10 → prove or disprove.

5. Prove that the set $\{0, 1, 2, 3, 4, 5, 6, 7\}$ forms a commutative ring with unity under addition and multiplication modulo 8 but that it does not form an integral domain.

6. Determine which of the following sets of 2 × 2 matrices form rings under matrix addition and multiplication. If a set forms a ring, determine whether it is an integral domain or a field:

(a) $\left\{S_1 = \begin{bmatrix} x & 0 \\ 0 & x \end{bmatrix} : x \in Z\right\}$;

(b) $S_2 = \left\{\begin{bmatrix} x & x \\ 0 & 0 \end{bmatrix} : x \in Q\right\}$;

(c) $S_3 = \left\{\begin{bmatrix} x & x \\ 0 & x \end{bmatrix} : x \in Z\right\}$;

(d) $S_4 = \left\{ \begin{bmatrix} x & y \\ 4y & x \end{bmatrix} : x \in Q \right\}$.

[Z is the set of all integers and Q is the set of rational numbers.]

7. Operations $*$ and $\wedge$ are defined on the set Z of all integers by

$$\left. \begin{array}{rcl} a * b & = & a + b - 1 \\ a \wedge b & = & a + b - ab \end{array} \right\} \forall a, b \in Z,$$

where the $+, -$ and multiplication used on the right-hand side are the usual addition, subtraction and multiplication of integers.

Show that $\{Z, *, \wedge\}$ is a commutative ring with a unity element, the zero element and unity element being integers 1 and 0, respectively.

Show also that if Z be replaced by R (set of all real numbers) then $\{R, *, \wedge\}$ is a field.

8. The set of matrices

$$\left\{ \begin{bmatrix} a + ib & c + id \\ -c + id & a - ib \end{bmatrix} : a, b, c, d \in R \right\}$$

forms a skew field (i.e., a division ring) but not a field under matrix addition and multiplication—prove this statement.

9. Prove that the field of rational numbers is a subfield of the field of real numbers under the usual addition and multiplication.

10. Prove that the set of real numbers of the form $a + b\sqrt{2}$, where a and b are rational numbers is a subfield of the field of real numbers.

B

1. If $a, b \in F$ (where F is a field under $+, \bullet$) and $a \neq 0$, then prove that $\exists$ a *unique solution* of $a \cdot x = b$ in F.

2. Prove that a ring R has no divisors of zero if and only if the cancellation laws hold.

3. If R is a ring, then $\forall a, b, c \in R$, prove that

   (a) $-(a + b) = -a - b$;

   (b) $a - (b + c) = (a - b) - c$;

   (c) $a \cdot (b - c) = a \cdot b - a \cdot c$;

   (d) $(b - c) \cdot a = b \cdot c - c \cdot a$.

4. If $\{R, +, \bullet\}$ be a ring and S is a non-empty subset of R, then prove that $\{S, +, \bullet\}$ is a **subring** if and only if $a - b \in S$ and $a \cdot b \in S$ whenever $a, b \in S$.

5. (a) Prove that in a ring R each of the equations

$$a + x = b \text{ and } y + a = b \text{ have unique solutions.}$$

(b) If in a ring R with unity, x has a multiplicative inverse prove that

$$y \in R, xy = 0 \Rightarrow y = 0.$$ [C.P. 1982]

6. Prove that the intersection of two subrings of a ring R is a subring of R.

7. Prove that the ring of residue classes modulo k (k is a composite number) has divisors of zero.

8. Prove that a commutative ring R is a field if and only if for $a, b \in R$ $(a \neq 0)$ there exists unique $x \in R$ such that $a \cdot x = b$.

[Compare **Theorem 8.5.6**]

9. Prove that the set S of all 2×2 matrices of the form $\begin{pmatrix} a & 0 \\ 0 & b \end{pmatrix}$, with a and b integers, forms a subring R of all 2×2 matrices having integers as elements.

Hints and Answers

A

2. (a) a ring; (b) not a ring; (c) an integral domain; (d) a field; (e) not a ring.

6. (a) an integral domain; (b) a field; (c) not a ring; (d) a ring.

B

3. (a) $(a+b)+(-a-b)=0$; prove this; hence the result.

(b) $$\begin{aligned} a-(b+c) &= a+\{-(b+c)\} = a+\{(-b)+(-c)\} \\ &= \{a+(-b)\}+(-c) = (a-b)-c. \end{aligned}$$

(c) $$\begin{aligned} a\cdot(b-c) &= a\cdot\{b+(-c)\} = a\cdot b + a\cdot(-c) \\ &= a\cdot b + \{-(a\cdot c)\} = a\cdot b - a\cdot c. \end{aligned}$$

4. Since S is a subring of R, if $a, b \in S$, then $-b \in S$ and $a-b \in S$; also $a\cdot b \in S$. Conversely, if $a-b \in S$, then in particular $a-a \in S$, i.e., $0 \in S$. If $a \in S$, then $0-a \in S$, i.e., $-a \in S$. Also for $a, b \in S$, $a+b = a-(-b) \in S$ ($\because\ -b \in S$). Thus, S is closed under addition. **Comm.** and **Ass.** properties hold in S ($\because$ they hold in the superset R). Thus S is a commutative group with regard to +. Also $a\cdot b \in S \Rightarrow \{S, \bullet\}$ is a semigroup since **Ass.** Law holds in the superset R (and hence in S). **Distributive** property holds in R and hence in S.

Chapter 9

Vector Space Over a Field

Vector Spaces	
1.	Concept of Vector Space over a Field: **Examples.**
2.	General Properties of Vector Spaces.
3.	Linear dependence and independence of a finite set of vectors.
4.	Vector Subspaces.
5.	Concepts of Generators and Basis of a finite dimensional vector space.
6.	**Problems** only on formation of basis of a vector space.

9.1 Introduction

We have, in the previous chapter, introduced algebraic structures like Groups, Rings, Integral Domains and Fields. Now we shall introduce another very important algebraic structure called in **Vector Space** (or a **Linear Space**) **V** over a field F, denoted by **V**(F).

It is advisable to recall the definition of a Field:

A field is an algebraic structure consisting of a non-empty set F together with two binary operations–addition (+) and multiplication (·) satisfying the following axioms:

Additive Composition:

A_1. Under addition **F is closed** (i.e., $\forall a, b \in F, a + b \in F$).

A_2. Addition is **Associative** (i.e., $\forall a, b, c \in F, a + (b + c) = (a + b) + c$.

A_3. $\exists$ an element 0 (Zero) $\in F$ s.t. $a + 0 = 0 + a = a, \forall a \in F$. 0 is called **additive identity of F.**

A_4. For every element $a \in F$, there exists an element $-a \in F$ (called **additive inverse of a**) s.t. $a + (-a) = (-a) + a = 0$.

A_5. Addition is **commutative**, i.e., $\forall a, b \in F, a + b = b + a$.

In other words, under addition $(+)$, F is an **Abelian Group.**

Multiplicative Composition:

M_1. Under multiplication $(\cdot)$ F **is closed** (i.e., $\forall a, b \in F, a \cdot b \in F$)
[Instead of $a \cdot b$ we may simply write ab.]

M_2. Multiplication is **associative** (i.e., $\forall a, b, c \in F, a \cdot (b \cdot c) = (a \cdot b) \cdot c$).

M_3. $\exists$ an element 1 (Unity) $\in F$, called **Multiplicative Identity**

$$\text{s.t. } \forall a \in F,\ a \cdot 1 = 1 \cdot a = a.$$

M_4. For every non-zero element $a (a \neq 0) \in F$, there exists an element $a^{-1} \in F$ (called **multiplicative inverse of a**) s.t. $a \cdot a^{-1} = a^{-1} \cdot a = 1$.

M_5. Multiplication is **commutative** (i.e., $\forall a, b \in F, a \cdot b = b \cdot a$).

D. Multiplication is **distributive** with respect to addition.

$$\text{(i.e., } \forall a, b, c \in F,\ a \cdot (b + c) = a \cdot b + a \cdot c.$$

Subfield. Let F be a field. A non-empty subset K of F is called a subfield of F if K is closed w.r.t. the operations of addition and multiplication in F, and K itself is a field for these operations.

▶ Example 9.1.1 *The set* **Q** *of all rational numbers, the set* **R** *of all real numbers and the set* **C** *of all complex numbers are all examples of Fields. In* **Q** *and* **R**, *0 is the additive identity element and 1 is the multiplicative identity but in* C, $0 + i0$ *and* $1 + i0$ *are the additive and the multiplicative identities respectively.*

Since $\mathbf{Q} \subset \mathbf{R}$ *and is a field the field of rational numbers is a subfield of the field of real numbers.*

Since $\mathbf{R} \subset \mathbf{C}$ *and* **R** *is a field therefore, the field of real numbers is a subfield of the field of complex numbers.*

Also the set of numbers of the form $a + b\sqrt{2}$ *where* a *and* b *are rational numbers, is also a field, a subfield of the field of real numbers.*

The following properties hold in a field: If F is a field and $a, b, c \in F$, then

(i) $a + b = a + c \Rightarrow b = c$

(ii) $a \cdot 0 = 0 \cdot a = 0$

(iii) $a \cdot (-b) = -(a \cdot b)$

(iv) $-(-a) = a$

(v) $(-a) \cdot (-b) = a \cdot b$

(vi) $a \cdot b = 0 \Rightarrow a = 0$ or $b = 0$ (or both $= 0$) (Field has no zero divisor)

(vii) $a \neq 0, a \cdot b = a \cdot c \Rightarrow b = c$.

9.2 Vector Spaces: Definitions and Examples

Definition 1. *Let* $\{F, +, \cdot\}$ *be a field. The elements of* F *will be called* **Scalars.**

Let **V** be non-empty set whose elements will be called **Vectors.**
Then **V** is a **vector space** over the field F if the following axioms hold:

V_1. Additive axioms

A binary operation $\oplus$, called *vector addition,* is defined in **V** and under this operation, **V** is an Abelian group, i.e.,

(i) **V** is closed under $\oplus$, i.e., $\forall \alpha, \beta \in V, (\alpha \oplus \beta) \in V$.

(ii) Vector addition is Associative, i.e., $\forall \alpha, \beta, \gamma \in V, \alpha \oplus (\beta \oplus \gamma) = (\alpha \oplus \beta) \oplus \gamma$.

(iii) $\exists$ an element $\theta \in V$ such that $\forall \alpha \in V$

$$\alpha \oplus \theta = \theta \oplus \alpha = \alpha.$$

This element $\theta \in V$ is called the *zero vector* or, *null vector.*

(iv) For every vector $\alpha \in V$, there exists a vector $(-\alpha) \in V$ such that

$$\alpha \oplus (-\alpha) = (-\alpha) \oplus \alpha = \theta.$$

This element $(-\alpha) \in V$ is called the *additive inverse* of α (or *negative of* α).

(v) $\forall \alpha, \beta \in V,\ \alpha \oplus \beta = \beta \oplus \alpha$ *(Commutative Property).*

V_2. Scalar Multiplication axioms

Another binary operation, called *Scalar Multiplication,* is defined thus: For each scalar $a \in F$ and for any vector $\alpha \in V$ there corresponds a unique vector $a\alpha \in V$. Instead of $a\alpha$ we may write αa. This scalar multiplication obeys the following axioms:

(i) **V** is closed under scalar multiplication.

(ii) $\forall a \in F$ and for any two vectors $\alpha, \beta \in V, a(\alpha \oplus \beta) = a\alpha \oplus a\beta$.

(iii) $\forall a, b \in F$ and for any vector $\alpha \in V, (a + b)\alpha = a\alpha \oplus b\alpha$.

(iv) $\forall a, b \in F$ and for any vector $\alpha \in V, (ab)\alpha = a(b\alpha)$.

(v) $1\alpha = \alpha, \forall \alpha \in V$ and 1 is the multiplicative identity of the field F.

Observations.

1. The vector space **V** over the field F is usually denoted by $V(F)$. But if F is understood from the context we may simply say that **V** *is a vector space.*

2. If F is the field R of real numbers, then **V** is a *real vector space* or it is a vector space over R. Similarly, if F is the field C of complex numbers we speak of *complex vector space.* In this text we shall be mainly concerned with real **vector spaces.**

3. Note that we have used the symbol '$\oplus$' for *vector addition in* V and the usual symbol '+' for *addition in* F. Often we may use the same symbol + for **vector addition** and also for **scalar addition.** See **Ex. 9.2.3.**

   No confusion should arise because it will be clear from the context which addition we mean.

4. Since V is an abelian group, therefore, properties of an abelian group will hold in V. In particular, remember that in V,

   (a) **Left Cancellation Law** and **Right Cancellation Law** both hold:

   $$\alpha \oplus \beta = \alpha \oplus \gamma \Rightarrow \beta = \gamma \text{ and } \beta \oplus \alpha = \gamma \oplus \alpha \Rightarrow \beta = \gamma.$$

   (b) $\alpha \oplus \beta = \theta \Rightarrow \alpha = -\beta$ and $\beta = -\alpha$.

   (c) $\alpha - \beta$ will mean $\alpha \oplus (-\beta)$.

   (d) $-(\alpha \oplus \beta) = -\alpha - \beta$.

   (e) $-(-\alpha) = \alpha$.

   (f) $\alpha \oplus \beta = \alpha \Rightarrow \beta = \theta$.

   (g) The identity element θ in V is unique. We may also write θ_V instead of θ to stress upon the fact that the null vector $\in$ V.

   (h) The additive inverse of each vector is unique.

   (i) $\alpha \oplus \beta = \gamma \Rightarrow \alpha \oplus \beta - \gamma = \theta$.

Examples of Important Vector Spaces: For establishing a vector space students must verify all the axioms in details as shown below:

▶ Example 9.2.1 ***The vector space V_n of all ordered n-tuples over the real field* R.**

Solution: An element $\alpha = (x_1, x_2, x_3, \ldots, x_n)$ where x_i's are all real numbers, is called an *n-tuple over* R. In particular, $\alpha = (a, b)$ where a and b are real numbers, is called *an ordered pair* of real numbers. The totality of all such ordered pairs will form what we call the system of complex numbers.

We shall prove that the set of all ordered n-tuples of real numbers (denoted by V_n) will form **a real vector space.** Each ordered n-tuple will be called a **vector.**

We have,

$$V_n = \{(x_1, x_2, \ldots, x_n) : x_1, x_2, x_3, \ldots, x_n \in R\}.$$

We first define **Vector addition, Scalar multiplication** and equality of two elements (or Vectors) in $\mathbf{V_n}$.

Let $\alpha = (x_1, x_2, \ldots, x_n)$ and $\beta = (y_1, y_2, \ldots, y_n)$ be two vectors in $\mathbf{V_n}$. We define

(i) *Equality of two vectors:* $\alpha = \beta$ if and only if

$$x_1 = y_1,\ x_2 = y_2, \ldots, x_n = y_n.$$

(ii) *Addition of two vectors:* $\alpha \oplus \beta$ is defined by the n-tuple

$$(x_1 + y_1, x_2 + y_2, \ldots, x_n + y_n).$$

(iii) *Scalar multiplication:* For any real number a (Scalar a) we define

$$a\alpha = (ax_1, ax_2, \ldots, ax_n).$$

Now we check whether all the postulates of a vector space are satisfied in $\mathbf{V}_n$.

$\mathbf{V}_1$. (i) If $\alpha, \beta \in \mathbf{V}_n$, then $\alpha \oplus \beta \in \mathbf{V}_n$

[Since $x_1 + y_1, x_2 + y_2, \ldots, x_n + y_n$ are all real numbers, the n-tuple

$$(x_1 + y_1, x_2 + y_2, \ldots, x_n + y_n) \in \mathbf{V}_n].$$

(ii) If $\alpha = (x_1, x_2, \ldots, x_n), \beta = (y_1, y_2, \ldots, y_n)$ and $\gamma = (z_1, z_2, \ldots, z_n) \in \mathbf{V}_n$, clearly,

$$\begin{aligned}
\alpha \oplus (\beta \oplus \gamma) &= (x_1, x_2, \ldots, x_n) \oplus (y_1 + z_1, y_2 + z_2, \ldots, y_n + z_n) \\
&= \{x_1 + (y_1 + z_1), x_2 + (y_2 + z_2), \ldots, x_n + (y_n + z_n)\} \\
&= \{(x_1 + y_1) + z_1, (x_2 + y_2) + z_2, \ldots, (x_n + y_n) + z_n\} \\
&\qquad \text{(using associative property of addition in R)} \\
&= (x_1 + y_1, x_2 + y_2, \ldots, x_n + y_n) + (z_1, z_2, \ldots, z_n) \\
&= (\alpha \oplus \beta) \oplus \gamma
\end{aligned}$$

This proves the associative property of addition of vectors in $\mathbf{V}_n$.

(iii) $\exists$ an element $\theta = (0, 0, \ldots, 0) \in \mathbf{V}_n$ called *Null Vector* in $\mathbf{V}_n$.

If $\alpha = (x_1, x_2, \ldots, x_n), \in \mathbf{V}_n$, then it is easy to see that $\alpha \oplus \theta = \theta \oplus \alpha = \alpha$, i.e., θ is the *additive identity* in V_n.

(iv) Given, $\alpha = (x_1, x_2, \ldots, x_n) \in \mathbf{V}_n$, we see that $\exists$ an element

$$(-x_1, -x_2, \ldots, -x_n) \in \mathbf{V}_n$$

such that $\alpha \oplus (-x_1, -x_2, \ldots, -x_n) = (-x_1, -x_2, \ldots, -x_n) \oplus \alpha = (0, 0, 0 \cdots 0)$, i.e., $(-x_1, -x_2, \ldots, -x_n)$ is the *additive inverse* of $\alpha (= -\alpha)$.

(v)
$$\begin{aligned}
\alpha \oplus \beta &= (x_1 + y_1, x_2 + y_2, \ldots, x_n + y_n) \\
&= (y_1 + x_1, y_2 + x_2, \ldots, y_n + x_n) \\
&\qquad (\because \text{commutative property of addition holds in R}) \\
&= (y_1, y_2, \ldots, y_n) \oplus (x_1, x_2, \ldots, x_n) = \beta + \alpha.
\end{aligned}$$

Thus Commutative Law for vector addition holds in $\mathbf{V}_n$.

Thus $\mathbf{V}_n$ is an additive abelian group.

$\mathbf{V}_2$. (i) If $a \in R, \alpha = (x_1, x_2, \ldots x_n) \in R$, then according to our definition,

$$a\alpha = (ax_1, ax_2, \ldots, ax_n).$$

But clearly $a\alpha \in \mathbf{V}_n$ because each $ax_i \in R$.

(ii) Let $a \in R$ and $\alpha = (x_1, x_2, \ldots, x_n)$ and $\beta = (y_1, y_2, \ldots, y_n) \in V_n$.
Then $\alpha \oplus \beta = (x_1 + y_1, x_2 + y_2, \ldots, x_n + y_n)$

$$\begin{aligned}
\text{and } a(\alpha \oplus \beta) &= a(x_1 + y_1, x_2 + y_2, \ldots, x_n + y_n) \\
&= \{a(x_1 + y_1), a(x_2 + y_2), \ldots, a(x_n + y_n)\} \\
&= (ax_1 + ay_1, ax_2 + ay_2, \ldots, ax_n + ay_n) \\
&\qquad [\because \text{ in } R, a(x_i + y_i) = ax_i + ay_i] \\
&= (ax_1, ax_2, \ldots, ax_n) \oplus (ay_1, ay_2, \ldots, ay_n) \\
&= a(x_1, x_2, \ldots, x_n) \oplus a(y_1, y_2, \ldots, y_n) \\
&= a\alpha \oplus a\beta.
\end{aligned}$$

(iii) If $a, b \in R$ and $\alpha = (x_1, x_2, \ldots, x_n) \in V_n$, then $(a + b) \in R$ and

$$\begin{aligned}
(a + b)\alpha &= (a + b)(x_1, x_2, \ldots, x_n) \\
&= ([a + b]x_1, [a + b]x_2, \ldots, [a + b]x_n) \\
&= [ax_1 + bx_1, ax_2 + bx_2, \ldots, ax_n + bx_n] \\
&= (ax_1, ax_2, \ldots, ax_n) \oplus (bx_1, bx_2, \ldots, bx_n) \\
&= a(x_1, x_2, \ldots, x_n) \oplus b(x_1, x_2, \ldots, x_n) \\
&= a\alpha \oplus b\alpha.
\end{aligned}$$

(iv) Let $a, b \in R$ and $\alpha = (x_1, x_2, \ldots, x_n) \in V_n$.
Then in R the product $ab \in R$.

$$\begin{aligned}
\therefore ab\alpha &= ab(x_1, x_2, \ldots, x_n) \\
&= (abx_1, abx_2, \ldots, abx_n) \\
&= \{a(bx_1), a(bx_2), \ldots, a(bx_n)\} \\
&= a(bx_1, bx_2, \ldots, bx_n) \\
&= a(b\alpha).
\end{aligned}$$

(v) Clearly, $1\alpha = 1(x_1, x_2, \ldots, x_n) = (x_1, x_2, \ldots, x_n) = \alpha$.
Thus all the axioms V_1 and V_2 of a vector space are satisfied in V_n over R; i.e., V_n is a real vector space.

Note 1. *We shall very often write* R^n *in place of* V_n. R^n *can be considered as the Cartesian product* $R \times R \times \cdots \times R$.

Thus note $R^2 = \{(a, b) : a, b \in R\}, R^3 = \{(a, b, c) : a, b, c \in R\}$ are real vector spaces.

Note that the field may be any arbitrary field F (not always the field of real numbers). We then write $V_n(F)$, the vector space of all ordered n-tuples $(a_1, a_2, \ldots, a_n)$ where a's $\in F$.

▶ Example 9.2.2 *V is the set of all* $m \times n$ *matrices with their elements as real numbers and F is the real field R. Prove that* $V(R)$ *is a real vector space.*

Here the vectors are all $m \times n$ matrices.

Vector addition $\oplus$ is the usual addition of two $m \times n$ matrices.

Scalar multiplication is the multiplication of an $m \times n$ matrix by a real number.

Null matrix of order $m \times n$ is the null vector θ and is the additive identity here.

Now it is easy to check that **V** is an abelian group under ordinary addition of two $m \times n$ matrices. For scalar multiplication also it is easy to check the axioms (i)—(v) under **V2.**

Observations: Suppose, if the elements of matrices were **rational numbers** and F is the field R of real numbers then **V will not be a vector space over R.** See that if the scalar is $\sqrt{3}$ and α is an $m \times n$ matrix whose elements are rational numbers then $\sqrt{3}\alpha \notin V$, i.e., under *scalar multiplication* **V** *is not closed;* as such $\mathbf{V}(R)$ is not a vector space in this case.

▶ Example 9.2.3 *Prove that the set of all polynomials over any field* F *is a vector space.*

Solution: Let F[x] denote the set of all polynomials in x over a field F. F may be a complex field or a real field or it may be any arbitrary field.

Then F[x] is a vector space over the field F with respect to addition of two polynomials (*vector addition* $\oplus$) and multiplication of a polynomial by a constant polynomial (i.e., by an element of F) (*Scalar multiplication*).

Let
$$\begin{aligned} f(x) &= a_0 + a_1x + a_2x^2 + \cdots \\ g(x) &= b_0 + b_1x + b_2x^2 + \cdots \\ h(x) &= c_0 + c_1x + c_2x^2 + \cdots \end{aligned}$$
be any three polynomials of F[x], the coefficients belonging to F.

Each one is **a vector** $\in$ F[x].

We define:

Equality: $f(x) = g(x)$ if and only if $a_i = b_i$ for all $i = 0, 1, 2, 3, \ldots$

Addition: $f(x) \oplus g(x) = (a_0 + b_0) + (a_1 + b_1)x + (a_2 + b_2)x^2 + \cdots$

Scalar multiplication: If $k \in F$ then $kf(x) = ka_0 + (ka_1)x + (ka_2)x^2 + \cdots$

Now the axiom $\mathbf{V}_1$ and $\mathbf{V}_2$ can be verified as shown below:

$\mathbf{V}_1$. Additive axioms

> Note that hence forward in all discussions that follow we have used the same symbol + both vector **addition** and **scalar addition.** the context will make it clear which addition we actually mean.

(i) Since $a_0 + b_0, a_1 + b_1, a_2 + b_2, \ldots$ are all elements of F,

therefore, $f(x) + g(x) \in F[x]$ (*Closure Property*).

(ii) $[f(x) + g(x)] + h(x)$
$$= [(a_0 + b_0) + (a_1 + b_1)x + (a_2 + b_2)x^2 + \cdots] + c_0 + c_1x + c_2x^2 + \cdots$$
$$= \{(a_0 + b_0) + c_0\} + \{(a_1 + b_1) + c_1\}x + \{(a_2 + b_2) + c_2\}x^2 + \cdots$$
$$= \{a_0 + (b_0 + c_0)\} + \{a_1 + (b_1 + c_1)\}x + \{a_2 + (b_2 + c_2)\}x^2 + \cdots$$
(Using associativity of addition in F)
$$= a_0 + a_1x + a_2x^2 + \cdots + (b_0 + c_0) + (b_1 + c_1)x + (b_2 + c_2)x^2 + \cdots$$
$$= f(x) + [g(x) + h(x)]$$ (*Associative Property*)

(iii) *Existence of identity* in F[x]· $\exists$ a polynomial $\theta(x) \in F[x]$, where

$$\theta(x) = 0 + 0x + 0x^2 + \cdots$$

Clearly, $f(x) + \theta(x) = \theta(x) + f(x) = f(x)$, for any polynomial $f(x)$ in F[x], i.e., $\theta(x)$ is the identity element of F[x].

(iv) **Existence of Additive Inverse:**

Let $f(x) = a_0 + a_1x + a_2x^2 + \cdots \in F[x]$

Then $-f(x) = -a_0 - a_1x - a_2x^2 + \cdots$ also $\in F(x)$

and $f(x) + [-f(x)] = 0 + 0x + 0x^2 + \cdots = \theta(x)$

and $[-f(x)] + f(x)$ is also equal to $\theta(x)$.

$\therefore -f(x)$ is the additive inverse of $f(x)$ in F[x].

(v)
$$\begin{aligned} f(x) + g(x) &= (a_0 + b_0) + (a_1 + b_1)x + (a_2 + b_2)x^2 + \cdots \\ &= (b_0 + a_0) + (b_1 + a_1)x + (b_2 + a_2)x^2 + \cdots \\ &\quad \text{(Using Commutative law of addition in F)} \\ &= (b_0 + b_1x + b_2x^2 + \cdots) + (a_0 + a_1x + a_2x^2 + \cdots) \\ &= g(x) + f(x) \ (\textit{Commutative Property}). \end{aligned}$$

Thus F[x] is an **abelian group** with respect to addition of polynomials.

V_2. Scalar Multiplication axioms

(i) If $k \in F$, we have defined for any polynomial $f(x) \in F[x]$,

$$kf(x) = ka_0 + (ka_1)x + (ka_2)x^2 + \cdots (a_is \in F),$$

since $ka_0, ka_1, ka_2, \ldots$ are all elements of F, therefore, $kf(x) \in F[x]$, i.e., F[x] is closed under scalar multiplication.

(ii) If $k \in F$, then $\forall f(x), g(x) \in F[x]$,

$$\begin{aligned} k[f(x) + g(x)] &= k[(a_0 + b_0) + (a_1 + b_1)x + (a_2 + b_2)x^2 + \cdots] \\ &= k(a_0 + b_0) + k(a_1 + b_1)x + k(a_2 + b_2)x^2 + \cdots \\ &= (ka_0 + kb_0) + (ka_1 + kb_1)x + (ka_2 + kb_2)x^2 + \cdots \\ &\quad \text{(Distributive Law of F)} \\ &= (ka_0 + (ka_1)x + (ka_2)x^2 + \cdots) \\ &\quad + (kb_0 + (kb_1)x + (kb_2)x^2 + \cdots) \\ &= kf(x) + kg(x). \end{aligned}$$

(iii) If $k_1, k_2 \in F$, then $\forall f(x) \in F[x]$,

$$\begin{aligned} (k_1 + k_2)f(x) &= (k_1 + k_2)a_0 + (k_1 + k_2)a_1x + (k_1 + k_2)a_2x^2 + \cdots \\ &= (k_1a_0 + k_2a_0) + (k_1a_1 + k_2a_1)x + (k_1a_2 + k_2a_2)x^2 + \cdots \\ &= (k_1a_0 + k_1a_1x + k_2a_2x^2 + \cdots) \\ &\quad + (k_2a_0 + k_2a_1x + k_2a_2x^2 + \cdots) \\ &= k_1f(x) + k_2g(x). \end{aligned}$$

(iv) If $k_1, k_2 \in F$, then $\forall f(x) \in F[x]$

$$\begin{aligned} (k_1k_2)f(x) &= (k_1k_2)a_0 + (k_1k_2)a_1x + (k_1k_2)a_2x^2 + (k_1k_2)a_3x^3 + \cdots \\ &= k_1[k_2a_0 + k_2a_1x + k_2a_2x^2 + k_2a_3x^3 + \cdots] \\ &= k_1[k_2f(x)]. \end{aligned}$$

(v) If 1 is the multiplication identity of F and $f(x) \in F(x)$, then

$$\begin{aligned} 1f(x) &= (1a_0) + (1a_1)x + (1a_2)x^2 + \cdots \\ &= a_0 + a_1x + a_2x^2 + \cdots \\ &= f(x). \end{aligned}$$

Hence, F[x] is a vector space over the field F.

► Example 9.2.4 *Let S be any non-empty set and F be any arbitrary field. Let V be the set of all mappings from S to F :*

$$V = \{f : f : S \to F\}.$$

The elements of S will be called vectors and those of F are scalars. Vector Addition and Scalar Multiplication are defined thus:

$$\begin{aligned} (f+g)(x) &= f(x) + g(x), \text{ where } x \in S \text{ (Vector Addition)} \\ (cf)(x) &= cf(x), \text{ where } x \in S \text{ (Scalar Multiplication)} \end{aligned}$$

where f and g are vectors $\in$ **V** *and C* **is a Scalar** $\in$ F.
Then, prove that **V** *is a vector space over the field* F.

Solution: We have $\forall x \in S, (f+g)x = f(x) + g(x)$.
Since $f(x), g(x)$ both belong to the field F, therefore, their sum $\in$ F.
Thus $f + g$ is also a mapping from S to F, i.e., $f + g \in \mathbf{V}\ \forall f, g \in \mathbf{V}$.

$\mathbf{V_1}$. (i) That is **V** is *closed under addition.*

(ii) *Addition is associative:*

$$\begin{aligned} [(f+g)+h](x) &= (f+g)(x) + h(x) \text{ [By def.]} \\ &= [f(x)+g(x)] + h(x) \text{ [By def.]} \\ &= f(x) + [g(x)+h(x)] \\ &\quad \text{(Since } f(x), g(x), h(x) \in F \text{ and addition is associative in F.)} \\ &= f(x) + (g+h)(x) \\ &= [f+(g+h)](x) \\ \therefore\ (f+g)+h &= f+(g+h) \end{aligned}$$

(iii) *Existence of additive identity:*

We define a mapping $\theta : S \to F$ such that $\theta(x) = 0\ \forall x \in S$, where 0 is the additive identity of the field F. We shall call θ, the **Zero mapping.**

We have $(f+\theta)(x) = f(x) + \theta(x) = f(x) + 0 = f(x)$

$\therefore$ $f + \theta = f$. Similarly, $\theta + f = f$, i.e., θ is the additive identity of V.

(iv) *Existence of additive inverse:*

For any element $f \in V$, another element $-f \in V$ is difined by

$$(-f)(x) = -[f(x)], \forall x \in S.$$

We have, $[f + (-f)](x) = f(x) + (-f)(x)$
$= f(x) + [-f(x)]$
$= f(x) - f(x) = \theta$, i.e., $f + (-f) = \theta$.

Similarly, $(-f) + f = \theta$, i.e., $-f$ is the additive inverse of f in V.

(v) *Commutative Property:*

We have, $(f + g)(x) = f(x) + g(x)$
$= g(x) + f(x)$ ($\because$ in F addition is commutative)
$= (g + f)x$
$\therefore f + g = g + f$.

Thus V is an abelian group with respect to addition.

V_2. If $c \in F$ and $f \in V$, then our definition of scalar multiplication is

$$(cf)(x) = cf(x).$$

Then $c \in F$ and $f(x) \in F$, therefore, $cf(x) \in F$,
i.e., cf is a mapping from S to F or $cf \in V$, for all $f \in V$ and for all $c \in F$.

(i) V is closed under scalar multiplication.

(ii) If $c \in F$ and $f, g \in V$,

$$[c(f + g)]x = c[f + g](x) = c[f(x) + g(x)] = cf(x) + cg(x)$$
$$= (cf)(x) + (cg)(x) = [cf + cg](x),$$
$$\text{i.e., } c(f + g) = cf + cg.$$

(iii) If $c_1, c_2 \in F$ and $f \in V$, then

$$[(c_1 + c_2)f](x) = (c_1 + c_2)f(x) = c_1 f(x) + c_2 f(x)$$
$$= (c_1 f)(x) + c_2 f(x) = (c_1 f + c_2 f)(x)$$
$$\therefore (c_1 + c_2)f = c_1 f + c_2 f.$$

(iv) If $c_1, c_2 \in F$ and $f \in V$, then it is easy to check $(c_1 c_2)f = c_1(c_2 f)$.

(v) If 1 is the multiplicative identity of F and $f \in V$, then

$$(1f)(x) = 1f(x) = f(x)$$
$$\therefore 1f = f.$$

$\therefore$ all the axioms for scalar multiplication are satisfied.

Hence, V is a vector space over F.

▶ Example 9.2.5 *The vector space of all real-valued continuous functions defined in some interval* $[a, b]$.

Solution: Let V denote the set of all real valued continuous functions defined on $[a, b]$. Then V is a vector space over the field R or real numbers with

Vector addition defined by $(f + g)(x) = f(x) + g(x), \forall x \in [a, b]$ and $\forall f, g \in V$.

Scalar multiplication defined by $(\alpha f)(x) = \alpha f(x)\ \forall \alpha \in R$ and $\forall f \in V$.

Note that if f and g are two real valued continuous function in $[a, b]$, then $f + g$ is also a real valued continuous function in $[a, b]$ and αf is also a real valued continuous function in $[a, b]$, where α is any real number. Thus under vector addition and scalar multiplication defined as above V is closed. Now verify that V is an abelian group under vector addition and the following axioms of scalar multiplication hold: (Proceed as in the previous examples.)

(i) If $\alpha \in R$ and $f, g \in V$, then $\alpha(f + g) = \alpha f + \alpha g$.

(ii) If $\alpha, \beta \in R$ and $f \in V$, then $(\alpha + \beta)f = \alpha f + \beta f$.

(iii) If $\alpha, \beta \in R$ and $f \in V$, then $(\alpha\beta)f = \alpha(\beta f)$.

(iv) Also $1f = f$ where 1 is the multiplicative identity of R and f is any element of V.

Hence, V is a vector space over R.

▶ Example 9.2.6 *Let* G *be a field and* F *is a subfield of* G. *Then prove that* G *itself is a vector space over* F.

Solution: Since G is a field, $(G, +)$ is an abelian group. The elements of G are *Vectors* and the elements of F are *Scalars* in this case. Scalar multiplication is multiplication of elements of F by elements of G and they are all defined. If 1 is the multiplicative identity of G, then 1 is also the multiplicative identity of F (subfield). Other axioms of scalar multiplication can be easily verified. Hence, G becomes a vector space over F.

Note 2. *Let* **C** *be the field of complex numbers and* **R**, *the field of real numbers which is a subfield of* **C**.

Then **C** is a vector space over **R**.

But see that **R** is not a vector space over **C**.

(In fact **R** is not closed under scalar multiplication, e.g., $5 \in$ **R** and $2 + 3i \in$ **C**, $(2 + 3i)5 \notin$ **R**.)

▶ Example 9.2.7 *Let* $V = \mathbf{R}^2 = \{(x, y) : x, y \in \mathbf{R}\}$ *and let* F *be the field of real numbers, i.e.,* $F = \mathbf{R}$. *Show that*

(i) *if we define*

Vector addition: $(x_1, y_1) + (x_2 + y_2) = (x_1 + x_2, y_1 + y_2)$

and *Scalar multiplication:* $c(x, y) = (cx, y)$,

then **V** *is* NOT *a vector space over* F.

But (ii) *if we define*

Vector addition: $(x_1, y_1) + (x_2, y_2) = (x_1 + x_2, y_1 + y_2)$

and *Scalar multiplication:* $c(x, y) = (cx, cy)$,

then **V** *is a vector space over* F.

Or, (iii) *if we define*

Vector addition: $(x_1, y_1) + (x_2, y_2) = (3y_1 + 3y_2, -x_1 - x_2)$

and *Scalar multiplication:* $c(x, y) = (3cy, -cx)$,

then **V** *is* NOT *a vector space over* F.

Solution:

(i) See that in this case the axiom $(c + d)(x, y) \neq c(x, y) + d(x, y)$.

$$\text{L.H.S.} = \{(c + d)x, y\} = (cx + dx, y)$$
$$\text{R.H.S.} = (cx, y) + (dx, y) = (cx + dx, 2y);$$

hence $V(\mathbf{R})$ is not a vector space.

(ii) This is a special case of $V_n(\mathbf{R})_n = \mathbf{R}$. See **Example 9.2** above.

(iii) Try yourself.

9.3 General Properties of Vector Spaces

Let **V** be a vector space over a field F.

We shall, in the following discussions, denote the Null Vector of $V(F)$ by θ and the Additive Identity of F by 0. Important properties of $V(F)$ are listed below:

I. $a\theta = \theta, \ \forall a \in F$.

II. $0\alpha = \theta, \ \forall \alpha \in V$.

III. $a(-\alpha) = -(a\alpha), \ \forall a \in F, \ \forall \alpha \in V$.

IV. $(-a)\alpha = -(a\alpha), \ \forall a \in F, \ \forall \alpha \in V$.

V. $a(\alpha - \beta) = a\alpha - a\beta, \ \forall a \in F, \ \forall \alpha, \beta \in V$.

VI. $a\alpha = \theta \Rightarrow a = 0$ or, $\alpha = \theta, \ \forall a \in F, \ \forall \alpha \in V$.

All these properties follow directly from the definition of vector space.

Proof.

I. We have, $a\theta = a(\theta + \theta)$ $[\because \theta = \theta + \theta]$
$= a\theta + a\theta$

Since $a\theta \in V$,

$$\theta + a\theta = a\theta$$
$$\therefore \theta + a\theta = a\theta + a\theta$$
$$\Rightarrow \theta = a\theta$$

[Using right cancellation law in V, which can be used as V is an abelian group with respect to addition.]

II. $0\alpha = (0+0)\alpha\ [\because 0 = 0+0]$
$= 0\alpha + 0\alpha.$

Since $0\alpha \in V$, $\theta + 0\alpha = 0\alpha$.

Hence $\theta + 0\alpha = 0\alpha + 0\alpha$
$\Rightarrow \theta = 0\alpha$ (Right cancellation law in V)

III. We have,

$$a[\alpha + (-\alpha)] = a\alpha + a(-\alpha)$$
$$\Rightarrow a\theta = a\alpha + a(-\alpha)$$
$$\Rightarrow \theta = a\alpha + a(-\alpha)\ (\because\ a\theta = \theta)$$

$\Rightarrow$ $a(-\alpha)$ is the additive inverse of $a\alpha$

$$\Rightarrow a(-\alpha) = -(a\alpha).$$

IV. We have,

$$[a + (-a)]\alpha = a\alpha + (-a)\alpha$$
$$\Rightarrow 0\alpha = a\alpha + (-a)\alpha$$
$$\Rightarrow \theta = a\alpha + (-a)\alpha$$

$\Rightarrow$ $(-a)\alpha$ is the additive inverse of $a\alpha$

$$\Rightarrow (-a)\alpha = -(a\alpha).$$

V. $a(\alpha - \beta) = a[\alpha + (-\beta)] = a\alpha + a(-\beta)$
$= a\alpha + [-(a\beta)]\ [\because a(-\beta) = -(a\beta)]$
$= a\alpha - a\beta$

VI. Let $a\alpha = \theta$ and $a \neq 0$. Then a^{-1} exists in F $(\because\ a \neq 0)$

$$\therefore a\alpha = \theta \Rightarrow a^{-1}(a\alpha) = a^{-1}\theta$$
$$\Rightarrow (a^{-1}a)\alpha = \theta$$
$$\Rightarrow 1\alpha = \theta$$
$$\Rightarrow \alpha = \theta.$$

Thus

$$a \neq 0,\ a\alpha = \theta \Rightarrow \alpha = \theta. \qquad (1)$$

Again, let $a\alpha = \theta$ and $\alpha \neq \theta$. To prove $a = 0$. If possible, $a \neq 0$, then a^{-1} exists.

$$\therefore a^{-1}(a\alpha) = a^{-1}\theta \Rightarrow (a^{-1}a)\alpha = \theta \Rightarrow 1\alpha = \theta \Rightarrow \alpha = \theta.$$

Thus we get a contradiction that α must be a zero vector (whereas we assumed $\alpha \neq \theta$). Therefore, a must be equal to 0,

$$\text{i.e., } \alpha \neq \theta, \ a\alpha = \theta \Rightarrow a = 0. \tag{2}$$

(1) & (2) together give the property $a\alpha = \theta \Rightarrow a = 0$ or $\alpha = \theta$.

Cancellation Laws

(i) If $a, b \in F$ and α is a non-zero of V, then $a\alpha = b\alpha \Rightarrow a = b$.

(ii) If $\alpha, \beta \in V$ and a is a non-zero element of F, then $a\alpha = a, \beta \Rightarrow \alpha = \beta$.

Proof.

(i) We have,

$$\begin{aligned} & a\alpha = b\alpha \ (\alpha \neq \theta) \\ \Rightarrow \ & a\alpha - b\alpha = \theta \\ \Rightarrow \ & (a - b)\alpha = \theta \\ \Rightarrow \ & a - b = 0 \text{ since } \alpha \neq \theta. \ \textbf{[Property VI]} \\ \Rightarrow \ & a = b. \end{aligned}$$

(ii) We have,

$$\begin{aligned} & a\alpha \ a\beta \ (a \neq 0) \\ \Rightarrow \ & a\alpha - a\beta = \theta \\ \Rightarrow \ & a(-\alpha - \beta) = \theta \\ \Rightarrow \ & \alpha - \beta = \theta \text{ since } a \neq 0 \ \textbf{[Property VI]} \\ \Rightarrow \ & \alpha = \beta. \end{aligned}$$

9.4 Linearly Dependent and Linearly Independent Set of Vectors

Definition 1. **Linear Combination.** *Let $V(F)$ be a vector space. If $\alpha_1, \alpha_2, \ldots, \alpha_k \in V$ then the expression $a_1\alpha_1 + a_2\alpha_2 + \cdots + a_k\alpha_k$ where $a_1, a_2, \ldots, a_k \in F$ is called a linear combination of the set of vector $\alpha_1, \alpha_2, \ldots, \alpha_k$.*

Definition 2. *Let V be a vector space over a field F. Let $S = \{\alpha_1, \alpha_2, \ldots, \alpha_k\}$ be a finite set of vectors of V.*

Then the set S of K vectors is said to be ***linearly dependent*** *(L.D.) if there exist scalars $a_1, a_2, a_3, \cdots, a_k \in F$,* NOT ***all of them zero,*** *(some of them may be zero) such that the linear combination*

$$a_1\alpha_1 + a_2\alpha_2 + a_3\alpha_3 + \cdots + a_k\alpha_k = \theta.$$

Otherwise the set S is **linearly independent**.

Definition 3. *Let V be a vector space over a field F. Let $S = \{\alpha_1, \alpha_2, \ldots, \alpha_k\}$ be a finite set of vectors of V. Then the set S of K vectors is said to be* ***linearly independent*** *(L.I.) if* ***every linear combination*** *of these K vectors is a null vector* ***only when*** *each scalar coefficient is zero.*

i.e., $a_1\alpha_1 + a_2\alpha_2 + \cdots + a_k\alpha_k = \theta$, where $a_1, a_2, \ldots, a_k \in F$, is true only when each a_i is zero.

Note 1. *An infinite set of vectors of* $V(F)$ *is said to be* ***linearly independent*** *if its every finite subset is linearly independent, otherwise it is* ***linearly dependent.***

We take examples from real vector spaces of n-tuples, i.e., $\mathbf{V}_n(\mathbf{R})$ or $\mathbf{R}^n$.

► Example 9.4.1 *In the real vector space of n-tuples* $V_n(R)$ *there are n unit vectors:*

$$\begin{aligned} e_1 &= (1,0,0,0,\ldots,0) \\ e_2 &= (0,1,0,0,\ldots,0) \\ e_3 &= (0,0,1,0,\ldots,0) \\ \ldots &\ldots \ldots \ldots \ldots \ldots \ldots \\ e_n &= (0,0,0,0,\ldots,1). \end{aligned}$$

Prove that they are linearly independent.

Solution: Let $a_1, a_2, a_3, \ldots, a_n$ be n real numbers (n scalars).
Then the linear combination

$$\begin{aligned} & a_1e_1 + a_2e_2 + a_3e_3 + \cdots + a_ne_3 = \theta \\ \Rightarrow\quad & a_1(1,0,0,\ldots,0) + a_2(0,1,0,0,\ldots,0) + \cdots + a_n(0,0,0,\ldots 1) = (0,0,0,\ldots 0) \\ \Rightarrow\quad & (a_1,0,0,\ldots,0) + (0,a_2,0,0,\ldots,0) + \cdots + (0,0,0,\ldots,a_n) = (0,0,0,0,\ldots,0) \\ \Rightarrow\quad & (a_1,a_2,a_3,\ldots,a_n) = (0,0,0,\ldots,0) \\ \Rightarrow\quad & a_1 = 0, a_2 = 0, a_3 = 0, \ldots, a_n = 0. \end{aligned}$$

∴ the set $(e_1, e_2, \ldots, e_n)$ of unit vectors of $\mathbf{V}_n(\mathbf{R})$ are L.I.

Particular cases: the set $\{e_1 = (1,0)$ and $e_2 = (0,1)\}$ is L.I. in $\mathbf{V}_2(\mathbf{R})$ or in $\mathbf{R}^2$.

$\{e_1 = (1,0,0), e_2 = (0,1,0)$ and $e_3 = (0,0,1)\}$ is L.I. in $\mathbf{V}_3(\mathbf{R})$ or in $\mathbf{R}^3$.

► Example 9.4.2 *Let* $S = \{(1,2,4),(1,0,0),(0,1,0),(0,0,1)\}$ *be a subset of* $\mathbf{V}_3(\mathbf{R})$ *or* $\mathbf{R}^3$. *Show that S is L.D.*

Solution: Let a_1, a_2, a_3, a_4 are 4 real numbers and we consider

$$\begin{aligned} & a_1(1,2,4) + a_2(1,0,0) + a_3(0,1,0) + a_4(0,0,1) = (0,0,0) \\ \Rightarrow\quad & (a_1 + a_2, 2a_1 + a_3, 4a_1 + a_4) = (0,0,0) \\ \Rightarrow\quad & a_1 + a_2 = 0, 2a_1 + a_3 = 0, 4a_1 + a_4 = 0 \\ \Rightarrow\quad & a_2 = -1, a_3 = -2a_1, a_4 = -4a_1. \end{aligned}$$

So the scalars $a_1 = 1, a_2 = -1, a_3 = -2, a_4 = -4$ will make the linear combination = zero vector (the scalar coefficients are not all zero).
∴ the Set S is L.D.

► Example 9.4.3 *Prove that the subset* $\{(1,2,0),(0,3,1),(-1,0,1)\}$ *of* $\mathbf{R}^3$ *is linearly independent whereas the subset* $\{(1,2,1),(3,1,5),(3,-4,7)\}$ *of* $\mathbf{R}^3$ *is linearly dependent.*

Solution: (i) Let a, b, c be three real numbers (scalars).

$$\begin{aligned} & a(1,2,0)+b(0,3,1)+c(-1,0,1)=(0,0,0) \\ \Rightarrow\ & (a,2a,0)+(0,3b,b)+(-c,0,c)=(0,0,0) \\ \Rightarrow\ & (a-c,2a+3b,b+c)=(0,0,0) \\ \Rightarrow\ & a-c=0, 2a+3b=0, b+c=0 \end{aligned}$$

whence $a = c, b = -c$ and $2a + 3b = 0$ gives $c = 0$ and hence $a = 0, b = 0$.

Otherwise Coefficient determinant $\begin{vmatrix} 1 & 0 & -1 \\ 2 & 3 & 0 \\ 0 & 1 & 1 \end{vmatrix} = 3 - 2 = 1 (\neq 0)$.

$\therefore$ the system of 3 homogeneous equations

$$a - c = 0, 2a + 3b = 0, b + c = 0$$

has unique solution, namely $a = 0, b = 0, c = 0$.

Thus the subset $\{(1,2,0),(0,3,1),(-1,0,1)\}$ is *linearly independent.*

(ii) **Again,** for the subset $\{(1,2,1),(3,1,5),(3,-4,7)\}$, any linear combination

$$a(1,2,1)+b(3,1,5)+c(3,-4,7)=(0,0,0)$$

$$\begin{aligned} \text{if } a+3b+3c &= 0 \quad (1)\\ 2a+b-4c &= 0 \quad (2)\\ a+5b+7c &= 0 \quad (3) \end{aligned}$$

Multiplying (1) by 2, we get,

$$\begin{aligned} 2a+6b+6c &= 0 \\ \text{and (2)} \quad 2a+b-4c &= 0. \end{aligned}$$

Hence, on subtraction, $5b + 10c = 0$, i.e., $b = -2c$.

Multiplying (2) by 5, we get

$$\begin{aligned} 10a+5b-20c &= 0 \\ \text{and (3)} \quad a+5b+7c &= 0. \end{aligned}$$

Hence, on subtraction, $9a - 27c = 0$, i.e., $a = 3c$, i.e., with $c = 1, b = -2, a = 3$ (not all zero) the linear combination becomes a zero vector.

Hence, the subset $\{(1,2,1),(3,1,5),(3,-4,7)\}$ is *linearly dependent.*

▶ Example 9.4.4 (i) *Show that the vectors* $(1,1,0,0),(0,1,-1,0),(0,0,0,3)$ *in* $\mathbf{R}^4$ *are Linearly Independent.*

(ii) *Show that in the vector space* $F[x]$ *of polynomials over the real field the set of vector* $\{1, x, x - x^2\}$ *is L.I.*

Solution:

(i) Let a, b, c be 3 real numbers such that

$$a(1,1,0,0)+b(0,1,-1,0)+c(0,0,0,3)=(0,0,0,0)$$

Then

$$\begin{aligned} a+0b+0c &= 0 \\ a+b+0c &= 0 \\ 0a-b+0c &= 0 \\ 0a+0b+3c &= 0. \end{aligned}$$

The only solution is $a=0, b=0, c=0$.
Hence, the vectors are L.I. (Linearly Independent).

(ii) The null vector in $F[x]$ is the *zero polynomial.*
Let a, b, c be any three real numbers such that

$$a\cdot 1+b\cdot x+c(x-x^2)=0, \text{ (i.e., zero polynomial.)}$$

Then,

$$\begin{aligned} & a+(b+c)x-cx^2=0+0x+0x^2 \\ \Rightarrow\quad & a=0, b+c=0, -c=0 \\ \Rightarrow\quad & a=0, b=0, c=0. \end{aligned}$$

$\therefore$ the set of vector $\{1, x, x-x^2\}$ is linearly independent.

▶ Example 9.4.5 *Prove that in a vector space* $\mathbf{V}(F)$ *any set of vectors which contains the null vector* θ *is linearly dependent.*

Solution: Let $S=\{\alpha_1, \alpha_2, \ldots, \alpha_{r-1}, \theta, \alpha_{r+1}, \ldots, \alpha_n\}$ be a set of n vectors of $\mathbf{V}(F)$ containing the null vector θ. To prove that S is L.D.

Clearly,

$$0\alpha_1+0\alpha_2+\cdots+0\alpha_{r-1}+1\theta+0\alpha_{r+1}+\cdots+0\alpha_n=\theta.$$

Thus at least one scalar coefficient is $1(\neq 0)$ but the linear combination is zero vector, therefore, by definition the set S is L.D.

Two Important Theorems

Let $\mathbf{V}(F)$ be a vector space over a field F.

Theorem 9.4.1. *Any subset of a linearly independent set of vectors of* $\mathbf{V}(F)$ *is linearly independent.*

Proof. Let $\{\alpha_1, \alpha_2, \ldots, \alpha_k\}$ be any given set of linearly independent vectors in $\mathbf{V}(F)$.

Let us consider any subset of this given set. To be more specific let $\{\alpha_1, \alpha_2, \ldots, \alpha_r\}$ where $r < k$, be a subset of the given set. We shall prove that this set of vectors is also L.I. If not, let the set $\{\alpha_1, \alpha_2, \ldots, \alpha_r\}$ be Linearly Dependent (L.D). Then the linear combination

$$a_1\alpha_1 + a_2\alpha_2 + \cdots a_j\alpha_j + \cdots + a_r\alpha_r = \theta \text{ (zero vector of } \mathbf{V}(F)),$$

where at least one of the scalar coefficients, say $a_j \neq 0$.

We can then write

$$a_1\alpha_1 + a_2\alpha_2 + \cdots + a_j\alpha_j + \cdots + a_r\alpha_r + 0\alpha_{r+1} + 0\alpha_{r+2} + \cdots + 0\alpha_k = \theta$$

where at least one scalar coefficient $a_j \neq 0$. This contradicts the statement that the given set of vectors is L.I.

We could have started with any subset of the given set and would arrive at the same conclusion. The theorem is thus proved.

Theorem 9.4.2. *Any superset of a given set of linearly dependent vectors is linearly dependent.*

Proof. Let $S = \{\alpha_1, \alpha_2, \ldots, \alpha_k\}$ be a L.D. set of vectors of $\mathbf{V}(F)$,

$$\text{so that } a_1\alpha_1 + a_2\alpha_2 + \cdots + a_k\alpha_k = \theta \text{ [zero vector of } \mathbf{V}(F)] \qquad (1)$$

where all the scalar coefficients are not zero (at least one scalar coefficient is not zero).

Next consider the set $T = \{\alpha_1, \alpha_2, \ldots, \alpha_k, \beta\}$ which is a super set of S containing one more vector β than S. Using (1) we can write

$$a_1\alpha_1 + a_2\alpha_2 + \cdots + a_k\alpha_k + 0\beta = \theta,$$

where all the scalar coefficients are not zero.

$\therefore$ the set T is also L.D.

We could take a bigger superset of S containing more than one additional vector than S and we would arrive at the same result. Thus the theorem is proved.

The following results should be remembered:

1. Any subset of a vector space $\mathbf{V}(F)$ is either L.D. or L.I.

2. A set containing only the null vector θ of $\mathbf{V}(F)$ is L.D.

   [See that $a\theta = \theta$ for any non-zero scalar coeff. a.]

3. A set containing the single non-zero vector is L.I.

   [See that $0\alpha = \theta$ for any vector $\alpha \in \mathbf{V}(F)$]

4. A set having one of the vectors as Null Vector is L.D.

   [See *Example* 9.4.5.]

5. If two vectors are L.D. then one of them is a scalar multiple of the other

$$[a\alpha_1 + b\alpha_2 = \theta, a, b \text{ not both zero; hence, etc.}]$$

6. Let $V(F)$ be a vector space. If $\alpha_1, \alpha_2, \ldots, \alpha_n$ are NON-ZERO vectors of V, then either *they are L.I.* or else, *some* α_k, $2 \leq k \leq n$ *is a linear combination of the preceding ones.*

9.5 Vector Subspaces

Definition 1. *Let V be a vector space over a field F and let* $W \subseteq V$. *Then W is called a subspace of V if W itself is a vector space over the same field F with respect to the vector addition and scalar multiplication in V.*

Theorem 9.5.1. *The necessary and sufficient condition for a non-empty subset W of a vector space V(F) to become a subspace of V is that for any two scalars* $a, b \in F$ *and for any two vectors* $\alpha, \beta \in W$, $a\alpha + b\beta \in W$.

Proof. The condition is necessary.

Assume that W is a subspace of V.

$\therefore$ W is closed under scalar multiplication and vector addition defined in $V(F)$.

Therefore, if $a \in F, \alpha \in W, a\alpha \in W$ and $b \in F, \beta \in W, b\beta \in W$.

$\therefore$ $a\alpha + b\beta \in W$. Hence, the condition is necessary.

The condition is sufficient. Suppose that W is a non-empty subset of V. Assume that if $a, b \in F$ and $\alpha, \beta \in W$, then $a\alpha + b\beta \in W$.

To prove W is a subspace of V.

(i) Taking, in particular, $a = 1, b = 1$, we see that if $\alpha, \beta \in W$, then

$$1\alpha + 1\beta \in W \Rightarrow \alpha + \beta \in W \; (\because \; \alpha \in W \Rightarrow \alpha \in V \text{ and } 1\alpha = \alpha \text{ in } V).$$

Thus W is *closed under vector addition.*

(ii) Taking $a = -1, b = 0$, we see that if $\alpha \in W$, then

$$\begin{aligned} & (-1)\alpha + 0\alpha \in W, \quad [\text{in place of } \beta \text{ we have taken } \alpha] \\ \text{i.e.,} \quad & -(1\alpha) + 0\alpha \in W \; [0\alpha = \theta, \text{ where } \theta \text{ is the null vector of } V] \\ \text{i.e.,} \quad & -\alpha + \theta \in W, \\ \text{i.e.,} \quad & -\alpha \in W. \end{aligned}$$

Thus the *additive inverse* of each element of $W \in W$.

(iii) Taking $a = 0, b = 0$, we see that if $\alpha \in W$, then

$$0\alpha + 0\alpha \in W, \text{ i.e., } \theta + \theta \in W, \text{ i.e., } \theta \in W.$$

Thus the null vector of V belongs to W. It will become the *null vector* of W.

(iv) Since the elements of W are also the elements of V, therefore vector addition will be associative as well as *commutative in* W.

Thus the given condition leads us to conclude that W is an abelian group with respect to vector addition.

Next, taking $\beta = \theta$, we see that if $a, b \in F$ and $\alpha \in W$, then

$$a\alpha + b\theta \in W, \text{ i.e., } a\alpha + \theta \in W, \text{ i.e., } a\alpha \in W.$$

Thus W is closed under scalar multiplication.

The other postulates of Scalar multiplication will hold in W because they hold in V of which W is a subset.

Hence, $W(F)$ is a subspace of $V(F)$.

Observations: Let $V(F)$ be a vector space of V. Then V itself is a subset of $V(F)$ and $\{\theta\}$ is also a subset of $V(F)$. Both are subspaces of V. They are known as **improper subspaces** of V. Other subspaces of V are called **proper subspaces** of V.

The subspace of V consisting of **zero vector** only is called the **zero subspace.**

In the following discussions, we shall take $\mathbf{R}^n$ as the real vector space of all n tuples

$$(x_1, x_2, \ldots, x_n), x_i \in \mathbf{R}.$$

▶ Example 9.5.1 *$\{(x_1, x_2, 0) \in \mathbf{R}^3; x_1, x_2$ are real numbers$\}$ is a subspace of the real vector space* $\mathbf{R}^3$.

Solution: Let $\alpha = (x_1, x_2, 0)$ and $\beta = (y_1, y_2, 0) \in W$, where x_1, x_2, y_1, y_2 are real numbers.

If a, b, be any two real numbers, then

$$\begin{aligned} a\alpha + b\beta &= a(x_1, x_2, 0) + b(y_1, y_2, 0) \\ &= (ax_1, ax_2, 0) + (by_1, by_2, 0) \\ &= (ax_1 + by_1, ax_2 + by_2, 0), \end{aligned}$$

where $ax_1 + by_1$ and $ax_2 + by_2$ are also real numbers.

Thus $a\alpha + b\beta \in W$. Hence, W is a subspace of $\mathbf{R}^3$.

▶ Example 9.5.2 *Verify that*

(i) $W = \{(x, 2y, 3z) : x, y, z$ *are* **real numbers**$\}$ *is a subspace of* $\mathbf{R}^3$.

$$\begin{aligned} [a\alpha + b\beta &= a(x_1, 2y_1, 3z_1) + b(x_2, 2y_2, 3z_2) \\ &= (ax_1, 2ay_1, 3az_1) + (bx_2, 2by_2, 3bz_2) \\ &= (ax_1 + bx_2, 2ay_1 + 2by_2, 3az_1 + 3bz_2) \in W] \end{aligned}$$

(ii) $W = \{(x, x, x) : x$ *is real numbers*$\}$ *is subspace of* $\mathbf{R}^3$.

$$[a\alpha + b\beta = a(x_1, x_1, x_1) + b(x_2, x_2, x_2) \in W\text{– How?}]$$

(iii) $W = \{(x, y, z) : x, y, z$ *are rational numbers*$\}$ *is not a subspace of* $\mathbf{R}^3$.

[Take $\alpha = (2,3,4), a = \sqrt{5}, a\alpha = (2\sqrt{5}, 3\sqrt{5}, 4\sqrt{5}) \notin W$. Thus W is not closed under scalar multiplication. Hence, etc.]

Theorem 9.5.2. *The intersection of any two subspaces W_1 and W_2 of a vector space $\mathbf{V}(F)$ is also a subspace of $\mathbf{V}(F)$.*

Proof. Let θ be the null vector of $\mathbf{V}(F)$. It is also the null vector of W_1 or W_2. Therefore, $\theta \in W_1 \cap W_2$, i.e., $W_1 \cap W_2$ is not an empty set.

Let $\alpha, \beta \in W_1 \cap W_2$ and $a, b \in F$.

Now $\alpha \in W_1 \cap W_2 \Rightarrow \alpha \in W_1$ and $\alpha \in W_2$.

Similarly $\beta \in W_1$ and $\beta \in W_2$.

Since W_1 is a subspace of $\mathbf{V}$, $a\alpha + b\beta \in W_1$ and since W_2 is a subspace of $\mathbf{V}$,

$$a\alpha + b\beta \in W_2.$$

Hence, $a\alpha + b\beta \in W_1 \cap W_2, \forall a, b \in F$ and $\alpha, \beta \in W_1 \cap W_2$.

$\therefore$ by sufficiency condition $W_1 \cap W_2$ is a subspace of $\mathbf{V}$.

Important note

The union of two subspaces of $\mathbf{V}(F)$ need not always be a subspace of $\mathbf{V}(F)$. e.g., if $W_1 = \{(0,0,z) : z \text{ is a real number}\}$ and $W_2 = \{(0,y,0) : y \text{ is a real number}\}$, then both are subspaces of $\mathbf{R}^3$ but not necessarily $W_1 \cup W_2$ is a subspace of $\mathbf{R}^3$.

See that $(0,0,4) \in W_1$ and $(0,3,0) \in W_2$ but $(0,0,4) + (0,3,0) = (0,3,4)$ does not belong to W_1 nor does it belong to W_2. Thus $W_1 \cup W_2$ is not closed under vector addition. Hence, $W_1 \cup W_2$ is not a subspace of $\mathbf{R}^3$.

We, however, assert here (without giving the proof) that

The Union of two subspaces is a subspace if and only if one of the two subspaces is a subset of the other.

[Let $W_1 \subseteq W_2$ or $W_2 \subseteq W_1$. Then $W_1 \cup W_2$ is either W_2 or W_1. But W_1, W_2 are subspaces and therefore, $W_1 \cup W_2$ is also a subspace. The proof of our assertion will be complete only when we prove the converse: If $W_1 \cup W_2$ is a subspace, then either $W_1 \subseteq W_2$ or $W_2 \subseteq W_1$. We leave out the proof in the present text.]

What is meant by "Smallest subspace" containing a subset of $\mathbf{V}(F)$?

Let $\mathbf{V}(F)$ be a vector space over a field F.

Let S be any subset of $\mathbf{V}$.

If W is a subspace of $\mathbf{V}$ containing S, and W itself is contained in every subspace of $\mathbf{V}$ containing S, then W is called the *smallest subspace* of $\mathbf{V}$ containing S. It is also called the subspace of $\mathbf{V}$ *generated* by S or *spanned* by S. The symbol $\{S\}$ is often used to mean the smallest subspace of $\mathbf{V}$ containing S. It can be easily seen that the intersection of all the subspaces of $\mathbf{V}(F)$ containing S is the subspace of $\mathbf{V}(F)$ generated by S.

If $\{S\} = \mathbf{V}$, then we say that "$\mathbf{V}$ is *spanned* by S."

Linear span

Let $\mathbf{V}(F)$ be a vector space and S be any non-empty subset of $\mathbf{V}$. Then by linear span of S, denoted by $L(S)$, we mean the set of all linear combinations of finite sets of elements of S.

Thus $L(S) = \{a_1\alpha_1 + a_2\alpha_2 + \cdots + a_n\alpha_n / \alpha_1, \alpha_2, \ldots \alpha_n$ is any arbitrary finite subset of S and $\{a_1, a_2, \ldots, a_n\}$ is any n arbitrary finite subset of F$\}$.

9.6 Basis and Dimension of a Vector Space

Definition 1. **Definition of a Basis.** *A subset* S *of a vector space* V(F) *is called a basis of the vector space* V(F) *if*

(i) S *contains vectors which are L.I. and*

(ii) S *generates* V(F), *i.e.,* L(S) = V, *i.e., each vector* $\alpha \in V$ *is a linear combination of a finite number of elements of* S.

▶ Example 9.6.1 *Prove that the set* $S = \{(1,0),(0,1)\}$ *is a basis of* $\mathbf{R}^2$.

Solution: First we show that S is a linearly independent set of vectors.

If a, b are real numbers, then

$$\begin{aligned} & a(1,0)+b(0,1)=(0,0) \\ \Rightarrow\quad & (a,0)+(0,b)=(0,0) \\ \Rightarrow\quad & (a,b)=(0,0) \\ \Rightarrow\quad & a=0 \text{ and } b=0, \end{aligned}$$

i.e., S consists of linearly independent vectors. (1)

Now we prove $L(S) = \mathbf{R}^2$. Clearly $L(S) \subseteq \mathbf{R}^2$.

We prove $\mathbf{R}^2 \subseteq L(S)$, i.e., any vector $(x_1, x_2) \in \mathbf{R}^2$ is a linear combination of the element of S.

We can write

$$(x_1,x_2) = x_1(1,0) + x_2(0,1). \tag{2}$$

Hence, by (1) and (2) S is a basis of R^2. We call this basis $\{(1,0),(0,1)\}$ as the **Standard Basis of** $\mathbf{R}^2$.

More generally

$$S = \{e_1 = (1,0,0,\ldots,0), e_2 = (0,1,0,\ldots,0), \ldots e_n = (0,0,\ldots,1),$$

where $e_1, e_2, \ldots, e_n \in \mathbf{R}^n$ is a **basis** (**standard basis**) of the set $\mathbf{R}^n$ of all n-tuples.

Here also we first show that S consists of linearly independent set of vectors:

If $a_1, a_2, \ldots, a_n$ are real numbers, then

$$\begin{aligned} & a_1e_1 + a_2e_2 + \cdots + a_ne_n = \theta = (0,0,0,\ldots,0) \\ \Rightarrow\quad & (a_1, a_2, \ldots, a_n) = (0,0,0,\ldots,0) \\ \Rightarrow\quad & a_1 = 0, a_2 = 0, \ldots, a_n = 0, \end{aligned}$$

i.e., S is a linearly independent set of vectors. (1)

Any vector $\mathbf{R}^n$, say $(x_1, x_2, x_3, \ldots, x_n)$ is a linear combination of the elements of S; thus

$$\begin{aligned} (x_1,x_2,\ldots,x_n) &= x_1(1,0,0,\ldots,0) + x_2(0,1,0,\ldots,0) + \cdots + x_n(0,0,\ldots,1) \\ &= x_1e_1 + x_2e_2 + \cdots + x_ne_n, \end{aligned}$$

i.e., $\mathbf{R}^n \subseteq L(S)$ and obviously $L(S) \leq \mathbf{R}^n$.

$$\therefore\ L(S) = \mathbf{R}^n. \tag{2}$$

Hence, by (1) and (2) S is a basis (**known as standard Basis**) of $\mathbf{R}^n$.

▶ Example 9.6.2 *Prove that the infinite set in the vector space* F[x] *of all polynomials over the field* F, $S = \{1, x, x^2, \ldots, x^n, \ldots\}$ *is linearly independent set of vectors. Now prove that* S *is a basis of* F[x].

Solution: Here S contains infinite number of elements.

Let $S_1 = \{x^{m_1}, x^{m_2}, \ldots, x^{m_n}\}$ be any finite subset of S having n vectors $\in F[x]$, where $m_1, m_2, \ldots, m_n$ are some non-negative integers. Let $a_1, a_2, \ldots, a_n \in F$ be n scalars such that

$$\begin{aligned} a_1x^{m_1} + a_2x^{m_2} + \cdots + a_nx^{m_n} &= \underset{\sim}{0} \text{ (zero polynomial)} \\ \therefore\ a_1 = 0, a_2 = 0, \ldots, a_n &= 0, \end{aligned}$$

i.e., every finite subset of S is linearly independent.

Therefore

$$S \text{ is linearly independent.} \tag{1}$$

Now we show that S spans F[x], i.e., any polynomial $\in F[x]$ can be expressed as a linear combination of a finite number of vectors $\in S$.

Let $f(x) = a_0 + a_1x + a_2x^2 + \cdots + a_mx^m$ be a polynomial of degree m.

Then

$$f(x) = a_0(1) + a_1(x) + a_2(x^2) + \cdots + a_m(x^m). \tag{2}$$

Hence, S is a basis of F[x], [$\because$ (1) and (2) are both true.]

Observations: We see that $\mathbf{R}^n$ has a finite basis, namely, $\{e_1, e_2, \ldots, e_n\}$ but F[x] has an infinite basis $\{1, x, x^2, \ldots, x^n, \ldots\}$. Clearly F[x] has no finite basis [if we take any finite set S of polynomials, then we can find a polynomial of degree greater than each of the terms of S—and such a polynomial cannot be expressed as a linear combination of elements of S].

What is a **Finite dimensional vector space**?

The vector space $\mathbf{V}(F)$ over a field F is called *Finite-dimensional if there exists a finite subset* S of $\mathbf{V}$ such that $\mathbf{V} = L(S)$.

e.g., $\mathbf{R}^n$ is a finite-dimensional vector space.

On the other hand a vector space which cannot be generated by a finite set of elements of the space is called an *Infinite dimensional* vector space; e.g., F[x] is infinite dimensional.

We now assert that (i) *There exists a basis for each finite dimensional vector space.* When we defined finite dimensional vector space we needed a subset S containing finite number of vectors of $\mathbf{V}$ such that $L(S) = \mathbf{V}$. Now it can be proved that for such a set S, either S is L.I. or a proper subset of S must be L.I. so that there always exists a basis for a finite dimensional vector space.

(ii) *If a vector space* $\mathbf{V}(F)$ *is finite dimensional there can exist more than one basis but any two basis of* $\mathbf{V}$ *must have the same number of elements.*

e.g., It can be easily checked that

$$\begin{aligned} S_1 &= \{(1,0,0), (0,1,0), (0,0,1)\} \\ \text{and } S_2 &= \{(1,0,0), (1,1,0), (1,1,1)\} \end{aligned}$$

are both basis of $\mathbf{R}^3$. See that S_1 and S_2 have the same number of elements, namely three.

Definition 2. *The number of elements in any basis of a finite dimensional vector space which is constant for all the bases of a vector space is called the dimension of the vector space denoted by* dim **V**.

Thus $\mathbf{R}^3$ *is a three-dimensional vector space,* $\mathbf{R}^n$ *is an n-dimensional vector space, etc., (i.e.,* $\dim \mathbf{R}^3 = 3, \dim \mathbf{R}^n = n$*, etc.)*

(iii) Let $S = \{\alpha_1, \alpha_2, \ldots, \alpha_n\}$ be a basis of a finite dimensional vector space $\mathbf{V}(F)$ of dimension n. Then every element $\alpha \in \mathbf{V}$ can be uniquely expressed as

$$\alpha = c_1\alpha_1 + c_2\alpha_2 + \cdots + c_n\alpha_n \ (c_i \in F).$$

We shall assume the truths of the following important properties of finite dimensional vector spaces:

A. Extension Theorem: *Let S be a linearly independent subset of a finite dimensional vector space* $\mathbf{V}(F)$. *Then S is either a basis of* **V** *or S will form a part of basis of* **V**, *(i.e., S can be extended to form a basis of* **V***).*

B. Replacement Theorem: If $\{\alpha_1, \alpha_2, \ldots, \alpha_n\}$ be a basis of a vector space **V** over a field F and a vector $\beta \neq \theta$ (i.e., non-null vector)$\in \mathbf{V}$ is expressed as

$$\beta = c_1\alpha_1 + c_2\alpha_2 + \cdots + c_j\alpha_j + \cdots + c_n\alpha_n (c_i \in F)$$

with one of the coefficients, say c_j, associated **with the vector** α_j **not zero,** then we can form a new basis by replacing α_j by β, i.e., the new basis will be

$$\{\alpha_1, \alpha_2, \ldots, \alpha_j, \beta, \alpha_{j+1}, \ldots, \alpha_n\}.$$

C. If a set S of n vectors of a finite dimension vector space $\mathbf{V}(F)$ of dimensional n, generates $\mathbf{V}(F)$, then S is a basis of **V**.

D. If $\mathbf{V}(F)$ be a finite dimensional vector space of dimension n, then **any set of** n **linearly independent** vectors in **V** forms a basis of **V**. This is an **important method** of determining whether a given set S of vector is a basis or not. See **Example 9.6.3(ii).**

Examples on Basis and Dimension

▶ Example 9.6.3 *Show that in* $\mathbf{R}^3$,

$$\begin{aligned} S_1 &= \{(1,1,2),(1,2,5),(5,3,4)\} \text{ is not a basis} \\ \text{but } S_2 &= \{(1,2,1),(2,1,0),(1,-1,2)\} \text{ is a basis.} \end{aligned}$$

Solution: (i) The elements of S_1 are $(1,1,2)$, $(1,2,5)$ and $(5,3,4)$. A linear combination of these three elements of S is

$$a(1,1,2) + b(1,2,5) + c(5,3,4).$$

It is equal to the null vector of $\mathbf{R}^3$, i.e., equals to $(0,0,0)$ if

$$\left.\begin{aligned} a + b + 5c &= 0 \\ a + 2b + 3c &= 0 \\ 2a + 5b + 4c &= 0 \end{aligned}\right\}. \qquad (1)$$

See that

$$\begin{vmatrix} 1 & 1 & 5 \\ 1 & 2 & 3 \\ 2 & 5 & 4 \end{vmatrix} = (8-15)-(4-6)+5(5-4) = 0.$$

$\therefore$ the system of homogeneous equations (1) has solutions other than the trivial solution $(0,0,0)$. In fact, if we solve we shall get $b = 2c, a = -7c$. Thus $c = 1, a = -7, b = 2$ is a solution of the system, i.e., $\exists$ scalars (not all zero) such that

$$a(1,1,2) + b(1,2,5) + c(5,3,4) = (0,0,0),$$

i.e., the set S_1 is linearly dependent (not linearly independent) and hence S_1 **cannot be a basis of** $\mathbf{R}^3$.

(ii) The elements of S_2 are $(1,2,1), (2,1,0)$ and $(1,-1,2)$.

A linear combination

$$a(1,2,1) + b(2,1,0) + c(1,-1,2) = (0,0,0)$$

gives

$$\left.\begin{aligned} a + 2b + c &= 0 \\ 2a + b - c &= 0 \\ a + 0b + 2c &= 0 \end{aligned}\right\}. \tag{2}$$

See that

$$\begin{vmatrix} 1 & 2 & 1 \\ 2 & 1 & -1 \\ 1 & 0 & 2 \end{vmatrix} = 2 - 2(4+1) + 1(-1) = -9 (\neq 0).$$

$\therefore$ the system of homogeneous equations (2) has unique solution, namely $a = 0$, $b = 0$, $c = 0$. This solution can also be easily checked by ordinary method of solving the system of equations (2). Thus the set $S_2 \subseteq \mathbf{R}^3$ is linearly independent.

We know that the set $\{(1,0,0),\ (0,1,0),\ (0,0,1)\}$ is a basis of $\mathbf{R}^3$. Therefore, $\dim \mathbf{R}^3 = 3$. Now we have proved S_2 is L.I.

$\therefore$ by property **D**, S_2 is a basis of **V**.

► Example 9.6.4 *Let* **V** *be the vector space of all* 2×2 *matrices over the field* F. *Let*

$$S = \left\{ \begin{pmatrix} 1 & 0 \\ 0 & 0 \end{pmatrix}, \begin{pmatrix} 0 & 1 \\ 0 & 0 \end{pmatrix}, \begin{pmatrix} 0 & 0 \\ 1 & 0 \end{pmatrix}, \begin{pmatrix} 0 & 0 \\ 0 & 1 \end{pmatrix} \right\}$$

be a subset of **V**. *Prove that* S *is a basis of* **V**.

Solution: We prove that S is linearly independent. We have,

$$a\begin{pmatrix} 1 & 0 \\ 0 & 0 \end{pmatrix} + b\begin{pmatrix} 0 & 1 \\ 0 & 0 \end{pmatrix} + c\begin{pmatrix} 0 & 0 \\ 1 & 0 \end{pmatrix} + d\begin{pmatrix} 0 & 0 \\ 0 & 1 \end{pmatrix} = \begin{pmatrix} 0 & 0 \\ 0 & 0 \end{pmatrix} \quad (a,b,c,d \in F).$$

This gives

$$\begin{pmatrix} a & b \\ c & d \end{pmatrix} = \begin{pmatrix} 0 & 0 \\ 0 & 0 \end{pmatrix} \quad \text{or, } a = b = c = d = 0.$$

Also any arbitrary element $\begin{pmatrix} x & y \\ z & w \end{pmatrix}$ where $x, y, z, w \in F$, can be expressed as

$$x\begin{pmatrix} 1 & 0 \\ 0 & 0 \end{pmatrix} + y\begin{pmatrix} 0 & 1 \\ 0 & 0 \end{pmatrix} + z\begin{pmatrix} 0 & 0 \\ 1 & 0 \end{pmatrix} + w\begin{pmatrix} 0 & 0 \\ 0 & 1 \end{pmatrix}, \text{ i.e., } L(S) = \mathbf{V}.$$

$\therefore$ S is a basis of **V**. Since the basis S contains 4 elements, $\dim \mathbf{V} = 4$.

▶ Example 9.6.5 *In the real vector space* $\mathbf{R}^3$, *prove that the set*

$$S = \{(3,5,1), (1,1,2), (2,-1,0)\}$$

is linearly independent. Write down a basis of $\mathbf{R}^3$. *What is the dimension of* $\mathbf{R}^3$?

Solution: To prove S is L.I. consider

$$a(3,5,1) + b(1,1,2) + c(2,-1,0) = (0,0,0)$$

and obtain $a = b = c = 0$. Therefore, the set S forms a basis of $\mathbf{R}^3$, $\dim \mathbf{R}^3 = 3$. Write the explanations in details as in Ex. **9.6.3** above.

▶ Example 9.6.6 *Find a basis for* $\mathbf{R}^3$ *that contains the vectors* $(2,4,0)$ *and* $(2,6,2)$.

Solution: $\dim \mathbf{R}^3 = 3$. The standard basis for $\mathbf{R}^3$ is

$$\{e_1 = (1,0,0),\ e_2 = (0,1,0),\ e_3 = (0,0,1)\}.$$

Let

$$\begin{aligned} \alpha &= (2,4,0) = 2e_1 + 4e_2 + 0e_3 \\ \text{and } \beta &= (2,6,2) = 2e_1 + 6e_2 + 2e_3. \end{aligned}$$

Using Replacement Theorem, we can replace e_1 in the standard basis by α. Then (α, e_2, e_3) is a new basis for $\mathbf{R}^3$.

Now we express $\beta = c_1\alpha + c_2e_2 + c_3e_3$,

$$\begin{aligned} \text{i.e.,}\quad & (2,6,2) = c_1(2,4,0) + c_2(0,1,0) + c_3(0,0,1) \\ \Rightarrow\quad & 2c_1 = 2, \text{ i.e., } c_1 = 1, 4c_1 + c_2 = 6, \\ \text{i.e.}\quad & c_2 = 2, c_3 = 2. \\ \therefore\quad & \beta = \alpha + 2e_2 + 2e_3. \end{aligned}$$

Again, using replacement theorem, we can replace e_3 by β.

$$\therefore \{\alpha, e_2, \beta\} \text{ is a basis for } \mathbf{R}^3 \text{ which contains } \alpha, \beta.$$

Try in a similar manner:

▶ Example 9.6.7 *Find a basis of the real vector space* $\mathbf{R}^3$ *containing the vectors* $(1,1,2)$ *and* $(3,5,2)$. **[C.P. 2003]**

▶ Example 9.6.8 *Given a basis of* $\mathbf{R}^3$ *as*

$$B = \{(1,1,1), (1,1,0), (1,0,0)\}$$

we can express any vector, say $(1,7,3)$, *in terms of the elements of* B.

Solution: $(1,7,3) = a(1,1,1) + b(1,1,0) + c(1,0,0)$

$$\Rightarrow a+b+c=1,\ a+b=7,\ a=3 \text{ whence } b=4, c=-6.$$

Thus, $(1,7,3) = 3(1,1,1) + 4(1,1,0) - 6(1,0,0)$.

These coefficients $3,4,-6$ are called the **coordinates of the vector** $(1,7,3)$ **relative to the basis** B.

9.7 Solved Problems on Vector Spaces

▶ Example 9.7.1 *Show that the set S of vectors $(1,0,0),(0,2,1),(1,1,0)$ form a basis of the real vector space R^3.* **[C.H. 1995]**

Solution: Consider the linear combination

$$c_1(1,0,0) + c_2(0,2,1) + c_3(1,1,0) = (0,0,0)$$

(c_1, c_2, c_3 are real constants)

$$\Rightarrow \quad c_1 + c_3 = 0;\ 2c_2 + c_3 = 0;\ c_2 = 0$$

$$\Rightarrow \quad c_1 = 0;\ c_2 = 0;\ c_3 = 0,$$

i.e., set of the three given vectors is L.I.

Any vector $(a,b,c) \in R^3$ can be expressed as a linear combination of the three vectors of S, i.e.,

$$(a,b,c) = \lambda(1,0,0) + \mu(0,2,1) + \nu(1,1,0)$$

$$\Rightarrow \quad \lambda + \nu = a,\ 2\mu + \nu = b,\ \mu = c$$

$$\Rightarrow \quad \mu = c,\ \nu = b - 2c,\ \lambda = a - b + 2c.$$

Thus S is L.I. and any vector of R^3 can be expressed as a linear combination of three vectors of S (i.e., S generates R^3). Hence, S is a basis of R^3.

Otherwise. The linear combination

$$c_1(1,0,0) + c_2(0,2,1) + c_3(1,1,0) = (0,0,0) \ (c_1, c_2, c_3 \text{ are scalars})$$

$$\left.\begin{aligned} \text{gives } c_1 + c_3 = 0 \quad &\text{or,} \quad c_1 + 0\cdot c_2 + c_3 = 0 \\ 2c_2 + c_3 = 0 \quad &\text{or,} \quad 0c_1 + 2c_2 + c_3 = 0 \\ c_2 = 0 \quad &\text{or,} \quad 0c_1 + c_2 + 0c_3 = 0 \end{aligned}\right\}.$$

The coefficient determinant of this homogeneous system of there equations is

$$\begin{vmatrix} 1 & 0 & 1 \\ 0 & 2 & 1 \\ 0 & 1 & 0 \end{vmatrix} = -1\ (\neq 0)$$

Therefore, by Cramer's rule there exists unique trivial solution

$$c_1 = 0,\ c_2 = 0,\ c_3 = 0,$$

i.e., S is linearly independent.

Since R^3 is a real vector space of dimension 3 and a subset S of R^3 is L.I., therefore, $S = \{(1,0,0),(0,2,1),(1,1,0)\}$ form a basis of R^3.

▶ **Example 9.7.2** *Show that the set of vectors* $\{(1,0,0),\ (0,1,1),\ (1,1,1)\}$ *in the real vector space* R^3 *are linearly dependent.* **[C.P. 1995]**

Solution: The relation

$$c_1(1,0,0)+c_2(0,1,1)+c_3(1,1,1)=(0,0,0)$$
$$\Rightarrow \quad c_1+c_3=0,\ c_2+c_3=0$$
$$\Rightarrow \quad c_1=-c_3=c_2.$$

We can take $c_1=c_2=1,\ c_3=-1$. Then

$$1(1,0,0)+1(0,1,1)-1(1,1,1)=(0,0,0).$$

Not all the coeff. are zero.

$\therefore$ the set $\{(1,0,0),\ (0,1,1),\ (1,1,1)\}$ is **Linearly dependent.**

▶ **Example 9.7.3** *Are the vectors* $(1,-2,0),\ (3,0,-2),\ (0,-1,-5)$ *linearly independent in the real vector space* R^3? **[C.P. 1996]**

Try yourself. **[*Ans.* Yes, Linearly independent.]**

▶ **Example 9.7.4** *Do the vectors* $\{(1,1,2),\ (3,5,2)$ *and* $(1,0,0)\}$ *form a L.I. subset of* R^3? *Justify.* **[C.P. 1997]**

Solution: The linear combination

$$a(1,1,2)+b(3,5,2)+c(1,0,0)=(0,0,0)$$
$$\text{if} \quad a+3b+c=0,\ a+5b=0,\ 2a+2b=0.$$

The coefficient determinant

$$=\begin{vmatrix} 1 & 3 & 1 \\ 1 & 5 & 0 \\ 2 & 2 & 0 \end{vmatrix}=2-10=-8\ (\neq 0).$$

$\therefore$ the three equations when solved, have only a trivial solution:

$$a=0,\ b=0,\ c=0, \text{ i.e., the given set is L.I.}$$

▶ **Example 9.7.5** *If* $\{x_1,x_2,x_3\}$ *be a L.I. set of vectors in a vector space V, show that the set of vectors* $\{x_1+x_2;\ x_1-x_2;\ x_1-2x_2+x_3\}$ *is also L.I.* **[C.P. 1996]**

Solution: Let a, b, c be scalars such that

$$a(x_1+x_2)+b(x_1-x_2)+c(x_1-2x_2+x_3)=0 \tag{1}$$
$$\text{or,} \quad (a+b+c)x_1+(a-b-2c)x_2+cx_3=0. \tag{2}$$

Given: x_1,x_2,x_3 are L.I.

Therefore, (2)

$$\Rightarrow a+b+c=0,\ a-b-2c=0,\ c=0.$$

The only solution of these equations $c=0,\ a=0,\ b=0$.

Therefore, by (1) the set of vectors $x_1+x_2,\ x_1-x_2,\ x_1-2x_2+x_3$ are linearly independent.

▶ Example 9.7.6 (a) *Define a basis of a vector space.* [C.P. 1997]

(b) *Show that the vectors* $(1,2,1), (2,1,0)$ *and* $(1,-1,2)$ *form a basis of* R^3.

Solution: (a) A subset S of a vector space V(F) over a field F, is said to be a basis of V(F), if

(i) S consists of linearly independent vectors.

(ii) S generates V(F), i.e., each vector in V is a linear combination of a finite number of elements of S.

(b) Show that the three vectors $(1,2,1)$, $(2,1,0)$ and $(1,-1,2)$ form a linearly independent set of R^3. Hence they form a basis of R^3. **See Property D.**

▶ Example 9.7.7 *Express a vector* $(7,11) \in R^2$ *as a linear combination of the vectors* $(2,3)$ *and* $(3,5)$. [C.P. 1997]

Solution: $(7,11) = a(2,3) + b(3,5) \Rightarrow 7 = 2a + 3b,\ 11 = 3a + 56.$
Solving we get, $a = 2$, $b - 1$.
$\therefore$ we can express $(7,11) = 2(2,3) + 1(3,5)$.

▶ Example 9.7.8 *Determine* K *so that the vectors* $(1,-1,2)$, $(0,K,3)$ *and* $(-1,2,3)$ *are linearly dependent.*

Solution: $c_1(1,-1,2) + c_2(0,K,3) + c_3(-1,2,3) = (0,0,0)$

$$\begin{aligned} \Rightarrow c_1 + 0c_2 - c_3 &= 0 \\ -c_1 + Kc_2 + 2c_3 &= 0 \\ 2c_1 + 3c_2 + 3c_3 &= 0. \end{aligned}$$

If the determinant

$$\begin{vmatrix} 1 & 0 & -1 \\ -1 & K & 2 \\ 2 & 3 & 3 \end{vmatrix} = 0,$$

then the set of homogeneous equations has non-trivial solutions and the set of vectors will then be Linearly Dependent (L.D).
determinant $= (3K-6) - (-3-2K) = 0$, i.e., $5K - 3 = 0$, or, $K = \frac{3}{5}$.

▶ Example 9.7.9 *Show that the vectors* $(2,1,0)$, $(1,1,0)$, $(4,2,0)$ *of* R^3 *are not linearly independent.*

Try yourself.

▶ Example 9.7.10 *Show that* $B = \{(1,2,1), (0,1,0), (0,0,1)\}$ *is a basis of* R^3. *Express the vector* $(1,2,3) \in R^3$ *as a linear combination of the basis* B.

[*Ans.* **Second part:** $(1,2,3) = a(1,2,1) + b(0,1,0) + c(0,0,1) \Rightarrow a = 1,\ 2a + b = 2,\ a + c = 3$ $\Rightarrow a = 1,\ b = 0,\ c = 2.$]

▶ Example 9.7.11 *Do the vectors* $(1,1,0), (1,0,1)$ *and* $(0,1,1)$ *form a basis for the vector space* R^3 *over the field of real numbers? Justify.*

Solution: Consider the linear combination

$$\begin{aligned} c_1(1,1,0)+c_2(1,0,1)+c_3(0,1,1) &= (0,0,0) \\ \text{if } c_1+c_2=0, c_1+c_3=0, c_2+c_3 &= 0, \\ \text{i.e., if } c_1=c_2=c_3 &= 0. \end{aligned}$$

Hence, the given vectors are L.I. Since R^3 is a real vector space of dimension 3, and a subset of three vector is L.I., therefore, they form a basis for R^3.

▶ Example 9.7.12 *Show that the subset* $S = \{(2,4,0), (0,1,0), (2,6,2)\}$ *of the real vector space* R^3 *is a basis of* R^3. [C.P. 2001]

Solution: The linear combination

$$a(2,4,0)+b(0,1,0)+c(2,6,2)=(0,0,0)$$

if $2a+2c=0$, $4a+b+6c=0$, $2c=0 \Rightarrow c=0, a=0, b=0$, i.e., three vectors of R^3 (a vector space of dimension three) are L.I. Hence they form a basis of R^3. See Property D.

▶ Example 9.7.13 *Find a basis of the real vector space* R^3 *containing the vectors* $(1,1,2)$ *and* $(3,5,2)$. [C.P. 2003]

Solution: R^3 is a vector space of dimension 3.

The standard basis for R^3 is $\{e_1,e_2,e_3\}$ where

$$e_1=(1,0,0),\ e_2=(0,1,0),\ e_3=(0,0,1).$$

Let $\alpha=(1,1,2), \beta=(3,5,2)$.

Now $\alpha=(1,1,2)=1(1,0,0)+1(0,1,0)+2(0,0,1)$.

By Replacement Theorem, α can replace e_1 or e_2 or e_3 (since in each case the coefficient $\neq 0$) in the basis (e_1,e_2,e_3). Suppose we replace e_1 by α.

Then (α,e_2,e_3) is a basis of R^3.

$$\begin{aligned} \text{Let } \beta=(3,5,2) &= c_1\alpha+c_2e_2+c_3e_3 \\ &= c_1(1,1,2)+c_2(0,1,0)+c_3(0,0,1) \\ \Rightarrow 3 &= c_1,\ c_1+c_2=5,\ 2c_1+c_3=2 \\ \Rightarrow c_1 &= 3,\ c_2=2,\ c_3=-4 \\ \therefore \beta &= 3\alpha+2e_2-4e_3. \end{aligned}$$

Again, by Replacement Theorem β can replace e_2 or e_3. Suppose we replace e_2 by β. Then $\{\alpha,\beta,e_3\}$ form a new basis for R^3 containing the vectors α,β.

▶ Example 9.7.14 *Find a basis and the dimension of the subspace* W *of* R^3 *where*

$$W=\{(x,y,z)\in R^3 \text{ such that } x+y+z=0\}.$$

Solution: Let $\alpha=(a,b,c)$ be any arbitrary vector of W.

Then $a+b+c=0$. We write

$$\alpha=(a,b,-a-b)=(a,b,-a)+(a,b,-b)=a(1,0,-1)+b(0,1,-1)$$

Any arbitrary vector of W can be expressed as a linear combination of $\{(1,0,-1),(0,1,-1)\}$, i.e., W is generated by the set of vector $(1,0,-1)$ and $(0,1,-1)$. Moreover $(1,0,-1)$ and $(0,1,-1)$ are L.I.

Since

$$\begin{aligned} & c_1(1,0,-1)+c_2(0,1,-1)=(0,0,0) \\ \Rightarrow\quad & c_1=0,\ c_2=0,\ -c_1-c_2=0, \text{ i.e., } c_1=0,\ c_2=0. \end{aligned}$$

$\therefore$ the set $\{(1,0,-1),(0,1,-1)\}$ is a basis of W and clearly dimension of $W=2$.

Exercises on Chapter 9

1. Let V be the set of all pairs (x,y) of real numbers and F be the field of real numbers. Define

$$\begin{aligned} (a,b)+(c,d) &= (a+c,0) \\ c(x,y) &= (cx,0) \end{aligned}$$

Show that V, with these operations, is not a vector space over F.

2. Is the set of all **non-zero** polynomials in x of degree ≤ 2 a vector space? [*Ans.* No]

3. Is the set of **all polynomials** in x of degree ≤ 2 a vector space? [*Ans.* Yes]

4. Show that the complex field $\mathbf{C}$ is a vector space over the field $\mathbf{R}$ of real numbers.

5. The set of ordered triads $(a_1,a_2,0)$ where $a_1,a_2\in\mathbf{R}$ is a subspace of $\mathbf{R}^3$. Prove.

6. Let V be a vector space of all polynomials in x over the real field $\mathbf{R}$. Let W be a subset of V consisting of all polynomials of degree $\le n$. Prove that W is a subspace of V.

7. Prove that the set of all solutions (x,y,z) satisfying $x+y+2z=0$ is a subspace of $\mathbf{R}^3$.

8. If $x,y\in R$, prove that the set of all triads $(x+2y,y,-x+3y)$ is a subspace of $\mathbf{R}^3$.

9. $W=\{(x,y,z): x,y,z$ are rational numbers$\}$. Is W a subspace of real vector space $V_3(\mathbf{R})$? [*Ans.* No]

   What can you say about $W=\{(x,2y,3z): x,y,z\in\mathbf{R}\}$? [*Ans.* Yes]

10. The set of vector $\alpha=(x,y)\in\mathbf{R}^2$ for which $x^2=y^2$ is a subspace of $\mathbf{R}^2$. True or False?

11. Show that the set of vectors
$$S = \{(1,2,4),\ (1,0,0),\ (0,1,0),\ (0,0,1)\}$$
is linearly dependent in $\mathbf{R}^3$, but the set $T = \{(1,2,0),\ (0,3,1),\ (-1,0,1)\}$ is linearly independent.

12. If $a_1b_2 - a_2b_1 = 0$, then and then only the vector (a_1, a_2) and $(b_1, b_2) \in \mathbf{R}^2$ are linearly dependent.

13. If α, β, γ are linearly dependent vector of $\mathbf{V}(F)$ where F is any subfield of the field of complex numbers, then prove that $\{\alpha+\beta,\ \beta+\gamma,\ \gamma+\alpha\}$ is also linearly dependent.

14. Show that the set $\{1, x, 1 + x + x^2\}$ is a set of linearly independent vector in the vector space of all polynomials over a real field.

15. In the vector space $\mathbf{R}^3$ let $\alpha = (1,2,1)$, $\beta = (3,1,5)$ and $\gamma = (3,-4,7)$. Show that the subspaces spanned by $S = \{\alpha, \beta\}$ and $T = \{\alpha, \beta, \gamma\}$ are the same, i.e., $L(T) = L(S)$.

[Prove $L(S) \subseteq L(T)$ and then $L(T) \subseteq L(S)$. The result will follow.]

16. Show that

(a) $\{(1,1,-1),(2,-3,5),(-2,1,4)\}$ of $\mathbf{R}^3$ are *L.I.*

(b) $\{(1,1,2,4),(2,-1,-5,2),(1,-1,-4,0),(2,1,1,6)\}$ of $\mathbf{R}^4$ are *L.I.*

17. Show that the vector $\alpha_1 = (1,0,-1)$, $\alpha_2 = (1,2,1)$, $\alpha_3 = (0,-3,2)$ form a basis of $\mathbf{R}^3$. Express each of the standard basis vector, namely $(1,0,0)$, $(0,1,0)$, $(0,0,$ as a linear combination of $\alpha_1, \alpha_2, \alpha_3$.

18. Select a basis, if any such exists, of $\mathbf{R}^3$ from the set $\{\alpha_1, \alpha_2, \alpha_3, \alpha_4\}$ where
$$\alpha_1 = (1,-3,2),\ \alpha_2 = (2,4,1),\ \alpha_3 = (3,1,3),\ \alpha_4 = (1,1,1).$$

[$\{\alpha_1, \alpha_2, \alpha_3\}$ cannot form a basis but $\{\alpha_1, \alpha_2, \alpha_4\}$ can.]

19. Show that the vectors

(a) $\{(2,1,0),(1,1,0),(4,2,0)\}$ form a basis of $\mathbf{R}^3$.

(b) $\{(1,1,2),(1,2,5),(5,3,4)\}$ do not form a basis of $\mathbf{R}^3$.

20. Find the coordinate vector of $(3,5,-2) \in R^3$ relative to the basis
$$B = \{(1,2,3),(2,0,1),(1,-1,0)\}.$$

21. Find a basis for $\mathbf{R}^3$ containing the vector/set of vectors given below:

(a) $\{(2,4,2)\}$;

(b) $\{(2,4,2),(3,6,2)\}$;

(c) $(1,1,2),(3,5,2)$. [C.P. 2003]

22. Prove that the set $S = \{(1,1,0),(1,0,1),(0,1,1)\}$ is a basis set for $\mathbf{R}^3$. Show that the vector $(1,1,1)$ may replace any one of the set S and form a new basis but the same is not true for the vector $(3,1,2)$.

23. Show that the vector $\{(1,0,0),(0,2,1),(1,1,0)\}$ form a basis of the real vector space R^3. **[C.P. 1995]**

[(i) $$a(1,0,0)+b(0,2,1)+c(1,1,0)=(0,0,0)$$
$$\Rightarrow \quad a+c=0, 2b+c=0, b=0.$$
These give $a=0, b=0, c=0$, i.e., the three vectors are L.I.

(ii) Any vector $(\alpha,\beta,\gamma) \in R^3$ can be expressed as a linear combinations of the three given vectors.

$$(\alpha,\beta,\gamma)=\lambda(1,0,0)+\mu(0,2,1)+\nu(1,1,0)$$
$$\Rightarrow \quad \alpha=\lambda+\nu, \beta=2\mu+\nu, \gamma=\mu$$
$$\Rightarrow \quad \mu=\gamma; \nu=\beta-2\mu=\beta-2\gamma; \lambda=\alpha-\nu=\alpha-\beta+2\gamma$$

(i) and (ii) together imply that the three vectors form a basis of $\mathbf{R}^3$.]

Chapter 10

Real Quadratic Form

Real Quadratic Form
1. Real Quadratic Form defined.
2. Associated Symmetric Matrix.
3. **Value class:** Definite, Semi-definite and Indefinite: Direct determination.
4. Eigen values of Associated Matrix: Leading Principal Minors: Rank, Signature of the Normal Form Criterion.

10.1 Real Quadratic Form: Definition and Examples

1. Definition. A Homogeneous polynomial of the form

$$\sum_{j=1}^{n}\sum_{i=1}^{n} a_{ij}x_i x_j \quad \left(\begin{array}{l} a_{ij} \text{ are real numbers and } a_{ij} = a_{ji} \end{array}\right)$$

is called a **real quadratic form** in n variables $x_1, x_2, \ldots, x_n$. We shall, however, restrict ourselves to not more than three variables.

Examples 10.1.1

(i) $3x^2 + 7xy + 5y^2$ is a real quadratic form in two variables x and y.

(ii) $x^2 - y^2 + 2z^2$ is a real quadratic form in the variables x, y and z.

(iii) $2x_1^2 + 2x_2^2 + 5x_3^2 - 4x_2x_3 - 6x_3x_1 + 7x_1x_2$ is a real quadratic form in three variables x_1, x_2, x_3.

2. If $Q(x_1, x_2, x_3, \ldots, x_n) = \sum_{j=1}^{n}\sum_{i=1}^{n} a_{ij}x_ix_j$ (a_{ij}'s are reals) be a real quadratic form in n variables $x_1, x_2, x_3, \ldots, x_n$, then there exists **a unique symmetric square**

matrix B of order n, such that

$$Q = X^TBX,$$

$$\text{where } X = \begin{bmatrix} x_1 \\ x_2 \\ \vdots \\ x_n \end{bmatrix} \text{ and } X^T\text{, its transpose } (x_1, x_2, \ldots, x_n).$$

This matrix B is called the **matrix of the quadratic form** Q.

3. Consider **Example 10.1.1(i)** given above, namely the quadratic form

$$3x^2 + 7xy + 5y^2 = 3x\cdot x + \tfrac{7}{2}x\cdot y + \tfrac{7}{2}y\cdot x + 5y\cdot y$$

We take the matrix

$$B = \begin{bmatrix} 3 & \frac{7}{2} \\ \frac{7}{2} & 5 \end{bmatrix} \quad \left\{ \text{cf. } \begin{matrix} & x & y \\ x & x.x & x.y \\ y & y.x & y.y \end{matrix} \right.$$

$$\text{and } X = \begin{bmatrix} x \\ y \end{bmatrix}, X^T = (\, x \quad y \,).$$

Then

$$X^TB = (\, x \quad y \,)\begin{pmatrix} 3 & \frac{7}{2} \\ \frac{7}{2} & 5 \end{pmatrix} = (\, 3x + \tfrac{7}{2}y \quad \tfrac{7}{2}x + 5y \,).$$

$$\begin{aligned} \therefore X^TBX &= \left(3x + \frac{7}{2}y \quad \frac{7}{2}x + 5y\right)\begin{bmatrix} x \\ y \end{bmatrix} \\ &= \left(3x^2 + \frac{7}{2}xy + \frac{7}{2}xy + 5y^2\right) \\ &= (3x^2 + 7xy + 5y^2). \end{aligned}$$

We then write the quadratic form as $X^TBX = 3x^2 + 7xy + 5y^2$ where

$$B = \begin{pmatrix} 3 & \frac{7}{2} \\ \frac{7}{2} & 5 \end{pmatrix} \text{ is the **matrix of the quadratic form.**}$$

4. Examples 10.1.2 *Write down a symmetric matrix* B *for each of the following quadratic forms and verify that they can be written as* X^TBX.

(i) $x^2 - 18xy + 5y^2$: Write $x^2 - 18xy + 5y^2 = x\cdot x - 9x\cdot y - 9y\cdot x + 5yy$

$$\text{Take } B = \begin{matrix} & x & y \\ x & \big[\, 1 & -9 \\ y & -9 & 5 \,\big] \end{matrix} \text{ and } X = \begin{bmatrix} x \\ y \end{bmatrix}.$$

Check that the given quadratic form is

$$X^TBX = (\, x \quad y \,)\begin{pmatrix} 1 & -9 \\ -9 & 5 \end{pmatrix}\begin{bmatrix} x \\ y \end{bmatrix}.$$

(ii) $x_1^2 + x_2^2 + x_3^2 - x_1x_2 - x_2x_3 - x_3x_1$.
We write the given expression in the form:

$$\left(x_1 \cdot x_1 - \frac{1}{2}x_1 \cdot x_2 - \frac{1}{2}x_1 \cdot x_3\right) + \left(-\frac{1}{2}x_2 \cdot x_1 + x_2 \cdot x_2 - \frac{1}{2}x_2 \cdot x_3\right)$$
$$+ \left(-\frac{1}{2}x_3 \cdot x_1 - \frac{1}{2}x \cdot x_2 + x_3 \cdot x_3\right).$$

We then take the matrix
$$B = \begin{matrix} & x_1 & x_2 & x_3 \\ x_1 & & & \\ x_2 & & & \\ x_3 & & & \end{matrix} \begin{pmatrix} 1 & -1/2 & -1/2 \\ -1/2 & 1 & -1/2 \\ -1/2 & -1/2 & 1 \end{pmatrix}.$$

Then the given quadratic form can be easily written as

$$X^T BX \text{ where } X = \begin{bmatrix} x_1 \\ x_2 \\ x_3 \end{bmatrix} \quad \textbf{(Check)}.$$

(iii) $x^2 + 2y^2 - 5z^2 - xy + 4yz - 3zx$.
This expression can be written as

$$\left(x \cdot x - \frac{1}{2}x \cdot y - \frac{3}{2}xz\right) + \left(-\frac{1}{2}y \cdot x + 2y \cdot y + 2y \cdot z\right)$$
$$+ \left(-\frac{3}{2}z \cdot x + 2z \cdot y - 5z \cdot z\right).$$

We take the matrix
$$B = \begin{matrix} & x & y & z \\ x & & & \\ x & & & \\ x & & & \end{matrix} \begin{pmatrix} 1 & -1/2 & -3/2 \\ -1/2 & 2 & 2 \\ -3/2 & 2 & -5 \end{pmatrix}.$$

Now see that if $X = \begin{bmatrix} x \\ y \\ z \end{bmatrix}$ so that $X^T = (\,x \quad y \quad z\,)$.

Then

$$X^T B = (\,x \quad y \quad z\,) \begin{pmatrix} 1 & -\frac{1}{2} & -\frac{3}{2} \\ -\frac{1}{2} & 2 & 2 \\ -\frac{3}{2} & 2 & -5 \end{pmatrix}$$

$$= \left(\,x - \tfrac{1}{2}y - \tfrac{3}{2}z \quad -\tfrac{1}{2}x + 2y + 2z \quad -\tfrac{3}{2}x + 2y - 5z\,\right)$$

$$\therefore\ X^T BX = \left(\,x - \tfrac{1}{2}y - \tfrac{3}{2}z \quad -\tfrac{1}{2}x + 2y + 2z \quad -\tfrac{3}{2}x + 2y - 5z\,\right) \begin{pmatrix} x \\ y \\ z \end{pmatrix}$$

$$= \left(\,x^2 - \tfrac{1}{2}xy - \tfrac{3}{2}xz - \tfrac{1}{2}yx + 2y^2 + 2yz - \tfrac{3}{2}xz + 2yz - 5z^2\,\right)$$

$$\text{or, } X^T BX = x^2 + 2y^2 - 5z^2 - xy - 3xz + 4yz.$$

5. *What is the quadratic form corresponding to the symmetric matrix:*

$$B = \begin{bmatrix} 1 & 2 & 3 \\ 2 & 0 & 3 \\ 3 & 3 & 1 \end{bmatrix}?$$

Solution. Let $X = \begin{bmatrix} x \\ y \\ z \end{bmatrix}$. Then $X^T BX$ is the quadratic form corresponding to the given matrix B.

We have

$$X^T B = \begin{pmatrix} x & y & z \end{pmatrix} \begin{pmatrix} 1 & 2 & 3 \\ 2 & 0 & 3 \\ 3 & 3 & 1 \end{pmatrix}$$

$$= \begin{pmatrix} x + 2y + 3z & 2x + 3z & 3x + 3y + z \end{pmatrix}$$

$$\therefore X^T BX = \begin{pmatrix} x + 2y + 3z & 2x + 3z & 3x + 3y + z \end{pmatrix} \begin{pmatrix} x \\ y \\ z \end{pmatrix}$$

$$= \left[x(x + 2y + 3z) + y(2x + 3z) + z(3x + 3y + z) \right]$$

$$\therefore X^T BX = x^2 + z^2 + 4xy + 6xz + 6yz \text{ (Reqd. Result)}$$

6. *Write down the quadratic form corresponding to the diagonal matrix*

$$A = \begin{bmatrix} \lambda_1 & 0 & 0 \\ 0 & \lambda_2 & 0 \\ 0 & 0 & \lambda_3 \end{bmatrix}.$$

Let $X = \begin{pmatrix} x_1 \\ x_2 \\ x_3 \end{pmatrix}$ so that $X^T = \begin{pmatrix} x_1 & x_2 & x_3 \end{pmatrix}$.

$$\therefore X^T A = \begin{pmatrix} x_1 & x_2 & x_3 \end{pmatrix} \begin{pmatrix} \lambda_1 & 0 & 0 \\ 0 & \lambda_2 & 0 \\ 0 & 0 & \lambda_3 \end{pmatrix}$$

$$= \begin{pmatrix} \lambda_1 x_1 & \lambda_2 x_2 & \lambda_3 x_3 \end{pmatrix}$$

Hence

$$X^T AX = \begin{pmatrix} \lambda_1 x_1 & \lambda_2 x_2 & \lambda_3 x_3 \end{pmatrix} \begin{bmatrix} x_1 \\ x_2 \\ x_3 \end{bmatrix}$$

$$= \left(\lambda_1 x_1^2 + \lambda_2 x_2^2 + \lambda_3 x_3^2 \right).$$

We write the quadratic form as $X^T AX = \lambda_1 x_1^2 + \lambda_2 x_2^2 + \lambda_3 x_3^2$.

7. *What is the associated matrix of the quadratic form $x^2 - 2y^2 + 3z^2$?*

The given quadratic form can be written as

$$\begin{aligned}(x\cdot 1x + x\cdot 0y + x\cdot 0z) \;&+\; (y\cdot 0x - y\cdot 2y + y\cdot 0z)\\ &+\; (z\cdot 0x + z\cdot 0y + z\cdot 3z).\end{aligned}$$

Take the matrix

$$B = \begin{matrix} & x & y & z \\ x & 1 & 0 & 0 \\ y & 0 & -2 & 0 \\ z & 0 & 0 & 3 \end{matrix}.$$

Then $X^TBX = x^2 - 2y^2 + 3z^2$.
So the required associated matrix is the diagonal matrix

$$B = \begin{pmatrix} 1 & 0 & 0 \\ 0 & -2 & 0 \\ 0 & 0 & 3 \end{pmatrix}.$$

10.2 Value Class of a Real Quadratic Form: Definite, Semi-definite and Indefinite

Definitions: Let $Q = X^TBX$ be a real quadratic form in n variables $x_1, x_2, \ldots, x_n$ with associated symmetric matrix B. Then the Q is said to be

(i) Positive definite if $Q \geq 0$ for all real values of the variables $x_1, x_2, \ldots, x_n$ and $Q = 0$ iff $X = 0$ (i.e., $Q = 0 \Rightarrow x_1 = x_2 = \cdots = x_n = 0$).

▶ Example 10.2.1 *The quadratic form* $x^2 - 4xy + 5y^2$ *in two variables is* **positive definite** *because it can be written as*

$$(x - 2y)^2 + y^2$$

which is ≥ 0 *for all real values of* x, y **and**

$$\begin{aligned}(x - 2y)^2 + y^2 = 0 \;&\Rightarrow\; x - 2y = 0 \textit{ and } y = 0\\ &\Rightarrow\; x = 0 \textit{ and } y = 0.\end{aligned}$$

▶ Example 10.2.2 *The quadratic form* $x_1^2 + x_2^2 + x_3^2$ *in three variables* x_1, x_2, x_3 *is clearly* **positive definite.**

(ii) Negative definite if $Q \leq 0$ for all real values of the variables $x_1, x_2, \ldots, x_n$ and $Q = 0$ if and only if $x_1 = x_2 = \cdots = x_n = 0$.

▶ Example 10.2.3 $-x^2 - 2y^2 - 3x_3^2$ *is a negative definite quadratic form in three variables* x, y, z.

(iii) Positive semi-definite if $Q \geq 0$ for all real values of the variables $x_1, x_2, \ldots, x_n$ and $Q = 0$ for some non-zero real vector X, i.e., $Q = 0$ for some real values of the variables $x_1, x_2, \ldots, x_n$, **not all zero.**

▶ Example 10.2.4 *The real quadratic form* $x^2 + y^2 + 2z^2 - 2xz - 2yz$ *is* **positive semi-definite** *because it can be written in the form* $(x - z)^2 + (y - z)^2$ *which is* ≥ 0, *for all real values of* x, y, z *but is zero for non-zero values also, for example* $x = y = z = 1$.

► Example 10.2.5 *The real quadratic form $x^2 + y^2 + 0z^2$ in three variables x, y, z is positive semi-definite. It is non-negative for all real values of x, y, z and it is zero for values $x = y = 0$ and $z = 3$, which are not all zero.*

Clearly it is **positive definite** *in two variables x and y. (Why?)*

(iv) Negative semi-definite. If $Q \le 0$ for all real values of the variables x_1, $x_2, \ldots, x_n$ and $Q = 0$ for some values of the variables $x_1, x_2, \ldots, x_n$, **not all zero.**

► Example 10.2.6 *The quadratic form $Q = -x^2 - y^2 - 0z^2$ in three variables x, y, z is* **negative semi-definite.**

(v) Indefinite if Q takes **positive** as well as **negative** values for real values of the variables $x_1, x_2, \ldots, x_n$.

► Example 10.2.7 *The quadratic form $x^2 - y^2 + z^2$ in three variables is indefinite. It takes positive value 1 when $x = 1, y = 1, z = 1$ and it takes negative value -1 when $x = 0, y = 1, z = 0$.*

Note 1. (i) *The aforesaid five classes of real quadratic forms are mutually exclusive and they are known as the*

Value classes of real quadratic forms

Every real quadratic form must belong to one and only one value class.

(ii) *A form which is either* positive definite *or* negative definite *is called* ***definite quadratic form.***

A form which is either positive semi-definite *or* negative semi-definite *is called* ***semi-definite quadratic form.***

On similar lines, can make the following **Classification of Real Symmetric Matrices:**

A real symmetric matrix A is said to be

definite, semi-definite or indefinite

if the corresponding quadratic form $X^T AX$ is respectively

definite, semi-definite or indefinite.

10.3 Value Class of a Real Quadratic Form by Direct Calculation

► Example 10.3.1 *By direct calculation justify that the real quadratic form in three variables $Q \equiv x^2 + 2y^2 + 4z^2 + 2xy - 4yz - 2zx$ is positive definite.*

Solution: Here $Q = (x + y - z)^2 + (y - z)^2 + 2z^2$.

$\therefore$ $Q \ge 0$ for all real values of x, y, z and $Q = 0$ when

$$x + y - z = 0,\ y - z = 0,\ z = 0$$
$$\Rightarrow z = 0,\ y = 0,\ x = 0.$$

Therefore, Q is **positive definite.**
Also we can state that the associated symmetric matrix

$$A = \begin{pmatrix} 1 & 1 & -1 \\ 1 & 2 & -2 \\ -1 & -2 & 4 \end{pmatrix} \text{ is positive definite.}$$

▶ Example 10.3.2 *The real quadratic form Q in three variables x_1, x_2, x_3, given by*

$$Q \equiv 2x_1^2 + x_2^2 + 2x_3^2 + 2x_1x_2 - 2x_2x_3 - 4x_3x_1$$

is positive semi-definite.

Solution: $Q = (x_1 + x_2 - x_3)^2 + (x_1 - x_3)^2 \geq 0$ for all real values of x_1, x_2, x_3. Also Q may be zero for $x_1 = 1, x_2 = 0, x_3 = 1$ where x_1, x_2, x_3 are not all zero.
∴ Q is positive semi-definite and also the associated matrix

$$\begin{pmatrix} 2 & 1 & -2 \\ 1 & 1 & 1 \\ -2 & 1 & 2 \end{pmatrix} \text{ is positive semi-definite.}$$

▶ Example 10.3.3 *Let*

$$\begin{aligned} Q(x, y, z) &= x^2 + y^2 + 2xy + 2yz + 2zx \\ &= (x + y + z)^2 - z^2. \end{aligned}$$

Q *is* **indefinite** *because*

$$\begin{aligned} & Q > 0 \textit{ for some } x, y, z \text{ (e.g., } x = 1, y = 1, z = 0) \\ \textit{and} \quad & Q < 0 \textit{ for some } x, y, z \text{ (e.g., } x = 0, y = 1, z = 1). \end{aligned}$$

10.4 Criterions for the Value Class of a Real Quadratic Form

We have four criteria for the determination of value class of a real quadratic form. We shall state the criterions without proof:

First criterion

By Direct calculation (as shown by examples in Art. 10.3)

Second Criterion

By finding the eigen values of the associated matrix. (The concept of Eigen value of a square matrix has been explained in **Chapter 11)**

Theorem 10.4.1. *A real quadratic form X^TAX in n variables is*

(i) **positive definite** *if and only if all the eigen values of A are positive*

(ii) **negative definite** *if and only if all the eigen values of A are negative*

(iii) **positive semi-definite** *if and only if all the eigen values of* A *are* ≥ 0 *and at least one eigen value of* A *is zero*

(iv) **negative semi-definite** *if and only if all the eigen values of* A *are* ≥ 0 *and at least one eigen value of* A *is zero*

(v) **indefinite** *if and only if* A *has positive as well as negative eigen values.*

▶ Example 10.4.1 *Write the associated matrix* A *of the quadratic form*

$$Q = 6x^2 + 35y^2 + 11z^2 + 4zx.$$

Find the eigen values of A *and hence determine the value class of* Q.

Solution: The associated matrix A of the given quadratic form Q is

$$A = \begin{bmatrix} 6 & 0 & 2 \\ 0 & 35 & 0 \\ 2 & 0 & 11 \end{bmatrix}.$$

The characteristic equation of A is

$$\begin{vmatrix} 6-\lambda & 0 & 2 \\ 0 & 35-\lambda & 0 \\ 2 & 0 & 11-\lambda \end{vmatrix} = 0$$

$$\begin{aligned} \text{or,} \quad & (6-\lambda)(35-\lambda)(11-\lambda) - 2(70-2\lambda) = 0 \\ \text{or,} \quad & (35-\lambda)\{(6-\lambda)(11-\lambda) - 4\} = 0 \\ \text{or,} \quad & (35-\lambda)(\lambda^2 - 17\lambda + 62) = 0 \\ & \lambda = 35, \ \frac{17+\sqrt{41}}{2}, \ \frac{17-\sqrt{41}}{2}. \end{aligned}$$

Since the eigen values of A are all positive, therefore, the quadratic form is positive definite.

Third criterion

Positive definite if the leading principal minors of the associated matrix A are all positive.

Leading principal minors of a square matrix:

Let $A = \begin{pmatrix} a_{11} & a_{12} & a_{13} \\ a_{21} & a_{22} & a_{23} \\ a_{31} & a_{32} & a_{33} \end{pmatrix}$ **be a square matrix of order three.**

Then

$$A_1 = \begin{vmatrix} a_{11} \end{vmatrix}, \ A_2 = \begin{vmatrix} a_{11} & a_{12} \\ a_{21} & a_{22} \end{vmatrix}, \ A_3 = \begin{vmatrix} a_{11} & a_{12} & a_{13} \\ a_{21} & a_{22} & a_{23} \\ a_{31} & a_{32} & a_{33} \end{vmatrix}$$

are called **leading principal minors** of A.

Theorem 10.4.2. *A real quadratic form* $X^T AX$ *is* **positive definite** *if and only if the leading principal minors of* A *are all positive.*

A real quadratic form $X^T AX$ *is* **negative definite** *if and only if the leading principal minors of* A *are* **negative** *and* **positive alternately.**

In case the leading principal minors of A *are, say* **positive, zero** *and* **negative,** *then the quadratic form is* **indefinite.**

▶ **Example 10.4.2** *Let* $Q = 18x^2 + 18y^2 + 18z^2 - 12xy - 12yz - 12zx.$

Solution: Here the associated matrix A of Q is

$$A = \begin{vmatrix} 18 & -6 & -6 \\ -6 & 18 & -6 \\ -6 & -6 & 18 \end{vmatrix}$$

The leading principal minors of A are

$$|\,18\,|, \begin{vmatrix} 18 & -6 \\ -6 & 18 \end{vmatrix}, \begin{vmatrix} 18 & -6 & -6 \\ -6 & 18 & -6 \\ -6 & -6 & 18 \end{vmatrix}$$

or, 18, 288, 216 × 16 (all positive)

∴ Q is **positive definite.**

▶ **Example 10.4.3** *Let* $Q = 2x^2 + 2y^2 + 2z^2 + 4xy - 4xz + 4yz.$

Solution: Here the associated matrix A of Q is

$$A = \begin{vmatrix} 2 & 2 & -2 \\ 2 & 2 & 2 \\ -2 & 2 & 2 \end{vmatrix}.$$

The leading principal minors of A are

$$|\,2\,|, \begin{vmatrix} 2 & 2 \\ 2 & 2 \end{vmatrix} = 0, \begin{vmatrix} 2 & 2 & -2 \\ 2 & 2 & 2 \\ -2 & 2 & 2 \end{vmatrix} = -32.$$

They are not all positive, not also alternately positive, negative.
Here the quadratic form is **indefinite.**

Fourth Criterion

Rank of a Real Quadratic Form

The **rank of a real quadratic** form $X^T AX$ is defined by the rank of the associated real symmetric matrix A.

If $X^T AX$ is a quadratic form of rank r, then there exists a non-singular matrix P which will reduce the quadratic form $X^T AX$ to a sum of r square terms. (See Example 10.4.4 below).

Working Rule

We shall transform the symmetric matrix A to diagonal form by applying **congruent** operations (will be explained through examples). Then the application of the corresponding column operations to the unit matrix **I** will give a non-singular matrix P such that $P^T AP$ = a diagonal matrix.

► Example 10.4.4 *Determine a non-singular matrix* P *such that* $P^T AP$ *is a diagonal matrix, where*

$$A = \begin{bmatrix} 6 & -2 & 2 \\ -2 & 3 & -1 \\ 2 & -1 & 3 \end{bmatrix}$$

What is the quadratic form corresponding to this symmetric matrix A? What is the non-singular transformation corresponding to the matrix P *which will reduce the quadratic form to the sum of three square terms?*

Solution: In order to reduce the matrix A to the diagonal form **we shall apply** congruent operations (i.e., *row operations, followed by exactly similar column operations*).

We write,

$$A = \mathbf{I}\ A\ \mathbf{I}$$

$$\begin{bmatrix} 6 & -2 & 2 \\ -2 & 3 & -1 \\ 2 & -1 & 3 \end{bmatrix} = \begin{bmatrix} 1 & 0 & 0 \\ 0 & 1 & 0 \\ 0 & 0 & 1 \end{bmatrix} A \begin{bmatrix} 1 & 0 & 0 \\ 0 & 1 & 0 \\ 0 & 0 & 1 \end{bmatrix}. \quad (1)$$

We have written on the right side **I A I.**

The row operations will be applied on the **pre-factor I** and the column operations will be applied on the **post-factor I.**

[In fact, there is no need of actually applying row-operations on the **pre-factor I,** because at any stage, the matrix obtained by applying row operations on **pre-factor I** will be the **transpose** of the matrix obtained by applying **column operations** on **post-factor I**]

Perform the congruent operations:

$$\left.\begin{aligned} r_2' \text{ (i.e., new row}_2) &= r_2 + \tfrac{1}{3}r_1 \\ r_3' \text{ (i.e., new row}_3) &= r_3 - \tfrac{1}{3}r_1 \end{aligned}\right\}. \quad \text{(I)}$$

$$\left.\begin{aligned} c_2' \text{ (i.e., new col}_2) &= c_2 + \tfrac{1}{3}c_1 \\ c_3' \text{ (i.e., new col}_3) &= c_3 - \tfrac{1}{3}c_1 \end{aligned}\right\}. \quad \text{(II)}$$

$$\begin{bmatrix} 6 & 0 & 0 \\ 0 & 7/3 & -1/3 \\ 0 & -1/3 & 7/3 \end{bmatrix} = \begin{bmatrix} 1 & 0 & 0 \\ 1/3 & 1 & 0 \\ -1/3 & 0 & 1 \end{bmatrix} A \begin{bmatrix} 1 & 1/3 & -1/3 \\ 0 & 1 & 0 \\ 0 & 0 & 1 \end{bmatrix}$$

First perform Col. operations II and then the row operations I (left matrix)

Now transpose (right matrix → middle matrix)

First perform Col. operations II on the post factor I (right matrix)

Next perform another set of congruent operations:

$$r_3' = r_3 + \tfrac{1}{7}r_2; \qquad \text{(III)}$$

$$c_3' = c_3 + \tfrac{1}{7}c_2. \qquad \text{(IV)}$$

$$\begin{bmatrix} 6 & 0 & 0 \\ 0 & 7/3 & 0 \\ 0 & 0 & 16/3 \end{bmatrix} = \overset{P^T}{\begin{bmatrix} 1 & 0 & 0 \\ 1/3 & 1 & 0 \\ -2/7 & 1/7 & 1 \end{bmatrix}} A \overset{P}{\begin{bmatrix} 1 & 1/3 & -2/7 \\ 0 & 1 & 1/7 \\ 0 & 0 & 1 \end{bmatrix}}$$

First perform Col. operations IV and then row operations III — Then transpose — First term IV operations

Thus we obtain a non-singular matrix

$$P = \begin{bmatrix} 1 & 1/3 & -2/7 \\ 0 & 1 & 1/7 \\ 0 & 0 & 1 \end{bmatrix}$$

such that $P^TAP =$ diagonal matrix $\{6, 7/3, 16/7\}$.

Now see that the quadratic form corresponding to the symmetric matrix A is

$$X^TAX = 6x_1^2 + 3x_2^2 + 3x_3^2 - 4x_1x_2 + 4x_1x_3 - 2x_2x_3. \qquad (2)$$

The non-singular transformation corresponding to the matrix P is given by $X = PY$, i.e.,

$$\begin{bmatrix} x_1 \\ x_2 \\ x_3 \end{bmatrix} = \begin{bmatrix} 1 & 1/3 & -2/7 \\ 0 & 1 & 1/7 \\ 0 & 0 & 1 \end{bmatrix} \begin{bmatrix} y_1 \\ y_2 \\ y_3 \end{bmatrix}$$

which gives

$$\left.\begin{aligned} x_1 &= y_1 + \tfrac{1}{3}y_2 - \tfrac{2}{7}y_3 \\ x_2 &= y^2 + \tfrac{1}{7}y_3 \\ x_3 &= y_3 \end{aligned}\right\}. \qquad (3)$$

The transformation (3) will reduce the quadratic form (2) to the diagonal form

$$(PY)^TA(PY) = 6y_1^2 + \frac{7}{3}y_2^2 + \frac{16}{7}y_3^2.$$

When the rank of the Quadratic form X^TAX is 3, we have observed that the transformation $X = PY$ (P, as given above) reduces to the form which is the sum of three square terms, the coefficient being 6, $\frac{7}{3}$, $\frac{16}{7}$. We could perform further operations ($r_1' = \frac{1}{\sqrt{6}}r_1$, $c_1' = \frac{1}{\sqrt{6}}c_1$; $r_2' = \frac{\sqrt{3}}{7}r_2$, $c_2' = \sqrt{\frac{3}{7}}c_2$ and $r_3' = \sqrt{\frac{7}{16}}r_3$, $c_3' = \sqrt{\frac{7}{16}}c_3$) and the new P would be

$$\begin{bmatrix} a & \frac{1}{3}b & -\frac{2}{7}c \\ 0 & b & \frac{1}{7}c \\ 0 & 0 & c \end{bmatrix} = P' \text{ (say)}$$

where $a = \frac{1}{\sqrt{6}}$, $b = \sqrt{\frac{7}{16}}$, $c = \sqrt{\frac{7}{16}}$, then the transformation $X = P'Y$ will reduce to the Normal Form or Canonical Form $y_1^2 + y_2^2 + y_3^2$, the coefficients being 1, 1, 1.

Important Definitions

If X^TAX be a real Quadratic form of rank r in n variables, then there exists a non-singular transformation $X = PY$ which transforms X^TAX to the form:

$$\begin{aligned} X^TAX &= (PY)^TA(PY) = y_1^2 + y_2^2 + \cdots + y_r^2 - y_{p+1}^2 - y_{p+2}^2 - \cdots - y_r^2 \\ &= \text{Sum and Difference of the squares of the new variables.} \end{aligned}$$

This expression is called the Canonical or Normal form of the given Quadratic form.
p = the number of positive terms in the normal form of X^TAX = index of the Quadratic form
s = the excess of the number of positive terms over the number of negative terms in the normal form = signature of the quadratic form,

i.e., $s = p - (r - p) = 2p - r.$

Theorem 10.4.3. *Suppose r is the* **rank** *and s is the* **signature** *of a real quadratic form $Q = X^TAX$ in n variables. Then*

(i) Q *is* **positive definite** *if and only if* $s = r = n$.
(ii) Q *is* **negative definite** *if and only if* $-s = r = n$.
(iii) Q *is* **positive semi-definite** *if and only if* $s = r < n$.
(iv) Q *is* **negative semi-definite** *if and only if* $-s = r < n$.
and (v) Q *is* **indefinite** *if and only if* $|s| \neq r$.

We assume the truth of this theorem without a formal proof.

Now in **Example 10.4.4** above.

Rank of the matrix $A = r = 3$ = Rank of the quadratic from X^TAX.

See that the diagonal form of the Quadratic form $= 6y_1^2 + \frac{7}{3}y_2^2 + \frac{16}{7}y_3^2$ (or its Normal form $y_1^2 + y_2^2 + y_3^2$).

$\therefore$ **Index** of the quadratic form = number of positive terms of the normal form = 3.

Hence **signature** $(s) = 2p - r = 2 \times 3 - 3 = 3$.

Thus in **Example 10.4.4,** $s = r = n = 3$ (n is the number of variables).

$\therefore$ Here the quadratic form is **positive definite.**

We give a few more examples:

▶ Example 10.4.5 *Reduce the quadratic form in 3 variables x, y, z : $X^TAX = 2x^2 + y^2 - 3z^2 - 8yz - 4zx + 12xy$ to canonical form (P). Find its rank and signature. What is the linear transformation ($X = PY$) which brings about the normal reduction? What is its value class by* **Theorem 10.4.3?**

Solution: The matrix A of the quadratic form $A = \begin{bmatrix} 2 & 6 & -2 \\ 6 & 1 & -4 \\ -2 & -4 & -3 \end{bmatrix}$.

We write $A = I\ A\ I$,

$$\text{i.e.,}\quad \begin{bmatrix} 2 & 6 & -2 \\ 6 & 1 & -4 \\ -2 & -4 & -3 \end{bmatrix} = \begin{bmatrix} 1 & 0 & 0 \\ 0 & 1 & 0 \\ 0 & 0 & 1 \end{bmatrix} A \begin{bmatrix} 1 & 0 & 0 \\ 0 & 1 & 0 \\ 0 & 0 & 1 \end{bmatrix}.$$

We shall reduce A to a diagonal matrix by applying congruence operation on its. Perform the operations

$$\left.\begin{aligned} r_2' &= r_2 - 3r_1 \\ \text{and then } r_3' &= r_3 - r_1 \end{aligned}\right\} \quad \left.\begin{aligned} c_2' &= c_2 - 3c_1 \\ c_3' &= c_3 - c_1 \end{aligned}\right\}$$

$$\begin{bmatrix} 2 & 0 & 0 \\ 0 & -17 & 2 \\ 0 & 2 & -5 \end{bmatrix} = \begin{bmatrix} 1 & 0 & 0 \\ -3 & 1 & 0 \\ 1 & 0 & 1 \end{bmatrix} A \begin{bmatrix} 1 & -3 & 1 \\ 0 & 1 & 0 \\ 0 & 0 & 1 \end{bmatrix}.$$

Now perform the operations $r_3' \to r_3 + \frac{2}{17}r_2,\ c_3' \to c_3 + \frac{2}{17}c_2$

$$\begin{bmatrix} 2 & 0 & 0 \\ 0 & -17 & 0 \\ 0 & 0 & -81/17 \end{bmatrix} = \begin{bmatrix} 1 & 0 & 0 \\ -3 & 1 & 0 \\ 11/17 & 2/17 & 1 \end{bmatrix} A \begin{bmatrix} 1 & -3 & 11/17 \\ 0 & 1 & 2/17 \\ 0 & 0 & 1 \end{bmatrix}.$$

Thus we obtain a non-singular matrix

$$P = \begin{bmatrix} 1 & -3 & 11/7 \\ 0 & 1 & 2/7 \\ 0 & 0 & 1 \end{bmatrix}.$$

such that $P^T AP = \text{diag}\{2, -17, -81/17\}$.

The non-singular transformation corresponding to matrix P is given by $X = PY$, i.e.,

$$\begin{bmatrix} x \\ y \\ z \end{bmatrix} = \begin{bmatrix} 1 & -3 & 11/7 \\ 0 & 1 & 2/7 \\ 0 & 0 & 1 \end{bmatrix} \begin{bmatrix} y_1 \\ y_2 \\ y_3 \end{bmatrix}$$

which is same as

$$\begin{aligned} x &= y_1 - 3y_2 + \tfrac{11}{7}y_3 \\ y &= y_2 + \tfrac{2}{7}y_3 \\ z &= y_3 \end{aligned}$$

This transformation will reduce the given quadratic form to the **diagonal form**

$$(PY)^T A(PY) = 2y_1^2 - 17y_2^2 - \frac{81}{17}y_3^2.$$

With further transformations we could find P' s.t. the quadratic form reduces to $y_1^2 - y_2^2 - y_3^2$.

The rank of the quadratic form = 3 = rank of the matrix A = r.

Index of the quadratic form = number of positive terms = 1 = p.

Signature = s = 2p − r = 2.1 − 3 = −1. Number of variables = 3 = n.

The quadratic form is **indefinite because** $|s| \neq r$.

What are the four criterion to determine the value class of a real quadratic form?
First criterion: Direct calculation (See Examples 10.2.1–10.2.7).
Also see Examples 10.3.1–10.3.3.
Second criterion: By finding eigen values of the associated matrix (See Example 10.4.1).
Third criterion: Leading Principal Minors of the matrix (See Examples 10.4.2–10.4.3).
Fourth criterion: Rank-signature criterion (See Examples 10.4.4–10.4.5).

Exercises on Chapter 10

1. Write down the matrix of each of the following real quadratic forms:

(a) $x^2 - 18xy + 5y^2$;

(b) $x^2 + 2y^2 + 3z^2 + 4xy + 5yz + 6zx$;

(c) $ax^2 + by^2 + cz^2 + du^2 + cv^2$;

(d) $x^2 + y^2 + z^2 + 4xy + 2xz + 4yz$.

2. Write down the quadratic forms corresponding to the following matrices:

(a) $\begin{bmatrix} 2 & 1 & 5 \\ 1 & 3 & -2 \\ 5 & -2 & 4 \end{bmatrix}$; **(b)** $\begin{bmatrix} 0 & 1 & 2 & 3 \\ 1 & 2 & 3 & 4 \\ 2 & 3 & 4 & 5 \\ 3 & 4 & 5 & 6 \end{bmatrix}$.

3. Reduce each of the following quadratic forms in three variables to canonical form (or diagonal form) and then find its rank and signature. Also write down in each the linear transformation which brings about the reduction:

(a) $x_1^2 - 2x_2^2 + 3x_3^2 - 4x_2x_3 + 6x_3x_1$;

(b) $6x^2 + 3y^2 + 14z^2 + 4yz + 18zx + 4xy$;

(c) $x_1^2 + 2x^2 + 2x_3^2 - 2x_1x_2 - 2x_2x_3 + x_3x_1$.

4. Prove that the quadratic form $6x^2 + 3y^2 + 14z^2 + 4yz + 18zx + 4xy$ in three variables is positive definite (by fourth criterion, i.e., by find the rank and signature).

5. Use the criterion of leading principal minors of the associated matrix of the following quadratic form to find the value class:

(a) $6x^2 + 3y^2 + 3z^2 - 4xy - 2yz + 4yz$;

(b) $6x_1^2 + 49x_2^2 + 5x_3^2 - 82x_2x_3 + 20x_3x_1 - 4x_1x_2$.

6. Write the matrix A of the quadratic form $6x^2 + 35y^2 + 11y^2 + 4zx$. find the eigen values of A and hence determine the value class of this quadratic form.

Answers

1. (a) $\begin{bmatrix} 1 & -9 \\ -9 & 5 \end{bmatrix}$;

(b) $\begin{bmatrix} 1 & 2 & 3 \\ 2 & 2 & 5/2 \\ 3 & 5/2 & 3 \end{bmatrix}$;

(c) $\begin{bmatrix} a & 0 & 0 & 0 & 0 \\ 0 & b & 0 & 0 & 0 \\ 0 & 0 & c & 0 & 0 \\ 0 & 0 & 0 & d & 0 \\ 0 & 0 & 0 & 0 & e \end{bmatrix}$ or we may write it as: diag $[a \ b \ c \ d \ e]$;

(d) $\begin{bmatrix} 1 & 2 & 1 \\ 2 & 1 & 2 \\ 1 & 2 & 1 \end{bmatrix}$.

2. (a) $2x^2 + 3y^2 + 4z^2 + 2xy + 10xz - 4yz$;

(b) $2y^2 + 4z^2 + 6u^2 + 2xy + 4xz + 6xu + 6yz + 8yu + 10zu$.

3. (a) $y_1^2 - 2y_2^2 - 4y_3^2$; rank $r = 3$; signatures $= -1$; $x_1 = y_1 - 3y_3$; $x_2 = y_2 - y_3$; $x_3 = y_3$.

(b) $y_1^2 + \frac{7}{3}y_2^2 + \frac{1}{14}y_3^2$.

Chapter 11

Characteristic Equation of a Square Matrix: Cayley-Hamilton Theorem

	Characteristic Equation: Eigen Values and Eigen Vectors
1.	Characteristic equation of a square matrix.
2.	Eigen values and Eigen vectors (*Problems*).
3.	*Cayley-Hamilton Theorem:* Its applications.

11.1 Introduction: Eigen values of a square matrix defined

Let $A = (a_{ij})_{n \times n}$ be a square matrix of order n. We shall assume that the elements of A are all real numbers. Let I be the unit square matrix of order n (i.e., all the elements of the main diagonal are 1 and the other elements are all zero).

The matrix $A - \lambda I$, where λ is a scalar, is called the **characteristic matrix** of A.

The determinant $|A - \lambda I|$ is a **polynomial** in λ of degree n. This polynomial is known as **characteristic polynomial** of A.

The equation $|A - \lambda I| = 0$ is called the **characteristic equation** of the **square matrix** A.

The roots of this equation are called the **characteristic roots or eigen values** of A.

Corresponding to each eigen value we can find one vector, called **eigen vector** (**To be discussed in more details at a subsequent article**).

The definitions given above will be more clear when we shall consider an example:

▶ **Example 11.1.1** *Find the eigen values of the following third order square matrix*

$$A = \begin{bmatrix} 1 & 2 & 3 \\ 0 & -4 & 2 \\ 0 & 0 & 7 \end{bmatrix}_{3 \times 3}.$$

Solution: The characteristic polynomial of A is

$$\begin{aligned}\det(A-\lambda I) &= \begin{bmatrix} 1-\lambda & 2 & 3 \\ 0 & -4-\lambda & 2 \\ 0 & 0 & 7-\lambda \end{bmatrix} \\ &= (1-\lambda)(-4-\lambda)(7-\lambda).\end{aligned}$$

∴ the characteristic equation is $(1-\lambda)(-4-\lambda)(7-\lambda)=0$.

The roots of the above cubic equation in λ are 1, −4 and 7. By definition, these three values are the eigen values of A.

Exercises on Chapter 11(I)

1. Determine the characteristic roots of each of the following square matrices over the field of complex numbers:

(a) $\begin{pmatrix} 0 & 1 & 2 \\ 1 & 0 & -1 \\ 2 & -1 & 0 \end{pmatrix}$; (c) $\begin{pmatrix} 1 & -1 & -1 \\ 1 & -1 & 0 \\ 1 & 0 & -1 \end{pmatrix}$; (e) $\begin{pmatrix} -9 & 4 & 4 \\ -8 & 3 & 4 \\ -16 & 8 & 7 \end{pmatrix}$;

(b) $\begin{pmatrix} 1 & 2 \\ -1 & 3 \end{pmatrix}$; (d) $\begin{pmatrix} 1 & -1 & 4 \\ 0 & 3 & 7 \\ 0 & 0 & 5 \end{pmatrix}$; (f) $\begin{pmatrix} 0 & 1 & 0 & 0 \\ 0 & 0 & 1 & 0 \\ 0 & 0 & 0 & 1 \\ 1 & 0 & 0 & 0 \end{pmatrix}$.

[*Ans.* (a) $2, -1 \pm \sqrt{3}$; (b) $4 \pm \sqrt{-4} = 4 \pm 2i$; (c) $-1, i, -i$;]
(d) $1, 3, 5$; (e) $-1, -1, 3$, (f) $1, -1, i, -i$.]

2. Verify that the matrices

$$\begin{pmatrix} b & c & a \\ c & a & b \\ a & b & c \end{pmatrix}, \begin{pmatrix} c & a & b \\ a & b & c \\ b & c & a \end{pmatrix}, \begin{pmatrix} a & b & c \\ b & c & a \\ c & a & b \end{pmatrix}$$

have the same *characteristic equation*.

3. Show that the matrices

$$A = \begin{pmatrix} 0 & h & g \\ h & 0 & f \\ g & f & 0 \end{pmatrix}, B = \begin{pmatrix} 0 & f & h \\ f & 0 & g \\ h & g & 0 \end{pmatrix}, C = \begin{pmatrix} 0 & g & f \\ g & 0 & h \\ f & h & 0 \end{pmatrix}$$

have the same *characteristic equation*, $\lambda^3 - (f^2+g^2+h^2)\lambda - 2fgh = 0$.

4. Show that the matrices A and A^t (Transpose of A) have the same characteristic roots.

5. Show that 0 is a characteristic root of a matrix if and only if the matrix is singular.

[*Hints*: $\lambda = 0$ satisfies $|A - \lambda I| = 0 \Rightarrow |A| = 0$, i.e., A is singular. Conversely take the steps backwards.]

6. Show that the characteristic roots of a triangular matrix are just the diagonal elements of the matrix.

[*Hints*: Let $A = \begin{bmatrix} a_{11} & a_{12} & a_{13} & \cdots & a_{in} \\ 0 & a_{22} & a_{23} & \cdots & a_{2n} \\ \cdots & \cdots & \cdots & \cdots & \cdots \\ 0 & 0 & 0 & \cdots & a_{nn} \end{bmatrix}$ be a triangular matrix of order n.

Now see that $\det(A - \lambda I) = (a_{11} - \lambda)(a_{22} - \lambda) \cdots (a_{nn} - \lambda)$; hence, etc.]

11.2 Cayley-Hamilton Theorem

Statement. *Every square matrix satisfies its characteristic equation.*

More clearly, *let* $|A - \lambda I| = (-1)^n[\lambda^n + a_1 x^{n-1} + \cdots + a_r\lambda^{n-r} + \cdots + a_n$ *be the characteristic polynomial of a* $n \times n$ *matrix A.*

Then the matrix equation

$$X^n + a_1X^{n-1} + a_2X^{n-2} + \cdots + a_rX^{n-r} + \cdots + a_nI = O.$$

is satisfied by A.

We first verify this statement with the help of a given matrix:

▶ Example 11.2.1 *Let* $A = \begin{pmatrix} 1 & 2 & 1 \\ -1 & 0 & 3 \\ 2 & -1 & 1 \end{pmatrix}$ *be a square matrix of order 3. We first find its characteristic equation.*

Solution: $\det(A - \lambda I) = 0$, i.e., $\begin{vmatrix} 1-\lambda & 2 & 1 \\ -1 & 0-\lambda & 3 \\ 2 & -1 & 1-\lambda \end{vmatrix} = 0$

or, $(1-\lambda)\begin{vmatrix} 0-\lambda & 3 \\ -1 & 1-\lambda \end{vmatrix} - 2\begin{vmatrix} -1 & 3 \\ 2 & 1-\lambda \end{vmatrix} + 1\begin{vmatrix} -1 & 0-\lambda \\ 2 & -1 \end{vmatrix} = 0$

or, $(1-\lambda)\{-\lambda(1-\lambda)+3\} - 2\{-1(1-\lambda)-6\} + 1\{1-2(-\lambda)\} = 0$

or, $(1-\lambda)(-\lambda+\lambda^2+3) - 2(\lambda-7) + (1+2\lambda) = 0$

or, $-\lambda+\lambda^2+3+\lambda^2-\lambda^3-3\lambda-2\lambda+14+1+2\lambda = 0$

or, $-\lambda^3+2\lambda^2-4\lambda+18 = 0.$

The characteristic equation can, therefore, be written as

$$\lambda^3 - 2\lambda^2 + 4\lambda - 18 = 0. \tag{1}$$

Cayley-Hamilton theorem asserts that this equation is satisfied by A, i.e.,

$$A^3 - 2A^2 + 4A - 18I = O. \tag{2}$$

We now verify that the assertion (2) is True.

We see that

$$A^2 = A \cdot A = \begin{pmatrix} 1 & 2 & 1 \\ -1 & 0 & 3 \\ 2 & -1 & 1 \end{pmatrix} \cdot \begin{pmatrix} 1 & 2 & 1 \\ -1 & 0 & 3 \\ 2 & -1 & 1 \end{pmatrix} = \begin{pmatrix} 1 & 1 & 8 \\ 5 & -5 & 2 \\ 5 & 3 & 0 \end{pmatrix}$$

$$A^3 = A^2 \cdot A = \begin{pmatrix} 1 & 1 & 8 \\ 5 & -5 & 2 \\ 5 & 3 & 0 \end{pmatrix}\begin{pmatrix} 1 & 2 & 1 \\ -1 & 0 & 3 \\ 2 & -1 & 1 \end{pmatrix} = \begin{pmatrix} 16 & -6 & 12 \\ 14 & 8 & -8 \\ 2 & 10 & 14 \end{pmatrix}$$

so that

$$A^3 - 2A^2 + 4A - 18I$$

$$= \begin{pmatrix} 16 & -6 & 12 \\ 14 & 8 & -8 \\ 2 & 10 & 14 \end{pmatrix} - 2\begin{pmatrix} 1 & 1 & 8 \\ 5 & -5 & 2 \\ 5 & 3 & 0 \end{pmatrix} + 4\begin{pmatrix} 1 & 2 & 1 \\ -1 & 0 & 3 \\ 2 & -1 & 1 \end{pmatrix} - 18\begin{pmatrix} 1 & 0 & 0 \\ 0 & 1 & 0 \\ 0 & 0 & 1 \end{pmatrix}$$

$$= \begin{pmatrix} 16-2+4-18 & -6-2+8+0 & 12-16+4+0 \\ 14-10-4+0 & 8+10+0-18 & -8-4+12+0 \\ 2-10+8+0 & 10-6-4+0 & 14+0+4-18 \end{pmatrix}$$

$$= \begin{pmatrix} 0 & 0 & 0 \\ 0 & 0 & 0 \\ 0 & 0 & 0 \end{pmatrix} = \bigcirc.$$

Hence, the matrix A satisfies the characteristic equation (1).

▶ Example 11.2.2 *Verify that the matrix*

$\begin{pmatrix} 1 & 2 \\ -1 & 3 \end{pmatrix}$ *satisfies its characteristic equation.*

Solution: For the matrix $A = \begin{pmatrix} 1 & 2 \\ -1 & 3 \end{pmatrix}$, we have the characteristic equation:

$$\det(A - \lambda I) = 0, \quad \text{or,} \quad \begin{vmatrix} 1-\lambda & 2 \\ -1 & 3-\lambda \end{vmatrix} = 0$$

$$\text{or,} \quad (1-\lambda)(3-\lambda) + 2 = 0$$

$$\text{or,} \quad \lambda^2 - 4\lambda + 5 = 0. \tag{1}$$

Cayley-Hamilton theorem asserts that the matrix A satisfies this equation, i.e.,

$$A^2 - 4A + 5I = \bigcirc \tag{2}$$

We verify that the assertion (2) is True.

Now

$$A^2 = \begin{pmatrix} 1 & 2 \\ -1 & 3 \end{pmatrix} \cdot \begin{pmatrix} 1 & 2 \\ -1 & 3 \end{pmatrix} = \begin{pmatrix} -1 & 8 \\ -4 & 7 \end{pmatrix}$$

$$\therefore A^2 - 4A + 5I = \begin{pmatrix} -1 & 8 \\ -4 & 7 \end{pmatrix} - 4\begin{pmatrix} 1 & 2 \\ -1 & 3 \end{pmatrix} + \begin{pmatrix} 5 & 0 \\ 0 & 5 \end{pmatrix}$$

$$= \begin{pmatrix} -1-4+5 & 8-8+0 \\ -4+4+0 & 7-12+5 \end{pmatrix} = \begin{pmatrix} 0 & 0 \\ 0 & 0 \end{pmatrix} = \bigcirc.$$

Hence, the matrix A satisfies the characteristic equation (1).

▶ Example 11.2.3 *Verify that the matrix*

(i) $A = \begin{pmatrix} 1 & 2 & 1 \\ 0 & 1 & -1 \\ 3 & -1 & 1 \end{pmatrix}$; (ii) $B = \begin{pmatrix} 1 & 0 & 2 \\ 0 & -1 & 1 \\ 0 & 1 & 0 \end{pmatrix}$

satisfy their respective characteristic equations. [C.P. 2003]

Try yourself. Proceed as in Ex. 11.2.2 above.

Proof of Cayley-Hamilton theorem:[1]

Statement. Given at the beginning of the article.

Proof. Since the elements of $(A-\lambda I)$ are at the most of first degree in λ, the elements of $\text{adj}(A-A\lambda)$ are at the most of degree $n-1$ in λ. Therefore, we can write

$$\text{adj}(A-\lambda I) = B_0\lambda^{n-1} + B_1\lambda^{n-2} + B_2\lambda^{n-3} + \cdots + B_{n-1},$$

where $B_0, B_1, B_2, \ldots, B_{n-1}$ are $n \times n$ matrices, their elements being functions of a_{ij}'s.

Since

$$(A-\lambda I)\cdot \text{adj}(A-\lambda I) = |A-\lambda I|I,$$

therefore

$$(A-\lambda I)(B_0\lambda^{n-1}+B_1\lambda^{n-2}+\cdots+B_{n-1}) = (-1)^n[\lambda^n + a_1\lambda^{n-1} + a_2\lambda^{n-2} + \cdots + a_n]I$$

Comparing the coefficients of like powers of λ, we obtain

$$\left.\begin{array}{rcl} -IB_0 & = & (-1)^n I \\ AB_0 - IB_1 & = & (-1)^n a_1 I \\ AB_1 - IB_2 & = & (-1)^n a_2 I \\ \cdots\cdots\cdots & \cdots & \cdots\cdots\cdots \\ AB_{n-1} & = & (-1)^n a_n I \end{array}\right\}$$

Pre-multiplying the above equations by $A^n, A^{n-1}, A^{n-2}, \ldots, I$ respectively and adding, we obtain

$$O = (-1)^n[A_n + a_1A_{n-1} + a_2A_{n-2} + \cdots + a_nI].$$

Thus $A^n + a_1A^{n-1} + a_2A^{n-2} + \cdots + a_nI = O$.

This establishes the theorem.

11.3 Applications of Cayley-Hamilton Theorem

▶ Example 11.3.1 *If* $A = \begin{pmatrix} 3 & 1 \\ -1 & 2 \end{pmatrix}$, *express* $2A^5 - 3A^4 + A^2 - 4I$ *as a linear polynomial in* A.

Solution: The characteristic equation of A is

$$\begin{vmatrix} 3-\lambda & 1 \\ -1 & 2-\lambda \end{vmatrix} = 0, \quad \text{or,} \quad (3-\lambda)(2-\lambda)+1=0$$

$$\text{or,} \quad \lambda^2 - 5\lambda + 7 = 0.$$

By Cayley-Hamilton theorem we can write

$$A^2 - 5A + 7I = O. \tag{1}$$

[1]The proof is not included in the C.U. B.Sc. general course syllabus.

Given expression

$$2A^5 - 3A^4 + A^2 - 4I = 2A^3(A^2-5A+7I) + 7A^2(A^2-5A+7I) + 21A(A^2-5A+7I) + 57(A^2-5A+7I) + 138A - 403I.$$

Using (1) we easily see that the given expression = 138A − 403I, which is a linear polynomial in A.

▶ Example 11.3.2 *Let* $A = \begin{pmatrix} 1 & 2 \\ -1 & 3 \end{pmatrix}$. *Express* $A^6 - 4A^5 + 8A^4 - 12A^3 + 14A^2$ *in the form* $-4A + 3I$ (*which is a linear polynomial in* A).

Try yourself.

▶ Example 11.3.3 ***(How to find the inverse of a square matrix by using Cayley-Hamilton theorem).***

Find the characteristic equation of the matrix

$$A = \begin{pmatrix} 1 & 3 & 2 \\ 0 & 2 & 1 \\ 0 & 5 & 3 \end{pmatrix}$$ *and hence compute its inverse.*

Solution: The characteristic equation of A is given by

$$\begin{vmatrix} 1-\lambda & 3 & 2 \\ 0 & 2-\lambda & 1 \\ 0 & 5 & 3-\lambda \end{vmatrix} = 0$$

$$\text{or,}\quad (1-\lambda)\{(2-\lambda)(3-\lambda)-5\} = 0$$

$$\text{or,}\quad (1-\lambda)(2-\lambda)(3-\lambda) - 5(1-\lambda) = 0$$

$$\text{or,}\quad (\lambda^2 - 3\lambda + 2)(3-\lambda) - 5(1-\lambda) = 0$$

$$\text{or,}\quad -\lambda^3 + 6\lambda^2 - 11\lambda + 6 - 5 + 5\lambda = 0$$

$$\text{or,}\quad \lambda^3 - 6\lambda^2 + 6\lambda - 1 = 0.$$

By Cayley-Hamilton theorem

$$A^3 - 6A^2 + 6A - I = O,$$

$$\text{or,}\quad A(A^2 - 6A + 6I) = I.$$

Hence

$$A^{-1} = A^2 - 6A + 6I$$

$$= \begin{pmatrix} 1 & 3 & 2 \\ 0 & 2 & 1 \\ 0 & 5 & 3 \end{pmatrix}\begin{pmatrix} 1 & 3 & 2 \\ 0 & 2 & 1 \\ 0 & 5 & 3 \end{pmatrix} - 6\begin{pmatrix} 1 & 3 & 2 \\ 0 & 2 & 1 \\ 0 & 5 & 3 \end{pmatrix} + 6\begin{pmatrix} 1 & 0 & 0 \\ 0 & 1 & 0 \\ 0 & 0 & 1 \end{pmatrix}$$

$$= \begin{pmatrix} 1 & 19 & 11 \\ 0 & 9 & 5 \\ 0 & 25 & 14 \end{pmatrix} - \begin{pmatrix} 6 & 18 & 12 \\ 0 & 12 & 6 \\ 0 & 30 & 18 \end{pmatrix} + \begin{pmatrix} 6 & 0 & 0 \\ 0 & 6 & 0 \\ 0 & 0 & 6 \end{pmatrix}$$

$$= \begin{pmatrix} 1 & 1 & -1 \\ 0 & 3 & -1 \\ 0 & -5 & 2 \end{pmatrix}.$$

▶ Example 11.3.4 *Find the characteristic equation of the matrix*

$$A = \begin{pmatrix} 1 & 2 & 3 \\ 2 & 3 & 4 \\ 1 & 0 & -1 \end{pmatrix} \text{ and hence compute } A^3.$$

Solution: The characteristic equation of A is given by

$$\begin{vmatrix} 1-\lambda & 2 & 3 \\ 2 & 3-\lambda & 4 \\ 1 & 0 & -1-\lambda \end{vmatrix} = 0$$

$$\begin{aligned}
\text{or, } & (1-\lambda)(3-\lambda)(-1-\lambda) - 2(-2-2\lambda-4) + 3\{0-3+\lambda\} = 0 \\
\text{or, } & \{-\lambda^3 + 3\lambda^2 + \lambda - 3\} + 12 + 4\lambda - 9 + 3\lambda = 0 \\
\text{or, } & -\lambda^3 + 3\lambda^2 + 8\lambda = 0 \\
\text{or, } & \lambda^3 - 3\lambda^2 - 8\lambda = 0.
\end{aligned}$$

By Cayley-Hamilton theorem, $A^3 - 3A^2 - 8A = \bigcirc$

$$\begin{aligned}
\text{or, } A^3 &= 3A^2 + 8A \\
&= 3\begin{pmatrix} 1 & 2 & 3 \\ 2 & 3 & 4 \\ 1 & 0 & -1 \end{pmatrix} \cdot \begin{pmatrix} 1 & 2 & 3 \\ 2 & 3 & 4 \\ 1 & 0 & -1 \end{pmatrix} + 8\begin{pmatrix} 1 & 2 & 3 \\ 2 & 3 & 4 \\ 1 & 0 & -1 \end{pmatrix} \\
&= 3\begin{pmatrix} 8 & 8 & 8 \\ 12 & 13 & 14 \\ 0 & 2 & 4 \end{pmatrix} + 8\begin{pmatrix} 1 & 2 & 3 \\ 2 & 3 & 4 \\ 1 & 0 & -1 \end{pmatrix}, \\
\text{i.e., } A^3 &= \begin{pmatrix} 32 & 40 & 48 \\ 52 & 63 & 74 \\ 8 & 6 & 4 \end{pmatrix}.
\end{aligned}$$

▶ Example 11.3.5 **A Very Important Problem:**

If $A = \begin{pmatrix} 1 & 0 & 0 \\ 1 & 0 & 1 \\ 0 & 1 & 0 \end{pmatrix}$, *show that for every integer* $n \geq 3$, $A^n = A^{n-2} + A^2 - I$. *Hence, determine* A^{50}.

Solution: For the matrix A, the characteristic polynomial is given by

$$\begin{vmatrix} 1-\lambda & 0 & 0 \\ 1 & 0-\lambda & 1 \\ 0 & 1 & 0-\lambda \end{vmatrix} = (1-\lambda)(\lambda^2 - 1) = -\lambda^3 + \lambda^2 + \lambda - 1.$$

∴ the characteristic equation of A is

$$\lambda^3 - \lambda^2 - \lambda + 1 = 0.$$

By Cayley-Hamilton theorem

$$\begin{aligned}
& A^3 - A^2 - A + I = 0 \\
\text{or, } & A(A^2 - I) = A^2 - I \qquad (1)
\end{aligned}$$

Pre-multiplying (1) throughout by A^{K-3}

$$\begin{aligned} A^{K-3}\cdot A(A^2-I) &= A^{K-3}(A^2-I) \\ \text{or, } A^{K-2}(A^2-I) &= A^{K-3}(A^2-I) \end{aligned} \tag{2}$$

Put $k = n, n-1, n-2, \dots, 3$ in succession in (2), we thus get

$$\left.\begin{array}{rrcl} (k=n) & A^{n-2}(A^2-I) & = & A^{n-3}(A^2-I) \\ (k=n-1) & A^{n-3}(A^2-I) & = & A^{n-4}(A^2-I) \\ (k=n-2) & A^{n-4}(A^2-I) & = & A^{n-5}(A^2-I) \\ \cdots\cdots\cdots & \cdots\cdots\cdots\cdots & \cdots & \cdots\cdots\cdots\cdots \\ (k=3) & A(A^2-I) & = & A^2-I \end{array}\right\}. \tag{3}$$

From (3) we can easily write

$$\begin{aligned} A^{n-2}(A^2-I) &= A^2-I \\ \text{or, } A^n - A^{n-2} &= A^2 - I \ (n \geq 3). \end{aligned} \tag{4}$$

$$A^n = A^{n-2} + A^2 - I \ (n \geq 3).$$

This proves the first part of the problem.
Put $n = 50, 48, 46, \dots 4$ successively in (4), we have

$$\begin{aligned} A^{50} - A^{48} &= A^2 - I \\ A^{48} - A^{46} &= A^2 - I \\ &\cdots\cdots\cdots\cdots \\ A^4 - A^2 &= A^2 - I \\ \hline \text{(on addition)} \quad A^{50} - A^2 &= 24(A^2 - I) \text{ How 24?} \end{aligned} \tag{5}$$

$$\begin{aligned} \therefore\ A^{50} &= 25A^2 - 24I \\ &= 25\begin{pmatrix} 1 & 0 & 0 \\ 1 & 0 & 1 \\ 0 & 1 & 0 \end{pmatrix}\begin{pmatrix} 1 & 0 & 0 \\ 1 & 0 & 1 \\ 0 & 1 & 0 \end{pmatrix} - 24\begin{pmatrix} 1 & 0 & 0 \\ 0 & 1 & 0 \\ 0 & 0 & 1 \end{pmatrix} \\ &= 25\begin{pmatrix} 1 & 0 & 0 \\ 1 & 1 & 0 \\ 1 & 0 & 1 \end{pmatrix} - 24\begin{pmatrix} 1 & 0 & 0 \\ 0 & 1 & 0 \\ 0 & 0 & 1 \end{pmatrix} \\ &= \begin{pmatrix} 1 & 0 & 0 \\ 25 & 1 & 0 \\ 25 & 0 & 1 \end{pmatrix} \text{ (Reqd. Answer)} \end{aligned}$$

▶ Example 11.3.6 *Evaluate* $A^5 - 27A^3 + 65A^2$, *where*

$$A = \begin{pmatrix} 0 & 0 & 1 \\ 3 & 1 & 0 \\ -2 & 1 & 4 \end{pmatrix}.$$

Solution: The characteristic equation of the matrix A is given by

$$\begin{vmatrix} 0-\lambda & 0 & 1 \\ 3 & 1-\lambda & 0 \\ -2 & 1 & 4-\lambda \end{vmatrix} = 0$$

$$\text{or,}\quad -\lambda(1-\lambda)(4-\lambda) + 1\{3+2-2\lambda\} = 0$$

$$\text{or,}\quad -\lambda^3 + 5\lambda^2 - 4\lambda + 5 - 2\lambda = 0$$

$$\text{or,}\quad \lambda^3 - 5x^2 + 6\lambda - 5 = 0.$$

By Cayley-Hamilton theorem we can write

$$A^3 - 5A^2 + 6A - 5I = O. \tag{1}$$

Multiplying by A^2 we get

$$A^5 - 5A^4 + 6A^3 - 5A^2 = 0, \tag{2}$$

$$\begin{aligned}
\text{i.e., } A^5 &= 5A^4 - 6A^3 + 5A^2 \\
&= A \cdot 5A^3 - 6A^3 + 5A^2 \\
&= 5A(5A^2 - 6A + 5I) - 6A^3 + 5A^2, \text{ using (1) we replace } 5A^3 \\
&= 19A^3 - 25A^2 + 25A.
\end{aligned}$$

$$\begin{aligned}
\therefore\ A^5 - 27A^3 + 65A^2 &= 19A^3 - 25A^2 + 25A - 27A^3 + 65A^2 \\
&= -8A^3 + 40A^2 + 25A \\
&= -8(5A^2 - 6A + 5I) + 40A^2 + 25A, \text{ using (1)} \\
&= 73A - 40I \\
&= 73\begin{pmatrix} 0 & 0 & 1 \\ 3 & 1 & 0 \\ -2 & 1 & 4 \end{pmatrix} - 40\begin{pmatrix} 1 & 0 & 0 \\ 0 & 1 & 0 \\ 0 & 0 & 1 \end{pmatrix} \\
&= \begin{pmatrix} -40 & 0 & 73 \\ 219 & 33 & 0 \\ -146 & 73 & 252 \end{pmatrix} \text{ (Reqd. Answer).}
\end{aligned}$$

Exercises on Chapter 11(II)

1. Verify Cayley-Hamilton theorem for the matrix

$$A = \begin{pmatrix} 1 & 2 & 1 \\ 1 & -1 & 1 \\ 2 & 3 & -1 \end{pmatrix}.$$

Now compute A^{-1}. [*Ans.* $A^{-1} = \frac{1}{9}\begin{pmatrix} -2 & 5 & 3 \\ 3 & -3 & 0 \\ 5 & 1 & -3 \end{pmatrix}$]

2. Using Example 11.3.5, evaluate A^{100}, where $A = \begin{pmatrix} 1 & 0 & 0 \\ 1 & 0 & 1 \\ 0 & 1 & 0 \end{pmatrix}$.

[*Ans.* $\begin{pmatrix} 1 & 0 & 0 \\ 50 & 1 & 0 \\ 50 & 0 & 1 \end{pmatrix}$]

3. Using Cayley-Hamilton theorem obtain A^{-1}, where $A = \begin{pmatrix} 1 & 2 & 1 \\ 0 & 1 & -1 \\ 3 & -1 & 1 \end{pmatrix}$.

[*Ans.* $\begin{pmatrix} 0 & \frac{1}{3} & \frac{1}{3} \\ \frac{1}{3} & \frac{2}{9} & -\frac{1}{9} \\ \frac{1}{3} & -\frac{7}{9} & -\frac{1}{9} \end{pmatrix}$]

4. What are the eigen values (or characteristic roots) of the matrix

$$A = \begin{pmatrix} 0 & 1 & 2 \\ 1 & 0 & -1 \\ 2 & -1 & 0 \end{pmatrix}?$$

[*Ans.* $2, -1 \pm \sqrt{3}$]

5. Find the eigen value of the matrix $A = \begin{bmatrix} 1 & 4 \\ 2 & 3 \end{bmatrix}$ and verify Cayley-Hamilton theorem. Find A^{-1} and express

$$A^5 - 4A^4 - 7A^3 + 11A^2 - A - 10I$$

as a linear polynomial in A. [*Ans.* Reqd. linear polynomial is $A + 5I$]

11.4 Eigen Values and their Corresponding Eigen Vectors

We have already defined eigen values of a square matrix A of order n as the roots of λ in

$$\det(A - \lambda I) = 0.$$

If λ is an eigen value of A, then the matrix $A - \lambda I$ is singular. Therefore, there exists a non-zero vector X such that

$$(A - \lambda I)X = 0 \text{ or, } AX = \lambda X.$$

Such a vector X is called the **eigen vector** of A corresponding to the eigen value λ.

▶ Example 11.4.1 *Determine the eigen values and the corresponding eigen vectors of the matrix*

$$A = \begin{bmatrix} 5 & 4 \\ 1 & 2 \end{bmatrix}.$$

Solution: The characteristic equation of A is

$$|A-\lambda I|=0, \quad \text{or,} \quad \begin{vmatrix} 5-\lambda & 4 \\ 1 & 2-\lambda \end{vmatrix}=0$$

$$\text{or,} \quad (5-\lambda)(2-\lambda)-4=0$$

$$\text{or,} \quad \lambda^2-7\lambda+6=0 \ (\textbf{characteristic equation of A})$$

$$\text{or,} \quad (\lambda-6)(\lambda-1)=0,$$

i.e., $\lambda_1=6, \lambda_2=1$ are the **characteristic roots** or **eigen values** of the given matrix.

The eigen vectors $X_1 = \begin{bmatrix} x_1 \\ x_2 \end{bmatrix}$ of A corresponding to the eigen value $\lambda_1 = 6$ are given by

$$AX_1 = 6X_1$$

$$\text{or,} \quad \begin{bmatrix} 5 & 4 \\ 1 & 2 \end{bmatrix}\begin{bmatrix} x_1 \\ x_2 \end{bmatrix} = 6\begin{bmatrix} x_1 \\ x_2 \end{bmatrix}$$

$$\text{or,} \quad \left.\begin{aligned} 5x_1+4x_2 &= 6x_1 \\ x_1+2x_2 &= 6x_2 \end{aligned}\right\}, \text{ i.e.,} \quad \left.\begin{aligned} x_1-4x_2 &= 0 \\ -x_1+4x_2 &= 0 \end{aligned}\right\}$$

The coefficient matrix of these equation has rank 1.
These equations reduce to the single equation

$$x_1-4x_2=0.$$

Obviously $x_2=1, x_1=4$ is a solution of this equation, i.e.,

$$X_1 = \begin{bmatrix} 4 \\ 1 \end{bmatrix}$$

is an eigen vector of A corresponding to the eigen value $\lambda_1=6$.

The set of ALL eigen vectors of A corresponding to the eigen value 6 is given by

$$c_1X_1 = c_1\begin{bmatrix} 4 \\ 1 \end{bmatrix},$$

where c_1 is any non-zero scalar.

The eigen vector X_2 of A corresponding to the eigen value 1 is given by the non-zero solutions of the equation

$$AX_2 = IX_2, \quad \text{or,} \quad \begin{bmatrix} 5 & 4 \\ 1 & 2 \end{bmatrix}\begin{bmatrix} x_1 \\ x_2 \end{bmatrix} = \begin{bmatrix} x_1 \\ x_2 \end{bmatrix}$$

$$\text{or,} \quad 5x_1+4x_2=x_1 \text{ and } x_1+2x_2=x_2.$$

Here also we get a single equation $x_2=-x_1$. Let us take $x_1=1$, then $x_2=-1$ and the vector $X_2 = \begin{bmatrix} 1 \\ -1 \end{bmatrix}$.

The set of ALL eigen vectors of A corresponding to the eigen value $\lambda_2=1$ is

$$c_2X_2 = c_2\begin{bmatrix} 1 \\ -1 \end{bmatrix},$$

where c_2 is any non-zero scalar.

▶ Example 11.4.2 *Find the eigen values and eigen vectors of the matrix*

$$A = \begin{bmatrix} -2 & 4 \\ 4 & 4 \end{bmatrix}.$$

Exactly similar to Example 11.4.1. [C.P. 2003]

▶ Example 11.4.3 *Determine the eigen values and their corresponding eigen vectors of the matrix*

$$A = \begin{pmatrix} 8 & -6 & 2 \\ -6 & 7 & -4 \\ 2 & -4 & 3 \end{pmatrix}.$$

Solution: The characteristic equation of the matrix A is

$$|A - \lambda I| = 0$$

$$\text{or,} \quad \begin{vmatrix} 8-\lambda & -6 & 2 \\ -6 & 7-\lambda & -4 \\ 2 & -4 & 3-\lambda \end{vmatrix} = 0$$

$$\text{or,} \quad \lambda^3 - 18\lambda^2 + 45\lambda = 0$$

$$\text{or,} \quad \lambda(\lambda - 3)(\lambda - 15) = 0.$$

$\therefore$ the three eigen values of A are $\lambda_1 = 0$, $\lambda_2 = 3$, $\lambda_3 = 15$.

(i) The eigen vector $X_1 = \begin{bmatrix} x_1 \\ x_2 \\ x_3 \end{bmatrix}$ of A corresponding to $\lambda_1 = 0$ is given by the non-zero solutions of the equation

$$(A - 0I)X_1 = 0$$

$$\text{or,} \quad \begin{pmatrix} 8 & -6 & 2 \\ -6 & 7 & -4 \\ 2 & -4 & 3 \end{pmatrix}\begin{pmatrix} x_1 \\ x_2 \\ x_3 \end{pmatrix} = \begin{pmatrix} 0 \\ 0 \\ 0 \end{pmatrix}$$

$$\text{or,} \quad \left.\begin{array}{r} 8x_1 - 6x_2 + 2x_3 = 0 \\ -6x_1 + 7x_2 - 4x_3 = 0 \\ 2x_1 - 4x_2 + 3x_3 = 0 \end{array}\right\}$$

These equations reduce to

$$\left.\begin{array}{r} x_2 = x_3 \\ 2x_1 - 4x_2 + 3x_3 = 0 \end{array}\right\}.$$

Taking $x_2 = x_3 = 1$, we get $x_1 = \frac{1}{2}$. So

$$X_1 = \begin{bmatrix} \frac{1}{2} \\ 1 \\ 1 \end{bmatrix}$$

is an eigen vector of A corresponding to the eigen value 0.

The set of ALL eigen vectors of A corresponding to the eigen value 0 is given by

$$c_1X_1 = c_1\begin{bmatrix} \frac{1}{2} \\ 1 \\ 1 \end{bmatrix},$$

where c_1 is a non-zero scalar.

(ii) The eigen vector $X_2 = \begin{bmatrix} x_1 \\ x_2 \\ x_3 \end{bmatrix}$ corresponding to the eigen value $X_2 = 3$ is given by the non-zero solutions of

$$(A - 3I)X_2 = O$$

$$\text{or,}\quad \begin{pmatrix} 5 & -6 & 2 \\ -6 & 4 & -4 \\ 2 & -4 & 0 \end{pmatrix}\begin{pmatrix} x_1 \\ x_2 \\ x_3 \end{pmatrix} = \begin{pmatrix} 0 \\ 0 \\ 0 \end{pmatrix}$$

$$\text{or,}\quad \left.\begin{aligned} 5x_1 - 6x_2 + 2x_3 &= 0 \\ -6x_1 + 4x_2 - 4x_3 &= 0 \\ 2x_1 - 4x_2 + 0x_3 &= 0 \end{aligned}\right\} \Rightarrow \left.\begin{aligned} x_1 &= 2x_2 \\ 5x_1 - 6x_2 + 2x_3 &= 0 \end{aligned}\right\}$$

$$\Rightarrow x_1 = 2x_2 \text{ and } x_3 = -2x_2.$$

Taking $x_2 = 1, x_1 = 2, x_3 = -2$ and the eigen vector $X_2 = \begin{bmatrix} 2 \\ 1 \\ -2 \end{bmatrix}$.

The set of ALL eigen vectors of A corresponding to the eigen value $\lambda_2 = 3$ is

$$c_2X_2 = c_2\begin{bmatrix} 2 \\ 1 \\ -2 \end{bmatrix},$$

where c_2 is a non-zero scalar.

(iii) The eigen vector $X_3 = \begin{bmatrix} x_1 \\ x_2 \\ x_3 \end{bmatrix}$ corresponding to $\lambda_3 = 15$ are given by the non-zero solutions of the equation

$$(A - 15I)X_3 = O$$

$$\text{or,}\quad \begin{bmatrix} -7 & -6 & 2 \\ -6 & -8 & -4 \\ 2 & -4 & -12 \end{bmatrix}\begin{bmatrix} x_1 \\ x_2 \\ x_3 \end{bmatrix} = \begin{bmatrix} 0 \\ 0 \\ 0 \end{bmatrix}$$

$$\text{or,}\quad \left.\begin{aligned} -7x_1 - 6x_2 + 2x_3 &= 0 \\ -6x_1 - 8x_2 - 4x_3 &= 0 \\ 2x_1 - 4x_2 - 12x_3 &= 0 \end{aligned}\right\} \Rightarrow \left.\begin{aligned} x_2 &= -2x_3 \\ 2x_1 - 4x_2 - 12x_3 &= 0 \end{aligned}\right\}$$

$$\Rightarrow \left.\begin{aligned} x_2 &= -2x_3 \\ x_1 &= 2x_3 \end{aligned}\right\}.$$

Taking $x_3 = 1$, $x_1 = 2$, $x_2 = -2$ and the eigen vector

$$X_3 = \begin{bmatrix} 2 \\ -2 \\ 1 \end{bmatrix}.$$

The set of ALL eigen vectors of A corresponding to the eigen value $\lambda_3 = 15$ is

$$c_3X_3 = c_3\begin{bmatrix} 2 \\ -2 \\ 1 \end{bmatrix},$$

where c_3 is a non-zero scalar.

▶ **Example 11.4.4** *Determine the eigen values and the corresponding eigen vectors of the matrix*

$$A = \begin{pmatrix} 6 & -2 & 2 \\ -2 & 3 & -1 \\ 2 & -1 & 3 \end{pmatrix}$$

Solution: The characteristic equation of A is

$$\begin{vmatrix} 6-\lambda & -2 & 2 \\ -2 & 3-\lambda & -1 \\ 2 & -1 & 3-\lambda \end{vmatrix} = 0$$

$$\begin{aligned}
\text{or,}\quad & (6-\lambda)\{(3-\lambda)^2-1\}+2(-4+2\lambda)+2(-4+2\lambda)=0 \\
\text{or,}\quad & (6-\lambda)(x^2-6\lambda+8)+8\lambda-16=0 \\
\text{or,}\quad & (6-\lambda)(\lambda-4)(\lambda-2)+8(\lambda-2)=0 \\
\text{or,}\quad & (\lambda-2)\{(6-\lambda)(\lambda-4)+8\}=0 \\
\text{or,}\quad & (\lambda-2)(-\lambda^2+10\lambda-16)=0 \\
\text{or,}\quad & (\lambda-2)(\lambda^2-10\lambda+16)=0, \\
\text{i.e.,}\quad & (\lambda-2)^2(\lambda-8)=0, \\
\text{i.e.,}\quad & \lambda = 2,\ 2,\ 8 \text{ (the two eigen values are equal).}
\end{aligned}$$

The eigen vectors of A corresponding to the eigen value 8 are given by the non-zero solutions of

$$(A-8I)X_1 = 0,$$

$$\text{i.e.,}\quad \begin{pmatrix} 6-8 & -2 & 2 \\ -2 & 3-8 & -1 \\ 2 & -1 & 3-8 \end{pmatrix}\begin{pmatrix} x_1 \\ x_2 \\ x_3 \end{pmatrix} = 0,$$

$$\text{i.e.,}\quad -2x_1-2x_2+2x_3=0, -2x_1-5x_2-x_3=0$$
$$\text{and } 2x_1-x_2-5x_3=0.$$

These equations give $x_2 = -x_3$ and $2x_1 - x_2 - 5x_3 = 0$. Taking $x_3 = 1$, we get $x_2 = -1$ and $x_1 = 2$. The eigen vector $X_1 = \begin{bmatrix} x_1 \\ x_2 \\ x_3 \end{bmatrix} = \begin{bmatrix} 2 \\ -1 \\ 1 \end{bmatrix}$ and all eigen vectors of A corresponding to the eigen value 8 is given by c $\begin{bmatrix} 2 \\ -1 \\ 1 \end{bmatrix}$, where c is a non-zero scalar.

The eigen vectors of A corresponding to eigen value 2 are given by non-zero solutions of

$$(A-2I)X_2 = O,$$

$$\text{i.e.,}\quad \begin{pmatrix} 6-2 & -2 & 2 \\ -2 & 1 & -1 \\ 2 & -1 & 1 \end{pmatrix}\begin{pmatrix} x_1 \\ x_2 \\ x_3 \end{pmatrix} = 0$$

$$\text{or,}\quad 4x_1 - 2x_2 + 2x_3 = 0, -2x_1 + x_2 - x_3 = 0$$

$$\text{and } 2x_1 - x_2 + x_3 = 0.$$

These equations give a single equation

$$2x_1 - x_2 + x_3 = 0.$$

We can take two linearly independent solutions of this equation, say

$$X_2 = \begin{bmatrix} -1 \\ 0 \\ 2 \end{bmatrix} X_3 = \begin{bmatrix} 1 \\ 2 \\ 0 \end{bmatrix}$$

$c_1X_2 + c_2X_3$ gives all the eigen vectors of A corresponding to the eigen value 2, c_1, c_2 being real numbers (not both zero).

Exercises on Chapter 11(III)

Find the eigen values and the corresponding eigen vectors of the following matrices:

1. $\begin{pmatrix} 1 & 3 \\ 4 & 5 \end{pmatrix}$.

2. $\begin{pmatrix} 4 & 0 & 0 \\ 0 & 5 & 0 \\ 0 & 0 & 6 \end{pmatrix}$;

3. $\begin{pmatrix} 1 & -1 & 1 \\ 2 & -2 & 4 \\ 3 & -3 & 6 \end{pmatrix}$;

4. $\begin{pmatrix} 3 & 2 & 4 \\ 2 & 0 & 2 \\ 4 & 2 & 3 \end{pmatrix}$;

5. $\begin{bmatrix} 8 & -6 & 2 \\ -6 & 7 & -4 \\ 2 & -4 & 3 \end{bmatrix}$;

6. $\begin{pmatrix} 1 & -1 & 2 \\ 0 & 1 & 0 \\ 1 & 2 & 1 \end{pmatrix}$.

[**Answers:**

1. **−1**; $c\begin{bmatrix} 3 \\ -2 \end{bmatrix}$; 7; $c\begin{bmatrix} 1 \\ 2 \end{bmatrix}$

2. **4**; $c\begin{bmatrix} 1 \\ 0 \\ 0 \end{bmatrix}$; 5; $c\begin{bmatrix} 0 \\ 1 \\ 0 \end{bmatrix}$; 6; $c\begin{bmatrix} 0 \\ 0 \\ 1 \end{bmatrix}$

3. **5**; $c\begin{bmatrix} 1 \\ 2 \\ 3 \end{bmatrix}$; **0, 0** → linearly independent eigen vectors are $c_1\begin{bmatrix} 1 \\ 1 \\ 0 \end{bmatrix} + c_2\begin{bmatrix} 0 \\ 2 \\ 1 \end{bmatrix}$

4. **8**; $c\begin{bmatrix} 2 \\ 1 \\ 2 \end{bmatrix}$; **−1, −1** linearly → independent eigen vectors are $c_1\begin{bmatrix} 0 \\ 2 \\ -1 \end{bmatrix} + c_2\begin{bmatrix} 1 \\ 0 \\ -1 \end{bmatrix}$

5. **0**; $c\begin{bmatrix} 1 \\ 2 \\ 2 \end{bmatrix}$; 3; $c\begin{bmatrix} 2 \\ 1 \\ -2 \end{bmatrix}$; 15; $c\begin{bmatrix} 2 \\ -2 \\ 1 \end{bmatrix}$

6. **1**; $c\begin{bmatrix} -4 \\ 2 \\ 1 \end{bmatrix}$; $1+\sqrt{2}$; $c\begin{bmatrix} 2 \\ 0 \\ 2 \end{bmatrix}$; $1-\sqrt{2}$; $c\begin{bmatrix} -\sqrt{2} \\ 0 \\ 1 \end{bmatrix}$.]

i.e. [illegible] $= 0$

$$-4x_1 - 2x_2 + 2x_3 = 0,\quad 2x_1 + x_2 - x_3 = 0$$

$$2x_1 + x_2 - x_3 = 0$$

1. [illegible]
2. [illegible]
3. [illegible]
4. [illegible]
5. [illegible]
6. [illegible]

[Answers:

1. −1, [illegible]
2. 4, [illegible]

5. 0, [illegible]; 3, [illegible]; 15, [illegible]

Calcutta University Questions

B.Sc. General Course (NEW Syllabus)

2005

Group A: Classical Algebra

(25 Marks)

Answer **Questions No. 1** and any **two** form the rest.

1. (a) Answer any one question: 2

   i. On the complex plane, let $P(z)$ be a variable point such that $|z + 3i| = 4$. Find the locus of P.

   ii. If the roots of the equation $x^3 - px^2 + qx - r = 0$ are in G.P. Show that $p^3 r = q^3$.

   iii. Find the inverse of the matrix $A = \begin{bmatrix} 5 & 3 \\ -2 & 2 \end{bmatrix}$.

   (b) Answer any question: 3

   i. Show that every skew-symmetric determinant of order three is zero.

   ii. Find the condition that the equation $x^3 + px^2 + qx + r = 0$ may have two roots equal but of opposite signs.

   iii. Let A, B, C be three matrices such that $A \neq 0$ and $AB = AC$, where 0 is the null matrix. Does it imply $B = C$? Justify your answer by an example. 5 + 5

2. (a) If $x = \cos\alpha + i\sin\alpha$, $y = \cos\beta + i\sin\beta$, $z = \cos\gamma + i\sin\gamma$ and $x + y + z = xyz$, prove that $\cos(\beta - \gamma) + \cos(\gamma - \alpha) + \cos(\alpha - \beta) = 1$.

   (b) If $\log\sin(\theta + i\phi) = \alpha + i\beta$, prove that $2e^{2\alpha} = \cosh 2\phi - \cos 2\theta$, where θ, ϕ, α, β are real numbers.

3. (a) If α, β, γ are the roots of the equation $x^3 + 3x + 1 = 0$, find the equation whose roots are $\frac{\alpha}{\beta} + \frac{\beta}{\alpha}, \frac{\beta}{\gamma} + \frac{\gamma}{\beta}, \frac{\gamma}{\alpha} + \frac{\alpha}{\beta}$ and hence find the value of $\sum\left(\frac{\alpha}{\beta} + \frac{\beta}{\gamma}\right)$. 4 + 1

   (b) Solve the equation $x^3 - 6x^2 - 6x - 7 = 0$ by Cardan's method. 5

4. (a) If the adjugate of $\Delta = \begin{vmatrix} a & h & g \\ h & b & f \\ g & f & c \end{vmatrix}$ is $\Delta - \begin{vmatrix} A & H & G \\ H & B & F \\ G & F & C \end{vmatrix}$, then prove that $\dfrac{BC - F^2}{a} = \dfrac{CA - G^2}{b} = \dfrac{AB - H^2}{c} = \Delta$. 5

   (b) Express $\Delta = \begin{vmatrix} 0 & (a-b)^2 & (a-c)^2 \\ (b-a)^2 & 0 & (b-c)^2 \\ (c-a)^2 & (c-b)^2 & 0 \end{vmatrix}$ as the product of two determinants, and hence show that $\Delta = 2(b-c)^2(c-a)^2(a-b)^2$. 3 + 2

5. (a) If $A + I = \begin{bmatrix} 1 & 3 & 4 \\ -1 & 1 & 3 \\ -2 & -3 & 1 \end{bmatrix}$, evaluate $(A + I)(A - I)$, where I represents the 3×3 identity matrix. 4

(b) Find the inverse of the matrix $\begin{bmatrix} 2 & 1 & 1 \\ 1 & -1 & 0 \\ 2 & 1 & -1 \end{bmatrix}$ and using it solve the following system of equations:

$$\left.\begin{aligned} 2x + y + z &= 5 \\ x - y &= 0 \\ 2x + y + z &= 1 \end{aligned}\right\}.$$ 3 + 1

(c) Find the rank of the matrix $\begin{bmatrix} 1 & 1 & 2 \\ 3 & 5 & 2 \\ 4 & 8 & 0 \end{bmatrix}$. 2

Group B: 25 Marks (Modern Algebra)

[Answer *Question No.* 6 and *any two* from the rest]

6. Examine the validity (*with proper justifications*) of the following statements:

(a) Answer *any one* question: 2

i. Let $f : X \to Y$ and $g : Y \to Z$ be two injective mappings. Then their composition $g \circ f$ is injective.

ii. For any element a of a group $(G, \bullet)$, $(a^{-1})^{-1} = a$.

iii. The unit circle $S^1 = \{(x, y) \in R^2 : x^2 + y^2 = 1\}$ in R^2 is a subspace of the vector space R^3 (over R).

(b) Answer *any one* question: 3

i. Let $f : R^2 \to R^2$ be the map defined by $f(x, y) = (x, 0)$.
If $A = \{(x, y) \in R^2 : x - y = 0\}$ and $B = \{(x, y) \in R^2 : x - y = 1\}$ find $f(A \cap B) = f(A) \cap f(B)$.

ii. The subset $\{(1, 0, 0), (0, 1, 0), (0, -1, 0)\}$ of R^3 is linearly dependent in the vector space R^3 (over R).

iii. The basis of the vector space R^3 over R in unique.

7. **(a)** Let $G = \{1, \omega, \omega^2\}$ be the set of three cube roots of unity. Show that—

i. G is a group under usual multiplication of complex numbers.

ii. G has no proper such group.

(b) Let G be a group and $H = \{y \in G : xy = yx \ \forall x \in G\}$. Show that H is a commutative subgroup of G. If G is abelian, find H. 5 + 5

8. **(a)** Let R be a ring. Prove that for all $x, y, z \in R$,

i. $(-x)y = -xy$;

ii. $(-x)(-y) = xy$;

iii. $x(y - z) = xy - xz$. 5

(b) Let $M = \left\{ \begin{bmatrix} a & b \\ 2b & a \end{bmatrix} ; a, b \text{ are rational} \right\}$. Prove that under usual matrix addition and multiplication, M is a field. 5

9. **(a)** Explain the following concepts with an example in each case:

i. Vector space;
ii. Generators of a finite dimensional vector space.
iii. Basis of a finite dimensional vector space. 6

(b) Examine whether the set $S = \{(1,0,1), (0,1,1), (1,1,0)\}$ form a basis of R^3 over R. 4

10. (a) State Cayley-Hamilton Theorem. Illustrate the theorem with an example. 2 + 3

(b) Find all the eigen value of the matrix $A = \begin{bmatrix} 1 & -1 & 2 \\ 2 & -2 & 4 \\ 3 & -3 & 6 \end{bmatrix}$ 3

(c) Find the symmetric matrix corresponding to the real quadratic form: $x^2 + y^2 + z^2 + 4xy + 2xz + 4yz$. 2

1997
B.Sc. Pass Course
First Paper
Group A: Short Answer Questions

1. (a) Answer *any three* questions: $3 \times 2 = 6$

i. If $\omega^n + \omega^{2n} + 1 = 0$, where ω is an imaginary cube root of unity and n an integer, is n divisible by 3?

ii. Express $\dfrac{i^5 + i^9 + i^{14}}{i^6 + i^{11} + i^{13}}$ $(i = \sqrt{-1})$ in the form $a + ib$.

iii. Find the remainder, by Synthetic Division, when the polynomial $3x^4 - 4x^3 + 2x^2 - 9x + 1$ is divided by $2x + 1$.

iv. Find the equation whose roots are equal in magnitude but opposite in sign to the roots of $x^4 + 3x^3 - 7x^2 + 2x + 1 = 0$.

v. If A, B, C are matrices of appropriate orders with $AB = AC$, then does it imply that $B = C$? Give an example in support of your conclusion.

vi. Without expanding show that

$$\begin{vmatrix} 0 & (x-y)^3 & (x-z)^3 \\ (y-x)^3 & 0 & (y-z)^3 \\ (z-x)^3 & (z-y)^3 & 0 \end{vmatrix} = 0.$$

(b) Answer *any two* questions: $2 \times 2 = 4$

i. Write down the subsets of the set $\{x, y, z\}$.

ii. Is $f : R \to R$, given by $f(x) = \log x$, $x \in R$, a mapping where R is the set of all real numbers?

iii. If the binary operation '$\circ$' be defined on I, the set of all integers by $a \circ b = a + b + 1$, $a, b \in I$, find the identity element w.r.t. $\circ$.

iv. In the above question (iii), find the inverse of a with respect to $\circ$.

v. Express the vector $(7, 11)$ as a linear combination of the vectors $(2, 3)$ and $(3, 5)$.

Group B

[Answer *Question No.* 7 and *any two* from the rest]

3. (a) If the amplitude of $\frac{z-1}{z+1}$ is $\frac{\pi}{4}$ where $z = x + iy$, show that z lies on a circle with centre at i. What is its radius?

(b) Express $\frac{(\cos\theta + i\sin\theta)^6}{(\sin\theta + i\cos\theta)^6}$ in the form $a + ib$.

(c) Show that all the values of i^i are real and are in G.P. $4+4+4=12$

4. (a) Find the cubic in x which vanishes for $x = 1$ and $x = -2$ and shall have the values 4 and 8 for $x = -1$ and $x = 2$ respectively.

(b) Show that the roots of $\frac{1}{x-a} + \frac{1}{x-b} + \frac{1}{x-c} = \frac{1}{x}$, where $a > b > c > 0$ are all real.

(c) Solve the equation $x^3 - 15x - 126 = 0$ by Cardan's method. $4+4+4=12$

5. (a) Find the matrices A and B where $A + B = 2B^t$ and $3A + 2B = I_3$, where B^t denotes the transpose of B and I_3 is the identity matrix of order 3.

(b) Find the rank of the matrix $\begin{pmatrix} 1 & 1 & 2 \\ 3 & 5 & 2 \\ 4 & 8 & 0 \end{pmatrix}$.

(c) If

$$A = \begin{pmatrix} 1 & -1 & 0 \\ 2 & 3 & 4 \\ 0 & 1 & 2 \end{pmatrix} \text{ and } C = \begin{pmatrix} 2 & 2 & -4 \\ -4 & 2 & -4 \\ 2 & -1 & 5 \end{pmatrix},$$

then verify that $AC = CA = 6I_3$ and use this result to solve the system of equations $x - y = 3$, $2x + 3y + 4z = 17$, $y + 2z = 7$. $4+4+4=12$

6. (a) Show that $\begin{vmatrix} a+b+c & -c & -b \\ -c & a+b+c & -a \\ -b & -a & a+b+c \end{vmatrix} = 2(b+c)(c+a)(a+b)$.

(b) If $ax + by + cz = 1$, $cx + ay + bz = 0$, $bx + cy + az = 0$, then show that $\begin{vmatrix} x & y & z \\ z & x & y \\ y & z & x \end{vmatrix}$ is the reciprocal of $\begin{vmatrix} a & c & b \\ b & a & c \\ c & b & a \end{vmatrix}$.

(c) Solve the following system of equations by Cramer's Rule:

$$2x - y = 3,\ 3y - 2z = 5 \text{ and } 2z - x = -4.$$

$4+4+4=12$

7. Answer *any four* of the following questions: $4 \times 4 = 16$

(a) If $A = \{5, 6, 7, 8, 9\}$, $B = \{2, 4, 6, 8, 10, 12\}$, $C = \{3, 6, 9, 12\}$, then verify that

$$A - (B \cup C) = (A - B) \cap (A - C).$$

(b) Show that the mapping $f : N \to N$ (N being the set of all natural numbers) defined by $f(n) = n + 2$, $n \in N$ is one-one but not onto.

(c) Show that the set G of all numbers of the form 2^n, where n is an integer is an abelian group w.r.t. ordinary multiplication.

(d) Prove that the set of all integers of the form $2n$ where n is any integer is a commutative ring w.r.t. addition and multiplication but not a field.

(e) Show that the set of all 2×2 real matrices $\begin{pmatrix} c & b \\ c & d \end{pmatrix}$ with $ad - bc = 1$ is a subgroup of the group of all 2×2 non-singular real matrices.

(f) Define a basis of a vector space. Do the vectors $(1, 1, 2)$, $(3, 5, 2)$ and $(1, 0, 0)$ form a linearly independent subset of R^3? Justify.

1998

Group A: Short Answer Questions

1. (a) Answer *any three* questions: $3 \times 2 = 6$

   i. Obtain an equation whose roots are twice the roots of the equation
   $$x^3 + 3x^2 + 4x + 5 = 0.$$

   ii. By Euler's exponential values of $\sin Z$ and $\cos Z$, where Z is a complex number, establish that $\cos 2Z = 1 - 2\sin^2 Z$.

   iii. If α, β, γ are the roots of the equation $ax^3 + bx^2 + cx + d = 0$, find the value of $\Sigma\alpha^2$, where $a \neq 0$.

   iv. Justify the statement: 'If two rows of a determinant are identical, then the determinant vanishes'.

   v. Cite one example to show that matrix multiplication is, in general, not commutative.

   vi. If both A and B are 2×2 real skew symmetric matrices, show that $AB = BA$.

   (b) Answer *any two* questions: $2 \times 2 = 4$

   i. If A and B are two subsets of a set S, find $(A \cap B')' \cup (A' \cap B)'$ in the simplest form, where A' denotes the complement of A.

   ii. In a group $(G, *)$, show that $(a * b)^{-1} = b^{-1} * a^{-1}$, where $a, b \in G$.

   iii. Give one example to show that union of two subgroups of a group may not be a subgroup of the same group.

   iv. Determine k so that the vectors $(1, -1, 2)$, $(0, k, 3)$ and $(-1, 2, 3)$ are linearly dependent.

Group B

[Answer *Question No.* 7 and *any two* from the rest]

3. (a) Prove De Moivre's theorem for positive integral index. $4 + 4 + 4 = 12$

   (b) Find all the values of $(i - 1)^{1/5}$.

   (c) If $\log \sin(\theta + i\phi) = \alpha + i\beta$, prove that
   $$2e^{2\alpha} = \cosh 2\phi - \cos 2\theta \ (\theta, \phi, \alpha, \beta \text{ are all real.})$$

4. Answer *any three:* $4 + 4 + 4 = 12$

   (a) State Decartes' rule of sign and apply it to prove that the equation $x^7 - 2x^4 + 3x^3 - 1 = 0$, has at least four imaginary roots.

(b) The roots of the equation $x^3 - 3x^2 + kx + 3 = 0$ are in arithmetic progression. Find the value of k and solve the equation.

(c) The roots of the equation $x^3 - px^2 + qx - r = 0$ are α, β, γ. Prove that

$$(\beta+\gamma)(\gamma+\alpha)(\alpha+\beta) = pq - r.$$

(d) Apply Cardan's method to solve the equation $x^3 - 24x + 72 = 0$.

5. (a) Considering a general third order determinant show that its square can be put in the form

$$\begin{vmatrix} A & H & G \\ H & B & F \\ G & F & C \end{vmatrix}.$$

(b) Evaluate $\begin{vmatrix} 0 & 1 & 1 \\ 1 & 0 & 1 \\ 1 & 1 & 0 \end{vmatrix}$ and hence or otherwise establish that

$$\begin{vmatrix} y+z & z+x & x+y \\ z+x & x+y & y+z \\ x+y & y+z & z+x \end{vmatrix} = 2\begin{vmatrix} x & y & z \\ y & z & x \\ z & x & y \end{vmatrix}.$$

(c) Define adjoint of a determinant. If Δ' be the adjoint of a third order determinant $\Delta(\neq 0)$, prove that $\Delta' = \Delta^2$. $4+4+4=12$

6. (a) If $A + I_3 = \begin{bmatrix} 1 & 3 & 4 \\ -1 & 1 & 3 \\ -2 & -3 & 1 \end{bmatrix}$, evaluate $(A+I_3)(A-I_3)$, where I_3 represents 3×3 identity matrix.

(b) Find the inverse of the matrix $\begin{bmatrix} 1 & 1 & 2 \\ 2 & -1 & 3 \\ 3 & -1 & -1 \end{bmatrix}$ and using it solve the equations

$$x + y + 2z = 4, 2x - y + 3z = 9, 3x - y - z = 2.$$

(c) Find the rank of the matrix $\begin{bmatrix} 3 & 4 & -6 \\ 2 & -1 & 7 \\ 1 & -2 & 8 \end{bmatrix}$. $4+4+4=12$

7. Answer *any four* questions: $4 \times 4 = 16$

(a) If $A = \{a, b, c, d\}$ and $B = \{a, b, c\}$ be two sets, determine $(A \times B) \cap (B \times A)$ and $(A \times B) - (B \times A)$.

(b) Let S be the set of all integers and f, g, h are three mappings from S to S defined by $f(n) = n^2$, $g(n) = |n|$ and $h(n) = n + 1$ for any $n \in S$. Show that (i) only one among f, g, h is one to one and (ii) $fg \neq gf$, where $(fg)(s) = f(g(s))$, $s \in S$.

(c) In the set Q of all rational numbers an operation $*$ is defined as follows: For any two elements $a, b \in Q$, $a * b = a + b + ab$. Is $(Q, *)$ a group? Justify your answer.

(d) Prove that the set of all 2×2 real matrices of the form $\begin{bmatrix} a & b \\ 0 & c \end{bmatrix}$ is a subring of the ring of all 2×2 real matrices under matrix addition and multiplication.

(e) Prove that the commutative group of all real numbers of the form $a + b\sqrt{2}(a, b$ are rational numbers) under usual addition is a field under usual multiplication.

(f) Show that the vectors $(1, 2, 1), (2, 1, 0)$ and $(1, -1, 2)$ form a basis of R^3.

1999

Group A: Short Answer Questions

1. **(a)** Answer *any three* questions: $3 \times 2 = 6$

i. Find the equation each of whose roots is greater by 2 than the roots of the equation $x^4 + 8x^3 + x - 5 = 0$.

ii. Find the smallest positive integer n, if $\left(\frac{1+i}{1-i}\right)^n = 1$.

iii. Is it possible that all the roots of the equation $2x^3 - 11x^2 + 28x - 24 = 0$ be complex? Justify your answer.

iv. Without expanding find the value of the determinant

$$\begin{vmatrix} 7 & 12 & -3 \\ 9 & 14 & -1 \\ 8 & 13 & -2 \end{vmatrix}.$$

v. What is the rank of unit matrix of order two? Justify your answer.

vi. If $A = \begin{bmatrix} 0 & \sqrt{5} \\ -\sqrt{5} & 0 \end{bmatrix}$, then show that A^2 is symmetric.

(b) Answer *any two* questions: $2 \times 2 = 4$

i. Write down all the subsets of the set $S = \{p, q, r\}$, what is the power set of the set S?

ii. Show whether the mapping $f : Z \to Z$, defined by $f(x) = |x|, x \in Z$, and Z is the set of all integers is onto or not? Justify your answer.

iii. Show that the set $x = \{1, 0, -1\}$ does not form a group under usual addition.

iv. Define a ring with unity element. Give an example of a ring without unity.

v. Show that the vectors $(2, 1, 0), (1, 1, 0), (4, 2, 0)$ of R^3 are not linearly independent.

Group B

[Answer *Question No.* 7 and any two from the rest]

3. **(a)** State De Moivre's theorem and applying it find the formula for $\cos 3\theta$.

(b) If $\sin\alpha + \sin\beta + \sin\gamma = \cos\alpha + \cos\beta + \cos\gamma = 0$, then prove that

$$\cos^2\alpha + \cos^2\beta + \cos^2\gamma = \sin^2\alpha + \sin^2\beta + \sin^2\gamma = \tfrac{3}{2}.$$

(c) If $\sin^{-1}(u+iv)=a+ib$, then prove that $1+u^2+v^2=\sin^2 a+\cosh^2 b$.

$4+4+4=12$

4. (a) If the roots of the equation $x^3+px^2+qx+r=0$ be α,β,γ, then find the equation whose roots are $\beta\gamma-\alpha^2, \gamma\alpha-\beta^2, \alpha\beta-\gamma^2$.

(b) Show that the number of real roots of the equation $x^n-1=0$ is two or one according as n is even or odd.

(c) Solve the equation by Cardan's method: $x^3-6x-9=0$. $4+4+4=12$

5. (a) Show that

$$\begin{vmatrix} 2ab & a^2 & b^2 \\ a^2 & b^2 & 2 \\ b^2 & 2ab & a^2 \end{vmatrix} = -(a^3+b^3)^2.$$

(b) If $1,\omega,\omega^2$ are cube roots of unity, prove that $a+b\omega+c\omega^2$ is a factor of $\begin{vmatrix} a & b & c \\ b & c & a \\ c & a & b \end{vmatrix}$. Find also other factors.

(c) Solve by Crammer's rule: $4+4+4=12$

$$\begin{aligned} x-3z &= 1 \\ 2x-y-4z &= 2 \\ y+z &= 4. \end{aligned}$$

6. (a) If $A=\begin{bmatrix} 1 & 0 & 0 \\ -1 & -2 & -1 \\ -2 & 3 & -2 \end{bmatrix}$, then show that $A+A^T$ is symmetric and $A-A^T$ is skew symmetric and then express A as the sum of a symmetric and a skew symmetric matrix.

(b) If $I=\begin{bmatrix} 1 & 0 \\ 0 & 1 \end{bmatrix}$ and $E=\begin{bmatrix} 0 & 1 \\ 0 & 0 \end{bmatrix}$, prove that $(2I-3E)^3=8I+36E$.

(c) Show that the matrix $A=\begin{bmatrix} 1/3 & 2/3 & 2/3 \\ 2/3 & 1/3 & -2/3 \\ -2/3 & 2/3 & -1/3 \end{bmatrix}$ is an orthogonal.

Hence, find A^{-1}. $4+4+4=12$

7. Answer *any four* questions: $4\times 4=16$

(a) For the three sets $A=\{p,q,r\}$, $B=\{s,t,u\}$, $C=\{s,u\}$, verify

$$A\times(B-C)=(A\times B)-(A\times C).$$

(b) A mapping $f:Z\to N$ is defined as follows:

$$\begin{aligned} f(n) &= 2n \text{ for all } n\geq 1 \\ &= -2n+1 \text{ for all } n\leq 0. \end{aligned}$$

Determine whether the mapping is one to one and onto.
(Z=the set of all integers, N=the set of all natural numbers)

(c) On the set of integers Z, the binary operation $*$ is defined as $a*b=a+b-2$ for all $a,b\in Z$. Show that $<Z,*>$ is a group.

(d) i. Show that if $(G, *)$ be a commutative group, then $(a * b)^2 = a^2 * b^2; a, b \in G$.

ii. Prove that in a ring $(R, +, \bullet)$ with the unity element 1, $(-1) \cdot a = -a$, $a \in R$.

(e) Considering the set of all matrices

$$M = \left\{ \begin{pmatrix} a & b \\ 2b & a \end{pmatrix} \Big/ a, b \text{ are rational} \right\}$$

as a ring under matrix addition and matrix multiplication, show that M forms a field.

(f) Show that $B = \{(1,2,1), (0,1,0), (0,0,1)\}$ is a basis of R^3. Express the vector $(1,2,3) \in R^3$ as a linear combination of the basis B.

B.Sc. General Course

2000

Group A: Short Answer Questions

1. (a) Answer *any three* questions: $3 \times 2 = 6$

i. Show that $(\sin\theta + i\cos\theta)^n = \cos n\left(\frac{\pi}{2} - \theta\right) + i \sin n\left(\frac{\pi}{2} - \theta\right)$ when n is an integer.

ii. For two complex numbers z_1, z_2 prove that

$$|z_1 + z_2|^2 + |z_1 - z_2|^2 = 2\left(|z_1|^2 + |z_2|^2\right).$$

iii. The equation $3x^5 - 4x^2 + 8 = 0$ has at least two imaginary roots; justify.

iv. Give an example of a third order skew-symmetric determinant and find its value.

v. Let A, B be matrices such that $AB = 0$, where 0 is the null matrix. Does it imply that $A = 0$ or $B = 0$? Give an example in support of your conclusion.

(b) Answer *any three* questions: $3 \times 2 = 6$

i. If $f : R^+ \to R^+$ be defined by $f(x) = x^2 + 3, x \in R^+$ where R^+ is the set of all positive real numbers, find f^{-1}.

ii. Verify that $A - (B \cup C) = (A - B) \cap (A - C)$ where

$$A = \{1,2,5\},\ B = \{2,4,6\},\ C = \{5,7\}.$$

iii. $S = \{1, i, -1, -i\}$, where $i^4 = 1$, is a group with respect to usual multiplication. Show that it is a cyclic group.

iv. Give an example of a ring which is not a field.

v. Find the eigen values of the matrix

$$\begin{bmatrix} 3 & 0 & 3 \\ 0 & 3 & 0 \\ 3 & 0 & 3 \end{bmatrix}.$$

Group B

[Answer *any two* questions]

2. (a) Show that $i \log \dfrac{x-i}{x+i} = \pi - \tan^{-1} x$. $4+3+5=12$

(b) If $\tan \log(x+iy) = a+ib$, where $a^2+b^2 \neq 1$, prove that

$$\tan \log(x^2+y^2) = \frac{2a}{1-a^2-b^2}.$$

(c) Use De Moivre's theorem to show that

$$\alpha^n + \beta^n = 2^{\frac{n}{2}+1} \cdot \cos \frac{n\pi}{4}$$

where α, β are the roots of $x^2-2x+2=0$ and n is a positive integer.

3. (a) If the roots of the equation $x^4+ax^3+bx^2+cx+d=0$ are in G.P., show that $c^2=a^2d$.

(b) If $\alpha, \beta, \gamma, \delta$ be the roots of $x^4-3x^3+4x^2-5x+6=0$, find the value of

$$(\alpha^2+3)(\beta^2+3)(\gamma^2+3)(\delta^2+3).$$

(c) Form the equation whose roots are squares of the roots of $x^3+px^2+qx+r=0$. $4+5+3=12$

4. (a) Prove without expanding that $\begin{vmatrix} 1 & a & a^2-bc \\ 1 & b & b^2-ca \\ 1 & c & c^2-ab \end{vmatrix} = 0.$

(b) If the adjugate of $\Delta = \begin{vmatrix} a & h & g \\ h & b & f \\ g & f & c \end{vmatrix}$ is $D' = \begin{vmatrix} A & H & G \\ H & B & F \\ G & F & C \end{vmatrix}$, then prove that

$$\frac{BC-F^2}{a} = \frac{CA-G^2}{b} = \frac{AB-H^2}{c} = \Delta.$$

(c) Define adjoint of a determinant. If Δ^1 be the adjoint of a third order determinant $\Delta(\Delta \neq 0)$, prove that $\Delta^1 = \Delta^2$. $3+4+5=12$

5. (a) Find A and B when $A+B=2B^T$ and $3A+2B=I_3$, where I_3 represents 3×3 identity matrix.

(b) If $A = \begin{bmatrix} 2 & -1 \\ 1 & 3 \end{bmatrix}$ and $A^2-5A+7I=0$ where I is the 2×2 unit matrix and 0 is the null matrix, then show that

$$A^{-1} = \frac{1}{7}\begin{bmatrix} 3 & 1 \\ -1 & 2 \end{bmatrix}.$$

(c) Solve, by the matrix method, the equations $x+y+z=8$, $x-y+2z=6$, $3x+5y-7z=14$. $4+4+4=12$

Group C

[Answer *any three* questions]

6. **(a)** Verify that $(A \cap B)' = A' \cup B'$ where

$$A = \{2,3,4\},\ B = \{3,4,5\},\ U = \{1,2,3,4,5,6,7,30\},$$

A' and B' are respectively complements of A and B in U. Also verify that $A \times B \neq B \times A$.

(b) Two mappings $f : \mathbf{R} \to \mathbf{R}$ and $g : \mathbf{R} \to \mathbf{R}$ (|**R** is the set of real numbers) are defined by

$$f(x) = x^2 \text{ and } g(x) = \left\{ \begin{array}{ll} x-1, & x \geq 0 \\ -x, & x < 0 \end{array} \right\}$$

Find the mappings $f \circ g$ and $g \circ f$.

(c) Give an example of a mapping $f : Z \to Z$ (Z=Set of all integers) which is injective but not surjective. $5+4+3=12$

7. **(a)** Prove that the set M of all 2×2 non-singular matrices forms a non-commutative group under usual matrix multiplication.

(b) Prove that a group in G of below statements.

i. Prove that $(ab)^{-1} = b^{-1}a^{-1}, ab \in G$.

ii. If $a^2 = e$ for all $a \in G$, prove that G is commutative.

(c) Prove that the set of all even integers, under usual addition, is a subgroup of the group of integers. $4+5+3=12$

8. **(a)** Prove that the set of all 2×2 real matrices of the form $\begin{pmatrix} a & 0 \\ 0 & b \end{pmatrix}$ is a subring of the ring of all 2×2 real matrices under usual matrix addition and multiplication.

(b) Show that the ring of matrix

$$\left\{ \begin{pmatrix} a & b \\ 2b & a \end{pmatrix} \Big/ a, b \in R \right\}$$

does not form a field. (R=Set of all real numbers).

(c) Determine k so that the vectors $(1,3,1)$, $(2,k,0)$ and $(0,4,1)$ are linearly dependent in R^3. $4+5+3=12$

9. **(a)** Show that $4xy + 2yz - 4zx - 6x^2 - 3y^2 - 3z^2$ is negative definite.

(b) Do the vectors $(1,1,0), (1,0,1)$ and $(0,1,1)$ form a basis of the vector space R^3 over the field of real numbers? Justify.

(c) State Cayley-Hamilton theorem and use it to compute the inverse of the matrix

$$\begin{pmatrix} 2 & 1 \\ 3 & 5 \end{pmatrix}.$$

$4+3+5=12$

2001

Group A: Short Answer Questions

1. (a) Answer *any three* questions: $2 \times 3 = 6$

   i. Show that the points $3-4i$, $-3+4i$ and $6-8i$ in the Argand plane are collinear.

   ii. Show that a root of $x^3 - 4x + 2 = 0$ lies between 1 and 2.

   iii. Show that $\beta - \gamma$ is a factor of the determinant

$$\Delta = \begin{vmatrix} 1 & 1 & 1 \\ \alpha & \beta & \gamma \\ \alpha^2 & \beta^2 & \gamma^2 \end{vmatrix}.$$

   iv. Show that the matrix $A = \begin{pmatrix} \frac{1}{\sqrt{2}} & -\frac{1}{\sqrt{2}} \\ \frac{1}{\sqrt{2}} & \frac{1}{\sqrt{2}} \end{pmatrix}$ is an orthogonal matrix.

   v. Write the matrix equation for the system of equations:

$$\begin{aligned} x + y + z &= 3 \\ 2x - 3y + 4z &= 17 \\ x + 6y - z &= -8. \end{aligned}$$

(b) Answer *any three* questions: $2 \times 3 = 6$

   i. If S (the universal set)$=\{1,2,3,4,5,6,7\}$ and A, B are its two subsets given by $A = \{1,5,6\}$, $B = \{5,6,7\}$, verify that $(A \cap B)' = A' \cup B'$ where '′' denotes complement in S.

   ii. Two mappings $f : R \to R^+$ and $g : R \to R^+$ are defined as follows:

$$\begin{aligned} f(x) &= x^2 + 1, \\ g(x) &= x, \end{aligned}$$

   Determine the composite mappings $f \circ g$ and $g \circ f$. (R is the set of all real numbers)

   iii. Given that the set of all 2×2 matrices of the form $\begin{pmatrix} x & y \\ 0 & z \end{pmatrix}$, where x, y, z are integers, forms a ring with respect to matrix addition and matrix multiplication. Is it a commutative ring? Justify your answer.

   iv. Show that the vectors $(2,4,0)$, $(0,1,0)$ and $(2,6,2)$ are linearly independent in the real vector space R^3.

   v. Write the associated matrix for the quadratic form $3x^2 + 7xy - y^2$.

Group B

[Answer *any two* questions]

2. (a) If the ratio $\dfrac{z-i}{z-1}$ is purely imaginary, where z is any complex number, then show that z lies on a circle whose centre is $\frac{1}{2}(1+i)$ and radius is $\frac{1}{\sqrt{2}}$.

(b) If $2\cos\alpha = m + \frac{1}{m}, 2\cos\beta = n + \frac{1}{n}$ and $2\cos\gamma = p + \frac{1}{p}$, then prove that

$$2\cos(\alpha+\beta+\gamma) = mnp + \frac{1}{mnp}.$$

(c) Prove that $\sin\left(i\log\frac{a-ib}{a+ib}\right) = \frac{2ab}{a^2+b^2}$, where a and b are real numbers. $4+4+4=12$

3. (a) If $2+i3$ be one of the roots of the equation $x^4 - 2x^3 + 6x^2 + 22x + 13 = 0$, find other roots.

(b) If α, β, γ be the roots of equation $x^3 + px^2 + qx + r = 0$, form the equation whose roots are $\frac{\alpha}{\beta+\gamma}, \frac{\beta}{\gamma+\alpha}, \frac{\gamma}{\alpha+\beta}$.

(c) Solve the cubic equation $x^3 - 12x + 65 = 0$ by Cardan's method.

$4+4+4=12$

4. (a) Define a symmetric determinant

If $\Delta = \begin{vmatrix} a_1 & b_1 & c_1 \\ a_2 & b_2 & c_2 \\ a_3 & b_3 & c_3 \end{vmatrix}$, show that Δ^2 is a symmetric determinant.

(b) Show that, $\begin{vmatrix} 1+a_1 & a_2 & a_3 \\ a_1 & 1+a_2 & a_3 \\ a_1 & a_2 & 1+a_3 \end{vmatrix} = 1 + a_1 + a_2 + a_3.$

(c) Solve by Crammer's rule:

$$\begin{aligned} x - y + 2z &= 1 \\ x + y + z &= 2 \\ 2x - y + z &= 5. \end{aligned}$$

$4+4+4=12$

5. (a) Express $\begin{pmatrix} 3 & 2 & -6 \\ 0 & -1 & 4 \\ 5 & -2 & 0 \end{pmatrix}$ as the sum of a symmetric and a skew symmetric matrix.

(b) If $A = \begin{pmatrix} 1 & 2 \\ 3 & 5 \end{pmatrix}$, show that $A \cdot \frac{\text{adj}\,A}{|A|} = \frac{\text{adj}\,A}{|A|} \cdot A = I_2$, where I_2 is the identity matrix of order 2. From this relation obtain the necessary condition for the existence of inverse of a square matrix.

(c) Find the rank of the matrix $A = \begin{pmatrix} 1 & 2 & 3 \\ 2 & 1 & 0 \\ 0 & 1 & 2 \end{pmatrix}$. $4+4+4=12$

Group C

[Answer *any two* questions]

6. (a) i. Write down all the subsets of the set $A = \{a, b, c\}$.

ii. If $A = \{1,2\}, B = \{1,2,3\}$, find $(A \times B) \cap (B \times A)$. $2+2$

(b) A mapping $f : R \to R$ is defined as follows:

$$f(x) = \begin{cases} 1, & \text{if } x \text{ is rational} \\ -1, & \text{if } x \text{ is irrational} \end{cases}$$

R is the set of all real numbers.
Find the range of f. What is the value of $f(\sqrt{7})$?
Is the mapping f one-one and onto? Give reason. $1+1+(1+1)$

(c) Define inverse mapping of $f : A \to B$, where A and B are two non-empty sets.
A mapping $f : R \to R$ (R is the set of all real numbers) is defined as follows: $f(x) = x^2 + 1, x \in R$; find $f^{-1}(10)$. $2+2$

7. (a) Define subgroup of a group.
Show that the set $H = \{1, -1\}$ is a subgroup of the multiplicative group $G = \{1, -1, i, -i\}$, where $i = \sqrt{-1}$. $1+3$

(b) Find the inverse of the permutation

$$p = \begin{pmatrix} 1 & 2 & 3 & 4 & 5 & 6 \\ 5 & 3 & 4 & 6 & 1 & 2 \end{pmatrix}.$$

Find whether the permutation p is even or odd. $2+2$

(c) Show that the set Z of all integers forms a cyclic group under the operation addition. What are its generators? $3+1$

8. (a) Show that the set $\{(a + b\sqrt{2})$, where a and b are integers$\}$ forms a commutative ring with the identity element, with respect to addition and multiplication. 5

(b) Define a field. Assuming that the set Z of all integers forms a commutative ring under addition and multiplication, show that Z is not a field. $2+2$

(c) Show that in a ring $(R, +, \bullet)$,

$$(-a) \bullet b = a \bullet (-b) = -(a \bullet b); a, b \in R. \quad 3$$

9. (a) Show that the subset $S = \{2, 4, 0), (0, 1, 0), (2, 6, 2)$ of the real vector space R^3 is a basis of R^3. 4

(b) Determine the eigen values of $A = \begin{bmatrix} 1 & 3 \\ 4 & 5 \end{bmatrix}$ and the eigen vectors corresponding to the positive eigen value of A. $2+2$

(c) State Clayey-Hamilton theorem and verify the theorem for the matrix

$$A = \begin{bmatrix} 3 & -1 \\ 2 & 4 \end{bmatrix}. \quad 1+3$$

2002

Group A: Short Answer Questions

1. (a) Answer *any three* questions: $2 \times 3 = 6$

   i. On the complex plane, let $P(z)$ be a variable point such that $|z-2i| = 3$. Find the locus of P.

   ii. Find the conditions so that $x^2 + 3px + q$ may have a factor of the form $(x-a)^2$.

   iii. Show that the equation $27x^4 - 48x^2 - 12x + 13 = 0$ has two positive roots in $(0,1)$ and $(1,2)$.

   iv. Show that every skew-symmetric determinant of order three is zero.

   v. Show that the matrix $A = \frac{1}{3}\begin{pmatrix} 1 & -2 & 2 \\ -2 & 1 & 2 \\ -2 & -2 & -1 \end{pmatrix}$ is orthogonal.

   (b) Answer *any three* questions: $2 \times 3 = 6$

   i. If $A = \{3, 5, 6, 8\}$, $B = \{-3, 0, 6\}$, find $(A \cup B - (A \cap B)$.

   ii. If $f : R \to R$ is defined by $f(x) = x^2 + 5, x \in R$, where R is the set of all real numbers, find $f^{-1}(6)$.

   iii. If e be the identity element in a group G and $a * a = e$, $a \in G$, then show that G is abelian.

   iv. Show that the ring of all even integers is not a field.

   v. Find the eigen values of the matrix $\begin{pmatrix} 2 & -2 \\ 1 & 1 \end{pmatrix}$.

Group B

[Answer *any two* questions]

2. (a) Find the cube roots of (-1) using De Moivre's theorem.

   (b) Express $\sin^4\theta\cos^4\theta$ in terms of sines and cosines of multiples of θ.

   (c) If $x + iy = \dfrac{1}{2 + \cos\theta + i\sin\theta}$, where x, y, θ are real, show that when θ varies, the point (x, y) moves on a circle whose centre is the point $(2/3, 0)$. $4 + 4 + 4 = 12$

3. (a) Find the equation of the fourth degree with rational coefficients, one root of which is $\sqrt{2} + i\sqrt{3}$.

   (b) Solve the equation $x^3 - 9x^2 + 39x - 126 = 0$.

   (c) If $\alpha_1, \alpha_2, \alpha_3$ be the roots of the equation $x^3 + x + 1 = 0$, then prove that $(\alpha_1^2 + 1)(\alpha_2^2 + 1)(\alpha_3^2 + 1) = 1$. $5 + 4 + 3 = 12$

4. (a) Find the value of

$$\begin{vmatrix} 0 & (\alpha-\beta)^2 & (\alpha-\gamma)^2 \\ (\beta-\alpha)^2 & 0 & (\beta-\gamma)^2 \\ (\gamma-\alpha)^2 & (\gamma-\beta)^2 & 0 \end{vmatrix}$$

by expressing it as the product of two determinants.

(b) Without expanding the determinant, prove that

$$\begin{vmatrix} b+c & c+a & a+b \\ q+r & r+p & p+q \\ y+z & z+x & x+y \end{vmatrix} = 2\begin{vmatrix} a & b & c \\ p & q & r \\ x & y & z \end{vmatrix}.$$

(c) If Δ' denotes the adjugate of a non-zero determinant Δ of third order, then show that $\Delta' = \Delta^2$. $4+4+4=12$

5. (a) Show that every square matrix can be expressed as a sum of a symmetric matrix and a skew symmetric matrix.

(b) If $A = \begin{pmatrix} 1 & 2 & 2 \\ 2 & 1 & 2 \\ 2 & 2 & 1 \end{pmatrix}$, then show that $A^2 - 4A - 5I = 0$, where I is the 3×3 unit matrix.

(c) Solve the equations by matrix method $x+y+z=4$, $2x-y+3z=1$, $3x+2y-z=1$. $4+4+4=12$

Group C

[Answer *any two* questions]

6. (a) In an examination of 100 students, 70 passed in Mathematics, 65 passed in Physics and 55 passed in Chemistry. Of these students, 50 passed in Mathematics and Physics, 45 passed in Mathematics and Chemistry, 40 passed in Physics and Chemistry and 35 passed in all the three subjects.

i. How many students passed in exactly two of the three subjects?
ii. How many students passed in exactly one of the three subjects?
iii. How many students failed in all the three subjects?

(b) Let N be the set of natural numbers. Show that the mapping $f: N \to N$, defined by $f(x) = 2x+4$ for all $x \in N$, is an one to one mapping but not onto mapping.

(c) If two mappings $f: R \to R$ and $g: R \to R$ are defined as follows: $f: x \to x^2, g: x \to x-2$, then show that $f \circ g \neq g \circ f$. $5+4+3=12$

7. (a) Show that the set of non-zero rational numbers forms an abelian group under multiplication.

(b) Prove that the intersection of any two subgroups of a group is a subgroup of the group.

(c) Let $(G, *)$ be a group. For any two elements $a, b \in G$, prove that (i) $(a^{-1})^{-1} = a$, and (ii) $(a * b)^{-1} = b^{-1} * a^{-1}$. $4+4+4=12$

8. (a) Prove that the set of all even integers forms a commutative ring.

(b) Show that the set of all 2×2 real matrices of the form $\begin{pmatrix} x & y \\ -y & x \end{pmatrix}$ forms a field with respect to matrix addition and multiplication.

(c) Define a cyclic group. Show that the group $G = \{-1, 1, i, -i\}$ where $i^2 = -1$ is a finite cyclic group with respect to multiplication.

$4 + 4 + 4 = 12$

9. (a) Show that $\alpha_1 = (1, 2, 1), \alpha_2 = (2, 1, 0), \alpha_3 = (1, -1, 2)$ form a basis of the vector space R^3 over the field of real numbers.

(b) If λ be an eigen value of a real orthogonal matrix A, prove that $1/\lambda$ is also an eigen value of A.

(c) Show that $5x^2 + y^2 + 14z^2 - 4yz - 10zx$ is positive definite.

$4 + 4 + 4 = 12$

2003

Group A: Short Answer Questions

1. (a) Answer *any three* questions: $2 \times 3 = 6$

i. If $\omega (\neq 1)$ be a cube root of unity, prove that

$$(1 - \omega)(1 - \omega^2)(1 - \omega^4)(1 - \omega^8) = 9.$$

ii. Express $3x^3 - 4x^2 + 5x + 6$ as a polynomial in $(x + 1)$.

iii. Find the condition that the equation $x^3 + px^2 + qr + r = 0$ may have two roots equal but of opposite signs.

iv. Prove, without expanding, that the determinant

$$\begin{vmatrix} 1 & a & b+c \\ 1 & b & c+a \\ 1 & c & a+b \end{vmatrix} = 0.$$

v. By an example show that the matrix multiplication is, in general, not commutative.

(b) Answer *any three* questions: $2 \times 3 = 6$

i. If $A = \{5, 6, 7, 8, 9\}$, $B = \{2, 4, 6, 8, 10, 12\}$, $C = \{3, 6, 9, 12\}$, then verify that

$$A \cup (B \cap C) = (A \cup B) \cap (A \cup C).$$

ii. Two mappings $f : R \to R$ and $g : R \to R$ are defined as follows:

$$f(x) = 4x - 1, \; g(x) = x^2 + 2.$$

Determine the composite mapping $f \circ g$ where $(f \circ g)(x) = f(g(x))$. (R is the set of all real numbers.)

iii. Write down the subsets of the set $\{1, 2, 3\}$.

iv. Show that the set $S = \{-3, -2, -1, 0, 1, 2, 3\}$ is not a group with respect to operation addition.

v. Show that the vectors $(1, 5, 2)$, $(1, 1, 0)$ and $(0, 0, 1)$ are linearly independent in the real vector space R^3.

Group B

[Answer *any two* questions]

2. **(a)** If z_1 and z_2 be a pair of conjugate complex numbers and z_3 and z_4 be another pair of conjugate complex numbers, then show that

$$\text{amp.}\left(\frac{z_1}{z_4}\right) = \text{amp.}\left(\frac{z_3}{z_2}\right).$$

(b) If $x + \frac{1}{x} = 2\cos\frac{\pi}{7}$, then show that $x^7 + \frac{1}{x^7} = -2$.

(c) Show that $\tan\left(i \log\frac{a - ib}{a + ib}\right) = \frac{2ab}{a^2 - b^2}$. $4 + 4 + 4 = 12$

3. **(a)** Solve the equation $4x^3 + 16x^2 - 9x - 36 = 0$ where the sum of two roots is zero.

(b) Show that the roots of equation

$$\frac{1}{x-a} + \frac{1}{x-b} + \frac{1}{x-c} = \frac{1}{x},$$

where $a > b > c > 0$ are real.

(c) Solve the equation $x^3 + 12x - 12 = 0$ by Cardon's method.

$4 + 4 + 4 = 12$

4. **(a)** Prove that

$$\begin{vmatrix} x^2+y^2+1 & x^2+2y^2+3 & x^2+3y^2+4 \\ y^2+2 & 2y^2+6 & 3y^2+8 \\ y^2+1 & 2y^2+3 & 3y^2+4 \end{vmatrix} = x^2y^2.$$

(b) Solve by Crammer's rule

$$\begin{aligned} 3x - 2y + z &= -1 \\ -x + y + 7z &= 1 \\ 4x - 3y - 2z &= -2. \end{aligned}$$

(c) If ω be a cube root of unity, show that $(a + b\omega + c\omega^2)$ is a factor of the determinant

$$\begin{vmatrix} a & b & c \\ b & c & a \\ c & a & b \end{vmatrix}.$$

Hence, show that the determinant is equal to $-(a^3 + b^3 + c^3 - 3abc)$.

$4 + 4 + 4 = 12$

5. **(a)** If $A = \begin{bmatrix} 0 & 1 \\ 1 & 1 \end{bmatrix}$ and $B = \begin{bmatrix} 0 & -1 \\ 1 & 0 \end{bmatrix}$ be two matrices, then show that

$$(A + B)(A - B) \neq A^2 - B^2.$$

(b) For the matrix $A = \begin{bmatrix} 2 & -1 \\ 1 & 3 \end{bmatrix}$, show that $A^2 - 5A + 7I = 0$, where I is the 2×2 unit matrix and 0 is the null matrix. Hence, show that

$$A^{-1} = \frac{1}{7}\begin{bmatrix} 3 & 1 \\ -1 & 2 \end{bmatrix}$$

(c) Find the rank of the matrix $\begin{pmatrix} 0 & 1 & 2 \\ 3 & 5 & 2 \\ 4 & 8 & 0 \end{pmatrix}$ $4+4+4=12$

Group C

[Answer *any two* questions]

6. **(a)** If A and B be two sets, then prove that $A \cup B = (A - B) \cup B$

(b) If $A = \{1, 2\}, B = \{2, 3\}$, and $C = \{3, 4\}$ are three given sets, then find the following sets: (i) $A \times B) \cup (A \times C)$; (ii) $A \times B) \cap (A \times C)$.

(c) Show that the map $f : Q \to Q, f(x) = 3x + 2$ is one-one and onto, where Q is the set of rational numbers and find f^{-1}. $4+4+4=12$

7. **(a)** Show that the set

$$M = \left\{ \begin{bmatrix} 1 & 0 \\ 0 & 1 \end{bmatrix}, \begin{bmatrix} 0 & 1 \\ -1 & 0 \end{bmatrix}, \begin{bmatrix} -1 & 0 \\ 0 & -1 \end{bmatrix}, \begin{bmatrix} 0 & -1 \\ 1 & 0 \end{bmatrix} \right\}$$

forms an Abelian group under matrix multiplication.

(b) Show that $\begin{pmatrix} 1 & 2 & 3 & 4 & 5 & 6 \\ 4 & 3 & 1 & 2 & 6 & 5 \end{pmatrix}$ is an even permutation and

$\begin{pmatrix} 1 & 2 & 3 & 4 & 5 & 6 & 7 \\ 4 & 3 & 1 & 2 & 6 & 7 & 5 \end{pmatrix}$ is an odd permutation.

(c) **i.** If $a^2 = e$ for all $a \in G$, prove that G is a commutative group where e is the identity element of the group G.

ii. Prove that a group contains only one identity element.

$4+4+(2+2)=12$

8. **(a)** Show that the ring of the following set of matrices

$$\left\{ \begin{pmatrix} a & b \\ 2b & a \end{pmatrix} \Big/ a, b \in R \right\}$$

is not a field. (R=set of all real numbers)

(b) If Z be the set of all integers, show that $(2Z, +, \cdot)$ is a commutative ring with respect to addition and multiplication.

(c) If I is the identity element in a ring $(R, +, \cdot)$, then prove that

i. $(-1)a = -a = a(-1), a \in R$;

ii. $(-1)(-1) = 1$. $4+4+(2+2)=12$

9. (a) Find a basis of the real vector space R^3 containing the vectors $(1,1,2)$ and $(3,5,2)$.

(b) Find the eigen values and eigen vectors of the matrix $A = \begin{bmatrix} -2 & 4 \\ 4 & 4 \end{bmatrix}$.

(c) State Cayley-Hamilton theorem and verify the theorem for the matrix

$$A = \begin{bmatrix} 1 & 0 & 2 \\ 0 & -1 & 1 \\ 0 & 1 & 0 \end{bmatrix}.$$

$4+4+4=12$

2004

Group A: Short Answer Questions

1. (a) Answer *any three* questions: $2 \times 3 = 6$

i. Let $\omega^n + \omega^{2n} + 1 = 0$, where $\omega^3 = 1$ and n be an integer. Test whether n is divisible by 3.

ii. Solve the equation $x^3 - 3x^2 - 6x + 8 = 0$, given that the roots are in A.P.

iii. If $\alpha, \beta\gamma$ be the roots of equation $x^3 + px + q = 0$, find the value of $\sum \frac{1}{\beta+\gamma}$.

iv. Prove, without expanding, that $\begin{vmatrix} 5 & 1 & 2 \\ 7 & 3 & 6 \\ 9 & 7 & 14 \end{vmatrix} = 0$.

v. What is the rank of unit matrix of order 2? Justify your answer.

(b) Answer *any three* questions $2 \times 3 = 6$

i. Four the sets $A = \{1,2,3\}$ and $B = \{4,5\}$, verify that $A \times B = B \times A$.

ii. Let the function $f : R \to R$ be defined by $f(x) = 2x + 3$, $x \in R$ (R is the set of real numbers).
If $A = \{x : 1 \leq x \leq 2\}$, find $f(A)$.

iii. Show that $\{1, \omega, \omega^2\}$ is a group under multiplication.

iv. Find the given values of the matrix $\begin{pmatrix} 1 & -1 & 0 \\ 1 & 2 & -1 \\ 3 & 2 & -2 \end{pmatrix}$.

v. Determine k so that the vectors $(1,-1,2)$, $(0,k,3)$ and $(-1,2,3)$ are linearly dependent.

Group B

[Answer *any two* questions]

2. (a) If the amplitude of $\frac{z-1}{z+1}$ is $\frac{\pi}{4}$, where $z = x + iy, i = \sqrt{-1}$, show that z his on a circle with centre at i. What is its radius?

(b) If $\tan(x+iy) = a + ib$, prove that $a^2 + b^2 + 2a \cot 2x = 1$.

(c) Use De Moivre's theorem, to show that $\alpha^n + \beta^n = 2^{\frac{n}{2}+1} \cos \frac{n\pi}{4}$, where α, β are the roots of $x^2 - 2x + 2 = 0$. $4+4+4$

3. (a) Find the condition that the roots of the equation $x^3 - px^2 + qx - r = 0$ will be in Arithmetical Progress.

(b) Solve the equation $x^4 - 3x^3 - 5x^2 + 9x - 2 = 0$, given $2 - \sqrt{3}$ is one of its roots.

(c) The roots of the equation $x^3 - px^2 + qx - r = 0$ are α, β, γ. Find the value of $(\beta + \gamma)(\gamma + \alpha)(\alpha + \beta)$.

4 + 4 + 4

4. **(a)** Show that $\begin{vmatrix} a^2 & ab & b^2 \\ b^2 & a^2 & 2ab \\ 2ab & b^2 & a^2 \end{vmatrix}$ is a perfect square.

(b) If A_i, B_i, C_i are the cofactors of a_i, b_i, c_i $(i = 1, 2, 3)$ in $\Delta = \begin{vmatrix} a_1 & b_1 & c_1 \\ a_2 & b_2 & c_2 \\ a_3 & b_3 & c_3 \end{vmatrix}$,

show that $\begin{vmatrix} B_1 + C_1 & C_1 + A_1 & A_1 + B_1 \\ B_2 + C_2 & C_2 + A_2 & A_2 + B_2 \\ B_3 + C_3 & C_3 + A_3 & A_3 + B_3 \end{vmatrix} = 2\Delta^2$.

(c) Prove that $\begin{vmatrix} a & b & c \\ b & c & a \\ c & a & b \end{vmatrix}^2 = \begin{vmatrix} 2bc - a^2 & c^2 & b^2 \\ c^2 & 2ca - b^2 & a^2 \\ b^2 & a^2 & 2ab - c^2 \end{vmatrix} - 0.$

4 + 4 + 4

5. **(a)** If $A = \begin{bmatrix} 1 & 2 \\ 2 & 4 \end{bmatrix}$, $B = \begin{bmatrix} 4 & -2 \\ -2 & 1 \end{bmatrix}$, $C = \begin{bmatrix} 1 & 2 \\ 3 & 4 \end{bmatrix}$, show that—

   i. $(A - B)(A + B) = A^2 - B^2$;
   ii. $(A - C)(A + C) \neq A^2 - C^2$.

(b) If $A = \begin{bmatrix} 1 & 1 & 2 \\ 2 & -1 & 3 \end{bmatrix}$, $B = \begin{bmatrix} 1 & -2 \\ -2 & 0 \\ 1 & 1 \end{bmatrix}$ show that $B'A' = (AB)'$, where X' denotes the transpose of X.

(c) Solve the equations by matrix method:

$$\left.\begin{aligned} x + y + z &= 4 \\ 2x - y + 3z &= 1 \\ 3x + 2y - z &= 1 \end{aligned}\right\}.$$

4 + 4 + 4

Group C

[Answer *any three* questions].
(R denotes the Set of Real Numbers)

6. **(a)** If $A = \{5, 6, 7, 8, 9\}$, $B = \{2, 4, 6, 8, 10, 12\}$ and $C = \{3, 6, 9, 12\}$ then verify that $A - (B \cup C) = (A - B) \cap (A - C)$.

(b) Let $A = \{x : x \text{ is an integer and } 1 \leq x \leq 10\}$ and $B = \{x : x \text{ is a prime integer less than } 10\}$. Verify $A \cap B = B$.

(c) Show that the function $f : R \to R$ given by $f(x) = |x|$, $x \in R$ is neither injective nor surjective.

4 + 4 + 4

7. **(a)** Let H be the set of all real matrices $\left\{ \begin{pmatrix} a & b \\ c & d \end{pmatrix} : ad - bc = 1 \right\}$. Prove that H form a non-commutative group under matrix multiplication.

(b) In a group (G, o), show that

i. $a \circ b = a \circ c \Rightarrow b = c$;

ii. $(a \circ b)^{-1} = b^{-1} \circ a^{-1}$, $a, b, c \in G$.

(c) Define a subgroup of a group. Let Q^* be the group of all non-zero rational numbers under usual multiplication. Is Z^+, a subgroup of Q^*? (Z^+ = set of all positive integers). Justify your answer. 4+4+4

8. (a) Let $\{Z, +, \bullet\}$ be a ring with unity. Show that $(2Z, +, \bullet)$ is a subring of $(Z, +, \bullet)$. Does the subring contain the unity? (Z = set of all integers).

(b) Show that the set of matrices of the form $\begin{pmatrix} a & b \\ 0 & 0 \end{pmatrix}$, where $a, b \in R$ form a ring under matrix addition and multiplication.

(c) Let $S = \{a + b\sqrt{2} : a, b \text{ are rational numbers}\}$ be a ring with respect to usual addition and multiplication. Show that S is a Field. 4+4+4

9. (a) Show that the set $S = \{(1, 0, 1), (0, 1, 1), (1, 1, 0)\}$ is a basis of the vector space R^3.

(b) Show that the quadratic form $5x^2 + y^2 + 5z^2 + 4xy - 8xy - 4yz$ is positive definite.

(c) Find the eigen values and eigen vector of the matrix $A = \begin{pmatrix} 2 & -1 \\ -1 & 2 \end{pmatrix}$.

4+4+4

Additional Topics

1. Sequences and Series of Real Numbers (Chapter 12)
2. Linear Difference Equations (Chapter 13)
3. Inequalities (Chapter 14)
4. Simple continued Fractions (Chapter 15)

Chapter 12

Sequences and Series of Real Numbers

Section I: Sequence of Real Numbers

12.1 Introduction

Our main objective is to study the theory of Infinite series, an important and a very useful branch of mathematical analysis. For its proper understanding a knowledge of sequences is essential. In the present chapter, therefore, we introduce the concept of a sequence and the idea of its convergence.

12.2 Definition of a Sequence

Definition: *Suppose that we have a set of real numbers x_n, one associated with each positive integer n. Such a set,* **arranged in order of increasing** n : $\{x_1, x_2, x_3, \ldots, x_n, \ldots\}$ *is called a* **sequence** *of real numbers. We write the sequence as* $\{x_n\}$. $x_1, x_2, x_3, \ldots$ *are respectively called first, second, third* $\cdots$ *terms of the sequence* $\{x_n\}$.

Examples (**with remarks, where necessary**)

► Example 12.2.1 *The positive integers themselves, arranged in order of magnitude, form a sequence* $\{n\} = \{1, 2, 3, \ldots, n, \ldots\}$.

► Example 12.2.2 *The reciprocals of positive integers form what we call a* **harmonic sequence:** $\left\{\frac{1}{n}\right\} = \left\{1, \frac{1}{2}, \frac{1}{3}, \ldots, \frac{1}{n}, \ldots\right\}$.

► Example 12.2.3 *It is not necessary that the numbers x_n should be all different:*

(i) $\{1, 1, 1, \ldots, \ldots\}$ *is n perfectly good sequence.*

We may write this sequence as $\{x_n\}$, *where* $x_n = 1, \forall n$.

(ii) *Similarly,* $\{x_n\}$, *where* $x_n = (-1)^{n-1}$ *gives* $\{1, -1, 1, -1, \ldots\}$.

(iii) $\{x_n\}$, *where* $x_n = \frac{1}{2}\{1 - (-1)^n\}$ *gives* $\{1, 0, 1, 0, \ldots\}$.

► Example 12.2.4 *There may not be any known 'explicit formula' for writing x_n. We just have an x_n associated with each and every n : e.g.,*

(i) $\{x_n\} = \{3, 1, 4, 1, 5, 9, 2, \ldots\}$ *(the successive integers in the decimal representation for π—there is no formula for x_n).*

(ii) $\{x_n\} = \{2, 3, 5, 7, 11, 13, \ldots\}$ *(the prime numbers).*

▶ Example 12.2.5 *x_n is, of course, a function of n: we might have written $x(n)$. Conversely, when a function $f(n)$ defined for positive integers n is given, then the numbers $\{f(1), f(2), \ldots, f(n), \ldots\}$ constitute a sequence.*

Take, for instance, $x_n = \frac{n-1}{2n} = f(n)$. Then this sequence is $\{0, \frac{1}{4}, \frac{1}{3}, \frac{3}{8}, \frac{2}{5}, \frac{5}{12}, \ldots\}$

Exercises. *Write explicitly the terms of the sequence $\{x_n\}$, where*

(i) $x_n = 1 + \left(-\frac{1}{2}\right)^n$; (ii) $x_n = \dfrac{\sin(n\pi/2)}{n}$; (iii) $x_n = \dfrac{3n^2}{(n-1)(n-2)}$.

Note 1. *In the last case x_1 and x_2 do not exist but x_n is defined for all other values of n; therefore, the sequence $\{x_n\}$ is said to exist even if a finite number of terms does not exist.*

But $\{\operatorname{cosec} n\pi\}$ or $\{\sec n\pi/2\}$ are not sequences because the terms are not defined for infinitely many n.

▶ Example 12.2.6 *A sequence may be written down from a recurrence relation:*

$$\{1, 1, 2, 3, 5, 8, 13, \ldots\} \text{ [Fibonacci's Sequence].}$$

Here $x_1 = 1, x_2 = 1$; after that every $x_n = x_{n-1} + x_{n-2} (n \geq 3)$.

12.3 Bounded and Unbounded Sequences

Definition: *A sequence $\{x_n\}$ is said to be* **bounded above** *if $\exists$ a real number U such that $x_n \leq U$, for all positive integers n. U is then called* **an upper bound** *of $\{x_n\}$. If we cannot find such an U, we say that $\{x_n\}$ is unbounded above.*

Definition: *A sequence $\{x_n\}$ is said to be* **bounded below** *if $\exists$ a real number L such that $x_n \geq L$, for all positive integers n. L is then called* **a lower bound** *of $\{x_n\}$. When we cannot find such an L, we say that $\{x_n\}$ is unbounded below.*

Definition: *A sequence $\{x_n\}$ is said to be* **bounded** *if it is bounded both above and below, i.e., in this case $\exists$ two numbers L, U such that $L \leq x_n \leq U$, for all n. This last condition may also be stated as: $\exists$ a number $M(> 0)$ such that $|x_n| \leq M$*, for all n. (M = greater of $|L|$ and $|U|$).

When a sequence is not bounded, we say that it is **unbounded.**

Examples

1. $\{-1, 1, -3, 1, -5, 1, -7, \ldots\}$ is such that $x_n \leq 1, \forall n$. Hence, this sequence is *bounded above.*

2. $\{1, 2, 3, \ldots, n, \ldots\}$ is *unbounded above.*

3. $\{1, \frac{1}{2}, 3, \frac{1}{4}, 5, \frac{1}{6}, \ldots\}$ is such that $x_n \geq 0, \forall n$. Hence, this sequence is *bounded below.*

4. $\{-1, -2, -3, \ldots, -n, \ldots\}$ is *bounded below.*

5. $\{x_n\}$, where $x_n = 1 + \frac{1}{n}$. This sequence is $\left\{1 + \frac{1}{1}, 1 + \frac{1}{2}, 1 + \frac{1}{3}, \ldots, 1 + \frac{1}{n}, \ldots\right\}$. Hence, $1 \leq x_n \leq 2 \, \forall n$. $\therefore$ $\{x_n\}$ is *bounded.*

6. $\{-1, +1, -1, +1, \ldots (-1)^n, \ldots\}$ is such that $|x_n| \leq 1 \, \forall n$. Hence, it is also *bounded.*

7. $\{-1, 2, -3, 4, \ldots, n(-1)^n, \ldots\}$ is unbounded, for, given any number G, however large it may be $\exists$ n such that

$$|x_n| = |n(-1)^n| = n > G.$$

Exercises for Practice

1. Give an example of a sequence which is

   (a) Bounded above, but not below.
   (b) Bounded below, but not above.
   (c) Unbounded both above and below.

2. Prove that a constant sequence is always *bounded.*

3. Verify whether the following sequences are bounded or not:

   (a) $\left\{\dfrac{n}{1+n}\right\}$; (b) $\left\{\dfrac{n+1}{n}\right\}$; (c) $\{\sin n\}$; (d) $\{\tan n\}$.

12.4 Limit of a Sequence

In a sequence $\{x_n\}$ we are more interested about the behaviour of x_n for very large values of n : what the "**very distant**" members of the sequence are like. For instance, the sequence $\{x_n\} = \left\{0, \frac{1}{2}, \frac{2}{3}, \frac{3}{4}, \ldots, \left(1 - \frac{1}{n}\right), \ldots\right\}$ evidently has all such terms very close to 1; again, the distant members of the sequence $\{y_n\} = \left\{-1, \frac{1}{2}, -\frac{1}{8}, \frac{1}{4}, \ldots, (-1)^n/n, \ldots\right\}$ are very close to 0.

We say: x_n tends to 1 and y_n tends to 0; 1 and 0 are respectively the **limits** of the two sequences.

Observe that the sequences $\{0, 2, 0, 2, \ldots, 1 + (-1)^n, \ldots\}$ and $\{-1, 2, -3, 4, \ldots, (-1)^n n, \ldots\}$ have no such property: We say that they have **no limits.**

Our present aim is to say *precisely* what is meant by the *limit* of a sequence $\{x_n\}$.

Definition of limit of a sequence

Let $\{x_n\}$ be a sequence of real numbers and let l be a fixed real number. We say that 'x_n tends to l' or 'x_n has a limit l' if the following condition is satisfied:

Given any positive number ϵ *(no matter, how small)* $\exists$ *a positive integer, say* N *(which depends on* ϵ*) such that whenever* $n > N$, *we always have* $|x_n - l| < \epsilon$, [**or,** $|x_n - l| < \epsilon$ for **all** $n > N$].

In such a case we write **either** $x_n \to l$ as $n \to \infty$, **or,** $\lim\limits_{n \to \infty} x_n = l$.

Geometrical representation

Let OX be the axis of real numbers (Fig. 12.1).

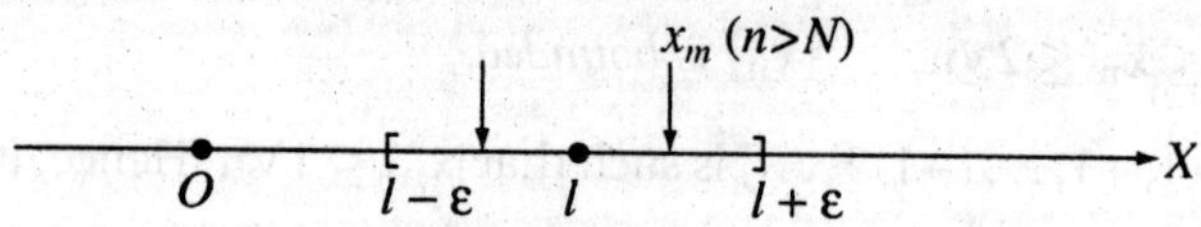

Fig. 12.1

We mark points x_n for various values of n and also mark the number l. If $x_n \to l$, then after some stage (corresponding to N-th stage in the above definition) **all members** of $\{x_n\} \in$ the open interval $(l-\epsilon, l+\epsilon)$.

Observations. The definition speaks of **an integer** N with a certain property. Clearly N may not be unique, indeed it is not so; any integer > N will also satisfy that property. What is important is that $\exists$ N for every possible choice of ϵ. There is no need to find the **best** value of N (by which we mean the least value satisfying the property); it is enough to find **some** n for every ϵ.

A very useful phrase: A proposition P_n is true **for all sufficiently large** n**:** this means that $\exists$ N such that P_n is true $\forall n > N$. Our definition of $x_n \to l$ may then be written

"$x_n \to l$, if given any $\epsilon > 0$, $|x_n - l| < \epsilon$ for all sufficiently large n"

Examples:

► Example 12.4.1 $\{\frac{1}{n}\}$ (*Harmonic sequence*): *We shall prove that* $\frac{1}{n} \to 0$ *as* $n \to \infty$.

Let $\epsilon (> 0)$ be given. Now $|\frac{1}{n} - 0| < \epsilon$ will be true if $\frac{1}{n} < \epsilon$, i.e., if $n > \frac{1}{\epsilon}$.

Take any integer $N > \frac{1}{\epsilon}$. Then $\forall n > N$, we find $|\frac{1}{n} - 0| < \epsilon$.

$$\therefore \lim_{n\to\infty} \frac{1}{n} = 0.$$

► Example 12.4.2 $\{\frac{2n+1}{n+1}\} \to 2$ *as* $n \to \infty$.

Let $\epsilon > 0$ be given. Then $|\frac{2n+1}{n+1} - 2| < \epsilon$ will be true if $|\frac{-1}{n+1}| < \epsilon$, i.e., if $\frac{1}{n+1} < \epsilon$, i.e., if $n > \frac{1}{\epsilon} - 1$.

Take an integer $N > \frac{1}{\epsilon} - 1$ (e.g., $\epsilon = 0{\cdot}01$, take, say $N = 102$).

Then $|x_n - 2| < \epsilon \; \forall n > N$, $\left(x_n = \frac{2n+1}{n+1}\right)$.

$$\therefore x_n \to 2 \text{ as } n \to \infty \text{ or, } \lim_{n\to\infty} x_n = 2.$$

Exercises for Practice

Try in a similar manner to establish

(i) $\dfrac{3n-1}{n+2} \to 3$ as $n \to \infty$; (ii) $\dfrac{n+1}{2n+1} \to \dfrac{1}{2}$ as $n \to \infty$.

(iii) $\dfrac{n-1}{2n} \to \dfrac{1}{2}$ as $n \to \infty$; (iv) $\dfrac{n^2+1}{2n^2+3} \to \dfrac{1}{2}$ as $n \to \infty$.

Next question arises: Can a sequence tend to have more than one value?

Theorem 12.4.1. *A sequence* $\{x_n\}$ *can have* **at most** *one limit.*

Proof. Suppose, if possible, $\{x_n\}$ tends to l and also to m, where $l \neq m$.
We take $\epsilon =$ a positive number $\frac{1}{2}|l-m|$.
$\therefore$ by definition $\exists$ a positive integer N_1 such that $\forall n > N_1$

$$|x_n - l| < \tfrac{1}{2}|l-m|; \qquad (1)$$

at the same time $\exists$ a positive integer N_2 such that $\forall n > N_2$

$$|x_n - m| < \tfrac{1}{2}|l-m|. \qquad (2)$$

We choose now a positive integer N which is greater than both N_1 and N_2. With such an N, both the inequalities (1) and (2) must be true. Thus $\forall n > N$

$$|l-m| = |l - x_n + x_n - m| \leq |l - x_n| + |x_n - m| < \tfrac{1}{2}|l-m| + \tfrac{1}{2}|l-m|.$$

This is a contradiction. Hence, our supposition that $x_n \to l$ as well as $x_n \to m$ $(l \neq m)$ was not true, i.e., x_n cannot tend to two distinct limits; hence the theorem.

12.5 Convergent Sequences

Definition: *A sequence* $\{x_n\}$ *is called a* **convergent sequence** *if* $\exists$ *a finite real number* l *such that* $\lim\limits_{n\to\infty} x_n = l$.
We then say that $\{x_n\}$ **converges to** l.
If a sequence does not have such a finite limit, then it is called a **divergent sequence.**

▶ Example 12.5.1 *Thus consider the sequence* $\left\{\frac{1}{n}\right\}$. *We have proved that* $\frac{1}{n} \to 0$ *as* $n \to 0$ *(0 is a finite real number); hence* $\left\{\frac{1}{n}\right\}$ *is a convergent sequence.*

▶ Example 12.5.2 *Let* $x_n = (-1)^n$ *: the sequence is* $\{-1, 1, -1, 1, -1, \ldots\}$.

We shall prove that it is not convergent.
Assume that $x_n \to$ a finite l. Then, given $\epsilon = \frac{1}{2}$, say, $\exists$ a positive integer N such that $\forall n \geq N$

$$|x_n - l| < \tfrac{1}{2}, \text{ i.e., } |(-1)^n - l| < \tfrac{1}{2}.$$

Take $n = 2k$ (but $> N$) and $n = 2k+1 (> N)$. We then have

$$|1 - l| < \tfrac{1}{2} \text{ and also } |-1-l| < \tfrac{1}{2};$$

i.e., $2 = |1-(-1)| = |1 - l + l - (-l)| \leq |l-1| + |l+1| < \frac{1}{2} + \frac{1}{2} = 1$.
This gives a contradiction. Hence, $\{x_n\}$ is not convergent.

Theorem 12.5.1. *A convergent sequence is bounded.*

Proof. Let $\{x_n\}$ be a convergent sequence with limit l.
Then corresponding to a particular $\epsilon = l$, say, $\exists$ a positive integer N such that $\forall n > N$, $|x_n - l| < 1$, i.e., $l - 1 < x_n < l + 1$.
Let

$$\begin{aligned} k &= \min\{x_1, x_2, \ldots, x_N, |l| - 1\} \\ \text{and } K &= \max\{x_1, x_2, \ldots, x_N, |l| + 1\}. \end{aligned}$$

Obviously $k < x_n < K \ \forall n$. $\therefore$ $\{x_n\}$ is bounded.

Corollary 12.5.1 *If a sequence is unbounded, then it is not convergent.*

Theorem 12.5.2. *Suppose $x_n \to l$ as $n \to \infty$. Then $|x_n| \to |l|$ as $n \to \infty$. The converse is, in general, not true. Only when $l = 0$, the converse is true.*

Proof. Given any $\epsilon > 0, \exists$ a positive integer N such that $|x_n - l| < \epsilon \, \forall n > N$.

But we have the inequality $||x_n| - |l|| \le |x_n - l|$;

Consequently $||x_n| - |l|| < \epsilon, \forall n > N$.

This proves $|x_n| \to |l|$ as $n \to \infty$.

Next, consider the sequence $y_n = (-1)^n$. We have $|y_n| = 1$ and hence $|y_n| \to 1$ as $n \to \infty$. But $\{y_n\}$ is not convergent (proved in **Ex. 12.5.2** above).

Now see that $x_n \to 0$ as $n \to \infty$ if and only if $|x_n| \to 0$ as $n \to \infty$.

Theorem 12.5.3. *Suppose that $x_n \to l$ as $n \to \infty$ and let $l > 0$. Then there is a certain positive number K such that $x_n > K$ for all sufficiently large n (i.e., $\exists$ a positive integer N such that $x_n > K \, \forall n > N$).*

Proof. Since $l > 0$ we may take $\frac{1}{2}l$ as a suitable ϵ in the definition of a limit. Thus, corresponding to $l/2 \, \exists$ a positive integer N such that $|x_n - l| < l/2 \, \forall n > N$, i.e., $l/2 < x < 3l/2, \forall n > N$.

In particular, $x > l/2 \, \forall n > N$ (thus $\exists K = l/2$).

Significance of this theorem: When a sequence of numbers has a positive limit the numbers themselves must eventually be positive, and in fact, ultimately stay a certain distance away from zero.

Theorem 12.5.4. *Suppose $x_n > 0 \, \forall n$ and that $x_n \to l$ as $n \to \infty$. Then l is either positive or $l = 0$ (i.e., $l \ge 0$).*

[**Important.** $x_n > 0$ does not always imply $l > 0$, it may be $l = 0$]

Proof. $x_n \to l$ as $n \to \infty$ implies that given any $\epsilon > 0, \exists$ a positive integer N such that $|x_n - l| < \epsilon \forall n > N$; that is, $l - \epsilon < x_n < l + \epsilon, \forall n > N$.

In particular, $x_n < l + \epsilon, \forall n > N$

$\therefore l > x_n - \epsilon$ and hence $> -\epsilon$.

$\therefore l$ exceeds all negative numbers and so $l \ge 0$.

Example. $x_n = 1/n > 0 \, \forall n$, but the limit is not > 0; in fact, the limit $= 0$. More general result is the following:

Theorem 12.5.5. *Suppose $x_n \to l, y_n \to m$ as $n \to \infty$ and that $x_n < y_n \, \forall n$. Then $l \le m$.*

Proof. Given any $\epsilon > 0$, $\exists$ a suitable positive integer N such that

$$\left.\begin{array}{r} l - \epsilon < x_n < l + \epsilon \\ \text{and} \quad m - \epsilon < y_n < m + \epsilon \end{array}\right\} \forall n > N$$

Then for $n > N$, $l - \epsilon < x_n < y_n < m + \epsilon$, so that $l - m < 2\epsilon$, or, $m - l > -2\epsilon$.

It then follows that $m - l \ge 0$ for ϵ is arbitrary. $\quad \therefore l \le m$.

Significance: $x_n < y_n \Rightarrow l \le m$. Thus with limit operations taken on both sides of an inequality, a strong inequality ($<$) becomes weak ($\le$).

Theorem 12.5.6. *Suppose* $x_n \le y_n \le z_n (\forall_n)$ *and suppose that* $x_n \to l$ *as* $n \to \infty$ *and also that* $z_n \to l$ *as* $n \to \infty$. *Then* $y_n \to l$ *as* $n \to \infty$.

Proof. Since $x_n \to l, z_n \to l$ as $n \to \infty$, given any $\epsilon > 0 \exists$ a suitable positive integer N such that $\forall_n > N$, $l - \epsilon < x_n < l + \epsilon$ and $l - \epsilon < z_n < l + \epsilon$.

Given: $x_n \le y_n \le z_n\ \forall n$. $\therefore\ \forall_n > N, l - \epsilon < x_n \le y_n \le z_n < l + \epsilon$.

Hence, $\forall_n > N, l - \epsilon < y_n < l + \epsilon$ or $|y_n - l| < \epsilon$

That is, $y_n \to l$ as $n \to \infty$.

Note that, we have proved two things: (i) $\lim\limits_{n\to\infty} y_n$ exists and (ii) that limit is l.

12.6 Limits of Combination of Sequences

Limit of a Sum of Two Sequences

Theorem 12.6.1. *If* $\lim\limits_{n\to\infty} x_n = l$ *and* $\lim\limits_{n\to\infty} y_n = m$, *then*

$$\lim_{n\to\infty} (x_n + y_n) = \lim_{n\to\infty} x_n + \lim_{n\to\infty} y_n = l + m.$$

Proof. Let $\epsilon > 0$ be given. Then corresponding to the positive number $\epsilon/2 \exists$ a **suitable positive integer** N such that $|x_n - l| < \epsilon/2$ and $|y_n - m| < \epsilon/2$, $\forall_n > N$.

[Usually it will happen that $\exists\ N_1, N_2$ such that $|x_n - l| < \epsilon/2\ \forall_n > N_1$ and $|y_n - m| < \epsilon/2\ \forall_n > N_2$.

By suitable positive integer we actually mean $[N = \max(N_1, N_2)]$]

$$\begin{aligned} \therefore\ |(x_n + y_n) - (l + m)| &= |(x_n - l) + (y_n - m)| \\ &\le |x_n - l| + |y_n - m| \\ &< \epsilon/2 + \epsilon/2 = \epsilon, \forall_n > N. \\ \therefore\ \lim_{n\to\infty} (x_n + y_n) &= l + m. \end{aligned}$$

Limit of Difference of Two Sequences

Corollary. Theorem: $x_n \to l$ and $y_n \to m \Rightarrow x_n - y_n \to l - m$.

Limit of Product of Two Sequences

Theorem 12.6.2. *If* $\lim\limits_{n\to\infty} x_n = l$ *and* $\lim\limits_{n\to\infty} y_n = m$, *then* $\lim\limits_{n\to\infty} (x_n y_n) = lm$, *i.e., limit of a product = product of the limits if the limits exist separately.*

Proof. Let us set down two positive numbers ϵ and k of which ϵ is any arbitrary positive number and k is a certain definite positive number which we shall specify later. Then ϵ/k is an arbitrary positive number.

Since $x_n \to l$, $y_n \to m$ as $n \to \infty$ it follows that corresponding to $\epsilon/k \exists$ a *suitable positive integer,* N such that $|x_n - l| < \epsilon/k$ and $|y_n - m| < \epsilon/k$, $\forall_n > N$.

Then,

$$\begin{aligned} |x_n y_n - lm| &= |x_n(y_n - m) + m(x_n - l)| \\ &\le |x_n|\,|y_n - m| + |m|\,|x_n - l| \\ &< \frac{\epsilon}{k}\{|x_n| + |m|\}, \forall_n > N. \end{aligned}$$

Since a convergent sequence is always bounded, $\{x_n\}$ must be bounded. That is $\exists$ a positive number K such that $|x_n| < K, \forall_n$.

$\therefore \forall_n > N$, we may take $|x_n y_n - lm| < \frac{\epsilon}{k}\{K + |m|\}$.

Now if k be so chosen from the beginning that $k > K + |m|$, $\left(\text{i.e., } \frac{K+|m|}{k} < 1\right)$, then we have $\forall_n > N, |x_n y_n - lm| < \epsilon$ and hence

$$\lim_{n\to\infty} x_n y_n = lm = \lim_{n\to\infty} x_n \times \lim_{n\to\infty} y_n.$$

Corollary 12.6.1 *If* $\lim_{n\to\infty} x_n = l$, $\lim_{n\to\infty} y_n = m$ *and* $\lim_{n\to\infty} z_n = t$, *then*

$$\lim_{n\to\infty} (x_n y_n z_n) = l\,m\,t.$$

In fact, the law is true for the product of a finite number of sequences.

Limit of the Quotient of Two Sequences

Theorem 12.6.3. *If* $\lim_{n\to\infty} x_n = l$ *and* $\lim_{n\to\infty} y_n = m$ $(m \neq 0)$, *then*

$$\lim_{n\to\infty} \frac{x_n}{y_n} = \frac{\lim_{n\to\infty} x_n}{\lim_{n\to\infty} y_n} = \frac{l}{m}.$$

Proof. As in the previous theorem we set down two positive numbers ϵ and k, of which ϵ is arbitrary and k is some fixed positive number to be specified later. Then ϵ/k is an arbitrary positive number. Since $x_n \to l$ and $y_n \to m$ as $n \to \infty$, corresponding to $\epsilon/k\,\exists$ a suitable positive integer N_1 such that

$$|x_n - l| < \epsilon/k \text{ and } |y_n - m| < \epsilon/k,\ \forall n > N_1.$$

Then

$$\left|\frac{x_n}{y_n} - \frac{l}{m}\right| = \left|\frac{m x_n - l y_n}{m y_n}\right| = \left|\frac{m(x_n - l) - l(y_n - m)}{m y_n}\right|$$
$$\leq \frac{|m|\,|x_n - l| + |l|\,|y_n - m|}{|m|\,|y_n|}$$
$$< \frac{\{|m| + |l|\}}{|m|\,|y_n|} \cdot \frac{\epsilon}{k},\ \forall n > N_1.$$

Now since $y_n \to m (m \neq 0)$ as $n \to \infty$, we have $|y_n| \to |m|$ when $n \to \infty$ (see **Theorem 12.5.2.**)

Let $0 < H < |m|$. Then $|y_n| > H$ for $n >$ a certain positive integer N_2.

Now if $N = \max\ (N_1, N_2)$, then $\forall n > N$.

$$\left|\frac{x_n}{y_n} - \frac{l}{m}\right| < \frac{\epsilon}{k} \frac{\{|l| + |m|\}}{|m|} \frac{1}{H},\ \forall n > N.$$

If to begin with we would choose k so that $\frac{|l|+|m|}{k|m|H} < 1$, i.e., if we would choose $k > \frac{|l|+|m|}{|m|H}$, then $\forall n > N$, we get

$$\left|\frac{x_n}{y_n} - \frac{l}{m}\right| < \epsilon. \quad \therefore \quad \lim_{n\to\infty} \frac{x_n}{y_n} = \frac{l}{m}\ (m \neq 0).$$

Corollary 12.6.2 First prove as in the above theorem:

If $\lim\limits_{n\to\infty} y_n = m$ and $m \neq 0$, then $\lim\limits_{n\to\infty} \frac{1}{y_n} = \frac{1}{m}$.

Then deduce: if $\lim\limits_{n\to\infty} x_n = l$ and $\lim\limits_{n\to\infty} y_n = m\ (\neq 0)$, then using Product Rule $\lim\limits_{n\to\infty} \frac{x_n}{y_n} = \frac{l}{m}$.

▶ Example 12.6.1 $\lim\limits_{n\to\infty}\left(\dfrac{3n^2+2n+1}{4n^2+5n+12}\right) = \lim\limits_{n\to\infty}\left|\dfrac{3+\frac{2}{n}+\frac{1}{n^2}}{4+\frac{5}{n}+\frac{12}{n^2}}\right| = \lim\limits_{n\to\infty}\left(\dfrac{x_n}{y_n}\right)$, *say*

$$\begin{aligned}\lim_{n\to\infty} x_n &= \lim_{n\to\infty}\left(3+\frac{2}{n}+\frac{1}{n^2}\right) = \lim_{n\to\infty} 3 + \lim_{n\to\infty}\frac{2}{n} + \lim_{n\to\infty}\frac{1}{n^2}\\ &= 3 + 2.0 + 0 = 3.\end{aligned}$$

Similarly, $\lim\limits_{n\to\infty} y_n = 4\ (\neq 0)$. $\therefore$ reqd. limit $= \dfrac{\lim\limits_{n\to\infty} x_n}{\lim\limits_{n\to\infty} y_n} = \dfrac{3}{4}$.

Exercises: Try in a similar manner

(i) $\lim\limits_{n\to\infty} \dfrac{1}{1+n^2} = 0$; (ii) $\lim\limits_{n\to\infty} \dfrac{(3n^2+5)(7n^3+9)}{5+7n^2+8n^5} = \dfrac{21}{8}$;

(iii) $\lim\limits_{n\to\infty} \dfrac{n^2+3n}{2n^2+n-1} = \dfrac{1}{2}$; (iv) $\lim\limits_{n\to\infty}\left(\dfrac{1}{n^2}+\dfrac{2}{n^2}+\cdots+\dfrac{n}{n^2}\right) = \dfrac{1}{2}$.

12.7 Non-Convergent Sequences

Definition: *A sequence* $\{x_n\}$ *is said to tend to* $+\infty$ *(or, diverges to* $+\infty$*) whenever, given any positive number* K *(no matter how large)* $\exists$ *a positive integer* N *such that* $\forall n \geq N$, $x_n > K$.

In this case, we write $\lim\limits_{n\to\infty} x_n = +\infty$ *(or* $x_n \to \infty$ *as* $n \to \infty$*).*

Geometrically, this means that however a point K may be chosen, all sequence points x_n (except perhaps a finite number of them) remain beyond it on the right.

▶ Example 12.7.1 *Let* $x_n = n^2$. *The sequence is* $\{1^2, 2^2, 3^2, \ldots, n^2, \ldots\}$.

Then, given any $k > 0$, $n^2 > k\,\forall n > \sqrt{k}$.
So choosing N, a positive integer $\geq \sqrt{k}$, we can say $n^2 > k$ whenever $n > N$.
$\therefore$ a sequence $\{x_n = n^2\}$ diverges to $+\infty$.

Definition: *A sequence* $\{x_n\}$ *is said to tend to* $-\infty$ *(or diverges to* $-\infty$*) if the sequence* $\{-x_1, -x_2, -x_3, \ldots\ldots, -x_n, \ldots\}$ *tends to* $+\infty$.

$\therefore$ a sequence $\{x_n\}$ tends to $-\infty$ if and only if the following condition is satisfied:
Given any positive number K (no matter how large) $\exists$ a positive integer N such that $\forall n > N$, $x_n < -K$.

In geometrical language, '$x_n \to -\infty$' means that however a point be chosen on the real line, all the points x_n, except at most a finite number of them, say $x_1, x_2, \ldots, x_n$, remain beyond it on the left.

▶ Example 12.7.2 $\{-n^2\}$ *diverges to* $-\infty$, *since* $-n^2 < -k\,\forall n > N$, *where* $N = \sqrt{K}$.

Definition: *A sequence which does not converge to a finite limit, nor does it diverge to* $+\infty$ *or* $-\infty$ *will be called an* **oscillatory sequence.**

If an oscillatory sequence $\{x_n\}$ be such that $\exists$ a fixed positive number C such that $|x_n| < C$ for all n, then we say that the sequence **oscillates finitely,** otherwise it is **oscillating infinitely.** Oscillating sequences are also called **improperly divergent sequences.** Thus we have the following scheme:

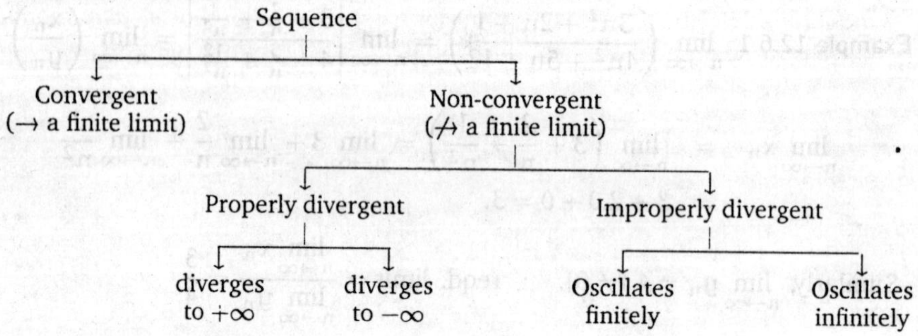

▶ Example 12.7.3 *We have already seen that* $\{(-1)^n\}$ *is not convergent. Further it is bounded. It does not clearly diverge to* $+\infty$ *or* $-\infty$.

$\therefore$ the sequence $\{(-1)^n\}$ *oscillates finitely.*

▶ Example 12.7.4 *Consider the sequence* $\{-1, +2, -3, +4, -5, +6, -7, \ldots\}$.

The sequence is unbounded (both above and below); it cannot, therefore, tend to a finite limit. It cannot diverge to $+\infty$ because we cannot say that $x_n >$ a certain given positive K **for every** $n >$ some positive integer N (though there may be many values of n for which $x_n > K$). What is important in the definition is that the inequality should be true for **every** $n > N$ (**not that**) *for an unlimited number of values of* n, **not also** *for* n *as large as we please.* Similarly the sequence does not diverge to $-\infty$.

Hence, the sequence *oscillates infinitely.*

Try in a similar manner

▶ Example 12.7.5 $\left\{x_n = \frac{(-1)^n(2n-1)}{n}\right\}$ *oscillates (between* -2 *and* $+2$*) finitely, but* $\{x_n\}$

$$\text{where } x_n = \tfrac{1}{2}n, \quad \text{if } n \text{ be even,}$$
$$= 0, \quad \text{if } n \text{ be odd,}$$

oscillates infinitely.

12.8 Monotone Sequences

Definition: *A sequence* $\{x_n\}$ *is said to be monotone increasing* (**m.i.**) *if* $x_n \le x_{n+1}$ *for every* n, **e.g.,** $\{1, 2, 3, \ldots, n, n+1, \ldots\}$ *and* $\{2, 2^2, 2^3, \ldots, 2^n, 2^{n+1}, \ldots\}$ *are examples of monotone increasing sequences.*

Definition: *A sequence* $\{x_n\}$ *is said to be monotone decreasing* (**m.d.**) *if* $x_n \ge x_{n+1}$ for every n, **e.g.,** $\left\{1, \frac{1}{2}, \frac{1}{3}, \ldots, \frac{1}{n}, \frac{1}{n+1} \cdots\right\}$ *is a monotone decreasing sequence.*

Observations.

1. A sequence need not be monotone (either increasing or decreasing): e.g., $\{1, 0, 1, 0, 1, 0, \ldots\}$ is neither **m.i.** nor **m.d.**

2. If $\{x_n\}$ is **m.i.**, then $\{-x_n\}$ is **m.d.**

3. If in the definition of m.i. sequence we replace the condition $x_n \leq x_{n+1}$ by a strong inequality, $x_n < x_{n+1}\ \forall n$, then the sequence is said to be *strictly monotone increasing;* similar definition ($x_n > x_{n+1} \forall n$) for *strictly monotone decreasing* sequence.

A monotone sequence has a definite behaviour—*either* it tends to a finite limit **convergent** *or,* it is **properly divergent** (tends to $+\infty$ or tends to $-\infty$). A monotone sequence can **never oscillate.**

We shall take this property as a **fundamental axiom:**

(i) **A monotone sequence, if bounded, must be convergent.**
If the sequence is **m.i.** and bounded above, it converges to a value which is known as its *exact upper bound.* If the sequence is **m.d.** and bounded below, it converges to a value which is called its *exact lower bound.*

(ii) **A monotone sequence, if unbounded, must be properly divergent.**
If the sequence is **m.i.** and unbounded above, then it must diverge to $+\infty$. If the sequence is **m.d.** and unbounded below, then it must diverge to $-\infty$.

12.9 Important Examples on Sequences

▶ Example 12.9.1 *Consider the sequence* $\{x_n\}$, *where* $x_n = \left(1 + \frac{1}{n}\right)^n$.

We shall prove that this sequence is monotone increasing and bounded and hence convergent.

Solution:

Step 1. *The sequence is monotone increasing.*

For a particular n, by Binomial Theorem, we get

$$\begin{aligned}
x_n &= \left(1+\frac{1}{n}\right)^n = 1 + n\cdot\frac{1}{n} + \frac{n(n-1)}{2!}\left(\frac{1}{n}\right)^2 \\
&\quad + \frac{n(n-1)(n-2)}{3!}\left(\frac{1}{n}\right)^3 + \cdots + \frac{n(n-1)\cdots 3\cdot 2\cdot 1}{n!}\left(\frac{1}{n}\right)^n. \\
&= 1 + 1 + \frac{1}{2!}\left(1-\frac{1}{n}\right) + \frac{1}{3!}\left(1-\frac{1}{n}\right)\left(1-\frac{2}{n}\right) \\
&\quad + \cdots + \frac{1}{n!}\left(1-\frac{1}{n}\right)\left(1-\frac{2}{n}\right)\cdots\left(1-\frac{n-1}{n}\right).
\end{aligned}$$

Each term on the right is positive and further positive terms are introduced as n is increased; and hence $x_{n-1} > x_n, \forall n$.

$\therefore$ $\{x_n\}$ is a *monotone increasing sequence* (in fact, a strictly **m.i.** sequence).

Step 2. *The sequence is bounded (both above and below).*

Clearly, $x_n \geq 2$, for $\forall n$, i.e., 2 is a lower bound. We shall prove that **3 is an upper bound.**

$$\begin{aligned} x_n &< 1+1+\frac{1}{2!}+\frac{1}{3!}+\cdots+\frac{1}{n!}(n \geq 2) \; [\because 1-\tfrac{1}{n}<1, 1-\tfrac{2}{n}<1, \text{etc.}] \\ &< 1+1+\frac{1}{2}+\frac{1}{2^2}+\cdots+\frac{1}{2^{n-1}} \; [\text{See that, } 3! = 3\cdot 2\cdot 1 > 2\cdot 2, \text{etc.}] \\ &= 1+\frac{1-\left(\frac{1}{2}\right)^n}{1-\frac{1}{2}} = 1+2-2\cdot\left(\frac{1}{2}\right)^n < 3, \forall n. \end{aligned}$$

$\because \{x_n\}$ is bounded above and 3 is an *upper bound.*

Then we have $2 \leq x_n < 3, \forall n$, i.e., the sequence is bounded (both above and below).

Conclusion. Since $\{x_n\}$ is monotone increasing as well as bounded it is convergent and it converges to a limit $l \leq 3$.

We denote this limit by e, thus

$$\lim_{n\to\infty}\left(1+\frac{1}{n}\right)^n = e \text{ (in fact, } 2 < e < 3)$$

Corollary 12.9.1 $\lim_{n\to\infty} \left(1-\frac{1}{n}\right)^n = e^{-1}$.

▶ Example 12.9.2 *To prove* $\lim_{n\to\infty} y_n = e$, *where* y_n *is defined by*

$$y_n = 1+\frac{1}{1!}+\frac{1}{2!}+\frac{1}{3!}+\cdots+\frac{1}{n!}.$$

Solution:

Step 3. *Clearly* $y_{n+1}-y_n = \frac{1}{n+1} > 0 \; \forall n$; *hence* $\{y_n\}$ *is* ***m.i.***

Also from the previous **Example** $2 < y_n < 3, \; \forall n$.

$\because \{y_n\}$ being **m.i.** and bounded, must converge to a limit e' (say), where $e' \leq 3$.

Step 4. *To prove* $e = e'$.

From **Example 12.9.1.** $x_n = \left(1+\frac{1}{n}\right)^n \leq 1+\frac{1}{1!}+\frac{1}{2!}+\cdots+\frac{1}{n!} = y_n$.

$$\therefore \lim_{n\to\infty} x_n \leq \lim_{n\to\infty} y_n, \text{ i.e., } e \leq e'. \tag{3}$$

We now take a positive integer $m < n$ and consider first $(m+1)$ terms. Since the terms are all positive it is obvious that

$$1+1+\frac{1}{2!}\left(1-\frac{1}{n}\right)+\cdots+\frac{1}{m!}\left(1-\frac{1}{n}\right)\left(1-\frac{2}{n}\right)\cdots\left(1-\frac{m-1}{n}\right) < x_n.$$

[Compare Ex. 12.9.1]

Let m remain fixed and we make $n \to \infty$, then this inequality gives

$$y_m = 1 + 1 + \frac{1}{2!} + \cdots + \frac{1}{m!} \le \lim_{n\to\infty} x_n = e,$$

for all values of m.

$$\therefore \text{ as } m \to \infty, \lim_{m\to\infty} y_m \le e, \text{ i.e., } e' \le e. \tag{4}$$

From equations (3) and (4) we conclude that $e = e'$.

Note 1. *The sequences $\{x_n\}$ and $\{y_n\}$ are such that every member is a rational number; but what is interesting is that they have a common limit e which can be proved to be not a rational number: an approximate value of $e = 2{\cdot}7182818285\cdots$*

Remember. If a sequence $\{x_n\}$ converges to the limit 0, then we say that the sequence $\{x_n\}$ is a **null sequence.**

▶ Example 12.9.3 *If $|x| < 1$, then $x^n \to 0$ as $n \to \infty$, i.e., $\{x^n, |x| < 1\}$ is a null sequence.*

Solution: If $x = 0$, then every member of $\{x^n\}$ is 0 and as such $x^n \to 0$ as $n \to \infty$. So, let $x \neq 0$.

Let $\frac{1}{|x|} = 1 + y (y > 0)$ $(\because |x| < 1)$.

$$\therefore |x^n - 0| = |x|^n = \frac{1}{(1+y)^n} = \frac{1}{1 + ny + \frac{n(n-1)}{2!}y^2 + \cdots + y^n} < \frac{1}{ny}.$$

Thus if $\epsilon > 0$ be given, $|x^n - 0| < \epsilon$ if $\frac{1}{ny} < \epsilon$.

$$\therefore \forall n > N \text{ where } N = \frac{1}{\epsilon y}, |x^n - 0| < \epsilon.$$

Thus $x^n \to 0$ as $n \to \infty$ (provided $|x| < 1$).

Corollary 12.9.2 $\left\{\frac{1}{2^n}\right\}, \left\{\frac{1}{3^n}\right\}, \left\{\frac{1}{10^n}\right\}$, *etc. are all null sequences.*

▶ Example 12.9.4 *To discuss the behaviour of $\{x^n\}$, where x is any real number.*

Solution:

Case 1. *If $-1 < x < 1$, then $x^n \to 0$ (proved before).*

Case 2. *If $x = 1$, all terms of the sequence being $1, x^n \to 1$ as $n \to \infty$; in fact, $|x^n - 1| = |1 - 1| = 0 <$ any ϵ $\forall n$.*

Case 3. *If $x > 1, x^n \to +\infty$ as $n \to \infty$.*

Take $x = 1 + y (y > 0)$. Then $x^n = (1+y)^n > 1 + ny$.

$\therefore$ $x^n > K$ (where K is any arbitrary positive number), if $n > \frac{K-1}{y}$; choose a positive integer $N > \frac{K-1}{y}$.

Then $\forall n > N, x^n > K$. Hence, $x^n \to \infty$ as $n \to \infty$.

Case 4. *If* $x < -1, \{x^n\}$ *oscillates infinitely.*

Case 5. *If* $x = -1$, *the terms are alternately* -1 *and* $+1$. *The sequence oscillates finitely* -1 *and* $+1$.

▶ Example 12.9.5 $x^{1/n} \to 1$ *as* $n \to \infty$ $(x > 0)$.

Case 1. $x = 1$, *every* $x^{1/n} = 1$ *and hence* $x^{1/n} \to 1$ *as* $n \to \infty$.

Case 2. $x > 1$. *Here* $x^{1/n} > 1$ *and then we can write* $x^{1/n} = 1 + y_n (y_n > 0)$ *and hence* $x = (1+y_n)^n = 1 + ny_n + n\frac{(n-1)}{2!}y_n^2 + \cdots + y_n^n > 1 + ny_n$.

$\therefore\ y_n < \frac{x-1}{n}$, *i.e.*, $0 < x^{1/n} - 1 < \frac{x-1}{n}$.

$\therefore\ |x^{1/n} - 1| < \epsilon\, \forall n > N$, where N is a positive integer $> \frac{x-1}{\epsilon}$.

Case 3. *When* $0 < x < 1, x^{1/n} < 1$ *and we then write* $x^{1/n} = \frac{1}{1+y_n}$ $(y_n > 0)$.

Then $x = \frac{1}{(1+y_n)^n} < \frac{1}{1+ny_n}$ so that $y_n < \frac{\frac{1}{x}-1}{n}$.

$\therefore\ 0 < y_n < \epsilon \forall n > N$ where N is a positive integer $> \frac{\frac{1}{x}-\epsilon}{\epsilon}$, i.e., $y_n \to 0$ as $n \to \infty$ and consequently $x^{1/n} = \frac{1}{1+y_n} \to 1$ as $n \to \infty$.

Alternative method

Let $y = 1/x > 1$ $(\because\ x < 1)$

$$\therefore\quad y^{1/n} = \frac{1}{x^{1/n}};\ \lim_{n\to\infty} y^{1/n} = 1 \text{ (by Case 2)}$$

$$\therefore\quad \lim_{n\to\infty} x^{1/n} = \frac{1}{\lim\limits_{n\to\infty} y^{1/n}} = 1.$$

▶ Example 12.9.6 $\lim\limits_{n\to\infty} n^{1/n} = 1$.

Solution: Let $n^{1/n} = 1 + x_n$, so that $x_n > 0$.

Then $n = (1+x_n)^n = 1 + nx_n + \frac{n(n-1)}{2!}x_n^2 + \cdots + (x_n)^n (n \geq 2)$.

$\therefore\ n > \frac{1}{2}n(n-1)x_n^2$, or, $x_n^2 < \frac{2}{n-1}$.

Hence, $1 < n^{1/n} < 1 + \sqrt{\frac{2}{n-1}}$ and $|n^{1/n} - 1| < \epsilon\ \forall n$ such that $\sqrt{\frac{2}{n-1}} < \epsilon$, i.e., $\forall n > N$ where N is a positive integer $> 1 + \frac{2}{\epsilon^2}$.

$$\therefore\ \lim_{n\to\infty} n^{1/n} = 1.$$

▶ Example 12.9.7 $\{nx^n\}, |x| < 1$ *is a null sequence.*

Solution: Let $|x| = \frac{1}{1+y} (y > 0)$

$$
\begin{aligned}
|nx^n| &= \frac{n}{(1+y)^n} = \frac{n}{1+ny+\frac{n(n-1)}{2!}y^2+\cdots+y^n} \\
&< \frac{n}{\frac{n(n-1)}{2!}y^2} \text{ (retaining only the third term)} \\
|n^n| &< \frac{2}{(n-1)y^2} < \epsilon \text{ if } n > 1 + \frac{2}{\epsilon y^2} \\
\therefore \lim_{n\to\infty} nx^n &= 0\ (|x| < 1).
\end{aligned}
$$

▶ Example 12.9.8 $x_n = \sqrt{n+1} - \sqrt{n}$. *To prove* $\lim_{n\to\infty} x_n = 0$.

Solution:

$$
\begin{aligned}
x_n &= \sqrt{n+1} - \sqrt{n} = \frac{(\sqrt{n+1}-\sqrt{n})(\sqrt{n+1}+\sqrt{n})}{\sqrt{n+1}+\sqrt{n}} \\
&= \frac{1}{\sqrt{n+1}+\sqrt{n}} < \frac{1}{2\sqrt{n}}.
\end{aligned}
$$

$\therefore$ $x_n <$ a given $\epsilon(> 0)$ if $\frac{1}{2\sqrt{n}} < \epsilon$, i.e., if $n > \frac{1}{4\epsilon^2}$.
$\therefore$ $|x_n - 0| < \epsilon \forall n > N$, where N is a positive integer $> \frac{1}{4\epsilon^2}$.
Hence, $\lim_{n\to\infty} x_n = 0$.

▶ Example 12.9.9 *If* $\{x_n\}$ *be a sequence such that*

$$\lim_{n\to\infty} \left|\frac{x_{n+1}}{x_n}\right| = l, 0 \le l < 1, \textit{ then } \lim_{n\to\infty} x_n = 0.$$

Solution: Choose $\epsilon > 0$ such that $l + \epsilon = k$, say, where $0 < k < 1$.

Then there is a positive integer N such that $\forall n \ge N$

$$\left|\frac{x_{n+1}}{x_n}\right| < k, \text{ i.e., } \left|\frac{x_{N+1}}{x_N}\right| < k, \left|\frac{x_{N+2}}{x_{N+1}}\right| < k, \ldots \left|\frac{x_n}{x_{n-1}}\right| < k (n > N)$$

and hence

$$\left|\frac{x_{N+1}}{x_N} \cdot \frac{x_{N+2}}{x_{N+1}} \cdots \frac{x_n}{x_{n-1}}\right| < k^{n-N}, \text{ i.e., } |x_n| < k^n \times \frac{x_N}{k^N} \to 0 \text{ as } n \to \infty.$$

Note. $0 < k < 1, k^n \to 0$ as $n \to \infty$ and hence $\frac{x_N}{k^N} \cdot k^n \to 0$ as $n \to \infty$, x_N/k^N being a fixed number. $\therefore$ $x_n \to 0$ as $n \to \infty$.

Applications

(i) $\lim_{n\to\infty} \frac{m(m-1)(m-2)\cdots(m-n+1)}{n!} x^n = 0 (|x| < 1)$

Solution: Call $s_n = \dfrac{m(m-1)(m-2)\cdots(m-n+1)}{n!} x^n$, then

$$s_{n+1} = \frac{m(m-1)\cdots(m-n)}{(n+1)!} x^{n+1}$$

Thus $\left|\frac{s_{n+1}}{s_n}\right| = \left|\frac{m-n}{n+1}\right||x| = \left|\frac{\frac{m}{n}-1}{1+\frac{1}{n}}\right||x| \to |x|$ as $n \to \infty$.

$\therefore$ $s_n \to 0$ as $n \to \infty$ ($\because$ $|x| < 1$, given).

(ii) $\lim\limits_{n\to\infty} \frac{x^n}{n!} = 0$; $\lim\limits_{n\to\infty} \frac{x^n}{n} = 0$ ($|x| \leq 1$)

See that in the first case

$$\left|\frac{x^{n+1}}{(n+1)!}\bigg/\frac{x^n}{n!}\right| = \frac{1}{n+1}|x| \to 0,$$

$\forall x$ as $n \to \infty$; hence $\lim\limits_{n\to\infty} \frac{x^n}{n!} = 0$.

Try similarly the second case.

12.10 Cauchy Sequences

A sequence $\{x_n\}$ is said to be a **Cauchy sequence** if for any given $\epsilon > 0 \exists$ a positive integer N such that for all $m, n > N$, $|x_m - x_n| < \epsilon$.

▶ Example 12.10.1 *Consider the constant sequence* $\{a, a, a, \ldots\} = \{x_n\}$, *say.*

Here $x_n = a \, \forall n$. Let $\epsilon > 0$ be given. Take $N = 1$, then

$$m, n > N = 1 \Rightarrow |x_m - x_n| = |a - a| = 0 < \epsilon.$$

$\therefore$ every constant sequence is a *Cauchy sequence.*

▶ Example 12.10.2 *Consider* $\left\{\frac{1}{n}\right\}$. *Let* $\epsilon > 0$ *be given. Take a positive integer* $N > 1/\epsilon$, *we see that* $m, n > N \Rightarrow \frac{1}{m}, \frac{1}{n} < \epsilon$, *hence* $\frac{1}{m} - \frac{1}{n} < \epsilon$, *or,* $\frac{1}{n} - \frac{1}{m} < \epsilon$.

Thus $\left|\frac{1}{m} - \frac{1}{n}\right| < \epsilon \, \forall n, m > N$. $\therefore$ $\left\{\frac{1}{n}\right\}$ is a Cauchy sequence.

Theorem 12.10.1. *Every convergent sequence is a Cauchy sequence.*

Proof. Let $\{x_n\}$ be a given sequence having l as its limit.

Let $\epsilon > 0$ be given. Then $\exists$ a positive integer N such that $|x_n - l| < \epsilon/2 \, \forall n > N$.

$\therefore$ $|x_n - x_m| = |x_n - l + l - x_m| \leq |x_n - l| + |l - x_m| < \epsilon/2 + \epsilon/2 = \epsilon$, whenever $m, n > N$.

$\therefore$ $\{x_n\}$ is a *Cauchy sequence.*

Observations: The converse of the theorem is also true. The proof is, however, beyond the scope of this book.

This theorem and its converse together can be stated in the following language known as *Cauchy's General Principle of Convergence:* [C.P. 1980]

A necessary and sufficient condition that a sequence $\{x_n\}$ converges to a finite limit is that, for any arbitrary choice of $\epsilon(\epsilon > 0)$, we can always find an integer N (depending on ϵ) such that

$|x_m - x_n| < \epsilon$, *whenever* m *and* n *are two indices, both greater than* N.

▶ Example 12.10.3 *Consider the sequence defined by*

$$x_n = 1 - \tfrac{1}{2} + \tfrac{1}{3} - \tfrac{1}{4} + \cdots + (-1)^{n-1}\tfrac{1}{n}.$$

Choosing $m > N$, we have

$$\begin{aligned}|x_m - x_n| &= \frac{1}{n+1} - \frac{1}{n+2} + \frac{1}{n+3} - \cdots \frac{1}{m} \\ &= \frac{1}{n+1} - \left(\frac{1}{n+2} - \frac{1}{n+3}\right) - \left(\frac{1}{n+4} - \frac{1}{n+5}\right) - \cdots - \\ &< \frac{1}{n+1}\end{aligned}$$

Now if N be so chosen that $N > \frac{1}{\epsilon} - 1$, then $|x_m - x_n| < \epsilon$, whenever $m, n > N$. Hence, the sequence converges.

▶ Example 12.10.4 $\{x_n\}$, *where* $x_n = 1 + \frac{1}{2} + \frac{1}{3} + \frac{1}{4} + \cdots + \frac{1}{n}$.

Choose $m = 2n$, we have

$$x_m - x_n = \frac{1}{n+1} + \frac{1}{n+2} + \cdots + \frac{1}{n+n} > n \cdot \frac{1}{2n} = \frac{1}{2}.$$

$\therefore$ $x_m - x_n$ **is not less than any positive number that we can name.**

$\therefore$ by Cauchy's principle $\{x_n\}$ does not converge. Since $\{x_n\}$ is **m.i.** and not convergent, hence it diverges to $+\infty$.

Exercises on Chapter 12(I)

1. Write the first four terms of each of the following sequences:

   (a) $\left\{\frac{2^{n-1}}{n^2+1}\right\}$; (b) $\left\{\frac{x^{n+1}}{(n+1)!}\right\}$; (c) $\left\{\frac{4}{3}\left(1 - \frac{1}{2^n}\right)\right\}$; (d) $\left(1 + \frac{1}{n}\right)^n$; (e) $\left\{\frac{\sin n\pi/2}{n}\right\}$.

2. Write an nth term of each of the following sequences:

   (a) $\left\{\frac{1}{2}, \frac{3}{4}, \frac{5}{6}, \frac{7}{8}, \ldots\right\}$; (b) $\left\{\frac{4}{3\cdot 5}, \frac{5}{4\cdot 6}, \frac{6}{5\cdot 7}, \frac{7}{6\cdot 8}, \ldots\right\}$.

3. Two students were asked to write the nth term of a sequence whose first four terms are 1, 16, 81, 256 and then to write the 5th term of the sequence. One student wrote $x_n = n^4$ and obtained $x_5 = 625$. Another student wrote

   $$x_n = 10n^3 - 35n^2 + 50n - 24 \text{ and obtained } x_5 = 601.$$

   Which student gave the correct nth term?

4. Use your definition of limit of a sequence to establish

   (a) $\left\{\frac{n}{n+1}\right\}$ converges to 1; $\left\{\frac{n^2}{2n^2+1}\right\}$ converges to $\frac{1}{2}$. **[C.P. 1980]**

(b) $\left\{1, 0, -\frac{1}{3}, 0, \frac{1}{5}, -\frac{1}{7}, \ldots, \frac{\sin n\pi/2}{n}, \ldots\right\}$ tends to 0.

(c) $\left\{\frac{3n^2+n}{n^2+3n+1}\right\}$ tends to the limit 3.

(d) $\left\{\sqrt[3]{n-2}\right\}$ diverges to ∞.

5. Find a suitable positive integer N such that

(a) $\left|\frac{2n+1}{n+1} - 2\right| < 0{\cdot}0001, \ \forall n > N.$

(b) $\left|\frac{2n+5}{6n-11} - \frac{1}{3}\right| < 0{\cdot}001, \ \forall n > N.$

(c) $|x_n - 2| < 0{\cdot}001, \ \forall n > N$, where $x_n = 2 + \left(-\frac{1}{2}\right)^n$.

6. Discuss whether the following sequences are bounded or not, monotone or not; then state whether the limit exists or not:

(a) $\{3, 3{\cdot}3, 3{\cdot}33, 3{\cdot}333, \ldots\}$

(b) $\{0{\cdot}2, 0{\cdot}23, 0{\cdot}233, 0{\cdot}2333, \ldots\}$

(c) $\{1, -2, 3, -4, 5, -6, \ldots\}$

(d) $\{-1, -1, -2, -2, -3, -3, \ldots\}$

(e) $\{x_n\}$, where $x_n = \frac{1}{n+1} + \frac{1}{n+2} + \frac{1}{n+3} + \cdots + \frac{1}{2n}$.

[*Hints*: (e) $x_{n+1} - x_n > 0$ verify; also see that $x_n < \frac{n}{n+1} = 1 - \frac{1}{n+1}$; $\forall n$ hence bdd.]

7. Consider the behaviour of the following sequences with respect to convergence or divergence:

(a) $\{(-1)^n\}$;

(b) $\left\{\frac{100,000}{n} + (-1)^n\right\}$;

(c) $\left\{100,000(-1)^n + \frac{1}{n}\right\}$;

(d) $\{(-1)^n(n+1)\}$; [C.P. 1980]

(e) $\left\{(-1)^n \frac{n^2+1}{n}\right\}$;

(f) $\left\{(-1)^n \frac{2n-1}{n}\right\}$.

8. Verify that (Limit Theorems may be used, when necessary):

(a) $\lim_{n\to\infty} \frac{6n^3+2n+1}{n^3+n^2} = 6$;

(b) $\lim_{n\to\infty} \left(\sqrt{n+1} - \sqrt{n}\right)\left(\sqrt{n+\frac{1}{2}}\right) = \frac{1}{2}$;

(c) $\lim_{n\to\infty} \left(\sqrt[3]{n+1} - \sqrt[3]{n}\right) = 0$;

(d) $\lim\limits_{n\to\infty}\left(\dfrac{1}{n^2}+\dfrac{2}{n^2}+\dfrac{3}{n^2}+\cdots+\dfrac{n}{n^2}\right)=\dfrac{1}{2}$;

(e) $\lim\limits_{n\to\infty}\left(\sqrt{n^2+n}-n\right)=\dfrac{1}{2}$;

(f) $\lim\limits_{n\to\infty}\left(\dfrac{1}{\sqrt{n}}+\dfrac{1}{\sqrt{n+1}}+\dfrac{1}{\sqrt{n+2}}+\cdots+\dfrac{1}{\sqrt{2n}}\right)=\infty$;

(g) $\lim\limits_{n\to\infty}\left(\dfrac{1}{\sqrt{n^2+1}}+\dfrac{1}{\sqrt{n^2+2}}+\cdots+\dfrac{1}{\sqrt{n^2+n}}\right)=1$.

[*Hints*: (f) $x_n > \frac{n+1}{\sqrt{2n}}$ and hence also $> \frac{n}{\sqrt{2n}} > K, \forall n > 2K^2$.

(g) $\frac{n}{\sqrt{n^2+n}} \le x_n \le \frac{n}{\sqrt{n^2+1}}$; extreme limits each$=1$; hence $\lim\limits_{n\to\infty} x_n = 1$.]

9. **(a)** Write the sequence

$$\left\{\sqrt{2},\ \sqrt{2+\sqrt{2}},\ \sqrt{2+\sqrt{2+\sqrt{2}}},\ \cdots\right\}$$

in the form of the recurrence relation $x_1=\sqrt{2}, x_n=\sqrt{2+x_{n-1}}\ (n>1)$. Verify: $x_{n+1}^2-x_n^2=x_n-x_{n-1}$. Hence, prove that $\{x_n\}$ is **m.i.** Now from $x_{n+1}>x_n$ deduce that $\{x_n\}$ is bounded above. Finally find the limit of the sequence.

(b) Prove that the sequence $\left\{\sqrt{7}, \sqrt{7+\sqrt{7}}, \sqrt{7+\sqrt{7+\sqrt{7}}}, \cdots\cdots\right\}$ converges to the positive root of the quadratic equation $t^2-t-7=0$.

(c) Prove that the sequence $\left\{\sqrt{2}, \sqrt{2\sqrt{2}}, \sqrt{2\sqrt{2\sqrt{2}}}, \cdots\right\}$ converges to the limit 2.

10. If $x_n \to l$ as $\lim\limits_{n\to\infty}$, prove that the sequence $\{y_n\}$, where

$$y_n=\frac{x_1+x_2+\cdots+x_n}{n}$$

also conveges to l. Apply this principle to prove

$$\lim_{n\to\infty}\frac{1+\frac{1}{2}+\frac{1}{3}+\cdots+\frac{1}{n}}{n}=0$$

and $$\lim_{n\to\infty}\frac{1+2^{1/2}+3^{1/3}+\cdots+n^{1/n}}{n^3}=1.$$

11. Use Cauchy's principle to prove that

$$\left\{1+\frac{1}{2}+\frac{1}{3}+\cdots+\frac{1}{n}\right\} \text{ diverges to } +\infty.$$

12. Determine a positive integer m such that

$$\left|\frac{n^2+n+1}{3n^2+1}-\frac{1}{3}\right|<0\cdot 1, \text{ for } \forall n \ge m.$$

Replacing $0\cdot 1$ by $0\cdot 001$, find the corresponding m.

13. Discuss the convergence or divergence of $\left\{\dfrac{n!}{n^n}\right\}$. [C.P. 1982]

Answers

1. (a) $\frac{1}{2}, \frac{2}{5}, \frac{4}{10}, \frac{8}{17}$; (c) $\frac{2}{3}, 1, \frac{7}{6}, \frac{5}{4}$; (e) $1, 0, -\frac{1}{3}, 0$.
(b) $\frac{x^2}{2!}, \frac{x^3}{3!}, \frac{x^4}{4!}, \frac{x^5}{5!}$; (d) $2, \frac{9}{4}, \frac{64}{27}, \frac{625}{256}$;

2. (a) $\frac{2n-1}{2n}$; (b) $\frac{n+3}{(n+2)(n+4)}$.

3. Both correct. Merely giving a finite number of terms at the beginning cannot define a **unique** nth term.

5. (a) $N = 9999$; (b) $N = 1446$; (c) $N = 19$.

6. (a), (b) bdd., **m.i.**; hence convgt. (c) Not bdd., Not monotone; limit does not exist. (d) bdd. above that but below; **m.d.**; diverges to $-\infty$. (e) **m.i.**, **bdd.**; convgt.

7. (a), (b) Osc. finitely (-1 and $+1$); (c) Osc. finitely ($-100,000$ and $+100,000$); (d) and (e) Osc. infinitely; (f) Osc. finitely.

12. $m = 4 : m = 333$.

13. $x_{n+1} = \frac{(n+1)!}{(n+1)^{n+1}} = \frac{n!}{n^n} \cdot \left(\frac{n}{n+1}\right)^n = x_n \left(\frac{n}{n+1}\right)^n$

Since $\left(\frac{n}{n+1}\right)^n < 1$, $\therefore$ $x_{n+1} < x_n$, i.e., $\{x_n\}$ is *m.d.*

Since $x_n > 0$ $\forall n$, the sequence is bounded below; thus $\lim\limits_{n\to\infty} x_n$ exists $= l$ (say) ≥ 0.

Since $\left(\frac{n}{n+1}\right)^n = \frac{1}{(1+\frac{1}{n})^n} < \frac{1}{1+n\cdot\frac{1}{n}} = \frac{1}{2}$.

$\therefore$ $\lim\limits_{n\to\infty} x_{n+1} \leq \frac{1}{2} \lim x_n$, i.e., $l \leq \frac{1}{2}l$, i.e., $l \leq 0$.

Thus $l \geq 0$ and $l \leq 0 \Rightarrow l = 0$.

Section II: Infinite Series

12.11 Introduction

What is an infinite series? Our answer is the following:

Let $\{u_1, u_2, u_3, \ldots, u_n, \ldots\}$ be any infinite sequence.

We shall call the *symbol*

$$u_1 + u_2 + u_3 + \cdots + u_n \cdots, \quad (1)$$

an *infinite series*; u_n is called its *nth term* (or *general term*). We shall often write,

$$\sum_{n=1}^{\infty} u_n \text{ or simply, } \sum u_n$$

for the infinite series **(1)**.

When does an infinite series converge?

For this, we shall construct another sequence $\{s_n\}$ defined by the following rule:

$$s_1 = u, \; s_2 = u_1 + u_2, \; \cdots, \; s_n = u_1 + u_2 + \cdots + u_n = \sum_{r=1}^{n} u_r, \text{ etc.}$$

This sequence is called the *sequence of partial sums* of the series **(1)**; s_n is called the *nth partial sum* of the series **(1)**.

If this sequence $\{s_n\}$ tends to a finite limit s (say) as $n \to \infty$, then we say that the *infinite series* Σu_n *converges* and that it *converges to* s. s is known as the '*sum of the series*'. When Σu_n converges to s in this sense, then we may write

$$s = \Sigma u_n = u_1 + u_2 + \cdots + u_n + \cdots$$

We shall make a more general statement:

An infinite series $\sum_{n=1}^{\infty} u_n$ *is convergent, divergent or oscillating according as the sequence of its partial sums is convergent, divergent or oscillating.*

The importance of the *Infinite Series* will be appreciated if we quote the following lines of *Sokolnikoff's:*

It is difficult to conceive of a single mathematical topic that occupies a more prominent place in Applied Mathematics than the subject of infinite series.

The scope of this book is, however, very much limited. We shall mostly consider infinite series in which the terms are *never negative*; we call such a series *infinite series of positive terms.*

Our aim will be mainly to determine (either *by using the above definition or with the help of some standard tests*) whether a given series is convergent or divergent. Towards the end, we have given the proofs of the tests we have applied, together with a very short discussion on an **extremely important idea**, namely, **absolute convergence.**

12.12 Illustrative Examples

▶ Example 12.12.1 *The following infinite series is known as* **geometric series.**

$$\sum_{n=1}^{\infty} r^{n-1} = 1 + r + r^2 + r^3 + \cdots + r^{n-1} + \cdots$$

As a particular case, consider $\sum_{n=1}^{\infty} \left(\frac{1}{2}\right)^{n-1}$. This series is *convergent*, for the sequence $\{S_n\}$ of its partial sums tends to a finite limit. See that

$$\begin{aligned} S_n &= 1+\frac{1}{2}+\frac{1}{2^2}+\cdots+\left(\frac{1}{2}\right)^{n-1} \\ &= \frac{1-(\frac{1}{2})^n}{1-\frac{1}{2}} = 2-\left(\frac{1}{2}\right)^{n-1} \to 2 \text{ as } n\to\infty. \end{aligned}$$

We say, in this case, that $\sum\limits_{n=1}^{\infty}(\frac{1}{2})^{n-1}$ is *convergent* and that this infinite series has a *sum* 2 (also called *sum to infinity*). We write

$$2 = 1+\frac{1}{2}+\frac{1}{2^2}+\frac{1}{2^3}+\cdots+\frac{1}{2^{n-1}}+\cdots$$

[**Note.** $|S_n-2| = |-(\frac{1}{2})^{n-1}| = (\frac{1}{2})^{n-1} <$ any $\epsilon\ (>0)\ \forall n > 1+\frac{\log 1/\epsilon}{\log 2}$]

In fact, $\sum\limits_{n=1}^{\infty} r^{n-1}$ converges whenever $-1<r<1$ and when convergent, the sum of the series $= \frac{1}{1-r}$.

▶ Example 12.12.2 $\displaystyle\sum_{n=1}^{\infty}\frac{1}{n(n+1)} = \frac{1}{1\cdot 2}+\frac{1}{2\cdot 3}+\frac{1}{3\cdot 4}+\cdots+\frac{1}{n(n+1)}+\cdots$

Here $u_n = \frac{1}{n}-\frac{1}{n+1}$; consequently

$$\begin{aligned} S_n &= u_1+u_2+u_3+\cdots+u_n \\ &= \left(\frac{1}{1}-\frac{1}{2}\right)+\left(\frac{1}{2}-\frac{1}{3}\right)+\left(\frac{1}{3}-\frac{1}{4}\right)+\cdots+\left(\frac{1}{n}-\frac{1}{n+1}\right) \\ &= 1-\frac{1}{n+1} \to 1 \text{ as } n\to\infty. \end{aligned}$$

∴ the infinite series $\sum\limits_{n=1}^{\infty}\frac{1}{n(n+1)}$ is convergent and that its sum is **1**.

▶ Example 12.12.3 $\displaystyle\sum_{n=1}^{\infty}\left(-\frac{3}{2}\right)^n = -\frac{3}{2}+\frac{9}{4}-\frac{27}{8}+\cdots+(-1)^n\frac{3^n}{2^n}+\cdots$

This series, however, is not convergent and it has no **sum**. See that

$$\begin{aligned} s_n &= -\frac{3}{2}+\frac{9}{4}-\frac{27}{8}+\cdots+(-1)^n\frac{3^n}{2^n} \\ &= -\frac{3}{2}\times\frac{1-(-3/2)^n}{1-(-3/2)} = +\frac{3}{5}\left\{\left(-\frac{3}{2}\right)^n-1\right\}. \end{aligned}$$

∴ $\lim\limits_{n\to\infty}(-\frac{3}{2})^n$ does not exist (in fact, this sequence oscillates).

∴ s_n does not tend to a finite limit and as such the infinite series $\sum\limits_{n=1}^{\infty}(-3/2)^n$ *does not converge.* We say that it *diverges.*

[**It is not always essential to point out whether the series is oscillating or diverges to ±∞; our main point is to find out whether the series is convergent or not; if it does not converge we shall say that it diverges.**]

12.13 A Basic Theorem for Convergence or Divergence of an Infinite Series

Theorem 12.13.1. *Cauchy's principle as applied to infinite series.* Σu_n *converges* **if and only if,** *corresponding to any pre-assigned positive number* ϵ *(no matter how small) there exists a positive integer* N *such that for all* $m \geq n \geq N$ *we have*

$$|u_{n+1} + u_{n+2} + u_{n+3} + \cdots + u_m| < \epsilon.$$

Otherwise

We may give an equivalent statement: Σu_n converges *if and only if* corresponding to any pre-assigned positive number ϵ there exists a positive integer N such that,

$$\text{for all } n \geq N \text{ and for all } p = 0, 1, 2, 3, \ldots$$

we have $|u_{n+1} + u_{n+2} + \cdots + u_{n+p}| < \epsilon$.

Proof. By definition, the infinite series Σu_n converges *if and only if* the sequence $\{s_n\}$ of its partial sums converges.

Now by Cauchy's principle applied on the sequence we see that Σu_n converges *if and only if,* given any $\epsilon > 0, \exists$ a positive integer N, such that for all $m \geq n \geq N$, $|s_m - s_n| < \epsilon$, i.e., $|u_{n+1} + u_{n+2} + \cdots + u_m| < \epsilon$.

An Important Corollary

The terms $\{u_n\}$ *of a convergent series form a null sequence.*

That is, if $\sum_{k=1}^{\infty} u_k$ converges, then $u_k \to 0$ as $k \to \infty$.

Proof. Take $m = n + 1$ in the result of the previous theorem.

For a convergent series, $|s_{n+1} - s_n|$ should be less than any pre-assigned positive number $\epsilon \; \forall n \geq N$,

$$\begin{aligned} &\text{i.e.,} \quad |u_{n+1}| < \epsilon, \forall n \geq N, \\ &\text{i.e.,} \quad u_{n+1} \to 0 \text{ as } n \to \infty, \\ &\text{i.e.,} \quad u_k \to 0 \text{ as } k \to \infty (k = n + 1). \end{aligned}$$

Note carefully that $u_n \to 0$ is a **necessary condition** for the convergence of

$$u_1 + u_2 + u_3 + \cdots + u_n + \cdots$$

But this is not a **sufficient condition:** For, we shall presently prove that the series (*harmonic series*) $1 + \frac{1}{2} + \frac{1}{3} + \cdots + \frac{1}{n} + \cdots$ diverges. But its nth term $u_n = \frac{1}{n} \to 0$ as $n \to \infty$.

What we have proved is this: [**a negative test**]

If in an infinite series $u_1 + u_2 + \cdots + u_n + \cdots$ *we find that* u_n *does not tend to zero as* $n \to \infty$ *(i.e.,* $u_n \not\to 0$ *as* $n \to \infty$*), then the series is certainly not convergent:*

e.g., *the series* $\sum_{n=1}^{\infty} \frac{n^n}{n!}$ *does not certainly converge;*

$$\text{for, its } n\text{th term} = \frac{n^n}{n!} = \frac{n}{1} \cdot \frac{n}{2} \cdot \frac{n}{3} \cdots \frac{n}{n-1} \cdot \frac{n}{n} > 1 \text{ whenever } n > 1.$$

$\therefore$ it cannot tend to zero as $n \to \infty$.

Thus the series does not converge

Applications

▶ Example 12.13.1 *The series* $1+\frac{1}{2}+\frac{1}{3}+\cdots+\frac{1}{n}+\cdots$ *diverges.*

Solution: Here $s_n = 1+\frac{1}{2}+\frac{1}{3}+\ldots+\frac{1}{n}$.
For any n, we see that

$$|s_{2n}-s_n| = \frac{1}{n+1}+\frac{1}{n+2}+\cdots\ldots+\frac{1}{2n}$$
$$> \frac{1}{2n}+\frac{1}{2n}+\cdots+\frac{1}{2n}, \text{ i.e., } > n\cdot\frac{1}{2n}, \text{ i.e., } > \frac{1}{2}.$$

Thus no n can be found such that $|s_m - s_n| (m = 2n)$ can be made less any ϵ (if we choose $\epsilon < \frac{1}{2}$), i.e., Cauchy's criterion is not satisfied. Hence, the series cannot converge.

▶ Example 12.13.2 *The series* $\frac{1}{1!}+\frac{1}{2!}+\frac{1}{3!}+\cdots+\frac{1}{n!}+\cdots$ *is a convergent series.*

Solution: Note here $\frac{1}{n!} = \frac{1}{1\cdot 2\cdot 3\cdots n} < \frac{1}{2\cdot 2\cdot 2\ldots 2} = \frac{1}{2^{n-1}}$
If $m > n$,

$$|s_m - s_n| = \frac{1}{(n+1)!}+\frac{1}{(n+2)!}+\cdots+\frac{1}{m!} < \frac{1}{2n}+\frac{1}{2^{n+1}}+\cdots+\frac{1}{2^{m-1}}$$
$$= \frac{1}{2^n}\frac{1-(\frac{1}{2})^{m-n}}{1-\frac{1}{2}}, \text{ i.e., } < 2\cdot\frac{1}{2^n}, \forall m > n.$$

The right side can be made $< \epsilon$ for $n > N$.
Hence, Cauchy's criterion is satisfied and the series converges.

▶ Example 12.13.3 $1-\frac{1}{2}+\frac{1}{3}-\frac{1}{4}+\frac{1}{5}-\frac{1}{6}+\cdots+(-1)^{n-1}\frac{1}{n}+\cdots$ *converges.*

Solution:

$$|s_m - s_n| = \frac{1}{n+1}-\frac{1}{n+2}+\frac{1}{n+3}-\ldots$$
$$= \frac{1}{n+1}-\left(\frac{1}{n+2}-\frac{1}{n+3}\right)-\ldots$$
$$< \frac{1}{n+1}, \text{ which is } < \epsilon, \forall n > \frac{1}{\epsilon}-1 = N.$$

∴ the series converges by Cauchy's criterion.
[This series is called the **alternating harmonic series.**]

12.14 Positive Series: Criterion of Convergence

If every $u_n \geq 0$, then the series Σu_n is called a positive series. For such series the following criterion holds:

Theorem 12.14.1. *A positive series* $\sum_{n=1}^{\infty} u_n$ *converges if the sequence* $\{s_n\}$ *of its partial sums be bounded; otherwise it will diverge to* $+\infty$.

Note 1. ***Thus a positive series cannot oscillate***—it will either converge (when $\{s_n\}$ is bounded) or it will diverge to $+\infty$.

Proof. For a positive series, the sequence $\{s_n\}$, where $s_n = u_1 + u_2 + \cdots + u_n$ is always monotone increasing (because $s_{n+1} - s_n = u_{n+1} \geq 0$) and hence if $\{s_n\}$ is bounded above, then it must converge; if unbounded above then $s_n \to \infty$ as $n \to \infty$.

Accordingly, Σu_n converges if $\{s_n\}$ is bounded; otherwise Σu_n diverges to $+\infty$.

Applications

► Example 12.14.1 *The geometric series* $1 + r + r^2 + \cdots + r^{n-1} + \cdots$ $(r > 0)$ *converges if* $0 < r < 1$, *diverges to* $+\infty$ *if* $r > 1$.

Solution: $S_n = 1 + r + r^2 + \cdots + r^{n-1}$; $\{S_n\}$ is **m.i.** $(\because\ r > 0)$

Again, when $0 < r < 1$

$$S_n = \frac{1 - r^n}{1 - r} < \frac{1}{1 - r}, \text{ for all } n,$$

i.e., $\{S_n\}$ is bounded above.

Accordingly the series converges when $0 < r < 1$.

But, when $r > 1$,

$$S_n = \frac{r^n - 1}{r - 1}$$

can exceed any number for large n, i.e., $\{S_n\}$ is unbounded above.

Hence, the series diverges to $+\infty$, if $r > 1$.

► Example 12.14.2 **The hyperharmonic series** $\sum\limits_{n=1}^{\infty} \frac{1}{n^p}$ **(also known as p-series).** *This series converges* **only when** $p > 1$; *for other values of* p *the series does not converge.*

Case 1. *If* $p \leq 0$, $\lim\limits_{n \to \infty} \frac{1}{n^p} \neq 0$ *and hence the series cannot converge.*

Case 2. $p = 1$, *we have the familiar* harmonic series $1 + \frac{1}{2} + \frac{1}{3} + \frac{1}{4} + \cdots + \frac{1}{n} + \cdots$

We have already proved (by using Cauchy's criterion) that this series diverges. We give another proof (based on the theorem stated earlier in this article).

We first consider the sum of 2^n terms

$$\begin{aligned}
S_{2^n} &= 1 + \frac{1}{2} + \frac{1}{3} + \frac{1}{4} + \frac{1}{5} + \cdots + \frac{1}{2^n}. \\
&= 1 + \frac{1}{2} + \left(\frac{1}{3} + \frac{1}{4}\right) + \left(\frac{1}{5} + \frac{1}{6} + \frac{1}{7} + \frac{1}{8}\right) + \cdots \\
&\quad + \left(\frac{1}{2^{n-1} + 1} + \frac{1}{2^{n-1} + 2} + \cdots + \frac{1}{2^n}\right) \\
&\geq 1 + \frac{1}{2} + \left(\frac{1}{4} + \frac{1}{4}\right) + \left(\frac{1}{8} + \frac{1}{8} + \frac{1}{8} + \frac{1}{8}\right) + \cdots \\
&\quad + \left(\frac{1}{2^n} + \frac{1}{2^n} + \cdots + \frac{1}{2^n}\right) \\
&= 1 + \frac{1}{2} + 2 \cdot \frac{1}{4} + 2^2 \cdot \frac{1}{8} + \cdots + 2^{n-1} \cdot \frac{1}{2^n} = 1 + \frac{n}{2}.
\end{aligned}$$

Given any $G > 0$, $S_{2^n} > G$ for $n > 2G - 2$.

Taking $m > 2^n$, we have $S_m > S_{2_n} > G$ for $m > 2^{2G-2}$, i.e., $\{S_m\}$ is unbounded above.

$\therefore$ the positive series $\Sigma \frac{1}{n}$ diverges to $+\infty$.

Case 3. $0 < p < 1$. *Here* $n^p < n$, *i.e.*, $\frac{1}{n^p} > \frac{1}{n}$.

Then

$$\begin{aligned} S_{2_n} &= 1 + \frac{1}{2^p} + \left(\frac{1}{3^p} + \frac{1}{4^p}\right) + \left(\frac{1}{5^p} + \frac{1}{6^p} + \frac{1}{7^p} + \frac{1}{8^p}\right) + \cdots \\ &\quad + \left(\frac{1}{(2^{n-1}+1)^p} + \cdots + \frac{1}{(2^n)^p}\right) \\ &\geq 1 + \frac{1}{2} + \left(\frac{1}{3} + \frac{1}{4}\right) + \left(\frac{1}{5} + \frac{1}{6} + \frac{1}{7} + \frac{1}{8}\right) + \cdots \\ &\quad + \left(\frac{1}{2^{n-1}+1} + \cdots + \frac{1}{2^n}\right) \\ &\geq 1 + \frac{n}{2} \text{ (as in Case 2).} \end{aligned}$$

Same arguments as in Case 2 will lead to the conclusion that the series

$$\sum_{n=1}^{\infty} \frac{1}{n^p} \text{ diverges if } 0 < p < 1.$$

Case 4. ***Suppose*** $p > 1$.

$$\begin{aligned} S_{2^n-1} &= 1 + \left(\frac{1}{2^p} + \frac{1}{3^p}\right) + \left(\frac{1}{4^p} + \cdots + \frac{1}{7^p}\right) + \cdots \\ &\quad + \left(\frac{1}{(2^{n-1})^p} + \cdots + \frac{1}{(2^n-1)^p}\right) \\ &\leq 1 + \left(\frac{1}{2^p} + \frac{1}{2^p}\right) + \left(\frac{1}{4^p} + \cdots + \frac{1}{4^p}\right) + \cdots \\ &\quad + \left(\frac{1}{(2^{n-1})^p} + \cdots + \frac{1}{(2^{n-1})^p}\right) \\ &= 1 + 2 \cdot \frac{1}{2^p} + 4 \cdot \frac{1}{4^p} + 8 \cdot \frac{1}{8^p} + \cdots + 2^{n-1} \frac{1}{(2^{n-1})^p} \\ &= 1 + \frac{1}{2^{p-1}} + \frac{1}{(2^{p-1})^2} + \frac{1}{(2^{-1})^3} + \cdots + \frac{1}{(2^{p-1})^{n-1}} \\ &= \frac{1 - \left(\frac{1}{2^{p-1}}\right)^n}{1 - \frac{1}{2^{p-1}}} \textit{ (summing the finite geometric series)} \\ \therefore\ S_{2^n-1} &< \frac{1}{1 - \frac{1}{2^{p-1}}} = \textit{a constant quantity, for all } n. \end{aligned}$$

Now, for any positive integer m, $\exists$ a positive integer n such that $2^{n-1} > m$.

$\therefore$ $S_m < S_{2^n-1} < \dfrac{1}{1 - \frac{1}{2^{p-1}}}$ = a constant quantity, i.e., $\{S_m\}$ is bounded above.

Consequently the positive series $\sum_{n=1}^{\infty} \frac{1}{n^p}$ $(p > 1)$ **converges.**

► Example 12.14.3 *The series* $\sum_{n=1}^{\infty} \frac{1}{(n-1)!} = 1 + \frac{1}{1!} + \frac{1}{2!} + \cdots + \frac{1}{(n-1)!} + \cdots$

Solution: Here

$$\begin{aligned} s_n &= 1+1+\frac{1}{2}+\frac{1}{2\cdot 3}+\frac{1}{2\cdot 3\cdot 4}+\cdots+\frac{1}{2\cdot 3\cdot 4\cdots(n-1)} \\ &< 1+1+\frac{1}{2}+\frac{1}{2\cdot 3}+\frac{1}{3\cdot 4}+\cdots+\frac{1}{(n-2)(n-1)} \quad (n>4) \\ \therefore\ s_n &< 1+1+\frac{1}{2}+\left(\frac{1}{2}-\frac{1}{3}\right)+\left(\frac{1}{3}-\frac{1}{4}\right)+\cdots \\ &\quad +\left(\frac{1}{n-2}-\frac{1}{n-1}\right) \quad (n>4), \\ \text{i.e.,} &< 1+1+\frac{1}{2}+\frac{1}{2}-\frac{1}{n-1} = 3-\frac{1}{n-1} < 3. \end{aligned}$$

But $s_1 < s_2 < s_3 < \cdots < s_n < s_{n+1} < \cdots$

$\therefore \{s_n\}$ is **m.i.** and bounded; hence convergent;

Accordingly, the series **converges.**

12.15 General Theorems on Convergence

Theorem 12.15.1. *If the infinite series* $\sum_{n=1}^{\infty} u_n$ *is convergent with sum to infinity* s, *then the infinite series.*

$u_{m+1} + u_{m+2} + u_{m+3} + \cdots = \sum_{r=m+1}^{\infty} u_r$ (*where* m *is a given positive integer*) *is also convergent and its sum to infinity is* $s - (u_1 + u_2 + \cdots + u_m)$.

Proof. Sum of n terms of the latter series

$$\begin{aligned} &= u_{m+1}+u_{m+2}+u_{m+3}+\cdots+u_{m+n} \\ &= s_{m+n}-s_m, \text{ where } s_r = u_1+u_2+\cdots+u_r. \end{aligned}$$

Since $\sum_{n=1}^{\infty} u_n$ converges, $\lim_{n\to\infty} s_{m+n}$ converges to s; hence etc.

Corollary 12.15.1 *If* $u_1+u_2+u_3+\cdots+u_n+\cdots$ *diverges, so also does* $u_{m+1}+u_{m+2}+\cdots+u_{m+n}+\cdots$ ($m =$ *a given positive integer*).

What we mean is: In discussing the convergence or divergence of a series, any finite number of terms at the beginning may be left out.

Theorem 12.15.2. *If the infinite series* $u_1 + u_2 + u_3 + \cdots$ *is convergent with sum to infinity* s, *and if* c *is any constant, then* $cu_1 + cu_2 + cu_3 + \cdots$ *is convergent with sum to infinity* cs.

Similarly, if $u_1 + u_2 + u_3 + \cdots$ diverges, so also does $cu_1 + cu_2 + cu_3 + \cdots$ unless $c = 0$.

The proof is very simple: we leave out as an exercise for the students.

We shall assume the truth of the following theorem:

Theorem 12.15.3. *A positive series converges to the same sum in whatever order the terms are taken.*

Another useful theorem is the following:

Theorem 12.15.4. *If* $u_1 + u_2 + u_3 \cdots$ *and* $v_1 + v_2 + v_3 + \cdots$ *be both convergent with sums to infinity* s *and* t, *then the series*

$$(au_1 + bv_1) + (au_2 + bv_2) + (au_3 + bv_3) + \cdots$$

is also convergent with sum to infinity $as + bt$ (a, b, *are constants*).

[See that $\lim\limits_{n\to\infty} [(au_1 + bv_1) + (au_2 + bv_2) + \cdots + (au_n + bv_n)]$

$$= a \lim_{n\to\infty} [u_1 + u_2 + \cdots + u_n] + b \lim_{n\to\infty} [v_1 + v_2 + \cdots + v_n] = as + bt.]$$

12.16 Tests for Convergence: Series of Positive Terms

We shall now give a few tests for convergence which are applicable only to positive series. Our intention is to make our students familiar with the statements of these tests and their careful applications. The curious readers will find the proofs of some of these tests at the end of the exercises of this chapter.

Test 1. Comparison Test for Convergence

If
$$u_1 + u_2 + u_3 + \cdots + u_n + \cdots = \Sigma u_n,$$
and
$$v_1 + v_2 + v_3 + \cdots + v_n + \cdots = \Sigma v_n$$

be two series of positive terms and if Σv_n *is known to be convergent, then* Σu_n *is convergent if either*

(i) $u_n \le v_n$, *for all values of* $n >$ *a fixed integer* m;

or (ii) $\lim\limits_{n\to\infty} \frac{u_n}{v_n} =$ *a finite quantity* l (*say*). (*it may be even zero*).

[More generally, condition (i) can be stated as:

$u_n \le cv_n$ for all $n > m$, c being a fixed positive quantity.]

▶ Example 12.16.1 *Prove, by comparison test, that the series*

$$1 + \frac{2}{1!} + \frac{2^2}{2!} + \frac{2^3}{3!} + \frac{2^4}{4!} + \frac{2^5}{5!} + \cdots \text{ is convergent.}$$

Solution: (i) We compare this series with the series

$$1 + \frac{2}{1!} + \frac{2^2}{2!} + \frac{2^2}{2!}\left(\frac{2}{3}\right) + \frac{2^2}{2!}\left(\frac{2}{3}\right)^2 + \frac{2^2}{2!}\left(\frac{2}{3}\right)^3 + \cdots$$

From the 5th term onwards, each term of the given series (call it Σu_n) is less than the corresponding term of the series Σv_n, say, chosen for comparison, that is, $u_n < v_n$ for $n > 4$.

But if we omit the first two terms of Σv_n, we have a G.P. with common ratio 2/3, which is, therefore, convergent.

$\therefore$ by comparison test, Σu_n is convergent.

Or, (ii) **Limit form:**

$$u_n = \frac{2^n}{n!}, \; v_n = \frac{2^2}{2!}\left(\frac{2}{3}\right)^{n-1}$$

$$\lim_{n\to\infty} \frac{u_n}{v_n} = \lim_{n\to\infty} \frac{2^n}{n!} \times \frac{2!}{2^2} \times \left(\frac{3}{2}\right)^{n-1}$$

$$= \lim_{n\to\infty} \frac{1}{3}\frac{3^n}{n!} = 0 \text{ (a finite quantity).}$$

Since Σv_n is convergent, it follows by limit form of comparison test, Σu_n is convergent.

Note 1. *In discussing convergence we may leave out a finite number of terms at the beginning; in the present problem we have left out the first term of Σu_n and first two terms of Σv_n.]*

We have used here: $\lim_{n\to\infty} \frac{x^n}{n!} = 0$, *for all values of* x. **[See Ex. 12.9.9, Applications (ii)]**

This follows from the fact that $\lim_{n\to\infty} \left|\frac{x_{n+1}}{x_n}\right| = l, 0 \le l < 1 \Rightarrow \lim_{n\to\infty} x_n = 0.$

Test 2. Comparison Test for Divergence

If $u_1 + u_2 + u_3 + \cdots + u_n + \cdots = \Sigma u_n$,
and $v_1 + v_2 + v_3 + \cdots + v_n + \cdots = \Sigma v_n$

be two series of positive terms and if Σv_n is divergent, then Σu_n diverges if either

(i) $u_n \ge v_n$, *for all values of $n >$ a fixed integer m;*

or, (ii) $\lim_{n\to\infty} \frac{u_n}{v_n}$ = *a finite quantity l (say).*

[More generally, condition (i) can be stated as: $u_n \ge cv_n$ **for all $n > m$, c being a positive constant.**]

▶ Example 12.16.2 *Prove (by comparison test) that the series*

$$\frac{1}{1\cdot 3} + \frac{2}{3\cdot 5} + \frac{3}{5\cdot 7} + \frac{4}{7\cdot 9} + \cdots \text{ is divergent.}$$

Solution: (i) The given series is Σu_n where

$$u_n = \frac{n}{(2n-1)(2n+1)}. \quad \therefore u_n > \frac{n}{2n\cdot 3n} = \frac{1}{6n}.$$

Compare Σu_n with Σv_n, where $v_n = 1/n$.

Since $u_n > \frac{1}{6}v_n$ and since Σv_n diverges, it follows that Σu_n **is divergent.**
(ii) **Limit form:**

$$u_n = \frac{n}{(2n-1)(2n+1)};\ \text{take } v_n = \frac{n}{n^2} = \frac{1}{n}$$

$$\lim_{n\to\infty}\frac{u_n}{v_n} = \lim_{n\to\infty}\frac{n}{(2n-1)(2n+1)}\times\frac{n}{1}$$

$$= \lim_{n\to\infty}\frac{1}{\left(2-\frac{1}{n}\right)\left(2+\frac{1}{n}\right)} = \frac{1}{4}\ \text{(a finite quantity).}$$

Now Σv_n is divergent ($\because$ it is the harmonic series which is known to be divergent); consequently Σu_n also diverges.

Some observations on comparison test

The utility of comparison tests depends on having at our disposal standard series which are known to be convergent or divergent.

In most cases we shall use either the Geometric series $\sum_{n=1}^{\infty} r^{n-1}$ **which converges only when** $0 < r < 1$ **or the p-series** $\sum_{n=1}^{\infty}\frac{1}{n^p}$ **which converges if** $p > 1$**, but diverges if** $p \leq 1$**.**

A practical hint: *If* $u_n = \frac{f(n)}{\varphi(n)}$ *where* $f(n)$ *and* $\varphi(n)$ *are polynomials in* n*, then take* $v_n = \frac{n^k}{n^l}$ *where* n^k *is the highest power of* n *in* $f(n)$ *and* n^l *is the highest power of* n *in* $\varphi(n)$.

▶ Example 12.16.3 *Suppose* $u_n = \frac{2n+3}{n(n+1)(n+3)}$; *take* $v_n = \frac{n}{n^3} = \frac{1}{n^2}$.

Now see that

$$\frac{u_n}{v_n} = \frac{2n+3}{n(n+1)(n+3)}\times\frac{n^2}{1} = \frac{n(2n+3)}{(n+1)(n+3)}$$

$$= \frac{2+3/n}{(1+1/n)(1+3/n)}\ \text{and}\ \frac{u_n}{v_n}\to \text{a finite limit 2 as } n\to\infty.$$

$\therefore$ by limit form of comparison test, $\sum u_n$ converges because $\sum\frac{1}{n^2}$ is convergent (it is a p-series with $p = 2 > 1$).

Another important example

▶ Example 12.16.4 *Suppose* $u_n = \sqrt[3]{n^3+1} - n$.

Then write

$$u_n = n\sqrt[3]{1+(1/n)^3} - n = n\left[\left(1+\frac{1}{n^3}\right)^{1/3} - 1\right]$$

$$= n\left[1+\frac{1}{3}\frac{1}{n^3}+\frac{\frac{1}{3}\left(\frac{1}{3}-1\right)}{2!}\frac{1}{n^6}+\cdots-1\right]$$

$$= \frac{1}{3}\frac{1}{n^2} + \text{higher powers of } \frac{1}{n}.$$

We take $v_n = \frac{1}{n^2}$ and then see that $\frac{u_n}{v_n} \to \frac{1}{3}$ as $n \to \infty$.
Now since Σv_n converges, Σu_n also converges.

Test 3. D′ Alembert's Ratio test for Convergence

The infinite series of positive terms $u_1 + u_2 + u_3 + \cdots + u_n + \cdots = \Sigma u_n$ *is convergent if either*

(i) $\frac{u_{n+1}}{u_n} < k < 1$, *whenever* $n \geq m$ (m *being a fixed positive integer*) *and* k *is a positive constant,*

or, (ii) $\lim_{n\to\infty} \frac{u_{n+1}}{u_n} = l$, *where* $0 < l < 1$.

Test 4. D′ Alembert's Ratio test for Divergence

The infinite series of positive terms $u_1 + u_2 + u_3 + \cdots + u_n + \cdots = \Sigma u_n$ *is divergent if either*

(i) $\frac{u_{n+1}}{u_n} \geq 1$, *whenever* $n \geq m$ (m *being a fixed positive integer*)

or, (ii) $\lim_{n\to\infty} \frac{u_{n+1}}{u_n} = l$, *where* $l > 1$.

No conclusion can be made about convergence or divergence of the series Σu_n, if $\lim_{n\to\infty} \frac{u_{n+1}}{u_n} = 1$. The series may or may not converge. Other test or the definition of convergence is to be used in this case: **See that**

$\Sigma \frac{1}{n}$ is known to be divergent, but here

$$\lim_{n\to\infty} \frac{u_{n+1}}{u_n} = \lim_{n\to\infty} \frac{1/(n+1)}{1/n} = \lim_{n\to\infty} \frac{n}{n+1} = 1$$

and $\Sigma \frac{1}{n^2}$ is known to be convergent, but **here also**

$$\lim_{n\to\infty} \frac{u_{n+1}}{u_n} = \lim_{h\to\infty} \frac{1/(n+1)^2}{1/n^2} = \lim_{n\to\infty} \frac{n^2}{(n+1)^2} = 1.$$

$\therefore$ just from $\lim_{n\to\infty} \frac{u_{n+1}}{u_n} = 1$, no conclusion should be made.

▶ Example 12.16.5 *The series* $\frac{1}{2} + \frac{2}{2^2} + \frac{3}{2^3} + \cdots + \frac{n}{2^n} \cdots$ *is convergent.*

Solution: Here take $u_n = \frac{n}{2^n}$. Then

$$\begin{aligned} \lim_{n\to\infty} \frac{u_{n+1}}{u_n} &= \lim_{n\to\infty} \frac{n+1}{2^{n+1}} \Big/ \frac{n}{2^n} = \lim_{n\to\infty} \frac{n+1}{2^{n+1}} \times \frac{2^n}{n} \\ &= \frac{1}{2} \lim_{n\to\infty} \left(1 + \frac{1}{n}\right) = \frac{1}{2}, \end{aligned}$$

which is positive and less than 1.

$$\therefore \ \Sigma u_n \text{ converges by ratio test.}$$

▶ Example 12.16.6 *Prove that the series* $x + \frac{1}{2}x^2 + \frac{1}{3}x^3 + \frac{1}{4}x^4 + \cdots$ ($x \geq 0$) *is convergent for* $0 \leq x < 1$ *and divergent for* $x \geq 1$.

Solution:

$$u_n = \frac{x^n}{n};\ u_{n+1} = \frac{x^{n+1}}{n+1};\ \frac{u_{n+1}}{u_n} = \frac{x^{n+1}}{n+1} \times \frac{n}{x^n} = \frac{1}{1+1/n}x$$

$$\therefore \lim_{n\to\infty} \frac{u_{n+1}}{u_n} = x.$$

Hence, by Ratio Test,

if $0 < x < 1$, $\sum u_n$ converges,

if $x > 1$, $\sum u_n$ diverges,

if $x = 1$, $\lim_{u\to\infty} \frac{u_{n+1}}{u_n} = 1$, and hence there is no conclusion from the ratio test.

The series then becomes $1 + \frac{1}{2} + \frac{1}{3} + \frac{1}{4} + \cdots$ (**harmonic series**), which we have proved to be divergent.

If $x = 0$, each term is 0, and the series is then also convergent. *Hence, the problem.*

Test 5. Cauchy's Root test

The series of positive terms

$$u_1 + u_2 + u_3 + \cdots + u_n + \cdots = \Sigma u_n$$

converges if $u_n^{1/n} \le k < 1$ *for all values of* $n >$ *some fixed positive integer* m, *but diverges if* $u_n^{1/n} \ge 1$, *for infinitely many* n.

Limit form: *The series of positive terms*

$$u_1 + u_2 + u_3 + \cdots + u_n + \cdots = \sum u_n$$

converges if $\lim_{n\to\infty} u_n^{1/n} = l$, *where* $l < 1$, *diverges if* $\lim_{n\to\infty} u_n^{1/n} = l$, *where* $l > 1$.

But if this limit $\left(\lim_{n\to\infty} u_n^{1/n}\right)$ *is equal to* 1, *no conclusion can be made about the convergence or divergence of* Σu_n. *The series may or may not converge.*

▶ Example 12.16.7 *The series* $\frac{1}{3} + \left(\frac{2}{5}\right)^2 + \left(\frac{3}{7}\right)^3 + \cdots + \left(\frac{n}{2n+1}\right)^n + \cdots$ *converges.*

We use root test:

$$u_n^{1/n} = \left\{\left(\frac{n}{2n+1}\right)^n\right\}^{\frac{1}{n}} = \frac{n}{2n+1} = \frac{1}{2+1/n} \to \frac{1}{2} \text{ as } n \to \infty.$$

This limit is finite and less than one; hence, by root test we conclude that the series converges.

Observations: In practice the root and ratio tests are among the most useful tests for deciding convergence or divergence of a series. However that ratio test is often easier to apply than the root test, for u_{n+1}/u_n is often a simpler function of n than $u_n^{1/n}$. But, remember, sometimes root test gives a more wide conclusion whereas ratio test fails to do so. One can be sure of getting the same result by *ratio test* if the root test shows the convergence of a series.

► Example 12.16.8 *Take a series whose* $u_n = 2^{-n-(-1)^n}$. **This series converges by root test,** *because* $u_n^{1/n} = 2^{-1-\frac{(-1)^n}{n}} \to 2^{-1}$ (*a finite limit* < 1) (*as* $n \to \infty$).

But see that

$$\frac{u_{n+1}}{u_n} = \frac{2^{-n-1-(-1)^{n+1}}}{2^{-n-(-1)^n}} = 2^{-1+(-1)^n - (-1)^{n+1}}$$

$\therefore$ $u_{n+1}/u_n = 2$, if n be even; $= 1/8$, if n be odd.
$\therefore$ **no conclusion by ratio test.**

► Example 12.16.9 $\sum_{n=1}^{\infty} \frac{x^n}{n} \cdot \frac{u_{n+1}}{u_n} = \frac{x^{n+1}}{n+1} \times \frac{n}{x^n} = \frac{1}{1+1/n} \cdot x \to x$ *as* $n \to \infty$.

$\therefore$ Σu_n converges if $0 < x < 1$, diverges when $x > 1$.
When $x = 1$, the ratio test is inconclusive, but then the series is harmonic and divergent.

The root test gives **the same results:** $\sqrt[n]{u_n} = \sqrt[n]{\frac{x^n}{n}} = \frac{x}{\sqrt[n]{n}} \to x$
($\because$ $\sqrt[n]{n} \to 1$ as $n \to \infty$, **see Ex. 12.9.6).**

Test 6. Raabe's Test

[To be applied if while using ratio test it is found that $\lim u_{n+1}/u_n = 1$]
For a series of positive terms

(i) *if* $u_n/u_{n+1} \geq 1 + \alpha/n$, *where* $\alpha > 1$, *then* Σu_n *converges,*

(ii) *if* $u_n/u_{n+1} \leq 1 + 1/n$, *then* Σu_n *diverges.*

[*The inequalities may hold for all* n *or for all* $n >$ *some fixed integer* m. *In* (i) α *must be a constant.*]

Limit form: *Let* Σu_n *be a series of positive terms.*
Let $R_n = n\left(\frac{u_n}{u_{n+1}} - 1\right)$.
If $\lim_{n\to\infty} R_n = l$, *then for* $l > 1$, Σu_n *converges and for* $l < 1$, *the series* $n \to \infty$ *diverges; no conclusion if* $l = 1$.
[Proof not given in this text.]

► Example 12.16.10 *The series* $1 + \frac{1}{2\cdot 3} + \frac{1\cdot 3}{2\cdot 4\cdot 5} + \frac{1\cdot 3\cdot 6}{2\cdot 4\cdot 6\cdot 7} + \cdots$

Here,

$$u_n = \frac{1\cdot 3\cdot 5\cdot 7\cdots(2n-1)}{2\cdot 4\cdot 6\cdot 8\cdots 2n} \cdot \frac{1}{2n+1}$$ [We leave out the first term of the given series.]

$$u_{n+1} = \frac{1\cdot 3\cdot 5\cdot 7\cdots(2n-1)(2n+1)}{2\cdot 4\cdot 6\cdot 8\cdots 2n(2n+2)} \cdot \frac{1}{2n+3}$$

$$\therefore \frac{u_n}{u_{n+1}} = \frac{(2n+2)(2n+3)}{(2n+1)^2} = \frac{4n^2+10n+6}{4n^2+4+1}$$

[Note that $\frac{u_{n+1}}{u_n} \to 1$ as $n \to \infty$; hence no conclusion by Ratio Test.]

$$\therefore R_n = n\left(\frac{u_n}{u_{n+1}} - 1\right) = n \times \frac{6n+5}{4n^2+4n+1}$$

$$= \frac{6+5/n}{4+4/n+1/n^2}, \text{ i.e., } R_n \to \tfrac{3}{4} \text{ as } n \to \infty$$

$\therefore \lim_{n\to\infty} R_n > 1$; hence by Raabe's test the **series converges.**

▶ Example 12.16.11 *To consider the convergence or divergence of the series*

$$x + \frac{1}{2}\frac{x^3}{3} + \frac{1\cdot 3}{2\cdot 4}\frac{x^5}{5} + \frac{1\cdot 3\cdot 5}{2\cdot 4\cdot 6}\frac{x^7}{7} + \cdots \quad (x > 0)$$

We first consider the *ratio test:*

$$\frac{u_{n+1}}{u_n} = \frac{(2n+1)^2}{(2n+2)(2n+3)} \cdot x^2 = \frac{4n^2+4n+1}{4n^2+10n+6} x^2 \to x^2 \text{ as } n \to \infty.$$

$\therefore$ by ratio test the series converges if $0 < x < 1$, diverges if $x > 1$.
For $x = 1$, $\lim_{n\to\infty} \frac{u_{n+1}}{u_n} = 1$. In that case we shall use Raabe's Test.
See **Example 12.6.10** above. The series converges for $x = 1$.
$\therefore$ the conclusion is: the series converges if $0 < x \leq 1$, diverges if $x > 1$.

Exercises on Chapter 12(II)

A

Direct applications

1. For each of the following series obtain the nth partial sum and then decide whether it is convergent or not. If convergent, find the sum of the series:

(a) $\frac{1}{1\cdot 2\cdot 3} + \frac{1}{2\cdot 3\cdot 4} + \frac{1}{3\cdot 4\cdot 5} + \cdots + \frac{1}{n(n-1)(n+2)} + \cdots$ [1/4]

(b) $\frac{1}{1\cdot 3} + \frac{1}{3\cdot 5} + \frac{1}{5\cdot 7} + \frac{1}{7\cdot 9} + \cdots + \frac{1}{(2n-1)((2n+1)} + \cdots$ [1/2]

(c) $1 + \frac{2}{3} + \left(\frac{2}{3}\right)^2 + \left(\frac{2}{3}\right)^3 + \cdots + \left(\frac{2}{3}\right)^{n-1} + \cdots$ [3]

2. (a) Prove that the series $\sum_{n=1}^{\infty} \frac{n}{n+1}$ cannot converge.

(b) Give an example of a series, whose nth term tends to zero as $n \to \infty$, but the series itself does not converge.

3. (a) Using **Cauchy's criterion** for the convergence and divergence of a real sequence prove that

$\sum_{n=1}^{\infty} \frac{1}{n}$ diverges; but $\sum_{n=1}^{\infty} \frac{1}{n!}$ and the series $\sum_{n=1}^{\infty} (-1)^{n-1}\frac{1}{n}$ are both convergent.

(b) Prove that $1+\frac{1}{\sqrt{2}}+\frac{1}{\sqrt{3}}+\frac{1}{\sqrt{4}}+\cdots+\frac{1}{\sqrt{n}}+\cdots$ is divergent.

4. Use *Comparison Test* to investigate the convergence or divergence of the following series:

(a) $\frac{1}{1\cdot 2}+\frac{1}{3\cdot 4}+\cdots+\frac{1}{(2n-1)2n}+\cdots$ [Convgt.]

(b) $\sum_{n=2}^{\infty}\frac{1}{\sqrt{n(n-1)}}$ [Divgt.]

(c) $\sum_{n=1}^{\infty}u_n$, where $u_n=\sqrt{n^2+1}-n$ [Divgt.]

(d) $\frac{1}{1\cdot 2\cdot 3}+\frac{3}{2\cdot 3\cdot 4}+\frac{5}{3\cdot 4\cdot 5}+\cdots;\ \sum_{n=1}^{\infty}\frac{n+1}{n^6}$ [both Convgt.] [C.P. 1980]

(e) $\sum_{n=1}^{\infty}\frac{x^{n-1}}{1+x^n}\ (x>0)$ [$x<1$, Convgt.; $x\geq 1$, Divgt.]

[$x<1, x^n\to 0, u_n/x^{n-1}\to 1$. Σx^{n-1} converges. $\therefore$ Σu_n converges.
$x>1, x^{-n}\to 0, u_n\to 1/x$. $\therefore$ Σu_n diverges.
$x=1, u_n=\frac{1}{2}$ and the series diverges.]

5. Use *D'Alembert's Ratio test* to test the convergence or divergence of the following series:

(a) $\frac{1^2}{2}+\frac{2^2}{2^2}+\frac{3^2}{2^3}+\frac{4^2}{2^4}+\frac{5^2}{2^5}+\cdots+\frac{n^2}{2^n}+\cdots$ [Convgt.]

(b) $\sum_{n=0}^{\infty}\frac{x^n}{n!}=1+\frac{x}{\lfloor 1}+\frac{x^2}{\lfloor 2}+\frac{x^3}{\lfloor 3}+\cdots+\frac{x^n}{\lfloor n}+\cdots(x>0)$ [Convgt. $\forall x$]

(c) $\frac{1}{4}+\frac{1\cdot 3}{4\cdot 7}+\frac{1\cdot 3\cdot 5}{4\cdot 7\cdot 10}+\frac{1\cdot 3\cdot 5\cdot 7}{4\cdot 7\cdot 10\cdot 13}+\cdots+\frac{1\cdot 3\cdot 5\cdot 7\cdots(2n-1)}{4\cdot 7\cdot 10\cdot 13\cdots(3n+1)}+\cdots$ [Convgt.]

(d) $\sum_{n=2}^{\infty}\frac{n^2-1}{n^2+1}x^n(x>o)$ [$0<x<1$ Convgt.; $x\geq 1$ Divgt.]

(e) $1+\frac{\sqrt{2}-1}{1!}+\frac{(\sqrt{2}-1)^2}{2!}+\frac{(\sqrt{2}-1)^3}{3!}+\cdots$ [Convgt.]

6. Use *Cauchy's Root test* to decide about the convergence or divergence of the following series:

(a) $\frac{1}{2}+\frac{1}{3^2}+\frac{1}{4^3}+\frac{1}{5^4}+\cdots+\frac{1}{(n+1)^n}+\cdots$ [Convgt.]

(b) $1+\frac{2^2}{2^2}+\frac{2^3}{3^3}+\frac{2^4}{4^4}+\cdots+\frac{2^n}{n^n}+\cdots$ [Convgt.]

(c) $\sum_{n=1}^{\infty} u_n$, where $u_n = \frac{n^{n^2}}{(1+n)^{n^2}}$ $\left[\sqrt[n]{u_n} \to \frac{1}{e}\right]$ [Convgt.]

(d) $\sum_{n=1}^{\infty} \frac{(n+1)^n x^n}{n^{n+1}}, (x > 0)$ [Convgt. if $0 < x < 1$, Divgt. if $x \geq 1$]

(e) $\left(\frac{2^2}{1^2} - \frac{2}{1}\right)^{-1} + \left(\frac{3^3}{2^3} - \frac{3}{2}\right)^{-2} + \left(\frac{4^4}{3^4} - \frac{4}{3}\right)^{-3} + \cdots + \left[\frac{(n+1)^{n+1}}{n^{n+1}} - \frac{n+1}{n}\right]^{-n} \cdots$ [Convgt.]

7. (a) Use *Raabe's test* to prove that the series

$$1 + \frac{1}{2 \cdot 3} + \frac{1 \cdot 3}{2 \cdot 4} \cdot \frac{1}{5} + \frac{1 \cdot 3 \cdot 5}{2 \cdot 4 \cdot 6} \cdot \frac{1}{7} + \cdots \text{ is convergent.}$$

(b) Use *Ratio test* to prove that the series

$$x + \frac{1}{2}\frac{x^3}{3} + \frac{1 \cdot 3}{2 \cdot 4}\frac{x^5}{5} + \frac{1 \cdot 3 \cdot 5}{2 \cdot 4 \cdot 6}\frac{x^7}{7} + \cdots (x > 0)$$

converges if $0 < x < 1$ and diverges if $x > 1$.

Use *Raabe's test* to show that the series also converges when $x = 1$.

(c) Determine whether the series

$$\sum_{n=1}^{\infty} \frac{1 \cdot 3 \cdot 5 \cdots (2n-1)}{2 \cdot 4 \cdot 6 \cdots 2n} = \frac{1}{2} + \frac{1 \cdot 3}{2 \cdot 4} + \frac{1 \cdot 3 \cdot 5}{2 \cdot 4 \cdot 6} + \cdots$$

is convergent or divergent. [Divgt.]

(d) Verify that *Raabe's test* also fails to give any information about the convergence or divergence of the series

$$\left(\frac{2}{3}\right)^2 + \left(\frac{2 \cdot 4}{3 \cdot 5}\right)^2 + \left(\frac{2 \cdot 4 \cdot 6}{3 \cdot 5 \cdot 7}\right)^2 + \left(\frac{2 \cdot 4 \cdot 6 \cdot 8}{3 \cdot 5 \cdot 7 \cdot 9}\right)^2 + \cdots$$

(e) Prove that $\sum \frac{4 \cdot 7 \cdot 10 \cdots (3n+1)}{5 \cdot 8 \cdot 11 \cdots (3n+2)}$ is divergent.

B

Miscellaneous

1. Test the following series for convergence or divergence:

(a) $\sum_{n=1}^{\infty} \left\{\left(\frac{n+1}{n}\right)^{n+1} - \frac{n+1}{n}\right\}^{-n}$ [Convgt.]

(b) $\frac{1}{1 \cdot 2} + \frac{1}{2 \cdot 3} + \frac{1}{3 \cdot 4} + \cdots + \frac{1}{n(n+1)} + \cdots$ [Convgt.]

(c) $1 + \frac{x}{2} + \frac{x^2}{5} + \frac{x^3}{10} + \cdots + \frac{x^n}{n^2+1} \cdots (x > 0)$

[Convgt. if $0 < x \le 1$; Divgt. if $x > 1$]

(d) $\frac{x}{1\cdot 2} + \frac{x^2}{2\cdot 3} + \frac{x^3}{3\cdot 4} + \frac{x^4}{4\cdot 5} \cdots (x \ge 0)$

[Convgt. if $0 \le x \le 1$; Divgt. if $x > 1$]

(e) $\frac{1}{1} + \frac{1}{1\cdot 3} + \frac{1}{1\cdot 3\cdot 5} + \cdots$

(f) $\frac{1}{1+2^{-1}} + \frac{2}{1+2^{-2}} + \frac{3}{1+2^{-3}} + \cdots$ [Divgt.]

(g) $1 + \frac{2!}{2^2} + \frac{3!}{3^3} + \frac{4!}{4^4} + \cdots + \frac{n!}{n^n} + \cdots$ [Convgt.] [$u_n < 2/n^2$]

(h) $x^2 + \frac{2^2}{3\cdot 4}x^4 + \frac{2^2\cdot 4^2}{3\cdot 4\cdot 5\cdot 6}x^6 + \cdots$ (take $x > 0$) [$0 < x \le 1$ Convgt.; $x > 1$ Divgt.]

(i) $1 + \frac{2^p}{2!} + \frac{3^p}{3!} + \frac{4^p}{4!} + \cdots$ (p is a positive constant) [Convgt.]

(j) $1 + \frac{2}{1!} + \frac{2^2}{2!} + \frac{2^3}{3!} + \cdots$ [Convgt.]

(k) $\sum_{n=1}^{\infty} \frac{x^n}{n^2}\cdot\ (x > 0)$ [$0 < x \le 1$ Convgt.; $x > 1$ Divgt.] [C.P. 1980]

(l) $x + 2x^2 + 3x^3 + 4x^4 + \cdots\ (x > 0)$ [$0 < x < 1$ Convgt.; $x \ge 1$ Divgt.]

(m) $\frac{1}{2^2} + \frac{2^2}{3^3} + \frac{3^3}{4^4} + \cdots + \frac{n^n}{(n+1)^{n+1}}$ [Divgt.]

(n) $1 + \frac{2\cdot 4}{1\cdot 3}x + \frac{2\cdot 4\cdot 6}{1\cdot 3\cdot 5}x^2 + \frac{2\cdot 4\cdot 6\cdot 8}{1\cdot 3\cdot 5\cdot 7}x^3 + \cdots\ (x > 0)$

[Convgt. $0 < x < 1$; Divgt. $x \ge 1$]

(o) $\frac{1}{\sqrt{1+1^2}} + \frac{1}{\sqrt{2(1+2^2)}} + \cdots + \frac{1}{\sqrt{n(1+n^2)}} + \cdots$ [Convgt.]

(p) $1 + 2x + 3^2\cdot\frac{x^2}{2!} + 4^3\cdot\frac{x^3}{3!} + 5^4\cdot\frac{x^4}{4!} + \cdots\ (x > 0)$

[Convgt. $0 < x < \frac{1}{e}$; Divgt. if $x > \frac{1}{e}$]

(q) $\sum_{n=2}^{\infty} \frac{\log n}{n^p}\ (p > 0)$ [Convgt. $p > 1$, diverges $0 < p \le 1$].

2. If Σu_n is a convergent series of positive terms, then prove that Σu_n^2 and $\Sigma u_n u_{n+1}$ are both convergent.

[$u_n \to 0 \Rightarrow 0 < u_n < 1\ \forall n \ge m$; $\therefore\ u_n^2 < u_n$ and $u_{n+1}u_n < u_n\cdot \forall n \ge m$; hence etc.]

3. (a) If $\Sigma u_n, \Sigma v_n$ be both convergent, $u_n > 0, v_n > 0$, prove that $\Sigma\sqrt{u_n v_n}$ is convergent.

(b) If Σu_n^2 is convergent, then prove that $\dfrac{\Sigma u_n}{n}$ is also convergent (take $u_n > 0$).

$$\left[\frac{u_n}{n} \leq \frac{1}{2}\left(u_n^2 + \frac{1}{n^2}\right)\right]$$

(c) Prove that if $a_n > 0$, and if one of the series $\Sigma a_n, \sum \dfrac{a_n}{1+a_n}$ is convergent (or divergent), then so is the other.

4. Use the idea $\dfrac{1}{a+nb} > \dfrac{1}{n(a+b)}$ $(a > 0, b > 0)$ to prove that the series $\displaystyle\sum_{n=0}^{\infty} \frac{1}{a+nb}$ is divergent.

5. **(a)** Use ratio test to prove that the series

$$1 + a + ab + a^2b + a^2b^2 + \cdots + a^n b^{n-1} + a^n b^n + \cdots \ (0 < a < b)$$

is convergent when $b < 1$; is divergent when $a > 1$.
Now see that root test gives also that the series converges when $ab < 1$ and diverges when $ab > 1$. What happens when $ab = 1$.

(b) If $0 < a < b < 1$, then using the root test prove that the series

$$a + b + a^2 + b^2 + a^3 + b^3 + \cdots$$

converges. Check whether the ratio test leads to the same conclusion.

6. Use the relation $\dfrac{1}{2}\left[\dfrac{1}{r(r+1} - \dfrac{1}{(r+1)(r+2)}\right] = \dfrac{1}{r(r+1)(r+2)}$ to prove that the series

$$\sum_{n=1}^{\infty} \frac{1}{n(n+1)(n+2)}$$

is convergent with sum to infinity $\dfrac{1}{2}$. What conclusion can you derive if you apply Ratio Test or Root Test?

Appendix

Proofs of Comparison Test, Ratio Test and Root Test: Concept of Absolute Convergence

1. Comparison Test: [Statements in Art **12.16.** Test **1, 2**].

For Convergence: Test 1:

(a) Let t be the sum to infinity of $v_1 + v_2 + v_3 + \cdots$
Then, any **nth partial sum** (i.e., for all n) $v_1 + v_2 + v_3 + \cdots + v_n < t$

$$\therefore\ u_{m+1} + u_{m+2} + \cdots + u_n \leq c(v_{m+1} + v_{m+2} + \cdots + v_n) < ct$$

for all values of $n > m$.

$\therefore\ u_{m+1} + u_{m+2} + \cdots + u_n + \cdots$ is convergent.

Adding a finite number of terms $u_1 + u_2 + \cdots + u_m$ (m is fixed, given). We see that the series $u_1 + u_2 + \cdots + u_m + u_{m+1} \cdots + u_n + \cdots$ is convergent.

(b) Limit form: Since $\lim\limits_{n\to\infty} \frac{u_n}{v_n}$ = a finite quantity $l (\geq 0)$, the value of $\frac{u_n}{v_n}$ can be made as near l as we please for **all** sufficiently large n.
$\therefore$ a fixed integer m exists such that $\frac{u_n}{v_n} < l + 1 = c$, say, whenever $n > m$.
$\therefore$ it follows from (a) that Σu_n is convergent.

For Divergence: Test 2:

(a) Since Σv_n is known to be divergent, and since m is a fixed integer,

$$v_{m+1} + v_{m+2} + \cdots + v_n$$

takes values as large as we please for **all sufficiently large** values of n.
But $u_{m+1} + u_{m+2} + \cdots + u_n \geq c(v_{m+1} + v_{m+2} + \cdots + v_n)$ $[c > 0]$.
$\therefore\ u_{m+1} + u_{m+2} + \cdots + u_n$ takes as large values as we please for all sufficiently large values of n.

$$\therefore \sum_{k=m+1}^{\infty} u_k \text{ diverges.}$$

Adding a finite number of terms, $\sum\limits_{n=1}^{\infty} u_n$ also diverges.

(b) Since $\lim\limits_{n\to\infty} \frac{u_n}{v_n} = l$, $(l > 0)$, the value of u_n/v_n is as near l as we please for **all sufficiently large values of n.**
$\therefore\ \frac{u_n}{v_n} > \frac{1}{2}l$ (say) for all sufficiently large values of n (i.e., say for all $n > m$).
$\therefore$ the result follows from the previous case.
Otherwise. If possible, let Σu_n converge. Then we have

$$v_n \leq \frac{1}{c} u_n \text{ for all } n > m.$$

$\therefore$ by **Test 1** Σv_n should be converge (**a contradiction**).

2. D'Alembert's Ratio Test: [Statements in Art. **12.16.** Test **3, 4.**] **[C.P. 1982]**

Test 3: [**For Convergence**]

(a) $u_{m+1} < k u_m; u_{m+2} < k u_{m+1} < k^2 u_m; u_{m+3} < k u_{m+2} < k^3 u_m$ and so on.
$\therefore$ each term of the series.

$$u_{m+1} + u_{m+2} + u_{m+3} + \cdots$$

is less than the corresponding term of an infinite geometric series

$$k u_m + k^2 u_m + k^3 u_m + \cdots$$

whose common ratio k is less than 1 and hence convergent.

$\therefore$ by comparison test, $\sum\limits_{k=m+1}^{\infty} u_k$ is convergent and adding a finite number (m) of terms $\sum\limits_{n=1}^{\infty} u_n$ must also be convergent.

(b) Limit form. Since $\lim\limits_{n\to\infty} \frac{u_{n+1}}{u_n} = l$ $(0 < l < 1)$, the value of u_{n+1}/u_n is as near l as we please for **all sufficiently large values of n.**

$\therefore$ if k be taken in such a way that $0 < l < k < 1$, then $u_{n+1}/u_n < k$, for all sufficiently large values of n (i.e., say for all $n \geq m$).

The result now follows from (a).

Test 4. [For Divergence]

(a) Here $u_{m+1} \geq u_m; u_{m+2} \geq u_{m+1} \geq u_m; u_{m+3} \geq u_{m+2} \geq u_m$; and so on.

$\therefore$ each term of the series $u_{m+1} + u_{m+2} + u_{m+3} + \cdots$ is greater than or equal to the corresponding term of the divergent series

$$u_m(1 + 1 + 1 + 1 + \cdots).$$

$\therefore$ by Comparison Test, the $\sum\limits_{k=m+1}^{\infty} u_k$ must diverge and consequently $\sum\limits_{n=1}^{\infty} u_n$ also diverges ($\because$ m is fixed).

(b) Limit form: Since $\lim\limits_{n\to\infty} \frac{u_{n+1}}{u_n} = l > 1$, the value of $\frac{u_{n+1}}{u_n}$ is as near l as we please for **all sufficiently large values of n.**

But $l > 1$. $\therefore$ $\frac{u_{n+1}}{u_n} \geq 1$ for all values of $n >$ some fixed integer m.

The result, therefore, follows from (a).

Otherwise. See that if $l > 1$, then $\frac{u_{n+1}}{u_n} \geq 1$ for all $n > m$ which implies that u_n cannot tend to zero as $n \to \infty$.

$\therefore$ Σu_n cannot converge; being a positive series it must, therefore, **diverge.**

3. Cauchy's Root Test: [Statements in Art. **12.16** Test 5.] **[C.P. 1980]**

For Convergence:

(a) Given $u_n^{1/n} \leq k < 1$, for all $n \geq m$.

$$\therefore \quad u_n \leq k^n, \text{ for all } n \geq m$$
$$\text{or,} \quad u_{m+1} + u_{m+2} + \cdots \leq k^{m+1} + k^{m+2} + \cdots$$

The right side is a geometric series with common ratio $k < 1$ and hence convergent.

$\therefore$ by Comparison Test $\sum\limits_{r=m+1}^{\infty} u_n$ converges. Adding a finite number of terms we see that $\sum\limits_{r=1}^{\infty} u_n$ converges.

(b) **Limit form.** If $\lim_{n\to\infty} u_n^{1/n} = l < 1$, then $u_n^{1/n}$ is as near l as we please for all sufficiently large values of n. If we take k, such that $l < k < l$, then $u_n^{1/n} < k$ for all sufficiently large n (i.e., for all $n >$ some fixed integer m).
The result then follows from (a).

For Divergence:

(a) If $u_n^{1/n} \geq 1$, for infinitely many n, then $u_n \geq 1$ for infinitely many n; then u_n cannot tend to zero as $n \to \infty$ and hence Σu_n cannot converge; being a positive series it has to diverge.

(b) **Limit form.** $\lim_{n\to\infty} u_n^{1/n} = l > 1$ implies that for infinitely many n, $u_n^{1/n}$ must remain greater than 1. The result, then follows from (a).

4. **Absolute Convergence and Conditional Convergence**

(a) Suppose that a series contains positive and negative terms. Let a new series be obtained by changing the sign of every negative term. This new series is thus a series of positive terms only. If this new series is convergent, then the original series (having positive and negative terms) is said to be **absolutely convergent,** e.g., the series $1 - \frac{1}{2^2} - \frac{1}{3^2} + \frac{1}{4^2} - \frac{1}{5^2} - \frac{1}{6^2} + \frac{1}{7^2} + \cdots$ is *absolutely convergent* because the series (taken all terms positively).

$$1 + \frac{1}{2^2} + \frac{1}{3^2} + \frac{1}{4^2} + \frac{1}{5^2} + \cdots \text{ is convergent.}$$

Our **definition** does not state that an absolutely convergent series Σu_n is itself convergent—the definition only states that Σu_n is absolutely convergent only when a **different series** $\Sigma|u_n|$ is convergent. But fortunately it can be proved that

A series which is absolutely convergent is also convergent.

We shall argue like this: If we can prove that the series

$$1 - \frac{1}{2^2} + \frac{1}{3^2} - \frac{1}{4^2} + \cdots$$

is *absolutely convergent,* then the series itself must be convergent (no other proof necessary).

(b) What is important to note is that it is possible that a series containing positive and negative terms is **convergent.**

$$\left(\text{e.g., } 1 - \frac{1}{2} + \frac{1}{3} - \frac{1}{4} + \cdots \text{ is convergent, proved earlier}\right)$$

but the series, taken all terms positively (here, $1 + \frac{1}{2} + \frac{1}{3} + \cdots$) is divergent. In this case, the original series is said to be **semi-convergent** or **conditionally convergent.** Thus the series $1 - \frac{1}{2} + \frac{1}{3} - \frac{1}{4} + \cdots$ is a conditionally convergent series.

(c) Given a series Σu_n containing positive and negative terms we construct a new series $\Sigma|u_n|$ and then use tests applicable to positive series. If we find that the series $\Sigma|u_n|$ is convergent, then we may just state that "Σu_n is absolutely convergent and *hence convergent*". But if $\Sigma|u_n|$ is divergent, then we may not sure about the convergence or divergence of Σu_n. We are then to study the series Σu_n itself by some other means.

e.g., $\sum_{n=0}^{\infty} \frac{x^n}{n!}$ *is convergent for all values of* x (*positive or negative*);

take $u_n = \frac{x^n}{n!}$.

We use ratio test on the series $\sum_{n=0}^{\infty} \left|\frac{x^n}{n!}\right|$.

Then

$$\left|\frac{u_{n+1}}{u_n}\right| = \left|\frac{x^{n+1}}{(n+1)!} \times \frac{n!}{x^n}\right| = \frac{1}{n+1}|x| \to 0\ (< 1).$$

The series $\Sigma|u_n|$ converges for all values of x_n.

$\therefore$ the conclusion is $\sum \frac{x^n}{n!}$ is absolutely convergent and hence convergent for all x.

(d) Alternating series: A series with alternately positive and negative terms is called an **alternating series,** e.g.,

$$u_1 - u_2 + u_3 - u_4 + \cdots \quad \text{(all } u_i\text{'s are positive)} \tag{1}$$

is an alternating series. For such series we have a test:

Leibnitz's Test of Convergence for an alternating series:

The alternating series (1) *converges if* (a) *the sequence* $\{u_n\}$ *is monotone decreasing* (m.d.) and (b) $u_n \to 0$ *as* $n \to \infty$.

See that the alternating series $1 - \frac{1}{2} + \frac{1}{3} - \frac{1}{4} + \cdots$ satisfies both the conditions (a) and (b) and hence by Leibntiz's Test, it converges.

Q. *What can you say about the series* $1 - \frac{1}{\sqrt{2}} + \frac{1}{\sqrt{3}} - \frac{1}{\sqrt{4}} + \frac{1}{\sqrt{5}} - \cdots$? *Is it absolutely convergent? Conditionally convergent?*

Chapter 13

Linear Difference Equations

13.1 Introduction

Since the time of Newton, finite differences have had a strong appeal to the Mathematicians. In this short note we shall describe a few techniques of solving some special type of difference equations—**linear difference equations with constant coefficients of order not more than two.**

13.2 Difference Operators

Let $u_1, u_2, u_3, \ldots, u_n, \ldots$ be a sequence of numbers.

(i) By the symbol Δu_n we mean the difference $u_{n+1} - u_n$ (e.g., $\Delta u_1 = u_2 - u_1$, $\Delta u_2 = u_3 - u_2$, etc.)

The symbol Δ is called **"Difference operator"**. The operator when applied to u_n produces $u_{n+1} - u_n$.

(ii) What happens if $\Delta u_1 = \Delta u_2 = \Delta u_3 = \cdots = \Delta u_n = 0$?

Clearly, then $u_1 = u_2 = \cdots = u_n$.

(iii) We define $\Delta^\circ u_n = u_n, \Delta^1 u_n = \Delta u_n = u_{n+1} - u_n$.

$$\begin{aligned} \text{Then } \Delta^2 u_n &= \Delta(\Delta u_n) = \Delta(u_{n+1} - u_n) \\ &= (u_{n+2} - u_{n+1}) - (u_{n+1} - u_n) \\ &= u_{n+2} - 2u_{n+1} + u_n \\ \Delta^3 u_n &= \Delta(\Delta^2 u_{n+2}) = \Delta(u_{n+2} - 2u_{n+1} + u_n) \\ &= [u_{n+3} - 2u_{n+2} + u_{n+1}] - [u_{n+2} - 2u_{n+1} + u_n] \\ &= u_{n+3} - 3u_{n+2} + 3u_{n+1} - u_n = 0. \end{aligned}$$

By induction we can prove that for a positive integer r,

$$\Delta^r u_n = u_{n+r} - {}^r c_1 u_{n+r-1} + {}^r c_2 u_{n+r-2} + \cdots + (-1)^r u_n.$$

[Assume the result for $n = k$, then deduce

$$\begin{aligned} \Delta^{k+1} u_n &= \Delta(\Delta^k u_n) = \Delta(u_{n+k} - {}^k c_1 u_{n+k-1} + \cdots + (-1)^k u_n) \\ &= u_{n+k+1} - {}^{k+1} c_1 u_{n+k} + \cdots + (-1)^{k+1} u_n. \end{aligned}$$

The result is found to be true for $r = 2$ or 3 and hence for $4, 5$, etc., i.e., the result is true for all positive integers r.]

(iv) If r and s are any **non-negative** integers, then

$$\Delta^{r+s}u_n = \Delta^r(\Delta^s u_n) = \Delta^s(\Delta^r u_n).$$

[See that: $\Delta^{r+s}u_n = \Delta(\Delta^{r+s-1}u_n) = \Delta\{\Delta(\Delta^{r+s-2}u_n)\} = \Delta^2(\Delta^{r+s-2}u_n)$ provided $r + s - 2 \geq 0$. Proceed Step by Step (i.e., by induction)].

(v) If $u_1, u_2, \ldots, u_n, \ldots$ and $v_1, v_2, \ldots, v_n, \ldots$ be two sequences, then

$$\Delta(u_n + v_n) = \Delta u_n + \Delta v_n.$$

[See that, left side $= (u_{n+1} + v_{n+1}) - (u_n + v_n)$
$= (u_{n+1} - u_n) + (v_{n+1} - v_n)$]

A point to note: Let $s_n = u_1 + u_2 + \cdots + u_n$

$$\Delta s_n = (u_1 + u_2 + \cdots + u_{n+1}) - (u_1 + u_2 + \cdots + u_n) = u_{n+1}$$

but $\Delta u_1 + \Delta u_2 + \cdots + \Delta u_n = (u_2 - u_1) + (u_3 - u_2) + \cdots + (u_{n+1} - u_n) = u_{n+1} - u_1$.

$\therefore\ \Delta s_n = \Delta u_1 + \Delta u_2 + \cdots + \Delta u_n$, only when $u_1 = 0$.

(vi) For a sequence, say $\{1^2, 2^2, 3^2, \ldots, n^2\}$, we usually construct a *Table of Differences* in the following manner:

u_n	Δu_n	$\Delta^2 u_n$	$\Delta^3 u_n$	$\Delta^4 u_n$	$\cdots$
1^2					
	3				
2^2		2			
	5		0		
3^2		2		0	
	7		0		
4^2		2		0	
	9		0		
5^2		2			
	11	$\vdots$			
6^2	$\vdots$	$\vdots$			
$\vdots$	$\vdots$				

Important to Remember: If u_x be a function of a variable x (*where x need not necessarily be a positive integer*), the meaning of Δu_x is

$$\Delta u_x = u_{x+1} - u_x,$$

(i.e., we make the convention that the independent variable x changes by finite jumps, each interval being equal and **equals to 1.**)

All that has been said about the symbol Δ still holds.

Illustration

▶ Example 13.2.1 $u_n = n^2$, $\Delta u_n = (n+1)^2 - n^2 = 2n+1$.
$\Delta^2 u_n = 2(n+1)+1-(2n+1) = 2$.
$\Delta^3 u_n = 0$, *i.e.*, $\Delta^r u_n = 0$, *for* $r \geq 3$.

▶ Example 13.2.2 $u_n = n^3$; *See that* $\Delta u_n = 3n^2 + 3n + 1, \Delta^2 u_n = 6n + 6$,

$$\Delta^3 u_n = 6, \Delta^4 u_n = 0, \textit{ i.e., } \Delta^r u_n = 0, \textit{ for } r \geq 4.$$

▶ Example 13.2.3 $u_x = 2^x; \Delta u_x = 2^{x+1} - 2^x = 2^x(2^1 - 1) = 2^x$.

$$\Delta^2 u_x = \Delta(\Delta u_x) = 2^{x+1} - 2^x = 2^x, \textit{ etc., i.e., } \Delta^r u_x = 2^x, \textit{ for all } r.$$

▶ Example 13.2.4

$$\begin{aligned}
u_n &= n(n-1)(n-2); \\
\Delta u_n &= (n+1)n(n-1) - n(n-1)(n-2) = 3n(n-1); \\
\Delta^2 u_n &= \Delta(\Delta u_n) = 3(n+1)n - 3n(n-1) = 6n; \\
\Delta^3 u_n &= 6(n+1) - 6n = 6. \\
\therefore \Delta^r u_n &= 0 \ (r \geq 4).
\end{aligned}$$

The reader should now work Exercises on Chapter 13, Nos. 1–4.

13.3 The Shift Operator $E \equiv 1 + \Delta$

(i) We define $(1+\Delta)^r u_n = (1+\Delta)\{(1+\Delta)^{r-1} u_n\}$, if $r > 1$.
By a simple inductive argument we can establish that

$$(1+\Delta)^r u_n = \left(1 + {}^rC_1\Delta + {}^rC_2\Delta^2 + \cdots + \Delta^r\right) u_n. \quad (1)$$

Now observe that

$$u_{n+1} = u_n + (u_{n+1} - u_n) = u_n + \Delta u_n = (1+\Delta)u_n.$$

We denote the operator $1+\Delta$ by E. Thus

$$Eu_n = u_{n+1} \ (E \text{ shifts } u_n \text{ to } u_{n+1})$$

Again,

$$\begin{aligned}
u_{n+1} &= (1+\Delta)u_n = (1+\Delta)^2 u_{n-1} \\
&= (1+\Delta)^3 u_{n-2} = \cdots = (1+\Delta)^n u_1. \\
\therefore u_{n+1} &= E^n u_1 = (1+\Delta)^n u_1 \\
&= (1 + {}^nC_1\Delta + {}^nC_2\Delta^2 + \cdots + \Delta^n)u_1. \quad (2)
\end{aligned}$$

This last result is known as *Newton's Difference Formula.*
Note, further that

$$E^r u_n = u_{n+r} \text{ (where } r \text{ is a positive integer).} \quad (3)$$

Even if u_x be a function of *any variable* x (not necessarily a positive integral variable), the operation E is defined by $Eu_x = u_{x+1}$.

Now observe that we may interpret the symbol $\Delta \equiv E - 1$. Then

$$\begin{aligned}\Delta^2 &= (E-1)^2, \Delta^3 = (E-1)^3, \ldots, \Delta^r = (E-1)^r, \qquad (4)\\ \Delta^2 u_x &= \Delta(\Delta u_x) = \Delta(u_{x+1} - u_x) = (u_{x+2} - u_{x+1}) - (u_{x+1} - u_x)\\ &= u_{x+2} - 2u_{x+1} + u_x = E^2 u_x - 2Eu_x + u_x \text{ [See (3)]}\\ &= (E^2 - 2E + 1)u_x = (E-1)^2 u_x; \text{ [others by induction]}\end{aligned}$$

13.4 The Inverse Operator Δ^{-1}

(i) For a sequence $\{u_n\}$ we can find another sequence $\{v_n\}$ such that $\Delta v_n = u_n$, for every n. We then write

$$v_n = \Delta^{-1} u_n \text{ (\textit{Anti-difference operator})}$$

To see this, it is enough to take

$$v_n = u_1 + u_2 + \cdots + u_{n-1}, (n \geq 2) \text{ and } v_1 = 0.$$

Then $\Delta v_1 = v_2 - v_1 = u_1 - 0 = u_1$ and for $n \geq 2$,

$$\Delta v_n = (u_1 + u_2 + \cdots + u_n) - (u_1 + u_2 + \cdots + u_{n-1}) = u_n.$$

(ii) In this connection the following result is useful to remember:

If $\{v_n\}$ and $\{w_n\}$ be two sequences such that $\Delta v_n = \Delta w_n$, for every n, then $v_n - w_n$ is independent of n.

[See that $\Delta(v_n - w_n) = 0$; hence, etc.]

Thus if $\{v_n\}$ be such that $\Delta v_n = \Delta u_n$, then $v_n = u_n + C$, where C is an arbitrary constant.

(iii) We introduce a symbol (called **Factorial**):

$$\begin{aligned}x^{(0)} &= 1; x^{(1)} = x; x^{(2)} = x(x-1);\\ x^{(3)} &= x(x-1)(x-2); \text{ etc.}\\ x^{(r)} &= x(x-1)(x-2)\cdots(x-\overline{r-1}).\end{aligned}$$

We can then verify

$$\begin{aligned}\Delta x^{(r)} &= x^{(r-1)} \cdot r\\ \Delta^2 x^{(r)} &= \Delta(\Delta x^{(r)}) = r(r-1)x^{(r-2)}, \text{ and so on.}\end{aligned}$$

See the analogy: $\frac{d}{dx}(x^n) = nx^{n-1}$ and $\Delta x^{(n)} = nx^{(n-1)}$.

We may also see that $\Delta\left(\frac{x^{(x)}}{n}\right) = x^{(n-1)}$ so that

$$\Delta^{-1} x^{(n-1)} = \frac{x^{(n)}}{n} + C.$$

This notation helps us to find $\Delta^{-1}x^2, \Delta^{-1}x^3$, etc.

We have

$$\begin{aligned} x^2 &= x(x-1)+x = x^{(2)}+x^{(1)} \\ \therefore\ \Delta^{-1}x^2 &= \Delta^{-1}x^{(2)}+\Delta^{-1}x^{(1)}+C = \frac{x^{(3)}}{3}+\frac{x^{(2)}}{2}+C \\ &= \frac{1}{3}x(x-1)(x-2)+\frac{1}{2}x(x-1)+C. \end{aligned}$$

Similarly,

$$\begin{aligned} x^3 &= x(x-1)(x-2)+3x(x-1)+x = x^{(3)}+3x^{(2)}+x^{(1)} \\ \therefore\ \Delta^{-1}x^3 &= \frac{x^{(4)}}{4}+3\frac{x^{(3)}}{3}+\frac{x^{(2)}}{2}+C. \end{aligned}$$

(iv) **An application in finding the sum to n terms**

To find $S_n = u_1+u_2+\cdots+u_n$, we construct a sequence $\{v_n\}$ such that $\Delta v_n = u_n$, for every n. Then

$$\begin{aligned} \Delta S_n &= (u_1+u_2+\cdots+u_{n+1})-(u_1+u_2+\cdots+u_n) = u_{n+1} = E^n u_1 \\ &= (1+\Delta)^n u_1 = (1+\Delta)^n \Delta v_1 = \Delta\{(1+\Delta)^n v_1\} \\ &= \Delta w_n \text{ (say)}. \end{aligned}$$

This implies $S_n = w_n + C$ where C is a constant independent of n.

[In fact, we may take $C = S_1 - w_1 = u_1-(1+\Delta)v_1$
$= u_1 - v_1 - \Delta v_1 = u_1 - v_1 - u_1 = -v_1$]

$\therefore$ for any n, $S_n - (1+\Delta)^n v_1 = -v_1$

$$\begin{aligned} \text{or, } S_n &= \{(1+\Delta)^n - 1\}v_1 = \{(1+\Delta)^n-1\}\Delta^{-1}u_1. \\ &= \{{}^nC_1\Delta + {}^nC_2\Delta^2 + {}^nC_3\Delta^3+\cdots+\Delta^n\}\Delta^{-1}u_1 \\ &= {}^nC_1 u_1 + {}^nC_2\Delta u_1 + {}^nC_3\Delta^2 u_1+\cdots+\Delta^{n-1}u_1. \end{aligned}$$

e.g., $1^2+2^2+3^2+\cdots+n^2 = S_n$

$$\begin{aligned} &= n\cdot 1 + \frac{n(n-1)}{2!}\cdot 3 + \frac{n(n-1)(n-2)}{3!}\cdot 2 \\ &= \frac{1}{6}n(n+1)(2n+1). \end{aligned}$$

[Using $u_1 = 1, \Delta u_1 = 3, \Delta^2 u_1 = 2, \Delta^3 u_1 = 0$, etc. **From Table of differences shown in Art. 13.2.**]

A formula to remember

$$\begin{aligned} S_n &= \{(1+\Delta)^n-1\}\Delta^{-1}u_1 \\ &= {}^nC_1u_1 + {}^nC_2\Delta u_1 + {}^nC_3\Delta^2 u_1 + \cdots + \Delta^{n-1}u_1. \end{aligned}$$

The reader should now work Exercises on Chapter 13 Nos. 5, 6, 7.

13.5 Finite Difference Equation

Definition 1. *A* ***difference equation*** *is an equation involving an independent variable* x, *the dependent variable* u_x *and the successive differences* $\Delta u_x, \Delta^2 u_x, \ldots$

Since the differences $\Delta u_x, \Delta^2 u_x$, etc. can be expressed in terms of the successive values of the dependent variable u_x, a difference equation is usually written involving the independent variable, dependent variable and its successive values.

We shall use 'n' in place of 'x' when the independent variable assumes only positive integral values.

The **order** of a difference equation in

$$u_x = \text{highest suffix of } u - \text{lowest suffix of } u.$$

e.g., $u_{x+2} - 2u_{x+1} - 3u_x = 0$ is a difference equation of **order 2.**

See that we can write this equation as

$$(E^2 - 2E - 3)u_x = 0 \text{ (using results like } u_{x+2} = E^2 u_x, u_{x+1} = Eu_x).$$

The highest power of the shift operator E also gives the *order of the equation.* Using $E = 1 + \Delta$ we see that the same equation could be written as

$$\Delta^2 u_x - 4u_x = 0.$$

Another example: $u_{x+2} - 3u_{x+1} + 5u_x = x^3$ is a difference equation of order 2. It could be written as

$$(E^2 - 3E + 5)u_x = x^3 \text{ or as, } \Delta^2 u_x - \Delta u_x + 3u_x = x^3.$$

Again, we can easily check that $u_x = 2^x + \left(-\frac{1}{2}\right)^x$ satisfies the equation

$$2u_{x+2} - 3u_{x+1} - 2u_x = 0.$$

We say, in this case, that $u_x = 2^x + \left(\frac{-1}{2}\right)^x$ is a solution of the given equation.

A solution containing arbitrary constants equal in number to the order of the equation is called its *general solution* or *complete primitive.* A *particular solution* is obtained by giving these constants special values.

[Compare Exercises on Chapter 13 Nos. 8–13]

13.6 Linear Difference Equation

An equation of the form

$$u_{x+n} + p_1 u_{x+n-1} + p_2 u_{x+n-2} + \cdots + p_n u_x = R(x),$$

where $p_1, p_2, \ldots, p_n$ and $R(x)$ are either functions of x or constants, is called a **linear** difference equation of order n. If $R(x) = 0$, the equation is called a *linear homogeneous equation of order* n; otherwise it is called a *linear non-homogeneous equation of order* n.

We shall discuss only those equations where the coefficients $p_1, p_2, \ldots, p_n$ are all constants. Moreover we shall limit ourselves to equations of *orders not more than two.*

A. Linear equations of first order (with constant coefficients)

▶ Example 13.6.1 **To solve:** $u_{x+1} - au_x = 0$ $(a \neq 1)$ (**or,** $(E - a)u_x = 0$)

Method of Solution

Multiplying by a^{-x-1} we get

$$a^{-(x+1)}u_{x+1} - a^{-x}u_x = 0 \Rightarrow \Delta(a^{-x}u_x) = 0$$

$$\therefore \quad a^{-x}u_x = C \text{ or, } u_x = Ca^x,$$

where C is either a constant or a periodic function of period 1.

[See that if C be such that $C = f(x)$ where $f(x+1) = f(x)$, then also $u_x = Ca^x$ satisfies $u_{x+1}/u_x = a$, i.e., satisfies the given equation.]

▶ Example 13.6.2 **To solve:** $u_{x+1} - u_x = 0$.

Here, we have $\Delta(u_x) = 0$, which implies $u_x - C$, where C is an arbitrary constant or a periodic function of period 1.

▶ Example 13.6.3 **To solve:** $u_{x+1} - u_x = Ca^x$ (*C is a constant or a periodic function of period* 1 *and* $a \neq 1$.)

We first observe $\Delta\left(\dfrac{Ca^x}{a-1}\right) = \dfrac{Ca^{x+1}}{a-1} - \dfrac{Ca^x}{a-1} = Ca^x (a \neq 1)$.

More generally, $\Delta\left(\dfrac{Ca^x}{a-1} + C'\right) = Ca^x$.

(C' is any arbitrary constant or a periodic function of period 1.)

$$\therefore \quad \frac{Ca^x}{a-1} + C' = \Delta^{-1}(Ca^x).$$

Now see that the given equation can be written as

$$\Delta(u_x) = Ca^x$$

$$\therefore \quad u_x = \Delta^{-1}(Ca^x) = \frac{Ca^x}{a-1} + C'.$$

▶ Example 13.6.4 **To solve:** $u_{x+1} - bu_x = Ca^x$, *where C is a constant or a periodic function of period* 1.

Two cases are to be considered:

Case 1. *When* $b \neq a$. *Multiply by* b^{-x-1}. *Then the equation becomes*

$$\begin{aligned} b^{-x-1}u_{x+1} - b^{-x}u_x &= Cb^{-1}(a^x \cdot b^{-x}) = Cb^{-1}(ab^{-1})^x, \\ \text{i.e., } \Delta(b^{-x}u_x) &= Cb^{-1}(ab^{-1})^x \\ \therefore b^{-x}u_x &= \frac{Cb^{-1}(ab^{-1})^x}{ab^{-1}-1} + C' \text{ (cf. Ex. 2).} \end{aligned}$$

(C' is any constant or a periodic function of period 1)

$$\therefore u_x = \frac{C\cdot b^x \cdot b^{-1}\cdot a^x \cdot b^{-x}}{ab^{-1}-1} + C'b^x = \frac{Ca^x}{a-b} + C'b^x.$$

Observations. The given equation can be written in terms of the shift operator E.

$$(E-b)u_x = Ca^x. \; (\because \; Eu_x = u_{x+1})$$

Note that the homogeneous equation $(E-b)u_x = 0$ has the *general solution* $C'b^x$ and check that $\frac{Ca^x}{a-b}$ is a *particular solution* of the given equation. Thus the general solution contains two parts:

$C'b^x$ (known as *Complementary Function*, C.F.)

$+ \quad \dfrac{Ca^x}{a-b}$ (known as *Particular Integral*, P. I.),

i.e., General solution = C.F. + P. I.

A rule to remember

$$\text{P. I.} = \frac{1}{E-b}CA^x = \frac{Ca^x}{a-b} \quad \textbf{(Replace E by a provided } a \neq b\textbf{)}.$$

Case 2. *When* $b = a$. *Multiply both sides by* a^{-x-1}. *The equation becomes*

$$a^{-x-1}(u_{x+1} - au_x) = Ca^x \cdot a^{-x-1}$$

or, $a^{-x-1}u_{x+1} - a^{-x}u_x = Ca^{-1}$

or, $\Delta(a^{-x}u_x) = Ca^{-1}\cdot 1$

$\therefore \quad a^{-x}u_x = Ca^{-1}\Delta^{-1}(1) = Ca^{-1}(x + C')$

or, $u_x = a^x(A + xB) \; [A = CC'a^{-1}, \; B = Ca^{-1}]$

Note 1. *In this case, the P. I.* $= B\cdot x\cdot a^x$, *C.F.* $= A\cdot a^x$.

In **Example 13.6.5**, below we give a systematic method of solving a linear difference equation of first order. The students are advised to follow similar steps in every problem:

▶ Example 13.6.5 **To solve:** $u_{x+1} - au_x = bx$.

Solution: We may write the equation as $(E-a)u_x = bx(E \equiv 1 + \Delta)$. The general solution consists of C.F. + P. I.

First step. To find the C.F. we find the general solution of the reduced equation:

$$u_{x+1} - au_x = 0.$$

Multiplying by a^{-x-1} we get

$$a^{-x-1}u_{x+1} - a^{-x}u_x = 0 \Rightarrow \Delta(a^{-x}u_x) = 0$$

$\therefore$ the reqd. C.F.$=Aa^x$ (A is an arbitrary constant or a periodic function of period 1.)

Second step. To find the P. I. we shall use the following operational method:

$$\begin{aligned}
\text{P. I.} &= \frac{1}{E-a}bx = \frac{1}{(1-a)+\Delta}bx \\
&= \frac{1}{1-a}\left\{1+\frac{\Delta}{1-a}\right\}^{-1} bx \\
&= \frac{1}{1-a}\left\{1-\frac{\Delta}{1-a}+\frac{\Delta^2}{(1-a)^2}+\cdots\right\}bx \\
&= \frac{1}{1-a}(bx)-\frac{1}{(1-a)^2}\Delta(bx)+\frac{1}{(1-a)^2}\Delta^2(bx)+\cdots \\
&= \frac{b}{1-a}x-\frac{b(x+1-x)}{(1-a)^2}+0+0+\cdots \\
&= \frac{bx}{1-a}-\frac{b}{(1-a)^2}.
\end{aligned}$$

$\therefore$ the general solution of the given equation is given by

$$u_x = A\cdot a^x + \frac{bx}{1-a} - \frac{b}{(1-a)^2} = A\cdot a^x - \frac{bx}{a-1} - \frac{b}{(a-1)^2}.$$

Note 2. $\frac{1}{E-1}x^2 = \frac{1}{\Delta}x^2 = \Delta^{-1}(x^2) = \frac{1}{3}x^{(3)} + \frac{1}{2}x^{(2)} + C$
$= \frac{1}{3}x(x-1)(x-2) + \frac{1}{2}x(x-1) + C.$

B. Linear difference equations of second order (with constant coefficients)

I. Homogeneous equations:

▶ Example 13.6.6 **To solve:** $u_{x+2} + au_{x+1} + bu_x = 0$, *i.e., to solve:*

$$(E^2 + aE + b)u_x = 0 \ (E = 1 + \Delta).$$

Solution: Let α, β be the roots of a quadratic equation

$$t^2 + at + b = 0$$

[This is called **Auxiliary Equation:** Such an equation can be obtained by replacing E by t.]

Then $\alpha + \beta = -a, \alpha\beta = b$.
Accordingly we may write the given equation as

$$\begin{aligned}
&(E^2 - (\alpha+\beta)E + \alpha\beta)u_x = 0 \\
\text{or,}\quad &(E-\alpha)(E-\beta)u_x = 0.
\end{aligned}$$

Case 1. *Suppose $\alpha \neq \beta$ (may be real or complex).*

Put $(E-\beta)u_x = v_x$. Then the equation reduces to

[See Ex. 13.6.1.] $(E-\alpha)v_x = 0$

$\therefore\ v_x = A\alpha^x$ [A is an arbitrary constant or a periodic function of period 1.]
Now, therefore, $(E-\beta)u_x = A\alpha^x$.

$$\text{[See Ex. 13.6.4]} \therefore\ u_x = B\cdot\beta^x + \frac{A\alpha^x}{\alpha-\beta} = C_1\alpha^x + C_2\beta^x, \text{ say.}$$

(C_1, C_2 are arbitrary constants or, periodic functions of period 1.)

Case 2. *Suppose $\alpha = \beta$. Then the equation is*

$$(E-\alpha)^2 u_x = 0.$$

As before, put $(E-\alpha)u_x = v_x$. Then we have

$$(E-\alpha)v_x = 0 \Rightarrow v_x = A\alpha^x.$$

Again, $(E-\alpha)u_x = v_x = A\alpha^x \Rightarrow u_{x+1} - \alpha u_x = A\alpha^x$

$$\begin{aligned} &\Rightarrow \alpha^{-x-1}u_{x+1} - \alpha^{-x}u_x = A\alpha^{-1} \\ &\Rightarrow \Delta(\alpha^{-x}u_x) = A\alpha^{-1} \\ &\therefore\ \alpha^{-x}u_x = A\alpha^{-1}x + B \\ &\therefore\ u_x = (C_1x + C_2)\alpha^x\ (A\alpha^{-1} = C_1, B = C_2) \end{aligned}$$

In particular, if $\alpha = p + iq, \beta = p - iq$ (p, q real), then

$$u_x = C_1(p+iq)^x + C_2(p-iq)^x = r^x(C_1\cos x\theta + C_2\sin x\theta)$$

where $p = r\cos\theta, q = r\sin\theta; r = +\sqrt{p^2+q^2}$ and θ is such that $\cos\theta = p/r$ and $\sin\theta = q/r$ (the proper signs of p and q need be considered).
[See that $C_1(p+iq)^x + C_2(p-iq)^x$

$$\begin{aligned} &= r^x[C_1(\cos x\theta + i\sin x\theta) + C_2(\cos x\theta - i\sin x\theta)], \\ &\qquad\qquad \text{using De Moivre's Theorem} \\ &= r^x[(C_1+C_2)\cos x\theta + (C_1 i + C_2 i)\sin x\theta] \\ &= r^x[C_1\cos x\theta + C_2\sin x\theta]. \end{aligned}$$

e.g., (i) $u_{x+2} - u_{x+1} - 6u_x = 0 \Rightarrow (E^2 - E - 6)u_x = 0$
$\Rightarrow (E-3)(E+2)u_x = 0$
$\Rightarrow u_x = C_1(-2)^x + C_2\cdot 3^x.$

(ii) $16u_{x+2} - 8u_{x+1} + u_x = 0 \Rightarrow (16E^2 - 8E + 1)u_x = 0$
$\Rightarrow 16\left(E-\frac{1}{4}\right)\left(E-\frac{1}{4}\right)u_x = 0$
$\Rightarrow u_x = (C_1 + C_2x)\left(\frac{1}{4}\right)^x.$

(iii) $u_{x+2} - 4u_{x+1} + 13u_x = 0 \Rightarrow (E^2 - 4E + 13)u_x = 0$
$\Rightarrow (E-\alpha)(E-\beta)u_x = 0,$
(where $\alpha = 2+3i, \beta = 2-3i$)
$\Rightarrow u_x = C_1(2+3i)^x + C_2(2-3i)^x$
$= r^x[C_1\cos x\theta + C_2\sin x\theta]$

where $r = \sqrt{2^2+3^2} = \sqrt{13}$ and θ is such that

$$\cos\theta = \frac{2}{\sqrt{13}} \text{ and } \sin\theta = \frac{3}{\sqrt{13}}.$$

II. Non-homogeneous Equations

$$u_{x+2} + pu_{x+1} + qu_x = R(x) \text{ or, } (E^2 + pE + q)u_x = R(x).$$

With regard to the non-homogeneous equations of second order (with constant coefficients) we have to make the following remarks:

1. General Solution = Complementary Function (C.F.) + Particular Integral (P. I.)

   C.F. ≡ General Solution of the reduced equation

   $$u_{x+2} + pu_{x+1} + qu_x = 0 \text{ (making right side=0).}$$

   P. I. = some particular solution of $u_{x+2} + pu_{x+1} + qu_x = R(x)$.

2. We shall ordinarily use the *operational method* for finding P. I.

3. If $R(x)$ is of the form x^m, then for a particular solution one may take a trial solution

   $$u_x = Ax^2 + Bx + C$$

   and then determine A, B, C from the given equation.

4. **Operational Methods of finding the P. I.**

   We shall use the functional symbol

   $$L(E) = E^2 + pE + q.$$

   The operator $L(E)u_x = E^2u_x + pEu_x + qu_x = u_{x+2} + pu_{x+1} + qu_x$.

In inverse operator $\frac{1}{L(E)}$ applied on $R(x)$, written as $\frac{1}{L(E)}R(x)$, requires to find a function $Q(x)$ such that $L(E)Q(x) = R(x)$.

For solving problems our readers may assume the following results:

(i) $\frac{1}{L(E)}Ka^x = \frac{1}{L(a)}Ka^x$, if $L(a) \neq 0$.

(ii) $\frac{1}{L(E)} \cdot a^xV(x) = a^x \cdot \frac{1}{L(aE)}V(x)$.

(iii) $\frac{1}{L(E)} \cdot x^m = \frac{1}{K}\{1 + f(\Delta)\}^{-1}x^m$, using $E \equiv 1 + \Delta$.

Illustrative Examples

► Example 13.6.7 **To solve:** $u_{x+2} - 16u_x = 9x^2$.

Solution: We write the given equation as $(E^2 - 16)u_x = 9x^2$.

The general solution is C.F. + P. I.

To find the C.F. we consider the reduced equation

$$(E^2 - 16)u_x = 0.$$

$$\text{That is, } (E+4)(E-4)u_x = 0.$$

$\therefore$ C.F.$=C_1 4^x + C_2(-4)^x$ {C_1, C_2 are constants or periodic functions of period 1.}

To find the P. I.

Method 1	Method 2
Assume $u_x = Ax^2 + Bx + C$ Then $u_{x+2} = A(x+2)^2 + B(x+2) + C$ From the given equation $\{A(x+2)^2 + B(x+2) + C\}$ $- 16\{Ax^2 + Bx + C\} = 9x^2$ Comparing the coefficients $-15A = 9,\ 4A - 15B = 0,$ $4A + 2B - 15C = 0$ which gives $A = -\frac{3}{5},\ B = -\frac{4}{25},\ C = -\frac{68}{375}$ $\therefore$ P. I. $= -\frac{3}{5}x^2 - \frac{4}{25}x - \frac{68}{375}$.	P. I. $= \frac{1}{E^2 - 16} 9x^2$ $= \frac{1}{(1+\Delta)^2 - 16} 9x^2$ $= -\frac{3}{5}\left[1 - \frac{2\Delta + \Delta^2}{15}\right]^{-1} x^2$ $= -\frac{3}{5}x^2 - \frac{2}{25}\Delta(x^2) - \frac{57}{1125}\Delta^2(x^2)$ $= -\frac{3}{5}\left[1 + \frac{2\Delta + \Delta^2}{15} + \frac{4\Delta^2}{225} + \cdots\right] x^2$ $= -\frac{3}{5}x^2 - \frac{2}{25}(2x+1) - \frac{57}{1125} \times 2$ $= -\frac{3}{5}x^2 - \frac{4}{25}x - \frac{68}{375}$

Hence, the general solution is

$$u_x = C_1 \cdot 4^x + C_2 \cdot (-4)^x - \frac{3}{5}x^2 - \frac{4}{25}x - \frac{68}{375}.$$

Note 3. *See that*

$$\Delta(x^2) = (x+1)^2 - x^2 = 2x + 1$$
$$\Delta^2(x^2) = \Delta(\Delta(x^2)) = \Delta(2x+1) = 2(x+1) + 1 - (2x+1) = 2.$$

▶ Example 13.6.8 **To solve:** $u_{x+2} - 7u_{x+1} + 10u_x = 12 \cdot 4^x$.

We write this equation as

$$(E^2 - 7E + 10)u_x = 12 \cdot 4^x$$

The general solution=C.F. + P. I.
To find the C.F. we find the general solution of the reduced equation

$$(E-5)(E-2)u_x = 0.$$

$\therefore$ C.F. $= A \cdot 2^x + B \cdot 5^x$ (A, B are arbitrary constants or periodic functions of period 1)

To find the P. I.

$$\text{P. I.} = \frac{1}{(E-5)(E-2)} 12 \cdot 4x = 12 \frac{1}{(4-5)(4-2)} 4^x = -6 \cdot 4^x$$

$$\left[\because \frac{1}{L(E)} a^x = \frac{1}{L(a)} \cdot a^x \text{ if } L(a) \neq 0)\right]$$

$\therefore$ the general solution $u_x = A \cdot 2^x + B \cdot 5^x - 6 \cdot 4^x$.
[In this case, also the P. I. can be obtained by assuming $u_x = A \cdot 4^x$ and substituting in the given equation, $A = -6$.]

▶ Example 13.6.9 **To solve:** $u_{x+2} - 7u_{x+1} + 12u_x = \cos x$.

C.F.$=C_1 \cdot 3^x + C_2 \cdot 4^x$ can be easily found out.
To find the P. I. one should proceed as:

$$\begin{aligned}
\text{P. I.} &= \frac{1}{E^2 - 7E + 12}\cos x \\
&= \text{real part of } \frac{1}{E^2 - 7E + 12}(e^i)^x \\
&= \text{real part of } \frac{1}{(e^i)^2 - 7e^i + 12}e^{ix} \left[\text{cf. } \frac{1}{L(E)}a^x = \frac{a^x}{L(a)}\right] \\
&= \text{real part of } \frac{1}{(\cos 2 - 7\cos 1 + 12) + i(\sin 2 - 7\sin 1)}(\cos x + i\sin x) \\
&= \text{the real part of } \frac{\{(\cos 2 - 7\cos 1 + 12) - i(\sin 2 - 7\sin 1)\}(\cos x + i\sin x)}{(\cos 2 - 7\cos 1 + 12)^2 + (\sin 2 - 7\sin 1)^2} \\
&= \frac{(\cos 2 - 7\cos 1 + 12)\cos x + (\sin 2 - 7\sin 1)\sin x}{(\cos 2 - 7\cos 1 + 12)^2 + (\sin 2 - 7\sin 1)^2}
\end{aligned}$$

The general solution u_x=C.F. + P. I. can then be written down.

A few problems with sequences

▶ Example 13.6.10 **(Fibonacci Sequence of numbers).** *The first term of a sequence is 1, the second is 2, and every other term is the sum of the two preceding terms. Find the nth term.* **[C.H. 1982]**

Let us denote the sequence $\{u_1, u_2, u_3, u_4, \ldots, u_{n-1}, u_n, u_{n+1}, \ldots\}$
Given $u_1 = 1$, $u_2 = 2$, $u_3 = 1+2 = 3$, $u_4 = 3+2 = 5$, etc.; in general, for $n \geq 3$,

$$u_n = u_{n-1} + u_{n-2} (u_1 = 1, u_2 = 2).$$

We have the difference equation

$$u_{n+2} - u_{n+1} - u_n = 0 \text{ or, } (E^2 - E - 1)u_n = 0.$$

The auxiliary equation $t^2 - t - 1 = 0$ has two roots $\frac{1}{2}(1 \pm \sqrt{5})$.

$$\therefore u_n = A\left\{\frac{1}{2}\left(1+\sqrt{5}\right)\right\}^n + B\left\{\frac{1}{2}\left(1-\sqrt{5}\right)\right\}^n.$$

Since $u_1 = 1, u_2 = 2$, putting $n = 1, 2$ successively we get

$$\begin{aligned}
1 &= A\left\{\frac{1}{2}(1+\sqrt{5})\right\} + B\left\{\frac{1}{2}(1-\sqrt{5})\right\}, \\
2 &= A\left\{\frac{1}{2}(1+\sqrt{5})\right\}^2 + B\left\{\frac{1}{2}(1-\sqrt{5})\right\}^2 \\
\text{whence, } A &= \frac{1}{2}\frac{\sqrt{5}+1}{\sqrt{5}} \text{ and } B = -\frac{1}{2}\frac{1-\sqrt{5}}{\sqrt{5}} \\
\therefore u_n &= \frac{1}{2^{n+1}}\frac{1}{\sqrt{5}}\left\{\left(1+\sqrt{5}\right)^{n+1} - \left(1-\sqrt{5}\right)^{n+1}\right\}.
\end{aligned}$$

▶ Example 13.6.11 **To solve:** $u_n - 5u_{n-1} + 6u_{n-2} = n^2 + 5^n + 2^n$.

Solution: The general solution of

$$\begin{aligned} & u_n - 5u_{n-1} + 6u_{n-2} = 0 \\ \text{or,}\quad & u_{n+2} - 5u_{n+1} + 6u_n = 0 \\ \text{or,}\quad & (E^2 - 5E + 6)u_n = 0 \\ \text{or,}\quad & (E-3)(E-2)u_n = 0. \end{aligned}$$

is $A\cdot 2^n + B\cdot 3^n$ **(This is the C.F.)**

(i) We now search for a *particular solution of*

$$u_n - 5u_{n-1} + 6u_{n-2} = n^2.$$

Assume as a trial solution: $u_n = a + bn + cn^2$, then

$$(a + bn + cn^2) - 5\{a + b(n-1) + c(n-1)^2\} + 6\{a + b(n-2) + c(n-2)^2\} = n^2.$$

Comparing the coefficients

$$2c = 1,\ 2b - 14c = 0,\ 2a - 7b + 19c = 0,$$

whence we obtain, $a = \frac{1}{2}\cdot 15,\ b = \frac{1}{2}\cdot 7,\ c = \frac{1}{2}$.

$\therefore\ \frac{1}{2}(n^2 + 7n + 15)$ is a particular solution.

(ii) Now we find a particular solution of

$$u_n - 5u_{n-1} + 6u_{n-2} = 5^n.$$

Assume, $u_n = a\cdot 5^n$, then $a(5^n - 5\cdot 5^{n-1} + 6\cdot 5^{n-2}) = 5^n$, which gives $a = 25/6$.

$\therefore$ a particular solution is $\frac{25}{6}\cdot 5^n = \frac{1}{6}\cdot 5^{n+2}$.

(iii) Next we search for a particular solution of $u_n - 5u_{n-1} + 6u_{n-2} = 2^n$. (Note here that in the right side we have 2^n and 2 is one of the roots of the auxiliary equation $t^2 - 5t + 6 = 0$)

We assume here as a trial solution: $u_n = an\cdot 2^n$, then the equation gives

$$an\cdot 2^n - 5a(n-1)\cdot 2^{n-1} + 6a(n-2)\cdot 2^{n-2} = 2^n \text{ whence } a = -2.$$

Thus a particular solution $= -2n\cdot 2^n = -n\cdot 2^{n+1}$.

Finally the general solution of the equation is

$$\begin{aligned} u_n &= \text{C.F.} + \text{P. I.} \\ &= A\cdot 2^n + B\cdot 3^n + \frac{1}{2}(n^2 + 7n + 15) + \frac{1}{6}\cdot 5^{n+2} - n\cdot 2^{n+1}, \\ \text{i.e., } u_n &= \frac{1}{2}(n^2 + 7n + 15) + \frac{1}{6}\cdot 5^{n+2} + (A - 2n)\cdot 2^n + B\cdot 3^n. \end{aligned}$$

Exercises on Chapter 13

1. Let $u_n = n^k$. Verify that

   (a) Δu_n is a polynomial in n of degree $k-1$.

   (b) $\Delta^2 u_n$ is a polynomial in n of degree $k-2$.

   (c) $\Delta^{k-1} u_n$ is of the form $an+b$ $(a, b$ constants$)$

   What can you say about $\Delta^k \cdot u_n, \Delta^m u_n (m > k)$?

2. If $S_n = u_1 + u_2 + \cdots + u_n$, prove that

$$\Delta^2 S_n = \Delta S_{n+1} - \Delta S_n.$$

[C.P. 1982]

3. Let $u_n = n^3 + n$. Obtain $u_1, u_2, u_3, u_4, u_5, u_6$.

   Construct a Table of successive differences $\Delta u_n, \Delta^2 u_n$, etc. Verify that $\Delta^4 u_n = 0$.

4. If $\Delta^3 u_n = an + b$ $(a, b$ are constants$)$, prove that Δu_n is of the form $cn^2 + dn + e$.

5. Establish Newton's Difference formula: $u_{n+1} = (1+\Delta)^n u_1$.
   Hence, deduce that if $\Delta^k u_n$ is constant (independent of n), then

$$\begin{aligned} u_{n+1} = u_1 &+ a_1 n + a_2 n(n-1) + a_3 n(n-1)(n-2) \\ &+ \cdots + a_k n(n-1)(n-2)\cdots(n-k+1), \end{aligned}$$

   where $a_p = \Delta^p u_1 / p!$

6. Using the formula $S_n = \{(1+\Delta)^n - 1\}\Delta^{-1} u_1$, prove that

   (a) $1^3 + 2^3 + 3^3 + \cdots + n^3 = \frac{1}{4} n^2 (n+1)^2$.

   (b) $\sum_{r=1}^{n} (r^2 + r + 1) = \frac{1}{3} n(n^2 + 3n + 5)$.

7. Find a cubic function of n which has values $1, -3, -1, 13$ when $n = 1, 2, 3, 4$ respectively.

   [*Hints*: Use $u_n = E^{n-1} u_1 = (1+\Delta)^{n-1} u_1$.]

   [*Ans.* $u_n = n^3 - 3n^2 - 2n + 5$]

8. Prove that $u_n = 3^n (A + Bn)$ satisfies the equation $u_{n+2} - 6u_{n+1} + 9u_n = 0$ $(A, B$ are constants$)$.

9. Prove that $u_r = A \cdot 2^{r-1} + B \cdot 3^{r-1}$ satisfies the equation $u_r - 5u_{r-1} + 6u_{r-2} = 0$.

10. Prove that $u_n = A + B \cdot 2^n - n$ satisfies the equation $u_{n+2} - 3u_{n+1} + 2u_n = 1$.

11. Write down the order of the following difference equations:

(a) $\Delta u_x = 0$,
(b) $\Delta^2 u_x = 0$;
(c) $\Delta^2 u_x + 8\Delta u_x + 3u_x = 0$;
(d) $(E^2 + 3E + 2)u_x = x^2$;
(e) $Eu_{x+2} + 6Eu_{x+1} + 5Eu_x = 0$.
(f) $\Delta^3 u_x + 2\Delta u_x + u_x = x + 3$
(g) $u_{x+4} - 5u_{x+2} + 6u_x = 0$.
(h) $u_{x+2} - 2u_{x+1} + u_x = 3x^3 + x^4$;
(i) $u_{x+1} - 8u_x = 0$.

12. (a) Eliminate A and B and form a difference equation from

$$u_n = A\cdot 2^n + B\cdot 3^n.$$

[Obtain $u_{n+1} = A\cdot 2^{n+1} + B\cdot 3^{n+1}$
$u_{n+2} = A\cdot 2^{n+2} + B\cdot 3^{n+2}$.

Eliminate A and B and the eliminant is

$$\begin{vmatrix} u_n & 1 & 1 \\ u_{n+1} & 2 & 3 \\ u_{n+2} & 4 & 9 \end{vmatrix} = 0,$$

i.e., $u_{n+2} - 5u_{n+1} + 6u_n = 0$ (an equation of order two)]

(b) Eliminate C_1 and C_2 to form a difference equation from

$$u_x = (C_1 + C_2 x)4^x;$$
$$u_n = \left(C_1 \cos\frac{n\pi}{3} + C_2 \sin\frac{n\pi}{3}\right)\cdot 2^n;$$
$$u_x = C_1\cdot 3^x + C_2(-1)^x.$$

13. Show that

(a) $u(x) = A\cos(2\pi x/3) + B\sin(2\pi/3)$ is a solution of

$$u(x+2) + u(x+1) + u(x) = 0.$$

(b) $u_x = C_1\cos\theta x + C_2\sin\theta x$ $(\sin\theta \neq 0)$ is the general solution of

$$u_{x+1} - 2u_x\cos\theta + u_{x-1} = 0.$$

(c) $u_x = C + x$ satisfies $u_{x+1} - u_x = 1$, but $u_x = C + \frac{1}{2}(x^2 - x)$ satisfies

$$u_{x+1} - u_x = x.$$

First-order difference equations:

14. *Solve:*

(a) $2u_{x+1} + 3u_x = 0$;
(b) $u_{x+1} - 2u_x = 0$;
(c) $u_{x+1} - 2u_x = 3^x$;
(d) $u_{x+1} - 3u_x = 2^x$;
(e) $3u_{x+1} - u_x = x$;
(f) $u_{x+1} - u_x = 1$;
(g) $u_{x+1} - 3u_x - x = 0$ (with $u_0 = 1$);

(h) $u_{x+1} - u_x = x^2$(with $u_0 = 1$); (j) $u_{x+1} - au_x = \sin bx$.

(i) $u_{x+1} - 2u_x = x^2 \cdot 2^x$;

Second-order difference equations:

15. *Solve:*

(a) $u_{x+2} - 8u_{x+2} + 15u_x = 0$;

(b) $u_{x+2} - 7u_{x+1} + 12u_x = 0$ (with $u_0 = 0, u_1 = 2$);

(c) $u_{x+2} + 2u_{x+1} + 4u_x = 0$;

(d) $u_{x+2} - 3u_{x+1} - 4u_x = 3^x$;

(e) $u_{x+2} - 5u_{x+1} + 6u_x = 5^x$;

(f) $u_{x+2} - 4u_x = 5 \cdot 3^x$;

(g) $u_{x+2} - 4u_{x+1} + 4u_x = 2^x$;

(h) $u_{x+2} + u_x = 5 \cdot 2^x$ (with $u_0 = 1, u_1 = 0$);

(i) $u_{x+2} - 4u_{x+1} + 4u_x = 3x + 2^x$ $\left(u_0 = 4, u_1 = -\frac{31}{4}\right)$;

(j) $u_{x+2} - 2u_{x+1} + u_x = 5 + 3x$;

(k) $(E^2 - 2E - 15)u_x = 32$;

(l) $u_x - u_{x-1} + 2u_{x-2} = x + 2^x$;

(m) $u_{x+2} - 2u_{x+1} + u_x = x^3$;.

(n) $u_{x+2} - u_{x+1} - 2u_x = x^2$;

(o) $u_{x+2} + u_x = \sin \pi x/2$.

(p) $\Delta^2 u_x + \Delta u_x = x + \sin x$;

(q) $u_{x+2} + u_x = \sin \frac{1}{2}x$;

(r) $y_{x+2} + a^2 y_x = \cos ax (a = 0, a = 1)$;

(s) $y_{x+2} - 7y_{x+1} + 12y_x = \cos x$;

(t) $y_{x+2} + y_{x+1} - 12y_x = 5^x + x^2 + 2x$.

16. If $2u_n - u_{n-1} = na$, show that

$$u_n = \frac{u_1}{2^{n-1}} + a\left\{(n-1) + \frac{1}{2^{n-2}}\right\}.$$

17. The first term of a sequence is a, the second is b and every other term is the arithmetic mean of the two preceding terms. Show that the nth term is

$$\frac{1}{3}(a+2b) - \frac{1}{3}(a-b)\left(-\frac{1}{2}\right)^{n-2}.$$

18. (a) Assume $u_n = an + b$ and then find a particular solution of

$$u_n - 5u_{n-1} + 6u_{n-2} = n.$$

(b) Assume $u_n = a\cdot 7^n$ and then find a particular solution of

$$u_n - 5u_{n-1} + 6u_{n-2} = 7^n.$$

(c) Assume $u_n = an^2 + bn + c$ and find a particular solution of

$$u_n - u_{n-1} - u_{n-2} = n^2.$$

(d) Assume $u_n = a\cos n\alpha + b\sin n\alpha$, find a particular solution of

$$u_n + u_{n-1} = \sin n\alpha.$$

In each case (a)–(d) write the general solution of the difference equations.

Applications

19. The seeds of a certain plant when one year old produce ten-fold and when two years old or more produce eighteen-fold. A seed is planted and every seed subsequently produced is planted as soon as it is produced. Prove that the number of grains at the end of the nth year is

$$\frac{1}{a}\left\{\left(\frac{11+a}{2}\right)^{n+1} - \left(\frac{11-a}{2}\right)^{n+1}\right\},\ a = 3\sqrt{17} \qquad \text{[CH. 1990]}$$

[u_n=number of grains at the end of the nth year.
Then $u_n = u_{n-1} + 18u_{n-2} + 10(u_{n-1} - u_{n-2})$.
Now solve for u_n with $u_0 = 1, u_1 = 11$.]

20. The national income equation is

$$u_{n+2} - 2au_{n+1} + au_n = I\ (0 < a < 1).$$

Assuming a and I constants, obtain u_n in the form:

$$u_n = C_1(\sqrt{a})^k \sin k\theta + C_2(\sqrt{a})^k \cos k\theta + \tfrac{I}{1-a}.$$

[See that when n increases sufficiently $\lim U_n = I/(1-a)$; this means that the national income oscillates above and below this limit during the approach.]

21. $u_n = n(n-1)(n-2)(n-3)(n-4)$. Prove $\Delta^5 u_n = 5!$.

22. Solve: $u_{n+3} + u_{n+2} - 8u_{n+1} - 12u_n = 0$. **[C.P. 1980]**
[Roots of A.E.$=-2,-2,3$; **Sol.** $u_n = (c_1 n + c_2)(-2)^n + c_3(3)^n$]

23. If u_r satisfies $u_r 4u_{r+1} + u_{r-2} = 0\ (2 \le r \le n)$, show that $u_r = \frac{A\sinh(n-r)\alpha}{\sinh n\alpha}$.
[Given $u_n = 0, u_0 = A$ and $\alpha = \log(2+\sqrt{3})$] **[C.H. 1980]**
[Eqn. is same as $u_{r+2} - 4u_{r+1} + u_r = 0$

or, $(E^2 - 4E + 1)u_r = 0$, A.E: $t^2 - 4t + 1 = 0$,

i.e., $t = 2 \pm \sqrt{3} = e^{\alpha}, e^{-\alpha}$ $(\because \alpha = \log(2+\sqrt{3}))$

$\therefore$ $u_r = C_1 e^{r\alpha} + C_2 e^{-r\alpha}$.

Since $u_0 = A$, $u_n = 0$.

We get $C_1 + C_2 = A$ and $C_1 e^{n\alpha} + C_2 e^{-n\alpha} = 0$.

Solving for C_1, C_2 and remembering $\frac{e^x - e^{-x}}{2}$ = sinh x we obtain the reqd. result.]

Exercises on Chapter 13: Answers

11. (a), (i) → order one; (f) → order 3; (g) → order 4; all others → order two.

12. (b) $u_{x+2} - 8u_{x+1} + 16u_x = 0$; $u_{n+2} - 2u_{n+1} + 4u_n = 0$; $u_{x+2} - 2u_{x+1} - 3u_x = 0$.

14. **(a)** $u_x = C(-\frac{3}{2})^x$;

(b) $u_x = C \cdot 2^x$;

(c) $u_x = C \cdot 2^x + 3^x$;

(d) $u_x = C \cdot 3^x - 2^x$;

(e) $u_x = C.\left(\frac{1}{3}\right)^x + \frac{1}{2}x - \frac{3}{2}$;

(f) $u_x = C + x$;

(g) $u_x = \frac{5}{4} \cdot 3^x - \frac{1}{2}x - \frac{1}{4}$;

(h) $u_x = \frac{1}{3}x^{(3)} + \frac{1}{2}x^{(2)} + 1$ (Factorial notations);

(i) $u_x = C \cdot 2^x + \frac{1}{3}x(x-1)(2x-1) \cdot 2^{x-2}$;

(j) $u_x = Ca^x + \frac{\sin b(x-1) - a \sin bx}{a^2 - 2a \cos b + 1}$.

15. **(a)** $u_x = C_1 \cdot 3^x + C_2 \cdot 5^x$;

(b) $u_x = 2 \cdot 4^x - 2 \cdot 3^x$;

(c) $u_x = 2^x(C_1 \cos\theta x + C_2 \sin\theta x)$; $\left(\theta = \frac{2}{3}\pi\right)$;

(d) $u_x = C_1 4^x + C_2(-1)^x - \frac{1}{4} \cdot 3^x$

(e) $u_x = C_1 \cdot 2^x + C_2 \cdot 3^x + \frac{1}{6}5^x$;

(f) $u_x = C_1 \cdot 2^x + C_2 \cdot (-2)^x + 3^x$;

(g) $u_x = (C_1 + C_2 x)2^x + 2^{x-3} \cdot x(x-1)$;

(h) $u_x = C_1 \cos\frac{\pi}{2}x + C_2 \sin\frac{\pi}{x} + 2^x$; $(C_1 = 0, C_2 = -2)$;

(i) $u_x = (4 - \frac{25}{2}x)2^x + 3x + 6 + \frac{1}{8}x^2 \cdot 2^x$;

(j) $u_x = C_1 + C_2 x + x^2 + \frac{1}{2}x^3$;

(k) $u_x = C_1 5^x + C_2(-3)^x - 2$;

(l) $u_x = \left(\sqrt{2}\right)^x(C_1 \cos\theta x + C_2 \sin\theta x) + \frac{1}{2}x + \frac{3}{4} + 2^x$; $\left(\theta = \tan^{-1}\sqrt{7}\right)$;

(m) $u_x = (C_1 + C_2 x) + \frac{1}{20}x^{(5)} + \frac{1}{4}x^{(4)} + \frac{1}{6}x^{(3)}$ [Factorial notations];

(n) $u_x = C_1\ (-1)^x + C \cdot 2^x - \left(\frac{1}{2}x^2 + x + \frac{1}{4} \cdot 5\right)$;

(o) $u_x = C_1 \cos\left(\frac{\pi x}{2} + C_2\right) - \frac{1}{2}x \sin(\frac{1}{2}\pi x)$;

(p) $u_x = C_1 \cos\theta x + C_2 \sin\theta x + \frac{1}{2}x^{(2)} - x - \frac{\cos(x - 3/2)}{2 \sin 1/2}$;

(q) $u_x = C_2 \cos \pi x/2 + C_2 \sin \pi x/2 +$ P. I.;

(r) $y_x = a^x\left(C_1 \cos\frac{\pi x}{2} + C_2 \sin\frac{\pi x}{2}\right) + \frac{a^2\cos ax + \cos a(x-2)}{1+2a^2\cos 2a + a^4}$; Put now $a = 0, a = 1$;

(s) $y_x = C_1 \cdot 3^x + C_2 \cdot 4^x + \frac{a\cos bx + b\sin x}{a^2+b^2}$; where $a = \cos 2 - 7\cos 1 + 12$ and $b = (\sin 2 - 7\sin 1)$;

(t) $y_x = C_1 \cdot 3^x + C_2(-4)^x - \frac{1}{10}x^2 - \frac{13}{50}x - \frac{16}{125} + \frac{1}{18}\cdot 5^x$.

18. (a) $A \cdot 2^n + B \cdot 3^n + \frac{1}{4}(2n+7)$;

(b) $A \cdot 2^n + B \cdot 3^n + \frac{1}{20} \cdot 7^{n+2}$;

(c) $A\left(\frac{1}{2}(1+\sqrt{5})\right)^n + B\left(\frac{1}{2}(1-\sqrt{5})\right)^n - (n^2 + 6n + 13)$;

(d) $C_1(-1)^n + \frac{\sin(n+1)\alpha + \sin n\alpha}{2(1+\cos\alpha)}$.

Chapter 14

Inequalities

14.1 Introduction

In this chapter we shall deal with real numbers only. Given two real numbers a and b, **either** $a > b$ (i.e., a is greater than b; this happens when $a - b$ is positive)

or, $a < b$ (i.e., a is less than b; this occurs when $a - b$ is negative).

or, $a = b$ (i.e., a is equal to b; this means $a - b = 0$).

The number 0 is neither positive nor negative. When a is a positive real number, we write $a > 0$.

When a is a negative real number, we write $a < 0$.

Remember. $a > b \Rightarrow a - b > 0; a < b \Rightarrow a - b < 0$. $a > b$ is same as $b < a$.

The symbols $>$ and $<$ are signs of inequalities and the symbol $=$ is the sign of equality.

The following properties will be constantly used:

1. If a is any real number, then $a \cdot a = a^2 \geq 0$; equality arises when $a = 0$.
2. $a > 0, b > 0$, then $ab > 0$.
   $a < 0, b > 0$, then $ab < 0$.
3. $a > b$ and $b > c \Rightarrow a > c$.
4. If $a > b$, then $a + c > b + c$ and $a - c < b - c$, for any real number c.
5. $a > b$ and $c > 0 \Rightarrow ac > bc$.
   $a > b$ and $c < 0 \Rightarrow ac < bc$.

Proofs are easy, e.g., to prove **5.**

$$ac - bc = c(a - b) = +\text{ve } (\because\ c > 0 \text{ and } a - b > 0) \Rightarrow ac > bc.$$

The students are advised to give similar proofs for other cases.

For most cases, we need to know the *Theorem of means*. *Cauchy's Inequality* is another very useful inequality. We shall be mainly concerned with these two types of inequalities. Other inequalities—*Weierstrass Inequality, Holder's Inequality, Minkowaski's Inequality, Tchebychef's Inequality*—will be introduced as examples only—just to get them known to our readers. We shall make two sections:

Section I contains problems involving Theorem of Means and Cauchy's Inequality.

Section II will contain some more involved inequalities like Holder's Inequality, Minkowski's Inequality, etc.

Section I

14.2 Two Important Inequalities

I. *If* $a_r > b_r > 0$ *for* $r = 1, 2, 3, \dots, n$, *then* $a_1a_2a_3 \cdots a_n > b_1b_2b_3 \cdots b_n$.

Proof. We know, if $a_1 > b_1 > 0$ and $a_2 > b_2 > 0$, then $a_1a_2 > b_1b_2 > 0$.

[For, $a_1a_2 - b_1b_2 = a_1(a_2 - b_2) + b_2(a_1 - b_1) > 0$ each term being positive]

Similarly, if $a_1 > b_1 > 0, a_2 > b_2 > 0$ and $a_3 > b_3 > 0$, then $a_1a_2 > b_1b_2 > 0$ and hence $(a_1a_2)a_3 > (b_1b_2)b_3 > 0$, i.e., $a_1a_2a_3 > b_1b_2b_3 > 0$.

Successive applications give $a_1a_2a_3 \cdots a_n > b_1b_2b_3 \cdots b_n > 0$.

[**To be more precise:** One may use the Principle of Mathematical Induction: For $n = 2$, the result is true (Sec. I).

Assume that the result is true for $n = m$, i.e., assume

$$a_1a_2 \cdots a_m > b_1b_2 \cdots b_m > 0.$$

Now for $a_{m+1} > b_{m+1} > 0$ we again use the result for $n = 2$ and obtain

$$(a_1a_2 \cdots a_m)a_{m+1} > (b_1b_2 \cdots b_m)b_{m+1} > 0,$$

i.e., the result becomes true for $n = m + 1$ if it is assumed to be true for $n = m$.

Hence, by induction, the result is true for all positive integers n.]

Corollary 14.2.1 *If* $a_r \geq b_r > 0$, *for* $r = 1, 2, 3, \dots n$, *then*

$$a_1a_2a_3 \cdots a_n \geq b_1b_2b_3 \cdots b_n > 0.$$

The equality sign holds if $a_1 = b_1, a_2 = b_2, \dots, a_n = b_n$.

An warning: This result is not necessarily true if the numbers are not positive, e.g., $3 > -2$ and $4 > -8$ but $12 \not> (-2) \cdot (-8)$.

II. *If* $a > b > 0$, *then* $a^n >$ *or* $< b^n$ *according as* n *is a positive integer or a negative integer.*

The result is still true if $n = p/q$ where q is a positive integer and p is any integer; only the convention is that $a^{p/q}$ will mean the positive qth root of a^p.

Proof.

(a) In the previous result (I) take $a_1 = a_2 = \cdots = a_n = a$ (say) and $b_1 = b_2 = \cdots = b_n = b$ (say). Then when $a > b > 0$, we have $a^n > b^n$.

OTHERWISE. Use the principle of Induction.

(b) **The case when n is a negative integer**

First observe that if $a > b > 0$, then $a^{-1} < b^{-1}$ [For $\frac{1}{a} - \frac{1}{b} = \frac{b-a}{ab} < 0$]

Now let $n = -m$, where m is a positive integer.

Since $a > b > 0$ and m is a positive integer, we have $a^m > b^m$

$$\therefore \quad \frac{1}{a^m} < \frac{1}{b^m} \text{ (by what we have observed before),}$$

$$\text{i.e.,} \quad a^{-m} < b^{-m} \text{ or, } a^n < b^n \ (n = \text{a negative integer}).$$

(c) Lastly $n = p/q$, where q is a positive integer and p is any integer (+ve or −ve).

We find that $a > b \Rightarrow a^{1/q} > b^{1/q}$, where $a^{1/q}$ and $b^{1/q}$ are the positive qth root of a and b respectively.

[If not then $a^{1/q}$ would be $\le b^{1/q}$
or, $(a^{1/q})^q$ would be $\le (b^{1/q})^q$
or, a would be $\le b$ (NOT TRUE)]

$$\therefore\ a > b \Rightarrow a^{1/q} > b^{1/q}$$
$$\Rightarrow (a^{1/q})^p > (b^{1/q})^p, \quad \text{if } p \text{ is a positive integer}$$
$$\text{or,} \quad (a^{1/q})^p < (b^{1/q})^p, \quad \text{if } p \text{ is a negative integer.}$$

Hence, $a^n > b^n$ or $a^n < b^n$ according as n is a positive fraction or a negative fraction.

14.3 Theorems of Means

Definition 1. *Let* $\{a_1, a_2, a_3, \ldots, a_n\}$ *be a set of* n *real positive numbers.*

Their Arithmetic Mean (A.M.) is defined by

$$A = \frac{a_1 + a_2 + \cdots + a_n}{n}.$$

Their Geometric Mean (G.M.) is defined by

$$G = \sqrt[n]{a_1 a_2 \cdots a_n}$$

Their Harmonic Mean (H.M.) is defined by

$$H = \frac{n}{\frac{1}{a_1} + \frac{1}{a_2} + \cdots + \frac{1}{a_n}}.$$

Theorem 14.3.1. $A \ge G \ge H$. *Equality occurs when* $a_1 = a_2 = a_3 = \cdots = a_n$.

Proof. (i) We first prove $A \ge G$ for a set of positive real numbers $\{a_1, a_2, \ldots, a_n\}$.

Case 1. *Let* $n = 2^m$ *(*m *is some positive integer).*

For two positive real numbers a_1, a_2 we can write

$$a_1 a_2 = \left(\frac{a_1 + a_2}{2}\right)^2 - \left(\frac{a_1 - a_2}{2}\right)^2$$
$$\le \left(\frac{a_1 + a_2}{2}\right)^2 \left[\because \left(\frac{a_1 - a_2}{2}\right)^2 \ge 0\right], \text{ i.e., } \frac{a_1 + a_2}{2} \ge \sqrt{a_1 a_2} \qquad (1)$$

and equality arises when $a_1 = a_2$, i.e., for $n = 2^1$ the theorem is TRUE.

Now assume that the theorem is TRUE for $n = 2^K$ numbers.

Let

$$a_1 a_2 a_3, \ldots, a_{2^{K+1}} \text{ be } 2^{K+1} \text{ given numbers.}$$

Take two sets of numbers with 2^K numbers in each set

$$\{a_1, a_2, \ldots, a_{2^k}\}, \{a_{2^k+1}, a_{2^k+2}, \ldots, a_{2^{k+1}}\}.$$

Then

$$\frac{a_1 + a_2 + \cdots + a_{2^k} + a_{2^k+1} + \cdots a_{2^{k+1}}}{2^{k+1}}$$

$$= \frac{1}{2}\left\{\frac{a_1 + a_2 + \cdots + a_{2^k}}{2^k} + \frac{a_{2^k+1} + a_{2^k+2} + \cdots + a_{2^{k+1}}}{2^k}\right\}$$

$$\geq \frac{1}{2}\left\{\sqrt[2^k]{a_1 \cdot a_2 \cdot a_3 \cdots a_{2^k}} + \sqrt[2^k]{a_{2^k+1} \cdot a_{2^k+2} \cdots a_{2^{k+1}}}\right\} \text{ [Using (1)]}$$

$$\geq \sqrt{(a_1 a_2 \cdots a_{2^k})^{\frac{1}{2^k}} \cdot (a_{2^k+1} \cdot a_{2^k+2} \cdots a_{2^{k+1}})^{\frac{1}{2^k}}} \text{ [Again using (1)]}$$

$$= (a_1 a_2 \cdots a_{2^k} \cdot a_{2^k+1} \cdots a_{2^{k+1}})^{\frac{1}{2^{k+1}}}.$$

Equality holds if and only if

$$a_1 = a_2 = \cdots = a_{2^k} = a_{2^k+1} = \cdots = a_{2^{k+1}}.$$

Thus if theorem holds for 2^k numbers, then it will be true for 2^{k+1} and it holds for 2^1. Therefore, by induction, the theorem must be true for all positive integers of the form 2^m.

Case 2. *Next suppose that we have n positive numbers $a_1, a_2, \ldots, a_n$ with geometric mean G. Let 2^m be any power of 2* **which is not less than n**. *Consider the set of 2^m numbers consisting of $a_1, a_2, \ldots, a_n$ and $G, G, \ldots$ taken $2^m - n$ times.*

Then

$$\frac{a_1 + a_2 + \cdots + a_n + (2^m - n)G}{2^m}$$

$$\geq \sqrt[2^m]{a_1 \cdot a_2 \cdots a_n \cdot G^{2^m - n}} \quad \text{(using Case 1)}$$

$$\text{or,} \quad \frac{a_1 + a_2 + \cdots + a_n + (2^m - n)G}{2^m}$$

$$\geq (G^n \cdot G^{2^m - n})^{\frac{1}{2^m}} = G. \text{ (Since } G^n = a_1 a_2 \cdots a_n)$$

with equality when $a_1 = a_2 = \cdots = a_n = G$.

$$\therefore \quad a_1 + a_2 + \cdots + a_n + (2^m - n)G \geq 2^m G$$

$$\text{or,} \quad \frac{a_1 + a_2 + \cdots + a_n}{n} \geq G, \text{ for any positive integer } n.$$

Thus the theorem is established for n positive numbers.

Corollary 14.3.1 *The arithmetic mean of n positive numbers,* NOT ALL EQUAL, *is greater than their geometric mean.* $A > G$ *if the set* $[a]$ *is not a* dull *set. (i.e., a set whose members are all equal.)*

(ii) Now *we prove:* $G \geq H$. Let $a_1, a_2, \ldots, a_n$ be n positive real numbers. Then $\frac{1}{a_1}, \frac{1}{a_2}, \ldots, \frac{1}{a_n}$ are also n positive real numbers.
Using A.M. $\geq$ G.M. on this latter set we obtain

$$\frac{\frac{1}{a_1} + \frac{1}{a_2} + \cdots + \frac{1}{a_n}}{n} \geq \left(\frac{1}{a_1 a_2 \cdots a_n}\right)^{1/n}$$

with equality when $1/a_1 = 1/a_2 = \cdots = 1/a_n$, i.e., when $a_1 = a_2 = \cdots = a_n$.
We write this result as

$$\frac{n}{\frac{1}{a_1} + \frac{1}{a_2} + \cdots + \frac{1}{a_n}} \leq (a_1 a_2 \cdots a_n)^{1/n},$$

i.e., H.M. $\leq$ G.M. (or $H \leq G$).

So, finally, from (i) and (ii) we can write $A \geq G \geq H$ with equality when the set is **dull.**

Weighted Means

Let $[a]$ be a set of n real positive numbers $a_1, a_2, a_3, \ldots, a_n$.

Let $[p]$ be a set of n positive rational numbers $p_1, p_2, \ldots, p_n$ associated with $a_1, a_2, \ldots, a_n$ respectively. Then the *weighted arithmetic mean* of set $[a]$ with associated weights $[p]$ is defined by

$$A(a,p) = \frac{\sum_{i=1}^{n} p_i a_i}{\sum_{i=1}^{n} p_i} = \frac{p_1 a_1 + p_2 a_2 + \cdots + p_n a_n}{p_1 + p_2 + \cdots + p_n}.$$

The *weighted geometric mean* of the set $[a]$ with associated weights $[p]$ is defined by

$$G(a,p) = \left(a_1^{p_1} a_2^{p_2} \cdots a_n^{p_n}\right)^{\frac{1}{p_1 + p_2 + \cdots + p_n}}.$$

The *weighted harmonic mean* of the set $[a]$ with associated weights $[p]$ is defined by

$$H(a,p) = \frac{\sum_{i=1}^{n} p_i}{\sum_{i=1}^{n} \frac{p_i}{a_i}} = \frac{p_1 + p_2 + \cdots + p_n}{p_1/a_1 + p_2/a_2 + \cdots + p_n/a_n}.$$

Theorem 14.3.2. **Generalised Theorem of Means** *states*

$$A(a,p) \geq G(a,p) \geq H(a,p).$$

Equality occurs when the members of the set $[a]$ are all equal, i.e.,

$$a_1 = a_2 = \cdots = a_n.$$

We prove: $A(a,p) \geq G(a,p)$.

Case 1. *Let $p_1, p_2, \ldots, p_n$ be n positive integers.*

Then we consider

p_1 quantities each equal to a_1,
p_2 quantities each equal to a_2,
....
p_n quantities each equal to a_n;

and use A.M. $\geq$ G.M.

$$\frac{(a_1 + a_1 + \cdots + a_1) + (a_2 + a_2 + \cdots + a_2) + \cdots + (a_n + a_n + \cdots + a_n)}{p_1 + p_2 + \cdots + p_n}$$

$$\geq \quad \left\{(a_1 \cdot a_1 \cdots a_1)(a_2 \cdot a_2 \cdots a_2) \cdots (a_n \cdot a_n \cdots a_n)\right\}^{\frac{1}{p_1+p_2+\cdots+p_n}},$$

i.e., $$\frac{p_1 a_1 + p_2 a_2 + \cdots + p_n a_n}{p_1 + p_2 + \cdots + p_n} \geq \left(a_1^{p_1} a_2^{p_2} \cdots a_n^{p_n}\right)^{\frac{1}{p_1+p_2+\cdots+p_n}},$$

i.e., $A(a, p) \geq G(a, p)$.

Clearly, equality occurs when $a_1 = a_2 = \cdots = a_n$.

Case 2. *Let $p_1, p_2, \ldots, p_n$ be n positive fractions and let d be the l.c.m. of their denominators.*

Then $p_1 d, p_2 d, \ldots, p_n d$ are n positive integers and applying the result of Case 1 we g

$$\frac{a_1(p_1 d) + a_2(p_2 d) + \cdots + a_n(p_n d)}{p_1 d + p_2 d + \cdots + p_n d} \geq \left(a_1^{p_1 d} \cdot a_2^{p_2 d} \cdots a_n^{p_n d}\right)^{\frac{1}{p_1 d + p_2 d + \cdots + p_n d}},$$

i.e., $$\frac{a_1 p_1 + a_2 p_2 + \cdots + a_n p_n}{p_1 + p_2 + \cdots + p_n} \geq \left\{(a_1^{p_1} \cdot a_2^{p_2} \cdots a_n^{p_n})^d\right\}^{\frac{1}{d(p_1+p_2+\cdots+p_n)}},$$

i.e., $\geq \left(a_1^{p_1} \cdot a_2^{p_2} \cdots a_n^{p_n}\right)^{\frac{1}{p_1+p_2+\cdots+p_n}}$, i.e., $A(a, p) \geq G(a, p)$.

Here also equality arises when $a_1 = a_2 = \cdots = a_n$.

Theorem 14.3.3. *Suppose, $a \neq 1$ is a positive real number and m is any rational number. Then*

(i) *when m does not lie between 0 and 1*, $a^m - 1 > m(a - 1)$.

(ii) *but if m lies between 0 and 1*, $a^m - 1 < m(a - 1)$.

Proof.

Case 1. *Given $a > 0$ and $a \neq 1$. Let m be a rational number greater than 1.*

Consider a^m and 1 associated with weights 1 and $m-1$ respectively. Then using weighted A.M. > weighted G.M., we get

$$\frac{a^m \cdot 1 + 1 \cdot (m-1)}{1 + m - 1} > \left\{(a^m)^1 \cdot (1)^{m-1}\right\}^{\frac{1}{1+m-1}}$$

or, $a^m + (m - 1) > ma$

or, $a^m - 1 > m(a - 1)$.

Case 2. *Let m be a rational number such that $m < 0$ so that $1 - m > 1$. Take $1/a > 0$ and $\neq 1$ and use Case 1.*

$$\left(\frac{1}{a}\right)^{1-m} - 1 > (1-m)\left(\frac{1}{a} - 1\right),$$

$$\text{or,}\quad \frac{a^m}{a} - 1 > (1-m)\frac{(1-a)}{a},$$

$$\text{or,}\quad a^m > (1-m)(1-a) + a = 1 - m + ma,$$

$$\text{or,}\quad a^m - 1 > m(a-1).$$

Case 3. *Given $a > 0$ and $a \neq 1$. Let $0 < m < 1$ (m rational). Consider two unequal positive numbers a and 1 associated respectively with weights m and $(1 - m)$. We get by using weighted A.M. > weighed G.M.*

$$\frac{a.m + 1.(1-m)}{m + 1 - m} > \left(a^m . 1^{1-m}\right)^{\frac{1}{m+1-m}},$$

$$\text{or,}\quad ma + 1 - m > a^m,$$

$$\text{or,}\quad m(a-1) > a^m - 1, \quad \text{i.e.,}\quad a^m - 1 < m(a-1).$$

This completely proves the theorem.

Theorem 14.3.4. **m-Theorem:** *If $[a] = \{a_1, a_2, \ldots, a_n\}$ be a set of n positive real numbers, not all equal, and m is any rational number, then*

$$\frac{a_1^m + a_2^m + \cdots + a_n^m}{n} > \text{or} < \left(\frac{a_1 + a_2 + \cdots + a_n}{n}\right)^m$$

according as m does not lie between 0 and 1 or m does lie between 0 and 1.

When $m = 0$ or $m = 1$ equality occurs.

Proof. Let us write

$$A = \frac{a_1 + a_2 + \cdots + a_n}{n}.$$

Then

$$\frac{a_1}{A}, \frac{a_2}{A}, \ldots, \frac{a_n}{A}$$

are all positive and not all of them are equal to 1. Then, by the Theorem **14.3.3**, proved above:

$$\left(\frac{a_1}{A}\right)^m - 1 > \text{or} < m\left(\frac{a_1}{A} - 1\right)$$

according as m does not lie between 0 and 1 or m does lie between 0 and 1.

Similarly,

$$\left(\frac{a_2}{A}\right)^m - 1 > \text{or} < m\left(\frac{a_2}{A} - 1\right)$$

according as m does not or does lie between 0 and 1 and so on.

$$\left(\frac{a_n}{A}\right)^m - 1 > \text{or} < m\left(\frac{a_n}{A} - 1\right)$$

according as m does not or does lie between 0 and 1.

Adding these inequalities we get according as m does not or does lie between

$$\left(\frac{a_1}{A}\right)^m + \left(\frac{a_2}{A}\right)^m + \cdots + \left(\frac{a_n}{A}\right)^m - n > m\left(\frac{a_1 + a_2 + \cdots + a_n}{A} - n\right) = 0,$$

i.e., $$\frac{a_1^m + a_2^m + \cdots + a_n^m}{A^m} > \text{or} < n$$

or, $$\frac{a_1^m + a_2^m + \cdots + a_n^m}{n} > \text{or} < \left(\frac{a_1 + a_2 + \cdots + a_n}{n}\right)^m.$$

This proves the m-theorem.

14.4 Illustrative Examples

▶ Example 14.4.1 *If a and b be two unequal positive real numbers, prove that*

$$\sqrt{ab} > \frac{2}{\frac{1}{a} + \frac{1}{b}} \text{ (G.M. > H.M.).}$$

Solution: The square of a non-zero real number is always positive. Hence, in this case

$$(a - b)^2 > 0 \text{ or, } (a + b)^2 - 4ab > 0$$

$$\text{or, } \quad ab(a + b)^2 > 4a^2b^2 \text{ or, } ab > \frac{4a^2b^2}{(a + b)^2}$$

$$\text{or, } \quad \sqrt{ab} > \frac{2ab}{a + b} \text{ (taking positive square roots)}$$

$$\text{or, } \quad \sqrt{ab} > \frac{2}{\frac{1}{a} + \frac{1}{b}}.$$

▶ Example 14.4.2 *If n is a positive integer > 1, then prove that*

(i) $n^n > 1 \cdot 3 \cdot 5 \ldots (2n - 1)$;

(ii) $(n!)^3 < n^n \left\{\frac{1}{2}(n + 1)\right\}^{2n}$.

Solution: Note that in both cases we have the strict inequality. Only when $n = 1$, they become equalities.

(i) Consider the set of n positive numbers (not all equal) $\{1 \cdot 3 \cdot 5 \ldots (2n - 1)\}$ and use A.M. > G.M.,

$$\text{i.e., } \quad \frac{1 + 3 + \cdots + (2n - 1)}{n} > \sqrt[n]{1 \cdot 3 \cdot 5 \ldots (2n - 1)}$$

$$\text{i.e., } \quad \left(\frac{n(1 + 2n - 1)}{2n}\right)^n > 1 \cdot 3 \cdot 5 \ldots (2n - 1)$$

$$\text{or, } \quad n^n > 1 \cdot 3 \cdot 5 \ldots (2n - 1).$$

(ii) Consider the set of n positive numbers (not all equal) $\{1^3 \cdot 2^3 \cdot 3^3 \ldots n^3\}$ and use A.M. > G.M.,

$$\text{i.e.,} \quad \frac{1^3 + 2^3 + 3^3 + \cdots + n^3}{n} > \sqrt[n]{1^3 \cdot 2^3 \cdot 3^3 \ldots n^3}$$

$$\text{or,} \quad \left(\frac{n(n+1)}{2}\right)^2 \cdot \frac{1}{n} > \sqrt[n]{1^3 \cdot 2^3 \cdot 3^3 \ldots n^3},$$

$$\text{i.e.,} \quad \left\{n\left(\frac{n+1}{2}\right)^2\right\}^n > (n!)^3$$

$$\text{or,} \quad (n!)^3 < n^n \left(\frac{n+1}{2}\right)^{2n}.$$

► Example 14.4.3 (*Use of m-Theorem*): *If a, b, c, d are four positive real numbers, then prove that*

$$\frac{3}{b+c+d} + \frac{3}{c+d+a} + \frac{3}{d+a+b} + \frac{3}{a+b+c} \geq \frac{16}{a+b+c+d}.$$

Solution: Let

$$a_1 = \frac{b+c+d}{3}, \quad a_2 = \frac{c+d+a}{3}, \quad a_3 = \frac{d+a+b}{3}, \quad a_n = \frac{a+b+c}{3}.$$

Use m-theorem with $m = -1$. That is

$$\frac{a_1^{-1} + a_2^{-1} + a_3^{-1} + a_4^{-1}}{4} > \left(\frac{a_1 + a_2 + a_3 + a_4}{4}\right)^{-1},$$

$$\text{i.e.,} \quad \frac{3}{b+c+d} + \frac{3}{c+d+a} + \frac{3}{d+a+b} + \frac{3}{a+b+c}$$

$$> \quad 4\left[\frac{\frac{b+c+d}{3} + \frac{c+d+a}{3} + \frac{d+a+b}{3} + \frac{a+b+c}{3}}{4}\right],$$

$$\text{i.e.,} \quad \frac{3}{b+c+d} + \frac{3}{c+d+a} + \frac{3}{d+a+b} + \frac{3}{a+b+c} > \frac{16}{a+b+c+d}.$$

Equality arises when $a_1 = a_2 = a_3 = a_4$, i.e., when $a = b = c = d$.

► Example 14.4.4 *If $a_1, a_2, \ldots, a_n$ are n positive real numbers and p, q are positive integers such that $p/q > 1$, then prove that*

$$\left(a_1^q + a_2^q + \cdots + a_n^q\right)^p < n^{p-q}\left(a_1^p + a_2^p + \cdots + a_n^p\right)^q.$$

Solution: In the m-theorem we take a_1^p in place of a_1, a_2^p in place of $a_2, \ldots, a_n^p$ in place of a_n and for m we take q/p, then since $0 < m < 1$, we have

$$\frac{(a_1^p)^{q/p} + (a_2^p)^{q/p} + \cdots + (a_n^p)^{q/p}}{n} < \left(\frac{a_1^p + a_2^p + \cdots + a_n^p}{n}\right)^{q/p}$$

$$\text{or,} \quad \left(\frac{a_1^q + a_2^q + \cdots + a_n^q}{n}\right)^p < \left(\frac{a_1^p + a_2^p + \cdots + a_n^p}{n}\right)^q$$

$$\text{or,} \quad \left(a_1^q + a_2^q + \cdots + a_n^q\right)^p < n^{p-q}\left(a_1^p + a_2^p + \cdots + a_n^p\right).$$

Applications to the problems of Maxima and Minima

Suppose $x, y, z, \ldots, w$ are n positive real variables and let k be a constant.

I. If the sum $x+y+z+\cdots+w$=the constant k, then the greatest value of the product $xyz\cdots w$ occurs when $x=y=z=\cdots=w=k/n$ and the greatest value is $(k/n)^n$.

[Since $(xyz\cdots w)^{1/n} \le \frac{x+y+z+\cdots+w}{n}$ it follows $xyz\cdots w \le \left(\frac{k}{n}\right)^n$.

$(\because x+y+z+\cdots+w=k)$.

Equality occurs when $x=y=z=\cdots=w=\frac{k}{n}$.]

II. If the product $xyz\cdots w = a$ constant c (say), then the least value of the sum $x+y+z+\cdots+w$ is $nc^{1/n}$; it occurs when $x=y=z=\cdots=w=(c)^{1/n}$.

III. If $x, y, z, \ldots, w$ be n positive real variables, then for any rational number $m(\neq 0, 1)$ we have

$$\frac{x^m+y^m+\cdots+w^m}{n} \ge \text{ or } \le \left(\frac{x+y+z+\cdots+w}{n}\right)^m$$

according as m does not or does lie between 0 and 1.

Equality occurs when $x=y=z=\cdots=w$.

Therefore, if $x+y+z+\cdots+w = a$ constant k and m is a rational number >1 or <0, then the **least value** of $x^m+y^m+\cdots+w^m$ is $\left(\frac{k}{n}\right)^m$ and it occurs when $x=y=z=\cdots=w=k/n$.

Also if $x+y+z+\cdots+w=k$ and m is a rational number lying between 0 and 1, then the **greatest value** of $x^m+y^m+\cdots+w^m$ is $n\cdot\left(\frac{k}{n}\right)^m$ and it occurs when $x=y=z=\cdots=w=k/n$.

IV. If $x^m+y^m+\cdots+w^m = a$ constant c, then according as the rational number $m>1$ or <0 and $0<m<1$ the greatest value and the least value of $x+y+z+\cdots+w = n^{1-1/m}k^{1/m}$ and they occur when $x=y=z=\cdots=w=\left(\frac{k}{n}\right)^{1/m}$.

▶ Example 14.4.5 *Find the greatest value of xyz for positive real variables, x, y, z subject to the condition $xy+yz+zx=12$.*

Solution: Take 3 quantities xy, yz, zx and use A.M. $\ge$ G.M.

$$\frac{xy+yz+zx}{3} \ge (xy\cdot yz\cdot zx)^{1/3}, \text{ or, } x^2y^2z^2 \le \left(\frac{xy+yz+zx}{3}\right)^3,$$

$$\text{i.e.,} \quad \le \left(\frac{12}{3}\right)^3 \quad (\because xy+yz+zx=12),$$

$$\text{i.e.,} \quad xyz \le 8.$$

$\therefore$ the greatest value of $xyz=8$; this occurs when $xy=yz=zx$, i.e., when $x=y=z=2$ ($\because xy+yz+zx=12 \Rightarrow 3x^2=12$ or $(x=2)$.

► Example 14.4.6 *If* $3x + 5y = 2$ *(x, y are positive real variables) find the greatest value of* x^2y^3. *What are then the values of x and y?*

Solution: Here take the quantities $\frac{3}{2}x$, $\frac{3}{2}x$, $\frac{5}{3}y$, $\frac{5}{3}y$, $\frac{5}{3}y$

(so that sum $= 3x + 5y$ and product is a const. x^2y^3).

Using A.M. ≥ G.M. we get

$$\frac{\frac{3}{2}x + \frac{3}{2}x + \frac{5}{3}y + \frac{5}{3}y + \frac{5}{3}y}{5} \geq \left(\frac{3}{2}x \cdot \frac{3}{2}x \cdot \frac{5}{3}y \cdot \frac{5}{3}y \cdot \frac{5}{3}y\right)^{1/5}.$$

Equality occurs when $\frac{3}{2}x = \frac{5}{3}y$ or, $x : y = 10 : 9$.
Thus

$$\frac{9}{4} \cdot \frac{125}{27} x^2y^3 \leq \left(\frac{3x + 5y}{5}\right)^5, \text{ i.e., } x^2y^3 \leq \frac{12}{125} \cdot \left(\frac{2}{5}\right)^5 = 2^7 \cdot 3 \cdot 5^{-8}.$$

∴ the greatest value of x^2y^3 is $2^7 \cdot 3 \cdot 5^{-8}$. This occurs when $x : y = 10 : 9$, i.e., when $x = 10k, y = 9k$ and $3x + 5y = 2$ gives $k = 2/75$, i.e., $x = 20/75 = 4/15$ and $y = 18/75$.

► Example 14.4.7 *Find the least value of* $\frac{1}{x} + \frac{1}{y} + \frac{1}{z}$ *for positive real variables* x, y, z *subject to the condition* $x + y + z = c$, *where c is a constant.*

Solution: Using m-Theorem (taking $m = -1$) on three quantities x, y, z we get

$$\frac{x^{-1} + y^{-1} + z^{-1}}{3} \geq \left(\frac{x + y + z}{3}\right)^{-1}.$$

Equality occurs when $x = y = z = c/3$ ($\because x + y + z = c$),

$$\text{i.e., } \frac{1}{x} + \frac{1}{y} + \frac{1}{z} \geq \frac{9}{x + y + z} = \frac{9}{c}.$$

∴ the least value of $\frac{1}{x} + \frac{1}{y} + \frac{1}{z}$ is $\frac{9}{c}$ which happens when $x = y = z = c/3$.

14.5 Cauchy-Schwarz Inequality

Cauchy-Schwarz Inequality is an important inequality. It is often called simply **Cauchy's Inequality.**

It arises from the fact

$$\begin{aligned}(ax + by)^2 - (a^2 + b^2)(x^2 + y^2) &= 2abxy - a^2y^2 - b^2x^2 \\ &= -(ay - bx)^2 \leq 0.\end{aligned}$$

$$\therefore (ax + by)^2 \leq (a^2 + b^2)(x^2 + y^2).$$

Equality arises when $ay = bx$, i.e., when $x/y = a/b$.

Theorem 14.5.1. *If $x_1, x_2, \ldots, x_n$ and $y_1, y_2, \ldots, y_n$ are any two sets of real numbers, then Cauchy's inequality states:*

$$(x_1y_1 + x_2y_2 + \cdots + x_ny_n)^2 \leq (x_1^2 + x_2^2 + \cdots + x_n^2)(y_1^2 + y_2^2 + \cdots + y_n^2)$$

with equality if and only if

$$\frac{x_1}{y} = \frac{x_2}{y_2} = \cdots = \frac{x_n}{y_n}, \text{ (i.e., iff the sets are proportional).}$$

Proof. $(x_1^2 + x_2^2 + \cdots + x_n^2)(y_1^2 + y_2^2 + \cdots + y_n^2) - (x_1y_1 + x_2y_2 + \cdots + x_ny_n)^2$

$$= \sum_{i<j} (x_i^2y_j^2 + x_j^2y_i^2 - 2x_ix_jy_iy_j)^2 = \sum_{i>j} (x_iy_j - x_jy_i)^2 \geq 0.$$

Equality occurs iff $x_iy_j = x_jy_i$, i.e., iff $\frac{x_i}{y_i} = \frac{x_j}{y_j}$.

Alternative Proof. We assume that not all x_i's or y_j's are zero because in that case both sides become zero (a *trivial case*).

Case 1. *If* $x_i = ky_i$ $(i = 1, 2, \ldots, n)$, *then*

$$(x_1^2 + x_2^2 + \cdots + x_n^2)(y_1^2 + y_2^2 + \cdots + y_n^2) = k^2(y_1^2 + y_2^2 + \cdots + y_n^2)^2$$

and $$(x_1y_1 + x_2y_2 + \cdots + x_ny_n)^2 = k^2(y_1^2 + y_2^2 + \cdots + y_n^2)^2.$$

$\therefore$ equality arises in this case.

Case 2. *Suppose the two sets* [x] *and* [y] *are* not proportional.

We consider the expression

$$(x_1 - ky_1)^2 + (x_2 - ky_2)^2 + \cdots + (x_n - ky_n)^2 \text{ (k is any real number).}$$

The expression $= Ak^2 - 2Bk + C$, where

$$A = \sum_{i=1}^{n} y_i^2, \; B = \sum_{i=1}^{n} x_iy_i \text{ and } C = \sum_{i=1}^{n} x_i^2.$$

The expression is clearly >0 or $=0$ for all real values of k. It comes zero if $x_i = ky_i$, i.e., the sets [x] and [y] are proportional, which is not the case here.

$\therefore$ $Ak^2 - 2Bk + C > 0$, for all real k, A being > 0.

Now $Ak^2 - 2Bk + C = A(k - \alpha)(k - \beta)$, where α, β are the roots of

$$Ak^2 - 2Bk + C = 0. \quad (1)$$

Since $A > 0$, we shall get $Ak^2 - 2Bk + C > 0 \; \forall$ real k, if and only if the roots of (1) are imaginary, [for otherwise $\exists$ some real k for which $Ak^2 - 2Bk + C \leq 0$] the condition for which is the Discriminant of (1) < 0, i.e., when $(-2B)^2 - 4AC < 0$ or, $B^2 < AC$.

Putting the values of B, A, C we get the required result.

Case 1 and Case 2 together give Cauchy's Inequality.

Corollary 14.5.1 *If there are three sets of* n *positive numbers* $[a] = \{a_1, a_2, \ldots, a_n\}$, $[b] = \{b_1, b_2, \ldots, b_n\}$, $[c] = \{c_1, c_2, \ldots, c_n\}$, *then*

$$(a_1b_1c_1 + a_2b_2c_2 + \cdots + a_nb_nc_n)^2 < (a_1^2 + a_2^2 + \cdots + a_n^2)(b_1^2 + b_2^2 + \cdots + b_n^2)(c_1^2 + c_2^2 + \cdots + c_n^2).$$

Using Cauchy's Inequality on the two sets $[a]$ and $[bc]$ we get

$$\{a_1(b_1c_1) + a_2(b_2c_2) + \cdots + a_n(b_nc_n)\}^2 \leq (a_1^2 + a_2^2 + \cdots + a_n^2)(b_1^2c_1^2 + b_2^2c_2^2 + \cdots + b_n^2c_n^2) \quad \text{(I)}$$

Now $(b_1^2 + b_2^2 + \cdots + b_n^2)(c_1^2 + c_2^2 + \cdots + c_n^2) = b_1^2c_1^2 + b_2^2c_2^2 + \cdots + b_n^2c_n^2 +$ other terms which are all positive.

$$\therefore\ b_1^2c_1^2 + b_2^2c_2^2 + \cdots + b_n^2c_n^2 < (b_1^2 + b_2^2 + \cdots + b_n^2)(c_1^2 + c_2^2 + \cdots + c_n^2).$$

Hence, from (I) it follows

$$(a_1b_1c_1 + a_2b_2c_2 + \cdots + a_nb_nc_n)^2 < (a_1^2 + a_2^2 + \cdots + a_n^2)(b_1^2 + b_2^2 + \cdots + b_n^2)(c_1^2 + c_2^2 + \cdots + c_n^2).$$

14.6 Some Standard Inequalities

I. If $a_1, a_2, \ldots, a_n$ are n positive real numbers, not all equal, then

$$\frac{a_1^{x+y} + a_2^{x+y} + \cdots + a_n^{x+y}}{n} \gtrless \frac{a_1^x + a_2^x + \cdots + a_n^x}{n} \cdot \frac{a_1^y + a_2^y + \cdots + a_n^y}{n}$$

according as x and y have the same or opposite signs.

Proof.

Case 1. *Let* x *and* y *have the same sign (both positive or both negative).*

Let r and s be any two of the integers $\{1, 2, 3, \ldots, n\}$.
Then $a_r^x - a_s^x$ and $a_r^y - a_s^y$ are both positive or both negative or both zero.
Hence,

$$(a_r^x - a_s^x)(a_r^y - a_s^y) \geq 0$$

or, $$a_r^x a_r^y + a_s^x a_s^y - a_r^x a_s^y - a_s^x a_r^y \geq 0,$$

i.e., $$a_r^{x+y} + a_s^{x+y} \geq a_r^x a_s^y + a_s^x a_r^y. \quad (1)$$

Since r and s are any two integers of the set $\{1, 2, 3, \ldots n\}$ we can write down ${}^nC_2 = \frac{n(n-1)}{2}$ relations of the type (1)
All these relations can not be equalities ($\because$ a's are not all equal)
We have on addition of ${}^nC_2$ inequalities a strict inequality, namely

$$\sum_{r \neq s} a_r^{x+y} + a_s^{x+y} > \sum a_r^x a_s^y. \quad (2)$$

In the above summation we have $\frac{n(n-1)}{2}$ terms, of which a_r^{x+y} occurs $(n-1)$ times and $a_r^x a_s^y$ occurs only once.

Hence, (2) can be written as

$$\begin{aligned} n\sum a_r^{x+y} &> \sum a_r^{x+y} + \sum a_r^x \sum a_s^y \ (r \neq s), \\ &= \sum a_r^x \cdot \sum a_r^y. \end{aligned}$$

The last result can be written as

$$\frac{\sum a_r^{x+y}}{n} > \frac{\sum a_r^x}{n} \cdot \frac{\sum a_r^y}{n} \text{ (Reqd. result with } x \text{ and } y \text{ of same sign)}$$

Case 2. *If x and y have opposite signs, then*

$$(a_r^x - a_r^y)(a_s^y - a_s^y) \leq 0$$

Now proceed exactly as in **Case 1.** We can easily deduce

$$\frac{\sum a_r^{x+y}}{n} < \frac{\sum a_r^x}{n} \cdot \frac{\sum a_r^y}{n} \text{ (Reqd. result with } x \text{ and } y \text{ of opposite signs)}$$

EXTENSION. If $a_1, a_2, \ldots, a_n$ are n positive real numbers, not all equal and if $x_1, x_2, \ldots, x_m$ are all positive or all negative, then

$$\frac{\sum a_r^{x_1+x_2+\cdots+x_m}}{n} > \text{ or } < \frac{\sum a_r^{x_1}}{n} \cdot \frac{\sum a_r^{x_2}}{n} \cdots \frac{\sum a_r^{x_m}}{n}.$$

Applications

▶ Example 14.6.1 *If a, b, c, d be positive real numbers, prove that*

$$(a+b+c+d)\left(\frac{1}{a}+\frac{1}{b}+\frac{1}{c}+\frac{1}{d}\right) > 16$$

Solution:

$$\frac{a^p+b^p+c^p+d^p}{4} \cdot \frac{a^q+b^q+c^q+d^q}{4} > \frac{a^{p+q}+b^{p+q}+c^{p+q}+d^{p+q}}{4}.$$

Let $p = 1, q = -1$ so that p, q are of at opposite signs.
Then

$$\frac{a+b+c+d}{4} \cdot \frac{a^{-1}+b^{-1}+c^{-1}+d^{-1}}{4} > 1$$

$$\Rightarrow \quad (a+b+c+d)\left(\frac{1}{a}+\frac{1}{b}+\frac{1}{c}+\frac{1}{d}\right) > 16.$$

II. If a be a positive number, not equal to 1, and x, y are positive rationals, then

$$\frac{a^x-1}{x} > \frac{a^y-1}{y} \text{ if } x > y.$$

Proof. a is positive and $\neq 1$. $a^x > 0$ and $\neq 1$. Consider two numbers a^x and 1 with associated positive rational numbers y and $x-y$ respectively.

Then

$$\frac{y \cdot a^x + (x-y) \cdot 1}{y + x - y} > \{(a^x)^y \cdot 1\}^{1/x}$$

$$\text{or,} \quad y \cdot a^x + x - y > x(a^y)$$

$$\text{or,} \quad y(a^x - 1) > x(a^y - 1)$$

$$\text{or,} \quad \frac{a^x - 1}{x} > \frac{a^y - 1}{y} \quad (\text{since } x > 0, y > 0).$$

III. If a and b are positive $a \neq b$ and x is any rational number $(x \neq 1)$, then

$$xa^{x-1}(a-b) > a^x - b^x > xb^{x-1}(a-b)$$

unless $0 < x < 1$, in which case $x^{ax-1}(a-b) < a^x - b^x < xb^{x-1}(a-b)$.

Proof. We see that, in either of the cases, the truth of one of the inequalities involves that of the other.

Thus if for all unequal values of a and b,

$$xa^{-x-1}(a-b) > a^x - b^x.$$

We may interchange a and b, so that

$$xb^{x-1}(b-a) > b^x - a^x,$$

and therefore, $a^x - b^x > xb^{x-1}(a-b)$.

Thus it is only necessary to prove one inequality in each case. Let $a/b = k$. Then

(i) If $x > 0, a^x - b^x \gtrless xb^{x-1}(a-b)$

according as $k^x - 1 \gtrless x(k-1)$

or, according as $\frac{k^x - 1}{x} \gtrless \frac{k-1}{1}$,

i.e., according as $x \gtrless 1$ (see II above).

(ii) If x is negative and equal to $-y$, then $xa^{x-1}(a-b) \gtrless a^x - b^x$

according as $-yk^{-y-1}(k-1) \gtrless k^{-y} - 1$,

i.e., according as $-y(k-1) \gtrless k - k^{y+1}$,

i.e., according as $k^{y+1} - 1 \gtrless k(y+1) - y - 1$,

i.e., according as $\frac{k^{y+1} - 1}{y+1} \gtrless \frac{k-1}{1}$,

i.e., according as $y + 1 \gtrless 1$ (see II above) or $y \gtrless 0$.

Now $y > 0$ and hence $xa^{x-1}(a-b) > a^x - b^x$.

This completes the proof of the required inequality.

PARTICULAR CASES. When $b = 1$, the second inequalities become

$$a^x - 1 > x(a-1), \text{ if } x < 0 \text{ or } > 1$$

$$a^x - 1 < x(a-1), \text{ if } 0 < x < 1.$$

Writing n for x and $1+x$ for a, it follows that if n is any rational number except 1 and $x > -1$ and $\neq 0$, then

$$(1+x)^n > 1 + nx \text{ if } n < 0 \text{ or } n > 1$$

$$(1+x)^n < 1 + nx \text{ if } 0 < n < 1$$

or again changing the sign of x, if $x < 1$ and $\neq 0$

$$(1-x)^n > 1 - nx \text{ if } n < 0 \text{ or } n > 1$$

$$(1-x)^n < 1 - nx \text{ if } 0 < n < 1.$$

Applications

▶ Example 14.6.2 *If $a_1, a_2, \ldots, a_n$ are n positive numbers and if s_n is their sum, then prove that*

$$1 + s_n + \frac{s_n^2}{\lfloor 2} + \cdots + \frac{s_n^n}{\lfloor n} - (1+a_1)(1+a_2)\cdots(1+a_n) \geq 0.$$

Solution: Denoting the b.h.s. by u_n we get, if $n > 1$.

$$u_n - u_{n-1} > \sum_{r=1}^{n} \frac{s_n^r - s_{n-1}^r}{\lfloor r} - a_n(1+a_1)(1+a_2)\cdots(1+a_{n-1}).$$

Now $s_n^r - s_{n-1}^r > r a_n s_{n-1}^{r-1}$, when $(r > 1)$, and when $r = 1$ this becomes an equality.

$$\therefore \frac{u_n - u_{n-1}}{a_n} > \sum_{r=1}^{n-1} \frac{s_{n-1}^{r-1}}{\lfloor r-1} - (1+a_1)(1+a_2)\cdots(1+a_{n-1}) > u_{n-1};$$

hence $u_n - u_{n-1} > a_n u_{n-1}$ and $u_n > (1+a_n)u_{n-1} > u_{n-1}$.

Now $u_1 > 0$, therefore, $u_n > 0$ for $n = 2, 3, 4 \cdots$

Exercises on Chapter 14

1. If a, b, c are three unequal positive real numbers, then prove that

(a) $a^3 + b^3 + c^3 > 3abc$;

(b) $(b+c)(c+a)(a+b) > 8abc$;

(c) $(a+b-c)(b+c-a)(c+a-b) < abc$, if $a+b > c$, $b+c > a$, $c+a > b$;

(d) $(b+c+a)(a^4+b^4+c^4) > 9a^2b^2c^2$.

[In each of these cases use A.M. > G.M., e.g., (b) $\frac{b+c}{2} > \sqrt{bc}$, $\frac{c+a}{2} > \sqrt{ca}$, $\frac{a+b}{2} > \sqrt{ab}$; multiply.]

2. Prove that

(a) $\left(\frac{n+1}{2}\right)^n > n!$ (n is a positive integer);

(b) $n(n+1)^3 < 8(1^3 + 2^3 + \cdots + n^3)$.

[(a) $\frac{1+2+3+\cdots+n}{n} > \sqrt[n]{1.2.3.\cdots n}$, i.e., $\frac{n(n+1)}{2n} > \sqrt[n]{n!}$; hence etc.;

(b) $\frac{1^3+2^3+3^3+\cdots+n^3}{n} > \left(\frac{1+2+3+\cdots+n}{n}\right)^3$, use m-theorem with $m = 3$.]

3. If $a_1, a_2, \ldots, a_n$ are n positive numbers, not all equal and if $a_1 + a_2 + \cdots + a_n = s$, then prove that

(a) $\dfrac{s}{s-a_1}+\dfrac{s}{s-a_2}+\cdots+\dfrac{s}{s-a_n}>\dfrac{n^2}{n-1}$.

(b) $\dfrac{a_1}{s-a_1}+\dfrac{a_2}{s-a_2}+\cdots+\dfrac{a_n}{s-a_n}>\dfrac{n}{n-1}$.

(c) $\left(\dfrac{s}{a_1}-1\right)^{a_1}\left(\dfrac{s}{a_2}-1\right)^{a_2}\cdots\left(\dfrac{s}{a_n}-1\right)^{a_n}<(n-1)^s$.

[(i) Use A.M. > G.M. on two sets $\{(s-a_1),(s-a_2),\ldots(s-a_n)\}$ and

$$\left\{\frac{s}{s-a_1},\frac{s}{s-a_2},\ldots,\frac{s}{s-a_n}\right\}$$

and then multiply.

(ii) Write in (i) $\dfrac{s}{s-a_1}=\dfrac{s-a_1+a_1}{s-a_1}=1+\dfrac{a_1}{s-a_1}$, etc.

(iii) Take the set $\left\{\dfrac{s}{a_1}-1,\dfrac{s}{a_2}-1,\dfrac{s}{a_3}-1,\ldots\dfrac{s}{a_n}-1\right\}$ associated with weights $\{a_1,a_2,\ldots a_n\}$ and use weighted A.M. > weighted G.M.]

4. If a, b, c are positive numbers, not all equal, prove that

$$\frac{9}{a+b+c}<\frac{2}{b+c}+\frac{2}{c+a}+\frac{2}{a+b}<\frac{1}{a}+\frac{1}{b}+\frac{1}{c}.$$

[Use A.M. > G.M. on the sets $\left\{\dfrac{1}{b+c},\dfrac{1}{c+a},\dfrac{1}{a+b}\right\}$ and $\{b+c,c+a,a+b\}$; multiply $\Rightarrow$ left side follows.

For the other part of the inequality, use m-theorem $(m=-1)$ on $\{b,c\}$

$$\frac{b^{-1}+c^{-1}}{2}>\left(\frac{b+c}{2}\right)^{-1}, \text{ i.e., } \frac{1}{b}+\frac{1}{c}>\frac{4}{b+c}.$$

Similarly obtain $\dfrac{1}{c}+\dfrac{1}{a}>\dfrac{4}{c+a}$, $\dfrac{1}{a}+\dfrac{1}{b}>\dfrac{4}{a+b}$ and then add.]

5. Prove that (a, b, c are positive real numbers.)

(a) $27(a^4+b^4+c^4)>(a+b+c)^4$; use m-theorem with $m=4$.

(b) a, b, c are in H.P. and n is a positive integer >1, then $a^n+c^n>2b^n$.

$$\left[\frac{a^n+c^n}{2}>\left(\frac{a+c}{2}\right)^n.\right. \qquad \text{(i)}$$

Now a, b, c are in H.P.; hence

$$\frac{a^{-1}+c^{-1}}{2}=\frac{1}{b}.$$

But

$$\frac{a^{-1}+c^{-1}}{2}>\left(\frac{a+c}{2}\right)^{-1}, \text{ or, } \frac{1}{b}>\left(\frac{a+c}{2}\right)^{-1}, \text{ or, } \frac{a+c}{2}>b$$

Now (i) gives $a^n+c^n>2b^n$.]

6. If a, b are positive variables subject to $a + b = 4$, prove that

$$\left(a+\frac{1}{a}\right)^2+\left(b+\frac{1}{b}\right)^2 \geq \frac{25}{2}.$$

$\left[\frac{a^2+b^2}{2} \geq \left(\frac{a+b}{2}\right)^2; \frac{a^{-2}+b^{-2}}{2}\left(\frac{a+b}{2}\right)^{-2}\right.$ and $a + b = 4$. Now write

$$\left(a+\frac{1}{a}\right)^2+\left(b+\frac{1}{b}\right)^2 = a^2+b^2+a^{-2}+b^{-2}+4$$

and use the above results.]

7. If $a_1 > a_2 > \cdots > a_n > a_{n+1}$, prove that

$$(a_1 - a_{n+1})^n > n^n(a_1 - a_2)(a_2 - a_3)\cdots(a_n - a_{n+1})$$

unless $a_1, a_2 \cdots a_n, a_{n+1}$ are in A.P.

[Write $a_1 - a_{n+1} = (a_1 - a_2) + (a_2 - a_3) + \cdots + (a_n - a_{n+1})$.

Use A.M. $\geq$ G.M. Equality occurs when $a_1 - a_2 = a_2 - a_3 = \cdots = a_n - a_{n+1}$, i.e., when $a_1, a_2 \cdots a_n, a_{n+1}$ are in A.P.]

8. If a, b, c, d are four real quantities such that $a + b + c + d = p$ where p is a constant, then prove that the greatest value of $(p - a)(p - b)(p - c)(p - d)$ is $\frac{81}{256}p^4$ when $a = b = c = d = p/4$.

9. If a, b, x are all positive, then prove that

$$\frac{a+x}{b+x} > \text{ or } < \frac{a}{b} \text{ according as } a < \text{ or } > b.$$

$\left[\frac{a+x}{b+x} \gtrless \frac{a}{b}\right.$ according as $ab + bx \gtrless ab + ax$, i.e., according as $b \gtrless a$ or $a \gtrless b$]

10. Using the result of **Example 9** above, prove that

(a) $\frac{1}{\sqrt{4n+1}} < \frac{3.7.11.\cdots(4n-1)}{5.9.13\cdots(4n+1)} < \sqrt{\frac{3}{4n+3}}$.

(b) $\frac{1}{2\sqrt{n+1}} < \frac{1.3.5.\cdots(2n-1)}{2.4.6\cdots 2n} < \frac{1}{\sqrt{2n+1}}$.

[(i) By **Example 9** above, $\frac{3}{5} < \frac{5}{7}, \frac{7}{9} < \frac{9}{11} \cdots \frac{4n-1}{4n+1} < \frac{4n+1}{4n+3}$.

Multiply: $\frac{3.7.11\cdots 4n-1}{5.9.13.\cdots 4n+1} < \frac{5}{7}.\frac{9}{11}\cdots\frac{4n+1}{4n+3}$, i.e., $u_n < \frac{5.9.13\cdots(4n+1)}{7.11.15\cdots(4n+3)}$

$$\therefore u_n \cdot u_n < \frac{5.9.13\cdots(4n+1)}{7.11.15\cdots(4n+3)}\cdot\frac{3.7.11\cdots(4n-1)}{5.9.13\cdots(4n+1)},$$

i.e., $< \dfrac{3}{4n+3}$ or, $u_n < \sqrt{\dfrac{3}{4n+3}}$ (2nd part)

Now $(4n+1)u_n = \dfrac{3}{1}\cdot\dfrac{7}{5}\cdot\dfrac{11}{9}\cdots\dfrac{4n-1}{4n-3}$.

Example 9 above gives $\dfrac{3}{1} > \dfrac{5}{3}, \dfrac{7}{5} > \dfrac{9}{7} \cdots \dfrac{4n-1}{4n-3} > \dfrac{4n+1}{4n-1}$

On multiplication we get

$$\frac{3.7.\cdots(4n-1)}{1.5.\cdots(4n-3)} > \frac{5.9\cdots(4n+1)}{3.7\cdots(4n-1)} \text{ or, } (4n+1)u_n > \frac{5.9\cdots(4n+1)}{3.7\cdots(4n-1)}.$$

$$\text{or,}\quad (4n+1)u_n\cdot u_n > \frac{5.9\cdots(4n+1)}{3.7\cdots(4n-1)}\cdot\frac{3\cdot7\cdot11\cdots(4n-1)}{5\cdot9\cdot13\cdots(4n+1)}$$

$$\text{or,}\quad (4n+1)u_n^2 > 1, \text{ i.e., } u_n > \frac{1}{\sqrt{4n+1}}$$

$$\therefore\quad \frac{1}{\sqrt{4n+1}} < u_n < \sqrt{\frac{3}{4n+3}}.]$$

11. Prove that

$$\lfloor\underline{1}\,\lfloor\underline{3}\,\lfloor\underline{5}\cdots\lfloor\underline{2n-1} > (\lfloor\underline{n})^n.$$

[Let a and b be two positive integers s.t. $a+b=2m$. Then $\lfloor\underline{a}\,\lfloor\underline{b}$ is least if

$$a = b = m.$$

In case $a+b=2m+1$, then $\lfloor\underline{a}\,\lfloor\underline{b}$ is least if $a = m$ and $b = m+1$.

For $\lfloor\underline{n-r}\,\lfloor\underline{r} > \lfloor\underline{n-(r+1)}\,\lfloor\underline{r+1}$ if $n > 2r+1$, i.e., if $r < f\dfrac{n-1}{2}$. Hence

$$\lfloor\underline{n-1}\,\lfloor\underline{1} > \lfloor\underline{n-2}\,\lfloor\underline{2} > \lfloor\underline{n-3}\,\lfloor\underline{3} > \cdots$$

In the present problem: take $a+b=2n$. Then the least value of $\lfloor\underline{2n-r}\,\lfloor\underline{r}$ is obtained when $2n-r=r$, i.e., when $r=n$, the least value being $\lfloor\underline{n}\,\lfloor\underline{n}$,

$$\begin{aligned} \text{i.e.,}\quad \lfloor\underline{n}\,\lfloor\underline{n} &< \lfloor\underline{2n-1}\,\lfloor\underline{1} \\ \lfloor\underline{n}\,\lfloor\underline{n} &< \lfloor\underline{2n-3}\,\lfloor\underline{3} \\ &\cdots\cdots\cdots\cdots \\ \lfloor\underline{n}\,\lfloor\underline{n} &< \lfloor\underline{3}\,\lfloor\underline{2n-3} \\ \lfloor\underline{n}\,\lfloor\underline{n} &< \lfloor\underline{1}\,\lfloor\underline{2n-1} \end{aligned}$$

Multiplying we get $(\lfloor\underline{n})^{2n} < \{\lfloor\underline{1}\,\lfloor\underline{3}\,\lfloor\underline{5}\cdots\lfloor\underline{2n-1}\}^2$; hence etc.]

12. Prove that

$$81abcd < (s-a)(s-b)(s-c)(s-d) < \frac{81}{256}s^4.$$

where a, b, c, d are four positive numbers, not all equal, and $s = a+b+c+d$.

[$s - a = b+c+d > 3\sqrt[3]{bcd}$; etc. Multiply and get the first inequality. For the second inequality,

$$\frac{(s-a)+(s-b)+(s-c)+(s-d)}{4} > \sqrt[4]{(s-a)(s-b)(s-c)(s-d)}]$$

13. If x, y, z are three positive rational numbers, prove that

$$\left(\frac{x^2+y^2+z^2}{x+y+z}\right)^{x+y+z} \geq x^x \cdot y^y \cdot z^z.$$

[(i) First suppose that x, y, z are positive integers and take x quantities, each to x; take y quantities each equal equal to y and take z quantities each equal to z and use A.M. $\geq$ G.M.

(ii) If x, y, z are positive fractions, then mx, my, mz are positive integers where m is the l.c.m. of the denominators of x, y, z. Now use (i).]

14. Show that if $a+b+c+d=3s$, a, b, c, d, $s-a$, $s-b$, $s-c$, $s-d$ are all positive, then prove that $abcd > 81(s-a)(s-b)(s-c)(s-d)$.

15. Find the greatest value of $x^\alpha y^\beta z^\gamma$ for different values of x, y, z subject to the condition $x+y+z$ is constant. (α, β, γ are any set of positive rational numbers.)

16. If a_1, a_2, a_3, a_4 are four distinct positive numbers and $S = a_1+a_2+a_3+a_4$, then prove that

$$\frac{S}{a_2+a_3+a_4}+\frac{S}{a_1+a_3+a_4}+\frac{S}{a_1+a_2+a_4}+\frac{S}{a_1+a_2+a_3} > \frac{16}{3}.$$

17. Let $s = a_1+a_2+\cdots+a_n$, where $a_1, a_2, \ldots a_n$ are n positive real numbers, not all equal. Prove that

$$(n-1)\left[\frac{1}{s-a_1}+\frac{1}{s-a_2}+\cdots+\frac{1}{s-a_n}\right] < \frac{1}{a_1}+\frac{1}{a_2}+\cdots\frac{1}{a_n}.$$

[Take $n-1$ real positive numbers, not all equal, $x_1, x_2 \cdots x_{n-1}$.
Using A.M. > G.M. we obtain

$$\frac{x_1+x_2+\cdots+x_{n-1}}{n-1} > \sqrt[n-1]{x_1x_2\cdots x_{n-1}}$$

and
$$\frac{\frac{1}{x_1}+\frac{1}{x_2}+\cdots+\frac{1}{x_{n-1}}}{n-1} > \sqrt[n-1]{\frac{1}{x_1x_2\cdots x_{n-1}}}.$$

$$\therefore \quad (x_1^{-1}+x_2^{-1}+\cdots+x_{n-1}^{-1}) > \frac{(n-1)^2}{x_1+x_2+\cdots+x_{n-1}}.$$

Let us replace the set $\{x_1, x_2, \ldots, x_{n-1}\}$ by each of the n sets that can be formed from the set $\{a_1, a_2, \ldots, a_n\}$ by omitting one term and then add the n inequalities. Since a's are not all equal, the strict inequality holds in at least one case. Hence

$$(n-1)\left(\frac{1}{a_1}+\frac{1}{a_2}+\cdots+\frac{1}{a_n}\right) > (n-1)^2\left(\frac{1}{s-a_1}+\frac{1}{s-a_2}+\cdots+\frac{1}{s-a_n}\right)$$

and the result follows.]

18. If $x > n$, then prove that

$$\left(1+\frac{x}{n+1}\right)^{n+1} \geq \left(1+\frac{x}{n}\right)^{n}.$$

[Consider $\{1, 1+\frac{x}{n}\}$ associated with weights 1 and n and use weighted A.M. $\geq$ weighted G.M. Equality occurs when $1 = 1 + \frac{x}{n}$, i.e., when $x = 0$]

19. If a, b, c, d, e are positive numbers, then prove that

$$\left(\frac{a+b+c+d+e}{5}\right)^5 \geq \left(\frac{a+b}{2}\right)^2 \left(\frac{c+d+e}{3}\right)^3.$$

20. If a, b, c, d are all positive numbers, then prove that

$$a^5 + b^5 + c^5 + d^5 \geq abcd(a+b+c+d).$$

$\left[\frac{\Sigma a^5}{4} > \frac{\Sigma a^4}{4}\cdot\frac{\Sigma a}{4}\right.$ and $\frac{\Sigma a^4}{4} > \sqrt[4]{a^4 \cdot b^4 \cdot c^4 \cdot d^4}$, i.e., $> abcd$. Hence, $\frac{\Sigma a^5}{4} > abcd\frac{\Sigma a}{4}$ and the result follows.]

21. If a be a positive number, $a \neq 1, x$ and y are positive rationals, then prove that

$$\frac{a^x - 1}{x} > \frac{a^y - 1}{y} \text{ if } x > y.$$

[$a^x > 0$ and $a^x \neq 1$. Take positive numbers a^x and 1 and associate positive rational weights y and $x - y$ respectively. Then

$$\frac{y \cdot a^x + (x-y)\cdot 1}{y + x - y} > \left[(a^x)^y \cdot 1^{x-y}\right]^{\frac{1}{y+x-y}}$$

Now, simplify.]

22. If the sum of any two of the three quantities a, b, c is greater than the third and if $x + y + z = 0$, prove that $a^2yz + b^2zx + c^2xy \leq 0$.

[(a) If $x = y = z = 0$, then the expression $a^2yz + b^2zx + c^2xy$ is clearly zero.

(b) If one of x, y, z is zero, say $x = 0$, then $x + y + z = 0$ gives $y + z = 0$, i.e., y and z are of opposite signs, i.e., $yz < 0$ and hence

$$a^2yz + b^2zx + c^2xy = a^2yz < 0.$$

(c) Suppose none of x, y, z is zero. Then $x + y + z = 0$ implies that the sign of one of x, y, z is opposite to the sum of the other two; say $x = -(y+z)$, where y and z are positive. Write

$$\begin{aligned} a^2yz + b^2zx + c^2xy &= a^2yz + x(b^2z + c^2y) \\ &= a^2yz + (-y-z)(b^2z + c^2y) \\ &= a^2yz - b^2yz - b^2z^2 - c^2y^2 - c^2yz \\ &= -[(cy - bz)^2 + yz\{(b+c)^2 - a^2\}] < 0 \end{aligned}$$

since $(cy - bz)^2 \geq 0$ and $yz > 0$ and $(b+c)^2 - a^2 > 0 (b + c > a)$]

23. If a is any positive number except 1, and x, y, z are rationals, no two of which are equal and which may be positive or negative, then prove that

$$a^x(y-z)+a^y(z-x)+a^z(x-y)>0.$$

[First suppose, x, y, z be integers and without loss of generality let $x>y>z$. Consider $a^x, a^x, \ldots, (y-z)$ times and $a^z, a^z, \ldots, (x-y)$ times and use A.M. > G.M. The result will follow. In case, x, y, z are fractions, then if d be the l.c.m. of their denominators, then dx, dy, dz are positive integers and use the previous case. The result will again follow.]

24. If x, y, z, w are in H.P., then prove that $x+w>y+z$.
[Let the reciprocals of x, y, z, w be $p-3q$, $p-q$, $p+q$, $p+3q$]

25. Prove that if a, b, c be three distinct positive numbers, then

$$\frac{b^2+c^2}{b+c}+\frac{c^2+a^2}{c+a}+\frac{a^2+b^2}{a+b}>a+b+c.$$

[use m-Theorem $(m=2)\dfrac{b^2+c^2}{2}>\left(\dfrac{b+c}{2}\right)^2$ or $\dfrac{b^2+c^2}{b+c}>\dfrac{1}{2}(b+c)$, etc.]

26. If a, b, c are positive and not all equal, then prove that

$$6abc<a^2(b+c)+b^2(c+a)+c^2(a+b)<2(a^3+b^3+c^3).$$

27. If a, b, c, d are all >1, then prove that

$$8(abcd+1)>(a+1)(b+1)(c+1)(d+1).$$

28. m-Theorem can be generalised thus:

If $a, b, \ldots k$ and $x, y, \ldots w$ are two sets of n rationals and these of the first set are not all equal to one another, then m being any rational.

$$\frac{xa^m+yb^m+\cdots+wk^m}{x+y+\cdots+w}\gtrless\left(\frac{xa+yb+\cdots+wk}{x+y+\cdots+w}\right)^m$$

according as m **does not** or **does** lie between 0 and 1. Deduce this theorem.

Hence, prove that $\dfrac{a^{-1}b+b^{-1}c+c^{-1}a}{b+c+a}\geq\dfrac{a+b+c}{ab+bc+ca}$.

Also prove that $3(a^2b+b^2c+c^2a)(ab^2+bc^2+ca^2)\geq abc(a+b+c)^3$.

29. Show that the expressions

$$a(a-b)(a-c)+b(b-c)(b-a)+c(c-a)(c-b)$$

and

$$a^2(a-b)(a-c)+b^2(b-c)(b-a)+c^2(c-a)(c-b)$$

are both positive.

30. Show that $\dfrac{(a+b)xy}{ay+bx}\leq\dfrac{ax+by}{a+b}$.

31. Prove that if $s=a+b+c+\cdots$ (n quantities)

$$\left(\frac{s-a}{n-1}\right)^a\left(\frac{s-b}{n-1}\right)^b\left(\frac{s-c}{n-1}\right)^c\cdots<\left(\frac{s}{n}\right)^s.$$

14.7 Weierstrass', Holder's and Minkowski's Inequalities

I. Weierstrass' Inequalities:

If $a_1, a_2, \ldots a_n$ are positive real numbers, all less than 1 and if $s = \sum_{i=1}^{n} a_i$, then

$$1 - s_n < (1 - a_1)(1 - a_2)\cdots(1 - a_n) < \frac{1}{1 + s_n}$$

$$\text{and} \quad 1 + s_n < (1 + a_1)(1 + a_2)\cdots(1 + a_n) < \frac{1}{1 - s_n}(s_n < 1).$$

Proof. $(1 - a_1)(1 - a_2) = 1 - (a_1 + a_2) + a_1 a_2 > 1 - (a_1 + a_2)$

$$\begin{aligned}(1 - a_1)(1 - a_2)(1 - a_3) &> \{1 - (a_1 + a_2)\}(1 - a_3)\\ &> \{1 - (a_1 + a_2 + a_3)\};\end{aligned}$$

continuing in this manner (or by using the Principle of Induction)

$$\begin{aligned}&(1 - a_1)(1 - a_2)(1 - a_3)\cdots(1 - a_n)\\ &> \{1 - (a_1 + a_2 + \cdots + a_n)\}, \text{ i.e., } > 1 - s_n.\end{aligned}$$

In the same way, we can deduce

$$(1 + a_1)(1 + a_2)(1 + a_3)\cdots(1 + a_n) > 1 + s_n.$$

Again, $(1 + a_1)(1 - a_1) = 1 - a_1^2 < 1$,

$$\text{i.e.,} \quad 1 - a_1 < \frac{1}{1 + a_1}; \text{ also } (1 + a_1) < \frac{1}{1 - a_1} \quad (\because\ 0 < a_1 < 1)$$

$$\therefore \quad (1 - a_1)(1 - a_2)\cdots(1 - a_n) < \frac{1}{(1 + a_1)(1 + a_2)\cdots(1 + a_n)} < \frac{1}{1 + s_n}.$$

Also

$$\begin{aligned}(1 + a_1)(1 + a_2)\cdots(1 + a_n) &< \frac{1}{(1 - a_1)(1 - a_2)\cdots(1 - a_n)}\\ &< \frac{1}{1 - s_n} \quad (s_n < 1 \text{ assumed}).\end{aligned}$$

II. Holder's Inequality

Let

$$[a] = [a_1, a_2, \ldots, a_n] \text{ and } [b] = [b_1, b_2, \ldots b_n]$$

be two sets of n positive numbers, which are not proportional. If α, β are positive and $\alpha + \beta = 1$, then

$$\sum_{r=1}^{n} a_r^{\alpha} b_r^{\beta} < \left(\sum_{r=1}^{n} a_r\right)^{\alpha} \left(\sum_{r=1}^{n} b_r\right)^{\beta}. \tag{A}$$

But if $\alpha > 1$ or if $\alpha < 0$, then

$$\sum_{r=1}^{n} a_r^{\alpha} b_r^{\beta} > \left(\sum_{r=1}^{n} a_r\right)^{\alpha} \left(\sum_{r=1}^{n} b_r\right)^{\beta}. \tag{B}$$

[α, β are supposed to be rational numbers.]

Proof. (i) First suppose that α and β are positive rational numbers.

Let $a_1 + a_2 + \cdots + a_n = A_n$ and $b_1 + b_2 + \cdots + b_n = B_n$.

Take two positive numbers

$$\frac{a_r}{A_n} \text{ and } \frac{b_r}{B_n} \text{ with associated weights } \alpha \text{ and } \beta.$$

Then using weighted A.M. $\geq$ weighted G.M., we get

$$\frac{\alpha\frac{a_r}{A_n} + \beta\frac{b_r}{B_n}}{\alpha + \beta} \geq \left\{\left(\frac{a_r}{A_n}\right)^{\alpha} \cdot \left(\frac{b_r}{B_n}\right)^{\beta}\right\}^{\frac{1}{\alpha+\beta}}$$

$$\therefore \left(\frac{a_r}{A_n}\right)^{\alpha} \cdot \left(\frac{b_r}{B_n}\right)^{\beta} < \alpha\frac{a_r}{A_n} + \beta\frac{b_r}{B_n} \text{ unless } \frac{a_r}{A_n} = \frac{b_r}{B_n}.$$

Adding the results given by $r = 1, 2, 3, \ldots, n$

$$\frac{\sum_{r=1}^{n} a_r^{\alpha} b_r^{\beta}}{A_n^{\alpha} B_n^{\beta}} < \alpha + \beta = 1.$$

Hence,

$$\sum_{r=1}^{n} a_r^{\alpha} b_r^{\beta} < \left(\sum_{r=1}^{n} a_r\right)^{\alpha} \left(\sum_{r=1}^{n} b_r\right)^{\beta}.$$

Note 1. *If* $\frac{a_r}{A_n} = \frac{b_r}{B_n}$, *for any particular value of* r *(not for all* r*) the corresponding inequality becomes an equality, but at least one of the inequalities will hold unless* $\frac{a_r}{A_n} = \frac{b_r}{B_n}$ *for all values of* r.

Hence, the inequality in **(A)** holds unless $[a]$ and $[b]$ are proportional sets, i.e., unless

$$\frac{a_1}{b_1} = \frac{a_2}{b_2} = \cdots = \frac{a_n}{b_n} = \frac{A_n}{B_n}.$$

(ii) Next suppose that $\alpha > 1$ which implies $\beta < 0$. We put $\alpha = \frac{1}{p}$ and $q = 1 - p$. This makes p and q both positive.

See that

$$\frac{\beta}{\alpha} = \frac{1-\alpha}{\alpha} = p - 1 = q.$$

Also put $a_r = c_r d_r$ and $b_r = (d_r)^{-\alpha/\beta}$.

This makes

$$a_r^{\alpha} b_r^{\beta} = (c_r d_r)^{\alpha} \cdot (d_r)^{-\frac{\alpha}{\beta} \cdot \beta} = c_r^{\alpha} = c_r^{\frac{1}{p}} \text{ and } d_r = (b_r)^{-\frac{\beta}{\alpha}} = b_r^{q}.$$

Now, in **(A)** we write $c_r^{\frac{1}{p}}$ for a_r, $d_r^{\frac{1}{q}}$ for b_r and p, q for α, β respectively. Then

$$\sum_{r=1}^{n}\left(c_r^{\frac{1}{p}}\right)^p \cdot \left(d_r^{\frac{1}{q}}\right)^q < \left(\sum_{r=1}^{n} c_r^{\frac{1}{p}}\right)^p \left(\sum_{r=1}^{n} d_r^{\frac{1}{q}}\right)^q,$$

$$\text{i.e.,} \quad \sum_{r=1}^{n} c_r d_r < \left(\sum_{r=1}^{n} c_r^{\frac{1}{p}}\right)^p \left(\sum_{r=1}^{n} d_r^{\frac{1}{q}}\right)^q$$

$$\text{or,} \quad \sum_{r=1}^{n} a_r < \left(\sum_{r=1}^{n} a_r^{\alpha} b^{\beta}\right)^{\frac{1}{\alpha}} e\cdot \left(\sum_{r=1}^{n} b_r\right)^{-\frac{\beta}{\alpha}}.$$

Taking αth power of both sides $(\alpha > 0)$ we get

$$\left(\sum_{r=1}^{n} a_r\right)^{\alpha} \left(\sum_{r=1}^{n} b_r\right)^{\beta} < \sum_{r=1}^{n} a_r^{\alpha} b_r^{\beta}.$$

By interchanging a, b and interchanging α, β it follows that **(B)** also holds for $\beta > 1$ which implies $\alpha < 0$.

The results **(A)** and **(B)** can be written as

$$\sum_{r=1}^{n} a_r b_r < \left(\sum_{r=1}^{n} a_r^{\frac{1}{\alpha}}\right)^{\alpha} \left(\sum_{r=1}^{n} b_r^{\frac{1}{\beta}}\right)^{\beta}, \quad \text{if } \alpha, \beta \text{ positive} \tag{C}$$

$$\text{and} \quad \sum_{r=1}^{n} a_r b_r > \left(\sum_{r=1}^{n} a_r^{\frac{1}{\alpha}}\right)^{\alpha} \left(\sum_{r=1}^{n} b_r^{\frac{1}{\beta}}\right)^{\beta}, \quad \text{if } \alpha > 1 \text{ or if } \alpha < 0. \tag{D}$$

[Instead of the sets $[a]$ and $[b]$, take the sets as $[a^{\frac{1}{\alpha}}]$ and $[b^{\frac{1}{\beta}}]$, not proportional and $\alpha + \beta = 1$]

III. Minkowski's Inequality

We shall introduce the following notations:

$$M_r(a) = \left\{\frac{a_1^r + a_2^r + \cdots\cdots + a_n^r}{n}\right\}^{\frac{1}{r}}$$

$$M_r(a, p) = \left(\frac{p_1 a_1^r + p_2 a_2^r + \cdots + p_n a_n^r}{p_1 + p_2 + \cdots + p_n}\right)^{\frac{1}{r}}.$$

In particular,

$$M_1(a) = \frac{a_1 + a_2 + \cdots + a_n}{n} = A(a) \text{ [Arithmetic Mean of the set } [a]]$$

$$M_1(a, p) = \frac{p_1 q_1 + p_2 q_2 + \cdots + p_n q_n}{p_1 + p_2 + \cdots + p_n} \text{ [Weighted A.M. of the set } [a]]$$

Minkowski's Inequality

Let $[a]$ and $[b]$ be two sets of n positive numbers. Suppose the sets are not proportional. Let $[s]$ be the set of n numbers

$$\{s_1 = a_1 + b_1, s_2 = a_2 + b_2, \ldots, s_n = a_n + b_n\}.$$

Then $M_r(s) < M_r(a) + M_r(b)$, if $r > 1$,

$$\text{i.e.,} \quad \left(\frac{s_1^r + s_2^r + \cdots + s_n^r}{n}\right)^{\frac{1}{r}} < \left(\frac{a_1^r + a_2^r + \cdots + a_n^r}{n}\right)^{\frac{1}{r}} + \left(\frac{b_1^r + b_2^r + \cdots + b_n^r}{n}\right)^{\frac{1}{r}} \quad (1)$$

and $\quad M_r(s > M_r(a) + M_r(b)$, if $r < 1, r \neq 0$ **(2)**

Proof. (i) First suppose $r > 1$. Let r' be such that $\frac{1}{r'} + \frac{1}{r} = 1$.

Put in **(C)** under the heading "Holder's Inequality"

$$\alpha = \frac{1}{r},\ \beta = \frac{1}{r'},\ b_i = (s_i)^{\frac{r}{r'}}$$

Then,

$$\sum_{i=1}^{n} a_i s_i^{r-1} = \sum_{i=1}^{n} a_i (s_i)^{\frac{r}{r'}} < \left(\sum_{i=1}^{n} a_i^r\right)^{\frac{1}{r}} \left(\sum_{i=1}^{n} s_i^r\right)^{\frac{1}{r'}}$$

unless $[a]$ and $[s]$ are proportional. Similarly

$$\sum_{i=1}^{n} b_i s_i^{r-1} < \left(\sum b_i^r\right)^{\frac{1}{r}} \left(\sum s_i^r\right)^{\frac{1}{r'}}$$

unless $[b]$ and $[s]$ are proportional.

Hence, by addition

$$\sum_{i=1}^{n} s_i^{r-1}(a_i + b_i) < \left(\sum s_i^r\right)^{\frac{1}{r'}} \left[\left(\sum_{i=1}^{n} a_i^r\right)^{\frac{1}{r}} + \left(\sum_{i=1}^{n} b_i^r\right)^{\frac{1}{r}}\right]$$

unless $[a]$ and $[b]$ are proportional.

Thus

$$\frac{\left(\sum_{i=1}^{n} s_i^r\right)}{\left(\sum_{i=1}^{n} s_i^r\right)^{\frac{1}{r'}}} < \left(\sum_{i=1}^{n} a_i^r\right)^{\frac{1}{r}} + \left(\sum_{i=1}^{n} b_i^r\right)^{\frac{1}{r}},$$

$$\text{i.e.,} \quad \left(\sum_{i=1}^{n} s_i^r\right)^{1-\frac{1}{r'}} < \left(\sum_{i=1}^{n} a_i^r\right)^{\frac{1}{r}} + \left(\sum_{i=1}^{n} b_i^r\right)^{\frac{1}{r}}.$$

$$\text{or,} \quad \left(\sum_{i=1}^{n} s_i^r\right)^{\frac{1}{r}} < \left(\sum_{i=1}^{n} a_i^r\right)^{\frac{1}{r}} + \left(\sum_{i=1}^{n} b_i^r\right)^{\frac{1}{r}}.$$

Dividing by n^r, we can write

$$M_r(s) < M_r(a) + M_r(b).$$

(ii) If $r < 1, r \neq 0$, then $\frac{1}{r} > 1$ or $\frac{1}{r} < 0$.

The result [D] under Holder's inequality gives opposite inequalities to those used in (i).

By repeated applications of Minkowaski's inequality, if the sets $[a], [b], \ldots, [l]$, each containing n positive numbers are not all proportional, and if

$$s_r = a_r + b_r + \cdots + l_r \text{ for } r = 1, 2, 3, \ldots, n,$$

then

$$M_r(s) < M_r(a) + M_r(b) + \cdots + M_r(l), \text{ if } r > 1$$

and $$M_r(s) > M_r(a) + M_r(b) + \cdots + M_r(l), \text{ if } r < 1, r \neq 0.$$

Applications

► Example 14.7.1 *If* $a^3 + b^3 + c^3 = 1$, *prove that* $(a+3)^3 + (b+4)^3 + (c+5)^3 < 343$.

Solution: Using Minkowski's inequality (with $r = 3 > 1$)

$$(a^3 + b^3 + c^3)^{1/3} + (3^3 + 4^3 + 5^3)^{1/3} > \{(a+3)^3 + (b+3)^3 + (c+3)^3\}^{1/3}$$

or, $$\{1 + (216)^{1/3}\}^3 > (a+3)^3 + (b+3)^3 + (c+3)^3;$$

hence the result.

► Example 14.7.2 $\sqrt{\{(a+b)^2 + (c+d)^2\}} \leq \sqrt{a^2 + b^2} + \sqrt{c^2 + d^2}$.

Solution: Use Minkowski's inequality with $r = 2 > 1$.

Geometrically $P_1P_2 \leq OP_1 + OP_2$.

IV. Convex Functions: Jensen's Inequalities

If for all values of $x_1, x_2 (x_1 \neq x_2)$ in the interval $a \leq x \leq b$

$$f\left(\frac{x_1 + x_2}{2}\right) < \frac{1}{2}\{f(x_1) + f(x_2)\},$$

then the function $f(x)$ is said to be *convex* in the interval.

We state here (without proof):

If the second derivative $f''(x)$ *exists and is positive in the interval, then* $f(x)$ *is necessarily convex.*

Jensen's Inequalities

If $f(x)$ *is convex in* $a \leq x \leq b$ *and if* $x_1, x_2, \ldots, x_n$ *are points in this interval and if they are not all equal, then*

$$f\left\{\frac{1}{n}(x_1 + x_2 + \cdots + x_n)\right\} < \frac{1}{n}\{f(x_1) + f(x_2) + \cdots + f(x_n)\}. \qquad (\alpha)$$

Moreover, if $p_1, p_2, \ldots, p_n$ *are positive and rational and if* $f(x)$ *is convex in* $a \leq x \leq b$, *then*

$$f\left(\frac{p_1x_1 + p_2x_2 + \cdots + p_nx_n}{p_1 + p_2 + \cdots + p_n}\right) < \frac{p_1f(x_1) + p_2f(x_2) + \cdots + p_nf(x_n)}{p_1 + p_2 + \cdots + p_n}. \quad (\beta)$$

▶ **Example 14.7.3** *If* a_1, a_2, x_1, x_2 *are positive and* $x_1 \neq x_2$, *prove that*

$$a_1x_1 \log x_1 + a_2x_2 \log x_2 > (a_1x_1 + a_2x_2) \log \frac{a_1x_1 + a_2x_2}{a_1 + a_2}.$$

Solution: If

$$\begin{aligned} f(x) &= x \log x \, (x > 0) \\ f'(x) &= 1 + \log x \\ \text{and } f''(x) &= 1/x > 0. \end{aligned}$$

$\therefore$ $f(x)$ is convex and hence from (β) above

$$\frac{a_1f(x_1) + a_2f(x_2)}{a_1 + a_2} > f\left\{\frac{a_1x_1 + a_2x_2}{a_1 + a_2}\right\},$$

$$\text{i.e., } \frac{a_1x_1 \log x_1 + a_2x_2 \log x_2}{a_1 + a_2} > \frac{a_1x_1 + a_2x_2}{a_1 + a_2} \log \frac{a_1x_1 + a_2x_2}{a_1 + a_2}.$$

Hence, the result.

Chapter 15

Simple Continued Fractions

15.1 Introduction

Definition 1. *Any expression of the form*

$$a_1 + \cfrac{b_2}{a_2 + \cfrac{b_3}{a_3 + \cfrac{b_4}{a_4 + \cdots}}}$$

where $a_1, a_2, a_3, \ldots$ *and* $b_2, b_3, b_4, \ldots$ *are real numbers is called a Continued Fraction.*

This expression is written in a compact form as

$$a_1 + \frac{b_2}{a_2+}\,\frac{b_3}{a_3+}\,\frac{b_4}{a_4+\cdots}$$

We shall consider a particular case: $b_2 = b_3 = b_4 = \cdots = 1$ and $a_1, a_2, a_3, \ldots$ are all positive integers. Then the continued fraction becomes

$$a_1 + \frac{1}{a_2+}\,\frac{1}{a_3+}\,\frac{1}{a_4+\cdots}$$

Such a continued fraction is called a *Simple Continued Fraction.* In this chapter we shall discuss only simple continued fractions (S.C.F.).

In the S.C.F. a_1 is called the *first quotient,* a_2 the *second quotient,* a_3 the *third quotient,* etc. When the number of quotients is finite, the continued fraction is said to be a *terminating continued fraction;* otherwise an *infinite continued fraction.*

A vulgar fraction can be expressed as a finite S.C.F., e.g.,

$$\frac{157}{68} = 2+\frac{21}{68} = 2+\cfrac{1}{\frac{68}{21}} = 2+\cfrac{1}{3+\frac{5}{21}}$$

$$= 2+\cfrac{1}{3+\cfrac{1}{\frac{21}{5}}} = 2+\cfrac{1}{3+\cfrac{1}{4+\frac{1}{5}}}$$

$$= 2+\frac{1}{3+}\,\frac{1}{4+}\,\frac{1}{5}.$$

$$\begin{array}{r|l} 68 & 157 \; (2 \\ & 136 \\ \hline & 21 \end{array} \qquad \begin{array}{r|l} 21 & 68 \; (3 \\ & 63 \\ \hline & 5 \end{array} \qquad \begin{array}{r|l} 5 & 21 \; (4 \\ & 20 \\ \hline & 1 \end{array}$$

Note that the quotients 2, 3, 4, 5 are the *successive quotients* when we obtain, by the usual method, the H.C.F. of 68 and 157.

In the discussions that follow we shall take a typical simple continued fraction in the form:

$$a_1 + \frac{1}{a_2+}\,\frac{1}{a_3+}\cdots+\frac{1}{a_n+}, \qquad (1)$$

where $a_1, a_2, \ldots, a_n, \ldots$ are all positive integers. (a_1 may be zero).

The fractions $a_1, a+\frac{1}{a_2}, a_1+\frac{1}{a_2+\frac{1}{a_3}}$, etc. are respectively called *first convergent, second convergent, third convergent,* etc. of the S.C.F. (1).

Suppose we write

$$x_1 = a_1 + \frac{1}{a_2+}\,\frac{1}{a_3+}\cdots$$

$$x_2 = a_2 + \frac{1}{a_3+}\cdots$$

$$x_3 = a_3 + \frac{1}{a_4+}\cdots$$

$$\cdots \quad \cdots \quad \cdots\cdots\cdots$$

$$x_n = a_n + \frac{1}{a_{n+1}+\cdots}\cdots$$

$$\cdots \quad \cdots \quad \cdots\cdots\cdots$$

Then x_1 is called the *first complete quotient* of the S.C.F. (1);

x_2 is the *second complete quotient,* and so on;
x_n is the nth *complete quotient* of the S.C.F. (1),

i.e., the continued fraction x_1 is obtained from its nth convergent by substituting the nth complete quotient x_n in place of the nth quotient a_n,

$$\text{i.e., } x_1 = a_1 + \frac{1}{a_2+}\,\frac{1}{a_3+}\cdots+\frac{1}{a_{n-1}+}\,\frac{1}{x_n}.$$

15.2 Law of Formation of Successive Convergents of a S.C.F.

Let

$$a_1 + \frac{1}{a_2+}\,\frac{1}{a_3+}\cdots\frac{1}{a_n+}\cdots$$

be a simple continued fraction (S.C.F.), where the quotients $a_1, a_2, a_3, \ldots, a_n \cdots$ are all positive integers.

The first three convergent are

$$\frac{a_1}{1}, \quad a+\frac{1}{a_2}, \qquad a_1+\frac{1}{a_2+}\frac{1}{a_3},$$

$$\text{i.e.,} \quad \frac{a_1}{1}, \ \frac{a_2a_1+1}{a_2}, \ \frac{a_3(a_1a_2+1)+a_1}{a_3a_2+1}.$$

We observe that the numerator of the third convergent

= third quotient $(a_3)\times$ numerator of the second convergent
+ numerator of the first convergent.

Exactly, similar rule for the denominator of the third convergent.

Suppose that the numerators of the first, the second, the third, $\cdots$ be denoted by $p_1, p_2, p_3, \ldots$ and the denominators be denoted by $q_1, q_2, q_3, \ldots$.

Suppose for the nth convergent the numerator p_n and the denominator q_n are formed by the same rule as we have stated for the third convergent, i.e., assumed that

$$p_n = a_n p_{n-1} + p_{n-2} \text{ and } q_n = a_n q_{n-1} + q_{n-2}. \tag{2}$$

Now it is clear that the $(n+1)$th convergent differs from the nth convergent only in having the quotient $a_n + \frac{1}{a_{n+1}}$ in the place of a_n; hence the $(n+1)$th convergent

$$= \frac{\left(a_n + \frac{1}{a_{n+1}}\right)p_{n-1} + p_{n-2}}{\left(a_n + \frac{1}{a_{n+1}}\right)q_{n-1} + q_{n-2}}$$

$$= \frac{a_{n+1}(a_n p_{n-1} + p_{n-2}) + p_{n-1}}{a_{n+1}(a_n q_{n-1} + q_{n-2}) + q_{n-1}}$$

$$= \frac{a_{n+1}p_n + p_{n-1}}{a_{n+1}q_n + q_{n-1}} \quad \text{[using (2) which we have assumed.]}$$

If, therefore, we put

$$p_{n+1} = a_{n+1}p_n + p_{n-1}, \; q_{n+1} = a_n q_n + q_{n-1}.$$

We see that the numerator and denominator of the $(n+1)$th convergent follow the law which was assumed to hold in case of nth convergent. That is the law holds for $(n+1)$th convergent if it is assumed to hold for nth convergent. But the law has been found to be true for the third convergent and hence it will be true for the fourth convergent and so on. By the principle of Mathematical Induction it holds for any nth convergent where n is a positive integer ≥ 3. Thus, the law of formation of successive convergents is

First Convergent$=a_1$

Second Convergent$=\dfrac{a_1a_2+1}{a_2}$;

nth Convergent$=p_n/q_n$, where

$$\left.\begin{aligned} p_n &= a_n p_{n-1} + p_{n-2} \\ q_n &= a_n q_{n-1} + q_{n-2} \end{aligned}\right\} n \geq 3.$$

With $p_1 = a_1, q_1 = 1$ and $p_2 = a_1a_2 + 1, q_2 = a_2$.

15.3 Properties of the Convergent (No. 1–No. 9)

We shall use the notations introduced in art. **15.2.**

1. $p_n q_{n-1} - p_{n-1} q_n = (-1)^n$.

Let us call $u_n = p_n q_{n-1} - p_{n-1} q_n$.

Then

$$\begin{aligned} u_n &= (a_n p_{n-1} + p_{n-2}) q_{n-1} - p_{n-1}(a_n q_{n-1} + q_{n-2}) \\ &= a_n p_{n-1} q_{n-1} + p_{n-2} q_{n-1} - a_n p_{n-1} q_{n-1} - p_{n-1} q_{n-2} \\ &= -(p_{n-1} q_{n-2} - p_{n-2} q_{n-1}) = -u_{n-1}. \end{aligned}$$

Similarly, $u_{n-1} = -u_{n-2}$ and so $u_n = (-1)^2 u_{n-2}$.

Again, $u_{n-2} = -u_{n-3}$ and so $u_n = (-1)^3 u_{n-3}$ and so on. We, therefore, can write $u_n = (-1)^{n-2} u_2 = (-1)^{n-2}(p_2 q_1 - p_1 q_2)$.

But we know $p_2 = a_1 a_2 + 1$, $q_2 = a_2$, $p_1 = a_1$, $q_1 = 1$.

$$\therefore \quad u_n = (-1)^{n-2}(a_1 a_2 + 1 - a_2 a_1) = (-1)^{n-2}(-1)^2,$$

$$\text{i.e.,} \quad p_n q_{n-1} - p_{n-1} q_n = (-1)^n.$$

DEDUCTIONS. (i) *Every convergent is in its lowest terms, also p_n is prime to p_{n-1} and q_n to q_{n-1}.*

For, if p_n and q_n had a common factor g other than 1, then g would divide ± 1,

$(\because\ p_n q_{n-1} - p_{n-1} q_n = (-1)^n)$ and this is not possible.

Similarly, there is no common factor between p_n and p_{n-1} or between q_n and q_{n-1}.

(ii) *The difference between two successive convergents:*

$$\frac{p_n}{q_n} - \frac{p_{n-1}}{q_{n-1}} = \frac{p_n q_{n-1} - p_{n-1} q_n}{q_n q_{n-1}} = \frac{(-1)^n}{q_n q_{n-1}}.$$

Again,

$$\begin{aligned} \frac{p_n}{q_n} - \frac{p_{n-2}}{q_{n-2}} &= \frac{p_n q_{n-2} - p_{n-2} q_n}{q_n q_{n-2}} \\ &= \frac{(a_n p_{n-1} + p_{n-2}) q_{n-2} - p_{n-2}(a_n q_{n-1} + q_{n-2})}{q_n q_{n-2}} \\ &= \frac{a_n (p_{n-1} q_{n-2} - q_{n-1} p_{n-2})}{q_n q_{n-2}} \\ &= \frac{a_n (-1)^{n-1}}{q_n q_{n-1}} \quad (\because\ p_{n-1} q_{n-2} - p_{n-2} q_{n-1} = (-1)^{n-1}) \end{aligned}$$

2. We see that $p_2 = a_1 a_2 + 1$ and $p_1 = a_1$ $(a_1, a_2, \ldots$ arc +ve integers).

$$\therefore\ p_2 > p_1.$$

For $n \geq 3$, $p_n - p_{n-1} = (a_n - 1) p_{n-1} + p_{n-2} \geq p_{n-2} \geq 1$ and so for the q's, since $q_2 \geq q_1$ $(q_2 = a_2, q_1 = 1)$.

Thus the conclusion: *The two sequences $\{p_n\}$ and $\{q_n\}$ are both monotone increasing. In case of an infinite S.C.F. they tend to infinity as $n \to \infty$.*

3. We have, for the Simple Continued Fraction, $a_1 + \frac{1}{a_2+}\frac{1}{a_3+}\cdots\frac{1}{a_n+}\cdots$

$$p_n = a_n p_{n-1} + p_{n-2},\quad q_n = a_n q_{n-1} + q_{n-2}.$$

$$\therefore\ \frac{p_n}{p_{n-1}} = a_n + \frac{p_{n-2}}{p_{n-1}} = a_n + \frac{1}{\frac{p_{n-1}}{p_{n-2}}} = a_n + \frac{1}{\frac{a_{n-1}p_{n-2}+p_{n-3}}{p_{n-2}}}$$

$$= a_n + \frac{1}{a_{n-1} + \frac{1}{p_{n-2}/p_{n-3}}} = a_n + \frac{1}{a_{n-1}+}\frac{1}{a_{n-2}+}\cdots\frac{1}{p_2/p_1}$$

$$= a_n + \frac{1}{a_{n-1}+}\frac{1}{a_{n-2}+}\cdots\frac{1}{\frac{a_1a_2+1}{a_1}} = a_n + \frac{1}{a_{n-1}+}\frac{1}{a_{n-2}+}\cdots\frac{1}{a_2+}\frac{1}{a_1}.$$

But

$$\frac{q_n}{q_{n-1}} = \frac{a_n q_{n-1} + q_{n-2}}{q_{n-1}} = a_n + \frac{1}{\frac{q_{n-1}}{q_{n-2}}}$$

$$= a_n + \frac{1}{a_{n-1}+}\cdots + \frac{1}{q_2/q_1}$$

$$= a_n + \frac{1}{a_{n-1}+}\frac{1}{a_{n-2}+}\cdots + \frac{1}{a_2/1}$$

$$= a_n + \frac{1}{a_{n-1}+}\frac{1}{a_{n-2}+}\cdots + \frac{1}{a_2}.$$

From these it follows that if p_r/q_r, p'_r/q'_r are respectively the rth convergent of the finite Simple Continued Fractions,

$$a_1 + \frac{1}{a_2+}\frac{1}{a_3+}\cdots\frac{1}{a_n} \text{ and } a_n + \frac{1}{a_{n-1}+}\frac{1}{a_{n-2}+}\cdots\frac{1}{a_1}$$

where in the second case, the quotients occur in the reverse order, then

$$p'_n = p_n,\ q'_n = p_{n-1},\ p'_{n-1} = q_n,\ q'_{n-1} = q_{n-1}.$$

For, clearly, by above,

$$\frac{p'_n}{q'_n} = \frac{p_n}{p_{n-1}} \text{ and } \frac{p'_{n-1}}{q'_{n-1}} = \frac{q_n}{q_{n-1}}$$

and they are in their lowest terms.

4. Let $a_1 + \frac{1}{a_2+}\frac{1}{a_3+}\cdots\frac{1}{a_n+}\cdots$ be a Simple Continued Fraction.

If nth convergent be p_n/q_n, then $p_nq_{n-1} - p_{n-1}q_n = (-1)^n$.

If p'/q' and p/q are the consecutive convergent, then the above relation implies $pq' - p'q = \pm 1$.

Conversely, let $pq' - p'q = \pm 1$ where $q' < q$.

Suppose the fraction p/q is expressed as a Simple Continued Fraction with **an even number** of quotients if the sign is + or with **an odd number** of quotients if the sign is −.

[**Remember:** Any rational fraction can be expressed as a Simple Continued Fraction which is terminating and the continued fraction can be arranged so as to have either an odd number or an even number of quotients. For the proof see **No. 9** below]

We can prove that under these conditions:

$$\frac{p'}{q'} \textbf{ is the convergent immediately preceding } \frac{p}{q}.$$

For if $\frac{p''}{q''}$ be the last but one convergent (last convergent, of course, will be p/q), then

$$pq'' - p''q = \pm 1 = pq' - p'q \text{ (by supposition)}$$

and hence $\quad p(q' - q'') = q(p' - p'')$.

Now p is prime to q ($\because\ pq' - p'q = \pm 1$); so q divides $q' - q''$ or else $q' = q''$.

Also $q' < q$ (by supposition) and $q'' \leq q$ ($\because\ p''/q''$ and p/q are consecutive convergent). Therefore, q cannot divide $q' - q''$ and hence $q' = q''$ and consequently $p' = p''$,

i.e., $p''/q'' = p'/q'$ is the convergent just preceding p/q. **(Proved)**

5. A Very Important Result

Definition 1. *Let*

$$a_1 + \frac{1}{a_2+}\frac{1}{a_3+}\cdots\frac{1}{a_n+}\cdots \tag{1}$$

be an infinite Simple Continued Fraction. Let p_n/q_n be its nth convergent where

$$p_n = a_n p_{n-1} + p_{n-2},\ q_n = a_n q_{n-1} + q_{n-2}.$$

*If $\lim_{n\to\infty} p_n/q_n$ = a finite real number x, then the Simple Continued Fraction (**1**) is said to be convergent and the value of the continued fraction is x.*

Theorem 15.3.1. *Every infinite simple continued fraction is convergent. Its value is greater than any odd convergent and is less than any even convergent.*

Proof. Let $u_n = p_n/q_n$ = nth Convergent of the Simple Continued Fraction **(1)**.

We first prove that

(i) *Every convergent lies between the two preceding convergents:*

Justification:

$$u_n - u_{n-1} = (-1)^n \frac{1}{q_n q_{n-1}} \text{ [See No. 1 (ii)]}$$

$$u_n - u_{n-2} = (-1)^{n-1} \frac{a_n}{q_n q_{n-2}}.$$

Since the values of $q_n, q_{n-1}, q_{n-2}, a_n$ are all positive, $(u_n - u_{n-1})$ and $(u_n - u_{n-2})$ are of opposite signs. Therefore, u_n lies between u_{n-1} and u_{n-2}.

Next we prove that

(ii) *The odd convergents form an increasing sequence, and the even convergents from a decreasing sequence and every odd convergent is less than any even convergent.*

We have $u_1 = a_1$ and $u_2 = a_1 + \frac{1}{a_2}$; therefore, $u_1 < u_2$.
By what we have observed before, $u_1 < u_3 < u_2$.
Again, $u_3 < u_2 \Rightarrow u_3 < u_4 < u_2$, $\therefore\ u_1 < u_3 < u_4 < u_2$.
Also $u_3 < u_4 \Rightarrow u_3 < u_5 < u_4$, $\therefore\ u_1 < u_3 < u_5 < u_4 < u_2$ and so on.
Finally, $u_1 < u_3 < u_5 < u_7 < \cdots < u_6 < u_4 < u_2$, i.e., $\{u_1, u_3, u_5, \ldots\}$ is an increasing sequence and $\{u_2, u_4, u_6, \ldots\}$ is a decreasing sequence.
Every member of the two sequences lie between $u_1 = a_1$ and $u_2 = a_1 + \frac{1}{a_2}$.
Hence, the two sequences $\{u_{2n-1}\}$ and $\{u_{2n}\}$ are both convergent; suppose they converge to l and m respectively.
Now

$$\begin{aligned} u_{2n} - u_{2n-1} &= \frac{p_{2n}}{q_{2n}} - \frac{p_{2n-1}}{q_{2n-1}} = \frac{p_{2n}q_{2n-1} - q_{2n}p_{2n-1}}{q_{2n}q_{2n-1}} \\ &= \frac{(-1)^{2n}}{q_{2n}q_{2n-1}} = \frac{1}{q_{2n}q_{2n-1}} \to 0 \text{ as } n \to \infty \\ \lim_{n\to\infty} u_{2n} &= \lim_{n\to\infty} u_{2n-1}, \text{ i.e., } l = m = x \text{ (say).} \end{aligned}$$

Clearly for every n, $u_{2n-1} < x < u_{2n}$. This proves the theorem completely.

6. *Each convergent is more close to the value of the Simple Continued Fraction than its preceding convergent,* i.e., To prove that the numerical difference between x, the value of the continued fraction and u_n is less than the numerical difference between x and u_{n-1}.

$$\text{or, } \left| x - \frac{p_n}{q_n} \right| < \left| x - \frac{p_{n-1}}{q_{n-1}} \right|.$$

Justification. Let

$$\begin{aligned} x &= a_1 + \frac{1}{a_2+} \cdots + \frac{1}{a_n+} \cdots \\ \text{and let } x' &= (n+1)\text{th complete quotient} \\ &= a_{n+1} + \frac{1}{a_{n+2}+} \cdots (n+1)\text{th convergent} \\ &= \frac{p_{n+1}}{q_{n+1}} = \frac{a_{n+1}p_n + p_{n-1}}{a_{n+1}q_n + q_{n-1}}. \end{aligned}$$

If in the expression for p_{n+1} we replace a_{n+1} by $(n+1)$th complete quotient x' and similarly in the expression for q_{n+1} we replace a_{n+1} by x', then we could replace p_{n+1}/q_{n+1} by the value of the continued fraction, namely x.
Thus,

$$x = \frac{x'p_n + p_{n-1}}{x'q_n + q_{n-1}}$$

$$\text{or, } x'(xq_n - p_n) = -xq_{n-1} + p_{n-1} = -(xq_{n-1} - p_{n-1})$$

$$\text{or, } x'q_n\left(x - \frac{p_n}{q_n}\right) = -q_{n-1}\left(x - \frac{p_{n-1}}{q_{n-1}}\right),$$

$$\text{i.e., } x - \frac{p_n}{q_n} = -\frac{q_{n-1}}{x'q_n}\left(x - \frac{p_{n-1}}{q_{n-1}}\right).$$

Now $q_n > q_{n-1}$ and $x' \geq 1$, therefore,

$$\left|x - \frac{p_n}{q_n}\right| < \left|x - \frac{p_{n-1}}{q_{n-1}}\right|;$$

i.e., x is nearer to p_n/q_n than to p_{n-1}/q_{n-1}.

7. *Any convergent p_n/q_n of a simple continued fraction is more close to the value of the continued fraction than a rational fraction p/q whose denominator $q < q_n$.*

Justification: Let x be the value of the Simple Continued Fraction

$$a_1 + \frac{1}{a_2+}\,\frac{1}{a_3+}\cdots+\frac{1}{a_n+}\cdots$$

If p_n/q_n is a certain convergent, then it is more close to x than p_{n-1}/q_{n-1} (Proved before).

To prove that a fraction p/q with $q < q_n$ cannot be closer to x than p_n/q_n.

If possible, let p/q with $q < q_n$ be more close to x than p_n/q_n; clearly then

$$\left|\frac{p}{q} - \frac{p_{n-1}}{q_{n-1}}\right| < \left|\frac{p_n}{q_n} - \frac{p_{n-1}}{q_{n-1}}\right|;$$

i.e., $|pq_{n-1} - p_{n-1}q| < \dfrac{q}{q_n}|p_nq_{n-1} - p_{n-1}q_n| < \dfrac{q}{q_n} < 1$ ($\because$ $q < q_n$),

i.e., a positive integer $|pq_{n-1} - p_{n-1}q| < 1$ (which is not possible)

$\therefore$ $\frac{p_n}{q_n}$ is more close to x than $\frac{p}{q}$ where $q < q_n$.

8. *If $\frac{p}{q}, \frac{p'}{q}$, be two consecutive convergent of a Simple Continued Fraction whose value is x, then $\frac{pp'}{qq'} >$ or $< x^2$ according as, $\frac{p}{q} >$ or $< \frac{p'}{q'}$.*

Justifications: Let k be the complete quotient corresponding to the convergent immediately succeeding p'/q'; then

$$x = \frac{kp' + p}{kq' + q}.$$

$$\frac{pp'}{qq'} - x^2 = \frac{pp'}{qq'} - \left(\frac{kp' + p}{kq' + q}\right)^2$$

$$= \frac{pp'(kq' + q)^2 - qq'(kp' + p)^2}{qq'(kq' + q)^2}$$

$$= \frac{(k^2p'q' - pq)(pq' - p'q)}{qq'(kq' + q)^2}.$$

The factor $k^2p'q' - pq$ is positive, since $p' > p$, $q' > q$ and $k > 1$ and the denominator is always positive.

Hence, $\frac{pp'}{qq'} >$ or $< x^2$ according as $pq' - p'q >$ or < 0, i.e., according as

$$\frac{p}{q} > \text{or} < \frac{p'}{q'}.$$

It also follows from previous arguments that the quantities $pq' - p'q$, $pp' - qq'x^2$, $p^2 - q^2x^2$, $q'^2x^2 - p'^2$ all are of the same sign.

9. (a) *Given a positive rational number we can find a terminating simple continued fraction corresponding to it. This Simple Continued Fraction can be so arranged as to have either an odd number of quotients or an even number of quotients.*

Let p/q be a positive rational number, where p and q are positive integers.

In the process of finding the H.C.F. of p and q let $a_1, a_2, a_3, \ldots$ be the quotients and $r_1, r_2, r_3, \ldots$ be the remainders, so that

$$\begin{aligned} p &= a_1 q + r_1, \\ q &= a_2 r_1 + r_2, \\ r_1 &= a_3 r_2 + r_3, \text{ etc. etc.} \end{aligned}$$

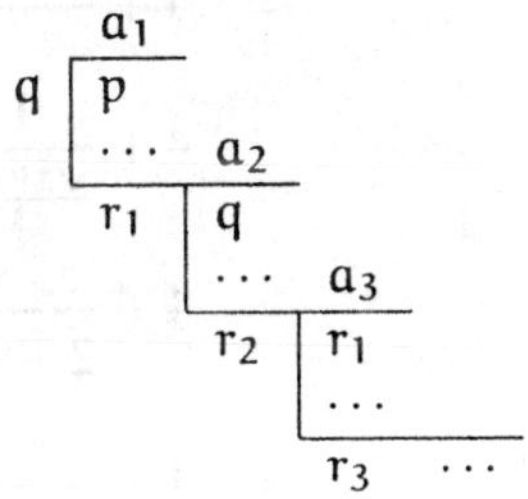

and therefore,

$$\frac{p}{q} = a_1 + \frac{r_1}{q} = a_1 + \frac{1}{q/r_1} = a_1 + \frac{1}{a_2 + r_2/r_1}$$

$$= a_1 + \frac{1}{a_2 + \frac{1}{r_1/r_2}} = a_1 + \frac{1}{a_2+}\,\frac{1}{a_3+}\,\frac{1}{r_2/r_3}.$$

The process terminates and if a_n is the last quotient, then

$$\frac{p}{q} = a_1 + \frac{1}{a_2+} \cdots + \frac{1}{a_n}.$$

If $a_n > 1$ we can write

$$\frac{1}{a_n} = \frac{1}{(a_n - 1)+}\,\frac{1}{1}.$$

If $a_n = 1$, then we can write

$$\frac{1}{a_{n-1}+}\,\frac{1}{a_n} = \frac{1}{a_{n-1} + 1}.$$

So we can always arrange so that the continued fraction has either an odd number or an even number of quotients, as may be required.

If $p < q$, then the first quotient is zero and we write $\frac{p}{q} = \frac{1}{q/p}$ and then proceed as before.

e.g., $$\frac{251}{802} = \frac{1}{\frac{802}{251}} = \frac{1}{3 + \frac{49}{251}} = \frac{1}{3} + \frac{1}{\frac{251}{49}}$$

$$= \frac{1}{3+}\,\frac{1}{5+}\,\frac{1}{8+}\,\frac{1}{6} \text{ (Even number of quotients)}$$

$$= \frac{1}{3+}\,\frac{1}{5+}\,\frac{1}{8+}\,\frac{1}{5+}\,\frac{1}{1} \text{ (Odd number of quotients)}$$

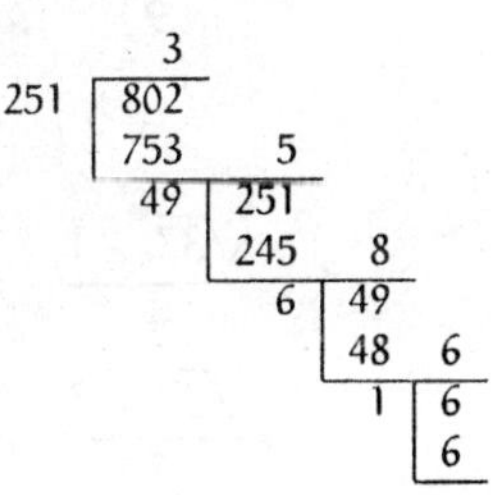

(b) *Any positive irrational number can be expressed as an infinite simple continued fraction and such representation is* **unique.**

Conversely, *the value of an infinite Simple Continued Fraction is an irrational number (for any rational number can be expressed as a finite Simple Continued Fraction)*

We shall take some specific examples and it will be observed that the method is perfectly general.

► **Example 15.3.1** *Express $\sqrt{19}$ as an infinite Simple Continued Fraction.*

Solution:

$$\begin{aligned}
\sqrt{19} &= 4+(\sqrt{19}-4) \ \text{[\textbf{Note:} 4 is the integral part of } \sqrt{19}] \\
&= 4+\frac{(\sqrt{19}-4)(\sqrt{19}+4)}{\sqrt{19}+4} = 4+\frac{3}{\sqrt{19}+4} \\
&= 4+\frac{1}{\frac{\sqrt{19}+4}{3}} = 4+\frac{1}{2+\frac{\sqrt{19}+4}{3}-2} \ \text{[2 is the integral part of } \tfrac{\sqrt{19}+4}{3}] \\
&= 4+\frac{1}{2+\frac{\sqrt{19}-2}{3}} = 4+\frac{1}{2+\frac{5}{\sqrt{19}+2}} = 4+\frac{1}{2+\frac{1}{\frac{\sqrt{19}+2}{5}}} \\
&= 4+\frac{1}{2+\frac{1}{1+\frac{\sqrt{19}-3}{5}}} = 4+\frac{1}{2+}\frac{1}{1+}\frac{1}{\frac{\sqrt{19}+3}{2}} = 4+\frac{1}{2+}\frac{1}{1+}\frac{1}{3+}\frac{1}{\frac{\sqrt{19}+3}{5}} \\
&= 4+\frac{1}{2+}\frac{1}{1+}\frac{1}{3+}\frac{1}{1+}\frac{1}{\frac{\sqrt{19}+2}{3}} = 4+\frac{1}{2+}\frac{1}{1+}\frac{1}{3+}\frac{1}{1+}\frac{1}{2+}\frac{1}{\sqrt{19}+4} \\
&= 4+\frac{1}{2+}\frac{1}{1+}\frac{1}{3+}\frac{1}{1+}\frac{1}{2+}\frac{1}{8+}\frac{1}{\frac{\sqrt{19}+4}{3}}.
\end{aligned}$$

After this the quotients 2, 1, 3, 1, 2, 8, recur. So we write

$$\sqrt{19} = 4+\frac{1}{\underset{*}{2+}}\frac{1}{1+}\frac{1}{3+}\frac{1}{1+}\frac{1}{2+}\frac{1}{\underset{*}{8+}}.$$

Such a continued fraction is called a **Recurring Continued Fraction**, the part marked with **two asteriks** recur—we call it the **Cycle**. The non-recurring elements, if exist, is called **acyclic part.**

► **Example 15.3.2** *Express $\frac{\sqrt{37}+8}{9}$ as a recurring simple continued fraction.*

$$\begin{aligned}
\frac{\sqrt{37}+8}{9} &= 1+\frac{\sqrt{37}+8}{9}-1 \left(1 \text{ is the integral part of } \tfrac{\sqrt{37}+8}{9}\right) \\
&= 1+\frac{\sqrt{37}-1}{9} = 1+\frac{4}{\sqrt{37}+1} \\
\frac{\sqrt{37}+1}{4} &= 1+\frac{\sqrt{37}+1}{4}-1 = 1+\frac{\sqrt{37}-3}{4} = 1+\frac{7}{\sqrt{37}+3} \\
{*}\frac{\sqrt{37}+3}{7} &= 1+\frac{\sqrt{37}+3}{7}-1 = 1+\frac{\sqrt{37}-4}{7} = 1+\frac{3}{\sqrt{37}+4} \\
\frac{\sqrt{37}+4}{3} &= 3+\frac{\sqrt{37}+4}{3}-3 = 3+\frac{\sqrt{37}-5}{3} = 3+\frac{4}{\sqrt{37}+5} \\
\frac{\sqrt{37}+5}{4} &= 2+\frac{\sqrt{37}+5}{4}-2 = 3+\frac{\sqrt{37}-3}{4} = 3+\frac{7}{\sqrt{37}+3} \\
\frac{\sqrt{37}+3}{7} &= 1+\text{, etc. (already occurred in } *)
\end{aligned}$$

The subsequent work consists of repetitions of the last three steps and hence

$$\begin{aligned}\frac{\sqrt{37}+8}{9} &= 1+\frac{1}{1+}\,\frac{1}{1+}\,\frac{1}{3+}\,\frac{1}{2+}\,\frac{1}{1+}\,\frac{1}{3+}\,\frac{1}{2+}\,\frac{1}{1+}\,\frac{1}{3+}\,\frac{1}{2+}\cdots\\ &= 1+\frac{1}{1+}\,\underset{*}{\frac{1}{1+}}\,\frac{1}{3+}\,\underset{*}{\frac{1}{2}},\end{aligned}$$

which is a recurring continued fraction with cyclic part

$\frac{1}{1+}\,\frac{1}{3+}\,\frac{1}{2+}$ and acyclic part $1+\frac{1}{1+}$.

15.4 Applications

I. Approximations

Theorem 15.4.1. *If e be the numerical value of the error, in representing the value* x *of a Simple Continued Fraction by its nth convergent* $u_n = p_n/q_n$, *then*

$$\frac{a_{n+2}}{q_n q_{n+2}} < e < \frac{1}{q_n q_{n+1}}, \qquad \text{(A)}$$

$$\textit{whence} \quad \frac{1}{q_n(q_n+q_{n+1})} < e < \frac{1}{q_n^2}. \qquad \text{(B)}$$

Proof. The proof depends on the fact:

The value x of an infinite Simple Continued Fraction is greater than any odd convergent and less than any even convergent. Also every convergent lies between the two preceding convergents.

So if $u_n = p_n/q_n$, then

$$u_n,\ u_{n+2},\ x,\ u_{n+1}$$

is either an increasing or a decreasing sequence.

Therefore,

$$|u_n - u_{n+2}| < |u_n - x| < |u_n - u_{n+1}|. \qquad \text{(C)}$$

It can be easily verified:

$$\begin{aligned} u_n - u_{n+2} &= \frac{p_n}{q_n} - \frac{p_{n+2}}{q_{n+2}} = \frac{p_n q_{n+2} - p_{n+2} q_n}{q_n q_{n+2}}\\ &= \frac{p_n(a_{n+2}q_{n+1}+q_n) - (a_{n+2}p_{n+1}+p_n)q_n}{q_n q_{n+2}}\\ &= -\frac{(a_{n+2})(p_{n+1}q_n - q_{n+1}p_n)}{q_n q_{n+2}}\\ &= (-1)^{n+2}\frac{a_{n+2}}{q_n q_{n+2}}\\ \text{so that } |u_n - u_{n+2}| &= \frac{a_{n+2}}{q_n q_{n+2}}. \end{aligned}$$

Similarly

$$|u_n - u_{n+1}| = \frac{1}{q_n q_{n+1}}.$$

The inequality (A) now follows at once ($\because$ $|u_r - x| = e$ from (C)).
Again, since, $q_{n+2} = a_{n+2}\, q_{n+1} + q_n$,

$$\therefore \frac{q_{n+2}}{a_{n+2}} = q_{n+1} + \frac{q_n}{a_{n+2}} \leq q_{n+1} + q_n.$$

Hence,

$$\frac{1}{q_n(q_{n+1}+q_n)} \leq \frac{a_{n+2}}{q_n q_{n+2}}.$$

Also $q_{n+1} > q_n$; hence

$$\frac{1}{q_n q_{n+1}} < \frac{1}{q_n^2}.$$

$$\therefore \frac{1}{q_n(q_{n+1}+q_n)} \leq \frac{a_{n+2}}{q_n q_{n+2}} < e < \frac{1}{q_n q_{n+1}} < \frac{1}{q_n{}^2}.$$

This gives the inequalities (B).

Points to remember

(i) Let a given number x be expressed as a Simple Continued Fraction of which p_r/q_r is the rth convergent. If p_n/q_n is taken the approximate value of the given Simple Continued Fraction, then the numerical value of the error is

$$e = |x - p_n/q_n| < \frac{1}{q_n q_{n+1}}.$$

In fact, we may take $e < \frac{1}{q_n{}^2}$.

Now $q_{n+1} = a_{n+1}q_n + q_{n-1}$; hence if a_{n+1} is large, p_n/q_n is a **good approximation.**

(ii) If we require an approximation with an error numerically less than a given fraction $\frac{1}{q}$ we shall calculate convergent until we find one p_n/q_n such that $q_n\, q_{n+1} \geq q$. The condition is satisfied of $q_n \geq \sqrt{q}$.

Solved Examples on Approximations

▶ Example 15.4.1 *Suppose we require to find an approximate value of π. Given*

$$\pi = 3 + \frac{1}{7+}\,\frac{1}{15+}\,\frac{1}{1+}\,\frac{1}{292+}\,\frac{1}{1+}\,\frac{1}{1+}\cdots$$

We find the successive convergents

$$\frac{3}{1}, \frac{22}{7}, \frac{333}{106}, \frac{355}{113}, \frac{103993}{33102}, \cdots$$

The 5th quotient 292 is comparatively large; so the 4th convergent $\frac{355}{113}$ is a good approximation of π and then the error is less than

$$\frac{1}{113 \times 33102} < \frac{1}{3740000} = 0.00000027 = 2.7 \times 10^{-7}.$$

► Example 15.4.2 *The mean tropical year is 365.2422642 days. It is required to find a practically convenient method of correcting the calendar, the year being taken as 365 days.*

Solution: $365.2422642 = 365 + 0.2422642$

$$0.2422642 = 0 + \frac{1}{4+}\,\frac{1}{7+}\,\frac{1}{1+}\,\frac{1}{4+}\,\frac{1}{1+}\,\frac{1}{6+}\,\frac{1}{1}\cdots$$

The successive convergents are

$$\frac{0}{1},\ \frac{1}{4},\ \frac{7}{29},\ \frac{8}{33},\ \frac{39}{161},\ \frac{47}{194},\ \frac{321}{1325}\cdots$$

Take 1/4 as a first approximation. This means that we may add 1 day for every four years (i.e., make every fourth year a leap year). But this is an over-correction.

However, if 8/33 is taken as an approximation, then it will amount to an addition of 24 days in every 99 years, i.e., make every fourth year a leap-year, except at the end of a century. This is also an over-correction.

► Example 15.4.3 *One metre is 39.37079 inches. Using the theory of continued fractions verify that 32 metres is approximately equal to 35 yards.*

Solution:

$$\text{One metre} = 39.37079 \times \frac{1}{36}\text{ yards} = 1.0936 \text{ yards}$$

$$\text{But } 1.0936 = 1 + \frac{1}{10+}\,\frac{1}{1+}\,\frac{1}{2+}\,\frac{1}{6+}\,\frac{1}{6}$$

```
          1
10000 ) 10936
        10000      10
          936 ) 10000
                  936
                  640
            1
     640 ) 936
           640     2
           296 ) 640
                 592
                  48
             6
      48 ) 296
           288     6
             8 ) 48
                 48
```

The successive convergent are

$$1,\ \frac{11}{10},\ \frac{12}{11},\ \frac{35}{32},\cdots$$

5th quotient is comparatively larger.

∴ 4th convergent may be taken as a good approximation.

∴ one metre $= \frac{35}{32}$ yards (nearly)

or, 32 metres is nearly equal to 35 yards.

II. Integral Solutions of $ax + by = c$.

► Example 15.4.4 *Find the integral solutions of* $137x - 106y = 4$.

Solution: We express $\frac{137}{106}$ as a simple continued fraction

$$1 + \frac{1}{3+}\,\frac{1}{2+}\,\frac{1}{2+}\,\frac{1}{1+}\,\frac{1}{1+}\,\frac{1}{2}$$

$\frac{p_6}{q_6}$ = Sixth convergent $= \frac{53}{41}$. Last convergent is the value $\frac{137}{106} = \frac{p_7}{q_7}$.

Use: $p_n q_{n-1} - p_{n-1} q_n = (-1)^n$.
Take $n = 7$, $p_7 q_6 - p_6 q_7 = (-1)^7 = -1$, i.e., $137.41 - 53.106 = -1$.

$$\therefore\ 137(-164) - 106(-212) = 4.$$

$\therefore$ the given equation is satisfied by $x = -164$ and $y = -212$.

Evidently it is also satisfied by $x = -164 + 106t, y = -212 + 137t$ and therefore, by $x = 48 + 106s$, $y = 62 + 137s$ (Putting $s = t - 2$) where s is any integer or zero.

General discussion: $ax + by = c$ (a, b, c are integers and a, b have no common factor)

Note that if h is the h.c.f. of a, b, then the equation can have no integral solution unless h is a factor of c and if h is a factor of c, then each side can be divided by h.

When a, b have no common factor, a solution $x = x_1, y = y_1$ may be found from the fraction a/b expressed as a Simple Continued Fraction as explained in the above example.

Also the more general solution $x = x_1 + bt$, $y = y_1 - at$ can then be written down.

▶ Example 15.4.5 *Solve in integers:* $11x - 9y = 4$.

[$x = x_1 = -16$, $y = y_1 = -20$; general solution $x = 9n + 2$, $y = 11n + 2$, n is any integer].

III. Problems on Recurring Continued Fractions

▶ Example 15.4.6 *Evaluate:* $\frac{1}{2+}\frac{1}{4+}\frac{1}{2+}\frac{1}{4+}\frac{1}{2+}\cdots$

Solution: Every infinite Simple Continued Fraction converges to a value x (say).

If p_r/q_r be the rth convergent, then $\lim_{r\to\infty} p_r/q_r = x$.

Here

$$\frac{p_{2n+2}}{q_{2n+2}} = \cfrac{1}{2 + \cfrac{1}{4 + p_{2n}/q_{2n}}}.$$

Now as $n \to \infty$, $\frac{p_{2n+2}}{q_{2n+2}} \to x$ and $\frac{p_{2n}}{q_{2n}} \to x$.

$$\therefore\quad x = \frac{1}{2 + \frac{1}{4+x}} = \frac{4+x}{2x+9}, \text{ or, } 2x^2 + 9x = 4 + x,$$

$$\text{or,}\quad 2x^2 + 8x - 4 = 0, \text{ or, } x^2 + 4x - 2 = 0,$$

$$\text{or,}\quad (x+2)^2 = 6, \text{ i.e., } x = \pm\sqrt{6} - 2.$$

But x must be positive (all elements of Simple Continued Fraction are positive), therefore, the value of $x = \sqrt{6} - 2$.

▶ Example 15.4.7 *Evaluate:* $1 + \frac{1}{2+}\frac{1}{3+}\frac{1}{2+}\frac{1}{3+}\frac{1}{2+}\cdots$.

Solution: Proceeding as in the previous example, if x be the value of the Simple Continued Fraction, then we can write

$$x-1=\frac{1}{2+}\frac{1}{3+}\frac{1}{2+}\frac{1}{3+}\cdots=\frac{1}{2+\frac{1}{3+(x-1)}}$$

$$\text{or,}\quad x-1=\frac{2+x}{2x+5},\ \text{or, } 2x^2+3x-5=x+2$$

$$\text{or,}\quad 2x^2+2x-7=0,\ \text{or, } x=\frac{-1\pm\sqrt{15}}{2}.$$

Since x is positive, the required value of $x=\frac{\sqrt{15}-1}{2}$.

▶ Example 15.4.8 *If*

$$x=a+\frac{1}{b+}\frac{1}{a+}\frac{1}{b+}\frac{1}{a+}\cdots\ \textit{and}\ y=b+\frac{1}{a+}\frac{1}{b+}\frac{1}{a+}\frac{1}{b+}\cdots$$

prove that $bx=ay$.

Solution: $x=a+\frac{1}{y}\Rightarrow xy=ay+1$ and $y=b+\frac{1}{x}\Rightarrow xy=bx+1$
Hence, $ay+1=bx+1$ **or,** $ay=bx$.

▶ Example 15.4.9 *Evaluate:*

$$2+\frac{1}{3+}\frac{1}{1+}\frac{1}{2+}\frac{1}{1+}\frac{1}{2+}\frac{1}{1+}\frac{1}{2+}\frac{1}{1+}\cdots$$

Solution: Let x be the value of the given Simple Continued Fraction and let

$$y=\frac{1}{1+}\frac{1}{2+}\frac{1}{1+}\frac{1}{2+}\frac{1}{1+}\cdots$$

So,

$$y=\frac{1}{1+\frac{1}{2+y}}=\frac{2+y}{3+y},\ \text{or, } y^2+2y-2=0\Rightarrow y=\sqrt{3}-1 (y>0).$$

$$\therefore\ x=2+\frac{1}{3+y}=\frac{7+2y}{3+y},\ \text{or, } x=\frac{7+2(\sqrt{3}-1)}{3+(\sqrt{3}-1)}=\frac{2\sqrt{3}+5}{\sqrt{3}+2}$$

$$=\frac{(2\sqrt{3}+5)(2-\sqrt{3})}{(2+\sqrt{3})(2-\sqrt{3})}=4-\sqrt{3}.$$

▶ Example 15.4.10 *If*

$$x=a+\frac{1}{b+}\frac{1}{c+}\frac{1}{a+}\frac{1}{b+}\frac{1}{c+}\cdots\ \textit{and}\ y=c+\frac{1}{b+}\frac{1}{a+}\frac{1}{c+}\frac{1}{b+}\frac{1}{a+}\cdots,$$

prove that $\frac{x}{y}=\frac{ab+1}{bc+1}$.

Solution:

$$x = a + \cfrac{1}{b + \cfrac{1}{c + \frac{1}{x}}} = a + \cfrac{1}{b + \frac{x}{cx+1}} = a + \frac{cx+1}{(bc+1)x+b},$$

i.e., $$x = \frac{(abc+a+c)x + (ab+1)}{(bc+1)x+b}$$

or, $$x^2(bc+1) - x(abc+a+c-b) - (ab+1) = 0, \qquad (1)$$

x is the positive root of this quadratic equation.

Replacing a by c and c by a, y must be the positive root of

$$y^2(ba+1) - x(cba+c+a-b) - (cb+1) = 0.$$

$\therefore\ -\frac{1}{y}$ satisfies the same equation as x and is the negative root.

Thus,

$$x \times \frac{1}{y} = -\text{ product of the roots of eqn. (1)}$$

$$= +\frac{(ab+1)}{bc+1}.$$

▶ Example 15.4.11 *If*

$$a + \frac{1}{a+}\frac{1}{a+}\frac{1}{a+}\ldots, \; b + \frac{1}{b+}\frac{1}{b+}\frac{1}{b+}\ldots \; \textit{and } c + \frac{1}{c+}\frac{1}{c+}\frac{1}{c+}\ldots$$

be in G.P. of common ratio r, *prove that* $c + \frac{1}{c} = \frac{a+c}{b}$.

Solution: Let

$$x = a + \frac{1}{a+}\cdots = a + \frac{1}{x},$$

$$y = b + \frac{1}{b+}\cdots = b + \frac{1}{y},$$

$$z = c + \frac{1}{c+}\cdots = c + \frac{1}{y},$$

i.e., $x^2 - ax - 1 = 0$, $y^2 - by - 1 = 0$, $z^2 - cz - 1 = 0$.

$$\Rightarrow \frac{x^2 - y^2}{y^2 - z^2} = \frac{ax - by}{by - cz} \qquad (2)$$

Given x, y, z are in G.P. of common ratio r

$$\therefore \frac{y}{x} = \frac{z}{y} = r, \text{ i.e., } y = rx, z = ry = r^2x.$$

Hence, (2) becomes

$$\frac{x^2 - r^2x^2}{r^2x^2 - r^2x^2} = \frac{ax - brx}{brx - cr^2x}, \text{ i.e., } \frac{1 - r^2}{r^2(1 - r^2)} = \frac{a - br}{r(b - cr)}$$

Since $r \neq 1$, we get

$$\frac{1}{r} = \frac{a-br}{b-cr} \text{ or, } b - cr = ar - br^2,$$

$$\text{i.e., } \quad (a+c)r = b(1+r^2) \text{ or, } \frac{a+c}{b} = \frac{1}{r} + r. \text{ (**Reqd. result**)}$$

15.5 Illustrative Examples

► Example 15.5.1 *Let* $a_1 + \frac{1}{a_2+}\frac{1}{a_3+}\cdots$ *be a Simple Continued Fraction. If* p_n/q_n *be its nth convergent prove that the difference between the first and nth convergent is numerically*

$$= \frac{1}{q_1q_2} - \frac{1}{q_2q_3} + \frac{1}{q_3q_4} - \cdots + \frac{(-1)^n}{q_{n-1}q_n}.$$

Solution:

$$\begin{aligned}
\frac{p_1}{q_1} - \frac{p_n}{q_n} &= \frac{p_1}{q_1} - \frac{p_2}{q_2} + \frac{p_2}{q_2} - \frac{p_3}{q_3} + \frac{p_3}{q_3} - \cdots + \frac{p_{n-1}}{q_{n-1}} - \frac{p_n}{q_n} \\
&= -\frac{p_2q_1 - p_1q_2}{q_1q_2} - \frac{p_3q_2 - p_2q_3}{q_2q_3} - \cdots - \frac{p_nq_{n-1} - p_{n-1}q_n}{q_{n-1}q_n} \\
&= -\frac{(-1)^2}{q_1q_2} - \frac{(-1)^3}{q_2q_3} \cdots - \frac{(-1)^n}{q_{n-1}q_n} \\
&= -\frac{1}{q_1q_2} + \frac{1}{q_2q_3} - \cdots - \frac{(-1)^n}{q_{n-1}q_n} \\
\therefore \left|\frac{p_1}{q_1} - \frac{p_n}{q_n}\right| &= \frac{1}{q_1q_2} - \frac{1}{q_2q_3} + \cdots + \frac{(-1)^n}{q_{n-1}q_n}.
\end{aligned}$$

► Example 15.5.2 *If* $\frac{M}{N}, \frac{P}{Q}, \frac{R}{S}$ *are the nth convergent, (n − 1)th convergent and (n − 2)th convergent to the continued fractions.*

$$\frac{1}{a_1+}\frac{1}{a_2+}\frac{1}{a_3+}\cdots, \quad \frac{1}{a_2+}\frac{1}{a_3+}\frac{1}{a_4+}\cdots, \quad \frac{1}{a_3+}\frac{1}{a_4+}\frac{1}{a_5+}\cdots$$

respectively, prove that $M = a_2P + R$, $N = (a_1a_2 + 1)P + a_1R$.

Solution: We have

$$\frac{M}{N} = \frac{1}{a_1+}\frac{P}{Q} = \frac{1}{a_1 + P/Q} = \frac{Q}{a_1Q + P}$$

$$\text{and } \frac{P}{Q} = \frac{1}{a_2+}\frac{R}{S} = \frac{1}{a_2 + R/S} = \frac{S}{a_2S + R}$$

But the fractions $\frac{M}{N}, \frac{P}{Q}, \frac{R}{S}$ are in their lowest terms; hence $\frac{Q}{a_1Q+P}$ and $\frac{S}{a_2S+R}$ are also in their lowest terms. [See Art. **15.3.** No. **1.** Deduction (i)]

$$\therefore M = Q, \; N = a_1Q + P, \; P = S, \; Q = a_2S + R.$$

Hence

$$\begin{aligned}
M &= a_2S + R = a_1P + R, \\
\text{and } N &= a_1(a_2S + R) + P = a_1(a_2P + R) + P = (a_1a_2 + 1)P + a_1R.
\end{aligned}$$

▶ **Example 15.5.3** *Prove that*

$$a\left(x_1+\frac{1}{ax_2+}\frac{1}{x_3+}\frac{1}{ax_4+}\cdots \text{ to } 2n \text{ quotients}\right)$$

$$= ax_1+\frac{1}{x_2+}\frac{1}{ax_3}\frac{1}{x_4+}\cdots \text{ to } 2n \text{ quotients.}$$

Solution:

$$\text{Left side} = ax_1+\frac{a}{ax_2+}\frac{1}{x_3+}\frac{1}{ax_4+}\cdots \text{ to } 2n \text{ quotients}$$

$$= ax_1+\frac{a/a}{\frac{ax_2}{a}+}\frac{1/a}{x_3+}\frac{1}{ax_4+}\cdots \text{ to } 2n \text{ quotients}$$

$$= ax_1+\frac{1}{x_2+}\frac{a/a}{ax_3+}\frac{a}{ax_4+}\cdots \text{ to } 2n \text{ quotients}$$

$$= ax_1+\frac{1}{x_2+}\frac{1}{ax_3+}\frac{a}{ax_4+}\cdots \text{ to } 2n \text{ quotients and so on}\cdots$$

$$= \textbf{R.H.S.}$$

▶ **Example 15.5.4** *If* p_n/q_n *be the nth convergent of the Simple Continued Fraction*

$$a_1+\frac{1}{a_2+}\frac{1}{a_3+}\cdots\frac{1}{a_n+}\cdots,$$

then prove that $p_{n+2}q_{n-2}\sim p_{n-2}q_{n+2}=a_{n+2}\cdot a_{n+1}-a_n+a_{n+2}+a_n$.

Solution: We have $p_{n-1}q_{n-2}-p_{n-2}q_{n-1}=(-1)^{n-1}$.

Again, $p_nq_{n-2}-p_{n-2}q_n$

$$= (a_np_{n-1}+p_{n-2})q_{n-2}-p_{n-2}(a_nq_{n-1}+q_{n-2})$$
$$= a_n(p_{n-1}q_{n-2}-p_{n-2}q_{n-1})=a_n(-1)^{n-1}.$$

Similarly, $p_{n+1}q_{n-2}-p_{n-2}q_{n+1}$

$$= (a_{n+1}p_n+p_{n-1})q_{n-2}-p_{n-2}(a_{n+1}q_n+q_{n-1})$$
$$= a_{n+1}(p_nq_{n-2}-p_{n-2}q_n)+(p_{n-1}q_{n-2}-p_{n-2}q_{n-1})$$
$$= a_{n+1}\cdot a_n(-1)^{n-1}+(-1)^{n-1}$$
$$= (-1)^{n-1}(a_{n+1}\cdot a_n+1).$$

Finally, $p_{n+2}q_{n-2}-p_{n-2}q_{n+2}$

$$= (a_{n+2}p_{n+1}+p_n)q_{n-2}-p_{n-2}(a_{n+2}q_{n+1}+q_n)$$
$$= a_{n+2}(p_{n+1}q_{n-2}-p_{n-2}q_{n+1})+(p_nq_{n-2}-p_{n-2}q_n)$$
$$= a_{n+2}\left[(-1)^{n-1}(a_{n+1}\cdot a_{n+1})\right]+a_n(-1)^{n-1}$$
$$= (-1)^{n-1}[a_{n+2}\cdot a_{n+1}\cdot a_n+a_{n+2}+a_n].$$

Hence, $|p_{n+2}q_{n-2}-p_{n-2}q_{n+2}|=a_{n+2}\cdot a_{n+1}\cdot a_n+a_{n+2}+a_n$.

▶ Example 15.5.5 *If* p_n/q_n *is the* n*th convergent of the Simple Continued Fraction*

$$a+\frac{1}{b+}\frac{1}{a+}\frac{1}{b+}\frac{1}{a+}\cdots$$

prove that

(i) $p_{2n}=q_{2n+1}$; (ii) $bp_{2n-1}=aq_{2n}$;
(iii) $p_{2n+2}-(ab+1)p_{2n}=aq_{2n}$; (iv) $q_{2n+2}-bp_{2n}=q_{2n}$.

Solution: We have the recurring S.C.F.

$$a+\frac{1}{b+}\frac{1}{a+}\frac{1}{b+}\frac{1}{a+}\cdots$$

$$\therefore \frac{p_{2n}}{q_{2n}} = a+\cfrac{1}{b+\cfrac{1}{\left(\frac{p_{2n-2}}{q_{2n-2}}\right)}} = a+\cfrac{1}{b+\frac{q_{2n-2}}{p_{2n-2}}}$$

$$= a+\frac{p_{2n-2}}{bp_{2n-2}+q_{2n-2}} = \frac{(ab+1)p_{2n-2}+aq_{2n-2}}{bp_{2n-2}+q_{2n-2}}.$$

Each convergent is in its lowest terms.
Hence,

$$\left.\begin{aligned} p_{2n} &= (ab+1)p_{2n-2}+aq_{2n-2} \\ \text{and } q_{2n} &= bp_{2n-2}+q_{2n-2} \end{aligned}\right\}. \tag{1}$$

Also $q_{2n}=bq_{2n-1}+q_{2n-2}$. $\therefore\ p_{2n-2}=q_{2n-1}$.
Consequently replacing n by $(n+1)$ we get $p_{2n}=q_{2n+1}$. **This proves (i).**
And $p_{2n}-aq_{2n}=\{(ab+1)p_{2n-2}+aq_{2n-2}\}-a(bp_{2n-2}+q_{2n-2})=p_{2n-2}$.
But $p_{2n}=bp_{2n-1}+p_{2n-2}$

$$\therefore\ bp_{2n-1}+p_{2n-2}-aq_{2n}=p_{2n-2}$$

Hence, $bp_{2n-1}=aq_{2n}$. **This proves (ii).**
Also we can write (1) in the form [replacing n by $(n+1)$]

$p_{2(n+1)}=(ab+1)p_{2(n+1)-2}+aq_{2(n+1)-2}$,
i.e., $p_{2n+2}=(ab+1)p_{2n}+aq_{2n}$. **This proves (iii).**
and $q_{2(n+1)}=bp_{2(n+1)-2}+a_{2(n+1)-2}$
or, $q_{2n+2}=bp_{2n}+q_{2n}$. **This proves (iv).**

▶ Example 15.5.6 *For the Simple Continued Fraction*

$$a+\frac{1}{b+}\frac{1}{c+}\frac{1}{a+}\frac{1}{b+}\frac{1}{c+}\cdots$$

prove that $p_{3n}q_3-p_3q_{3n}=q_{3n-3}$.

Solution:

$$\frac{p_{3n}}{q_{3n}} = a+\frac{1}{b+}\frac{1}{c+}\frac{1}{\left(\frac{p_{3n-3}}{q_{3n-3}}\right)} = a+\frac{1}{b+}\frac{p_{3n-3}}{cp_{3n-3}+q_{3n-3}}$$

$$= a+\frac{cp_{3n-3}+q_{3n-3}}{(bc+1)p_{3n-3}+bq_{3n-3}}$$

$$= \frac{(abc+a+c)p_{3n-3}+(ab+1)q_{3n-3}}{(bc+1)p_{3n-3}+bq_{3n-3}}$$

whence (each convergent is in its lowest terms)

$$\begin{aligned} p_{3n} &= (abc+a+c)p_{3n-3}+(ab+1)q_{3n-3} \\ q_{3n} &= (bc+1)p_{3n-3}+bq_{3n-3}. \end{aligned}$$

Also we have

$$\begin{aligned} \frac{p_3}{q_3} &= a+\frac{1}{b+\frac{1}{c}} = a+\frac{c}{bc+1} = \frac{abc+a+c}{bc+1}, \\ \text{i.e., } p_3 &= abc+a+c,\ q_3 = bc+1. \end{aligned}$$

$$\begin{aligned} \therefore\ p_{3n}q_3 - p_3q_{3n} &= (abc+a+c)p_{3n-3}\cdot(bc+1)+(ab+1)q_{3n-3}(bc+1) \\ &\quad -(abc+a+c)\cdot\{(bc+1)p_{3n-3}+bq_{3n-3}\} \\ &= \{(ab+1)(bc+1)-(abc+a+c)b\}q_{3n-3} \\ &= q_{3n-3}\ \textbf{(proved)}. \end{aligned}$$

▶ Example 15.5.7 (**A General Continued Fraction**)

Show that $\dfrac{n}{n+1} = \dfrac{1}{2-}\dfrac{1}{2-}\dfrac{1}{2-}\ldots$ *to n quotients where n is a positive integer.*

Solution:

$$\begin{aligned} \frac{n}{n+1} &= \frac{1}{\frac{n+1}{n}} = \frac{1}{2-\frac{n-1}{n}} \\ \frac{n-1}{n} &= \frac{1}{\frac{n}{n-1}} = \frac{1}{2-\frac{n-2}{n-1}} \\ \ldots\ldots &\quad \ldots\ \ldots\ldots\ldots\ldots \\ \frac{2}{3} &= \frac{1}{\frac{3}{2}} = \frac{1}{2-\frac{1}{2}} \end{aligned}$$

Hence, $\dfrac{n}{n+1} = \dfrac{1}{2-}\dfrac{1}{2-}\dfrac{1}{2-}\cdots$ to n quotients.

▶ Example 15.5.8 (*Assumed the knowledge of solution of a difference equation*)

Find the rth convergent of $1+\frac{1}{1+}\frac{1}{1+}\frac{1}{1+}\cdots$

Solution: Here $p_r = p_{r-1}+p_{r-2}$ and $q_r = q_{r-1}+q_{r-2}$.

But the solution of the difference equation

$$u_r - u_{r-1} - u_{r-2} = 0$$

is of the form $u_r = A\alpha^{r-1}+B\beta^{r-1}$, when α, β are the roots of $x^2-x-1=0$, i.e., $\frac{1}{2}(1\pm\sqrt{5})$.

$$\begin{aligned} \therefore\ p_r &= c\left\{\frac{1}{2}(1+\sqrt{5})\right\}^{r-1} + d\left\{\frac{1}{2}(1-\sqrt{5})\right\}^{r-1} \\ q_r &= g\left\{\frac{1}{2}(1+\sqrt{5})\right\}^{r-1} + h\left\{\frac{1}{2}(1-\sqrt{5})\right\}^{r-1}, \end{aligned}$$

in which c, d, g, h are constants whose values can be found from $p_1 = 1$, $p_2 = 2$, $q_1 = 1$, $q_2 = 1$.

Thus

$$c + d = 1,\ c\left(1+\sqrt{5}\right) + d\left(1-\sqrt{5}\right) = 4$$

$$\Rightarrow \quad c = \left\{\frac{1}{2}(1+\sqrt{5})\right\}^2 \Big/ \sqrt{5}$$

and $$d = \left\{\frac{1}{2}(1-\sqrt{5})\right\}^2 \Big/ \sqrt{5}$$

$$\therefore \quad p_r\sqrt{5} = \left\{\frac{1}{2}(1+\sqrt{5})\right\}^{r+1} - \left\{\frac{1}{2}(1-\sqrt{5})\right\}^{r+1}.$$

Similarly, obtain

$$q_r\sqrt{5} = \left\{\frac{1}{2}(1+\sqrt{5})\right\}^{r} - \left\{\frac{1}{2}(1-\sqrt{5})\right\}^{2}$$

$$\therefore \quad \frac{p_r}{q_r} = \frac{(1+\sqrt{5})^{r+1} - (1-\sqrt{5})^{r+1}}{2\{(1+\sqrt{5})^r - (1-\sqrt{5})^r\}}.$$

Exercises on Chapter 15

1. Calculate the successive convergent to the following simple continued fractions:

(a) $2 + \frac{1}{6+}\,\frac{1}{1+}\,\frac{1}{1+}\,\frac{1}{11+}\,\frac{1}{2}$;

(b) $5 + \frac{1}{4+}\,\frac{1}{3+}\,\frac{1}{2+}\,\frac{1}{1}$;

(c) $\frac{1}{2+}\,\frac{1}{2+}\,\frac{1}{3+}\,\frac{1}{1+}\,\frac{1}{4+}\,\frac{1}{2+}\,\frac{1}{6}$;

(d) $3 + \frac{1}{3+}\,\frac{1}{1+}\,\frac{1}{2+}\,\frac{1}{2+}\,\frac{1}{1+}\,\frac{1}{9}$. [*Ans.* (a) $\frac{2}{1}, \frac{13}{6}, \frac{15}{7}, \frac{28}{13}, \frac{323}{150}, \frac{674}{313}$; etc.]

2. Express the following quantities as continued fractions:

(a) $\frac{253}{179}$; **(c)** 0.37; **(e)** $\frac{1189}{3927}$;

(b) $\frac{832}{159}$; **(d)** 0.3029; **(f)** $\frac{753}{356}$.

[*Ans.* (e) $\frac{1}{3+}\,\frac{1}{3+}\,\frac{1}{3+}\,\frac{1}{3+}\,\frac{1}{3+}\,\frac{1}{3+}\,\frac{1}{3}$; etc.]

3. Express 1.079 as a Simple Continued Fraction with an odd number of convergent and find its fourth convergent. [*Ans.* $1 + \frac{1}{12+}\,\frac{1}{1+}\,\frac{1}{1+}\,\frac{1}{1+}\,\frac{1}{12+}\,\frac{1}{2}$; $\frac{27}{25}$]

4. Let $a_1 + \frac{1}{a_2+}\frac{1}{a_3+}\cdots$ be a Simple Continued Fraction and let p_n/q_n be its nth convergent. Prove that

(a) $\left(\frac{p_{n+2}}{p_n}-1\right)\left(1-\frac{p_{n-1}}{p_{n+1}}\right)=\left(\frac{q_{n+2}}{q_n}-1\right)\left(1-\frac{q_{n-1}}{q_{n+1}}\right).$

(b) $\frac{p_{n+1}-p_{n-1}}{q_{n+1}-q_{n-1}}=\frac{p_n}{q_n}.$

(c) $p_nq_{n-3}-q_np_{n-3}=(-1)^{n-2}(a_na_{n-1}+1)$

(d) $\frac{p_n^2+q_n^2}{p_{n-2}^2+q_{n-2}^2}=\frac{(p_np_{n-1}+q_nq_{n-1})^2+1}{(p_{n-1}p_{n-2}+q_{n-1}q_{n-2})^2+1}.$

(e) $p_nq_{n-4}-p_{n-4}q_n=(-1)^{n-1}(a_na_{n-1}a_{n-2}+a_n+a_{n-2}).$

5. If $\frac{n^4+n^2-1}{n^3+n^2+n+1}$ be expressed as a continued fraction, show that the quotients are $(n-1)$ and $(n+1)$ alternately, and find its successive convergent.

[*Ans.* Convergents are $\frac{n-1}{1}, \frac{n^2}{n+1}, \frac{n^3-n^2+n-1}{n^2}, \frac{n^4+n^2-1}{n^3+n^2+n+1}$]

6. In the continued fraction $\frac{1}{a+}\frac{1}{a+}\frac{1}{a+}\frac{1}{a+}\cdots$, show that

$$p_n^2+p_{n+1}^2=p_{n-1}p_{n+1}+p_np_{n+2} \text{ and } p_n=q_{n-1}.$$

7. If p_n/q_n be the nth convergent of the recurring Simple Continued Fraction

$$\frac{1}{a+}\frac{1}{b+}\frac{1}{a+}\frac{1}{b+}\frac{1}{a+}\frac{1}{b+}\cdots,$$

prove that $q_{2n}=p_{2n+1}$, $q_{2n-1}=\frac{a}{b}p_{2n}$.

Also deduce

$$p_{n+2}-(ab+2)p_n+p_{n-2}=0 \text{ and } q_{n+2}-(ab+2)q_n+q_{n-2}=0$$

8. If p_n/q_n is the nth convergent to

$$\frac{1}{a+}\frac{1}{a+}\frac{1}{a+}\frac{1}{a+}\cdots$$

prove that p_n and q_n are respectively the coefficients of x^n in the expansions of

$$\frac{x}{1-ax-x^2} \text{ and } \frac{ax+x^2}{1-ax-x^2}.$$

Hence, show that $p_n=q_{n-1}=\frac{\alpha^n-\beta^n}{\alpha-\beta}$, where α, β are the roots of the quadratic equations $t^2-at-1=0.$

[**Solution.** $p_n = ap_{n-1}+p_{n-2}$; $q_n = aq_{n-1}+q_{n-2}$; thus the numerators and denominators of the successive convergents are the coefficients of the terms of a recurring series whose scale of relation is $1 - ax - x^2$.

Let $S = p_1x + p_2x^2 + p_3x^3 + p_4x^4 + \cdots$

Then $(1 - ax - x^2)S = p_1x + (p_2 - ap_1)x^2$ (where $p_1 = 1, p_2 = a$).

Hence,

$$S = \frac{x}{1 - ax - x^2} = \frac{x}{(1 - \alpha x)(1 - \beta x)},$$

where α and β are the two roots of $t^2 - at - 1 = 0$

$$\therefore S = \frac{1}{\alpha - \beta}\left(\frac{1}{1 - \alpha x} - \frac{1}{1 - \beta x}\right).$$

$$\therefore p_n = \text{coefficient of } x^n \text{ in } S = \frac{\alpha^n - \beta^n}{\alpha - \beta}.$$

Similarly, if $S' = q_1x^2 + q_3x^3 + \cdots$, we find

$$S' = \frac{ax^2 + x^3}{1 - ax - x^2} = \frac{x}{1 - ax - x^2} - x,$$

hence $q_{n-1} = \dfrac{\alpha^n - \beta^n}{\alpha - \beta}$].

9. If p_n/q_n is the nth convergent to

$$\frac{1}{a+}\,\frac{1}{b+}\,\frac{1}{a+}\,\frac{1}{b+}\cdots,$$

show that p_n and q_n are respectively the coefficients of x^n in the expansions of

$$\frac{x + bx^2 - x^3}{1 - (ab + 2)x^2 + x^4} \text{ and } \frac{ax + (ab + 1)x^2 - x^4}{1 - (ab + 2)x^2 + x^4}.$$

Hence, show that $ap_{2n} = bq_{2n-1} = ab\dfrac{\alpha^n - \beta^n}{\alpha - \beta}$,

$$p_{2n+1} = q_{2n} = \frac{\alpha^{n+1} - \beta^{n+1} - (\alpha^n - \beta^n)}{\alpha - \beta}$$

where α, β are the values of x^2 in the equation $1 - (ab + 2)x^2 + x^4 = 0$.

10. If p_n/q_n and p'_n/q'_n are the nth convergent of

$$a_1 + \frac{1}{a_2+}\,\frac{1}{a_3+}\cdots + \frac{1}{a_n} \text{ and } a_n + \frac{1}{a_{n-1}+}\,\frac{1}{a_{n-2}+}\cdots + \frac{1}{a_2+}\,\frac{1}{a_1}$$

respectively, show that $p'_n = p_n, q'_n = p_{n-1}, p'_{n-1} = q_n$ and $q'_{n-1} = q_{n-1}$.

11. Verify that $\sqrt{11} = 3 + \dfrac{1}{3+}\,\dfrac{1}{6+}\,\dfrac{1}{3+}\,\dfrac{1}{6+}\,\dfrac{1}{3+}\,\dfrac{1}{x}$ and find the value of x. [*Ans.* $3 + \sqrt{11}$]

12. If p_r/q_r is the rth convergent of $a + \frac{1}{b+}\frac{1}{c+}\frac{1}{a+}\frac{1}{b+}\frac{1}{c+};\frac{1}{a+}\ldots$ prove that

$$p_{3n}q_3 - p_3q_{3n} = q_{3n-3}.$$

13. Find the rth convergent of $1 + \frac{1}{2+}\frac{1}{1+}\frac{1}{2+}\frac{1}{1+}\ldots$

[$p_{2r+1} - 4p_{2r-1} + p_{2r-2} = 0$;

$$\begin{aligned} p_{2r+1} &= a(2+\sqrt{3})^r + b(2-\sqrt{3})^r \\ &= \{a(1+\sqrt{3})^{2r} + b(1-\sqrt{3})^{2r}\}2^{-r}. \end{aligned}$$

Similarly for p_{2r}]

14. Using the theory of continued fractions find the least positive integral solution of $19x - 117y = 11$. [*Ans.* $x = 56, y = 9$]

15. Find the number of solutions in positive integers of $7x + 9y = 1000$. [*Ans.* 16]

[$x = 148 - 9n, y = 7n - 4, 0 < n < 17$]

16. Solve in integers: $13x + 17y = 8$. [*Ans.* $15 - 17n, 13n - 11$]

17. If p_r/q_r is rth convergent of $1 + \frac{1}{2+}\frac{1}{3+}\frac{1}{1+}\frac{1}{2+}\frac{1}{3+}\frac{1}{1+}\cdots$, then prove that

$$3q_{2n} = 2p_{2n} + p_{2n-2}.$$

18. Express the following as an infinite S.C.F.:

(i) $\sqrt{15}$; (ii) $\sqrt{26}$; (iii) $3 - \sqrt{5}$.

[*Ans.* (i) $3 + \frac{1^*}{1+}\frac{1^*}{6}$; (ii) $5 + \frac{1^{**}}{10}$; (iii) $\frac{1}{1+}\frac{1}{3+}\frac{1}{4}\frac{*}{}$]

19. Prove that

(a) $$\left(a + \frac{1}{b+}\frac{1}{a+}\frac{1}{b+}\frac{1}{a+}\cdots\right)\left(\frac{1}{b+}\frac{1}{a+}\frac{1}{b+}\frac{1}{a+}\cdots\right) = \frac{a}{b}.$$

(b) Prove that $$\frac{c + \frac{1}{1+}\frac{1}{b+}\frac{1}{a+}\frac{1}{1+}\cdots}{b + \frac{1}{1+}\frac{1}{a+}\frac{1}{b+}\frac{1}{1+}\cdots} = \frac{1+a}{1+b}.$$

20. **(a)** If

$$x = a + \frac{1}{b+}\frac{1}{b+}\frac{1}{a+}\frac{1}{a+}\cdots \text{ and } y = b + \frac{1}{a+}\frac{1}{a+}\frac{1}{b+}\frac{1}{b+}\cdots,$$

then prove that $(ab^2 + a + b)x - (a^2b + a + b)y = a^2 - b^2$.

(b) Prove that the difference between

$$\frac{1}{a+}\frac{1}{b+}\frac{1}{c+}\frac{1}{a+}\frac{1}{b+}\frac{1}{c+}\ldots \text{ and } \frac{1}{b+}\frac{1}{a+}\frac{1}{c+}\frac{1}{b+}\frac{1}{a+}\frac{1}{c+}\ldots$$

is equal to $\dfrac{a-b}{1+ab}$.

21. **(a)** If

$$x = a + \frac{1}{b+}\frac{1}{c+}\frac{1}{a+}\frac{1}{b+}\frac{1}{c+}\frac{1}{a+}\cdots;$$

$$y = b + \frac{1}{c+}\frac{1}{a+}\frac{1}{b+}\frac{1}{c+}\frac{1}{a+}\frac{1}{b+}\cdots;$$

$$z = c + \frac{1}{a+}\frac{1}{b+}\frac{1}{c+}\frac{1}{a+}\frac{1}{b+}\frac{1}{c+}\cdots;$$

prove that

$$xyz = t + \frac{1}{t+}\frac{1}{t+}\cdots,$$

where $t = a + b + c + abc$, and either $x - \frac{1}{y} = a$ or $x + \frac{x}{y} = \frac{c-b}{cb+1}$.

(b) If

$$x = \frac{1}{a+}\frac{1}{b+}\frac{1}{a+}\frac{1}{b+}\frac{1}{a+}\cdots$$

$$y = \frac{1}{2a+}\frac{1}{2b+}\frac{1}{2a+}\frac{1}{2b+}\frac{1}{2a+}\cdots$$

$$z = \frac{1}{3a+}\frac{1}{3b+}\frac{1}{3a+}\frac{1}{3b+}\frac{1}{3a+}\cdots,$$

then prove that $x(y^2 - z^2) + 2y(z^2 - x^2) + 3z(x^2 - y^2) = 0$.

22. **(a)** If p_r/q_r be the rth convergent of Simple Continued Fraction

$$\frac{1}{a+}\frac{1}{b+}\frac{1}{c+}\frac{1}{a+}\frac{1}{b+}\frac{1}{c+}\frac{1}{a+}\cdots,$$

prove that $p_{3n+3} = bp_{3n} + (bc+1)q_{3n}$.

(b) For the fraction

$$\frac{1}{a_1+}\frac{1}{a_2+}\cdots+\frac{1}{a_n+}\cdots,$$

prove that

$$p_n q_{n-3} - p_{n-3} q_n = (-1)^{n-1}(a_n a_{n-1} + 1)$$

and $$p_n q_{n-4} - p_{n-4} q_n = (-1)^{n-1}(q_n q_{n-1} q_{n-2} + q_n + q_{n-2})$$

23. If

$$a_1 + \frac{1}{a_2+}\cdots\frac{1}{a_{r-1}+}\frac{1}{a_r+}\frac{1}{a_{r-1}+}\cdots\frac{1}{a_2+}\frac{1}{a_1} = \frac{P}{Q}$$

and $\frac{P}{Q}$ is in its lowest terms, prove that $Q^2 - 1$ is divisible by P.

24. Let $p, q, p'q'$ be positive integers such that $q' < q$ and $pq' - p'q = 1$. If p/q can be expressed as a Simple Continued Fraction with even number of quotients. Prove that p'/q' is the convergent immediately before p/q.

25. For the fraction $F = \frac{1}{a+}\frac{1}{b+}\frac{1}{a+}\frac{1}{b+}$ .., prove that

$$q_n = p_{n-1} \text{ and } F = \frac{1}{2a}\left(-ab + \sqrt{ab(ab+4)}\right).$$

Also if α, β are the roots of $x^2 - (ab+2)x + 1 = 0$ and $d_n = \alpha^n - \beta^n$, then prove that $ap_{2n} = bq_{2n-1} = ab\frac{d_n}{d_1}$, $p_{2n+1} = q_{2n} = \frac{d_{n+1} - d_n}{d_1}$.

26. **(a)** Two scales of equal length are divided into 162 and 209 equal parts respectively. If their zero points be coincident, prove that the 31st of one nearly coincide with 40th division of the other.

(b) A lunar month consists of 29.53 days and a tropical year consists of 365.24 days. Show, by the theory of continued fractions, that there are nearly 99 lunar months in 8 tropical years and 235 lunar months in 19 tropical years.

27. **(a)** Find the limit to the error in taking $\frac{222}{203}$ yards as equivalent to a metre (Given that 1 metre=1.0936 yds.)

(b) Show that the error in taking the 4th convergent for $\sqrt{11}$ is less than 0.00005.

(c) Express $\sqrt{14}$ as a S.C.F. Show that $\frac{449}{120}$ differs from $\sqrt{14}$ by less than $\frac{1}{90000}$.

(d) For the fraction

$$\frac{1}{1+}\frac{1}{2+}\frac{1}{3+}\frac{1}{4+}\cdots$$

find a convergent which represents its value within 0.001.

28. Prove that

$$a + \frac{1}{b+}\frac{1}{2a+}\frac{1}{b+}\frac{1}{2a+}\cdots = \sqrt{\frac{a}{b}(ab+2)}.$$

Hence, deduce

$$\sqrt{a(a+1)} = a + \frac{1}{2+}\frac{1}{2a+}\frac{1}{2+}\frac{1}{2a+}\cdots$$

$$\sqrt{a(a+2)} = a + \frac{1}{1+}\frac{1}{2a+}\frac{1}{1+}\frac{1}{2a+}\cdots$$

$$\sqrt{a^2+2} = a + \frac{1}{a+}\frac{1}{2a+}\frac{1}{a+}\frac{1}{2a+}\frac{1}{a+}\frac{1}{2a+}\frac{1}{a+}\cdots$$

29. Prove that $\sqrt{a^2+1} = a + \frac{1}{2a+}\frac{1}{2a+}\frac{1}{2a+}\ldots$ and obtain the following

(a) $2(a^2+1)q_n = p_{n-1} + p_{n+1}$;

(b) $2p_n = q_{n-1} + q_{n+1}$;

(c) $p_n^2 - (a^2+1)q_n^2 = (-1)^n$.